Student Solutions Manual

Kevin Bodden   Randy Gallaher

Lewis and Clark Community College

Elementary and Intermediate Algebra

for College Students

■ Third Edition

Allen R. **Angel**

PEARSON

Prentice Hall

Upper Saddle River, NJ 07458

Vice President and Editorial Director, Mathematics: Christine Hoag
Executive Editor: Paul Murphy
Project Manager, Editorial: Dawn Nuttall
Editorial Assistant: Georgina Brown
Senior Managing Editor, Mathematics: Linda Behrens
Associate Managing Editor, Mathematics: Bayani Mendoza de Leon
Project Manager, Production: Traci Douglas
Supplement Cover Manager: Paul Gourhan
Supplement Cover Designer: Victoria Colotta
Operations Specialist: Ilene Kahn
Senior Operations Supervisor: Diane Peirano

© 2008 Pearson Education, Inc.
Pearson Prentice Hall
Pearson Education, Inc.
Upper Saddle River, NJ 07458

The author and publisher of this book have used their best efforts in preparing this book. These efforts include the development, research, and testing of the theories and programs to determine their effectiveness. The author and publisher make no warranty of any kind, expressed or implied, with regard to these programs or the documentation contained in this book. The author and publisher shall not be liable in any event for incidental or consequential damages in connection with, or arising out of, the furnishing, performance, or use of these programs.

Printed in the United States of America

10 9 8 7 6 5 4 3 2 1

ISBN 13: 978-0-13-614777-0 Standalone

ISBN 10:   0-13-614777-1 Standalone

ISBN 13: 978-0-13-614946-0 Component

ISBN 10:   0-13-614946-4 Component

Pearson Education Ltd., *London*
Pearson Education Australia Pty. Ltd., *Sydney*
Pearson Education Singapore, Pte. Ltd.
Pearson Education North Asia Ltd., *Hong Kong*
Pearson Education Canada, Inc., *Toronto*
Pearson Educación de Mexico, S.A. de C.V.
Pearson Education—Japan, *Tokyo*
Pearson Education Malaysia, Pte. Ltd.

Table of Contents

Chapter 1

Exercise Set 1.1

1 – 9. Answers will vary.

11. To prepare properly for this class, you need to do all the homework carefully and completely; preview the new material that is to be covered in class.

13. At least 2 hours of study and homework time for each hour of class time is generally recommended.

15. a. You need to do the homework in order to practice what was presented in class.

b. When you miss class, you miss important information. Therefore it is important that you attend class regularly.

17. Answers will vary.

Exercise Set 1.2

1. understand, translate, calculate, check, state answer

3. The symbol $\approx$ means "is approximately equal to."

5. Rank the data. The median is the value in the middle.

7. Divide the sum of the data by the number of pieces of data.

9. A mean average of 80 corresponds to a total of 800 points for the 10 quizzes. Walter's mean average of 79 corresponds to a total of 790 points for the 10 quizzes. Thus, he actually missed a B by 10 points.

11. a. $\dfrac{78+97+59+74+74}{5} = \dfrac{382}{5} = 76.4$

The mean grade is 76.4.

b. 59, 74, 74, 78, 97
The middle value is 74.
The median grade is 74.

13. a. $\dfrac{96.56+108.78+87.23+85.90+79.55+65.88}{6}$

$= \dfrac{523.90}{6} \approx 87.32$

The mean bill is about \$87.32.

b. \$65.88, \$79.55, \$85.90, \$87.23, \$96.56, \$108.78
The middle values are \$85.90 and \$87.23.
$\dfrac{85.90+87.23}{2} = \dfrac{173.13}{2} = 86.57$
The median bill is about \$86.57.

15.a. $\dfrac{10.63+10.67+10.68+10.83+11.6+11.76+11.87+12.18+12.8+12.91}{10} = \dfrac{115.93}{10} = 11.593$

The mean inches for rainfall for the 10 years is 11.593.

b. 10.63, 10.67, 10.68, 10.83, 11.6, 11.76, 11.87, 12.18, 12.8, 12.91
The middle values are 11.6 and 11.76.

$\dfrac{11.6+11.76}{2} = \dfrac{23.36}{2} = 11.68$
The median inches for rainfall for the 10 years is 11.68.

---

17. Barbara's earnings = 5% of sales
Barbara's earnings = 0.05(9400) = 470
Her week's earnings were \$470.

19. a. sales tax = 8% of price
sales tax = 0.08(16,700) = 1336
The sales tax was \$1,336.

b. Total cost = price + tax
Total cost = 16,700 + 1,336 = 18,036
The total cost was \$18,036.

21. operations performed = (number of operations in billions)(amount of time in seconds)

$$= (2.3)(0.7)$$
$$= 1.61 \text{ billion}$$

In 0.7 seconds, 1,610,000,000 operations can be performed.

---

23. a. time to use energy $= \dfrac{\text{kJ in hamburger}}{\text{kJ/min running}}$

$$= \frac{1550}{80}$$
$$= 19.375$$

It takes 19.375 minutes to use up the energy from a hamburger by running.

b. time to use energy $= \dfrac{\text{kJ in milkshake}}{\text{kJ/min walking}}$

$$= \frac{2200}{25}$$
$$= 88$$

It takes 88 minutes to use up the energy from a chocolate milkshake by walking.

c. time to use energy

$$= \frac{\text{kJ in glass of skim milk}}{\text{kJ/min cycling}}$$
$$= \frac{350}{35}$$
$$= 10$$

It takes 10 minutes to use up the energy from a glass of skim milk by cycling.

25. miles per gallon $= \dfrac{\text{number of miles}}{\text{number of gallons}}$

$$= \frac{16,935.4 - 16,741.3}{10.5}$$
$$= \frac{194.1}{10.5}$$
$$\approx 18.49$$

His car gets about 18.49 miles per gallon.

27. savings = local cost – mail order cost

local cost $= 425 + (0.08)(425)$

$$= 425 + 34$$
$$= 459$$

mail cost $= 4(62.30 + 6.20 + 8)$

$$= 4(76.50)$$
$$= 306$$

savings $= 459 - 306 = 153$

Eric saved $153.

29. A single green block should be placed on the 3 on the right. This is required for

$$-6 + (-4) + 2(-2) + 3(3) + 5$$
$$= -6 + (-4) + (-4) + 9 + 5$$
$$= -10 + (-4) + 9 + 5$$
$$= -14 + 9 + 5$$
$$= -5 + 5$$
$$= 0$$

31. a. gallons per year = 365(gallons per day)

$$= 365(11.25 \text{ gallons})$$
$$= 4106.25$$

There are 4106.25 gallons of water wasted each year.

b. additional money spent

= (cost)(gallons wasted)

$$= \frac{5.20}{1000 \text{ gallons}} \cdot 4106.25 \text{ gallons}$$
$$\approx 21.35$$

About $21.35 extra is spent because of the wasted water.

33. a. cost

$$= \text{deductible} + 20\%(\text{doctor bill} - \text{deductible})$$
$$= 150 + 0.20(365 - 150)$$
$$= 150 + 0.20(215)$$
$$= 150 + 43$$
$$= 193$$

Mel will be responsible for $193.

b. The insurance company would be responsible for the remainder of the bill which would be 365 – 193 = $172.

35. a. Hong Kong-China; 550

b. Mexico; 385

c. 550 – 385 = 165

37. a. 1992: 275,000
2004: 1,050,000

b. $1,050,000 - 275,000 = 775,000$

c. $\dfrac{1,050,000}{275,000} \approx 3.818$

≈ 3.82 times greater

39. a. 82% of 1.7 million $= 0.82(1.7 \text{ million})$
$= 1.394$ million

b. 15% of 1.7 million $= 0.15(1.7 \text{ million})$
$= 0.255$ million

c. 3% of 1.7 million $= = 0.03(1.7 \text{ million})$
$= 0.051$ million

41. a. $\text{mean} = \dfrac{\text{sum of grades}}{\text{number of exams}}$

$60 = \dfrac{50 + 59 + 67 + 80 + 56 + \text{last}}{6}$

$360 = 312 + \text{last}$

$\text{last} = 360 - 312$

$= 48$

Lamond needs at least a 48 on the last exam.

b. $70 = \dfrac{312 + \text{last}}{6}$

$420 = 312 + \text{last}$

$\text{last} = 420 - 312$

$= 108$

Lamond would need 108 points on the last exam, so he cannot get a C.

43. bachelor's degree
$\dfrac{49,900}{21,600} \approx 2.3$

45. Answers will vary.
One possible solution is:
50, 60, 70, 80, 90

$\text{mean} = \dfrac{50 + 60 + 70 + 80 + 90}{5} = \dfrac{350}{5} = 70$

Exercise Set 1.3

1. a. Variables are letters that represent numbers.

b. Letters often used to represent variables are x, y, and z.

3. a. The top number is the numerator.

b. The bottom number is the denominator.

5. Divide out factors that are common to both the numerator and the denominator.

7. a. The least common denominator is the smallest number divisible by the two denominators.

b. Answers will vary.

9. Part b) shows simplifying a fraction. In part a) common factors are divided out of two fractions.

11. Part a) is incorrect because you cannot divide out common factors when adding.

13. c) $\dfrac{4}{5} \cdot \dfrac{1}{4} = \dfrac{1}{5} \cdot \dfrac{1}{1} = \dfrac{1}{5}$. Divide out the common factor, 4. This process can be used only when multiplying fractions and so cannot be used for a) or b). Part d) becomes $\dfrac{4}{5} \cdot \dfrac{4}{1}$ so no common factor can be divided out.

15. Multiply numerators and multiply denominators.

17. Write fractions with a common denominator, add or subtract numerators, keep the common denominator.

19. Yes, it is simplified because the greatest common divisor of the numerator and denominator is 1.

21. Yes, it can be written as a mixed number.
$\dfrac{13}{5} = \dfrac{10}{5} + \dfrac{3}{5} = 2\dfrac{3}{5}$

23. The greatest common factor of 10 and 15 is 5.
$\dfrac{10}{15} = \dfrac{10 \div 5}{15 \div 5} = \dfrac{2}{3}$

25. The greatest common factor of 3 and 12 is 3.
$\dfrac{3}{12} = \dfrac{3 \div 3}{12 \div 3} = \dfrac{1}{4}$

27. The greatest common factor of 36 and 76 is 4.
$$\frac{36}{76} = \frac{36 \div 4}{76 \div 4} = \frac{9}{19}$$

29. The greatest common factor of 9 and 21 is 3.
$$\frac{9}{21} = \frac{9 \div 3}{21 \div 3} = \frac{3}{7}$$

31. 18 and 49 have no common factors other than 1. Therefore, the fraction is already simplified.

33. 12 and 25 have no common factors other than 1. Therefore, the fraction is already simplified.

35. $2\dfrac{13}{15} = \dfrac{30+13}{15} = \dfrac{43}{15}$

37. $7\dfrac{2}{3} = \dfrac{21+2}{3} = \dfrac{23}{3}$

39. $3\dfrac{5}{18} = \dfrac{54+5}{18} = \dfrac{59}{18}$

41. $9\dfrac{6}{17} = \dfrac{153+6}{17} = \dfrac{159}{17}$

43. $\dfrac{7}{4} = 1\dfrac{3}{4}$ because $7 \div 4 = 1$ R3

45. $\dfrac{13}{4} = 3\dfrac{1}{4}$ because $13 \div 4 = 3$ R 1

47. $\dfrac{32}{7} = 4\dfrac{4}{7}$ because $32 \div 7 = 4$ R 4

49. $\dfrac{86}{14} = 6\dfrac{2}{14} = 6\dfrac{1}{7}$ because $86 \div 14 = 6$ R 2

51. $\dfrac{2}{3} \cdot \dfrac{4}{5} = \dfrac{2 \cdot 4}{3 \cdot 5} = \dfrac{8}{15}$

53. $\dfrac{5}{12} \cdot \dfrac{4}{15} = \dfrac{\cancel{5}^{1}}{{}_{3}\cancel{12}} \cdot \dfrac{\cancel{4}^{1}}{\cancel{15}_{3}} = \dfrac{1 \cdot 1}{3 \cdot 3} = \dfrac{1}{9}$

55. $\dfrac{3}{4} \div \dfrac{1}{2} = \dfrac{3}{{}_{2}\cancel{4}} \cdot \dfrac{\cancel{2}^{1}}{1} = \dfrac{3}{2} \cdot \dfrac{1}{1} = \dfrac{3}{2}$ or $1\dfrac{1}{2}$

57. $\dfrac{3}{8} \div \dfrac{3}{4} = \dfrac{\cancel{3}^{1}}{{}_{2}\cancel{8}} \cdot \dfrac{\cancel{4}^{1}}{\cancel{3}_{1}} = \dfrac{1 \cdot 1}{2 \cdot 1} = \dfrac{1}{2}$

59. $\dfrac{10}{3} \div \dfrac{5}{9} = \dfrac{\cancel{10}^{2}}{{}_{1}\cancel{3}} \cdot \dfrac{\cancel{9}^{3}}{\cancel{5}_{1}} = \dfrac{2 \cdot 3}{1 \cdot 1} = \dfrac{6}{1} = 6$

61. $\dfrac{15}{4} \cdot \dfrac{2}{3} = \dfrac{\cancel{15}^{5}}{{}_{2}\cancel{4}} \cdot \dfrac{\cancel{2}^{1}}{\cancel{3}_{1}} = \dfrac{5 \cdot 1}{2 \cdot 1} = \dfrac{5}{2}$ or $2\dfrac{1}{2}$

63. $5\dfrac{3}{8} \div 1\dfrac{1}{4}$

$5\dfrac{3}{8} = \dfrac{40+3}{8} = \dfrac{43}{8}$

$1\dfrac{1}{4} = \dfrac{4+1}{4} = \dfrac{5}{4}$

$5\dfrac{3}{8} \div 1\dfrac{1}{4} = \dfrac{43}{8} \div \dfrac{5}{4} = \dfrac{43}{{}_{2}\cancel{8}} \cdot \dfrac{\cancel{4}^{1}}{5} = \dfrac{43 \cdot 1}{2 \cdot 5} = \dfrac{43}{10}$ or $4\dfrac{3}{10}$

65. $\dfrac{28}{13} \cdot \dfrac{2}{7} = \dfrac{\cancel{28}^{4}}{13} \cdot \dfrac{2}{\cancel{7}_{1}} = \dfrac{4 \cdot 2}{13 \cdot 1} = \dfrac{8}{13}$

67. $\dfrac{3}{8} + \dfrac{2}{8} = \dfrac{3+2}{8} = \dfrac{5}{8}$

69. $\dfrac{3}{14} - \dfrac{1}{14} = \dfrac{3-1}{14} = \dfrac{2}{14} = \dfrac{1}{7}$

71. $\dfrac{4}{5} + \dfrac{6}{15}$

$\dfrac{4}{5} = \dfrac{4}{5} \cdot \dfrac{3}{3} = \dfrac{12}{15}$

$\dfrac{4}{5} + \dfrac{6}{15} = \dfrac{12}{15} + \dfrac{6}{15} = \dfrac{12+6}{15} = \dfrac{18}{15} = \dfrac{6}{5}$ or $1\dfrac{1}{5}$

73. $\dfrac{8}{17} + \dfrac{2}{34}$

$\dfrac{8}{17} = \dfrac{8}{17} \cdot \dfrac{2}{2} = \dfrac{16}{34}$

$\dfrac{8}{17} + \dfrac{2}{34} = \dfrac{16}{34} + \dfrac{2}{34} = \dfrac{16+2}{34} = \dfrac{18}{34} = \dfrac{9}{17}$

75. $\dfrac{1}{3} + \dfrac{1}{4}$

$\dfrac{1}{3} = \dfrac{1}{3} \cdot \dfrac{4}{4} = \dfrac{4}{12}$

$\dfrac{1}{4} = \dfrac{1}{4} \cdot \dfrac{3}{3} = \dfrac{3}{12}$

$\dfrac{1}{3} + \dfrac{1}{4} = \dfrac{4}{12} + \dfrac{3}{12} = \dfrac{4+3}{12} = \dfrac{7}{12}$

77. $\dfrac{7}{12} - \dfrac{2}{9}$

$$\dfrac{7}{12} = \dfrac{7}{12} \cdot \dfrac{3}{3} = \dfrac{21}{36}$$

$$\dfrac{2}{9} = \dfrac{2}{9} \cdot \dfrac{4}{4} = \dfrac{8}{36}$$

$$\dfrac{7}{12} - \dfrac{2}{9} = \dfrac{21}{36} - \dfrac{8}{36} = \dfrac{21-8}{36} = \dfrac{13}{36}$$

79. $3\dfrac{1}{8} - \dfrac{5}{12}$

$$3\dfrac{1}{8} = \dfrac{24+1}{8} = \dfrac{25}{8} \cdot \dfrac{3}{3} = \dfrac{75}{24}$$

$$\dfrac{5}{12} = \dfrac{5}{12} \cdot \dfrac{2}{2} = \dfrac{10}{24}$$

$$3\dfrac{1}{8} - \dfrac{5}{12} = \dfrac{75}{24} - \dfrac{10}{24} = \dfrac{75-10}{24} = \dfrac{65}{24} \text{ or } 2\dfrac{17}{24}$$

81. $6\dfrac{1}{3} - 3\dfrac{1}{2}$

$$6\dfrac{1}{3} = \dfrac{18+1}{3} = \dfrac{19}{3} \cdot \dfrac{2}{2} = \dfrac{38}{6}$$

$$3\dfrac{1}{2} = \dfrac{6+1}{2} = \dfrac{7}{2} \cdot \dfrac{3}{3} = \dfrac{21}{6}$$

$$6\dfrac{1}{3} - 3\dfrac{1}{2} = \dfrac{38}{6} - \dfrac{21}{6} = \dfrac{38-21}{6} = \dfrac{17}{6} \text{ or } 2\dfrac{5}{6}$$

83. $9\dfrac{2}{5} - 6\dfrac{1}{2}$

$$9\dfrac{2}{5} = \dfrac{45+2}{5} = \dfrac{47}{5} \cdot \dfrac{2}{2} = \dfrac{94}{10}$$

$$6\dfrac{1}{2} = \dfrac{12+1}{2} = \dfrac{13}{2} \cdot \dfrac{5}{5} = \dfrac{65}{10}$$

$$9\dfrac{2}{5} - 6\dfrac{1}{2} = \dfrac{94}{10} - \dfrac{65}{10} = \dfrac{94-65}{10} = \dfrac{29}{10} \text{ or } 2\dfrac{9}{10}$$

85. $5\dfrac{9}{10} + 3\dfrac{1}{3}$

$$5\dfrac{9}{10} = \dfrac{50+9}{10} = \dfrac{59}{10} \cdot \dfrac{3}{3} = \dfrac{177}{30}$$

$$3\dfrac{1}{3} = \dfrac{9+1}{3} = \dfrac{10}{3} \cdot \dfrac{10}{10} = \dfrac{100}{30}$$

$$5\dfrac{9}{10} + 3\dfrac{1}{3} = \dfrac{177}{30} + \dfrac{100}{30} = \dfrac{177+100}{30} = \dfrac{277}{30} \text{ or } 9\dfrac{7}{30}$$

87. $\dfrac{5}{6} - \dfrac{3}{8}$

$$\dfrac{5}{6} \cdot \dfrac{4}{4} = \dfrac{20}{24}$$

$$\dfrac{3}{8} \cdot \dfrac{3}{3} = \dfrac{9}{24}$$

$$\dfrac{5}{6} - \dfrac{3}{8} = \dfrac{20}{24} - \dfrac{9}{24} = \dfrac{20-9}{24} = \dfrac{11}{24}$$

It is $\dfrac{11}{24}$ mile larger.

89. $55\dfrac{3}{16} - 46\dfrac{1}{4}$

$$55\dfrac{3}{16} = \dfrac{880+3}{16} = \dfrac{883}{16}$$

$$46\dfrac{1}{4} = \dfrac{184+1}{4} = \dfrac{185}{4} = \dfrac{185}{4} \cdot \dfrac{4}{4} = \dfrac{740}{16}$$

$$53\dfrac{3}{16} - 46\dfrac{1}{4} = \dfrac{883}{16} - \dfrac{740}{16} = \dfrac{143}{16} = 8\dfrac{5}{16}$$

Rebecca has grown $8\dfrac{5}{16}$ inches.

91. $1 - \dfrac{31}{50} = \dfrac{50}{50} - \dfrac{31}{50} = \dfrac{50-31}{50} = \dfrac{19}{50}$

The fraction of putts not made were $\dfrac{19}{50}$.

93. $1 - \dfrac{37}{40} = \dfrac{40}{40} - \dfrac{37}{40} = \dfrac{40-37}{40} = \dfrac{3}{40}$

There were $\dfrac{3}{40}$ people unemployed.

95. 15 feet $2\dfrac{1}{2}$ in. $- 3$ feet $3\dfrac{1}{4}$ in.

$= 14$ feet $14\dfrac{2}{4}$ in. $- 3$ feet $3\dfrac{1}{4}$ in.

$= 11$ feet $11\dfrac{1}{4}$ in.

Now convert to inches,

$$11(12) + 11\dfrac{1}{4} = 132 + 11\dfrac{1}{4} = 143\dfrac{1}{4} \text{ in.}$$

Now convert to feet.

$$143\dfrac{1}{4} \div 12 = \dfrac{572+1}{4} \cdot \dfrac{1}{12} = \dfrac{573}{48} \approx 11.94 \text{ft.}$$

97. $10\dfrac{1}{2} = \dfrac{21}{2}\cdot\dfrac{5}{5} = \dfrac{105}{10}$

$8\dfrac{2}{5} = \dfrac{42}{5}\cdot\dfrac{2}{2} = \dfrac{84}{10}$

$10\dfrac{1}{2} - 8\dfrac{2}{5} = \dfrac{105}{10} - \dfrac{84}{10} = \dfrac{105-84}{10} = \dfrac{21}{10} = 2\dfrac{1}{10}$

She improved by $2\dfrac{1}{10}$ minutes.

99. $3\dfrac{1}{8} = \dfrac{24+1}{8} = \dfrac{25}{8}$

$3\dfrac{1}{8} \div 2 = 3\dfrac{1}{8} \div \dfrac{2}{1} = \dfrac{25}{8}\cdot\dfrac{1}{2} = \dfrac{25}{16} \text{ or } 1\dfrac{9}{16}$

Each piece is $1\dfrac{9}{16}$ inches long.

101. $\dfrac{1}{16}\cdot 80 = \dfrac{1}{16}\cdot\dfrac{80}{1} = \dfrac{1}{1}\cdot\dfrac{5}{1} = 5$

Mr. Krisanda should be given 5 milligrams of the drug.

103. $15 \div \dfrac{3}{8} = \dfrac{15}{1}\cdot\dfrac{8}{3} = \dfrac{5}{1}\cdot\dfrac{8}{1} = \dfrac{5\cdot 8}{1\cdot 1} = \dfrac{40}{1} = 40$

Tierra can wash her hair 40 times.

105. $\dfrac{1}{4} + \dfrac{1}{4} + 1 = \dfrac{1}{4} + \dfrac{1}{4} + \dfrac{4}{4} = \dfrac{6}{4} = \dfrac{3}{2} \text{ or } 1\dfrac{1}{2}$

The total thickness is $1\dfrac{1}{2}$ inches.

107. $4\dfrac{2}{3} = \dfrac{12+2}{3} = \dfrac{14}{3}$

$28 \div \dfrac{14}{3} = \dfrac{28}{1}\cdot\dfrac{3}{14} = \dfrac{2}{1}\cdot\dfrac{3}{2} = \dfrac{6}{1} = 6$

There will be 6 whole strips of wood.

109. a. Total height of computer + monitor

$= 7\dfrac{1}{2} \text{ in.} + 14\dfrac{3}{8} \text{ in.}$

$7\dfrac{1}{2} = \dfrac{15}{2} = \dfrac{15}{2}\cdot\dfrac{4}{4} = \dfrac{60}{8}$

$14\dfrac{3}{8} = \dfrac{112+3}{8} = \dfrac{115}{8}$

$7\dfrac{1}{2} + 14\dfrac{3}{8} = \dfrac{60}{8} + \dfrac{115}{8} = \dfrac{175}{8} \text{ or } 21\dfrac{7}{8}$

Total height of computer and monitor is $21\dfrac{7}{8}$ inches, so there is sufficient room.

b. $22\dfrac{1}{2} = \dfrac{44+1}{2} = \dfrac{45}{2} = \dfrac{45}{2}\cdot\dfrac{4}{4} = \dfrac{180}{8}$

$22\dfrac{1}{2} - 21\dfrac{7}{8} = \dfrac{180}{8} - \dfrac{175}{8} = \dfrac{5}{8}$

There will be $\dfrac{5}{8}$ inch of extra height.

c. $22\dfrac{1}{2} = \dfrac{44+1}{2} = \dfrac{45}{2} = \dfrac{45}{2}\cdot\dfrac{2}{2} = \dfrac{90}{4}$

$26\dfrac{1}{2} = \dfrac{52+1}{2} = \dfrac{53}{2} = \dfrac{53}{2}\cdot\dfrac{2}{2} = \dfrac{106}{4}$

$2\dfrac{1}{2} = \dfrac{4+1}{2} = \dfrac{5}{2} = \dfrac{5}{2}\cdot\dfrac{2}{2} = \dfrac{10}{4}$

$1\dfrac{1}{4} = \dfrac{4+1}{4} = \dfrac{5}{4}$

$22\dfrac{1}{2} + 26\dfrac{1}{2} + 2\dfrac{1}{2} + 1\dfrac{1}{4}$

$= \dfrac{90}{4} + \dfrac{106}{4} + \dfrac{10}{4} + \dfrac{5}{4}$

$= \dfrac{211}{4} \text{ or } 52\dfrac{3}{4}$

The height of the desk is $52\dfrac{3}{4}$ in.

111. a. $\dfrac{*}{a} + \dfrac{?}{a} = \dfrac{*+?}{a}$

b. $\dfrac{\odot}{?} - \dfrac{\square}{?} = \dfrac{\odot - \square}{?}$

c. $\dfrac{\Delta}{\square} + \dfrac{4}{\square} = \dfrac{\Delta + 4}{\square}$

d. $\dfrac{x}{3} - \dfrac{2}{3} = \dfrac{x-2}{3}$

e. $\dfrac{12}{x} - \dfrac{4}{x} = \dfrac{12-4}{x} = \dfrac{8}{x}$

113. number of pills

$= \dfrac{(\text{mg per day})(\text{days per month})(\text{no. of months})}{\text{mg per pill}}$

$= \dfrac{(450)(30)(6)}{300}$

$= 270$

Dr. Muechler should prescribe 270 pills.

115. Answers will vary.

116. $\dfrac{9+8+15+32+16}{5} = \dfrac{80}{5} = 16$

The mean is 16.

117. In order, the values are: 8, 9, 15, 16, 32.
The median is 15.

118. Variables are letters used to represent numbers.

Exercise Set 1.4

1. A set is a collection of elements.

3. Answers will vary. One possible answer is:
An empty set is a set that contains no elements.

5. The set of whole numbers contains zero and all of the natural numbers. The set of natural numbers does not contain zero.

7. a. The natural number 7 is a whole number because it is a member of {0, 1, 2, 3, ...}.

   b. The natural number 7 is a rational number because it can be written as the quotient of two integers, $\dfrac{7}{1}$.

   c. All natural numbers are real numbers.

9. a. yes
   b. no
   c. no
   d. yes

11. The integers are {..., –3, –2, –1, 0, 1, 2, 3, ...}.

13. The whole numbers are {0, 1, 2, 3, ...}.

15. The negative integers are {..., –3, –2, –1}.

17. True; the whole numbers are {0, 1, 2, 3, ...}.

19. True; any number that can be represented on a real number line is a real number.

21. False; the integers are {..., –2, –1, 0, 1, 2, ...}.

23. False; $\sqrt{2}$ cannot be expressed as the quotient of two integers.

25. True; $-\dfrac{1}{5}$ is a quotient of two integers, $\dfrac{-1}{5}$.

27. True; 0 can be expressed as a quotient of two integers, $\dfrac{0}{1}$.

29. False; $4\dfrac{5}{8}$ is rational since it can be expressed as a quotient of two integers.

31. True; $-\sqrt{5}$ is irrational since it cannot be expressed exactly as a decimal number.

33. True, either   or { } is used.

35. False; irrational numbers are real but not rational.

37. True; any rational number can be represented on a real number line and is therefore real.

39. True; irrational numbers are real numbers which are not rational.

41. False; any negative irrational number is a counterexample.

43. True; the symbol $\mathbb{R}$ represents the set of real numbers.

45. False; every number greater than zero is positive but not necessarily an integer.

47. True; the integers are

$$\left\{ \underbrace{..., -2, -1,}_{\text{negative integers}} \underbrace{0}_{\text{zero}}, \underbrace{1, 2, ...}_{\text{positive integers}} \right\}.$$

49. a. 13 is a positive integer.
   b. –2 and 13 are rational numbers.
   c. –2 and 13 are real numbers.
   d. 13 is a whole number.

51. a. 3 and 77 are positive integers.
   b. 0, 3, and 77 are whole numbers
   c. 0, –2, 3, and 77 are integers.
   d. $-\dfrac{5}{7}$, 0, –2, 3, $6\dfrac{1}{4}$, 1.63, and 77 are rational numbers.
   e. $\sqrt{7}$ and $-\sqrt{3}$ are irrational numbers.
   f. $-\dfrac{5}{7}$, 0, –2, 3, $6\dfrac{1}{4}$, $\sqrt{7}$, $-\sqrt{3}$, 1.63, and 77 are real numbers.

For Exercises 53–63, answers will vary. One possible answer is given for each exercise.

53. 0, 1, 2

55. $-\sqrt{2},\ -\sqrt{3},\ -\sqrt{7}$

57. $-\dfrac{2}{3},\ \dfrac{1}{2},\ 6.3$

59. $-13, -5, -1$

61. $\sqrt{2},\ \sqrt{3},\ -\sqrt{5}$

63. $-7, 1, 5$

65. $\{8, 9, 10, 11, \ldots, 94\}$
$94 - 8 + 1 = 86 + 1 = 87$
The set has 87 elements.

67. **a.** $A = \{1, 3, 4, 5, 8\}$
 b. $B = \{2, 5, 6, 7, 8\}$
 c. A and $B = \{5, 8\}$
 d. A or $B = \{1, 2, 3, 4, 5, 6, 7, 8\}$

69. **a.** Set B continues beyond 4.
 b. Set A has 4 elements.
 c. Set B has an infinite number of elements.
 d. Set B is an infinite set.

71. **a.** There are an infinite number of fractions between any 2 numbers.
 b. There are an infinite number of fractions between any 2 numbers.

73. $5\dfrac{2}{5} = \dfrac{5 \cdot 5 + 2}{5} = \dfrac{25 + 2}{5} = \dfrac{27}{5}$

74. $\dfrac{16}{3} = 5\dfrac{1}{3}$ because $16 \div 3 = 5\ \text{R}\ 1$

75. $\dfrac{7}{8} - \dfrac{2}{3}$
$\dfrac{7}{8} = \dfrac{7}{8} \cdot \dfrac{3}{3} = \dfrac{21}{24}$
$\dfrac{2}{3} = \dfrac{2}{3} \cdot \dfrac{8}{8} = \dfrac{16}{24}$
$\dfrac{7}{8} - \dfrac{2}{3} = \dfrac{21}{24} - \dfrac{16}{24} = \dfrac{21 - 16}{24} = \dfrac{5}{24}$

76. $\dfrac{3}{5} \div 6\dfrac{3}{4}$

$6\dfrac{3}{4} = \dfrac{24 + 3}{4} = \dfrac{27}{4}$

$\dfrac{3}{5} \div 6\dfrac{3}{4} = \dfrac{3}{5} \div \dfrac{27}{4} = \dfrac{\overset{1}{\cancel{3}}}{5} \cdot \dfrac{4}{\underset{9}{\cancel{27}}} = \dfrac{1 \cdot 4}{5 \cdot 9} = \dfrac{4}{45}$

Exercise Set 1.5

1. **a.**
 b.
 c. −2 is greater than −4 because it is farther to the right on the number line.
 d. $-4 < -2$
 e. $-2 > -4$

3. **a.** 4 is 4 units from 0 on a number line.
 b. −4 is 4 units from 0 on a number line.
 c. 0 is 0 units from 0 on a number line.

5. Yes; for example, $5 > 3$ and $3 < 5$. Also, $-2 > -5$ and $-5 < -2$.

7. No, $-4 < -3$ but $|-4| > |-3|$.

9. No, $|-3| < |-4|$ but $-3 > -4$.

11. $|7| = 7$

13. $|-15| = 15$

15. $-|0| = 0$

17. $-|-5| = -(5) = -5$

19. $-|21| = -(21) = -21$

21. $6 > 2$; 6 is to the right of 2 on a number line.

23. $-4 < 0$; −4 is to the left of 0 on a number line.

25. $\dfrac{1}{2} > -\dfrac{2}{3}$; $\dfrac{1}{2}$ is to the right of $-\dfrac{2}{3}$ on a number line.

27. $0.7 < 0.8$; 0.7 is to the left of 0.8 on a number line.

29. $-\dfrac{1}{2} > -1$; $-\dfrac{1}{2}$ is to the right of -1 on a number line.

31. $-5 < 5$; -5 is to the left of 5 on a number line.

33. $-2.1 < -2$; -2.1 is to the left of -2 on a number line.

35. $\dfrac{4}{5} > -\dfrac{4}{5}$; $\dfrac{4}{5}$ is to the right of $-\dfrac{4}{5}$ on a number line.

37. $-\dfrac{3}{8} < \dfrac{3}{8}$; $-\dfrac{3}{8}$ is to the left of $\dfrac{3}{8}$ on a number line.

39. $0.49 > 0.43$; 0.49 is to the right of 0.43 on a number line.

41. $-0.086 > -0.095$; -0.086 is to the right of -0.095 on a number line.

43. $0.001 < 0.002$; 0.001 is to the left of 0.002 on a number line.

45. $\dfrac{5}{8} > 0.6$ because $\dfrac{5}{8} = 0.625$ and 0.625 is to the right of 0.6 on a number line.

47. $-\dfrac{4}{3} < -\dfrac{1}{3}$; $-\dfrac{4}{3}$ is to the left of $-\dfrac{1}{3}$ on a number line.

49. $-0.8 < -\dfrac{3}{5}$; -0.8 is to the left of -0.6 on a number line.

51. $0.3 < \dfrac{1}{3}$; 0.3 is to the left of $.333...$ on a number line.

53. $-\dfrac{17}{30} > -\dfrac{16}{20}$; $-\dfrac{34}{60}$ is to the right of $-\dfrac{48}{60}$ on a number line.

55. $-(-6) > -(-5)$; 6 is to the right of 5 on a number line.

57. $5 > |-2|$ since $|-2| = 2$

59. $\dfrac{3}{4} < |-4|$ since $|-4| = 4$

61. $|0| < |-4|$ since $|0| = 0$ and $|-4| = 4$

63. $4 < \left|-\dfrac{9}{2}\right|$ since $\left|-\dfrac{9}{2}\right| = \dfrac{9}{2}$ or $4\dfrac{1}{2}$

65. $\left|-\dfrac{4}{5}\right| < \left|-\dfrac{5}{4}\right|$ since $\left|-\dfrac{4}{5}\right| = \dfrac{4}{5} = \dfrac{16}{20}$ and $\left|-\dfrac{5}{4}\right| = \dfrac{5}{4} = \dfrac{25}{20}$

67. $|-4.6| = \left|-\dfrac{23}{5}\right|$ since $|-4.6| = 4.6$ and $\left|-\dfrac{23}{5}\right| = \dfrac{23}{5} = 4.6$

69. $\dfrac{2}{3} + \dfrac{2}{3} + \dfrac{2}{3} + \dfrac{2}{3} = 4 \cdot \dfrac{2}{3}$ since

$\dfrac{2}{3} + \dfrac{2}{3} + \dfrac{2}{3} + \dfrac{2}{3} = \dfrac{2+2+2+2}{3} = \dfrac{8}{3}$ and

$4 \cdot \dfrac{2}{3} = \dfrac{4}{1} \cdot \dfrac{2}{3} = \dfrac{8}{3}$

71. $\dfrac{1}{2} \cdot \dfrac{1}{2} < \dfrac{1}{2} \div \dfrac{1}{2}$ since $\dfrac{1}{2} \cdot \dfrac{1}{2} = \dfrac{1 \cdot 1}{2 \cdot 2} = \dfrac{1}{4}$ and

$\dfrac{1}{2} \div \dfrac{1}{2} = \dfrac{1}{2} \cdot \dfrac{2}{1} = \dfrac{1 \cdot 1}{1 \cdot 1} = 1$

73. $\dfrac{5}{8} - \dfrac{1}{2} < \dfrac{5}{8} \div \dfrac{1}{2}$ since $\dfrac{5}{8} - \dfrac{1}{2} = \dfrac{5}{8} - \dfrac{4}{8} = \dfrac{1}{8}$ and

$\dfrac{5}{8} \div \dfrac{1}{2} = \dfrac{5}{8} \cdot \dfrac{2}{1} = \dfrac{10}{8}$

75. $-|-1|, \dfrac{3}{7}, \dfrac{4}{9}, 0.46, |-5|$ because $-|-1| = -1$,

$\dfrac{3}{7} \approx 0.429$, $\dfrac{4}{9} = 0.444...$, and $|-5| = 5$.

77. $\dfrac{5}{12}, 0.6, \dfrac{2}{3}, \dfrac{19}{25}, |-2.6|$ because $\dfrac{5}{12} = 0.41666...$,

$\dfrac{2}{3} = 0.666...$, $\dfrac{19}{25} = 0.76$, and $|-2.6| = 2.6$.

79. 4 and -4 since $|4| = |-4| = 4$

For Exercises 81–87 answers will vary. One possible answer is given for each exercise.

81. There are no real numbers that are less than 4 and greater than 6.

83. Three numbers less than –2 and greater than –6 are –3, –4, –5.

85. Three numbers greater than –3 and greater than 3 are 4, 5, 6.

87. Three numbers greater than $|-2|$ and less than $|-6|$ are 3, 4, 5.

89. a. Between does not include endpoints.

    b. Three real numbers between 4 and 6 are 4.1, 5, and $5\frac{1}{2}$.

    c. No, 4 is an endpoint.

    d. Yes, 5 is greater than 4 and less than 6.

    e. True

91. a. dietary fiber and thiamin

    b. vitamin E, niacin, and riboflavin

93. The result of dividing a number by itself is 1. Thus, the result of dividing a number between 0 and 1 by itself is a number, 1, which is greater than the original number.

95. No, an absolute value of a number cannot be negative.

98. $2\frac{3}{5} + 3\frac{1}{3}$

$$2\frac{3}{5} = \frac{10+3}{5} = \frac{13}{5} \cdot \frac{3}{3} = \frac{39}{15}$$

$$3\frac{1}{3} = \frac{9+1}{3} = \frac{10}{3} \cdot \frac{5}{5} = \frac{50}{15}$$

$$2\frac{3}{5} + 3\frac{1}{3} = \frac{39}{15} + \frac{50}{15} = \frac{39+50}{15} = \frac{89}{15} = 5\frac{14}{15}$$

99. The set of integer numbers is $\{..., -3, -2, -1, 0, 1, 2, 3, ...\}$.

100. The set of whole numbers is $\{0, 1, 2, 3, ...\}$.

101. a. 5 is a natural number.

    b. 5 and 0 are whole numbers.

    c. 5, –2, and 0 are integers.

    d. $5, -2, 0, \frac{1}{3}, -\frac{5}{9}$, and 2.3 are rational numbers.

    e. $\sqrt{3}$ is an irrational number.

    f. $5, -2, 0, \frac{1}{3}, \sqrt{3}, -\frac{5}{9}$, and 2.3 are real numbers.

Mid-Chapter Test: 1.1-1.5

1. At least two hours of study and homework for each hour of class time is generally recommended.

2. a. The mean is
$$\frac{78.83 + 96.57 + 62.23 + 88.79 + 101.75 + 55.62}{6}$$
$$= \frac{483.78}{6} = \$80.63.$$

    b. To find the median place the numbers in order: 55.62, 62.23, 78.83, 88.79, 96.57, 101.75. Since there are an even amount of numbers, take the two in the middle and take their mean.
$$\frac{78.83 + 88.79}{2} = \frac{167.62}{2} = \$83.81.$$

3. New balance = Old balance + Deposits – Purchases
   = 652.70 + 230.75 – 3(19.62)
   = 652.70 + 230.75 – 58.86
   = 824.59
   Her new balance is $824.59.

4. a. Rental cost from Natwora's
   = 7.50(each 15-minute increment)
   = 7.5(16)
   = 120

   Rental cost for Gurney's
   =18(each 30-minute increment)
   =18(8)
   =144

   Natwora's is the better deal.

    b. 144–120 = 24
    You will save $24.

5. We must find out how many 1000 gallons was used. $\dfrac{33,700}{1000} = 33.7$

Water Bill = 1.85(number of 1000 gallons used)
$$= 1.85(33.7)$$
$$\approx 62.345$$

The water bill would be $62.35.

6. $\dfrac{3}{7} \cdot \dfrac{7}{18} = \dfrac{\cancel{3}^1}{\cancel{7}_1} \cdot \dfrac{\cancel{7}^1}{\cancel{18}_6} = \dfrac{1 \cdot 1}{1 \cdot 6} = \dfrac{1}{6}$

7. $\dfrac{9}{16} \div \dfrac{12}{13} = \dfrac{\cancel{9}^3}{16} \cdot \dfrac{13}{\cancel{12}_4} = \dfrac{3 \cdot 13}{16 \cdot 4} = \dfrac{39}{64}$

8. $\dfrac{5}{8} \cdot \dfrac{5}{5} = \dfrac{25}{40}$

$\dfrac{3}{5} \cdot \dfrac{8}{8} = \dfrac{24}{40}$

$\dfrac{5}{8} + \dfrac{3}{5} = \dfrac{25}{40} + \dfrac{24}{40} = \dfrac{25+24}{40} = \dfrac{49}{40} = 1\dfrac{9}{40}$

9. $6\dfrac{1}{4} = \dfrac{24+1}{4} = \dfrac{25}{4} \cdot \dfrac{5}{5} = \dfrac{125}{20}$

$3\dfrac{1}{5} = \dfrac{15+1}{5} = \dfrac{16}{5} \cdot \dfrac{4}{4} = \dfrac{64}{20}$

$6\dfrac{1}{4} - 3\dfrac{1}{5} = \dfrac{125}{20} - \dfrac{64}{20} = \dfrac{125-64}{20} = \dfrac{61}{20} = 3\dfrac{1}{20}$

10. $p = 2l + 2w = 2\left(14\dfrac{2}{3}\right) + 2\left(12\dfrac{1}{2}\right)$

$$= 2\left(\dfrac{44}{3}\right) + 2\left(\dfrac{25}{2}\right)$$

$$= \dfrac{88}{3} + \dfrac{50}{2}$$

$$= \left(\dfrac{88}{3} \cdot \dfrac{2}{2}\right) + \left(\dfrac{50}{2} \cdot \dfrac{3}{3}\right)$$

$$= \dfrac{176}{6} + \dfrac{150}{6}$$

$$= \dfrac{326}{6} = 54\dfrac{2}{6} = 54\dfrac{1}{3}$$

He will need $54\dfrac{1}{3}$ feet of fencing.

11. False

12. True

13. False

14. True

15. False

16. $-\left|-\dfrac{7}{10}\right| = -\dfrac{7}{10}$

17. $-0.005 > -0.006$ because -0.005 is to the right of -0.006 on the number line.

18. $\dfrac{7}{8} > \dfrac{5}{6}$ because $\dfrac{7}{8} = \dfrac{7}{8} \cdot \dfrac{3}{3} = \dfrac{21}{24}$ and $\dfrac{5}{6} = \dfrac{5}{6} \cdot \dfrac{4}{4} = \dfrac{20}{24}$.

19. $|-9| < |-12|$ because $|-9| = 9$ and $|-12| = 12$.

20. $\left|-\dfrac{3}{8}\right| = |-0.375|$ because $\left|-\dfrac{3}{8}\right| = \dfrac{3}{8} = 0.375$ and $|-0.375| = 0.375$.

Exercise Set 1.6

1. The 4 basic operations of arithmetic are addition, subtraction, multiplication, and division.

3. a. No; $-\dfrac{2}{3} + \dfrac{3}{2}$ does not sum to 0.

   b. The opposite of $-\dfrac{2}{3}$ is $\dfrac{2}{3}$ because

   $-\dfrac{2}{3} + \dfrac{2}{3} = 0$.

5. If we add two negative numbers, the sum is negative. When two numbers have the same sign, their sum also has the same sign.

7. Answers will vary.

9. a. He owed 162, a negative, and then paid a positive amount, 85, toward his debt. This would translate to $-162 + 85$.

   b. $-162 + 85$
   The numbers have different signs so find the difference between the absolute values.
   $|-162| - |85| = 162 - 85 = 77$
   $|-162|$ is greater so sum is negative.
   $-162 + 85 = -77$

   c. Since this is a negative amount, it is considered a debt.

11. Yes, it is correct.

11

13. The opposite of 9 is –9 since 9 + (–9) = 0.

15. The opposite of –28 is 28 since –28 + 28 = 0.

17. The opposite of 0 is 0 since 0 + 0 = 0.

19. The opposite of $\frac{5}{3}$ is $-\frac{5}{3}$ since $\frac{5}{3} + \left(-\frac{5}{3}\right) = 0$.

21. The opposite of $2\frac{3}{5}$ is $-2\frac{3}{5}$ since
$2\frac{3}{5} + \left(-2\frac{3}{5}\right) = 0$.

23. The opposite of 3.72 is –3.72 since
3.72 + (–3.72) = 0.

25. Numbers have same sign, so add absolute values.
$|5| + |6| = 5 + 6 = 11$
Numbers are positive so sum is positive.
5 + 6 = 11

27. Numbers have different signs so find difference between larger and smaller absolute values.
$|4| - |-3| = 4 - 3 = 1$. $|4|$ is greater than $|-3|$ so the sum is positive.
4 + (–3) = 1

29. Numbers have same sign, so add absolute values.
$|-4| + |-2| = 4 + 2 = 6$.
Numbers are negative, so sum is negative.
–4 + (–2) = –6

31. Numbers have different signs, so find difference between absolute values.
$|6| - |-6| = 6 - 6 = 0$
$6 + (-6) = 0$

33. Numbers have different signs, so find difference between absolute values.
$|-4| - |4| = 4 - 4 = 0$
–4 + 4 = 0

35. Numbers have same sign, so add absolute values. $|-8| + |-2| = 8 + 2 = 10$. Numbers are negative, so sum is negative. –8 + (–2) = –10

37. Numbers have different signs, so take difference between larger and smaller absolute values.
$|-7| - |3| = 7 - 3 = 4$. $|-7|$ is greater than $|3|$ so sum is negative. –7 + 3 = –4

39. Numbers have same sign, so add absolute values.
$|-8| + |-5| = 8 + 5 = 13$ $|-8| + |-5| = 8 + 5 = 13$
Numbers are negative, so sum is negative.
–8 + (–5) = –13

41. 0 + 0 = 0

43. –6 + 0 = –6

45. Numbers have different signs, so find difference between larger and smaller absolute values.
$|18| - |-9| = 18 - 9 = 9$. $|18|$ is greater than $|-9|$ so sum is positive. 18 + (–9) = 9

47. Numbers have same sign, so add absolute values.
$|-33| + |-31| = 33 + 31 = 64$. Numbers are negative, so sum is negative. –33 + (–31) = –64

49. Numbers have same sign, so add absolute values.
$|7| + |9| = 7 + 9 = 16$. Numbers are positive, so sum is positive. 7 + 9 = 16

51. Numbers have same sign, so add absolute values.
$|-8| + |-4| = 8 + 4 = 12$. Numbers are negative, so sum is negative. –8 + (–4) = –12

53. Numbers have different signs, so find difference between larger and smaller absolute values.
$|6| - |-3| = 6 - 3 = 3$. $|6|$ is greater than $|-3|$ so sum is positive. 6 + (–3) = 3

55. Numbers have different signs, so take difference between larger and smaller absolute values.
$|-19| + |13| = 19 - 13 = 6$. $|-19|$ is greater than $|13|$ so sum is negative. 13 + (–19) = –6

57. Numbers have different signs, so find difference between larger and smaller absolute values.
$|-200| - |180| = 200 - 180 = 20$. $|-200|$ is greater than $|180|$ so sum is negative.
180 + (–200) = –20

59. Numbers have same sign, so add absolute values.
$|-11| + |-20| = 11 + 20 = 31$. Numbers are negative, so sum is negative. –11 + (–20) = –31

61. Numbers have different signs, so find difference between larger and smaller absolute values.
$|-67| - |28| = 67 - 28 = 39$. $|-67|$ is greater than $|28|$ so sum is negative. –67 + 28 = –39

63. Numbers have different signs, so find difference between larger and smaller absolute values. $|184| - |-93| = 184 - 93 = 91$. $|184|$ is greater than $|-93|$ so sum is positive. $184 + (-93) = 91$

65. Numbers have different signs, so find difference between larger and smaller absolute values. $|-90.4| - |80.5| = 90.4 - 80.5 = 9.9$. $|-90.4|$ is greater than $|80.5|$ so sum is negative. $80.5 + (-90.4) = -9.9$

67. Numbers have same sign, so add absolute values. $|-124.7| + |-19.3| = 124.7 + 19.3 = 144.0$. Numbers are negative, so sum is negative. $-124.7 + (-19.3) = -144.0$

69. Numbers have same sign, so add absolute values. $|-123.56| + |-18.35| = 123.56 + 18.35 = 141.91$. Numbers are negative, so sum is negative. $-123.56 + (-18.35) = -141.91$

71. Numbers have different signs, so take difference between larger and smaller absolute values. $|-99.36| - |45.71| = 99.36 - 45.71 = 53.65$.
$|-99.36|$ is greater than $|45.71|$ so sum is negative.
$-99.36 + 45.71 = -53.65$

73. $\dfrac{3}{5} + \dfrac{1}{7} = \dfrac{21}{35} + \dfrac{5}{35} = \dfrac{21+5}{35} = \dfrac{26}{35}$

75. $\dfrac{5}{12} + \dfrac{6}{7} = \dfrac{35}{84} + \dfrac{72}{84} = \dfrac{35+72}{84} = \dfrac{107}{84}$ or $1\dfrac{23}{84}$

77. $-\dfrac{8}{11} + \dfrac{4}{5} = -\dfrac{40}{55} + \dfrac{44}{55}$
The numbers have different signs, so find the difference between the larger and smaller absolute values. $\left|\dfrac{44}{55}\right| - \left|-\dfrac{40}{55}\right| = \dfrac{44}{55} - \dfrac{40}{55} = \dfrac{4}{55}$.
Now, $\left|\dfrac{44}{55}\right|$ is greater than $\left|-\dfrac{40}{55}\right|$, so the sum is positive. $-\dfrac{8}{11} + \dfrac{4}{5} = -\dfrac{40}{55} + \dfrac{44}{55} = \dfrac{4}{55}$

79. $-\dfrac{7}{10} + \dfrac{11}{90} = -\dfrac{63}{90} + \dfrac{11}{90}$
The numbers have different signs, so find the difference between the larger and smaller absolute values. $\left|-\dfrac{63}{90}\right| - \left|\dfrac{11}{90}\right| = \dfrac{63}{90} - \dfrac{11}{90} = \dfrac{52}{90} = \dfrac{26}{45}$. Now, $\left|-\dfrac{63}{90}\right|$ is greater than $\left|\dfrac{11}{90}\right|$, so the sum is negative.
$-\dfrac{7}{10} + \dfrac{11}{90} = -\dfrac{63}{90} + \dfrac{11}{90} = -\dfrac{52}{90} = -\dfrac{26}{45}$

81. $-\dfrac{7}{30} + \left(-\dfrac{5}{6}\right) = -\dfrac{7}{30} + \left(-\dfrac{25}{30}\right)$
The numbers have the same signs, so add the absolute values.
$\left|-\dfrac{7}{30}\right| + \left|-\dfrac{25}{30}\right| = \dfrac{7}{30} + \dfrac{25}{30} = \dfrac{32}{30} = \dfrac{16}{15}$. The numbers are negative, so the sum is negative.
$-\dfrac{7}{30} + \left(-\dfrac{5}{6}\right) = -\dfrac{7}{30} + \left(-\dfrac{25}{30}\right) = -\dfrac{32}{30} = -\dfrac{16}{15}$ or $-1\dfrac{1}{15}$

83. $\dfrac{9}{25} + \left(-\dfrac{3}{50}\right) = \dfrac{18}{50} + \left(-\dfrac{3}{50}\right)$
The numbers have different signs, so find the difference between the larger and smaller absolute values. $\left|\dfrac{18}{50}\right| - \left|-\dfrac{3}{50}\right| = \dfrac{18}{50} - \dfrac{3}{50} = \dfrac{15}{50} = \dfrac{3}{10}$. Now, $\left|\dfrac{18}{50}\right|$ is greater than $\left|-\dfrac{3}{50}\right|$, so the sum is positive.
$\dfrac{9}{25} + \left(-\dfrac{3}{50}\right) = \dfrac{18}{50} + \left(-\dfrac{3}{50}\right) = \dfrac{15}{50} = \dfrac{3}{10}$

85. $-\dfrac{4}{5} + \left(-\dfrac{5}{75}\right) = -\dfrac{60}{75} + \left(-\dfrac{5}{75}\right)$
The numbers have the same signs, so add the absolute values.
$\left|-\dfrac{60}{75}\right| + \left|-\dfrac{5}{75}\right| = \dfrac{60}{75} + \dfrac{5}{75} = \dfrac{65}{75} = \dfrac{13}{15}$. The numbers are negative, so the sum is negative.
$-\dfrac{4}{5} + \left(-\dfrac{5}{75}\right) = -\dfrac{60}{75} + \left(-\dfrac{5}{75}\right) = -\dfrac{65}{75} = -\dfrac{13}{15}$

87. Numbers have different signs, so find difference between larger and smaller absolute values.

$$-\frac{9}{24}+\frac{5}{7}=-\frac{63}{168}+\frac{120}{168}=\left|\frac{120}{168}\right|-\left|-\frac{63}{168}\right|=$$

$$\frac{120}{168}-\frac{63}{168}=\frac{120-63}{168}=\frac{57}{168}=\frac{19}{56}$$

$\left|\frac{120}{168}\right|$ is greater than $\left|-\frac{63}{168}\right|$ so sum is positive.

$$-\frac{9}{24}+\frac{5}{7}=\frac{19}{56}$$

89. $-\dfrac{5}{12}+\left(-\dfrac{3}{10}\right)=-\dfrac{25}{60}+\left(-\dfrac{18}{60}\right)$

The numbers have the same signs, so add the

absolute values. $\left|-\dfrac{25}{60}\right|+\left|-\dfrac{18}{60}\right|=\dfrac{25}{60}+\dfrac{18}{60}=\dfrac{43}{60}$.

The numbers are negative, so the sum is

negative. $-\dfrac{5}{12}+\left(-\dfrac{3}{10}\right)=-\dfrac{25}{60}+\left(-\dfrac{18}{60}\right)=-\dfrac{43}{60}$

91. $-\dfrac{13}{14}+\left(-\dfrac{7}{42}\right)=-\dfrac{39}{42}+\left(-\dfrac{7}{42}\right)$

The numbers have the same signs, so add the absolute values.

$\left|-\dfrac{39}{42}\right|+\left|-\dfrac{7}{42}\right|=\dfrac{39}{42}+\dfrac{7}{42}=\dfrac{46}{42}=\dfrac{23}{21}$. The

numbers are negative, so the sum is negative.

$-\dfrac{13}{14}+\left(-\dfrac{7}{42}\right)=-\dfrac{39}{42}+\left(-\dfrac{7}{42}\right)=-\dfrac{46}{42}=-\dfrac{23}{21}$ or

$-1\dfrac{2}{21}$

93. a. Positive; $|587|$ is greater than $|-197|$ so sum will be positive.

b. $587 + (-197) = 390$

c. Yes; By part a) we expect a positive sum. The magnitude of the sum is the difference between the larger and smaller absolute values.

95. a. Negative; the sum of 2 negative numbers is always negative.

b. $-84+(-289)=-373$

c. Yes. The sum of 2 negative numbers should be (and is) a negative number with a larger absolute value.

97. a. Negative; $|-947|$ is greater than $|495|$ so sum will be negative.

b. $-947+495=-452$

c. Yes; by part a) we expect a negative sum. Magnitude of the sum is the difference between the larger and smaller absolute values.

99. a. Negative; the sum of 2 negative numbers is always negative.

b. $-496+(-804)=-1300$

c. Yes. The sum of 2 negative numbers should be (and is) a negative number with a larger absolute value.

101. a. Negative; $|-375|$ is greater than $|263|$ so sum will be negative.

b. $-375 + 263= -112$

c. Yes; by part a) we expect a negative sum. The magnitude of the sum is the difference between the larger and smaller absolute values.

103. a. Negative; the sum of 2 negative numbers is always negative.

b. $-1833+(-2047)=-3880$

c. Yes. The sum of 2 negative numbers should be (and is) a negative number with a larger absolute value.

105. a. Positive; $|3124|$ is greater than $|-2013|$ so sum will be positive.

b. $3124+(-2013)=1111$

c. Yes; by part a) we expect a positive sum. Magnitude of sum is difference between larger and smaller absolute values.

107. a. Negative; the sum of 2 negative numbers is always negative.

b. $-1025+(-1025)=-2050$

c. Yes. The sum of 2 negative numbers should be (and is) a negative number with a larger absolute value.

109. True; the sum of two negative numbers is always negative.

111. True; the sum of two positive numbers is always positive.

113. False; the sum has the sign of the number with the larger absolute value.

115. David's balance was –$94. His new balance can be found by adding. $-94 + (-183) = -277$
David owes the bank $277.

117. Total loss can be represented as $-18 + (-3)$. $|-18| + |-3| = 18 + 3 = 21$. The total loss in yardage is 21 yards.

119. The depth of the well can be found by adding $-27 + (-34) = -61$. The well is 61 feet deep.

121. The total height can be represented by 33,480 feet. The base can be represented by $-19,684$ feet. The height of the peak above sea level is $33,480 + (-19,684) = 13,796$ feet.

123. a. –12 million
The deficit is $12 million.
     b. 1999: –20 million; 2000; 8 million; 2001; 17 million
$-20 + 8 + 17 = 5$
From 1999-2001, there was a surplus of $5 million.

125. $(-4) + (-6) + (-12) = (-10) + (-12) = -22$

127. $29 + (-46) + 37 = (-17) + 37 = 20$

129. $(-12) + (-10) + 25 + (-3) = (-22) + 25 + (-3)$
$$= 3 + (-3)$$
$$= 0$$

131. $\dfrac{1}{2} + \left(-\dfrac{1}{3}\right) + \dfrac{1}{5} = \left(\dfrac{3}{6} - \dfrac{2}{6}\right) + \dfrac{1}{5}$
$$= \dfrac{1}{6} + \dfrac{1}{5}$$
$$= \dfrac{5}{30} + \dfrac{6}{30}$$
$$= \dfrac{11}{30}$$

133. $1 + 2 + 3 + \cdots + 10 = (1 + 10) + (2 + 9) + \cdots + (5 + 6)$
$$= 5(11)$$
$$= 55$$

135. $2\dfrac{3}{8} = \dfrac{16 + 3}{8} = \dfrac{19}{8}$

$\left(\dfrac{4}{7}\right)\left(2\dfrac{3}{8}\right) = \dfrac{\cancel{4}^{\,1}}{7} \cdot \dfrac{19}{\cancel{8}_{2}} = \dfrac{1 \cdot 19}{7 \cdot 2} = \dfrac{19}{14}$ or $1\dfrac{5}{14}$

136. $3 = \dfrac{3}{1} \cdot \dfrac{16}{16} = \dfrac{48}{16}$

$3 - \dfrac{5}{16} = \dfrac{48}{16} - \dfrac{5}{16} = \dfrac{48 - 5}{16} = \dfrac{43}{16}$ or $2\dfrac{11}{16}$

137. False, –0.25 is less than zero and not an integer.

138. $|-3| > 2$ since $|-3| = 3$

139. $8 < |-12|$ since $|-12| = 12$

Exercise Set 1.7

1. $2 - 7$

3. $☺ - ?$

5. a. To subtract b from a, add the opposite of b to a.
     b. $5 + (-14)$
     c. $5 + (-14) = -9$

7. a. $a - (-b) = a + b$
     b. $-4 - (-12) = -4 + 12$
     c. $-4 + 12 = 8$

9. a. $3 - (-6) + (-5) = 3 + 6 - 5$
     b. $3 + 6 - 5 = 9 - 5 = 4$

11. Yes it is correct.

13. $8 - (+2) = 8 + (-2) = 6$

15. $12 - 5 = 12 + (-5) = 7$

17. $8 - 9 = 8 + (-9) = -1$

19. $9 - (-3) = 9 + 3 = 12$

21. $-6 - 6 = -6 + (-6) = -12$

23. $0 - 7 = 0 + (-7) = -7$

25. $8 - 8 = 8 + (-8) = 0$

27. $-3 - 1 = -3 + (-1) = -4$

29. $-8 - (-5) = -8 + 5 = -3$

31. $6 - (-3) = 6 + 3 = 9$

33. $-9 - 11 = -9 + (-11) = -20$

35. $0 - (-9.8) = 0 + 9.8 = 9.8$

37. $-4.8 - (-5.1) = -4.8 + 5.1 = 0.3$

39. $14 - 7 = 14 + (-7) = 7$

41. $-8 - (-12) = -8 + 12 = 4$

43. $18 - (-4) = 18 + 4 = 22$

45. $-9 - 2 = -9 + (-2) = -11$

47. $-90.7 - 40.3 = -90.7 + (-40.3) = -131.0$

49. $-45 - 37 = -45 + (-37) = -82$

51. $70 - (-70) = 70 + 70 = 140$

53. $42.3 - 49.7 = 42.3 + (-49.7) = -7.4$

55. $-7.85 - (-3.92) = -7.85 + 3.92 = -3.93$

57. $4 - 15 = 4 + (-15) = -11$

59. $21 - 21 = 21 + (-21) = 0$

61. $-6.3 - (-12.4) = -6.3 + 12.4 = 6.1$

63. $10.3 - (-7.9) = 10.3 + 7.9 = 18.2$

65. $13 - 24 = 13 + (-24) = -11$

67. $-10.3 - 7.8 = -10.3 + (-7.8) = -18.1$

69.
$$\frac{5}{9} - \frac{3}{8} = \frac{5}{9} + \left(-\frac{3}{8}\right)$$
$$= \frac{40}{72} + \left(-\frac{27}{72}\right)$$
$$= \frac{40 + (-27)}{72}$$
$$= \frac{13}{72}$$

71.
$$\frac{8}{15} - \frac{7}{45} = \frac{8}{15} + \left(-\frac{7}{45}\right)$$
$$= \frac{24}{45} + \left(-\frac{7}{45}\right)$$
$$= \frac{24 + (-7)}{45}$$
$$= \frac{17}{45}$$

73.
$$-\frac{7}{10} - \frac{5}{12} = -\frac{7}{10} + \left(-\frac{5}{12}\right)$$
$$= -\frac{42}{60} + \left(-\frac{25}{60}\right)$$
$$= \frac{-42 + (-25)}{60}$$
$$= -\frac{67}{60}$$

75.
$$-\frac{4}{15} - \frac{3}{20} = -\frac{4}{15} + \left(-\frac{3}{20}\right)$$
$$= -\frac{16}{60} + \left(-\frac{9}{60}\right)$$
$$= \frac{-16 + (-9)}{60}$$
$$= -\frac{25}{60} = -\frac{5}{12}$$

77.
$$-\frac{7}{12} - \frac{5}{40} = -\frac{7}{12} + \left(-\frac{5}{40}\right)$$
$$= -\frac{70}{120} + \left(-\frac{15}{120}\right)$$
$$= \frac{-70 + (-15)}{120}$$
$$= -\frac{85}{120} = -\frac{17}{24}$$

79.
$$\frac{3}{8} - \frac{6}{48} = \frac{3}{8} + \left(-\frac{6}{48}\right)$$
$$= \frac{18}{48} + \left(-\frac{6}{48}\right)$$
$$= \frac{18 + (-6)}{48}$$
$$= \frac{12}{48} = \frac{1}{4}$$

81.
$$-\frac{4}{9} - \left(-\frac{3}{5}\right) = -\frac{4}{9} + \frac{3}{5}$$
$$= -\frac{20}{45} + \frac{27}{45}$$
$$= \frac{-20 + 27}{45}$$
$$= \frac{7}{45}$$

83.
$$\frac{3}{16} - \left(-\frac{5}{8}\right) = \frac{3}{16} + \frac{5}{8}$$
$$= \frac{3}{16} + \frac{10}{16}$$
$$= \frac{3 + 10}{16}$$
$$= \frac{13}{16}$$

85.
$$\frac{4}{7} - \frac{7}{9} = \frac{4}{7} + \left(-\frac{7}{9}\right)$$
$$= \frac{36}{63} + \left(-\frac{49}{63}\right)$$
$$= \frac{36 + (-49)}{63}$$
$$= -\frac{13}{63}$$

87.
$$-\frac{5}{12} - \left(-\frac{3}{10}\right) = -\frac{5}{12} + \frac{3}{10}$$
$$= -\frac{25}{60} + \frac{18}{60}$$
$$= \frac{-25 + 18}{60}$$
$$= -\frac{7}{60}$$

89. a. Positive; $378 - 279 = 378 + (-279)$
$|378|$ is greater than $|-279|$ so the sum will be positive.

b. $378 + (-279) = 99$

c. Yes; by part a) we expect a positive sum. The size of the sum is the difference between the absolute values of the 2 numbers.

91. a. Negative; $-482 - 137 = -482 + (-137)$
The sum of 2 negative numbers is always negative.

b. $-482 + (-137) = -619$

c. Yes. The sum of 2 negative numbers should be (and is) a negative number with a larger absolute value.

93. a. Positive; $843 - (-745) = 843 + 745$.
The sum of 2 positive numbers is always positive.

b. $843 + 745 = 1588$

c. Yes; by part a) we expect a positive answer. The size of the sum is the sum of the absolute values of the numbers.

95. a. Positive; $-408 - (-604) = -408 + 604$.
$|604|$ is greater than $|-408|$ so the sum will be positive.

b. $-408 + 604 = 196$

c. Yes; by part a) we expect a positive answer. The size of the answer is the difference between the larger and smaller absolute values.

97. a. Negative; $-1024 - (-576) = -1024 + 576$.
$|-1024|$ is greater than $|576|$ so the sum will be negative.

b. $-1024 + 576 = -448$

c. Yes; by part a) we expect a negative answer. The size of the answer is the difference between the larger and the smaller absolute values.

99. a. Positive; $165.7 - 49.6 = 165.7 + (-49.6)$.

$|165.7|$ is greater than $|-49.6|$ so the sum will be positive.

b. $165.7 + (-49.6) = 116.1$

c. Yes; by part a) we expect a negative answer. The size of the answer is the difference between the larger and the smaller absolute values.

101. a. Negative; $295 - 364 = 295 + (-364)$.

Since $|-364|$ is greater than $|295|$ the answer will be negative.

b. $295 + (-364) = -69$

c. Yes; by part a) we expect a negative answer. The size of the answer is the difference between the larger and the smaller absolute values.

103. a. Negative; $-1023 - 647 = -1023 + (-647)$. The sum of two negative numbers is always negative.

b. $-1023 + (-647) = -1670$

c. Yes. The sum of 2 negative numbers should be (and is) a negative number with a larger absolute value.

105. a. Zero; $-7.62 - (-7.62) = -7.62 + 7.62$. The sum of two opposite numbers is always zero.

b. $-7.62 + 7.62 = 0$

c. Yes; by part a) we expect zero.

107. $7 + 5 - (+8) = 7 + 5 + (-8) = 12 + (-8) = 4$

109. $-6 + (-6) + 6 = -12 + 6 = -6$

111. $-13 - (+5) + 3 = -13 + (-5) + 3 = -18 + 3 = -15$

113. $-9 - (-3) + 4 = -9 + 3 + 4 = -6 + 4 = -2$

115. $5 - (-9) + (-1) = 5 + 9 + (-1) = 14 + (-1) = 13$

117. $\begin{aligned}17 + (-8) - (+14) &= 17 + (-8) + (-14) \\ &= 9 + (-14) \\ &= -5\end{aligned}$

119. $-36 - 5 + 9 = -36 + (-5) + 9 = -41 + 9 = -32$

121. $-2 + 7 - 9 = -2 + 7 + (-9) = 5 + (-9) = -4$

123. $25 - 19 + 3 = 25 + (-19) + 3 = 6 + 3 = 9$

125. $\begin{aligned}-4 - 6 + 5 - 7 &= (-4) + (-6) + 5 + (-7) \\ &= -10 + 5 + (-7) \\ &= -5 + (-7) \\ &= -12\end{aligned}$

127. $\begin{aligned}17 + (-3) - 9 - (-7) &= 17 + (-3) + (-9) + 7 \\ &= 14 + (-9) + 7 \\ &= 5 + 7 \\ &= 12\end{aligned}$

129. $\begin{aligned}-9 + (-7) + (-5) - (-3) &= -9 + (-7) + (-5) + 3 \\ &= -16 + (-5) + 3 \\ &= -21 + 3 \\ &= -18\end{aligned}$

131. a. $300 - 343 = 300 + (-343) = -43$

They had 43 sweaters on back order.

b. $43 + 100 = 143$

They would need to order 143 sweaters.

133. $\begin{aligned}2\frac{1}{4} - \frac{3}{8} &= 2\frac{1}{4} + \left(-\frac{3}{8}\right) \\ &= \frac{9}{4} + \left(-\frac{3}{8}\right) \\ &= \frac{18}{8} + \left(-\frac{3}{8}\right) \\ &= \frac{18 + (-3)}{8} \\ &= \frac{15}{8} = 1\frac{7}{8}\end{aligned}$

After the second day $1\frac{7}{8}$ inches of water remains.

135. $44 - (-56) = 44 + 56 = 100$

Thus the temperature dropped 100°F.

137. a. $288 + (-7) = 281$

In 2006, his score was 281.

b. $-5 - (-2) = -5 + 2 = -3$

T. Clark had 3 less strokes than S. Cink.

139. $1-2+3-4+5-6+7-8+9-10$
$= (1-2)+(3-4)+(5-6)+(7-8)+(9-10)$
$= (-1)+(-1)+(-1)+(-1)+(-1)$
$= -5$

141. a. 8 units

b. $-3-(-11) = -3+11 = 8$

143. a. $3+2+2+1+1 = 9$
The ball travels 9 feet vertically.

b. $-3+2+(-2)+1+(-1) = -3$
The net distance is -3 feet.

144. The counting numbers are $\{1, 2, 3, \ldots\}$.

145. The set of rational numbers together with the set of irrational numbers forms the set of real numbers.

146. $|-3| > -5$ since $|-3| = 3$

147. $-|-9| < -|-5|$ since $-|-9| = -9$ and $-|-5| = -5$

148. $\dfrac{5}{6} - \dfrac{7}{8} = \dfrac{5}{6} + \left(-\dfrac{7}{8}\right)$

$\quad = \dfrac{20}{24} + \left(-\dfrac{21}{24}\right)$

$\quad = \dfrac{20+(-21)}{24}$

$\quad = -\dfrac{1}{24}$

Exercise Set 1.8

1. Like signs: product is positive. Unlike signs: product is negative.

3. When multiplying 3 or more real numbers, the product is positive if there is an even number of negative numbers and the product is negative if there is an odd number.

5. a. When a is a nonzero real number, $\dfrac{0}{a} = 0$.

b. When a is a nonzero real number, $\dfrac{a}{0}$ is undefined.

7. a. With $3-5$ you subtract, but with $3(-5)$ you multiply.

b. $3-5 = 3+(-5) = -2$
$3(-5) = -15$

9. a. With $x-y$ you subtract, but with $x(-y)$ you multiply.

b. $x-y = 5-(-2) = 5+2 = 7$

c. $x(-y) = 5[-(-2)] = 5\cdot 2 = 10$

d. $-x-y = -5-(-2) = -5+2 = -3$

11. The product $(8)(4)(-5)$ is negative since there is an odd number (1) of negatives.

13. The product $(-102)(-16)(24)(19)$ is positive since there is an even number of negatives.

15. The product $(-40)(-16)(30)(50)(-13)$ is negative since there is an odd number (3) of negatives.

17. Since the numbers have like signs, the product is positive. $(-5)(-4) = 20$

19. Since the number have unlike signs, the product is negative. $6(-3) = -18$

21. Since the numbers have like signs, the product is positive. $(-8)(-10) = 80$

23. Since the numbers have unlike signs, the product is negative. $-2.1(6) = -12.6$

25. Since the numbers have like signs, the product is positive. $6(7) = 42$

27. Since the numbers have unlike signs, the product is negative. $9(-9) = -81$

29. Since the numbers have like signs, the product is positive. $(-5)(-6) = 30$

31. Zero multiplied by any real number equals zero. $(-9(0)(-6) = 0(-6) = 0$

33. Since there is one negative number (an odd number), the product will be negative. $(21)(-1)(4) = (-21)(4) = -84$

35. Since there are three negative numbers (an odd number), the product will be negative. $-1(-3)(3)(-8) = 3(3)(-8) = 9(-8) = -72$

37. Since there are two negative numbers (an even number), the product will be positive.
$$(-4)(5)(-7)(1) = (-20)(-7)(1)$$
$$= (140)(1)$$
$$= 140$$

39. Zero multiplied by any real number equals zero.
$$(-1)(3)(0)(-7) = (-3)(0)(-7) = 0(-7) = 0$$

41. $\left(\dfrac{-1}{2}\right)\left(\dfrac{3}{5}\right) = \dfrac{(-1)(3)}{2 \cdot 5} = \dfrac{-3}{10} = -\dfrac{3}{10}$

43. $\left(\dfrac{-5}{9}\right)\left(\dfrac{-7}{15}\right) = \left(\dfrac{-\cancel{5}^{1}}{9}\right)\left(\dfrac{-7}{\cancel{15}_{3}}\right) = \dfrac{(-1)(-7)}{9 \cdot 3} = \dfrac{7}{27}$

45. $\left(\dfrac{6}{-3}\right)\left(\dfrac{4}{-2}\right) = (-2)(-2) = 4$

47. $\left(\dfrac{3}{4}\right)\left(\dfrac{-2}{15}\right) = \left(\dfrac{\cancel{3}^{1}}{{}_{2}\cancel{4}}\right)\left(\dfrac{-\cancel{2}^{1}}{\cancel{15}_{5}}\right)$
$$= \dfrac{(1)(-1)}{(2)(5)}$$
$$= \dfrac{-1}{10} = -\dfrac{1}{10}$$

49. Since the numbers have like signs, the quotient is positive. $\dfrac{42}{6} = 7$

51. Since the numbers have like signs, the quotient is positive. $-16 \div (-4) = \dfrac{-16}{-4} = 4$

53. Since the numbers have like signs, the quotient is positive. $\dfrac{-36}{-9} = 4$

55. Since the numbers have unlike signs, the quotient is negative. $\dfrac{36}{-2} = -18$

57. Since the numbers have like signs, the quotient is positive. $\dfrac{-19.8}{-2} = 9.9$

59. Since the numbers have unlike signs, the quotient is negative. $40/(-4) = \dfrac{40}{-4} = -10$

61. Since the numbers have unlike signs, the quotient is negative. $\dfrac{-66}{11} = -6$

63. Since the numbers have unlike signs, the quotient is negative. $\dfrac{48}{-6} = -8$

65. Since the numbers have like signs, the quotient is positive. $-64.8 \div (-4) = \dfrac{-64.8}{-4} = 16.2$

67. Zero divided by any nonzero number is zero.
$$\dfrac{0}{4} = 0$$

69. Since the numbers have unlike signs, the quotient is negative. $\dfrac{30.8}{-5.6} = -5.5$

71. Since the numbers have like signs, the quotient is positive. $\dfrac{-30}{-5} = 6$

73. $\dfrac{3}{12} \div \left(\dfrac{-5}{8}\right) = \dfrac{3}{12} \cdot \left(\dfrac{8}{-5}\right)$
$$= \dfrac{1}{{}_{1}\cancel{4}} \cdot \left(\dfrac{\cancel{8}^{2}}{-5}\right)$$
$$= \dfrac{1 \cdot 2}{1(-5)}$$
$$= \dfrac{2}{-5} = -\dfrac{2}{5}$$

75. $\dfrac{-5}{12} \div (-3) = \dfrac{-5}{12} \cdot \dfrac{1}{-3} = \dfrac{-5(1)}{12(-3)} = \dfrac{-5}{-36} = \dfrac{5}{36}$

77. $\dfrac{-15}{21} \div \left(\dfrac{-15}{21}\right) = \dfrac{-\cancel{15}}{{}_{1}\cancel{21}} \cdot \dfrac{\cancel{21}^{1}}{-\cancel{15}_{1}}$
$$= \dfrac{(-1)(1)}{(1)(-1)}$$
$$= \dfrac{-1}{-1}$$
$$= 1$$

79. $-12 \div \dfrac{5}{12} = \dfrac{-12}{1} \cdot \dfrac{12}{5}$

$= \dfrac{(-12)(12)}{(1)(5)}$

$= \dfrac{-144}{5}$

$= -\dfrac{144}{5}$

81. Since the numbers have unlike signs, the product is negative. $-4(8) = -32$

83. Since the numbers have like signs, the quotient is positive. $\dfrac{-100}{-5} = 20$

85. Since the numbers have unlike signs, the product is negative. $-7(2) = -14$

87. Since the numbers have unlike signs, the quotient is negative. $27.9 \div (-3) = \dfrac{27.9}{-3} = -9.3$

89. Since the numbers have unlike signs, the quotient is negative. $\dfrac{-100}{5} = -20$

91. Since the numbers have like signs, the quotient is positive. $\dfrac{-90}{-90} = 1$

93. Zero divided by any nonzero number is zero.
$0 \div 8.6 = \dfrac{0}{8.6} = 0$

95. Any nonzero number divided by zero is undefined. $\dfrac{5}{0}$ is undefined.

97. Zero divided by any nonzero number is zero.
$0 \div (-7) = \dfrac{0}{-7} = 0$

99. Any nonzero number divided by zero is undefined. $\dfrac{8}{0}$ is undefined.

101. a. Since the numbers have unlike signs, the product will be negative.
  b. $92(-38) = -3496$
  c. Yes; as expected the product is negative.

103. a. Since the numbers have unlike signs, the quotient will be negative.
  b. $-240 / 15 = \dfrac{-240}{15} = -16$
  c. Yes; as expected the quotient is negative.

105. a. Since the numbers have unlike signs, the quotient will be negative.
  b. $243 \div (-27) = \dfrac{243}{-27} = -9$
  c. Yes; as expected the quotient is negative.

107. a. Since the numbers have like signs, the product will be positive.
  b. $(-49)(-126) = 6174$
  c. Yes; as expected the product is positive.

109. a. The quotient will be zero; zero divided by any nonzero number is zero.
  b. $\dfrac{0}{5335} = 0$
  c. Yes; as expected the answer is zero.

111. a. Undefined; any nonzero number divided by 0 is undefined.
  b. $8.2 \div 0 = \dfrac{8.2}{0}$ is undefined.
  c. Yes; as expected the quotient is undefined.

113. a. Since the numbers have like signs, the quotient will be positive.
  b. $8 \div 2.5 = \dfrac{8}{2.5} = 3.2$
  c. Yes; as expected the quotient is positive.

115. a. Since there are two negative numbers (an even number), the product will be positive.
  b. $(-3.0)(4.2)(-18) = 226.8$
  c. Yes; as expected the product is positive.

117. False; the product of two numbers with like signs is a positive number

119. False; the quotient of two numbers with unlike signs is a negative number.

121. True; the quotient of two numbers with like signs is a positive number.

123. True

125. False; zero divided by 1 is zero.

127. True; any nonzero number divided by zero is undefined.

129. $3(-15) = -45$
The total loss was 45 yards.

131. a. $\dfrac{1}{5}(520) = \dfrac{520}{5} = 104$
She paid back \$104.

b. $-520 + 104 = -416$
Her new balance is $-\$416$.

133. Find out how much is left after giving the husbands each \$50.
$775.40 - (4 \cdot 50) = 775.40 - 200 = 575.40$
Now take the remainder and divide it by 4.
$\dfrac{575.40}{4} = 143.85$
Each woman receives \$143.85.

135. a. $5(-4) = -20$
Josue lost 20 points.

b. $100 - 20 = 80$
His test score is 80.

137. a. $220 - 50 = 170$
$60\% \text{ of } 170 = 0.6(170) = 102$
$75\% \text{ of } 170 = 0.75(170) = 127.5$
Target heart rate is 102 to 128 beats per minute.

b. Answers will vary.

139. $(-5)^3 = (-5)(-5)(-5) = 25(-5) = -125$

141. $1^{100} = 1$

143. The product $(-1)(-2)(-3)(-4)\cdots(-10)$ will be positive because there are an even number (10) of negative numbers.

146. $|-3.6| > |-2.7|$

147. $-\dfrac{7}{12} + \left(-\dfrac{1}{10}\right) = -\dfrac{35}{60} + \left(-\dfrac{6}{60}\right)$
$= \dfrac{-35 + (-6)}{60}$
$= -\dfrac{41}{60}$

148. $-20 - (-18) = -20 + 18 = -2$

149. $6 - 3 - 4 - 2 = 3 - 4 - 2 = -1 - 2 = -3$

150. $5 - (-2) + 3 - 7 = 5 + 2 + 3 - 7$
$= 7 + 3 - 7$
$- 10 - 7$
$= 3$

Exercise Set 1.9

1. In the expression a^b, a is the base and b is the exponent.

3. a. Every number has an understood exponent of 1.

b. In $5x^3 y^2 z$, 5 has exponent of 1, x has an exponent of 3, y has an exponent of 2, and z has an exponent of 1.

5. a. $y + y + y + y = 4y$

b. $y \cdot y \cdot y \cdot y = y^4$

7. The order of operations are parentheses, exponents, multiplication or division, then addition or subtraction.

9. No; $4 + 5 \times 2 = 4 + 10 = 14$, on a scientific calculator.

11. a. $15 - 10 \div 5 = 15 - 2 = 13$

b. $(15 - 10) \div 5 = 5 \div 5 = 1$

c. The keystrokes in b) are used since the fraction bar is a grouping symbol.

13. b. $\left[10 - (16 \div 4)\right]^2 - 6^3 = [10 - 4]^2 - 6^3$
$= 6^2 - 6^3$
$= 36 - 216$
$= -180$

15. b. When $x = 5$:
$$-4x^2 + 3x - 6 = -4(5)^2 + 3(5) - 6$$
$$= -4(25) + 3(5) - 6$$
$$= -100 + 15 - 6$$
$$= -85 - 6$$
$$= -91$$

17. $5^2 = 5 \cdot 5 = 25$

19. $1^7 = 1 \cdot 1 \cdot 1 \cdot 1 \cdot 1 \cdot 1 \cdot 1 = 1$

21. $-8^2 = -(8)(8) = -64$

23. $(-3)^2 = (-3)(-3) = 9$

25. $(-1)^3 = (-1)(-1)(-1) = -1$

27. $-10^2 = -(10)(10) = -100$

29. $(-9)^2 = (-9)(-9) = 81$

31. $3^3 = 3 \cdot 3 \cdot 3 = 27$

33. $(-4)^4 = (-4)(-4)(-4)(-4) = 256$

35. $-2^4 = -(2)(2)(2)(2) = -16$

37. $\left(\dfrac{3}{4}\right)^2 = \dfrac{3}{4} \cdot \dfrac{3}{4} = \dfrac{9}{16}$

39. $\left(-\dfrac{1}{2}\right)^5 = \left(-\dfrac{1}{2}\right)\left(-\dfrac{1}{2}\right)\left(-\dfrac{1}{2}\right)\left(-\dfrac{1}{2}\right)\left(-\dfrac{1}{2}\right) = -\dfrac{1}{32}$

41. $5^2 \cdot 3^2 = 5 \cdot 5 \cdot 3 \cdot 3 = 225$

43. $4^3 \cdot 3^2 = 4 \cdot 4 \cdot 4 \cdot 3 \cdot 3 = 576$

45. a. Positive; a positive number raised to any power is positive.

    b. $7^3 = 343$

    c. Yes; as expected the answer is positive.

47. a. Positive; a positive number raised to any power is positive.

    b. $6^4 = 1296$

    c. Yes; as expected the answer is positive.

49. a. Negative; a negative number raised to an odd power is negative.

    b. $(-3)^5 = -243$

    c. Yes; as expected the answer is negative.

51. a. Positive; a negative number raised to an even power is positive.

    b. $(-5)^4 = 625$

    c. Yes; as expected the answer is positive.

53. a. Negative; the opposite of a positive number is negative..

    b. $-(-9)^2 = -(-9)(-9) = -81$

    c. Yes; as expected the answer is negative.

55. a. Negative; $\left(\dfrac{3}{8}\right)^2$ is positive therefore,

    $-\left(\dfrac{3}{8}\right)^2$ is negative.

    b. $-\left(\dfrac{3}{8}\right)^2 = -0.140625$

    c. Yes; as expected the answer is negative.

57. $3 + 2 \cdot 6 = 3 + 12 = 15$

59. $6 - 6 + 8 = 0 + 8 = 8$

61. $-7 + 2 \cdot 6^2 - 8 = -7 + 2 \cdot 36 - 8$
$$= -7 + 72 - 8$$
$$= 65 - 8$$
$$= 57$$

63. $-3^3 + 27 = -27 + 27 = 0$

65. $(4-3) \cdot (5-1)^2 = (1) \cdot (4)^2 = 1 \cdot 16 = 16$

67. $3 \cdot 7 + 4 \cdot 2 = 21 + 8 = 29$

69. $5 - 2(7+5) = 5 - 2(12) = 5 - 24 = -19$

71. $-32 - 5(7-10)^2 = -32 - 5(-3)^2$
$$= -32 - 5(9)$$
$$= -32 - 45$$
$$= -77$$

73. $\dfrac{3}{4} + 2\left(\dfrac{1}{5}\right)^2 = \dfrac{3}{4} + 2\left(\dfrac{1}{25}\right)$

$\qquad = \dfrac{3}{4} + \dfrac{2}{25}$

$\qquad = \dfrac{75}{100} + \dfrac{8}{100}$

$\qquad = \dfrac{83}{100}$

75. $-4 + 3\left[-1 + \left(12 \div 2^2\right)\right] = -4 + 3\left[-1 + \left(12 \div 4\right)\right]$

$\qquad = -4 + 3\left[-1 + 3\right]$

$\qquad = -4 + 3\left[2\right]$

$\qquad = -4 + 6$

$\qquad = 2$

77. $\left(6 \div 3\right)^3 + 4^2 \div 8 = \left(2\right)^3 + 4^2 \div 8$

$\qquad = 8 + 16 \div 8$

$\qquad = 8 + 2$

$\qquad = 10$

79. $-7 - 48 \div 6 \cdot 2^2 + 5 = -7 - 48 \div 6 \cdot 4 + 5$

$\qquad = -7 - 8 \cdot 4 + 5$

$\qquad = -7 - 32 + 5$

$\qquad = -39 + 5$

$\qquad = -34$

81. $\left(9 \div 3\right) + 4\left(7 - 2\right)^2 = \left(9 \div 3\right) + 4\left(5\right)^2$

$\qquad = \left(9 \div 3\right) + 4\left(25\right)$

$\qquad = 3 + 100$

$\qquad = 103$

83. $\left[4 + \left(\left(5 - 2\right)^2 \div 3\right)^2\right]^2 = \left[4 + \left(\left(3\right)^2 \div 3\right)^2\right]^2$

$\qquad = \left[4 + \left(9 \div 3\right)^2\right]^2$

$\qquad = \left[4 + \left(3\right)^2\right]^2$

$\qquad = \left[4 + 9\right]^2$

$\qquad = \left(13\right)^2$

$\qquad = 169$

85. $\left(-3\right)^2 + 8 \div 2 = 9 + 8 \div 2$

$\qquad = 9 + 4$

$\qquad = 13$

87. $2\left[1.55 + 5\left(3.7\right)\right] - 3.35 = 2\left[1.55 + 18.5\right] - 3.35$

$\qquad = 2(20.05) - 3.35$

$\qquad = 40.1 - 3.35$

$\qquad = 36.75$

89. $\left(\dfrac{2}{5} + \dfrac{3}{8}\right) - \dfrac{3}{20} = \left(\dfrac{16}{40} + \dfrac{15}{40}\right) - \dfrac{3}{20}$

$\qquad = \dfrac{31}{40} - \dfrac{3}{20}$

$\qquad = \dfrac{31}{40} - \dfrac{6}{40}$

$\qquad = \dfrac{25}{40} = \dfrac{5}{8}$

91. $\dfrac{3}{4} - 4 \cdot \dfrac{5}{40} = \dfrac{3}{4} - \dfrac{4}{1} \cdot \dfrac{5}{40} = \dfrac{3}{4} - \dfrac{1}{2} = \dfrac{3}{4} - \dfrac{2}{4} = \dfrac{1}{4}$

93. $\dfrac{4}{5} + \dfrac{3}{4} \div \dfrac{1}{2} - \dfrac{2}{3} = \dfrac{4}{5} + \dfrac{3}{4} \cdot \dfrac{2}{1} - \dfrac{2}{3}$

$\qquad = \dfrac{4}{5} + \dfrac{3}{2} - \dfrac{2}{3}$

$\qquad = \dfrac{24}{30} + \dfrac{45}{30} - \dfrac{20}{30}$

$\qquad = \dfrac{49}{30}$

95. $\dfrac{-4 - \left[2\left(9 \div 3\right) - 5\right]}{6^2 - 3^2 \cdot 7} = \dfrac{-4 - \left[2\left(3\right) - 5\right]}{36 - 9 \cdot 7}$

$\qquad = \dfrac{-4 - \left[6 - 5\right]}{36 - 63}$

$\qquad = \dfrac{-4 - 1}{-27}$

$\qquad = \dfrac{-5}{-27} = \dfrac{5}{27}$

97. $\dfrac{-\left[4 - \left(6 - 12\right)^2\right]}{\left[\left(9 \div 3\right) + 4\right]^2 + 2^2} = \dfrac{-\left[4 - \left(-6\right)^2\right]}{\left(3 + 4\right)^2 + 4}$

$\qquad = \dfrac{-\left[4 - 36\right]}{7^2 + 4}$

$\qquad = \dfrac{-\left(-32\right)}{49 + 4}$

$\qquad = \dfrac{32}{53}$

99. $\left\{5-2\left[4-(6\div2)\right]^2\right\}^2 = \left\{5-2\left[4-3\right]^2\right\}^2$

$\qquad = \left\{5-2(1)^2\right\}^2$

$\qquad = \left\{5-2(1)\right\}^2$

$\qquad = \left\{5-2\right\}^2$

$\qquad = (3)^2$

$\qquad = 9$

101. $-\left\{4-\left[-3-(2-5)\right]^2\right\} = -\left\{4-\left[-3-(-3)\right]^2\right\}$

$\qquad = -\left\{4-\left[-3+3\right]^2\right\}$

$\qquad = -\left\{4-\left[0\right]^2\right\}$

$\qquad = -\left\{4-0\right\}$

$\qquad = -(4)$

$\qquad = -4$

103. $\left\{4-3\left[2-(9\div3)\right]^2\right\}^2 = \left\{4-3\left[2-3\right]^2\right\}^2$

$\qquad = \left\{4-3\left[-1\right]^2\right\}^2$

$\qquad = \left\{4-3\left[1\right]\right\}^2$

$\qquad = \left\{4-3\right\}^2$

$\qquad = 1^2$

$\qquad = 1$

105. Substitute 5 for x

a. $x^2 = 5^2 = 5\cdot5 = 25$

b. $-x^2 = -5^2 = -(5)(5) = -25$

c. $(-5)^2 = (-5)^2 = (-5)(-5) = 25$

107. Substitute -2 for x

a. $x^2 = (-2)^2 = (-2)(-2) = 4$

b. $-x^2 = -(-2)^2 = -(-2)(-2) = -(4) = -4$

c. $(-x)^2 = 2^2 = 2\cdot2 = 4$

109. Substitute 6 for x

a. $x^2 = 6^2 = 6\cdot6 = 36$

b. $-x^2 = -6^2 = -(6\cdot6) = -36$

c. $(-x)^2 = (-6)^2 = (-6)(-6) = 36$

111. Substitute $-\dfrac{1}{3}$ for x.

a. $x^2 = \left(-\dfrac{1}{3}\right)^2 = \left(-\dfrac{1}{3}\right)\left(-\dfrac{1}{3}\right) = \dfrac{1}{9}$

b. $-x^2 = -\left(-\dfrac{1}{3}\right)^2 = -\left(-\dfrac{1}{3}\right)\left(-\dfrac{1}{3}\right) = -\dfrac{1}{9}$

c. $(-x)^2 = \left(\dfrac{1}{3}\right)^2 = \left(\dfrac{1}{3}\right)\left(\dfrac{1}{3}\right) = \dfrac{1}{9}$

113. Substitute -2 for x in the expression.
$x+6 = -2+6 = 4$

115. Substitute 6 for z in the expression.
$-7z-3 = -7(6)-3 = -42-3 = -45$

117. Substitute -3 for a in the expression.
$a^2-6 = (-3)^2-6 = 9-6 = 3$

119. Substitute 2 for each p in the expression.
$3p^2-6p-4 = 3(2)^2-6(2)-4$

$\qquad = 3(4)-12-4$

$\qquad = 12-12-4$

$\qquad = 0-4$

$\qquad = -4$

121. Substitute -1 for each x in the expression.
$-4x^2-2x+1 = -4(-1)^2-2(-1)+1$

$\qquad = -4(1)-2(-1)+1$

$\qquad = -4+2+1$

$\qquad = -2+1$

$\qquad = -1$

123. Substitute $\dfrac{1}{2}$ for each x in the expression.

$-x^2-2x+5 = -(\tfrac{1}{2})^2-2(\tfrac{1}{2})+5$

$\qquad = -\dfrac{1}{4}-1+5$

$\qquad = -\dfrac{1}{4}-\dfrac{4}{4}+\dfrac{20}{4}$

$\qquad = -\dfrac{5}{4}+\dfrac{20}{4}$

$\qquad = \dfrac{15}{4} \text{ or } 3\dfrac{3}{4}$

125. Substitute 5 for each x in the expression.
$$4(3x+1)^2 - 6x = 4(3(5)+1)^2 - 6(5)$$
$$= 4(15+1)^2 - 30$$
$$= 4(16)^2 - 30$$
$$= 4(256) - 30$$
$$= 1024 - 30$$
$$= 994$$

127. Substitute -2 for r and -3 for s in the expression.
$$r^2 - s^2 = (-2)^2 - (-3)^2 = 4 - 9 = -5$$

129. Substitute 1 for x and -5 for y in the expression.
$$5(x-6y) + 3x - 7y = 5(1 - 6(-5)) + 3(1) - 7(-5)$$
$$= 5(1 - (-30)) + 3 + 35$$
$$= 5(31) + 3 + 35$$
$$= 155 + 3 + 35$$
$$= 158 + 35$$
$$= 193$$

131. Substitute -1 for x and -2 for y in the expression.
$$3(x-4)^2 - (3y-4)^2 = 3(-1-4)^2 - [3(-2)-4]^2$$
$$= 3(-5)^2 - (-6-4)^2$$
$$= 3(-5)^2 - (-10)^2$$
$$= 3(25) - 100$$
$$= 75 - 100$$
$$= -25$$

133. $6 \cdot 3$ Multiply 6 by 3

$(6 \cdot 3) - 4$ Subtract 4 from the product

$[(6 \cdot 3) - 4] - 2$ Subtract 2 from the difference

Evaluate:
$$[(6 \cdot 3) - 4] - 2 = [18 - 4] - 2 = 14 - 2 = 12$$

135. $10 \cdot 4$ Multiply 10 by 4

$(10 \cdot 4) + 9$ Add 9 to the product

$[(10 \cdot 4) + 9] - 6$ Subtract 6 from the sum

$\{[(10 \cdot 4) + 9] - 6\} \div 7$ Divide the difference by 7

Evaluate:
$$\{[(10 \cdot 4) + 9] - 6\} \div 7 = \{[40 + 9] - 6\} \div 7$$
$$= \{49 - 6\} \div 7$$
$$= 43 \div 7$$
$$= \frac{43}{7} \text{ or } 6\frac{1}{7}$$

137. $\dfrac{4}{5} + \dfrac{3}{7}$ Add $\dfrac{4}{5}$ to $\dfrac{3}{7}$

$\left(\dfrac{4}{5} + \dfrac{3}{7}\right) \cdot \dfrac{2}{3}$ Multiply the sum by $\dfrac{2}{3}$

Evaluate:
$$\left(\frac{4}{5} + \frac{3}{7}\right)\frac{2}{3} = \left(\frac{28}{35} + \frac{15}{35}\right) \cdot \frac{2}{3} = \left(\frac{43}{35}\right)\left(\frac{2}{3}\right) = \frac{86}{105}$$

139. $-\left(x^2\right) = -x^2$ is true for all real numbers.

141. When $t = 2.5$, $65t = 65(2.5) = 162.5$
The car travels 162.5 miles.

143. When $t = 2$,
$$-16t^2 + 48t + 70 = -16(2)^2 + 48(2) + 70$$
$$= -16(4) + 48(2) + 70$$
$$= -64 + 96 + 70$$
$$= 32 + 70$$
$$= 102$$
After 2 seconds the height will be 102 feet.

145. **a.** $2 \div 5^2 = 2 \div 25 = 0.08$

     b. $(2 \div 5)^2 = (0.4)^2 = 0.16$

147. When $R = 2$ and $T = 70$,
$$0.2R^2 + 0.003RT + 0.0001T^2$$
$$= 0.2(2)^2 + 0.003(2)(70) + 0.0001(70)^2$$
$$= 0.2(4) + 0.003(2)(70) + 0.0001(4900)$$
$$= 0.8 + 0.42 + 0.49$$
$$= 1.71$$
The growth is 1.71 inches.

149. $12 - (4 - 6) + 10 = 24$

155. **a.** There are 3 houses with 2 dogs.

     b.

Dogs	Number of Houses
0	4
1	5
2	3
3	1
4	1

c. $4(1) + 3(1) + 2(3) + 1(5) + 0(4)$
$= 4 + 3 + 6 + 5 + 0$
$= 7 + 6 + 5$
$= 13 + 5$
$= 18$
There are 18 dogs in all.

d. Number of dogs = 18
$$\text{mean} = \frac{\text{number of dogs}}{\text{number of houses}} = \frac{18}{14} \approx 1.29$$
There is a mean of 1.29 dogs per house.

156. $3 = \frac{6}{2} = \frac{1}{2} + \frac{5}{2} = \frac{1}{2} + \frac{20}{8}$
Cost $= \$2.40 + 20(0.20) = \$2.40 + \$4.00 = \$6.40$

157. $-\frac{7}{12} + \frac{4}{9} = \frac{-21}{36} + \frac{16}{36} = \frac{-21+16}{36} = -\frac{5}{36}$

158. $\left(\dfrac{-5}{7}\right) \div \left(\dfrac{-3}{14}\right) = \left(\dfrac{-5}{\,_1 7}\right) \cdot \left(\dfrac{\cancel{14}^{\,2}}{-3}\right)$

$\qquad = \dfrac{-5}{1} \cdot \dfrac{2}{-3}$

$\qquad = \dfrac{(-5)(2)}{(1)(-3)}$

$\qquad = \dfrac{10}{3} \text{ or } 3\dfrac{1}{3}$

Exercise Set 1.10

1. The commutative property of addition states that the sum of two numbers is the same regardless of the order in which they are added. One possible example is $3 + 4 = 4 + 3$.

3. The associative property of addition states that the sum of 3 numbers is the same regardless of the way the numbers are grouped. One possible example is $(2+3) + 4 = 2 + (3+4)$.

5. a. In $x + (y + z)$ the sum of y and z is added to x whereas in $x(y+z)$, x is multiplied by the sum.

b. When $x = 4, y = 5$, and $z = 6$,
$x + (y + z) = 4 + (5 + 6) = 4 + 11 = 15$.

c. When $x = 4, y = 5$, and $z = 6$,
$x(y + z) = 4(5 + 6) = 4(11) = 44$.

7. The associative property involves changing parentheses with one operation whereas the distributive property involves distributing a multiplication over an addition.

9. 0

11. a. -6 **b.** $\dfrac{1}{6}$

13. a. 3 **b.** $-\dfrac{1}{3}$

15. a. $-x$ **b.** $\dfrac{1}{x}$

17. a. -1.6 **b.** $\dfrac{1}{1.6}$ or 0.625

19. a. $-\dfrac{1}{5}$ **b.** 5

21. a. $\dfrac{5}{6}$ **b.** $-\dfrac{6}{5}$

23. Distributive property

25. Associative property of addition

27. Commutative property of multiplication

29. Associative property of multiplication

31. Distributive property

33. Identity property for multiplication

35. Inverse property for multiplication

37. $1 + (-4)$

39. $(-6 \cdot 4) \cdot 2$

41. $-2x - 2y$

43. $y \cdot x$

45. $3y + 4x$

47. $a + (b + 3)$

49. $3x + (4 + 6)$

51. $(m + n)3$

53. $4x + 4y + 12$

55. 0

57. $\dfrac{5}{2}n$

59. Yes; the order does not affect the outcome so the process is commutative.

61. Yes; the order does not affect the outcome so the process is commutative.

63. No; the order affects the outcome, so the process is not commutative.

65. Yes; the outcome is not affected by whether you do the first two items first or the last two first, so the process is associative.

67. No; the outcome is affected by whether you do the first two items first or the last two first, so the process is not associative.

69. No; the outcome is affected by whether you do the first two items first or the last two first, so the process is not associative.

71. In $(3+4)+x = x+(3+4)$ the $(3+4)$ is treated as one value.

73. This illustrates the commutative property of addition because the change is $3 + 5 = 5 + 3$.

75. No; it illustrates the associative property of addition since the grouping is changed.

77. $2\dfrac{3}{5}+\dfrac{2}{3}$

$2\dfrac{3}{5}=\dfrac{13}{5}=\dfrac{3}{3}\cdot\dfrac{13}{5}=\dfrac{39}{15}$

$\dfrac{2}{3}=\dfrac{2}{3}\cdot\dfrac{5}{5}=\dfrac{10}{15}$

$2\dfrac{3}{5}+\dfrac{2}{3}=\dfrac{39}{15}+\dfrac{10}{15}=\dfrac{49}{15}$ or $3\dfrac{4}{15}$

78. $3\dfrac{5}{8}-2\dfrac{3}{16}$

$3\dfrac{5}{8}=\dfrac{29}{8}=\dfrac{2}{2}\cdot\dfrac{29}{8}=\dfrac{58}{16}$

$2\dfrac{3}{16}=\dfrac{35}{16}$

$3\dfrac{5}{8}-2\dfrac{3}{16}=\dfrac{58}{16}-\dfrac{35}{16}=\dfrac{23}{16}$ or $1\dfrac{7}{16}$

79. $102.7+(-113.9)=-11.2$

80. $-\dfrac{7}{8}$

Chapter 1 Review Exercises

1. $13(12)-(32+29+36+31)=156-128$
$$=28$$
He had 28 doughnuts left over.

2. $1.05\big[1.05(500.00)\big]=1.05\big[525\big]=551.25$
In 2 years the goods will cost $551.25.

3. **a.** $899.99(.0825)=74.25$
The sales tax is $74.25.

   b. total cost = cost of laptop + sales tax
   total cost $= 899.99+74.25$
   $$=974.24$$
   The total cost of the laptop is $974.24.

4. $\big[30+12(25)\big]-300=\big[30+300\big]-300$
$$=330-300$$
$$=30$$
She can save $30.

5. **a.** mean $=\dfrac{75+79+86+88+64}{5}=\dfrac{392}{5}=78.4$
The mean grade is 78.4.

   b. 64, 75, 79, 86, 88
   The middle number is 79. The median grade is 79.

6. **a.** mean $=\dfrac{21+3+17+10+9+6}{6}=\dfrac{66}{6}=11$
The mean is 11.

   b. 3, 6, 9, 10, 17, 21
   The middle numbers are 9 and 10. Their average is $\dfrac{9+10}{2}=\dfrac{19}{2}=9.5$. The median is 9.5.

7. **a.** 30 minutes

   b. 27 minutes

28

8. a. $1.6(0.637) = 1.0192$

In 1967, 1.0192 million freshman applied to two or fewer colleges.

b. $2.1(0.247) = 0.5187$

In 2003, 0.5187 million freshman applied to six or more colleges.

9. $\dfrac{3}{5} \cdot \dfrac{5}{6} = \dfrac{\cancel{3}^{1}}{\cancel{5}_{1}} \cdot \dfrac{\cancel{5}^{1}}{\cancel{6}_{2}} = \dfrac{1 \cdot 1}{1 \cdot 2} = \dfrac{1}{2}$

10. $3\dfrac{5}{7} + 2\dfrac{1}{3} = \dfrac{26}{7} + \dfrac{7}{3}$

$= \dfrac{26}{7} \cdot \dfrac{3}{3} + \dfrac{7}{3} \cdot \dfrac{7}{7}$

$= \dfrac{78}{21} + \dfrac{49}{21}$

$= \dfrac{78 + 49}{21}$

$= \dfrac{127}{21}$ or $6\dfrac{1}{21}$

11. $\dfrac{5}{12} \div \dfrac{3}{5} = \dfrac{5}{12} \cdot \dfrac{5}{3} = \dfrac{5 \cdot 5}{12 \cdot 3} = \dfrac{25}{36}$

12. $\dfrac{5}{6} + \dfrac{1}{3} = \dfrac{5}{6} + \dfrac{1}{3} \cdot \dfrac{2}{2} = \dfrac{5}{6} + \dfrac{2}{6} = \dfrac{7}{6}$ or $1\dfrac{1}{6}$

13. $3\dfrac{1}{6} - 1\dfrac{1}{4} = \dfrac{19}{6} - \dfrac{5}{4}$

$= \dfrac{19}{6} \cdot \dfrac{2}{2} - \dfrac{5}{4} \cdot \dfrac{3}{3}$

$= \dfrac{38}{12} - \dfrac{15}{12}$

$= \dfrac{38 - 15}{12}$

$= \dfrac{23}{12}$ or $1\dfrac{11}{12}$

14. $7\dfrac{3}{8} \div \dfrac{5}{12} = 7\dfrac{3}{8} \cdot \dfrac{12}{5}$

$= \dfrac{59}{\cancel{8}_{2}} \cdot \dfrac{\cancel{12}^{3}}{5}$

$= \dfrac{59 \cdot 3}{2 \cdot 5}$

$= \dfrac{177}{10}$ or $17\dfrac{7}{10}$

15. The natural numbers are $\{1, 2, 3, \ldots\}$.

16. The whole numbers are $\{0, 1, 2, 3, \ldots\}$.

17. The integers are $\{\ldots, -3, -2, -1, 0, 1, 2, \ldots\}$.

18. The set of rational numbers is the set of all numbers which can be expressed as the quotient of two integers, denominator not zero.

19. a. 3 and 426 are positive integers.

b. 3, 0, and 426 are whole numbers.

c. 3, −5, −12, 0, and 426 are integers.

d. 3, −5, −12, 0, $\dfrac{1}{2}$, −0.62, 426, and $-3\dfrac{1}{4}$ are rational numbers.

e. $\sqrt{7}$ is an irrational number.

f. 3, −5, −12, 0, $\dfrac{1}{2}$, −0.62, $\sqrt{7}$, 426, and $-3\dfrac{1}{4}$ are real numbers.

20. a. 1 is a natural number.

b. 1 is a whole number.

c. −8 and −9 are negative numbers.

d. −8, −9, and 1 are integers.

e. −2.3, −8, −9, $1\dfrac{1}{2}$, 1, and $-\dfrac{3}{17}$ are rational numbers.

f. −2.3, −8, −9, $1\dfrac{1}{2}$, $\sqrt{2}$, $-\sqrt{2}$, 1, and $-\dfrac{3}{17}$ are real numbers.

21. $-7 < -5$; −7 is to the left of −5 on a number line.

22. $-2.6 > -3.6$; −2.6 is to the right of −3.6 on a number line.

23. $0.50 < 0.509$; 0.50 is to the left 0.509 on a number line.

24. $4.6 > 4.06$; 4.6 is to the right of 4.06 on a number line.

25. $-6.3 < -6.03$; −6.3 is to the left of −6.03 on a number line.

26. $5 > |-3|$ since $|-3|$ equals 3.

27. $\left|-\dfrac{9}{2}\right| = |-4.5|$ since $\left|-\dfrac{9}{2}\right| = |-4.5| = 4.5$.

28. $|-10| > |-7|$ since $|-10|$ equals 10 and $|-7|$ equals 7.

29. $-9 + (-5) = -14$

30. $-6 + 6 = 0$

31. $0 + (-3) = -3$

32. $-10 + 4 = -6$

33. $-8 - (-2) = -8 + 2 = -6$

34. $-2 - (-4) = -2 + 4 = 2$

35. $4 - (-4) = 4 + 4 = 8$

36. $12 - 12 = 12 + (-12) = 0$

37. $2 - 7 = 2 + (-7) = -5$

38. $7 - (-7) = 7 + 7 = 14$

39. $0 - (-4) = 0 + 4 = 4$

40. $-7 - 5 = -7 + (-5) = -12$

41. $\dfrac{4}{3} - \dfrac{3}{4} = \dfrac{16}{12} - \dfrac{9}{12} = \dfrac{16 - 9}{12} = \dfrac{7}{12}$

42. $\dfrac{1}{2} + \dfrac{3}{5} = \dfrac{5}{10} + \dfrac{6}{10} = \dfrac{5 + 6}{10} = \dfrac{11}{10}$

43. $\dfrac{5}{9} - \dfrac{3}{4} = \dfrac{20}{36} - \dfrac{27}{36} = \dfrac{20 - 27}{36} = -\dfrac{7}{36}$

44. $-\dfrac{5}{7} + \dfrac{3}{8} = -\dfrac{40}{56} + \dfrac{21}{56} = \dfrac{-40 + 21}{56} = -\dfrac{19}{56}$

45. $-\dfrac{5}{12} - \dfrac{5}{6} = -\dfrac{5}{12} - \dfrac{10}{12} = \dfrac{-5 - 10}{12} = -\dfrac{15}{12} = -\dfrac{5}{4}$

46. $-\dfrac{6}{7} + \dfrac{5}{12} = -\dfrac{72}{84} + \dfrac{35}{84} = \dfrac{-72 + 35}{84} = -\dfrac{37}{84}$

47. $\dfrac{2}{9} - \dfrac{3}{10} = \dfrac{20}{90} - \dfrac{27}{90} = \dfrac{20 - 27}{90} = -\dfrac{7}{90}$

48. $\dfrac{5}{12} - \left(-\dfrac{3}{5}\right) = \dfrac{25}{60} + \dfrac{36}{60} = \dfrac{25 + 36}{60} = \dfrac{61}{60}$ or $1\dfrac{1}{60}$

49. $9 - 4 + 3 = 5 + 3 = 8$

50. $-8 - 9 + 4 = -17 + 4 = -13$

51. $-5 - 4 - 3 = -9 - 3 = -12$

52. $-2 + (-3) - 2 = -5 - 2 = -7$

53. $7 - (+4) - (-3) = 7 - 4 + 3 = 3 + 3 = 6$

54. $6 - (-2) + 3 = 6 + 2 + 3 = 8 + 3 = 11$

55. Since the numbers have unlike signs, the product is negative; $7(-9) = -63$

56. Since the numbers have like signs, the product is positive; $(-8.2)(-3.1) = 25.42$

57. Since there are an odd number (3) of negatives the product is negative;
$(-4)(-5)(-6) = (20)(-6) = -120$

58. $\left(\dfrac{3}{5}\right)\left(\dfrac{-2}{7}\right) = \dfrac{3(-2)}{5 \cdot 7} = \dfrac{-6}{35} = -\dfrac{6}{35}$

59. $\left(\dfrac{10}{11}\right)\left(\dfrac{3}{-5}\right) = \dfrac{2}{11} \cdot \dfrac{3}{-1} = \dfrac{2 \cdot 3}{(11)(-1)} = \dfrac{6}{-11} = -\dfrac{6}{11}$

60. $\left(\dfrac{-5}{8}\right)\left(\dfrac{-3}{7}\right) = \dfrac{(-5)(-3)}{8 \cdot 7} = \dfrac{15}{56}$

61. Zero multiplied by any real number is zero.
$$0 \cdot \dfrac{4}{9} = 0$$

62. Since there are four negative numbers (an even number), the product is positive.
$$\begin{aligned}(-4)(-6)(-2)(-3) &= (24)(-2)(-3) \\ &= (-48)(-3) \\ &= 144\end{aligned}$$

63. Since the numbers have unlike signs, the quotient is negative; $15 \div (-3) = \dfrac{15}{-3} = -5$

64. Since the numbers have unlike signs, the quotient is negative.
$$12 \div (-2) = \dfrac{12}{-2} = -6$$

65. Since the numbers have unlike signs, the quotient is negative;
$$-14.72 \div 4.6 = \dfrac{-14.72}{4.6} = -3.2$$

66. Since the numbers have like signs, the quotient is positive: $-37.41 \div (-8.7) = 4.3$

67. Since the numbers have like signs, the quotient is positive: $-88 \div (-11) = 8$

68. $-4 \div \left(\dfrac{-4}{9}\right) = \dfrac{-4}{1} \cdot \dfrac{9}{-4} = \dfrac{-1}{1} \cdot \dfrac{9}{-1} = \dfrac{-9}{-1} = 9$

69. $\dfrac{28}{-3} \div \left(\dfrac{9}{-2}\right) = \left(\dfrac{28}{-3}\right) \cdot \left(\dfrac{-2}{9}\right) = \dfrac{-56}{-27} = \dfrac{56}{27}$

70. $\dfrac{14}{3} \div \left(\dfrac{-6}{5}\right) = \dfrac{\overset{7}{\cancel{14}}}{3} \cdot \left(\dfrac{5}{\underset{3}{\cancel{-6}}}\right) = \dfrac{7(5)}{3(-3)} = -\dfrac{35}{9}$

71. Zero divided by any nonzero number is zero;
$0 \div 5 = \dfrac{0}{5} = 0$

72. Zero divided by any nonzero number is zero;
$0 \div (-6) = \dfrac{0}{-6} = 0$

73. Any real number divided by zero is undefined;
$-12 \div 0 = \dfrac{-12}{0}$ is undefined.

74. Any real number divided by zero is undefined;
$-4 \div 0 = \dfrac{-4}{0}$ is undefined.

75. Any real number divided by zero is undefined;
$\dfrac{8.3}{0}$ is undefined

76. Zero divided by any nonzero number is zero;
$\dfrac{0}{-9.8} = 0$

77. $-5(3 - 8) = -5(-5) = 25$

78. $2(4 - 8) = 2(-4) = -8$

79. $(3 - 6) + 4 = -3 + 4 = 1$

80. $(-4 + 3) - (2 - 6) = (-1) - (-4) = -1 + 4 = 3$

81. $[6 + 3(-2)] - 6 = [6 + (-6)] - 6 = 0 - 6 = -6$

82. $(-5 - 3)(4) = [-5 + (-3)](4) = (-8)(4) = -32$

83. $[12 + (-4)] + (6 - 8) = 8 + (-2) = 6$

84. $9[3 + (-4)] + 5 = 9(-1) + 5 = -9 + 5 = -4$

85. $-4(-3) + [4 \div (-2)] = (12) + (-2) = 10$

86. $(-3 \cdot 4) \div (-2 \cdot 6) = -12 \div (-12) = 1$

87. $(-3)(-4) + 6 - 3 = 12 + 6 - 3 = 18 - 3 = 15$

88. $[-2(3) + 6] - 4 = [-6 + 6] - 4 = 0 - 4 = -4$

89. $-6^2 = -(6)(6) = -36$

90. $(-6)^2 = (-6)(-6) = 36$

91. $2^4 = (2)(2)(2)(2) = 16$

92. $(-3)^3 = (-3)(-3)(-3) = -27$

93. $(-1)^9 = (-1)(-1)(-1)(-1)(-1)(-1)(-1)(-1)(-1)$
$= -1$

94. $(-2)^5 = (-2)(-2)(-2)(-2)(-2) = -32$

95. $\left(\dfrac{-4}{5}\right)^2 = \left(\dfrac{-4}{5}\right)\left(\dfrac{-4}{5}\right) = \dfrac{16}{25}$

96. $\left(\dfrac{2}{5}\right)^3 = \left(\dfrac{2}{5}\right)\left(\dfrac{2}{5}\right)\left(\dfrac{2}{5}\right) = \dfrac{8}{125}$

97. $5^3 \cdot (-2)^2 = (5)(5)(5)(-2)(-2) = 500$

98. $(-2)^4 \left(\dfrac{1}{2}\right)^2 = (-2)(-2)(-2)(-2)\left(\dfrac{1}{2}\right)\left(\dfrac{1}{2}\right) = 4$

99. $\left(-\dfrac{2}{3}\right)^2 \cdot 3^3 = \left(-\dfrac{2}{3}\right)\left(-\dfrac{2}{3}\right)(3)(3)(3) = 12$

100. $(-4)^3 (-2)^2 = (-4)(-4)(-4)(-2)(-2) = -256$

101. $-5 + 3 \cdot 4 = -5 + 12 = 7$

102. $4 \cdot 6 + 4 \cdot 2 = 24 + 8 = 32$

103. $(3.7 - 4.1)^2 + 6.2 = (-0.4)^2 + 6.2$
$\qquad\qquad\qquad\quad\; = 0.16 + 6.2$
$\qquad\qquad\qquad\quad\; = 6.36$

104. $10 - 36 \div 4 \cdot 3 = 10 - 9 \cdot 3 = 10 - 27 = -17$

105. $6 - 3^2 \cdot 5 = 6 - 9 \cdot 5 = 6 - 45 = -39$

106. $\left[6.9 - (3 \cdot 5)\right] + 5.8 = \left[6.9 - 15\right] + 5.8$
$\qquad\qquad\qquad\qquad\quad\; = -8.1 + 5.8$
$\qquad\qquad\qquad\qquad\quad\; = -2.3$

107. $\dfrac{6^2 - 4 \cdot 3^2}{-\left[6 - (3 - 4)\right]} = \dfrac{36 - 4 \cdot 9}{-\left[6 - (-1)\right]}$
$\qquad\qquad\qquad\quad\; = \dfrac{36 - 36}{-7}$
$\qquad\qquad\qquad\quad\; = \dfrac{0}{-7} = 0$

108. $\dfrac{4 + 5^2 \div 5}{6 - (-3 + 2)} = \dfrac{4 + 25 \div 5}{6 - (-1)} = \dfrac{4 + 5}{7} = \dfrac{9}{7} \text{ or } 1\dfrac{2}{7}$

109. $3\left[9 - (4^2 + 3)\right] \cdot 2 = 3\left[9 - (16 + 3)\right] \cdot 2$
$\qquad\qquad\qquad\qquad\; = 3[9 - 19] \cdot 2$
$\qquad\qquad\qquad\qquad\; = 3(-10) \cdot 2$
$\qquad\qquad\qquad\qquad\; = -30 \cdot 2$
$\qquad\qquad\qquad\qquad\; = -60$

110. $(-3^2 + 4^2) + (3^2 \div 3) = (-9 + 16) + (9 \div 3)$
$\qquad\qquad\qquad\qquad\qquad\; = (7) + (3)$
$\qquad\qquad\qquad\qquad\qquad\; = 10$

111. $2^3 \div 4 + 6 \cdot 3 = 8 \div 4 + 6 \cdot 3 = 2 + 18 = 20$

112. $(4 \div 2)^4 + 4^2 \div 2^2 = (2)^4 + 16 \div 4$
$\qquad\qquad\qquad\qquad\; = 16 + 16 \div 4$
$\qquad\qquad\qquad\qquad\; = 16 + 4$
$\qquad\qquad\qquad\qquad\; = 20$

113. $\left(8 - 2^2\right)^2 - 4 \cdot 3 + 10 = (8 - 4)^2 - 4 \cdot 3 + 10$
$\qquad\qquad\qquad\qquad\qquad\; = (4)^2 - 4 \cdot 3 + 10$
$\qquad\qquad\qquad\qquad\qquad\; = 16 - 4 \cdot 3 + 10$
$\qquad\qquad\qquad\qquad\qquad\; = 16 - 12 + 10$
$\qquad\qquad\qquad\qquad\qquad\; = 4 + 10$
$\qquad\qquad\qquad\qquad\qquad\; = 14$

114. $4^3 \div 4^2 - 5(2 - 7) \div 5 = 4^3 \div 4^2 - 5(-5) \div 5$
$\qquad\qquad\qquad\qquad\qquad\quad\; = 64 \div 16 - 5(-5) \div 5$
$\qquad\qquad\qquad\qquad\qquad\quad\; = 4 - (-25) \div 5$
$\qquad\qquad\qquad\qquad\qquad\quad\; = 4 - (-5)$
$\qquad\qquad\qquad\qquad\qquad\quad\; = 4 + 5$
$\qquad\qquad\qquad\qquad\qquad\quad\; = 9$

115. $-\left\{-4\left[27 \div 3^2 - 2(4 - 2)\right]\right\}$
$\quad = -\left\{-4\left[27 \div 3^2 - 2(2)\right]\right\}$
$\quad = -\left\{-4\left[27 \div 9 - 2(2)\right]\right\}$
$\quad = -\left\{-4[3 - 4]\right\}$
$\quad = -\left\{-4[-1]\right\}$
$\quad = -\{4\}$
$\quad = -4$

116. $2\left\{4^3 - 6\left[4 - (2 - 4)\right] - 3\right\}$
$\quad = 2\left\{4^3 - 6\left[4 - (-2)\right] - 3\right\}$
$\quad = 2\left\{4^3 - 6[6] - 3\right\}$
$\quad = 2\left\{64 - 6[6] - 3\right\}$
$\quad = 2\left\{64 - 36 - 3\right\}$
$\quad = 2\{28 - 3\}$
$\quad = 2\{25\}$
$\quad = 50$

117. Substitute 4 for x;
$\quad 3x - 7 = 3(4) - 7 = 12 - 7 = 5$

118. Substitute -5 for x;
$\quad 6 - 4x = 6 - 4(-5) = 6 - (-20) = 6 + 20 = 26$

119. Substitute 6 for x;
$\quad 2x^2 - 5x + 3 = 2(6)^2 - 5(6) + 3$
$\qquad\qquad\qquad\; = 2(36) - 30 + 3$
$\qquad\qquad\qquad\; = 72 - 30 + 3$
$\qquad\qquad\qquad\; = 42 + 3$
$\qquad\qquad\qquad\; = 45$

120. Substitute -1 for y;
$$5y^2 + 3y - 2 = 5(-1)^2 + 3(-1) - 2$$
$$= 5(1) - 3 - 2$$
$$= 5 - 3 - 2$$
$$= 2 - 2$$
$$= 0$$

121. Substitute -2 for x;
$$-x^2 + 2x - 3 = -(-2)^2 + 2(-2) - 3$$
$$= -4 + (-4) - 3$$
$$= -8 - 3$$
$$= -11$$

122. Substitute 2 for x;
$$-x^2 + 2x - 3 = -2^2 + 2(2) - 3$$
$$= -4 + 4 - 3$$
$$= 0 - 3$$
$$= -3$$

123. Substitute 1 for x;
$$-3x^2 - 5x + 5 = -3(1)^2 - 5(1) + 5$$
$$= -3(1) - 5 + 5$$
$$= -3 - 5 + 5$$
$$= -8 + 5$$
$$= -3$$

124. Substitute -3 for x and -2 for y;
$$-x^2 - 8x - 12y = -(-3)^2 - 8(-3) - 12(-2)$$
$$= -9 - (-24) + 24$$
$$= -9 + 24 + 24$$
$$= 15 + 24$$
$$= 39$$

125. a. $278 + (-493) = -215$

    b. $|-493|$ is greater than $|278|$ so the sum should be (and is) negative.

126. a. $324 - (-29.6) = 324 + 29.6 = 353.6$

    b. The sum of two positive numbers is always positive. As expected, the answer is positive.

127. a. $\dfrac{-17.28}{6} = -2.88$

    b. Since the numbers have unlike signs, the quotient is negative, as expected.

128. a. $(-62)(-1.9) = 117.8$

    b. Since the numbers have like signs, the product is positive, as expected.

129. a. $(-4)^8 = 65,536$

    b. A negative number raised to an even power is positive. As expected, the answer is positive.

130. a. $-(4.2)^3 = -74.088$

    b. Since $(4.2)^3$ is positive, $-(4.2)^3$ should be (and is) negative.

131. Associative property of addition

132. Distributive property

133. Commutative property of addition

134. Commutative property of multiplication

135. Distributive property

136. Associative property of addition

137. Identity property of multiplication

138. Inverse property of addition

Chapter 1 Practice Test

1. a. $2(1.30) + 4.75 + 3(1.10)$
$= 2.60 + 4.75 + 3.30$
$= 7.35 + 3.30$
$= 10.65$
The bill is $10.65 before tax.

    b. $0.07(3.30) \approx 0.23$
The tax on the soda is $0.23.

    c. $10.65 + 0.23 = 10.88$
The total bill is $10.88.

    d. $50 - 10.88 = 39.12$
Her change will be $39.12.

2. $\dfrac{1,600,000}{643,500} \approx 2.48$

The price was about 2.5 times greater in the twelfth year compared to the first year.

3. a. $13.0 - 5.1 = 7.9$
        About 7.9 million listened to WRAB at this time.

    b. During this specific time, half the time KFUN had more than 8.8 million listeners and half the time KFUN had less than 8.8 million listeners.

4. a. 42 is a natural number.

    b. 42 and 0 are whole numbers.

    c. −6, 42, 0, −7, and −1 are integers.

    d. $-6, 42, -3\frac{1}{2}, 0, 6.52, \frac{5}{9}, -7$, and −1 are rational numbers.

    e. $\sqrt{5}$ is an irrational number.

    f. $-6, 42, -3\frac{1}{2}, 0, 6.52, \sqrt{5}, \frac{5}{9}, -7$, and −1 are real numbers.

5. $-9.9 < -9.09$; −9.9 is to the left of −9.09 on a number line.

6. $|-3| > |-2|$ since $|-3| = 3$ and $|-2| = 2$.

7. $-7 + (-8) = -15$

8. $-6 - 5 = -6 + (-5) = -11$

9. $15 - 12 - 17 = 3 - 17 = -14$

10. $(-4 + 6) - 3(-2) = (2) - (-6) = 2 + 6 = 8$

11. $(-4)(-3)(2)(-1) = (12)(2)(-1) = (24)(-1) = -24$

12. $\left(\dfrac{-2}{9}\right) \div \left(\dfrac{-7}{8}\right) = \dfrac{-2}{9} \cdot \dfrac{8}{-7} = \dfrac{-16}{-63} = \dfrac{16}{63}$

13. $\left(-18 \cdot \dfrac{1}{2}\right) \div 3 = \left(\dfrac{-\overset{9}{\cancel{18}}}{1} \cdot \dfrac{1}{\cancel{2}_1}\right) \div 3$

    $= \left(\dfrac{-9 \cdot 1}{1 \cdot 1}\right) \div 3$

    $= -9 \div 3$

    $= -3$

14. $-\dfrac{3}{8} - \dfrac{4}{7} = -\dfrac{21}{56} - \dfrac{32}{56} = \dfrac{-21 - 32}{56} = -\dfrac{53}{56}$

15. $-6(-2 - 3) \div 5 \cdot 2 = -6(-5) \div 5 \cdot 2$

    $= 30 \div 5 \cdot 2$

    $= 6 \cdot 2$

    $= 12$

16. $\left(-\dfrac{2}{3}\right)^5 = \left(-\dfrac{2}{3}\right)\left(-\dfrac{2}{3}\right)\left(-\dfrac{2}{3}\right)\left(-\dfrac{2}{3}\right)\left(-\dfrac{2}{3}\right) = -\dfrac{32}{243}$

17. $\left[6 + ((9 - 3)^2 \div 18)^2\right]^2 = \left[6 + ((6)^2 \div 18)^2\right]^2$

    $= \left[6 + (36 \div 18)^2\right]^2$

    $= \left[6 + (2)^2\right]^2$

    $= [6 + 4]^2$

    $= [10]^2$

    $= 100$

18. Answers will vary. One possibility follows:
    $-x^2$ means $(-x)^2$ and x^2 will always be positive for any nonzero value of x. Therefore, $-x^2$ will always be negative.

19. Substitute −3 for x;
    $5x^2 - 8 = 5(-3)^2 - 8 = 5(9) - 8 = 45 - 8 = 37$

20. Substitute −2 for each x;
    $-x^2 - 6x + 3 = -(-2)^2 - 6(-2) + 3$

    $= -4 - (-12) + 3$

    $= -4 + 12 + 3$

    $= 8 + 3$

    $= 11$

21. Substitute 3 for x and −2 for y;
    $6x - 3y^2 + 4 = 6(3) - 3(-2)^2 + 4$

    $= 6(3) - 3(4) + 4$

    $= 18 - 12 + 4$

    $= 6 + 4$

    $= 10$

22. Substitute 1 for x and −2 for y;
    $-x^2 + xy + y^2 = -(1)^2 + (1)(-2) + (-2)^2$

    $= -1 + (-2) + 4$

    $= -3 + 4$

    $= 1$

23. Commutative property of addition

24. Distributive property

25. Associative property of addition

Chapter 2

1. **a.** The terms of an expression are the parts that are added.

   b. The terms of $3x - 4y - 5$ are $3x$, $-4y$, and -5.

   c. The terms of $6xy + 3x - y - 9$ are $6xy$, $3x$, $-y$, and -9.

3. **a.** In $2x - 5$, the x is called the variable.

   b. In $2x - 5$, the -5 is called the constant.

   c. In $2x - 5$, the 2 is called the coefficient.

5. **a.** Yes, $5x$ and $-7x$ are like terms because they have the same variable with the same exponent.

   b. No, $7y$ and 2 are not like terms because the 2 does not have a y variable.

   c. Yes, $-3(t + 2)$ and $6(t + 2)$ are like terms because they have the same variable with the same exponent.

   d. Yes, $4pq$ and $-9pq$ are like terms because they have the same variables with the same exponents

7. **a.** The signs of all the terms inside the parentheses are changed when the parentheses are removed.

   b. $-(x - 8) = -x + 8$

9. $6x + 3x = 9x$

11. There are no like terms.
    $3x + 6$

13. $y + 3 + 4y = y + 4y + 3 = 5y + 3$

15. $\dfrac{3}{4} - \dfrac{6}{11} = \dfrac{33}{44} - \dfrac{24}{44} = \dfrac{9}{44}$
    $\dfrac{3}{4}a - \dfrac{6}{11}a = \dfrac{9}{44}a$

17. $2 - 6x + 5 = -6x + 2 + 5 = -6x + 7$

19. $-2w - 3w + 5 = -5w + 5$

21. $-x + 2 - x - 2 = -x - x + 2 - 2 = -2x$

23. $3 + 6x - 3 - 6x = 6x - 6x + 3 - 3 = 0$

25. $5 + 2x - 4x + 6 = 2x - 4x + 5 + 6 = -2x + 11$

27. $4r - 6 - 6r - 2 = 4r - 6r - 6 - 2 = -2r - 8$

29. $3x^2 - 9y^2 + 7x^2 - 5 - y^2 - 2$
    $= 3x^2 + 7x^2 - 9y^2 - y^2 - 5 - 2$
    $= 10y^2 - 10y^2 - 7$

31. $-2x + 4x - 3 = 2x - 3$

33. $b + 4 + \dfrac{3}{5} = b + \dfrac{20}{5} + \dfrac{3}{5} = b + \dfrac{23}{5}$

35. $5.1n + 6.42 - 4.3n = 5.1n - 4.3n + 6.42$
    $= 0.8n + 6.42$

37. There are no like terms.
    $\dfrac{1}{2}a + 3b + 1$

39. $13.4x + 1.2x + 8.3 = 14.6x + 8.3$

41. $-x^2 + 2x^2 + y = x^2 + y$

43. $2x - 7y - 5x + 2y = 2x - 5x - 7y + 2y = -3x - 5y$

45. $4 - 3n^2 + 9 - 2n = -3n^2 - 2n + 4 + 9$
    $= -3n^2 - 2n + 13$

47. $-19.36 + 40.02x + 12.25 - 18.3x$
    $= 40.02x - 18.3x - 19.36 + 12.25$
    $= 21.72x - 7.11$

49. $\dfrac{3}{5}x - 3 - \dfrac{7}{4}x - 2 = \dfrac{3}{5}x - \dfrac{7}{4}x - 3 - 2$
    $= \dfrac{12}{20}x - \dfrac{35}{20}x - 5$
    $= -\dfrac{23}{20}x - 5$

51. There are no like terms.
    $5w^3 + 2w^2 + w + 3$

53. $2z - 5z^3 - 2z^3 - z^2 = -5z^3 - 2z^3 - z^2 + 2z$
    $= -7z^3 - z^2 + 2z$

55. There are no like terms.
$6x^2 - 6xy + 3y^2$

57. $4a^2 - 3ab + 6ab + b^2 = 4a^2 + 3ab + b^2$

59. $5(x+2) = 5x + 5(2) = 5x + 10$

61. $5(x+4) = 5x + 5(4) = 5x + 20$

63. $3(x-6) = 3x + 3(-6) = 3x - 18$

65. $-\dfrac{1}{2}(2x-4) = -\dfrac{1}{2}\left[2x+(-4)\right]$
$$= -\dfrac{1}{2}(2x) + \left(-\dfrac{1}{2}\right)(-4)$$
$$= -x + 2$$

67. $1(-4+x) = 1(-4) + 1(x) = -4 + x = x - 4$

69. $\dfrac{4}{5}(s-5) = \dfrac{4}{5}s - \dfrac{4}{5}(5) = \dfrac{4}{5}s - 4$

71. $-0.3(3x+5) = -0.3(3x) + (-0.3)(5)$
$$= -0.9x + (-1.5)$$
$$= -0.9x - 1.5$$

73. $-\dfrac{1}{3}(3r-12) = -\dfrac{1}{3}(3r) + \left(-\dfrac{1}{3}\right)(-12) = -r + 4$

75. $0.7(2x+0.5) = 0.7(2x) + 0.7(0.5) = 1.4x + 0.35$

77. $-(-x+y) = -1(-x+y)$
$$= -1(-x) + (-1)(y)$$
$$= x + (-y)$$
$$= x - y$$

79. $-(2x+4y-8) = -1\left[2x+4y+(-8)\right]$
$$= -1(2x) + (-1)(4y) + (-1)(-8)$$
$$= -2x + (-4y) + 8$$
$$= -2x - 4y + 8$$

81. $1.1(3.1x - 5.2y + 2.8)$
$$= 1.1[3.1x + (-5.2y) + 2.8]$$
$$= (1.1)(3.1x) + (1.1)(-5.2y) + (1.1)(2.8)$$
$$= 3.41x + (-5.72y) + 3.08$$
$$= 3.41x - 5.72y + 3.08$$

83. $(2x-9y)5 = (2x)5 + (-9y)5 = 10x - 45y$

85. $(x+3y-9) = 1\left[x+3y+(-9)\right]$
$$= 1(x) + 1(3y) + (1)(-9)$$
$$= x + 3y + (-9) = x + 3y - 9$$

87. $-3(-x+2y+4) = -3(-x) + (-3)(2y) + (-3)(4)$
$$= 3x + (-6y) + (-12)$$
$$= 3x - 6y - 12$$

89. $5 - (3x+4) = 5 - 3x - 4$
$$= -3x + 5 - 4$$
$$= -3x + 1$$

91. $-2(3-x) + 7 = -6 + 2x + 7$
$$= 2x - 6 + 7$$
$$= 2x + 1$$

93. $6x + 2(4x+9) = 6x + 8x + 18 = 14x + 18$

95. $2(x-y) + 2x + 3 = 2x - 2y + 2x + 3$
$$= 2x + 2x - 2y + 3$$
$$= 4x - 2y + 3$$

97. $4(2c-3) - 3(c-4) = 8c - 12 - 3c + 12$
$$= 8c - 3c - 12 + 12$$
$$= 5c$$

99. $8x - (x-3) = 8x - x + 3 = 7x + 3$

101. $-\left(\dfrac{3}{4}x - \dfrac{1}{3}\right) + 2x = -\dfrac{3}{4}x + \dfrac{1}{3} + 2x$
$$= -\dfrac{3}{4}x + 2x + \dfrac{1}{3}$$
$$= -\dfrac{3}{4}x + \dfrac{8}{4}x + \dfrac{1}{3}$$
$$= \dfrac{5}{4}x + \dfrac{1}{3}$$

103. $\dfrac{2}{3}x + \dfrac{1}{2}(5x-4) = \dfrac{2}{3}x + \dfrac{5}{2}x - 2$
$$= \dfrac{4}{6}x + \dfrac{15}{6}x - 2$$
$$= \dfrac{19}{6}x - 2$$

105. $-(3s+4)-(s+2) = -3s-4-s-2$
$$= -3s-s-4-2$$
$$= -4s-6$$

107. $4(x-1)+2(3-x)-4 = 4x-4(1)+2(3)-2x-4$
$$= 4x-4+6-2x-4$$
$$= 4x-2x-4+6-4$$
$$= 2x-2$$

109. $4(m+3)-4m-12 = 4m+4(3)-4m-12$
$$= 4m+12-4m-12$$
$$= 4m-4m+12-12$$
$$= 0$$

111. $0.4-(y+5)+0.6-2 = 0.4-y-5+0.6-2$
$$= -y+0.4-5+0.6-2$$
$$= -y-6$$

113. $4+(3x-4)-5 = 4+3x-4-5$
$$= 3x+4-4-5$$
$$= 3x-5$$

115. $4(x+2)-3(x-4)-5 = 4x+4(2)-3x-3(-4)-5$
$$= 4x+8-3x+12-5$$
$$= 4x-3x+8+12-5$$
$$= x+15$$

117. $-0.2(6-x)-4(y+0.4)$
$$= -0.2(6)-0.2(-x)-4y-4(0.4)$$
$$= -1.2+0.2x-4y-1.6$$
$$= 0.2x-4y-1.2-1.6$$
$$= 0.2x-4y-2.8$$

119. $-6x+7y-(3+x)+(x+3)$
$$= -6x+7y-3-x+x+3$$
$$= -6x-x+x+7y-3+3$$
$$= -6x+7y$$

121. $\dfrac{1}{2}(x+3)+\dfrac{1}{3}(3x+6) = \dfrac{1}{2}x+\dfrac{3}{2}+\dfrac{3}{3}x+\dfrac{6}{3}$
$$= \dfrac{1}{2}x+\dfrac{3}{2}+x+2$$
$$= \dfrac{1}{2}x+x+\dfrac{3}{2}+2$$
$$= \dfrac{1}{2}x+\dfrac{2}{2}x+\dfrac{3}{2}+\dfrac{4}{2}$$
$$= \dfrac{3}{2}x+\dfrac{7}{2}$$

123. $\square + \ominus + \ominus + \square + \ominus = 2\square + 3\ominus$

125. $x+y+\Delta+\Delta+x+y+y = x+x+y+y+y+\Delta+\Delta$
$$= 2x+3y+2\Delta$$

127. $3\Delta+5\square-\Delta-3\square = 3\Delta-\Delta+5\square-3\square = 2\Delta+2\square$

129. $1\cdot18,\ 2\cdot9,\ 3\cdot6$
positive factors: 1, 2, 3, 6, 9, 18

131. $4x^2+5y^2+6(3x^2-5y^2)-4x+3$
$$= 4x^2+5y^2+18x^2-30y^2-4x+3$$
$$= 4x^2+18x^2-4x+5y^2-30y^2+3$$
$$= 22x^2-25y^2-4x+3$$

133. $2[3+4(x-5)]-[2-(x-3)]$
$$= 2[3+4x-20]-[2-x+3]$$
$$= 6+8x-40-2+x-3$$
$$= 8x+x+6-40-2-3$$
$$= 9x-39$$

135. $|-7| = 7$

136. $-|-16| = -(16) = -16$

137. $-4-3-(-6) = -4-3+6 = -7+6 = -1$

138. Answers will vary. The answer should include that the order is parentheses, exponents, multiplication and division from left to right, and addition and subtraction from left to right.

139. Substitute –1 for each x in the expression.
$$-x^2 + 5x - 6 = -(-1)^2 + 5(-1) - 6$$
$$= -1 + (-5) - 6$$
$$= -6 - 6$$
$$= -12$$

Exercise Set 2.2

1. An equation is a statement that shows two algebraic expressions are equal.

3. A solution to an equation may be checked by substituting the value in the equation and determining if it results in a true statement.

5. Equivalent equations are two or more equations with the same solution.

7. Subtract 2 from both sides of the equation to isolate the variable.

9. One example is $x + 2 = 1$.

11. All 3 equations have $x = 1$ as their solution.

13. Substitute 2 for $x = 2$.
$$4x - 3 = 5$$
$$4(2) - 3 = 5$$
$$8 - 3 = 5$$
$$5 = 5 \text{ True}$$
Since we obtain true statement, 2 is a solution.

15. Substitute –3 for $x, x = -3$.
$$2x - 5 = 5(x + 2)$$
$$2(-3) - 5 = 5[(-3) + 2]$$
$$-6 - 5 = 5(-1)$$
$$-11 = -5 \text{ False}$$
Since we obtain a false statement, –3 is not a solution.

17. Substitute –15 for $p, p = -15$.
$$2p - 5(p + 7) = 10$$
$$2(-15) - 5(-15 + 7) = 10$$
$$-30 - 5(-8) = 10$$
$$-30 + 40 = 10$$
$$10 = 10 \text{ True}$$
Since we obtain a true statement, –15 is a solution.

19. Substitute 3.4 for $x, x = 3.4$.
$$3(x + 2) - 3(x - 1) = 9$$
$$3(3.4 + 2) - 3(3.4 - 1) = 9$$
$$3(5.4) - 3(2.4) = 9$$
$$16.2 - 7.2 = 9$$
$$9 = 9 \text{ True}$$
Since we obtain a true statement, 3.4 is a solution.

21. Substitute $\frac{1}{2}$ for $x, x = \frac{1}{2}$.
$$4x - 4 = 2x - 2$$
$$4\left(\frac{1}{2}\right) - 4 = 2\left(\frac{1}{2}\right) - 2$$
$$2 - 4 = 1 - 2$$
$$-2 = -1 \text{ False}$$
Since we obtain a false statement, $\frac{1}{2}$ is not a solution.

23. Substitute $\frac{11}{2}$ for $x, x = \frac{11}{2}$.
$$3(x + 2) = 5(x - 1)$$
$$3\left(\frac{11}{2} + 2\right) = 5\left(\frac{11}{2} - 1\right)$$
$$3\left(\frac{15}{2}\right) = 5\left(\frac{9}{2}\right)$$
$$\frac{45}{2} = \frac{45}{2} \text{ True}$$
Since we obtain a true statement, $\frac{11}{2}$ is a solution.

25. $x + 2 = 7$
$$x + 2 - 2 = 7 - 2$$
$$x + 0 = 5$$
$$x = 5$$
Check: $x + 2 = 7$
$$5 + 2 = 7$$
$$7 = 7 \text{ True}$$

27. $-6 = x + 1$
$$-6 - 1 = x + 1 - 1$$
$$-7 = x + 0$$
$$-7 = x$$
Check: $-6 = x + 1$
$$-6 = -7 + 1$$
$$-6 = -6 \text{ True}$$

29. $x - 4 = -8$

$\quad x - 4 + 4 = -8 + 4$

$\quad\quad x + 0 = -4$

$\quad\quad\quad x = -4$

Check: $x - 4 = 8$

$\quad\quad\quad -4 - 4 = -8$

$\quad\quad\quad\quad -8 = -8$ True

31. $x + 9 = 52$

$\quad x + 9 - 9 = 52 - 9$

$\quad\quad x + 0 = 43$

$\quad\quad\quad x = 43$

Check: $x + 9 = 52$

$\quad\quad\quad 43 + 9 = 52$

$\quad\quad\quad\quad 52 = 52$ True

33. $-6 + w = 9$

$\quad -6 + 6 + w = 9 + 6$

$\quad\quad 0 + w = 15$

$\quad\quad\quad w = 15$

Check: $-6 + w = 9$

$\quad\quad\quad -6 + 15 = 9$

$\quad\quad\quad\quad 9 = 9$ True

35. $27 = x + 16$

$\quad 27 - 16 = x + 16 - 16$

$\quad\quad 11 = x + 0$

$\quad\quad 11 = x$

Check: $27 = x + 16$

$\quad\quad\quad 27 = 11 + 16$

$\quad\quad\quad 27 = 27$ True

37. $-18 = x - 14$

$\quad -18 + 14 = x - 14 + 14$

$\quad\quad -4 = x + 0$

$\quad\quad -4 = x$

Check: $-18 = x - 14$

$\quad\quad\quad -18 = -4 - 14$

$\quad\quad\quad -18 = -18$ True

39. $9 + x = 4$

$\quad 9 - 9 + x = 4 - 9$

$\quad\quad 0 + x = -5$

$\quad\quad\quad x = -5$

Check: $9 + x = 4$

$\quad\quad\quad 9 + (-5) = 4$

$\quad\quad\quad\quad 4 = 4$ True

41. $4 + x = -9$

$\quad 4 - 4 + x = -9 - 4$

$\quad\quad 0 + x = -13$

$\quad\quad\quad x = -13$

Check: $\quad 4 + x = -9$

$\quad\quad\quad 4 + (-13) = -9$

$\quad\quad\quad\quad -9 = -9$ True

43. $7 + r = -23$

$\quad 7 + r = -23$

$\quad 7 - 7 + r = -23 - 7$

$\quad\quad 0 + r = -30$

$\quad\quad\quad r = -30$

Check: $\quad 7 + r = -23$

$\quad\quad\quad 7 + (-30) = -23$

$\quad\quad\quad\quad -23 = -23$ True

45. $8 = 8 + v$

$\quad 8 - 8 = 8 - 8 + v$

$\quad\quad 0 = 0 + v$

$\quad\quad 0 = v$

Check: $8 = 8 + v$

$\quad\quad\quad 8 = 8 + 0$

$\quad\quad\quad 8 = 8$ True

47. $7 + x = -50$

$\quad 7 - 7 + x = -50 - 7$

$\quad\quad 0 + x = -57$

$\quad\quad\quad x = -57$

Check: $\quad 7 + x = -50$

$\quad\quad\quad 7 + (-57) = -50$

$\quad\quad\quad\quad -50 = -50$ True

49. $12 = 16 + x$

$\quad 12 - 16 = 16 - 16 + x$

$\quad\quad -4 = 0 + x$

$\quad\quad -4 = x$

Check: $12 = 16 + x$

$\quad\quad\quad 12 = 16 + (-4)$

$\quad\quad\quad 12 = 12$ True

51. $15 + x = -5$
$15 - 15 + x = -5 - 15$
$0 + x = -20$
$x = -20$
Check: $15 + x = -5$
$15 + (-20) = -5$
$-5 = -5$ True

53. $-15 + x = -15$
$-15 + 15 + x = -15 + 15$
$0 + x = 0$
$x = 0$
Check: $-15 + x = -15$
$-15 + 0 = -15$
$-15 = -15$ True

55. $5 = x - 12$
$5 + 12 = x - 12 + 12$
$17 = x + 0$
$17 = x$
Check: $5 = x - 12$
$5 = 17 - 12$
$5 = 5$ True

57. $-50 = x - 24$
$-50 + 24 = x - 24 + 24$
$-26 = x + 0$
$-26 = x$
Check: $-50 = x - 24$
$-50 = -26 - 24$
$-50 = -50$ True

59. $43 = 15 + p$
$43 - 15 = 15 - 15 + p$
$28 = 0 + p$
$28 = p$
Check: $43 = 15 + p$
$43 = 15 + 28$
$43 = 43$ True

61. $40.2 + x = -5.9$
$40.2 - 40.2 + x = -5.9 - 40.2$
$0 + x = -46.1$
$x = -46.1$
Check: $40.2 + x = -5.9$
$40.2 + (-46.1) = -5.9$
$-5.9 = -5.9$ True

63. $-37 + x = 9.5$
$-37 + 37 + x = 9.5 + 37$
$0 + x = 46.5$
$x = 46.5$
Check: $-37 + x = 9.5$
$-37 + 46.5 = 9.5$
$9.5 = 9.5$ True

65. $x - 8.77 = -17$
$x - 8.77 + 8.77 = -17 + 8.77$
$x + 0 = -8.23$
$x = -8.23$
Check: $x - 8.77 = -17$
$-8.23 - 8.77 = -17$
$-17 = -17$ True

67. $9.32 = x + 3.75$
$9.32 - 3.75 = x + 3.75 - 3.75$
$5.57 = x + 0$
$5.57 = x$
Check: $9.32 = x + 3.75$
$9.32 = 5.57 + 3.75$
$9.32 = 9.32$ True

69. No; there are no real numbers that can make $x + 1 = x + 2$.

71. $x - \Delta = \square$
$x - \Delta + \Delta = \square + \Delta$
$x = \square + \Delta$

73. $\odot = \square + \Delta$
$\odot - \Delta = \square + \Delta - \Delta$
$\odot - \Delta = \square$

76. $-\dfrac{7}{15} + \dfrac{5}{6} = -\dfrac{14}{30} + \dfrac{25}{30} = \dfrac{-14 + 25}{30} = \dfrac{11}{30}$

77. $-\dfrac{11}{12} + \left(-\dfrac{3}{8}\right) = -\dfrac{22}{24} + \left(-\dfrac{9}{24}\right) = \dfrac{-22 + (-9)}{24} = -\dfrac{31}{24}$

78. $4x + 3(x - 2) - 5x - 7 = 4x + 3x - 6 - 5x - 7$
$= 4x + 3x - 5x - 6 - 7$
$= 2x - 13$

79. $-(2t + 4) + 3(4t - 5) - 3t = -2t - 4 + 12t - 15 - 3t$
$= -2t + 12t - 3t - 4 - 15$
$= 7t - 19$

Exercise Set 2.3

1. Answers will vary. Answer should include that both sides of an equation can be multiplied by the same nonzero number without changing the solution to the equation.

3. **a.**
$$-x = a$$
$$-1x = a$$
$$(-1)(-1x) = (-1)a$$
$$1x = -a$$
$$x = -a$$

   b.
$$-x = 5$$
$$-1x = 5$$
$$(-1)(-1x) = (-1)5$$
$$1x = -5$$
$$x = -5$$

   c.
$$-x = -5$$
$$-1x = -5$$
$$(-1)(-1x) = (-1)(-5)$$
$$1x = 5$$
$$x = 5$$

5. Divide both sides by 3 to isolate the variable.

7. Multiply both sides by 2 because $2 \cdot \dfrac{x}{2} = x$.

9. $4x = 12$
$$\frac{4x}{4} = \frac{12}{4}$$
$$x = 3$$
Check: $4x = 12$
$$4(3) = 12$$
$$12 = 12 \ \text{True}$$

11. $\dfrac{x}{3} = 7$
$$3\left(\frac{x}{3}\right) = 3(7)$$
$$x = 21$$
Check: $\dfrac{x}{3} = 7$
$$\frac{21}{3} = 7$$
$$7 = 7 \ \text{True}$$

13. $-4x = 12$
$$\frac{-4x}{-4} = \frac{12}{-4}$$
$$x = -3$$
Check: $-4x = 12$
$$-4(-3) = 12$$
$$12 = 12 \ \text{True}$$

15. $\dfrac{x}{4} = -2$
$$4\left(\frac{x}{4}\right) = 4(-2)$$
$$x = 4(-2)$$
$$x = -8$$
Check: $\dfrac{x}{4} = -2$
$$\frac{-8}{4} = -2$$
$$-2 = -2 \ \text{True}$$

17. $\dfrac{x}{5} = 1$
$$5\left(\frac{x}{5}\right) = 5(1)$$
$$x = 5$$
Check: $\dfrac{x}{5} = 1$
$$\frac{5}{5} = 1$$
$$1 = 1 \ \text{True}$$

19. $-27n = 81$
$$\frac{-27n}{-27} = \frac{81}{-27}$$
$$n = -3$$
Check: $-27n = 81$
$$-27(-3) = 81$$
$$81 = 81 \ \text{True}$$

21. $-7 = 3r$
$$\frac{-7}{3} = \frac{3r}{3}$$
$$-\frac{7}{3} = r$$
Check: $-7 = 3r$
$$-7 = 3\left(-\frac{7}{3}\right)$$
$$-7 = -7 \ \text{True}$$

23.
$$-x = 13$$
$$-1x = 13$$
$$(-1)(-1x) = (-1)(13)$$
$$1x = -13$$
$$x = -13$$
Check: $-x = 13$
$$-(-13) = 13$$
$$13 = 13 \text{ True}$$

25.
$$-x = -8$$
$$-1x = -8$$
$$(-1)(-1x) = (-1)(-8)$$
$$1x = 8$$
$$x = 8$$
Check: $-x = -8$
$$-(8) = -8$$
$$-8 = -8 \text{ True}$$

27.
$$-\frac{w}{3} = -13$$
$$\frac{w}{-3} = -13$$
$$(-3)\left(\frac{w}{-3}\right) = (-3)(-13)$$
$$w = 39$$
Check: $-\frac{w}{3} = -13$
$$-\frac{39}{3} = -13$$
$$-13 = -13 \text{ True}$$

29.
$$4 = -12x$$
$$\frac{4}{-12} = \frac{-12x}{-12}$$
$$-\frac{1}{3} = x$$
Check: $4 = -12x$
$$4 = -12\left(-\frac{1}{3}\right)$$
$$4 = 4 \text{ True}$$

31.
$$-\frac{x}{3} = -2$$
$$\frac{x}{-3} = -2$$
$$(-3)\left(\frac{x}{-3}\right) = (-3)(-2)$$
$$x = 6$$
Check: $-\frac{x}{3} = -2$
$$-\frac{6}{3} = -2$$
$$-2 = -2 \text{ True}$$

33. $43t = 26$
$$\frac{43t}{43} = \frac{26}{43}$$
$$t = \frac{26}{43}$$
Check: $43t = 26$
$$43\left(\frac{26}{43}\right) = 26$$
$$26 = 26 \text{ True}$$

35. $-4.2x = -8.4$
$$\frac{-4.2x}{-4.2} = \frac{-8.4}{-4.2}$$
$$x = 2$$
Check: $-4.2x = -8.4$
$$-4.2(2) = -8.4$$
$$-8.4 = -8.4 \text{ True}$$

37. $3x = \frac{3}{5}$
$$\frac{1}{3} \cdot 3x = \frac{1}{3} \cdot \frac{3}{5}$$
$$x = \frac{1}{1} \cdot \frac{1}{5}$$
$$x = \frac{1 \cdot 1}{1 \cdot 5}$$
$$x = \frac{1}{5}$$
Check: $3x = \frac{3}{5}$
$$3\left(\frac{1}{5}\right) = \frac{3}{5}$$
$$\frac{3}{5} = \frac{3}{5} \text{ True}$$

39. $5x = -\dfrac{3}{8}$

$\dfrac{1}{5} \cdot 5x = \left(\dfrac{1}{5}\right) \cdot \left(-\dfrac{3}{8}\right)$

$x = \dfrac{1(-3)}{5(8)}$

$x = -\dfrac{3}{40}$

Check: $5x = -\dfrac{3}{8}$

$5\left(-\dfrac{3}{40}\right) = -\dfrac{3}{8}$

$-\dfrac{3}{8} = -\dfrac{3}{8}$ True

41. $15 = -\dfrac{x}{4}$

$15 = \dfrac{x}{-4}$

$(-4)(15) = (-4) \cdot \left(\dfrac{x}{-4}\right)$

$-60 = x$

Check: $15 = -\dfrac{x}{4}$

$15 = -\dfrac{(-60)}{4}$

$15 = 15$ True

43. $-\dfrac{b}{4} = -60$

$\dfrac{b}{-4} = -60$

$-4\left(\dfrac{b}{-4}\right) = (-4)(-60)$

$b = 240$

Check: $-\dfrac{b}{4} = -60$

$-\dfrac{240}{4} = -60$

$-60 = -60$ True

45. $\dfrac{x}{5} = -7$

$5\left(\dfrac{x}{5}\right) = 5(-7)$

$x = -35$

Check: $\dfrac{x}{5} = -7$

$\dfrac{-35}{5} = -7$

$-7 = -7$ True

47. $5 = \dfrac{x}{4}$

$4 \cdot 5 = 4\left(\dfrac{x}{4}\right)$

$20 = x$

Check: $5 = \dfrac{x}{4}$

$5 = \dfrac{20}{4}$

$5 = 5$ True

49. $\dfrac{3}{5}d = -30$

$\dfrac{5}{3} \cdot \dfrac{3}{5}d = \dfrac{5}{3}(-30)$

$d = -50$

Check: $\dfrac{3}{5}d = -30$

$\dfrac{3}{5}(-50) = -30$

$-30 = -30$ True

51. $\dfrac{y}{-2} = 0$

$(-2)\left(\dfrac{y}{-2}\right) = (-2)(0)$

$y = 0$

Check: $\dfrac{y}{-2} = 0$

$\dfrac{0}{-2} = 0$

$0 = 0$ True

53. $\dfrac{-7}{8}w = 0$

$\dfrac{8}{-7}\left(\dfrac{-7}{8}w\right) = \dfrac{8}{-7}\cdot 0$

$\qquad w = 0$

Check: $\dfrac{-7}{8}w = 0$

$\qquad \dfrac{-7}{8}(0) = 0$

$\qquad\qquad 0 = 0$ True

55. $\dfrac{1}{5}x = 4.5$

$5\left(\dfrac{1}{5}x\right) = 5(4.5)$

$\qquad x = 22.5$

Check: $\dfrac{1}{5}x = 4.5$

$\qquad \dfrac{1}{5}(22.5) = 4.5$

$\qquad\qquad 4.5 = 4.5$ True

57. $-4 = -\dfrac{2}{3}z$

$\left(-\dfrac{3}{2}\right)(-4) = \left(-\dfrac{3}{2}\right)\left(-\dfrac{2}{3}\right)z$

$\qquad 6 = z$

Check: $-4 = -\dfrac{2}{3}z$

$\qquad -4 = -\dfrac{2}{3}\cdot 6$

$\qquad -4 = -4$ True

59. $-1.4x = 28.28$

$\dfrac{-1.4x}{-1.4} = \dfrac{28.28}{-1.4}$

$\qquad x = -20.2$

Check: $\qquad -1.4x = 28.28$

$\qquad -1.4(-20.2) = 28.28$

$\qquad\qquad 28.28 = 28.28$ True

61. $-8x = -56$

$\qquad x = \dfrac{-56}{-8}$

$\qquad x = 7$

Check: $-8x = -56$

$\qquad -8(7) = -56$

$\qquad -56 = -56$ True

63. $\dfrac{2}{3}x = 6$

$\qquad x = \dfrac{3}{2}\cdot 6$

$\qquad x = 9$

Check: $\dfrac{2}{3}x = 6$

$\qquad \dfrac{2}{3}(9) = 6$

$\qquad\qquad 6 = 6$ True

65. a. In $5 + x = 10$, 5 is added to the variable, whereas in $5x = 10$, 5 is multiplied by the variable.

b. $\qquad 5 + x = 10$

$\qquad 5 + x - 5 = 10 - 5$

$\qquad\qquad x = 5$

c. $\quad 5x = 10$

$\qquad \dfrac{5x}{5} = \dfrac{10}{5}$

$\qquad x = 2$

67. Multiplying by $\dfrac{3}{2}$ is easier because the equation involves fractions.

$\qquad \dfrac{2}{3}x = 4$

$\left(\dfrac{3}{2}\right)\left(\dfrac{2}{3}\right)x = \left(\dfrac{3}{2}\right)\left(\dfrac{4}{1}\right)$

$\qquad x = \dfrac{12}{2}$

$\qquad x = 6$

69. Multiplying by $\dfrac{7}{3}$ is easier because the equation involves fractions.

$\qquad \dfrac{3}{7}x = \dfrac{4}{5}$

$\left(\dfrac{7}{3}\right)\dfrac{3}{7}x = \left(\dfrac{7}{3}\right)\dfrac{4}{5}$

$\qquad x = \dfrac{28}{15}$

71. a. ⊡

   b. Divide both sides of the equation by $\triangle$.

   c. $\boxdot = \dfrac{\smiley}{\triangle}$

73. $-8 - (-4) = -8 + 4 = -4$

74. $(-3)(-2)(5)(-1) = 6(5)(-1) = 30(-1) = -30$

75. $\begin{aligned} 4^2 - 2^3 \cdot 6 \div 3 + 6 &= 16 - 8 \cdot 6 \div 3 + 6 \\ &= 16 - 48 \div 3 + 6 \\ &= 16 - 16 + 6 \\ &= 0 + 6 \\ &= 6 \end{aligned}$

76. Associative property of addition

77. $\begin{aligned} -48 &= x + 9 \\ -48 - 9 &= x + 9 - 9 \\ -57 &= x + 0 \\ -57 &= x \end{aligned}$

Exercise Set 2.4

1. No; the variable x is on both sides of the equal sign.

3. $x = \dfrac{1}{3}$ because $1x = x$.

5. $x = -\dfrac{1}{2}$ because $-x = -1x$.

7. $x = \dfrac{4}{9}$ because $-x = -1x$.

9. You evaluate an expression. An expression cannot be solved. It is not true or false for given values; instead, it has a numerical value.

11. a. Answers will vary.

   b. Answers will vary.

13. a. Use the distributive property.
   Subtract 8 from both sides of the equation.
   Divide both sides of the equation by 6.

   b. $\begin{aligned} 2(3x + 4) &= -4 \\ 6x + 8 &= -4 \\ 6x + 8 - 8 &= -4 - 8 \\ 6x &= -12 \\ \dfrac{6x}{6} &= -\dfrac{12}{6} \\ x &= -2 \end{aligned}$

15. $\begin{aligned} 5x - 6 &= 19 \\ 5x - 6 + 6 &= 19 + 6 \\ 5x &= 25 \\ \dfrac{5x}{5} &= \dfrac{25}{5} \\ x &= 5 \end{aligned}$

17. $\begin{aligned} -4w - 5 &= 11 \\ -4w - 5 + 5 &= 11 + 5 \\ -4w &= 16 \\ \dfrac{-4w}{-4} &= \dfrac{16}{-4} \\ w &= -4 \end{aligned}$

19. $\begin{aligned} 3x + 6 &= 12 \\ 3x + 6 - 6 &= 12 - 6 \\ 3x &= 6 \\ \dfrac{3x}{3} &= \dfrac{6}{3} \\ x &= 2 \end{aligned}$

21. $\begin{aligned} 5x - 2 &= 10 \\ 5x - 2 + 2 &= 10 + 2 \\ 5x &= 12 \\ \dfrac{5x}{5} &= \dfrac{12}{5} \\ x &= \dfrac{12}{5} \end{aligned}$

23. $\begin{aligned} -5k - 4 &= -19 \\ -5k - 4 + 4 &= -19 + 4 \\ -5k &= -15 \\ \dfrac{-5k}{-5} &= \dfrac{-15}{-5} \\ k &= 3 \end{aligned}$

25.
$$12 - x = 9$$
$$12 - 12 - x = 9 - 12$$
$$-x = -3$$
$$(-1)(-x) = (-1)(-3)$$
$$x = 3$$

27.
$$8 + 3x = 19$$
$$8 - 8 + 3x = 19 - 8$$
$$3x = 11$$
$$\frac{3x}{3} = \frac{11}{3}$$
$$x = \frac{11}{3}$$

29.
$$16x + 5 = -14$$
$$16x + 5 - 5 = -14 - 5$$
$$16x = -19$$
$$\frac{16x}{16} = \frac{-19}{16}$$
$$x = -\frac{19}{16}$$

31.
$$-4.2 = 3x + 25.8$$
$$-4.2 - 25.8 = 3x + 25.8 - 25.8$$
$$-30 = 3x$$
$$\frac{-30}{3} = \frac{3x}{3}$$
$$-10 = x$$

33.
$$7r - 16 = -2$$
$$7r - 16 + 16 = -2 + 16$$
$$7r = 14$$
$$\frac{7r}{7} = \frac{14}{7}$$
$$r = 2$$

35.
$$60 = -5s + 9$$
$$60 - 9 = -5s + 9 - 9$$
$$51 = -5s$$
$$\frac{51}{-5} = \frac{-5s}{-5}$$
$$-\frac{51}{5} = s$$

37.
$$14 = 5x + 8 - 3x$$
$$14 = 5x - 3x + 8$$
$$14 = 2x + 8$$
$$14 - 8 = 2x + 8 - 8$$
$$6 = 2x$$
$$\frac{6}{2} = \frac{2x}{2}$$
$$3 = x$$

39.
$$2.3x - 9.34 = 6.3$$
$$2.3x - 9.34 + 9.34 = 6.3 + 9.34$$
$$2.3x = 15.64$$
$$\frac{2.3x}{2.3} = \frac{15.64}{2.3}$$
$$x = 6.8$$

41.
$$0.91y + 2.25 - 0.01y = 5.85$$
$$0.91y - 0.01y + 2.25 = 5.85$$
$$0.90y + 2.25 = 5.85$$
$$0.90y + 2.25 - 2.25 = 5.85 - 2.25$$
$$0.90y = 3.60$$
$$\frac{0.90y}{0.90} = \frac{3.60}{0.90}$$
$$y = 4$$

43.
$$28.8 = x + 1.40x$$
$$28.8 = 2.40x$$
$$\frac{28.8}{2.40} = \frac{2.40x}{2.40}$$
$$12 = x$$

45.
$$\frac{1}{7}(x + 6) = 4$$
$$7\left[\frac{1}{7}(x + 6)\right] = 7(4)$$
$$x + 6 = 28$$
$$x + 6 - 6 = 28 - 6$$
$$x = 22$$

47.
$$\frac{d + 3}{7} = 9$$
$$7\left(\frac{d + 3}{7}\right) = 7(9)$$
$$d + 3 = 63$$
$$d + 3 - 3 = 63 - 3$$
$$d = 60$$

49.

$$\frac{1}{3}(t-5) = -6$$

$$3\left[\frac{1}{3}(t-5)\right] = 3(-6)$$

$$t-5 = -18$$

$$t-5+5 = -18+5$$

$$t = -13$$

51.

$$\frac{3}{4}(x-5) = -12$$

$$\frac{4}{3}\left[\frac{3}{4}(x-5)\right] = \frac{4}{3}(-12)$$

$$x-5 = -16$$

$$x-5+5 = -16+5$$

$$x = -11$$

53.

$$\frac{x+4}{7} = \frac{3}{7}$$

$$7\left(\frac{x+4}{7}\right) = 7\left(\frac{3}{7}\right)$$

$$x+4 = 3$$

$$x+4-4 = 3-4$$

$$x = -1$$

55.

$$\frac{3}{4} = \frac{4m-5}{6}$$

$$12\left(\frac{3}{4}\right) = 12\left(\frac{4m-5}{6}\right)$$

$$9 = 2(4m-5)$$

$$9 = 8m-10$$

$$9+10 = 8m-10+10$$

$$19 = 8m$$

$$\frac{19}{8} = \frac{8m}{8}$$

$$\frac{19}{8} = m$$

57.

$$4(n+2) = 8$$

$$4n+8 = 8$$

$$4n+8-8 = 8-8$$

$$4n = 0$$

$$\frac{4n}{4} = \frac{0}{4}$$

$$n = 0$$

59.

$$-2(x-3) = 26$$

$$-2x+6 = 26$$

$$-2x+6-6 = 26-6$$

$$-2x = 20$$

$$\frac{-2x}{-2} = \frac{20}{-2}$$

$$x = -10$$

61.

$$-4 = -(x+5)$$

$$-4 = -x-5$$

$$-4+5 = -x-5+5$$

$$1 = -x$$

$$(-1)(1) = (-1)(-x)$$

$$-1 = x$$

63.

$$12 = 4(x-3)$$

$$12 = 4x-12$$

$$12+12 = 4x-12+12$$

$$24 = 4x$$

$$\frac{24}{4} = \frac{4x}{4}$$

$$6 = x$$

65.

$$2x-3(x+5) = 6$$

$$2x-3x-15 = 6$$

$$-x-15 = 6$$

$$-x-15+15 = 6+15$$

$$-x = 21$$

$$x = -21$$

67.

$$-3r-4(r+2) = 11$$

$$-3r-4r-8 = 11$$

$$-7r-8 = 11$$

$$-7r-8+8 = 11+8$$

$$-7r = 19$$

$$\frac{-7r}{-7} = \frac{19}{-7}$$

$$r = -\frac{19}{7}$$

69. $x - 3(2x + 3) = 11$

$x - 6x - 9 = 11$

$-5x - 9 = 11$

$-5x - 9 + 9 = 11 + 9$

$-5x = 20$

$\dfrac{-5x}{-5} = \dfrac{20}{-5}$

$x = -4$

71. $5x + 3x - 4x - 7 = 9$

$4x - 7 = 9$

$4x - 7 + 7 = 9 + 7$

$4x = 16$

$\dfrac{4x}{4} = \dfrac{16}{4}$

$x = 4$

73. $0.7(x - 3) = 1.4$

$0.7x - 2.1 = 1.4$

$0.7x - 2.1 + 2.1 = 1.4 + 2.1$

$0.7x = 3.5$

$\dfrac{0.7x}{0.7} = \dfrac{3.5}{0.7}$

$x = 5$

75. $2.5(4q - 3) = 0.5$

$10q - 7.5 = 0.5$

$10q - 7.5 + 7.5 = 0.5 + 7.5$

$10q = 8$

$\dfrac{10q}{10} = \dfrac{8}{10}$

$q = 0.8$

77. $3 - 2(x + 3) + 2 = 1$

$3 - 2x - 6 + 2 = 1$

$-2x + 3 - 6 + 2 = 1$

$-2x - 1 = 1$

$-2x - 1 + 1 = 1 + 1$

$-2x = 2$

$\dfrac{-2x}{-2} = \dfrac{2}{-2}$

$x = -1$

79. $1 + (x + 3) + 6x = 6$

$1 + x + 3 + 6x = 6$

$x + 6x + 1 + 3 = 6$

$7x + 4 = 6$

$7x + 4 - 4 = 6 - 4$

$7x = 2$

$\dfrac{7x}{7} = \dfrac{2}{7}$

$x = \dfrac{2}{7}$

81. $4.85 - 6.4x + 1.11 = 22.6$

$-6.4x + 4.85 + 1.11 = 22.6$

$-6.4x + 5.96 = 22.6$

$-6.4x + 5.96 - 5.96 = 22.6 - 5.96$

$-6.4x = 16.64$

$\dfrac{-6.4x}{-6.4} = \dfrac{16.64}{-6.4}$

$x = -2.6$

83. $7 = 8 - 5(m + 3)$

$7 = 8 - 5m - 15$

$7 = -5m - 7$

$7 + 7 = -5m - 7 + 7$

$14 = -5m$

$\dfrac{14}{-5} = \dfrac{-5m}{-5}$

$-\dfrac{14}{5} = m$

85. $10 = \dfrac{2s + 4}{5}$

$5(10) = 5\left(\dfrac{2s + 4}{5}\right)$

$50 = 2s + 4$

$50 - 4 = 2s + 4 - 4$

$46 = 2s$

$\dfrac{46}{2} = \dfrac{2s}{2}$

$23 = s$

87.
$$x + \frac{2}{3} = \frac{3}{5}$$
$$15\left(x + \frac{2}{3}\right) = 15\left(\frac{3}{5}\right)$$
$$15x + 10 = 9$$
$$15x + 10 - 10 = 9 - 10$$
$$15x = -1$$
$$\frac{15x}{15} = \frac{-1}{15}$$
$$x = -\frac{1}{15}$$

89.
$$\frac{r}{3} + 2r = 7$$
$$3\left(\frac{r}{3} + 2r\right) = 3(7)$$
$$r + 6r = 21$$
$$7r = 21$$
$$\frac{7r}{7} = \frac{21}{7}$$
$$r = 3$$

91.
$$\frac{3}{7} = \frac{3t}{4} + 1$$
$$28\left(\frac{3}{7}\right) = 28\left(\frac{3t}{4} + 1\right)$$
$$12 = 21t + 28$$
$$12 - 28 = 21t + 28 - 28$$
$$-16 = 21t$$
$$\frac{-16}{21} = \frac{21t}{21}$$
$$-\frac{16}{21} = t$$

93.
$$\frac{1}{2}r + \frac{1}{5}r = 7$$
$$10\left(\frac{1}{2}r + \frac{1}{5}r\right) = 10(7)$$
$$5r + 2r = 70$$
$$7r = 70$$
$$\frac{7r}{7} = \frac{70}{7}$$
$$r = 10$$

95.
$$\frac{2}{8} + \frac{3}{4} = \frac{w}{5}$$
$$\frac{1}{4} + \frac{3}{4} = \frac{w}{5}$$
$$\frac{4}{4} = \frac{w}{5}$$
$$1 = \frac{w}{5}$$
$$5(1) = 5\left(\frac{w}{5}\right)$$
$$5 = w$$

97.
$$\frac{1}{2}x + 4 = \frac{1}{6}$$
$$6\left(\frac{1}{2}x + 4\right) = 6\left(\frac{1}{6}\right)$$
$$3x + 24 = 1$$
$$3x + 24 - 24 = 1 - 24$$
$$3x = -23$$
$$\frac{3x}{3} = \frac{-23}{3}$$
$$x = -\frac{23}{3}$$

99.
$$\frac{4}{5}s - \frac{3}{4}s = \frac{1}{10}$$
$$20\left(\frac{4}{5}s - \frac{3}{4}s\right) = 20\left(\frac{1}{10}\right)$$
$$16s - 15s = 2$$
$$s = 2$$

101.
$$\frac{4}{9} = \frac{1}{3}(n - 7)$$
$$\frac{4}{9} = \frac{1}{3}n - \frac{7}{3}$$
$$9\left(\frac{4}{9}\right) = 9\left(\frac{1}{3}n - \frac{7}{3}\right)$$
$$4 = 3n - 21$$
$$4 + 21 = 3n - 21 + 21$$
$$25 = 3n$$
$$\frac{25}{3} = \frac{3n}{3}$$
$$\frac{25}{3} = n$$

103.
$$-\frac{3}{5} = -\frac{1}{9} - \frac{3}{4}x$$
$$180\left(-\frac{3}{5}\right) = 180\left(-\frac{1}{9} - \frac{3}{4}x\right)$$
$$-108 = -20 - 135x$$
$$-108 + 20 = -20 + 20 - 135x$$
$$-88 = -135x$$
$$\frac{-88}{-135} = \frac{-135x}{-135}$$
$$\frac{88}{135} = x$$

105. a. By subtracting first, you will not have to work with fractions.

   b.
$$3x + 2 = 11$$
$$3x + 2 - 2 = 11 - 2$$
$$3x = 9$$
$$\frac{3x}{3} = \frac{9}{3}$$
$$x = 3$$

107. $3(x-2) - (x+5) - 2(3-2x) = 18$
$$3x - 6 - x - 5 - 6 + 4x = 18$$
$$3x - x + 4x - 6 - 5 - 6 = 18$$
$$6x - 17 = 18$$
$$6x - 17 + 17 = 18 + 17$$
$$6x = 35$$
$$\frac{6x}{6} = \frac{35}{6}$$
$$x = \frac{35}{6}$$

109. $4[3 - 2(x+4)] - (x+3) = 13$
$$4(3 - 2x - 8) - x - 3 = 13$$
$$4(-2x - 5) - x - 3 = 13$$
$$-8x - 20 - x - 3 = 13$$
$$-9x - 23 = 13$$
$$-9x - 23 + 23 = 13 + 23$$
$$-9x = 36$$
$$\frac{-9x}{-9} = \frac{36}{-9}$$
$$x = -4$$

113. False. $\sqrt{2}$ is a real number but is irrational.

114. $\left[5(2-6) + 3(8 \div 4)^2\right]^2 = \left[5(-4) + 3(2)^2\right]^2$
$$= \left[5(-4) + 3(4)\right]^2$$
$$= \left[-20 + 12\right]^2$$
$$= \left[-8\right]^2$$
$$= 64$$

115. To solve an equation, we need to isolate the variable on one side of the equation.

116. To solve the equation, we divide both sides of the equation by –4.

Mid-Chapter Test: 2.1-2.4

1. $5x - 9y - 12 + 4y - 7x + 6$
$$= 5x - 7x - 9y + 4y - 12 + 6$$
$$= -2x - 5y - 6$$

2. $\frac{2}{5}x - 8 - \frac{3}{4}x + \frac{1}{2} = \frac{2}{5}x - \frac{3}{4}x - 8 + \frac{1}{2}$
$$= \frac{8}{20}x - \frac{15}{20}x - \frac{16}{2} + \frac{1}{2}$$
$$= -\frac{7}{20}x - \frac{15}{2}$$

3. $-4(2a - 3b + 6) = -8a + 12b - 24$

4. $1.6(2.1x - 3.4y - 5.2) = 3.36x - 5.44y - 8.32$

5. $5(t-3) - 3(t+7) - 2 = 5t - 15 - 3t - 21 - 2$
$$= 5t - 3t - 15 - 21 - 2$$
$$= 2t - 38$$

6. Substitute 2 for x in the following equation .
$$3(x - 4) = -2(x + 1)$$
$$3(2 - 4) = -2(2 + 1)$$
$$3(-2) = -2(3)$$
$$-6 = -6 \text{ True}$$
Since we obtained a true statement, 2 is a solution.

7. Substitute $\dfrac{2}{5}$ for p in the following equation.

$$7p - 3 = 2p - 5$$

$$7\left(\dfrac{2}{5}\right) - 3 = 2\left(\dfrac{2}{5}\right) - 5$$

$$\dfrac{14}{5} - \dfrac{15}{5} = \dfrac{4}{5} - \dfrac{25}{5}$$

$$-\dfrac{9}{5} = -\dfrac{21}{5} \quad \text{False}$$

Since we obtained a false statement, $\dfrac{2}{5}$ is not a solution.

8.
$$x - 5 = -9 \qquad \text{Check: } x - 5 = -9$$
$$x - 5 + 5 = -9 + 5 \qquad\qquad -4 - 5 = -9$$
$$x = -4 \qquad\qquad -9 = -9 \;\; \text{True}$$

9.
$$12 + x = -4 \qquad \text{Check: } \quad 12 + x = -4$$
$$12 - 12 + x = -4 - 12 \qquad 12 + (-16) = -4$$
$$x = -16 \qquad\qquad -4 = -4$$

10.
$$-16 = 7 + y \qquad \text{Check: } -16 = 7 + y$$
$$-16 - 7 = 7 - 7 + y \qquad -16 = 7 + (-23)$$
$$-23 = y \qquad\qquad -16 = -16 \;\; \text{True}$$

11. Multiply both sides of the equation by 4.

12.
$$6 = 12y \qquad \text{Check: } 6 = 12y$$
$$\dfrac{6}{12} = \dfrac{12y}{12} \qquad\qquad 6 = 12\left(\dfrac{1}{2}\right)$$
$$\dfrac{1}{2} = y \qquad\qquad 6 = 6 \;\; \text{True}$$

13.
$$\dfrac{x}{8} = 3 \qquad \text{Check: } \dfrac{x}{8} = 3$$
$$8\left(\dfrac{x}{8}\right) = 8(3) \qquad\qquad \dfrac{24}{8} = 3$$
$$x = 24 \qquad\qquad 3 = 3 \;\; \text{True}$$

14.
$$-\dfrac{x}{5} = -2 \qquad \text{Check: } -\dfrac{x}{5} = -2$$
$$\dfrac{x}{-5} = -2 \qquad\qquad -\dfrac{10}{5} = -2$$
$$-5\left(\dfrac{x}{-5}\right) = -5(-2) \qquad -2 = -2 \;\; \text{True}$$
$$x = 10$$

15.
$$-x = \dfrac{3}{7} \qquad \text{Check: } \quad -x = \dfrac{3}{7}$$
$$-1x = \dfrac{3}{7} \qquad\qquad -\left(-\dfrac{3}{7}\right) = \dfrac{3}{7}$$
$$-1(-1x) = -1\left(\dfrac{3}{7}\right) \qquad \dfrac{3}{7} = \dfrac{3}{7}$$
$$x = -\dfrac{3}{7}$$

16.
$$6x - 3 = 12$$
$$6x = 15$$
$$x = \dfrac{15}{6}$$
$$x = \dfrac{5}{2}$$

17.
$$-4 = -2w - 7$$
$$3 = -2w$$
$$-\dfrac{3}{2} = w$$

18.
$$\dfrac{3}{4} = \dfrac{4n - 1}{6}$$
$$12\left(\dfrac{3}{4}\right) = 12\left(\dfrac{4n - 1}{6}\right)$$
$$9 = 2(4n - 1)$$
$$9 = 8n - 2$$
$$11 = 8n$$
$$\dfrac{11}{8} = n$$

19.
$$-5(x + 4) - 7 = 3$$
$$-5x - 20 - 7 = 3$$
$$-5x - 27 = 3$$
$$-5x = 30$$
$$x = -6$$

20.
$$8 - 9(y + 4) + 6 = -2$$
$$8 - 9y - 36 + 6 = -2$$
$$-9y + 8 - 36 + 6 = -2$$
$$-9y - 22 = -2$$
$$-9y = 20$$
$$y = -\dfrac{20}{9}$$

Exercise Set 2.5

1. Answers will vary.

3. **a.** An identity is an equation that is true for infinitely many values of the variable.

   b. The solution is all real numbers.

5. The equation is an identity because both sides of the equation are identical.

7. An equation has no solution if it simplifies to a false statement.

9. **a.** Use the distributive property.
   Combine like terms.
   Subtract $5x$ from both sides of the equation.
   Subtract 6 from both sides of the equation.
   Divide both sides of the equation by 2.

   b.
   $$4x + 3(x + 2) = 5x - 10$$
   $$4x + 3x + 6 = 5x - 10$$
   $$7x + 6 = 5x - 10$$
   $$2x + 6 = -10$$
   $$2x = -16$$
   $$x = -8$$

11.
$$3x = -2x + 15$$
$$3x + 2x = -2x + 2x + 15$$
$$5x = 15$$
$$\frac{5x}{5} = \frac{15}{5}$$
$$x = 3$$

13.
$$-4x + 10 = 6x$$
$$-4x + 4x + 10 = 6x + 4x$$
$$10 = 10x$$
$$\frac{10}{10} = \frac{10x}{10}$$
$$1 = x$$

15.
$$5x + 3 = 6$$
$$5x + 3 - 3 = 6 - 3$$
$$5x = 3$$
$$\frac{5x}{5} = \frac{3}{5}$$
$$x = \frac{3}{5}$$

17.
$$21 - 6p = 3p - 2p$$
$$21 - 6p = p$$
$$21 - 6p + 6p = p + 6p$$
$$21 = 7p$$
$$\frac{21}{7} = \frac{7p}{7}$$
$$3 = p$$

19.
$$2x - 4 = 3x - 6$$
$$2x - 2x - 4 = 3x - 2x - 6$$
$$-4 = x - 6$$
$$-4 + 6 = x - 6 + 6$$
$$2 = x$$

21.
$$6 - 2y = 9 - 8y + 6y$$
$$6 - 2y = 9 - 2y$$
$$6 - 2y + 2y = 9 - 2y + 2y$$
$$6 = 9 \qquad \text{False}$$
Since a false statement is obtained, there is no solution.

23.
$$124.8 - 9.4x = 4.8x + 32.5$$
$$124.8 - 9.4x + 9.4x = 4.8x + 9.4x + 32.5$$
$$124.8 = 14.2x + 32.5$$
$$124.8 - 32.5 = 14.2x + 32.5 - 32.5$$
$$92.3 = 14.2x$$
$$\frac{92.3}{14.2} = \frac{14.2x}{14.2}$$
$$6.5 = x$$

25.
$$0.62x - .065 = 9.75 - 2.63x$$
$$0.62x + 2.63x - 0.65 = 9.75 - 2.63x + 2.63x$$
$$3.25x - 0.65 = 9.75$$
$$3.25x - 0.65 + 0.65 = 9.75 + 0.65$$
$$3.25x = 10.4$$
$$\frac{3.25x}{3.25} = \frac{10.4}{3.25}$$
$$x = 3.2$$

27.
$$5x + 3 = 2(x + 6)$$
$$5x + 3 = 2x + 12$$
$$5x - 2x + 3 = 2x - 2x + 12$$
$$3x + 3 = 12$$
$$3x + 3 - 3 = 12 - 3$$
$$3x = 9$$
$$\frac{3x}{3} = \frac{9}{3}$$
$$x = 3$$

29.
$$4y - 2 - 8y = 19 + 5y - 3$$
$$4y - 8y - 2 = 5y + 19 - 3$$
$$-4y - 2 = 5y + 16$$
$$-4y + 4y - 2 = 5y + 4y + 16$$
$$-2 = 9y - 16$$
$$-2 - 16 = 9y + 16 - 16$$
$$-18 = 9y$$
$$\frac{-18}{9} = \frac{9y}{9}$$
$$-2 = y$$

31.
$$2(x - 2) = 4x - 6 - 2x$$
$$2x - 4 = 2x - 6$$
$$2x - 2x - 4 = 2x - 2x - 6$$
$$-4 = -6 \text{ False}$$
Since a false statement is obtained, there is no solution.

33.
$$-(w + 2) = -6w + 32$$
$$-w - 2 = -6w + 32$$
$$-w + w - 2 = -6w + w + 32$$
$$-2 = -5w + 32$$
$$-2 - 32 = -5w + 32 - 32$$
$$-34 = -5w$$
$$\frac{-34}{-5} = \frac{-5w}{-5}$$
$$\frac{34}{5} = w$$

35.
$$-3(2t - 5) + 5 = 3t + 13$$
$$-6t + 15 + 5 = 3t + 13$$
$$-6t + 20 = 3t + 13$$
$$-6t + 6t + 20 = 3t + 6t + 13$$
$$20 = 9t + 13$$
$$20 - 13 = 9t + 13 - 13$$
$$7 = 9t$$
$$\frac{7}{9} = \frac{9t}{9}$$
$$\frac{7}{9} = t$$

37.
$$\frac{a}{5} = \frac{a - 3}{2}$$
$$10\left(\frac{a}{5}\right) = 10\left(\frac{a - 3}{2}\right)$$
$$2a = 5(a - 3)$$
$$2a = 5a - 15$$
$$2a - 5a = 5a - 5a - 15$$
$$-3a = -15$$
$$\frac{-3a}{-3} = \frac{-15}{-3}$$
$$a = 5$$

39.
$$\frac{n}{10} = 9 - \frac{n}{5}$$
$$10\left(\frac{n}{10}\right) = 10\left(9 - \frac{n}{5}\right)$$
$$n = 90 - 2n$$
$$n + 2n = 90 - 2n + 2n$$
$$3n = 90$$
$$\frac{3n}{3} = \frac{90}{3}$$
$$n = 30$$

41.
$$\frac{5}{2} - \frac{x}{3} = 3x$$
$$6\left(\frac{5}{2} - \frac{x}{3}\right) = 6(3x)$$
$$15 - 2x = 18x$$
$$15 - 2x + 2x = 18x + 2x$$
$$15 = 20x$$
$$\frac{15}{20} = \frac{20x}{20}$$
$$\frac{15}{20} = x$$
$$\frac{3}{4} = x$$

43.
$$\frac{5}{8} + \frac{1}{4}a = \frac{1}{2}a$$
$$8\left(\frac{5}{8} + \frac{1}{4}a\right) = 8\left(\frac{1}{2}a\right)$$
$$5 + 2a = 4a$$
$$5 + 2a - 2a = 4a - 2a$$
$$5 = 2a$$
$$\frac{5}{2} = \frac{2a}{2}$$
$$\frac{5}{2} = a$$

45.
$$0.1(x + 10) = 0.3x - 4$$
$$0.1x + 1 = 0.3x - 4$$
$$0.1x - 0.1x + 1 = 0.3x - 0.1 - 4$$
$$1 = 0.2x - 4$$
$$1 + 4 = 0.2x - 4 + 4$$
$$5 = 0.2x$$
$$\frac{5}{0.2} = \frac{0.2x}{0.2}$$
$$25 = x$$

47. $2(x + 4) = 4x + 3 - 2x + 5$
$$2x + 8 = 4x + 3 - 2x + 5$$
$$2x + 8 = 2x + 8$$
Since the left side of the equation is identical to the right side, the equation is true for all values of *x*. Thus the solution is all real numbers.

49.
$$5(3n + 3) = 2(5n - 4) + 6n$$
$$15n + 15 = 10n - 8 + 6n$$
$$15n + 15 = 16n - 8$$
$$15n - 15n + 15 = 16n - 15n - 8$$
$$15 = n - 8$$
$$15 + 8 = n - 8 + 8$$
$$23 = n$$

51.
$$-(3 - p) = -(2p + 3)$$
$$-3 + p = -2p - 3$$
$$-3 + p + 2p = -2p + 2p - 3$$
$$-3 + 3p = -3$$
$$-3 + 3 + 3p = -3 + 3$$
$$3p = 0$$
$$\frac{3p}{3} = \frac{0}{3}$$
$$p = 0$$

53. $-(x + 4) + 5 = 4x + 1 - 5x$
$$-x - 4 + 5 = 4x + 1 - 5x$$
$$-x + 1 = -x + 1$$
Since the left side of the equation is identical to the right side, the equation is true for all values of *x*. Thus the solution is all real numbers.

55.
$$35(2x - 1) = 7(x + 4) + 3x$$
$$70x - 35 = 7x + 28 + 3x$$
$$70x - 35 = 10x + 28$$
$$70x - 10x - 35 = 10x - 10x + 28$$
$$60x - 35 = 28$$
$$60x - 35 + 35 = 28 + 35$$
$$60x = 63$$
$$\frac{60x}{60} = \frac{63}{60}$$
$$x = \frac{21}{20}$$

57.
$$0.4(x+0.7)=0.6(x-4.2)$$
$$0.4x+0.28=0.6x-2.52$$
$$0.4x-0.4x+0.28=0.6x-0.4x-2.52$$
$$0.28=0.2x-2.52$$
$$0.28+2.52=0.2x-2.52+2.52$$
$$2.8=0.2x$$
$$\frac{2.8}{0.2}=\frac{0.2x}{0.2}$$
$$14=x$$

59.
$$\frac{3}{5}x-2=x+\frac{1}{3}$$
$$15\left(\frac{3}{5}x-2\right)=15\left(x+\frac{1}{3}\right)$$
$$9x-30=15x+5$$
$$9x-9x-30=15x-9x+5$$
$$-30=6x+5$$
$$-30-5=6x+5-5$$
$$-35=6x$$
$$\frac{-35}{6}=\frac{6x}{6}$$
$$-\frac{35}{6}=x$$

61.
$$\frac{y}{5}+2=3(y-4)$$
$$\frac{y}{5}+2=3y-12$$
$$\frac{y}{5}+2-2=3y-12-2$$
$$\frac{y}{5}=3y-14$$
$$5\left(\frac{y}{5}\right)=5(3y-14)$$
$$y=15y-70$$
$$y-15y=15y-15y-70$$
$$-14y=-70$$
$$\frac{-14y}{-14}=\frac{-70}{-14}$$
$$y=5$$

63.
$$12-3x+7x=-2(-5x+6)$$
$$12+4x=10x-12$$
$$12+4x-4x=10x-4x-12$$
$$12=6x-12$$
$$12+12=6x-12+12$$
$$24=6x$$
$$\frac{24}{6}=\frac{6x}{6}$$
$$4=x$$

65.
$$3(x-6)-4(3x+1)=x-22$$
$$3x-18-12x-4=x-22$$
$$-9x-22=x-22$$
$$-9x+9x-22=x+9x-22$$
$$-22=10x-22$$
$$-22+22=10x-22+22$$
$$0=10x$$
$$\frac{0}{10}=\frac{10x}{10}$$
$$0=x$$

67.
$$5+2x=6(x+1)-5(x-3)$$
$$5+2x=6x+6-5x+15$$
$$5+2x=x+21$$
$$5+2x-x=x-x+21$$
$$5+x=21$$
$$5-5+x=21-5$$
$$x=16$$

69.
$$7-(-y-5)=2(y+3)-6(y+1)$$
$$7+y+5=2y+6-6y-6$$
$$12+y=-4y$$
$$12+y-y=-4y-y$$
$$12=-5y$$
$$\frac{12}{-5}=\frac{-5y}{-5}$$
$$-\frac{12}{5}=y$$

71. $\dfrac{3}{5}(x-6) = \dfrac{2}{3}(3x-5)$

$\dfrac{3}{5}x - \dfrac{18}{5} = 2x - \dfrac{10}{3}$

$15\left(\dfrac{3}{5}x - \dfrac{18}{5}\right) = 15\left(2x - \dfrac{10}{3}\right)$

$9x - 54 = 30x - 50$

$9x - 9x - 54 = 30x - 9x - 50$

$-54 = 21x - 50$

$-54 + 50 = 21x - 50 + 50$

$-4 = 21x$

$\dfrac{-4}{21} = \dfrac{21x}{21}$

$-\dfrac{4}{21} = x$

73. $\dfrac{3(2r-5)}{5} = \dfrac{3r-6}{4}$

$\dfrac{6r-15}{5} = \dfrac{3r-6}{4}$

$20\left(\dfrac{6r-15}{5}\right) = 20\left(\dfrac{3r-6}{4}\right)$

$4(6r-15) = 5(3r-6)$

$24r - 60 = 15r - 30$

$24r - 15r - 60 = 15r - 15r - 30$

$9r - 60 = -30$

$9r - 60 + 60 = -30 + 60$

$9r = 30$

$\dfrac{9r}{9} = \dfrac{30}{9}$

$r = \dfrac{30}{9} = \dfrac{10}{3}$

75. $\dfrac{2}{7}(5x+4) = \dfrac{1}{2}(3x-4) + 1$

$\dfrac{10x}{7} + \dfrac{8}{7} = \dfrac{3x}{2} - 2 + 1$

$\dfrac{10x}{7} + \dfrac{8}{7} = \dfrac{3x}{2} - 1$

$14\left(\dfrac{10x}{7} + \dfrac{8}{7}\right) = 14\left(\dfrac{3x}{2} - 1\right)$

$20x + 16 = 21x - 14$

$20x - 20x + 16 = 21x - 20x - 14$

$16 = x - 14$

$16 + 14 = x - 14 + 14$

$30 = x$

77. $\dfrac{a-5}{2} = \dfrac{3a}{4} + \dfrac{a-25}{6}$

$12\left(\dfrac{a-5}{2}\right) = 12\left(\dfrac{3a}{4} + \dfrac{a-25}{6}\right)$

$6(a-5) = 9a + 2(a-25)$

$6a - 30 = 9a + 2a - 50$

$6a - 30 = 11a - 50$

$6a - 6a - 30 = 11a - 6a - 50$

$-30 = 5a - 50$

$-30 + 50 = 5a - 50 + 50$

$20 = 5a$

$\dfrac{20}{5} = \dfrac{5a}{5}$

$4 = a$

79. a. One example is $x + x + 1 = x + 2$.

   b. It has a single solution.

   c. Answers will vary. For equation given in part **a**):

$x + x + 1 = x + 2$

$2x + 1 = x + 2$

$2x - x + 1 = x - x + 2$

$x + 1 = 2$

$x + 1 - 1 = 2 - 1$

$x = 1$

81. a. One example is $x + x + 1 = 2x + 1$.

   b. Both sides simplify to the same expression.

   c. The solution is all real numbers.

83. a. One example is $x + x + 1 = 2x + 2$.

   b. It simplifies to a false statement.

   c. The solution is that there is no solution.

85. $5* - 1 = 4* + 5*$

$5* - 5* - 1 = 9* - 5*$

$-1 = 4*$

$\dfrac{-1}{4} = \dfrac{4*}{4}$

$-\dfrac{1}{4} = *$

87. $3☺ - 5 = 2☺ - 5 + ☺$

$3☺ - 5 = 3☺ - 5$

The left side of the equation is identical to the right side. The solution is all real numbers.

89. $4 - \left[5 - 3(x+2)\right] = x - 3$

$$4 - (5 - 3x - 6) = x - 3$$
$$4 - 5 + 3x + 6 = x - 3$$
$$3x + 5 = x - 3$$
$$3x - x + 5 = x - x - 3$$
$$2x + 5 = -3$$
$$2x + 5 - 5 = -3 - 5$$
$$2x = -8$$
$$\frac{2x}{2} = \frac{-8}{2}$$
$$x = -4$$

91. a. $|4| = 4$

    b. $|-7| = 7$

    c. $|0| = 0$

92. $\left(\dfrac{2}{3}\right)^5 \approx 0.131687243$

93. Factors are expressions that are multiplied together; terms are expressions that are added together.

94. $2(x-3) + 4x - (4-x) = 2x - 6 + 4x - 4 + x$
$$= 7x - 10$$

95. $2(x-3) + 4x - (4-x) = 0$
$$2x - 6 + 4x - 4 + x = 0$$
$$7x - 10 = 0$$
$$7x - 10 + 10 = 0 + 10$$
$$7x = 10$$
$$\frac{7x}{7} = \frac{10}{7}$$
$$x = \frac{10}{7}$$

96. $(x+4) - (4x-3) = 16$
$$x + 4 - 4x + 3 = 16$$
$$-3x + 7 = 16$$
$$-3x + 7 - 7 = 16 - 7$$
$$-3x = 9$$
$$\frac{-3x}{-3} = \frac{9}{-3}$$
$$x = -3$$

Exercise Set 2.6

1. A formula is an equation used to express a relationship mathematically.

3. The simple interest formula is:
$i = prt$ where i is interest, p is principle, r is the interest rate, and t is time.

5. $d = rt$; d represents distance, r represents rate and t represents time

7. No, π is an irrational number that is approximately equal to 3.14, but not exactly.

9. When you multiply a unit by the same unit, you get a square unit.

11. Substitute 60 for r and 4 for t.
$$d = rt$$
$$d = 60(4) = 240$$

13. Substitute 12 for l and 8 for w.
$$A = lw$$
$$A = 12(8) = 96$$

15. Substitute 2000 for p, 0.06 for r, and 3 for t.
$$i = prt$$
$$i = 2000(0.06)(3) = 360$$

17. Substitute 8 for l and 5 for w.
$$P = 2l + 2w$$
$$P = 2(8) + 2(5) = 16 + 10 = 26$$

19. Substitute 5 for r.
$$A = \pi r^2$$
$$A = \pi(5)^2 = 25\pi \approx 78.54$$

21. Substitute 72 for a, 81 for b, and 93 for c.
$$A = \frac{a+b+c}{3}$$
$$A = \frac{72+81+93}{3} = \frac{246}{3} = 82$$

23. Substitute 100 for x, 80 for m, and 10 for s.
$$z = \frac{x-m}{s}$$
$$z = \frac{100-80}{10} = \frac{20}{10} = 2$$

25. Substitute 28 for P and 6 for w.
$$P = 2l + 2w$$
$$28 = 2l + 2(6)$$
$$28 = 2l + 12$$
$$28 - 12 = 2l + 12 - 12$$
$$16 = 2l$$
$$\frac{16}{2} = \frac{2l}{2}$$
$$8 = l$$

27. Substitute 678.24 for V, and 6 for r.
$$V = \pi r^2 h$$
$$678.24 = \pi(6)^2 h$$
$$678.24 = 36\pi h$$
$$\frac{678.24}{36\pi} = \frac{36\pi h}{36\pi}$$
$$\frac{678.24}{36\pi} = h$$
$$6.00 \approx h$$

29. Substitute 24 for B and 61 for h.
$$B = \frac{703w}{h^2}$$
$$24 = \frac{703w}{(61)^2}$$
$$24(61)^2 = \frac{703w}{61^2}(61)^2$$
$$89,304 = 703w$$
$$\frac{89,304}{703} = \frac{703w}{703}$$
$$127.03 \approx w$$

31. Substitute 8 for r.
$$A = \pi r^2$$
$$A = \pi(4)^2 = 16\pi \approx 50.27 \text{ft}^2$$

33. Substitute 3 for h, 4 for b and 7 for d.
$$A = \frac{1}{2}h(b + d)$$
$$A = \frac{1}{2}(3)(4 + 7) = \frac{1}{2}(3)(11) = \frac{1}{2}(33) = 16.5 \text{ ft}^2$$

35. Substitute 4 for r and 9 for h.
$$V = \pi r^2 h$$
$$V = \pi(4)^2(9) = \pi(16)(9) = 144\pi \approx 452.39 \text{ cm}^3$$

37. Substitute 50 for F.
$$C = \frac{5}{9}(F - 32)$$
$$C = \frac{5}{9}(50 - 32) = \frac{5}{9}(18) = 10$$
The equivalent temperature is 10°C.

39. Substitute 25 for C.
$$F = \frac{9}{5}C + 32$$
$$F = \frac{9}{5}(25) + 32 = 45 + 32 = 77$$
The equivalent temperature is 77°F.

41. $P = \dfrac{KT}{V}$
$$P = \frac{(2)(20)}{1} = \frac{40}{1} = 40$$

43. $P = \dfrac{KT}{V}$
$$3 = \frac{(0.5)(30)}{V}$$
$$3 = \frac{15}{V}$$
$$3V = 15$$
$$V = \frac{15}{3} = 5$$

45. $A = lw$
$$\frac{A}{l} = \frac{lw}{l}$$
$$\frac{A}{l} = w$$

47. $d = rt$
$$\frac{d}{r} = \frac{rt}{r}$$
$$\frac{d}{r} = t$$

49. $i = prt$
$$\frac{i}{pr} = \frac{prt}{pr}$$
$$\frac{i}{pr} = t$$

51. $A = \dfrac{1}{2}bh$

$2A = 2\left(\dfrac{1}{2}bh\right)$

$2A = bh$

$\dfrac{2A}{h} = \dfrac{bh}{h}$

$\dfrac{2A}{h} = b$

53. $P = 2l + 2w$

$P - 2l = 2l - 2l + 2w$

$P - 2l = 2w$

$\dfrac{P - 2l}{2} = \dfrac{2w}{2}$

$\dfrac{P - 2l}{2} = w$

55. $3 - 2r = n$

$3 - 3 - 2r = n - 3$

$-2r = n - 3$

$\dfrac{-2r}{-2} = \dfrac{n - 3}{-2}$

$r = \dfrac{n - 3}{-2} = \dfrac{3 - n}{2}$

57. $y = mx + b$

$y - mx = mx - mx + b$

$y - mx = b$

59. $d = a + b + c$

$d - a = a - a + b + c$

$d - a = b + c$

$d - a - c = b + c - c$

$d - a - c = b$

61. $ax + by + c = 0$

$ax - ax + by + c = -ax$

$by + c = -ax$

$by + c - c = -ax - c$

$by = -ax - c$

$\dfrac{by}{b} = \dfrac{-ax - c}{b}$

$y = \dfrac{-ax - c}{b}$

63. $V = \dfrac{1}{3}\pi r^2 h$

$3V = 3\left(\dfrac{1}{3}\pi r^2 h\right)$

$3V = \pi r^2 h$

$\dfrac{3V}{\pi r^2} = \dfrac{\pi r^2 h}{\pi r^2}$

$\dfrac{3V}{\pi r^2} = h$

65. $A = \dfrac{m + d}{2}$

$2A = 2\left(\dfrac{m + d}{2}\right)$

$2A = m + d$

$2A - d = m + d - d$

$2A - d = m$

67. $2x + y = 5$

$2x - 2x + y = -2x + 5$

$y = -2x + 5$

69. $-3x + 3y = -15$

$-3x + 3x + 3y = 3x - 15$

$3y = 3x - 15$

$\dfrac{3y}{3} = \dfrac{3x - 15}{3}$

$y = \dfrac{3x}{3} + \dfrac{-15}{3}$

$y = x - 5$

71. $4x = 6y - 8$

$4x + 8 = 6y - 8 + 8$

$4x + 8 = 6y$

$\dfrac{4x + 8}{6} = \dfrac{6y}{6}$

$\dfrac{4x}{6} + \dfrac{8}{6} = y$

$\dfrac{2}{3}x + \dfrac{4}{3} = y$

$y = \dfrac{2}{3}x + \dfrac{4}{3}$

73. $5y = -10 + 3x$

$$\frac{5y}{5} = \frac{-10 + 3x}{5}$$

$$y = \frac{-10}{5} + \frac{3x}{5}$$

$$y = -2 + \frac{3}{5}x$$

$$y = \frac{3}{5}x - 2$$

75. $-6y = 15 - 3x$

$$\frac{-6y}{-6} = \frac{15 - 3x}{-6}$$

$$y = \frac{15}{-6} + \frac{-3x}{-6}$$

$$y = -\frac{5}{2} + \frac{1}{2}x$$

$$y = \frac{1}{2}x - \frac{5}{2}$$

77. $-8 = -x - 2y$

$$x - 8 = x - x - 2y$$

$$x - 8 = -2y$$

$$\frac{x - 8}{-2} = \frac{-2y}{-2}$$

$$\frac{-x}{2} + \frac{-8}{-2} = y$$

$$-\frac{1}{2}x + 4 = y$$

$$y = -\frac{1}{2}x + 4$$

79. $y + 3 = -\frac{1}{3}(x - 4)$

$$y + 3 = -\frac{1}{3}x + \frac{4}{3}$$

$$y + 3 - 3 = -\frac{1}{3}x + \frac{4}{3} - 3$$

$$y = -\frac{1}{3}x + \frac{4}{3} - \frac{9}{3}$$

$$y = -\frac{1}{3}x - \frac{5}{3}$$

81. $y - \frac{1}{5} = 2\left(x + \frac{1}{3}\right)$

$$y - \frac{1}{5} = 2x + \frac{2}{3}$$

$$y - \frac{1}{5} + \frac{1}{5} = 2x + \frac{2}{3} + \frac{1}{5}$$

$$y = 2x + \frac{10}{15} + \frac{3}{15}$$

$$y = 2x + \frac{13}{15}$$

83. $d = rt$

$$d = (2r)\left(\frac{1}{2}t\right)$$

$$d = 2 \cdot \frac{1}{2}rt$$

$$d = rt$$

The distance remains the same.

85. $A = s^2$

$$A = (2s)^2 = 4s^2$$

The area is 4 times as large as the original area.

87. Substitute $\frac{1}{2}s$ for r in the area formula.

$$A_{\text{circle}} = \pi r^2$$

$$= \pi\left(\frac{1}{2}s\right)^2$$

$$= \frac{1}{4}\pi s^2$$

$$= \frac{1}{4}(3.14)s^2$$

$$= 0.785s^2$$

$$A_{\text{square}} = s^2$$

The area of a square is larger because $1s^2 > 0.785s^2$.

89. $i = prt$

$$i = (6000)(0.08)(3) = 1440$$

He will pay $1440 interest.

91. $i = prt$

$450 = p(0.03)(3)$

$450 = 0.09p$

$\dfrac{450}{0.09} = \dfrac{0.09p}{0.09}$

$5000 = p$

She placed \$5000 in the savings account.

93. $d = rt$

$150 = r(3)$

$50 = r$

Her average speed was 50 mph.

95. $d = rt$

$d = (763.2)(0.01) = 7.632$

The car traveled 7.632 mi.

97. $A = lw$

$A = (8)(6) = 48$

The area of the screen is 48 sq. in.

99. $A = \dfrac{1}{2}bh$

$A = \dfrac{1}{2}(36)(31) = 558$

The area is 558 sq. in.

101. $C = 2\pi r$

$C = 2\pi(12) = 24\pi \approx 75.40$

The circumference of the pool is about 75.40 ft.

103. Total area = Area of top triangle + area of bottom triangle

Total Area $= 0.5b_1\,h_1 + 0.5b_2\,h_2$

Total Area $= 0.5(2)(1) + 0.5(2)(2)$

Total Area $= 1 + 2\ = 3$

The area of the kite is 3 square feet.

105. $A = \dfrac{1}{2}h(b+d)$

$A = \dfrac{1}{2}(2)(4+3) = \dfrac{1}{2}(2)(7) = (1)7 = 7$

The area of the sign is 7 sq. ft.

107. $C = 2\pi r$

$390 = 2\pi(r)$

$390 = 2\pi r$

$\dfrac{390}{2\pi} = \dfrac{2\pi r}{2\pi}$

$62.07 \approx r$

The radius of the roots is about 62.07 feet.

The diameter is twice the radius, so
$d = 2(62.07) = 124.14$
The diameter is about 124.1 ft.

109. The radius is half the diameter.

$V = \dfrac{4}{3}\pi r^3$

$V = \dfrac{4}{3}\pi\left(\dfrac{9}{2}\right)^3$

$V \approx 381.7$

The volume of the basketball is about 381.7 cu. in.

111. a. $B = \dfrac{703w}{h^2}$

b. 5 feet 3 inches $= 5(12)+3$

$\qquad\qquad\qquad = 60+3$

$\qquad\qquad\qquad = 63$ inches

$B = \dfrac{703(135)}{(63)^2} = \dfrac{94{,}905}{3969} \approx 23.91$

113. a. $V = lwh$

$V = (3x)(x)(6x-1)$

$\quad = 3x^2(6x-1)$

$\quad = 18x^3 - 3x^2$

b. $V = 18x^3 - 3x^2$

$V = 18(7)^3 - 3(7)^2 = 6174 - 147 = 6027$

Volume is 6027 cm^3.

c. $S = 2lw + 2lh + 2wh$

$S = 2(3x)(x) + 2(3x)(6x-1) + 2(x)(6x-1)$

$\quad = 6x^2 + 36x^2 - 6x + 12x^2 - 2x$

$\quad = 54x^2 - 8x$

d. $S = 54x^2 - 8x$

$S = 54(7)^2 - 8(7) = 2646 - 56 = 2590$

Surface area is 2590 cm^2.

115. $-\dfrac{4}{15}+\dfrac{2}{5}=-\dfrac{4}{15}+\dfrac{6}{15}=\dfrac{2}{15}$

116. $-6+7-4-3=1-4-3=-3-3=-6$

117. $\left[4\left(12\div 2^2-3\right)^2\right]^2=\left[4\left(12\div 4-3\right)^2\right]^2$
$$=\left[4\left(3-3\right)^2\right]^2$$
$$=\left[4\left(0\right)^2\right]^2$$
$$=\left[0\right]^2$$
$$=0$$

118. $\dfrac{r}{2}+2r=20$
$$2\left(\dfrac{r}{2}+2r\right)=2\left(20\right)$$
$$r+4r=40$$
$$5r=40$$
$$r=8$$

Exercise Set 2.7

1. A ratio is a quotient of two quantities.

3. The ratio of c to d can be written as c to d, $c{:}d$, and $\dfrac{c}{d}$.

5. To set up and solve a proportion, we need a given ratio and one of the two parts of a second ratio.

7. Yes, similar figures have the same shape but not necessarily the same size.

9. Yes; the terms in each ratio are in the same order.

11. No; The terms in each ratio are not in the same order.

13. $6{:}9=2{:}3$

15. $3{:}6=1{:}2$

17. Total grades $=6+4+9+3+2=24$
Ratio of total grades to D's $=24:3=8:1$

19. $7{:}4$

21. $5{:}15=1{:}3$

23. 3 hours $=3\times 60=180$ minutes
Ratio is $\dfrac{180}{30}=\dfrac{6}{1}$ or $6{:}1$.

25. 12 nickels is $12\cdot\left(\dfrac{1}{2}\right)=6$ dimes
Ratio is $7{:}6$.

27. Gear ratio $=\dfrac{\text{number of teeth on driving gear}}{\text{number of teeth on driven gear}}$
$$=\dfrac{40}{5}=\dfrac{8}{1}$$
Gear ratio is $8{:}1$.

29. **a.** $50{:}23$
   b. Since $50\div 23\approx 2.17$, $50{:}23\approx 2.17{:}1$.

31. **a.** $5.15{:}3.35$ or $1.03{:}0.67$
   b. Since $5.15\div 3.35\approx 1.54$, $5.15{:}3.35\approx 1.54{:}1$.

33. **a.** $19.2{:}2.2$ or $9.6{:}1.1$
   b. $6.7{:}3.1$

35. **a.** $40{:}32$ or $5{:}4$
   b. $15{:}11$

37. $\dfrac{x}{3}=\dfrac{20}{5}$
$$x\cdot 5=3\cdot 20$$
$$5x=60$$
$$\dfrac{5x}{5}=\dfrac{60}{5}$$
$$x=12$$

39. $\dfrac{5}{3}=\dfrac{75}{a}$
$$5\cdot a=3\cdot 75$$
$$5a=225$$
$$\dfrac{5a}{5}=\dfrac{225}{5}$$
$$a=45$$

41.
$$\frac{-7}{3} = \frac{21}{p}$$
$$-7 \cdot p = 3 \cdot 21$$
$$-7p = 63$$
$$\frac{-7p}{-7} = \frac{63}{-7}$$
$$p = -9$$

43.
$$\frac{15}{45} = \frac{x}{-6}$$
$$15 \cdot -6 = 45 \cdot x$$
$$-90 = 45x$$
$$\frac{-90}{45} = \frac{45x}{45}$$
$$-2 = x$$

45.
$$\frac{3}{z} = \frac{-1.5}{27}$$
$$3 \cdot 27 = z \cdot -1.5$$
$$81 = -1.5z$$
$$\frac{81}{-1.5} = \frac{-1.5z}{-1.5}$$
$$-54 = z$$

47.
$$\frac{9}{12} = \frac{x}{8}$$
$$9 \cdot 8 = 12 \cdot x$$
$$72 = 12x$$
$$\frac{72}{12} = \frac{12x}{12}$$
$$6 = x$$

49.
$$\frac{3}{12} = \frac{8}{x}$$
$$3x = 12 \cdot 8$$
$$3x = 96$$
$$\frac{3x}{3} = \frac{96}{3}$$
$$x = 32$$
Thus the side is 32 inches in length.

51.
$$\frac{4}{7} = \frac{9}{x}$$
$$4x = 7 \cdot 9$$
$$4x = 63$$
$$\frac{4x}{4} = \frac{63}{4}$$
$$x = 15.75$$
Thus the side is 15.75 inches in length.

53.
$$\frac{16}{12} = \frac{26}{x}$$
$$16x = (12)(26)$$
$$16x = 312$$
$$\frac{16x}{16} = \frac{312}{16}$$
$$x = 19.5$$
Thus the side is 19.5 inches in length.

55. Let x = number of loads one bottle can do.
$$\frac{4 \text{ fl ounces}}{1 \text{ load}} = \frac{100 \text{ fl ounces}}{x \text{ loads}}$$
$$\frac{4}{1} = \frac{100}{x}$$
$$4x = 100$$
$$x = 25$$
One bottle can do 25 loads.

57. Let x = number of miles that can be driven with a full tank.
$$\frac{19 \text{ miles}}{1 \text{ gallon}} = \frac{x}{14.2 \text{ gallons}}$$
$$\frac{19}{1} = \frac{x}{14.2}$$
$$x = (19)(14.2)$$
$$x = 269.8$$
It can travel 269.8 miles on a full tank.

59. Let x = length of model in feet.
$$\frac{1 \text{ foot model}}{20 \text{ foot train}} = \frac{x \text{ foot model}}{30 \text{ foot train}}$$
$$\frac{1}{20} = \frac{x}{30}$$
$$20x = 30$$
$$x = 1.5$$
The model should be 1.5 feet long.

61. Let x = number of teaspoons needed for sprayer.

$$\frac{3 \text{ teaspoons}}{1 \text{ gallon water}} = \frac{x \text{ teaspoons}}{8 \text{ gallons water}}$$

$$\frac{3}{1} = \frac{x}{8}$$

$$3 \cdot 8 = 1 \cdot x$$

$$24 = x$$

Thus 24 teaspoons are needed for the sprayer.

63. Let x = length of beak of blue heron in inches.

$$\frac{3.5 \text{ inches in photo}}{3.75 \text{ feet}} = \frac{0.4 \text{ inches in photo}}{x \text{ feet}}$$

$$\frac{3.5}{3.75} = \frac{0.4}{x}$$

$$3.5x = 3.75 \cdot 0.4$$

$$3.5x = 1.5$$

$$x \approx 0.43$$

It's beak is about 0.43 feet long.

65. Let x = number of cups needed.

$$\frac{1\frac{1}{2} \text{ cups}}{6 \text{ servings}} = \frac{x \text{ cups}}{15 \text{ servings}}$$

$$\frac{\frac{3}{2}}{6} = \frac{x}{15}$$

$$6x = 15 \cdot \frac{3}{2}$$

$$6x = 22.5$$

$$x = 3.75$$

3.75 cups of onions are needed.

67. Let x = length of the model bull in feet.

$$\frac{2.95 \text{ feet metal bull}}{1 \text{ feet real bull}} = \frac{28 \text{ feet metal bull}}{x \text{ feet real bull}}$$

$$\frac{2.95}{1} = \frac{28}{x}$$

$$2.95 \cdot x = 1 \cdot 28$$

$$2.95x = 28$$

$$x \approx 9.49$$

The model bull is about 9.49 feet long.

69. Let x = number of milliliters to be given.

$$\frac{1 \text{ milliliter}}{400 \text{ micrograms}} = \frac{x \text{ milliliter}}{220 \text{ micrograms}}$$

$$\frac{1}{400} = \frac{x}{220}$$

$$1 \cdot 220 = 400 \cdot x$$

$$220 = 400x$$

$$\frac{220}{400} = x$$

$$0.55 = x$$

Thus 0.55 milliliter should be given.

71. Let x = number of minutes to read entire book.

$$\frac{40 \text{ pages}}{30 \text{ minutes}} = \frac{760 \text{ pages}}{x \text{ minutes}}$$

$$\frac{40}{30} = \frac{760}{x}$$

$$40x = (30)(760)$$

$$40x = 22,800$$

$$x = \frac{22,800}{40} = 570$$

Thus it will take her 570 minutes or 9 hours 30 minutes to read the entire book.

73. Let x = number of children born with Prader-Willi Syndrome.

$$\frac{12,000 \text{ births}}{1 \text{ baby with syndrome}} = \frac{4,063,000 \text{ births}}{x \text{ babies with syndrome}}$$

$$\frac{12,000}{1} = \frac{4,063,000}{x}$$

$$12,000x = 4,063,000$$

$$x = \frac{4,063,000}{12,000} \approx 339$$

Thus, about 339 children were born with Prader-Willi Syndrome.

75.

$$\frac{12 \text{ inches}}{1 \text{ foot}} = \frac{78 \text{ inches}}{x \text{ feet}}$$

$$\frac{12}{1} = \frac{78}{x}$$

$$12x = 78$$

$$x = \frac{78}{12} = 6.5$$

Thus 78 inches equals 6.5 feet.

77. $\dfrac{9 \text{ square feet}}{1 \text{ square yard}} = \dfrac{26.1 \text{ square feet}}{x \text{ square yards}}$

$$\dfrac{9}{1} = \dfrac{26.1}{x}$$
$$9x = 26.1$$
$$x = \dfrac{26.1}{9} = 2.9$$

Thus 26.1 square feet equals 2.9 square yards.

79. $\dfrac{2.54 \text{ cm}}{1 \text{ inch}} = \dfrac{50.8 \text{ cm}}{x \text{ inches}}$

$$\dfrac{2.54}{1} = \dfrac{50.8}{x}$$
$$2.54x = 50.8$$
$$x = \dfrac{50.8}{2.54} = 20$$

Thus the length of the newborn is 20 inches.

81. Let x = number of home runs needed to break Bonds' record

$$\dfrac{73 \text{ home runs}}{162 \text{ games}} = \dfrac{x}{50 \text{games}}$$
$$\dfrac{73}{162} = \dfrac{x}{50}$$
$$162 \cdot x = 73 \cdot 50$$
$$162x = 3650$$
$$x = \dfrac{3650}{162} \approx 22.53$$

A player would need to hit 23 home runs.

83. Let x = amount of interest Jim would earn after 500 days.

$$\dfrac{\$110.52 \text{ interest}}{180 \text{ days}} = \dfrac{\$x \text{ interest}}{500 \text{ days}}$$
$$\dfrac{110.52}{180} = \dfrac{x}{500}$$
$$180x = 55,260$$
$$x = \dfrac{55,260}{180} = 307$$

He would earn $307 in interest.

85. Let x = the number of pesos she would receive in return for U.S. dollars.

$$\dfrac{\$1.00 \text{ U.S.}}{10.567 \text{ pesos}} = \dfrac{\$200 \text{ U.S.}}{x \text{ pesos}}$$
$$\dfrac{1}{10.567} = \dfrac{200}{x}$$
$$x = 2113.4$$

She will receive 2113.4 pesos.

87. The ratio of Mrs. Ruff's low density to high density cholesterol is $\dfrac{127}{60}$. If we divide 127 by 60 we obtain approximately 2.12. Thus Mrs. Ruff's ratio is approximately equivalent to 2.12:1. Therefore, her ratio is less than the desired 4:1 ratio.

89. In $\dfrac{a}{b} = \dfrac{c}{d}$, if b and d remain the same while a increases, then c increases because $ad = bc$. If a increases ad increases, so bc must increase by increasing c.

91. Let x = number of miles remaining on the life of each tire.
Inches remaining on the life of each tire:
$0.31 - 0.06 = 0.25$

$$\dfrac{0.03 \text{ inches}}{5000 \text{ miles}} = \dfrac{0.25 \text{ miles}}{x \text{ miles}}$$
$$\dfrac{0.03}{5000} = \dfrac{0.25}{x}$$
$$0.03x = 5000 \cdot 0.25$$
$$0.03x = 1250$$
$$x = \dfrac{1250}{0.03}$$
$$x \approx 41,667$$

The tires will last about 41,667 more miles.

93. Let x = number of cubic centimeters of fluid needed.

$$\dfrac{1}{40} = \dfrac{x}{25}$$
$$40x = 25$$
$$x = \dfrac{25}{40}$$
$$x = 0.625$$

0.625 cubic centimeters of fluid should be drawn up into a syringe.

96. Commutative property of addition

97. Associative property of multiplication

98. Distributive property

99.
$$3(4x-3)=6(2x+1)-15$$
$$12x-9=12x+6-15$$
$$12x-9=12x-9$$
$$12x-12x-9=12x-12x-9$$
$$-9=-9 \quad \text{True}$$
Since a true statement is obtained, the solution is all real numbers.

100.
$$y=mx+b$$
$$y-b=mx+b-b$$
$$y-b=mx$$
$$\frac{y-b}{x}=\frac{mx}{x}$$
$$\frac{y-b}{x}=m$$

Chapter 2 Review Exercises

1. $3(x+4)=3x+3(4)$
$$=3x+12$$

2. $5(x-2)=5[x+(-2)]$
$$=5x+5(-2)$$
$$=5x+(-10)$$
$$=5x-10$$

3. $-2(x+4)=-2x+(-2)(4)$
$$=-2x+(-8)$$
$$=-2x-8$$

4. $-(x+2)=-1(x+2)$
$$=(-1)(x)+(-1)(2)$$
$$=-x+(-2)$$
$$=-x-2$$

5. $-(m+3)=-1(m+3)$
$$=(-1)(m)+(-1)(3)$$
$$=-m-3$$

6. $-4(4-x)=-4[4+(-x)]$
$$=(-4)(4)+(-4)(-x)$$
$$=-16+4x$$

7. $5(5-p)=5[5+(-p)]$
$$=5(5)+5(-p)$$
$$=25+(-5p)$$
$$=25-5p$$

8. $6(4x-5)=6(4x)-6(5)$
$$=24x-30$$

9. $-5(5x-5)=-5[5x+(-5)]$
$$=-5(5x)+(-5)(-5)$$
$$=-25x+25$$

10. $4(-x+3)=4(-x)+4(3)$
$$=-4x+12$$

11. $\frac{1}{2}(2x+4)=\left(\frac{1}{2}\right)(2x)+\left(\frac{1}{2}\right)(4)$
$$=x+2$$

12. $-\frac{1}{3}(3+6y)=\left(-\frac{1}{3}\right)(3)+\left(-\frac{1}{3}\right)(6y)$
$$=-1+(-2y)$$
$$=-1-2y$$

13. $-(x+2y-z)=-1[x+2y+(-z)]$
$$=-1(x)+(-1)(2y)+(-1)(-z)$$
$$=-x+(-2y)+z$$
$$=-x-2y+z$$

14. $-3(2a-5b+7)=-3[2a+(-5b)+7]$
$$=-3(2a)+(-3)(-5b)+(-3)(7)$$
$$=-6a+15b+(-21)$$
$$=-6a+15b-21$$

15. $7x-3x=4x$

16. $5-3y+3=-3y+5+3=-3y+8$

17. $1+3x+2x=1+5x=5x+1$

18. $-2x-x+3y=-3x+3y$

19. $4m + 2n + 4m + 6n = 4m + 4m + 2n + 6n$

$\qquad\qquad\qquad\quad = 8m + 8n$

20. There are no like terms.

$9x + 3y + 2$ cannot be further simplified.

21. $6x - 2x + 3y + 6 = 4x + 3y + 6$

22. $x + 8x - 9x + 3 = 9x - 9x + 3 = 3$

23. $-4x^2 - 8x^2 + 3 = -12x^2 + 3$

24. $-2\left(3a^2 - 4\right) + 6a^2 - 8 = -6a^2 + 8 + 6a^2 - 8$

$\qquad\qquad\qquad\qquad\qquad = -6a^2 + 6a^2 + 8 - 8$

$\qquad\qquad\qquad\qquad\qquad = 0$

25. $2x + 3\left(x + 4\right) - 5 = 2x + 3x + 12 - 5 = 5x + 7$

26. $-4 + 2\left(3 - 2b\right) - b = -4 + 6 - 4b - b$

$\qquad\qquad\qquad\qquad = 2 - 5b$

$\qquad\qquad\qquad\qquad = -5b + 2$

27. $6 - \left(-7x + 6\right) - 7x = 6 + 7x - 6 - 7x$

$\qquad\qquad\qquad\qquad\quad = 7x - 7x + 6 - 6$

$\qquad\qquad\qquad\qquad\quad = 0$

28. $2\left(2x + 5\right) - 10 - 4 = 4x + 10 - 10 - 4$

$\qquad\qquad\qquad\qquad\quad = 4x - 4$

29. $-6\left(4 - 3x\right) - 18 + 4x = -24 + 18x - 18 + 4x$

$\qquad\qquad\qquad\qquad\qquad = 18x + 4x - 24 - 18$

$\qquad\qquad\qquad\qquad\qquad = 22x - 42$

30. $4y - 3\left(x + y\right) + 6x^2 = 4y - 3x - 3y + 6x^2$

$\qquad\qquad\qquad\qquad\qquad = 6x^2 - 3x + 4y - 3y$

$\qquad\qquad\qquad\qquad\qquad = 6x^2 - 3x + y$

31. $\dfrac{1}{4}d + 2 - \dfrac{3}{5}d + 5 = \dfrac{1}{4}d - \dfrac{3}{5}d + 2 + 5$

$\qquad\qquad\qquad\qquad = \dfrac{5}{20}d - \dfrac{12}{20}d + 7$

$\qquad\qquad\qquad\qquad = -\dfrac{7}{20}d + 7$

32. $3 - \left(x - y\right) + \left(x - y\right) = 3 - x + y + x - y$

$\qquad\qquad\qquad\qquad\qquad = -x + x + y - y + 3$

$\qquad\qquad\qquad\qquad\qquad = 3$

33. $\dfrac{5}{6}x - \dfrac{1}{3}\left(2x - 6\right) = \dfrac{5}{6}x - \dfrac{2}{3}x + 2$

$\qquad\qquad\qquad\qquad = \dfrac{5}{6}x - \dfrac{4}{6}x + 2$

$\qquad\qquad\qquad\qquad = \dfrac{1}{6}x + 2$

34. $\dfrac{2}{3} - \dfrac{1}{4}n - \dfrac{1}{3}\left(n + 2\right) = \dfrac{2}{3} - \dfrac{1}{4}n - \dfrac{1}{3}n - \dfrac{2}{3}$

$\qquad\qquad\qquad\qquad\qquad = -\dfrac{1}{4}n - \dfrac{1}{3}n + \dfrac{2}{3} - \dfrac{2}{3}$

$\qquad\qquad\qquad\qquad\qquad = -\dfrac{3}{12}n - \dfrac{4}{12}n + 0$

$\qquad\qquad\qquad\qquad\qquad = -\dfrac{7}{12}n$

35. $-3x = -3$

$\qquad \dfrac{-3x}{-3} = \dfrac{-3}{-3}$

$\qquad\quad x = 1$

36. $x + 6 = -7$

$\quad x + 6 - 6 = -7 - 6$

$\qquad\quad x = -13$

37. $x - 4 = 7$

$\quad x - 4 + 4 = 7 + 4$

$\qquad\quad x = 11$

38. $\dfrac{x}{3} = -9$

$\quad 3\left(\dfrac{x}{3}\right) = 3(-9)$

$\qquad\quad x = -27$

39. $5x + 1 = 12$

$\quad 5x + 1 - 1 = 12 - 1$

$\qquad\quad 5x = 11$

$\qquad \dfrac{5x}{5} = \dfrac{11}{5}$

$\qquad\quad x = \dfrac{11}{5}$

40.
$$14 = 3 + 2x$$
$$14 - 3 = 3 - 3 + 2x$$
$$11 = 2x$$
$$\frac{11}{2} = \frac{2x}{2}$$
$$\frac{11}{2} = x$$

41.
$$4c + 3 = -21$$
$$4c + 3 - 3 = -21 - 3$$
$$4c = -24$$
$$\frac{4c}{4} = \frac{-24}{4}$$
$$c = -6$$

42.
$$9 - 2a = 15$$
$$9 - 9 - 2a = 15 - 9$$
$$-2a = 6$$
$$\frac{-2a}{-2} = \frac{6}{-2}$$
$$a = -3$$

43.
$$-x = -12$$
$$-1x = -12$$
$$(-1)(-1x) = (-1)(-12)$$
$$1x = 12$$
$$x = 12$$

44.
$$3(x - 2) = 6$$
$$3x - 6 = 6$$
$$3x - 6 + 6 = 6 + 6$$
$$3x = 12$$
$$\frac{3x}{3} = \frac{12}{3}$$
$$x = 4$$

45.
$$-12 = 3(2x - 8)$$
$$-12 = 6x - 24$$
$$-12 + 24 = 6x - 24 + 24$$
$$12 = 6x$$
$$\frac{12}{6} = \frac{6x}{6}$$
$$2 = x$$

46.
$$4(6 + 2x) = 0$$
$$24 + 8x = 0$$
$$24 - 24 + 8x = 0 - 24$$
$$8x = -24$$
$$\frac{8x}{8} = \frac{-24}{8}$$
$$x = -3$$

47.
$$-6n + 2n + 6 = 0$$
$$-4n + 6 = 0$$
$$-4n + 6 - 6 = 0 - 6$$
$$-4n = -6$$
$$\frac{-4n}{-4} = \frac{-6}{-4}$$
$$n = \frac{6}{4} = \frac{3}{2}$$

48.
$$-3 = 3w - (4w + 6)$$
$$-3 = 3w - 4w - 6$$
$$-3 = -1w - 6$$
$$-3 + 6 = -1w - 6 + 6$$
$$3 = -1w$$
$$\frac{3}{-1} = \frac{-1w}{-1}$$
$$-3 = w$$

49.
$$6 - (2n + 3) - 4n = 6$$
$$6 - 2n - 3 - 4n = 6$$
$$6 - 3 - 2n - 4n = 6$$
$$3 - 6n = 6$$
$$3 - 3 - 6n = 6 - 3$$
$$-6n = 3$$
$$\frac{-6n}{-6} = \frac{3}{-6}$$
$$n = -\frac{3}{6} = -\frac{1}{2}$$

50.
$$4x + 6 - 7x + 9 = 18$$
$$-3x + 15 = 18$$
$$-3x + 15 - 15 = 18 - 15$$
$$-3x = 3$$
$$\frac{-3x}{-3} = \frac{3}{-3}$$
$$x = -1$$

51. $5+3(x-1)=3(x+1)-1$

$5+3x-3=3x+3-1$

$3x+2=3x+2$

$3x-3x+2=3x-3x+2$

$2=2$ True

The solution is all real numbers.

52. $8.4r-6.3=6.3+2.1r$

$8.4r-2.1r-6.3=6.3+2.1r-2.1r$

$6.3r-6.3=6.3$

$6.3r-6.3+6.3=6.3+6.3$

$6.3r=12.6$

$\dfrac{6.3r}{6.3}=\dfrac{12.6}{6.3}$

$r=2$

53. $19.6-21.3t=80.1-9.2t$

$19.6-21.3t+21.3t=80.1-9.2t+21.3t$

$19.6=80.1+12.1t$

$19.6-80.1=80.1-80.1+12.1t$

$-60.5=12.1t$

$\dfrac{-60.5}{12.1}=\dfrac{12.1t}{12.1}$

$-5=t$

54. $0.35(c-5)=0.45(c+4)$

$0.35c-1.75=0.45c+1.8$

$0.35c-0.35c-1.75=0.45c-0.35c+1.8$

$-1.75=0.10c+1.8$

$-1.75-1.8=0.10c+1.8-1.8$

$-3.55=0.10c$

$\dfrac{-3.55}{0.10}=\dfrac{0.10c}{0.10}$

$-35.5=c$

55. $0.2(x+6)=-0.3(2x-1)$

$0.2x+1.2=-0.6x+0.3$

$0.2x+0.6x+1.2=-0.6x+0.6x+0.3$

$0.8x+1.2=0.3$

$0.8x+1.2-1.2=0.3-1.2$

$0.8x=-0.9$

$\dfrac{0.8x}{0.8}=\dfrac{-0.9}{0.8}$

$x=-1.125$

56. $-2.3(x-8)=3.7(x+4)$

$-2.3x+18.4=3.7x+14.8$

$-2.3x+2.3x+18.4=3.7x+2.3x+14.8$

$18.4=6.0x+14.8$

$18.4-14.8=6.0x+14.8-14.8$

$3.6=6.0x$

$\dfrac{3.6}{6.0}=\dfrac{6.0x}{6.0}$

$0.6=x$

57. $\dfrac{p}{3}+2=\dfrac{1}{4}$

$12\left(\dfrac{p}{3}+2\right)=12\left(\dfrac{1}{4}\right)$

$4p+24=3$

$4p+24-24=3-24$

$4p=-21$

$\dfrac{4p}{4}=\dfrac{-21}{4}$

$p=-\dfrac{21}{4}$

58. $\dfrac{d}{6}+\dfrac{1}{7}=2$

$42\left(\dfrac{d}{6}+\dfrac{1}{7}\right)=42(2)$

$7d+6=84$

$7d+6-6=84-6$

$7d=78$

$\dfrac{7d}{7}=\dfrac{78}{7}$

$d=\dfrac{78}{7}$

59. $\dfrac{3}{5}(r-6)=3r$

$\dfrac{3}{5}r-\dfrac{18}{5}=3r$

$5\left(\dfrac{3}{5}r-\dfrac{18}{5}\right)=5(3r)$

$3r-18=15r$

$3r-3r-18=15r-3r$

$-18=12r$

$\dfrac{-18}{12}=\dfrac{12r}{12}$

$-\dfrac{3}{2}=r$

60.
$$\frac{2}{3}w = \frac{1}{7}(w-2)$$
$$\frac{2}{3}w = \frac{1}{7}w - \frac{2}{7}$$
$$21\left(\frac{2}{3}w\right) = 21\left(\frac{1}{7}w - \frac{2}{7}\right)$$
$$14w = 3w - 6$$
$$14w - 3w = 3w - 3w - 6$$
$$11w = -6$$
$$\frac{11w}{11} = \frac{-6}{11}$$
$$w = -\frac{6}{11}$$

61.
$$8x - 5 = -4x + 19$$
$$8x + 4x - 5 = -4x + 4x + 19$$
$$12x - 5 = 19$$
$$12x - 5 + 5 = 19 + 5$$
$$12x = 24$$
$$\frac{12x}{12} = \frac{24}{12}$$
$$x = 2$$

62.
$$-(w+2) = 2(3w-6)$$
$$-w - 2 = 6w - 12$$
$$-w - 2 + 12 = 6w - 12 + 12$$
$$-w + 10 = 6w$$
$$-w + w + 10 = 6w + w$$
$$10 = 7w$$
$$\frac{10}{7} = \frac{7w}{7}$$
$$\frac{10}{7} = w$$

63.
$$2x + 6 = 3x + 9 - 3$$
$$2x + 6 = 3x + 6$$
$$2x - 2x + 6 = 3x - 2x + 6$$
$$6 = x + 6$$
$$6 - 6 = x + 6 - 6$$
$$0 = x$$

64.
$$-5a + 3 = 2a + 10$$
$$-5a + 3 - 10 = 2a + 10 - 10$$
$$-5a - 7 = 2a$$
$$-5a + 5a - 7 = 2a + 5a$$
$$-7 = 7a$$
$$\frac{-7}{7} = \frac{7a}{7}$$
$$-1 = a$$

65.
$$5p - 2 = -2(-3p+6)$$
$$5p - 2 = 6p - 12$$
$$5p - 5p - 2 = 6p - 5p - 12$$
$$-2 = p - 12$$
$$-2 + 12 = p - 12 + 12$$
$$10 = p$$

66.
$$3x - 12x = 24 - 9x$$
$$-9x = 24 - 9x$$
$$-9x + 9x = 24 - 9x + 9x$$
$$0 = 24 \quad \text{False}$$
Since a false statement is obtained, there is no solution.

67.
$$4(2x-3) + 4 = 8x - 8$$
$$8x - 12 + 4 = 8x - 8$$
$$8x - 8 = 8x - 8$$
Since the equation is true for all values of x, the solution is all real numbers.

68.
$$4 - c - 2(4-3c) = 3(c-4)$$
$$4 - c - 8 + 6c = 3c - 12$$
$$5c - 4 = 3c - 12$$
$$5c - 3c - 4 = 3c - 3c - 12$$
$$2c - 4 = -12$$
$$2c - 4 + 4 = -12 + 4$$
$$2c = -8$$
$$\frac{2c}{2} = \frac{-8}{2}$$
$$c = -4$$

69.
$$2(x+7) = 6x+9-4x$$
$$2x+14 = 6x+9-4x$$
$$2x+14 = 2x+9$$
$$2x-2x+14 = 2x-2x+9$$
$$14 = 9 \quad \text{False}$$
Since a false statement is obtained, there is no solution.

70. $-5(3-4x) = -6+20x-9$
$$-15+20x = -6+20x-9$$
$$-15+20x = -15+20x$$
The statement is true for all values of x, thus the solution is all real numbers.

71. $4(x-3)-(x+5) = 0$
$$4x-12-x-5 = 0$$
$$3x-17 = 0$$
$$3x-17+17 = 0+17$$
$$3x = 17$$
$$\frac{3x}{3} = \frac{17}{3}$$
$$x = \frac{17}{3}$$

72. $-2(4-x) = 6(x+2)+3x$
$$-8+2x = 6x+12+3x$$
$$-8+2x = 9x+12$$
$$-8-12+2x = 9x+12-12$$
$$-20+2x = 9x$$
$$-20+2x-2x = 9x-2x$$
$$-20 = 7x$$
$$\frac{-20}{7} = \frac{7x}{7}$$
$$-\frac{20}{7} = x$$

73. $\dfrac{x+3}{2} = \dfrac{x}{2}$
$$2(x+3) = 2x$$
$$2x+6 = 2x$$
$$2x-2x+6 = 2x-2x$$
$$6 = 0 \quad \text{False}$$
Since a false statement is obtained, there is no solution.

74. $\dfrac{x}{6} = \dfrac{x-4}{2}$
$$2 \cdot x = 6(x-4)$$
$$2x = 6x-24$$
$$2x-6x = 6x-6x-24$$
$$-4x = -24$$
$$\frac{-4x}{-4} = \frac{-24}{-4}$$
$$x = 6$$

75. $\dfrac{1}{5}(3s+4) = \dfrac{1}{3}(2s-8)$
$$\frac{3}{5}s+\frac{4}{5} = \frac{2}{3}s-\frac{8}{3}$$
$$15\left(\frac{3}{5}s+\frac{4}{5}\right) = 15\left(\frac{2}{3}s-\frac{8}{3}\right)$$
$$9s+12 = 10s-40$$
$$9s-9s+12 = 10s-9s-40$$
$$12 = s-40$$
$$12+40 = s-40+40$$
$$52 = s$$

76. $\dfrac{2(2t-4)}{5} = \dfrac{3t+6}{4}-\dfrac{3}{2}$
$$\frac{4t-8}{5} = \frac{3t+6}{4}-\frac{3}{2}$$
$$20\left(\frac{4t-8}{5}\right) = 20\left(\frac{3t+6}{4}-\frac{3}{2}\right)$$
$$4(4t-8) = 5(3t+6)-30$$
$$16t-32 = 15t+30-30$$
$$16t-32 = 15t+0$$
$$16t-16t-32 = 15t-16t$$
$$-32 = -1t$$
$$\frac{-32}{-1} = \frac{-1t}{-1}$$
$$32 = t$$

77. $\dfrac{2}{5}(2-x) = \dfrac{1}{6}(-2x+2)$

$\dfrac{4}{5} - \dfrac{2}{5}x = -\dfrac{2}{6}x + \dfrac{2}{6}$

$\dfrac{4}{5} - \dfrac{2}{5}x = -\dfrac{1}{3}x + \dfrac{1}{3}$

$15\left(\dfrac{4}{5} - \dfrac{2}{5}x\right) = 15\left(-\dfrac{1}{3}x + \dfrac{1}{3}\right)$

$12 - 6x = -5x + 5$

$12 - 6x + 6x = -5x + 6x + 5$

$12 = x + 5$

$12 - 5 = x + 5 - 5$

$7 = x$

78. $\dfrac{x}{4} + \dfrac{x}{6} = \dfrac{1}{2}(x+3)$

$\dfrac{x}{4} + \dfrac{x}{6} = \dfrac{x}{2} + \dfrac{3}{2}$

$12\left(\dfrac{x}{4} + \dfrac{x}{6}\right) = 12\left(\dfrac{x}{2} + \dfrac{3}{2}\right)$

$3x + 2x = 6x + 18$

$5x = 6x + 18$

$5x - 6x = 6x - 6x + 18$

$-1x = 18$

$\dfrac{-1x}{-1} = \dfrac{18}{-1}$

$x = -18$

79. Substitute 7 for y, 2 for x, and 1 for b. Then solve for m.

$y = mx + b$

$7 = m(2) + 1$

$7 = 2m + 1$

$7 - 1 = 2m + 1 - 1$

$6 = 2m$

$\dfrac{6}{2} = \dfrac{2m}{2}$

$3 = m$

80. Substitute 12 for h, 3 for b, and 5 for d. Then solve for A.

$A = \dfrac{1}{2}h(b+d)$

$A = \dfrac{1}{2}(12)(3+5) = \dfrac{1}{2}(12)(8) = 48$

81. $A = \dfrac{1}{2}bh$

$A = \dfrac{1}{2}(8)(3) = 12 \text{ cm}^2$

82. $V = \dfrac{4}{3}\pi r^3$

$V = \dfrac{4}{3}\pi(2)^3 = \dfrac{4}{3}\pi(8) = \dfrac{32}{3}\pi \approx 33.51 \text{ in.}^3$

83. $\qquad P = 2l + 2w$

$P - 2w = 2l + 2w - 2w$

$P - 2w = 2l$

$\dfrac{P - 2w}{2} = \dfrac{2l}{2}$

$\dfrac{P - 2w}{2} = l$

84. $y - y_1 = m(x - x_1)$

$\dfrac{y - y_1}{x - x_1} = \dfrac{m(x - x_1)}{(x - x_1)}$

$\dfrac{y - y_1}{x - x_1} = m$

85. $\qquad -x + 3y = 2$

$-x + x + 3y = 2 + x$

$3y = x + 2$

$\dfrac{3y}{3} = \dfrac{x + 2}{3}$

$y = \dfrac{x}{3} + \dfrac{2}{3}$

$y = \dfrac{1}{3}x + \dfrac{2}{3}$

86. $d = rt$

$d = (61.7)(5) = 308.5$

He traveled 308.5 miles.

87. $A = lw$

$A = (20)(12) = 240$

The area of the flower garden is 240 sq. ft.

88. $V = \pi r^2 h$

$V = \pi(2)^2(2) = 8\pi \approx 25.13$

The volume of the tuna fish can is about 25.13 sq. in.

89. $12 : 20 = 3 : 5$

90. $80 \text{ ounces} = \dfrac{80}{16} = 5 \text{ pounds}$

The ratio of 80 ounces to 12 pounds is thus 5:12.

91. 4 minutes = 240 seconds

The ratio of 4 minutes to 40 seconds is $\dfrac{240}{40} = \dfrac{6}{1}$.

The ratio is 6:1.

92. $\dfrac{x}{4} = \dfrac{8}{16}$

$16 \cdot x = 8 \cdot 4$

$16x = 32$

$x = \dfrac{32}{16} = 2$

93. $\dfrac{5}{20} = \dfrac{x}{80}$

$20 \cdot x = 80 \cdot 5$

$20x = 400$

$x = \dfrac{400}{20} = 20$

94. $\dfrac{3}{x} = \dfrac{15}{45}$

$3 \cdot 45 = 15 \cdot x$

$135 = 15x$

$x = \dfrac{135}{15} = 9$

95. $\dfrac{20}{45} = \dfrac{15}{x}$

$20 \cdot x = 15 \cdot 45$

$20x = 675$

$x = \dfrac{675}{20} = \dfrac{135}{4}$

96. $\dfrac{6}{5} = \dfrac{-12}{x}$

$6 \cdot x = -12 \cdot 5$

$6x = -60$

$x = \dfrac{-60}{6} = -10$

97. $\dfrac{b}{6} = \dfrac{8}{-3}$

$-3 \cdot b = 6 \cdot 8$

$-3b = 48$

$x = \dfrac{48}{-3} = -16$

98. $\dfrac{-7}{9} = \dfrac{-12}{y}$

$-7 \cdot y = -12 \cdot 9$

$-7y = -108$

$y = \dfrac{-108}{-7}$

$y = \dfrac{108}{7}$

99. $\dfrac{x}{-15} = \dfrac{30}{-5}$

$-5 \cdot x = -15 \cdot 30$

$-5x = -450$

$x = \dfrac{-450}{-5} = 90$

100. $\dfrac{6}{8} = \dfrac{30}{x}$

$6 \cdot x = 8 \cdot 30$

$6x = 240$

$x = \dfrac{240}{6} = 40$

Therefore the length of the side is 40 in.

101. $\dfrac{7}{3.5} = \dfrac{2}{x}$

$7 \cdot x = 2 \cdot 3.5$

$7x = 7$

$x = \dfrac{7}{7} = 1$

Therefore the length of the side is 1 ft.

102. Let t = the time it will take in hours to make the boat trip.

$$\frac{40 \text{ miles}}{1.8 \text{ hours}} = \frac{140 \text{ miles}}{t \text{ hours}}$$

$$\frac{40}{1.8} = \frac{140}{t}$$

$$40 \cdot t = 1.8 \cdot 140$$

$$40t = 252$$

$$\frac{40t}{40} = \frac{252}{40}$$

$$t = 6.3$$

It will take the boat 6.3 hours to travel 140 miles.

103. Let x = number of dishes he can wash in 21 minutes.

$$\frac{12 \text{ dishes}}{3.5 \text{ minutes}} = \frac{x \text{ dishes}}{21 \text{ minutes}}$$

$$\frac{12}{3.5} = \frac{x}{21}$$

$$12 \cdot 21 = 3.5 \cdot x$$

$$252 = 3.5x$$

$$\frac{252}{3.5} = \frac{3.5x}{3.5}$$

$$72 = x$$

He can wash 72 dishes in 21 minutes.

104. Let x = number of pages that can be copied in 22 minutes.

$$\frac{1 \text{ minutes}}{20 \text{ pages}} = \frac{22 \text{ minutes}}{x \text{ pages}}$$

$$\frac{1}{20} = \frac{22}{x}$$

$$1 \cdot x = 22 \cdot 20$$

$$x = 440$$

440 pages can be copied in 22 minutes.

105. Let x = number of inches representing 380 miles.

$$\frac{60 \text{ miles}}{1 \text{ inch}} = \frac{380 \text{ miles}}{x \text{ inches}}$$

$$\frac{60}{1} = \frac{380}{x}$$

$$60 \cdot x = 380 \cdot 1$$

$$60x = 380$$

$$x = \frac{380}{60} = 6\frac{1}{3}$$

$6\frac{1}{3}$ inches on the map represent 380 miles.

106. Let x = size of actual car in feet

$$\frac{1 \text{ inch}}{1.5 \text{ feet}} = \frac{10.5 \text{ inches}}{x \text{ feet}}$$

$$\frac{1}{1.5} = \frac{10.5}{x}$$

$$1 \cdot x = 1.5 \cdot 10.5$$

$$x = 15.75$$

The size of the actual car is 15.75 ft.

107. Let x = the value of 1 peso in terms of U.S. dollars.

$$\frac{\$1 \text{ U.S.}}{9.165 \text{ pesos}} = \frac{x \text{ dollars}}{1 \text{ peso}}$$

$$\frac{1}{9.165} = \frac{x}{1}$$

$$9.165 \cdot x = 1 \cdot 1$$

$$9.165x = 1$$

$$x = \frac{1}{9.165} \approx 0.109$$

1 peso equals about $0.109.

108. Let x = number of bottles the machine can fill and cap in 2 minutes.

2 minutes = 120 seconds

$$\frac{50 \text{ seconds}}{80 \text{ bottles}} = \frac{120 \text{ seconds}}{x \text{ bottles}}$$

$$\frac{50}{80} = \frac{120}{x}$$

$$50 \cdot x = 80 \cdot 120$$

$$50x = 9600$$

$$x = \frac{9600}{50} = 192$$

The machine can fill and cap 192 bottles in 2 minutes.

Chapter 2 Practice Test

1. $-3(4 - 2x) = -3[4 + (-2x)]$
$$= -3(4) + (-3)(-2x)$$
$$= -12 + 6x \text{ or } 6x - 12$$

2. $-(x + 3y - 4) = -[x + 3y + (-4)]$
$$= -1[x + 3y + (-4)]$$
$$= (-1)(x) + (-1)(3y) + (-1)(-4)$$
$$= -x + (-3y) + 4$$
$$= -x - 3y + 4$$

3. $5x - 8x + 4 = -3x + 4$

4. $4 + 2x - 3x + 6 = 2x - 3x + 4 + 6 = -x + 10$

5. $-y - x - 4x - 6 = -x - 4x - y - 6 = -5x - y - 6$

6. $a - 2b + 6a - 6b - 3 = a + 6a - 2b - 6b - 3$
$$= 7a - 8b - 3$$

7. $2x^2 + 3 + 2(3x - 2) = 2x^2 + 3 + 6x - 4$
$$= 2x^2 + 6x + 3 - 4$$
$$= 2x^2 + 6x - 1$$

8.
$$2.4x - 3.9 = 3.3$$
$$2.4x - 3.9 + 3.9 = 3.3 + 3.9$$
$$2.4x = 7.2$$
$$\frac{2.4x}{2.4} = \frac{7.2}{2.4}$$
$$x = 3$$

9.
$$\frac{5}{6}(x - 2) = x - 3$$
$$\frac{5}{6}x - \frac{10}{6} = x - 3$$
$$6\left(\frac{5}{6}x - \frac{10}{6}\right) = 6(x - 3)$$
$$5x - 10 = 6x - 18$$
$$5x - 5x - 10 = 6x - 5x - 18$$
$$-10 = x - 18$$
$$-10 + 18 = x - 18 + 18$$
$$8 = x$$

10.
$$2x - 3(-2x + 4) = -13 + x$$
$$2x + 6x - 12 = -13 + x$$
$$8x - 12 = -13 + x$$
$$8x - 12 + 12 = -13 + 12 + x$$
$$8x = -1 + x$$
$$8x - x = -1 + x - x$$
$$7x = -1$$
$$\frac{7x}{7} = \frac{-1}{7}$$
$$x = -\frac{1}{7}$$

11.
$$3x - 4 - x = 2(x + 5)$$
$$3x - 4 - x = 2x + 10$$
$$2x - 4 = 2x + 10$$
$$2x - 2x - 4 = 2x - 2x + 10$$
$$-4 = 10 \text{ False}$$
Since a false statement is obtained, there is no solution.

12.
$$-3(2x + 3) = -2(3x + 1) - 7$$
$$-6x - 9 = -6x - 2 - 7$$
$$-6x - 9 = -6x - 9$$
Since the equation is true for all values of x, the solution is all real numbers.

13.
$$ax + by + c = 0$$
$$ax + by - by + c = 0 - by$$
$$ax + c = -by$$
$$ax + c - c = -by - c$$
$$ax = -by - c$$
$$\frac{ax}{a} = \frac{-by - c}{a}$$
$$x = \frac{-by - c}{a}$$

14.
$$-6x + 5y = -2$$
$$-6x + 6x + 5y = -2 + 6x$$
$$5y = 6x - 2$$
$$\frac{5y}{5} = \frac{6x - 2}{5}$$
$$y = \frac{6x}{5} - \frac{2}{5}$$
$$y = \frac{6}{5}x - \frac{2}{5}$$

15.
$$\frac{1}{7}(2x - 5) = \frac{3}{8}x - \frac{5}{7}$$
$$\frac{2}{7}x - \frac{5}{7} = \frac{3}{8}x - \frac{5}{7}$$
$$56\left(\frac{2}{7}x - \frac{5}{7}\right) = 56\left(\frac{3}{8}x - \frac{5}{7}\right)$$
$$16x - 40 = 21x - 40$$
$$16x - 16x - 40 = 21x - 16x - 40$$
$$-40 = 5x - 40$$
$$-40 + 40 = 5x - 40 + 40$$
$$0 = 5x$$
$$\frac{0}{5} = \frac{5x}{5}$$
$$0 = x$$

16.
$$\frac{9}{x} = \frac{3}{-15}$$
$$9(-15) = 3x$$
$$-135 = 3x$$
$$\frac{-135}{3} = x$$
$$-45 = x$$

17. a. An equation that has exactly one solution is a conditional equation.

b. An equation that has no solution is a contradiction.

c. An equation that has all real numbers as its solution is an identity.

18.
$$\frac{3}{4} = \frac{8}{x}$$
$$3x = 4 \cdot 8$$
$$3x = 32$$
$$x = \frac{32}{3}$$

The length of side x is $\frac{32}{3}$ feet or $10\frac{2}{3}$ feet.

19. Let r = the simple interest rate.
$$i = prt$$
$$80 = 2000 \cdot r \cdot 1$$
$$80 = 2000r$$
$$\frac{80}{2000} = \frac{2000r}{2000}$$
$$0.04 = r$$

The interest rate was 4%.

20. Let C = the circumference of the pie.
$$C = 2\pi r$$
$$C = 2\pi(4.5) = 9\pi \approx 28.27$$

The circumference of the pie is 28.27 in.

21. Let x = number of minutes it will take.
$$\frac{25 \text{ miles}}{35 \text{ minutes}} = \frac{125 \text{ miles}}{x \text{ minutes}}$$
$$\frac{25}{35} = \frac{125}{x}$$
$$25x = 35 \cdot 125$$
$$25x = 4375$$
$$x = \frac{4375}{25} = 175$$

It would take 175 minutes or 2 hours 55 minutes.

Chapter 2 Cumulative Review Test

1. $\dfrac{52}{15} \cdot \dfrac{10}{13} = \dfrac{\overset{4}{\cancel{52}}}{\underset{3}{\cancel{15}}} \cdot \dfrac{\overset{2}{\cancel{10}}}{\underset{1}{\cancel{13}}} = \dfrac{4 \cdot 2}{3 \cdot 1} = \dfrac{8}{3}$

2. $\dfrac{5}{24} \div \dfrac{2}{9} = \dfrac{5}{24} \cdot \dfrac{9}{2} = \dfrac{5}{8} \cdot \dfrac{3}{2} = \dfrac{5 \cdot 3}{8 \cdot 2} = \dfrac{15}{16}$

3. $|-2| > 1$ since $|-2| = 2$ and $2 > 1$.

4. $-5 - (-4) + 12 - 8 = -5 + 4 + 12 - 8$
$$= -1 + 12 - 8$$
$$= 11 - 8$$
$$= 3$$

5. $-7 - (-6) = -7 + 6 = -1$

6. $20 - 6 \div 3 \cdot 2 = 20 - 2 \cdot 2 = 20 - 4 = 16$

7. $3\left[6 - \left(4 - 3^2\right)\right] - 30 = 3\left[6 - (4 - 9)\right] - 30$
$$= 3\left[6 - (-5)\right] - 30$$
$$= 3\left[6 + 5\right] - 30$$
$$= 3(11) - 30$$
$$= 33 - 30$$
$$= 3$$

8. Substitute -2 for each x.
$$-2x^2 - 6x + 8 = -2(-2)^2 - 6(-2) + 8$$
$$= -2(4) - (-12) + 8$$
$$= -8 + 12 + 8$$
$$= 4 + 8$$
$$= 12$$

9. Distributive property

10. $8x + 2y + 4x - y = 8x + 4x + 2y - y$
$$= 12x + y$$

11. $9 - \dfrac{2}{3}x + 16 + \dfrac{3}{4}x = \dfrac{3}{4}x - \dfrac{2}{3}x + 9 + 16$
$$= \dfrac{9}{12}x - \dfrac{8}{12}x + 25$$
$$= \dfrac{1}{12}x + 25$$

12. $3x^2 + 5 + 4(2x - 7) = 3x^2 + 5 + 8x - 28$
$$= 3x^2 + 8x - 23$$

13. $7x + 3 = -4$
$7x + 3 - 3 = -4 - 3$
$7x = -7$
$\dfrac{7x}{7} = \dfrac{-7}{7}$
$x = -1$

14. $\dfrac{1}{4}x = -11$
$4\left(\dfrac{1}{4}x\right) = 4(-11)$
$x = -44$

15. $4(x - 2) = 5(x - 1) + 3x + 2$
$4x - 8 = 5x - 5 + 3x + 2$
$4x - 8 = 5x + 3x - 5 + 2$
$4x - 8 = 8x - 3$
$4x - 4x - 8 = 8x - 4x - 3$
$-8 = 4x - 3$
$-8 + 3 = 4x - 3 + 3$
$-5 = 4x$
$\dfrac{-5}{4} = \dfrac{4x}{4}$
$-\dfrac{5}{4} = x$

16. $\dfrac{3}{4}n - \dfrac{1}{5} = \dfrac{2}{3}n$
$60\left(\dfrac{3}{4}n - \dfrac{1}{5}\right) = 60\left(\dfrac{2}{3}n\right)$
$45n - 12 = 40n$
$45n - 45n - 12 = 40n - 45n$
$-12 = -5n$
$\dfrac{-12}{-5} = \dfrac{-5n}{-5}$
$\dfrac{12}{5} = n$

17. $A = \dfrac{a + b + c}{3}$
$3 \cdot A = 3 \cdot \left(\dfrac{a + b + c}{3}\right)$
$3A = a + b + c$
$3A - a = a - a + b + c$
$3A - a = b + c$
$3A - a - c = b + c - c$
$3A - a - c = b$

18. $\dfrac{40}{30} = \dfrac{3}{x}$
$40 \cdot x = 30 \cdot 3$
$40x = 90$
$\dfrac{40x}{40} = \dfrac{90}{40}$
$x = \dfrac{90}{40} = \dfrac{9}{4}$ or 2.25

19. Let A = the area of the trampoline.
$A = \pi r^2$
$A = \pi(11)^2 = 122\pi \approx 380.13$
The area of the trampoline is 380.13 sq. ft.

20. Let x = amount he earns after 8 hours.
$\dfrac{2 \text{ hours}}{\$10.50} = \dfrac{8 \text{ hours}}{x \text{ dollars}}$
$\dfrac{2}{10.5} = \dfrac{8}{x}$
$2x = (10.5)(8)$
$2x = 84$
$x = \dfrac{84}{2} = 42$
He earns $42 after 8 hours.

77

Chapter 3

Exercise Set 3.1

1. Added to, more than, increased by, and sum indicate the operation of addition.

3. Multiplied by, product of, twice, and three times indicate the operation of multiplication.

5. The cost is increased by 25% of the cost, so the expression needs to be $c + 0.25c$.

7. $25 - x$

9. Answers will vary. Some possible answers are *is, was, will be, yields, gives.*

11. $h + 4$

13. $a - 5$

15. $5h$

17. $2d$

19. $\frac{1}{2}a$

21. $r - 5$

23. $8 - m$

25. $2w + 8$

27. $5a - 4$

29. $\frac{1}{3}w - 7$

31. $x =$ Sonya's height

33. $x =$ the length of Jones Beach

35. $x =$ the number of medals Finland won

37. $x =$ the cost of the Chevy

39. $x =$ Teri's grade

41. $x =$ the amount that Kristen receives or $x =$ the amount Yvonne receives

43. $x =$ Don's weight or $x =$ Angela's weight

45. If $c =$ the cost of the chair, then $3c =$ the cost of the table.

47. If $a =$ the area of the kitchen, then $2a + 20 =$ the area of the living room.

49. If $w =$ the width of the rectangle, then $5w - 2 =$ the length of the rectangle.

51. If $w =$ the number of medals won by Sweden, then $38 - w =$ the number of medals won by Brazil.
If $w =$ the number of medals won by Brazil, then $38 - w =$ the number of medals won by Sweden.

53. If $g =$ George's age, then $\frac{1}{2}g + 2 =$ Mike's age.

55. If $m =$ the number of miles Jan walked, then $6.4 - m =$ the number of miles Edward walked.
If $m =$ the number of miles Edward walked, then $6.4 - m =$ the number of miles Jan walked.

57. $n + 8$

59. $\frac{1}{2}x$ or $\frac{x}{2}$

61. $2a - 1$

63. $2t - 30$

65. $1.2p + 20$

67. $80{,}000 - m$

69. $2r - 673$

71. $10x$

73. $100d$

75. $45 + 0.40x$

77. $s + 0.20s$

79. $e - 0.12e$

81. $c + 0.07c$

83. $m - 0.313m$

85. If $f =$ Frieda's weight, then $(f + 15) =$ Jennifer's weight. The sum of their weights is $f + (f + 15)$.

87. If $l =$ Luis' height, then $(2l - 1)$ is Armando's height. The difference in their heights is $(2l - 1) - l$.

89. If $n =$ the larger number, then $(3n - 40) =$ the smaller number. Their difference is $n - (3n - 40)$.

91. If $w =$ the younger child's weight, then $(2w - 3) =$ the weight of the older child. The sum of their weights is $w + (2w - 3)$.

93. If r = the area of Rhode Island, then $(479r + 462)$ = the area of Alaska. The sum of their areas is $r + (479r + 462)$.

95. If n = the cost of the bank stock, then $(n + 0.06n)$ = the price of Apple stock. Their sum is $n + (n + 0.06n)$.

97. If a = the assets of the bank in 2005, then $(a - 0.023a)$ = the assets of the bank in 2006. Their difference in assets is $a - (a - 0.023a)$.

99. Let x = one number, then $4x$ = second number. First number + second number = 20 $x + 4x = 20$

101. Let x = smaller integer, then $x + 1$ = larger consecutive integer. Smaller + larger = 41 $x + (x + 1) = 41$

103. Let x = the number. Twice the number decreased by 8 is 12. $2x - 8 = 12$

105. Let x = the number. One-fifth of the sum of the number and 10 is 150. $\frac{1}{5}(x + 10) = 150$

107. Let x = smaller consecutive even integer. Then $x + 2$ = larger consecutive even integer. smaller + twice larger = 22 $x + 2(x + 2) = 22$

109. $12.50h = 150$

111. $2.99x = 17.94$

113. $25q = 175$

115. Let a = Julie's age. Then Darla's age = $2a + 1$ Julie's age + Darla's age = 52 $a + (2a + 1) = 52$

117. Let s = the number of cards Saul owns. Then $(2s + 300)$ = the number of cards Jakob owns. Their difference is equal to 420. $(2s + 300) - s = 420$

119. Let s = the distance traveled by the Southern Pacific train. Then $(2s - 4)$ = the distance traveled by the Amtrak train. Their sum is equal to 890. $s + (2s - 4) = 890$

121. Let m = the number of miles Malik walked. Then $(3m - 2)$ = the number of miles Donna walked. Their sum is equal to 12.6 $m + (3m - 2) = 12.6$

123. Let c = the cost of the car. The new price is \$89,600 after a 2.3% increase. $c + 0.023c = 89,600$

125. Let p = the population of the town. The town decreased by 1.9% to 12,087. $p - 0.019p = 12,087$

127. Let c = the cost of the car. When 7% sales tax was added the car increased in price to \$32,600. $c + 0.07c = 32,600$

129. Let c = the cost of the meal. When the 15% tip was added the price increased to \$42.50. $c + 0.15c = 42.50$

131. a. 1 minute = 60 seconds
1 hour = 60 minutes = 3600 seconds
1 day = 24 hours
$\qquad$ = 1440 minutes
$\qquad$ = 86,400 seconds
$86,400d + 3600h + 60m + s$

b. $86,400d + 3600h + 60m + s$
$\quad = 86,400(4) + 3600(6) + 60(15) + 25$
$\quad = 368,125$ seconds

134. $3\left[(4 - 16) \div 2\right] + 5^2 - 3$
$= 3\left[(-12) \div 2\right] + 25 - 3$
$= 3(-6) + 25 - 3$
$= -18 + 25 - 3$
$= 7 - 3$
$= 4$

135. Substitute 40 for P and 5 for w.
$\quad P = 2l + 2w$
$40 = 2l + 2(5)$
$40 = 2l + 10$
$30 = 2l$
$15 = l$

136.
$$3x - 2y = 6$$
$$3x - 3x - 2y = -3x + 6$$
$$-2y = -3x + 6$$
$$\frac{-2y}{-2} = \frac{-3x + 6}{-2}$$
$$y = \frac{3x - 6}{2}$$
$$y = \frac{3}{2}x - 3$$

137.
$$\frac{3.6}{x} = \frac{10}{7}$$
$$3.6 \cdot 7 = 10 \cdot x$$
$$25.2 = 10x$$
$$\frac{25.2}{10} = \frac{10x}{10}$$
$$2.52 = x$$

138. First, to express the ratio, we need both quantites to be in the same units. Since there are 16 ounces in 1 pound, we can convert 4 pounds to 64 ounces. The ratio is $26:64$ or $13:32$ in lowest terms.

Exercise Set 3.2

1. Answers will vary.

3. Let n = a number
$$4n - 3 = 17$$
$$4n = 20$$
$$n = 5$$
The number is 5.

5. Let x = smaller integer, then $x + 1$ = next consecutive integer. Smaller number + larger number = 87.
$$x + (x + 1) = 87$$
$$2x + 1 = 87$$
$$2x = 86$$
$$x = 43$$
Smaller number = 43
Larger number = $x + 1 = 43 + 1 = 44$

7. Let x = smaller odd integer, then $x + 2$ = next consecutive odd integer. Sum of integers = 96.
$$x + (x + 2) = 96$$
$$2x + 2 = 96$$
$$2x = 94$$
$$x = 47$$
Smaller integer = 47
Larger integer = $47 + 2 = 49$

9. Let x = one number. Then $2x + 3$ = second number. First number + second number = 27
$$x + (2x + 3) = 27$$
$$3x + 3 = 27$$
$$3x = 24$$
$$x = 8$$
First number = 8
Second number = $2x + 3 = 2(8) + 3 = 19$

11. Let x = the smaller number. Then the larger number = $5x - 4$. Larger number – smaller number = 4
$$(5x - 4) - x = 4$$
$$5x - 4 - x = 4$$
$$4x - 4 = 4$$
$$4x = 8$$
$$x = 2$$
The smaller number is 2 and the larger number is $5(2) - 4 = 6$.

13. Let x = smaller integer, then larger integer = $2x - 8$. Larger integer – Smaller integer = 17
$$(2x - 8) - x = 17$$
$$2x - 8 - x = 17$$
$$x - 8 = 17$$
$$x = 25$$
Smaller number = 25
Larger number = $2x - 8 = 2(25) - 8 = 42$

15. Let x = the number of baseball cards given to Richey, then $3x$ = number of baseball cards given to Erin. Number of cards given to Richey + Number of cards given to Erin = 260
$$x + 3x = 260$$
$$4x = 260$$
$$4x = 260$$
$$x = 65$$
Grandma gave 65 baseball cards to Richey.

17. Let x = the number of hours it takes to build a horse, then $(2x + 1.4)$ = the number of hours it takes to attach the gloves to the horse.
Time to build horse + Time to attach gloves to horse = 32.6
$$x + (2x + 1.4) = 32.6$$
$$3x + 1.4 = 32.6$$
$$3x = 31.2$$
$$x = 10.4$$
$2x + 1.4 = 2(10.4) + 1.4 = 22.2$
It took him 22.2 hours to attach the gloves to the horse.

19. Let x = the number of weeks, then $6x$ = the amount she wishes to add to her collection over x weeks.
Amount started with in collection + Amount added each week to the collection over x weeks = Total number in collection
$$422 + 6x = 500$$
$$6x = 78$$
$$x = 13$$
It will take her about 13 weeks to get 500 frogs in her collection.

21. Let x = the number of years, then $250x$ = the number of employees retiring after x years.
Current amount of employees – Employees retiring after x years = Future total employees
$$4600 - 250x = 2200$$
$$-250x = -2400$$
$$x = 9.6$$
In 9.6 years the total employment will be 2200.

23. Let x = the average number of tornados in December, then $(11x - 16)$ = the average number of tornados in June.
Number of tornados in June – Number of tornados in December = 204
$$(11x - 16) - x = 204$$
$$11x - 16 - x = 204$$
$$10x - 16 = 204$$
$$10x = 220$$
$$x = 22$$
There was an average of 22 tornados in December and $11(22) - 16 = 226$ in June.

25. Let x = the average hourly wage paid to housekeepers in New Orleans, then $(2x + 1.46)$ = the average hourly wage paid to housekeepers in New York City.
Wage of housekeepers in New York – Wage of housekeepers in New Orleans = 8.10
$$(2x + 1.46) - x = 8.10$$
$$2x + 1.46 - x = 8.10$$
$$x + 1.46 = 8.10$$
$$x = 6.64$$
The average hourly wages paid to a housekeeper in New York City is $2(6.64) + 1.46 = \$14.74$.

27. Let x = the cost to produce a shirt in Northern China, then $(3x - 0.16)$ = the cost to produce a shirt in Mexico.

Cost to produce a shirt in Northern China + the cost to produce a shirt in Mexico = 3.28
$$x + (3x - 0.16) = 3.28$$
$$x + 3x - 0.16 = 3.28$$
$$4x - 0.16 = 3.28$$
$$4x = 3.44$$
$$x = 0.86$$
The cost to produce a shirt in Northern China is \$0.86 and the cost to produce a shirt in Mexico is $3(0.86) + 0.16 = \$2.42$.

29. Let x = number of gallons of gasoline that Luis can purchase.
(Cost per gallon) (Price per gallon) = Total cost
$$3.20x = 48$$
$$x = 15$$
Luis can purchase 15 gallons of gasoline.

31. Let x the number of copies made, then $0.02x$ = the cost to make x number of copies.
Cost of machine + Cost of copies made = Total cost
$$2100 + 0.02x = 2462$$
$$0.02x = 362$$
$$x = 18,100$$
In one year, 18,100 copies were made.

33. Let x = the number of On Demand movies watched, then $3.95x$ = the cost to watch x On Demand movies.
Cost for Cable + Additional fee for On Demand movies = 96.38
$72.68 + 3.95x = 96.38$
$\qquad 3.95x = 23.70$
$\qquad\qquad x = 6$
He watched 6 movies.

35. Let x = number of miles driven, then $20 + 0.25x$ = American rental fee and $35 + 0.15x$ = SavMor rental fee.
American rental fee = SavMor rental fee
$20 + 0.25x = 35 + 0.15x$
$20 + 0.10x = 35$
$\qquad 0.10x = 15$
$\qquad\qquad x = 150$
Driving 150 miles would result in both plans having the same total cost.

37. Let x = number of years until the salaries are the same.
yearly salary = base salary + (yearly increase) · (number of years)
yearly salary at Data Tech. = yearly salary at Nuteck
$40,000 + 2400x = 49,600 + 800x$
$40,000 + 1600x = 49,600$
$\qquad 1600x = 9600$
$\qquad\qquad x = 6$
It will take 6 years for the two salaries to be the same.

39. Let x = the number of pages, then $0.06x$ = the cost of printing x pages on the HP and $0.08x$ = the cost of printing x pages on the Lexmark.

Cost of the HP printer + Cost of printing x pages on the HP = Cost of the Lexmark printer + Cost of printing x pages on the Lexmark
$499 + 0.06x = 419 + 0.08x$
$\qquad 499 = 419 + 0.02x$
$\qquad\; 80 = 0.02x$
$\quad 4000 = x$
For the two printers to have the same cost, 4000 pages would have to be printed.

41. Let x = the number of envelopes used, then $0.39x$ = the mailing cost.
Printing cost + Mailing cost = Total cost
$600 + 0.39x = 1380$
$\qquad 0.39x = 780$
$\qquad\qquad x = 2000$
There were 2000 newsletters mailed.

43. Let x = the cost of the flight before tax, then $0.07x$ = the sales tax on the flight.
Cost of flight before tax + Sales tax on flight = Total cost
$x + 0.07x = 280$
$\quad 1.07x = 280$
$\qquad\; x = 261.68$
The cost of the flight before taxes was $261.68.

45. Let x = Zhen's present salary, then $x + 0.30x$ = Zhen's salary at his new job.
New salary = 30,200
$x + 0.30x = 30,200$
$\qquad 1.3x = 30,200$
$\qquad\; x = 23230.77$
Zhen's present salary is $23,230.77.

47. Let x = the number of bushels of oysters harvested in 2001, then $x - 0.93x$ = the number of bushels of oysters harvested in 2004.
The number of bushels of oysters harvested in 2004 = 26,000
$x - 0.93x = 26,000$
$\quad 0.07x = 26,000$
$\qquad\; x \approx 371,428.57$
About 371 thousand bushels were harvested in 2001.

49. Let x = total amount collected at door;
3000 + 3% of admission fees = total amount received.
$3000 + 0.03x = 3750$
$\qquad 0.03x = 750$
$\qquad\qquad x = \dfrac{7.50}{0.03} = 25,000$
The total amount collected at the door was $25,000.

51. Let x = average salary before wage cut.
(average salary before cut) − (decrease in salary) = average salary after wage cut
$x - 0.02x = 38,600$
$\quad 0.98x = 38,600$
$\qquad\; x \approx 39,387.76$
The average salary before the wage cut was $39,387.76.

53. Let x = the average salary of a graduate with a Bachelor's degree, then $x - 0.246x$ = the average salary of a graduate with an Associate's degree. Average salary of a graduate with an Associate's degree = 37,600
$$x - 0.246x = 37,600$$
$$0.754x = 37,600$$
$$x \approx 49,867.374$$
The average salary of a graduate with a bachelor's degree is about $49,867.

55. Let x = the amount of sales in dollars, then $600 + 0.02x$ = Plan 1 salary and $0.10x$ = Plan 2 salary.
Plan 1 salary = Plan 2 salary
$$600 + 0.02x = 0.10x$$
$$600 = 0.08x$$
$$7500 = x$$
Sales of $7,500 will result in the same salary from both plans.

57. Let x = the number of pages in the second edition, then $x - 0.04x$ = the number of pages in the third edition.
Number of pages in the third edition = 480
$$x - 0.04x = 480$$
$$0.96x = 480$$
$$x = 500$$
There were 500 pages in the second edition.

59. Let x = the amount of sales in dollars, then $400 + 0.02x$ = Plan 1 salary and $250 + 0.16x$ = Plan 2 salary.
Plan 1 salary = Plan 2 salary
$$400 + 0.02x = 250 + 0.16x$$
$$400 = 250 + 0.14x$$
$$150 = 0.14x$$
$$1071.43 = x$$
Sales of $1071.43 will result in the same salary from both plans.

61. Let x = maximum price of meal she can afford, then amount of tax = $0.07x$ and amount of tip = $0.15x$.
Price of meal + tax + tip = 30
$$x + 0.07x + 0.15x = 30$$
$$1.22x = 30$$
$$x \approx 24.59$$
The maximum price she can afford for a meal is $24.59.

63. Let x = the amount in dollars Phil's daughter receives, then $x + 0.25x$ = amount in dollars Phil's wife receives.
Daughter's share + Wife's share = $140,000
$$x + (x + 0.25x) = 140,000$$
$$2x + 0.25x = 140,000$$
$$2,25x = 140,000$$
$$x = 62,222.22$$
$$x + 0.25x = 62,222.22 + (0.25)(62,222.22)$$
$$= 77,777.78$$
Phil's wife will receive $77,777.78.

65. a. $\dfrac{74 + 88 + 76 + x}{4} = 80$

b. $\dfrac{74 + 88 + 76 + x}{4} = 80$
$$74 + 88 + 76 + x = 320$$
$$238 + x = 320$$
$$x = 82$$
Paul must receive an 82 on his fourth exam.

67. $4\left[(4-6) \div 2\right] + 3^2 - 1$
$$= 4\left[(-2) \div 2\right] + 9 - 1$$
$$= 4\left[-1\right] + 9 - 1$$
$$= -4 + 9 - 1$$
$$= 4$$

68. Commutative property of addition

69. $A = \dfrac{1}{2}bh$
$$2A = 2\left(\dfrac{1}{2}bh\right)$$
$$2A = bh$$
$$\dfrac{2A}{b} = \dfrac{bh}{b}$$
$$\dfrac{2A}{b} = h$$

70. $\dfrac{4.5}{6} = \dfrac{9}{x}$
$$4.5 \cdot x = 6 \cdot 9$$
$$4.5x = 54$$
$$x = 12$$

Mid-Chapter Test: 3.1-3.2

1. $6w$

2. $3h + 5$

3. $c + 0.20c$

4. Let m = number of miles driven, then $0.25m$ = total mileage cost. The total rental cost is $(40 + 0.25m)$.

5. $50n$

6. $25 - x$

7. $(c - 25)$ would be the cost of the item minus \$25. The cost of an item at a 25% off sale would be $(c - 0.25c)$.

8. x = the length of the Poison Dart frog

9. If p = the distance Pedro traveled, then $(4p + 6)$ = the distance Mary traveled.

10. Let v = the value of a car in 2006, then $(v - 0.18v)$ = the value of the car in 2005. Car value in 2006 – Car value in 2005 = $v - (v - 0.18v)$.

11. Let p = the original population of Cedar Oaks, then $0.12p$ = the amount of increase in the population.
Original population + the increase in population = 38,619
$p + 0.12p = 38,619$

12. Let x = the smaller integer and $(x + 2)$ = the larger consecutive odd integer.
Smaller integer + 3 times the larger is 26.
$x + 3(x + 2) = 26$

13. Let x = the smaller integer, then $(x + 1)$ = the larger consecutive integer.
Smaller integer + Larger integer = 93
$$x + (x + 1) = 93$$
$$x + x + 1 = 93$$
$$2x + 1 = 93$$
$$2x = 92$$
$$x = 46$$
The smaller integer is 46 and the larger is $(46 + 1) = 47$.

14. Let x = the smaller integer, then $(3x - 1)$ = the larger integer.
Larger integer – Smaller integer = 7
$$(3x - 1) - x = 7$$
$$3x - 1 - x = 7$$
$$2x - 1 = 7$$
$$2x = 8$$
$$x = 4$$
The smaller number is 4 and the larger is $3(4) - 1 = 11$.

15. Let x = the number of days of production, then $20x$ = the increased production.
Original production + Increase = 600
$$240 + 20x = 600$$
$$20x = 360$$
$$x = 18$$
It will take 18 days for production to reach 600 boxes daily.

16. Let h = the number of hours tennis is played, then $4h$ = the court cost at Dale's and $8h$ = the court cost at Abel's.
Total cost at Dale's = Total cost at Abel's
$$90 + 4h = 30 + 8h$$
$$90 = 30 + 4h$$
$$60 = 4h$$
$$15 = h$$
Kristina would have to play 15 hours of tennis a month for the total cost to be the same at both clubs.

17. Let c = the cost of the television, then $0.07c$ = the amount of sales tax.
Cost of the television + Sales tax = Total cost
$$c + 0.07c = 749$$
$$1.07c = 749$$
$$c = 700$$
The cost of the television before sales tax was \$700.

18. Let c = the number of clients that Betty has, then $(2c + 12)$ = the number of clients that Anita has.
Clients of Betty + Clients of Anita = 600
$$c + (2c + 12) = 600$$
$$c + 2c + 12 = 600$$
$$3c + 12 = 600$$
$$3c = 588$$
$$c = 196$$
Betty has 196 clients and Anita has $2(196) + 12 = 404$ clients.

19. Let m = the number of miles driven, then $0.18m$ = the mileage cost.
Truck cost + Mileage cost = Total cost
$36 + 0.18m = 45.36$
$$0.18m = 9.36$$
$$m = 52$$
There were 52 miles driven that day.

20. Let s = the amount of sales, then $0.08s$ = the commission from Plan 1 and $0.06s$ = the commission from Plan 2.
Total salary from Plan 1 = Total salary from Plan 2
$200 + 0.08s = 300 + 0.06s$
$$200 + 0.02s = 300$$
$$0.02s = 100$$
$$s = 5000$$
He must make \$5,000 in sales per week for the two plans to have the same total salary.

Exercise Set 3.3

1. $A = (2l) \cdot \left(\dfrac{w}{2}\right) = lw$

The area remains the same

3. $V = 2l \cdot 2w \cdot 2h = 8(lwh)$

The volume is eight times as great.

5. $A = \pi r^2 = \pi(3r)^2 = \pi(9r^2) = 9\pi r^2$
The area is nine times as great.

7. An isosceles triangle is a triangle with 2 equal sides.

9. The sum of the measures of the angles in a triangle is 180°.

11. Let x = the measure of the two equal angles, then $x + 42$ = the measure of the third angle.
Sum of the three angles = 180º
$x + x + (x + 42) = 180$
$\qquad 3x + 42 = 180$
$\qquad\qquad 3x = 138$
$\qquad\qquad\quad x = 46$
The two equal angles are each 46°. The third angle is $x + 42° = 46° + 42° = 88°$.

13. Let x = length of each side of the triangle, then $P = x + x + x = 3x$.
Perimeter = 34.5
$\qquad 3x = 34.5$
$\qquad\quad x = 11.5$
The length of each side is 11.5 inches.

15. Let x = measure of angle B. Then $2x + 21$ = measure of angle A.
Sum of the 2 angles = 90
$x + (2x + 21) = 90$
$\qquad 3x + 21 = 90$
$\qquad\qquad 3x = 69$
$\qquad\qquad\quad x = 23$
Measure of angle $A = 2(23) + 21 = 67°$
Measure of angle $B = 23°$

17. Let x = measure of angle A, then $3x - 8$ = measure of angle B.
Sum of the 2 angles = 180
$x + (3x - 8) = 180$
$\qquad 4x - 8 = 180$
$\qquad\quad 4x = 188$
$\qquad\qquad x = \dfrac{188}{4} = 47$
Measure of angle $A = 47°$
Measure of angle $B = 3(47) - 8 = 141 - 8 = 133°$

19. The two angles have equal measures.
$2x + 50 = 4x + 12$
$\qquad 38 = 2x$
$\qquad 19 = x$
$2x + 50 = 2(19) + 50 = 38 + 50 = 88º$
$4x + 12 = 4(19) + 12 = 76 + 12 = 88°$
Each angle measures 88°.

21. Let x = measure of smallest angle. Then second angle = $x + 10$ and third angle = $2x - 30$.
Sum of the 3 angles = 180
$x + (x + 10) + (2x - 30) = 180$
$\qquad\qquad 4x - 20 = 180$
$\qquad\qquad\quad 4x = 200$
$\qquad\qquad\qquad x = 50$
The first angle is 50°.
The second angle is $50 + 10 = 60°$.
The third angle is $2(50) - 30 = 70°$.

23. Let x = width of rectangle. Then $x + 6$ = length of rectangle.
$P = 2l + 2w$
$44 = 2(x + 6) + 2x$
$44 = 2x + 12 + 2x$
$44 = 4x + 12$
$32 = 4x$
$8 = x$
Width is 8 feet and length is $8 + 6 = 14$ feet.

25. Let x = width of tennis court.
Then $2x + 6$ = length of tennis court.
$$P = 2l + 2w$$
$$228 = 2(2x + 6) + 2x$$
$$228 = 4x + 12 + 2x$$
$$228 = 6x + 12$$
$$216 = 6x$$
$$36 = x$$
The width is 36 feet and the length is
$2(36) + 6 = 78$ feet.

27. Let x = measure of each smaller angle.
Then $3x - 20$ = measure of each larger angle.
(measure of the two smaller angles) + (measure of the two larger angles) = 360°
$$x + x + (3x - 20) + (3x - 20) = 360$$
$$8x - 40 = 360$$
$$8x = 400$$
$$x = 50$$
Each smaller angle is 50°. Each larger angle is
$3(50) - 20 = 130°$.

29. Let x = measure of each smaller angle, then $5x$ = measure of each larger angle.
(measure of the two smaller angles) + (measure of the two larger angles) = 360°
$$x + x + 5x + 5x = 360$$
$$12x = 360$$
$$x = 30$$
Each smaller angle is 30°.
Each larger angle is $5(30) = 150°$.

31. Let x = measure of the smallest angle.
Then $x + 10$ = measure of the second angle,
$2x + 14$ = measure of third angle, and
$x + 21$ = measure of fourth angle.
Sum of the four angles = 360°
$$x + (x + 10) + (2x + 14) + (x + 21) = 360$$
$$5x + 45 = 360$$
$$5x = 315$$
$$x = 63$$
Thus the angles are 63°, $63 + 10 = 73°$
$2(63) + 14 = 140°$ and $63 + 21 = 84°$.

33. Let x = width of bookcase shelf.
Then $x + 3$ = height of bookcase.
4 shelves + 2 sides = total lumber available.
$$4x + 2(x + 3) = 30$$
$$4x + 2x + 6 = 30$$
$$6x + 6 = 30$$
$$6x = 24$$
$$x = 4$$
The width of each shelf is 4 feet and the height is
$4 + 3 = 7$ feet.

35. Let x = length of a shelf.
Then $2x$ = height of bookcase.
4 shelves + 2 sides = total lumber available
$$4x + 2(2x) = 20$$
$$4x + 4x = 20$$
$$8x = 20$$
$$x = \frac{20}{8} = 2.5$$
The width of the bookcase is 2.5 feet. The height
of the bookcase is $2(2.5) = 5$ feet.

37. Let x = width of fenced in area.
Then $x + 5$ = length of fenced in area
Five "widths" + one "length" = total fencing
$$5x + (x + 5) = 71$$
$$6x + 5 = 71$$
$$6x = 66$$
$$x = 11$$
Width is 11 feet and length is $11 + 5 = 16$ feet.

39. $ac + ad + bc + bd$

41. $-|-6| < |-4|$ since $-|-6| = -6$ and $|-4| = 4$

42. $|-3| > -|3|$ since $|-3| = 3$ and $-|3| = -3$

43.
$$-8 - (-2) + (-4) = -8 + 2 + (-4)$$
$$= -6 + (-4)$$
$$= -10$$

44.
$$-7y + x - 3(x - 2) + 2y$$
$$= -7y + x - 3x + 6 + 2y$$
$$= x - 3x - 7y + 2y + 6$$
$$= -2x - 5y + 6$$

45.

$$6x + 3y = 9$$

$$6x - 6x + 3y = -6x + 9$$

$$3y = -6x + 9$$

$$\frac{3y}{3} = \frac{-6x + 9}{3}$$

$$y = \frac{-6x + 9}{3} \text{ or } y = -2x + 3$$

Exercise Set 3.4

1. Let t = time in hours it will take the ferries to be 6 miles apart.

Ferry	Rate	Time	Distance
Cat	34	t	$34t$
Bird	28	t	$28t$

$$34t - 28t = 6$$

$$6t = 6$$

$$t = 1$$

In 1 hour the two ferries will be 6 miles apart.

3. Let r = Abe's rate of speed..

Person	Rate	Time	Distance
Jodi	8	2	16
Abe	r	2	$2r$

$$16 - 2r = 4$$

$$-2r = -12$$

$$r = 6$$

Abe was traveling at 6 mph.

5. Let t be the time it takes for the planes to pass each other.

Plane	Rate	Time	Distance
Jet Blue	560	t	$560t$
SW	580	t	$580t$

$$560t + 580t = 821$$

$$1140t = 821$$

$$t \approx 0.72$$

It will take about 0.72 hours for the planes to pass each other.

7. Let t be the time it takes for Willie and Shanna to be 16.8 miles apart.

Person	Rate	Time	Distance
Willie	3	t	$3t$
Shanna	4	t	$4t$

$$3t + 4t = 16.8$$

$$7t = 16.8$$

$$t = 2.4$$

It will take 2.4 hours.

9. Let r be the second rate of the machine.

Machine	Rate	Time	Distance
First	60	7.2	432
Second	r	6.8	$6.8r$

$$432 + 6.8r = 908$$

$$6.8r = 476$$

$$r = 70$$

The second speed the machine was set at was 70 miles per hour.

11. Let t be the time it takes for the waves to be 80 miles apart.

Wave	Rate	Time	Distance
p-wave	3.6	t	$3.6t$
s-wave	1.8	t	$1.8t$

$$3.6t - 1.8t = 80$$

$$1.8t = 80$$

$$t \approx 44.4$$

It will take approximately 44.4 seconds.

13. Let r be the speed of the cutter coming from the east (westbound). Then $r + 5$ is the speed of the cutter coming from the west (eastbound).

Cutter	Rate	Time	Distance
Eastbound	$r + 5$	3	$3(r + 5)$
Westbound	r	3	$3r$

$$3(r + 5) + 3r = 225$$
$$3r + 15 + 3r = 225$$
$$6r = 210$$
$$r = 35$$

The speed of the westbound cutter is 35 mph and the speed of the eastbound cutter is 40 mph.

15. Let t be amount of time Samia walked for before she turned around.

Trip	Rate	Time	Distance
In	4	t	$4t$
Out	3.2	$t + 0.5$	$3.2(t + 0.5)$

$$4t + 3.2(t + 0.5) = 6$$
$$4t + 3.2t + 1.6 = 6$$
$$7.2t = 4.4$$
$$t \approx 0.61$$

Samia walked 0.61 hours before she turned around.

17. Let r be the rate of the slower crew. Then $r + 0.75$ is the rate of the faster crew.

Crew	Rate	Time	Distance
Slow	r	3.2	$3.2r$
Fast	$r + 0.75$	3.2	$3.2(r + 0.75)$

$$3.2r + 3.2(r + 0.75) = 12$$
$$3.2r + 3.2r + 2.4 = 12$$
$$6.4r = 9.6$$
$$r = 1.5$$

One crew works at a rate of 1.5 miles/day, and the other crew works at a rate of 2.25 miles/day.

19. Let r be the rate of *Apollo*. Then $r + 4$ is the rate of *Pythagoras*.

Boat	Rate	Time	Distance
Apollo	r	0.7	$0.7r$
Pythagoras	$r + 4$	0.7	$0.7(r + 4)$

$$0.7r + 0.7(r + 4) = 9.8$$
$$0.7r + 0.7r + 2.8 = 9.8$$
$$1.4r = 7$$
$$r = 5$$

The speed of *Apollo* is 5 mph, and the speed of *Pythagoras* is 9 mph.

21. Let t = time, in hours, Betty was traveling at 50 mph.

Speed	Rate	Time	Distance
Faster	70 mph	$t - 0.5$	$70(t - 0.5)$
Slower	50 mph	t	$50t$

$$50t - 70t + 35 = 5$$
$$-20t = -30$$
$$t = 1.5$$

Betty traveled for 1.5 hours at 50 mph.

23. **a.** Distance = rate × time = $(2.31)(1.04) \approx 2.4$ miles

 b. Distance = rate × time = $(23.0)(4.87) \approx 112.0$ miles

 c. Distance = rate × time = $(8.45)(3.1) \approx 26.2$ miles

 d. $2.4 + 112.0 + 26.2 = 140.6$ miles

 e. $1.04 + 4.87 + 3.10 = 9.01$ hours

25. Let x be the amount invested at 5%. Then $12,000 - x$ is the amount invested at 7%.

Principal	Rate	Time	Interest
x	5%	1	$0.05x$
$12,000 - x$	7%	1	$0.07(12,000 - x)$

$$0.05x + 0.07(12,000 - x) = 800$$
$$0.05x + 840 - 0.07x = 800$$
$$-0.02x = -40$$
$$x = 2000$$

They invested $2000 at 5% and $10,000 at 7%.

27. Let x be the amount invested at 6%. Then $6000 - x$ is the amount invested at 4%.

Principal	Rate	Time	Interest
x	6%	1	$0.06x$
$6000 - x$	4%	1	$0.04(6000 - x)$

$$0.06x = 0.04(6000 - x)$$
$$0.06x = 240 - 0.04x$$
$$0.10x = 240$$
$$x = 2400$$

She invested $2400 at 6% and $3600 at 4%.

29. Let x be the amount invested at 4%. Then $10,000 - x$ is the amount invested at 5%.

Principal	Rate	Time	Interest
x	4%	1	$0.04x$
$10,000 - x$	5%	1	$0.05(10,000 - x)$

$$0.05(10,000 - x) - 0.04x = 320$$
$$500 - 0.05x - 0.04x = 320$$
$$-0.09x = -180$$
$$x = 2000$$

She invested $2000 at 4% and $8000 at 5%.

31. Let t be the time, in months, during which Patricia paid $17.10 per month. Then $12 - t$ is the time during which she paid $18.40 per month.

Rate	Time	Amount
17.10	t	$17.10t$
18.40	$12 - t$	$18.40(12 - t)$

$$17.10t + 18.40(12 - t) = 207.80$$
$$17.10t + 220.80 - 18.40t = 207.80$$
$$-1.30t = -13$$
$$t = 10$$

She paid $17.10 for the first 10 months of the year, and paid $18.40 for the remainder of the year. The rate increase took effect in November.

33. Let x be the number of hours worked at Home Depot ($6.50 per hour). Then $18 - x$ is the number of hours worked at the veterinary clinic ($7.00 per hour).

Rate	Hours	Total
$6.50	x	$6.5x$
$7.00	$18 - x$	$7(18 - x)$

$$6.5x + 7(18 - x) = 122$$
$$6.5x + 126 - 7x = 122$$
$$-0.5x = -4$$
$$x = 8$$

Mihály worked 8 hours at Home Depot and 10 hours at the clinic.

35. Let x be the number of adults who went to the Baseball Hall of Fame. Then $2100 - x$ is the number of children who attended.

	Price	Number	Total
Adults	14.5	x	$14.50x$
Children	5.00	$2100 - x$	$5.00(2100 - x)$

$$14.5x + 5(2100 - x) = 21,900$$
$$14.5x + 10,500 - 5x = 21,900$$
$$9.5x = 11,400$$
$$x = 1200$$

There were 1200 adults at the Baseball Hall of Fame.

37. a. Let x be the number of shares of Nike. Then $5x$ is the number of shares of Kellogg.

Stock	Price	Shares	Total
Nike	78	x	$78x$
Kellogg	33	$5x$	$33(5x)$

$$78x + 33(5x) = 10,000$$
$$78x + 165x = 10,000$$
$$244x = 10,000$$
$$x \approx 40.98$$

Since only whole shares can be purchased, he will purchase 41 shares of Nike and 205 shares of Kellogg.

b. Mike spent $41(78) + 205(33) = \$9963$.

He has $\$10,000 - \$9963 = \$37$ left over.

39. Let x be the number of pounds of almonds. Then $30 - x$ is the number of pounds of walnuts.

Nut	Cost	Pounds	Total
Almond	6.40	x	$6.40x$
Walnut	6.80	$30 - x$	$6.8(30 - x)$
Mixture	6.65	30	199.50

$$6.4x + 6.8(30 - x) = 199.50$$
$$6.4x + 204 - 6.8x = 199.50$$
$$-0.4x = -4.5$$
$$x = 11.25$$

So, 11.25 pounds of almonds should be mixed with 18.75 pounds of walnuts.

41. Let x be the amount of topsoil in cubic yards.

Soil Type	Cost	Amount	Total
Strained	160	x	$160x$
Unstrained	120	$8 - x$	$120(8 - x)$
Mixture	150	8	$150(8)$

$$160x + 120(8 - x) = 150(8)$$
$$160x + 960 - 120x = 1200$$
$$40x + 960 = 1200$$
$$40x = 240$$
$$x = 6$$

He should mix 6 cubic yards of the strained with 2 cubic yards of the unstrained.

43. Let x be the cost per pound of the mixture.

Type	Cost	Pounds	Total
Good & Plenty	2.49	3	$2.49(3)$
Sweet Treats	2.89	5	$2.89(5)$
Mixture	x	8	$8x$

$$2.49(3) + 2.89(5) = 8x$$
$$7.47 + 14.45 = 8x$$
$$21.92 = 8x$$
$$2.74 = x$$

The mixture should sell for $2.74 per pound.

45. Let x be the percentage of alcohol in the mixture.

Percentage	Liters	Amount of Alcohol
12%	5	0.6
9%	2	0.18
$x\%$	7	$\left(\frac{x}{100}\right) \cdot 7$

$$\left(\frac{x}{100}\right) \cdot 7 = 0.6 + 0.18$$
$$0.07x = 0.78$$
$$x \approx 11.1$$

The alcohol content of the mixture is about 11.1%.

47. Let x be the amount of 12% sulfuric acid solution.

Solution	Strength	Liters	Amount
20%	0.20	1	0.20
12%	0.12	x	$0.12x$
Mixture	0.15	$x + 1$	$0.15(x + 1)$

$$0.20 + 0.12x = 0.15(x + 1)$$
$$0.20 + 0.12x = 0.15x + 0.15$$
$$-0.03x = -0.05$$
$$x = \frac{5}{3}$$

$1\frac{2}{3}$ liters of 12% sulfuric acid should be used.

49. Let x be the percentage of alcohol in the mixture.

Brand	Percentage	Ounces	Amount
Listerine	21.6%	6	$(0.216) \cdot 6$
Scope	15%	4	$(0.15) \cdot 4$
Mixture	x%	10	$\left(\frac{x}{100}\right) \cdot 10$

$$(0.216) \cdot 6 + (0.15) \cdot 4 = \left(\frac{x}{100}\right) \cdot 10$$
$$1.296 + 0.6 = \frac{x}{10}$$
$$1.896 = \frac{x}{10}$$
$$18.96 = x$$

The mixture has an 18.96% alcohol content.

51. Let x be the percent of orange juice in the new mixture.

Percent	Quarts	Orange Juice
12%	6	0.72
x	$6\frac{1}{2}$	$\frac{13}{2}x$

$$\frac{13}{2}x = 0.72$$
$$x \approx 0.111$$

The new mixture consists of about 11.1% orange juice.

53. Let x be the percentage of pure juice before mixing.

Solution	Percentage	Amount of Solution	Amt. of Juice
Concentrate	x	12	$12\left(\frac{x}{100}\right)$
Mixture	10%	48	4.8

$$12\left(\frac{x}{100}\right) = 4.8$$
$$0.12x = 4.8$$
$$x = 40$$

40% of the concentrate is pure juice.

55. Let x be the amount of the Clorox to be added to a 6 cups of of water.

Product	Percentage	Cups	Amount
Clorox	5.25%	x	$(0.0525)x$
Shock Treatment	10.5%	$6 - x$	$0.105(6 - x)$
Mixture	7.2%	6	$0.072(6)$

$$0.0525x + 0.105(6 - x) = 0.072(6)$$
$$0.0525x + 0.63 - 0.105x = 0.432$$
$$-0.0525x = -0.198$$
$$x \approx 3.77$$

The mixture needs about 3.77 cups Clorox and $6 - 3.77 = 2.23$ cups of shock treatment.

57. Let x be the number of pints of 12% alcohol.

Type	Percentage	Pints	Amount
12% concen.	12%	x	$0.12x$
5% concen.	5%	15	$0.05(15)$
Mixture	8%	$x + 15$	$0.08(x + 15)$

$$0.12x + 0.05(15) = 0.08(x + 15)$$
$$0.12x + 0.75 = 0.08x + 1.2$$
$$0.04x + 0.75 = 1.2$$
$$0.04x = 0.45$$
$$x = 11.25$$

The mixture should consist of 11.25 pints of the 12% alcohol.

59. The time it takes for the transport to make the trip is:

$$\text{Time} = \frac{\text{Distance}}{\text{Rate}} = \frac{1720}{370} \approx 4.65 \text{ hours}$$

The time it takes for the Hornets to make the trip is: $\text{Time} = \dfrac{\text{Distance}}{\text{Rate}} = \dfrac{1720}{900} \approx 1.91 \text{ hours}$

It takes the transport $4.65 - 1.91 = 2.74$ hours longer to make the trip. Since it needs to arrive 3 hours before the Hornets, it should leave about $2.74 + 3 = 5.74$ hours before them.

62. a.
$$2\frac{3}{4} \div 1\frac{5}{8} = \frac{11}{4} \div \frac{13}{8}$$
$$= \frac{11}{4} \cdot \frac{8}{13}$$
$$= \frac{22}{13} \text{ or } 1\frac{9}{13}$$

b.
$$2\frac{3}{4} + 1\frac{5}{8} = \frac{11}{4} + \frac{13}{8}$$
$$= \frac{22}{8} + \frac{13}{8}$$
$$= \frac{35}{8} \text{ or } 4\frac{3}{8}$$

63. $6(x-3) = 4x - 18 + 2x$

$\quad\ 6x - 18 = 6x - 18$

All real numbers are solutions.

64.
$$\frac{6}{x} = \frac{72}{9}$$
$$6 \cdot 9 = 72x$$
$$54 = 72x$$
$$x = \frac{54}{72} = \frac{3}{4} \text{ or } 0.75$$

65. Let $x =$ the first integer. Then, $x+1$ is the next integer.

$$x + (x+1) = 77$$
$$2x + 1 = 77$$
$$2x = 76$$
$$x = 38$$

The two integers are 38 and 39.

Chapter 3 Review Exercises

1. $3n + 7$

2. $1.2g$

3. $d - 0.25d$

4. $16y$

5. $200 - x$

6. Let $d =$ Dino's age, then $7d + 6 =$ Mario's age.

7. Let c be the number of robberies in 2005, then $(c - 0.12c)$ is the amount of robberies in 2006. The difference between the robberies is $c - (c - 0.12c)$.

8. Let n be the larger number, then $(3n - 24)$ is the smaller number. Their difference is 8. The equation is $n - (3n - 24) = 8$.

9. Let $x =$ the smaller number.
Then $x + 8 =$ the larger number.
Smaller number + larger number = 74
$$x + (x + 8) = 74$$
$$2x + 8 = 74$$
$$2x = 66$$
$$x = 33$$
The smaller number is 33 and the larger number is $33 + 8 = 41$.

10. Let $x =$ smaller integer.
Then $x + 1 =$ next consecutive integer.
Smaller number + larger number = 237
$$x + (x + 1) = 237$$
$$2x + 1 = 237$$
$$2x = 236$$
$$x = 118$$
The smaller number is 118 and the larger number is $118 + 1 = 119$.

11. Let $x =$ the smaller integer.
Then $5x + 3 =$ the larger integer
Larger number − smaller number = 31
$$(5x + 3) - x = 31$$
$$4x + 3 = 31$$
$$4x = 28$$
$$x = 7$$
The smaller number is 7 and the larger number is $5(7) + 3 = 38$.

12. Let x = cost of car before tax.
Then $0.07x$ = amount of tax.
Cost of car before tax + tax on car
= cost of car after tax
$$x + 0.07x = 23,260$$
$$1.07x = 23,260$$
$$x = \frac{23,260}{1.07} = 21,738.32$$
The cost of the car before tax is \$21,738.32.

13. Let x = the number of months, then $20x$ = the increase in production of bagels over x months.
Current production + Increase in production =
Future production
$$520 + 20x = 900$$
$$20x = 380$$
$$x = 19$$
It will take 19 months.

14. Let x = weekly dollar sales that would make total salaries from both companies the same.
The commission at present company = $0.03x$ and commission at new company = $0.08x$.
Salary + commission for present company
= salary + commission for new company
$$600 + 0.03x = 500 + 0.08x$$
$$100 + 0.03x = 0.08x$$
$$100 = 0.05x$$
$$2000 = x$$
Irene's weekly sales would have to be \$2000 for the total salaries from both companies to be the same.

15. Let x = original price of camcorder. Then $0.20x$ = reduction.
Original price –reduction = sale price
$$x - 0.20x = 495$$
$$0.8x = 495$$
$$x = 618.75$$
The original price of the camcorder was \$618.75.

16. Let h = the number of hours for both landscapers to charge the same price.
Two Brothers Nursery = ABC Nursery
$$400 + 45h = 200 + 65h$$
$$400 = 200 + 20h$$
$$200 = 20h$$
$$10 = h$$
It would take 10 hours of labor before the costs from both landscapers would be the same.

17. Let x be the median cost of a house in 2003.
Then $0.117x$ would be the amount of increase.
Median price in 2003 + Amount of increase =
Median price in 2005
$$x + 0.117x = 191,000$$
$$1.117x = 191,000$$
$$x = 170,993.73$$
The median cost in 2003 was about \$171,000.

18. Let x be the average annual refund in 1980.
Then $4x - 282$ is the average annual refund in 2004.
$$4x - 282 = 2454$$
$$4x = 2736$$
$$x = 684$$
The average annual refund in 1980 was \$684.

19. Let x = measure of the smallest angle. Then $x + 10$ = measure of second angle and $2x - 10$ = measure of third angle.
Sum of the three angles = 180°
$$x + (x + 10) + (2x - 10) = 180$$
$$4x = 180$$
$$x = 45$$
The angles are 45°, 45 + 10 = 55°, and $2(45) - 10 = 80°$.

20. Let x = measure of the smallest angle. Then
$x + 10$ = measure of second angle,
$5x$ = measure of third angle,
$4x + 20$ = measure of the fourth angle.
Sum of the four angles = 360°
$$x + (x + 10) + 5x + (4x + 20) = 360$$
$$11x + 30 = 360$$
$$11x = 330$$
$$x = \frac{330}{11} = 30$$
The angles are 30°, 30 + 10 = 40°.
$5(30) = 150°$, and $4(30) + 20 = 140°$.

21. Let w = width of garden. Then
$w + 4$ = length of garden.
$$P = 2l + 2w$$
$$70 = 2(w + 4) + 2w$$
$$70 = 4w + 8$$
$$62 = 4w$$
$$15.5 = w$$
The width is 15.5 feet and the length is 15.5 + 4 = 19.5 feet.

22. Let x = the width of the room. Then
$x + 30$ = the length of the room. The amount of string used is the perimeter of the room plus the wall separating the two rooms.

$P = 2l + 2w + w$
$P = 2l + 3w$
$310 = 2(x + 30) + 3x$
$310 = 2x + 60 + 3x$
$310 = 5x + 60$
$250 = 5x$
$50 = x$

The width of the room is 50 feet and the length is
$50 + 30 = 80$ feet.

23. Let a be the measure of each of the two smaller angles. Then each of the two larger angles is $3a$.
$a + a + 3a + 3a = 360$
$8a = 360$
$a = 45$
The angles measure 45°, 45°, and $3(45) = 135°$ and 135°.

24. Let h be the height of the bookcase. Then $2h$ would be its length.
Amount of lumber for 4 shelves + Amount of lumber for the sides = 20
$4(2h) + 2h = 20$
$8h + 2h = 20$
$10h = 20$
$h = 2$
The height of the bookcase is 2 feet and the length is 4 feet.

25. Let t be the time it takes for the joggers to be 4 kilometers apart.

Jogger	Rate	Time	Distance
Harold	8	t	$8t$
Susan	6	t	$6t$

$8t - 6t = 4$
$2t = 4$
$t = 2$
It takes the joggers 2 hours to be 4 kilometers apart.

26. Let t be the amount of time it takes for the trains to be 440 miles apart.

Train	Rate	Time	Distance
First	50	t	$50t$
Second	60	t	$60t$

$50t + 60t = 440$
$110t = 440$
$t = 4$
After 4 hours, the two trains will be 440 miles apart.

27. $\text{Rate} = \dfrac{\text{Distance}}{\text{Time}} = \dfrac{200 \text{ feet}}{22.73 \text{ seconds}} \approx 8.8$ ft/sec

The cars travel at about 8.8 ft/sec.

28. Let x be the amount invested at 8%. Then
$12,000 - x$ is the amount invested at $7\frac{1}{4}$%.

Principal	Rate	Time	Interest
x	8%	1	$0.08x$
$12,000 - x$	$7\frac{1}{4}$%	1	$0.0725(12,000 - x)$

$0.08x + 0.0725(12,000 - x) = 900$
$0.08x + 870 - 0.0725x = 900$
$0.0075x = 30$
$x = 4000$
Tatiana should invest $4000 at 8% and $8000 at $7\frac{1}{4}$%.

29. Let x be the amount invested at 3%. Then
$4,000 - x$ is the amount invested at 3.5%.

Principal	Rate	Time	Interest
x	3%	1	$0.03x$
$4,000 - x$	3.5%	1	$0.035(4,000 - x)$

$0.03x + 94.50 = 0.035(4,000 - x)$
$0.03x + 94.50 = 140 - .035x$
$0.065x = 45.5$
$x = 700$
Aimee invested $700 in the 3% account and $3300 in the 3.5% account.

30. Let x be the number of gallons of pure punch. Then $2 - x$ is the number of liters of the 5% acid solution.

Punch Solution	Strength	Gallons	Amount
98%	0.98	2	0.98(2)
100%	1	x	$1x$
98.5%	0.985	$x + 2$	0.985(x + 2)

$1.96 + x = 0.985(x + 2)$
$1.96 + x = 0.985x + 1.97$
$0.015x = 0.01$
$x \approx 0.67$

Marcie should add about 0.67 gallons of pure punch.

31. Let x be the number of small wind chimes sold.

Type	Price	Number Sold	Amount
Small	$8	x	$8x$
Large	$20	$30 - x$	$20(30 - x)$

$8x + 20(30 - x) = 492$
$8x + 600 - 20x = 492$
$-12x = -108$
$x = 9$

He sold 9 small and 21 large wind chimes

32. Let x be the number of liters of the 10% solution. Then $2 - x$ is the number of liters of the 5% acid solution.

Solution	Strength	Liters	Amount
10%	0.10	x	$0.10x$
5%	0.05	$2 - x$	$0.05(2 - x)$
Mixture	0.08	2	0.16

$0.10x + 0.05(2 - x) = 0.16$
$0.10x + 0.10 - 0.05x = 0.16$
$0.05x = 0.06$
$x = 1.2$

The chemist should mix 1.2 liters of 10% solution with 0.8 liters of 5% solution.

33. Let x = smaller odd integer. Then $x + 2$ = next consecutive odd integer.
Smaller number + larger number = 208
$x + (x + 2) = 208$
$2x + 2 = 208$
$2x = 206$
$x = 103$
The smaller number is 103 and the larger number is $103 + 2 = 105$.

34. Let x = cost of television before tax. Then amount of tax = $0.06x$. Cost of television before tax + tax on television = cost of television after tax.
$x + 0.06x = 477$
$1.06x = 477$
$x = \dfrac{477}{1.06} = 450$
The cost of the television before tax is $450.

35. Let x = his dollar sales. Then $0.05x$ = amount of commission.
Salary + commission = 900
$300 + 0.05x = 900$
$0.05x = 600$
$x = \dfrac{600}{0.05} = 12,000$
His sales last week were $12,000.

36. Let x = measure of the smallest angle. Then $x + 8$ = measure of second angle and $2x + 4$ = measure of third angle.
Sum of the three angles = 180°
$x + (x + 8) + (2x + 4) = 180$
$4x + 12 = 180$
$4x = 168$
$x = 42$
The angles are 42°, 42 + 8 = 50°, and 2(42) + 4 = 88°.

37. Let t = number of years. Then $25t$ = increase in employees over t years
Present number of employees + increase in employees = future number of employees
$427 + 25t = 627$
$25t = 200$
$t = \dfrac{200}{25} = 8$

It will take 8 years before they reach 627 employees.

38. Let x = measure of each smaller angle. Then
$x + 40$ = measure of each larger angle
(measure of the two smaller angles)
+(measure of the two larger angles) = 360°

$$x + x + (x + 40) + (x + 40) = 360$$
$$4x + 80 = 360$$
$$4x = 280$$
$$x = 70$$

Each of the smaller angles is 70° and each of the two larger angles is $70 + 40 = 110°$.

39. Let x = number of copies that would result in both centers charging the same. Then charge for copies at Copy King = $0.04x$ and charge for copies at King Kopie = $0.03x$
Monthly fee + charge for copies at Copy King = monthly fee + charge for copies at King Kopie.
$$20 + 0.04x = 25 + 0.03x$$
$$0.04x = 5 + 0.03x$$
$$0.01x = 5$$
$$x = \frac{5}{0.01} = 500$$

500 copies would result in both centers charging the same.

40. Let r = Jim's swim rate.

Person	Rate	Time	Distance
Rita	1	0.5	0.5
Jim	r	0.5	$0.5r$

$$0.5 - 0.5r = 0.2$$
$$-0.5r = -0.3$$
$$r = 0.6$$
Jim swims at 0.6 mph.

41. Let x be the amount of $3.50 per pound ground beef. Then $80 - x$ is the amount of $4.10 per pound of ground beef.

Ground Beef	Price	Amount	Total
$3.50	3.50	x	$3.50x$
$4.10	4.10	$80 - x$	$4.10(80 - x)$
Mixture	3.65	80	292

$$3.50x + 4.10(80 - x) = 292$$
$$3.50x + 328 - 4.10x = 292$$
$$-0.60x = -36$$
$$x = 60$$

The butcher mixed 60 lbs of $3.50 per pound ground beef with 20 lbs of $4.10 per pound ground beef.

42. Let x = the rate the older brother travels. Then $x + 5$ = the rate the younger brother travels.

Brother	Rate	Time	Distance
Younger	$x + 5$	2	$2(x + 5)$
Older	x	2	$2x$

Younger brother's distance + older brother's distance = 230 miles.
$$2(x + 5) + 2x = 230$$
$$2x + 10 + 2x = 230$$
$$4x + 10 = 230$$
$$4x = 220$$
$$x = 55$$

The older brother travels at 55 miles per hour and the younger brother travels at $55 + 5 = 60$ miles per hour.

43. Let x = the number of liters of 30% solution.

Percent	Liters	Amount
30%	x	$0.30x$
12%	2	$(0.12)(2)$
15%	$x + 2$	$0.15(x + 2)$

$$0.30x + (0.12)(2) = 0.15(x + 2)$$
$$0.30x + 0.24 = 0.15x + 0.30$$
$$0.15x + 0.24 = 0.30$$
$$0.15x = 0.06$$
$$x = 0.4$$

0.4 liters of the 30% acid solution need to be added.

44. Let w be the width of the yard. Then $1.5w$ would be the width of the yard.
3 widths + 2 lengths = 96
$$3w + 2(1.5w) = 96$$
$$3w + 3w = 96$$
$$6w = 96$$
$$w = 16$$

The width of the yard is 16 ft. and the length is $1.5(16) = 24$ ft.

45. Let x be the number of liters of 8% solution.

Percent	Liters	Amount
3%	6	$0.03(6)$
8%	x	$0.08x$
4%	$6 + x$	$0.04(6 + x)$

$$0.03(6) + 0.08x = 0.04(6 + x)$$
$$0.18 + 0.08x = 0.24 + 0.04x$$
$$0.18 + 0.04x = 0.24$$
$$0.04x = 0.06$$
$$x = 1.5$$

There was 1.5 liters of the 8% solution used in the mixture.

Chapter 3 Practice Test

1. $500 - n$

2. $2w + 6000$

3. $60t$

4. $c + 0.06c$

5. If n = the number of packages of orange, then $7n - 105$ = number of packages of peppermint.

6. If x = the number of men, then $600 - x$ = the number of women.
If x = the number of women, then $600 - x$ = the number of men.

7. If n = the smaller number, then $2n - 1$ is the larger number. The larger number minus the smaller number can be represented by $(2n - 1) - n$.

8. If n = the smaller bottle, then $n + 18$ is the larger bottle. The sum of the amounts can be represented by $n + (n + 18)$.

9. Let c = the cost of the peanuts, then $c + 0.84c$ is the cost of the deluxe can. Their difference is $(c + 0.84c) - c$.

10. Let x = smaller integer.
Then $2x - 10$ = larger integer.
Smaller number + larger number = 158
$$x + (2x - 10) = 158$$
$$3x - 10 = 158$$
$$3x = 168$$
$$x = 56$$
The smaller number is 56 and the larger number is $2(56) - 10 = 102$

11. Let x = smallest integer.
Then $x + 2$ is the consecutive odd integer.
Sum of the two integers = 33
$$x + 4(x + 2) = 33$$
$$x + 4x + 8 = 33$$
$$5x + 8 = 33$$
$$5x = 25$$
$$x = 5$$
The integers are 5 and $5 + 2 = 7$

12. Let x = one number.
Then $5x - 12$ is the other number.
Sum of the two numbers = 42
$$x + (5x - 12) = 42$$
$$x + 5x - 12 = 42$$
$$6x - 12 = 42$$
$$6x = 54$$
$$x = 9$$
The integers are 9 and $5(9) - 12 = 33$.

13. Let c = the cost of the furniture before tax
Tax amount = (cost) · (tax rate) = $0.06c$
Total cost = cost before tax + tax amount
$$2650 = c + 0.06c$$
$$2640 = 1.06c$$
$$\frac{2650}{1.06} = \frac{1.06c}{1.06}$$
$$2500 = c$$
The cost of the furniture before tax was $2500.

14. Let x = price of most expensive meal he can order. Then $0.15x$ = amount of the tip.
$$x + 0.15x = 40$$
$$1.15x = 40$$
$$x = 34.78$$
The price of the most expensive meal he can order is $34.78.

15. Let x = the amount of money Peter invested.
Then $2x$ = the amount of money Julie invested.
Then $2x$ = the amount of profit Julie receives.
Peter's profit + Julie's profit = Total profit
$$x + 2x = 120,000$$
$$3x = 120,000$$
$$x = 40,000$$
Peter will receive $40,000 and Julie receives $2(\$40,000) = \$80,000$.

16. Let x = the number of times the plow is needed for the costs to be equal.
Elizabeth's charge = $80 + 5x$
Jon charge = $50 + 10x$
The charges are equal when:
$$80 + 5x = 50 + 10x$$
$$30 + 5x = 10x$$
$$30 = 5x$$
$$6 = x$$
The snow would need to be plowed 6 times for the costs to be the same.

17. Let x = number of pages to be printed for the total cost of both printers to be equal.
Then $0.01x$ = cost of printing x pages on the Delta printer,
and $0.03x$ = cost of printing x pages on the TexMar printer.
Delta = TexMar
$$499 + 0.01x = 350 + 0.03x$$
$$499 = 0.02x + 350$$
$$149 = 0.02x$$
$$7450 = x$$
It would take about 7450 pages to be printed for the costs to be the same.

18. Let x = length of smallest side.
Then $x + 15$ = length of second side and $2x$ = length of third side.
Sum of the three sides = perimeter
$$x + (x + 15) + 2x = 75$$
$$4x + 15 = 75$$
$$4x = 60$$
$$x = 15$$
The three sides are 15 inches,
$15 + 15 = 30$ inches, and $2(15) = 30$ inches.

19. Let w = width of flag
Then $2w - 4$ = length of flag.
$2l + 2w$ = perimeter
$$2(2w - 4) + 2w = 28$$
$$4w - 8 + 2w = 28$$
$$6w - 8 = 28$$
$$6w = 36$$
$$w = 6$$
The width is 6 feet and the length is 8 feet.

20. Let x = the measure of the two smaller angles, then $(2x + 3)$ = the measure of the two larger angles.
$$x + x + (2x + 3) + (2x + 3) = 360$$
$$x + x + 2x + 3 + 2x + 3 = 360$$
$$6x + 6 = 360$$
$$6x = 354$$
$$x = 59$$
The angles measure 59°, 59°, $2(59) + 3 = 121°$, 121°.

21. Let x = the rate Harlene digs.
Then $x + 0.2$ is the rate Ellis digs.

Name	Rate	Time	Distance
Harlene	x	84	$84x$
Ellis	$x + 0.2$	84	$84(x + 0.2)$

Distance Harlene digs + distance Ellis digs
= total length of trench

$$84x + 84(x + 0.2) = 67.2$$
$$84x + 84x + 16.8 = 67.2$$
$$168x = 50.4$$
$$x = 0.3$$

Harlene digs at 0.3 feet per minute and Ellis digs at $0.3 + 2 = 0.5$ feet per minute.

22. Let x = the rate Bonnie is running.

Name	Rate	Time	Distance
Alice	8	2	16
Bonnie	x	2	$2x$

$$16 - 2x = 4$$
$$-2x = -12$$
$$x = 6$$

Bonnie is running at 6 mph.

23. Let x be the number of pounds of Jelly Belly candy. Then $3 - x$ is the number of pounds of Kits candy.

Type	Cost	Pounds	Total
Jelly Belly	$2.20	x	$2.20(x)$
Kits	$2.75	$3 - x$	$2.75(3 - x)$
Mixture	$2.40	3	$2.40(3)$

$$2.20x + 2.75(3 - x) = 2.40(3)$$
$$2.2x + 8.25 - 2.75x = 7.2$$
$$-0.55x + 8.25 = 7.2$$
$$-0.55x = -1.05$$
$$x \approx 1.91$$

The mixture should contain about 1.91 pounds of Jelly Belly candy and about 1.09 pounds of Kits candy.

24. Let x = amount of 20% salt solution to be added.

Percent	Liters	Amount
20%	x	$0.20x$
40%	60	$(0.40)(60)$
35%	$x + 60$	$0.35(x + 60)$

$$0.20x + (0.40)(60) = 0.35(x + 60)$$
$$0.20x + 24 = 0.35x + 21$$
$$-0.15x + 24 = 21$$
$$-0.15x = -3$$
$$x = 20$$

20 liters of 20% solution must be added.

25. Let x = number of liters of 8% solution.

Percent	Liters	Amount
8%	x	$0.08x$
5%	$3 - x$	$0.05(3 - x)$
6%	3	$0.06(3)$

$$0.08x + 0.05(3 - x) = 0.06(3)$$
$$0.08x + 0.15 - 0.05x = 0.18$$
$$0.03x + 0.15 = 0.18$$
$$0.03x = 0.03$$
$$x = 1$$

There will be 1 liter of 8% solution and $3 - 1 = 2$ liters of 5% solution.

Chapter 3 Cumulative Review Test

1. 40% of $40,000 per year
$$(0.40)(40,000) = 16,000$$
Emily receives $16,000 in social security.

2. a. $38.7\% - 4.7\% = 34.0\%$

b. 4.7% of 1.3 million
$$(0.047)(1.3) \approx 0.06 \text{ million or } 60,000$$

3. a. $\dfrac{5 + 6 + 8 + 12 + 5}{5} = 7.2$

The mean level was 7.2 parts per million.

b. Carbon dioxide levels in order: 5, 5, 6, 8, 12
The median is 6 parts per million.

4. $\dfrac{5}{12} \div \dfrac{3}{4} = \dfrac{5}{12} \cdot \dfrac{4}{3} = \dfrac{20}{36} = \dfrac{5}{9}$

5. $\dfrac{2}{3} - \dfrac{1}{8} = \dfrac{2 \cdot 8}{3 \cdot 8} - \dfrac{1 \cdot 3}{8 \cdot 3} = \dfrac{16}{24} - \dfrac{3}{24} = \dfrac{13}{24}$

$\dfrac{2}{3}$ inch is $\dfrac{13}{24}$ of an inch greater than $\dfrac{1}{8}$ inch.

6. a. $\{1, 2, 3, 4, \ldots\}$

b. $\{0, 1, 2, 3, \ldots\}$

c. A rational number is a quotient of two integers, denominator not 0.

7. a. $\left|-4\right| = 4$

b. $\left|-5\right| = 5$ and $\left|-3\right| = 3$. Since $5 > 3, \left|-5\right| > \left|-3\right|$.

8. $2 - 6^2 \div 2 \cdot 2 = 2 - 36 \div 2 \cdot 2$
$= 2 - 18 \cdot 2$
$= 2 - 36$
$= -34$

9. $4(2x - 3) - 2(3x + 5) - 6$
$= 8x - 12 - 6x - 10 - 6$
$= 2x - 28$

10. $5x - 6 = x + 14$
$5x - 6 - x = x - x + 14$
$4x - 6 = 14$
$4x - 6 + 6 = 14 + 6$
$4x = 20$
$\dfrac{4x}{4} = \dfrac{20}{4}$
$x = 5$

11. $6r = 2(r + 3) - (r + 5)$
$6r = 2r + 6 - r - 5$
$6r = r + 1$
$6r - r = r - r + 1$
$5r = 1$
$\dfrac{5r}{5} = \dfrac{1}{5}$
$r = \dfrac{1}{5}$

12. $2(x + 5) = 3(2x - 4) - 4x$
$2x + 10 = 6x - 12 - 4x$
$2x + 10 = 2x - 12$
$2x - 2x + 10 = 2x - 2x - 12$
$10 = -12$ False
The equation has no solution.

13. $\dfrac{4.8}{x} = -\dfrac{3}{5}$
$5(4.8) = -3x$
$24 = -3x$
$-8 = x$

14. Substitute 6 for r.
$A = \pi r^2$
$A = \pi (6)^2 = 36\pi \approx 113.10$

15. a. $4x + 8y = 16$
$4x - 4x + 8y = 16 - 4x$
$8y = 16 - 4x$
$\dfrac{8y}{8} = \dfrac{16 - 4x}{8}$
$y = \dfrac{16}{8} - \dfrac{4x}{8}$
$y = 2 - \dfrac{1}{2}x$
$y = -\dfrac{1}{2}x + 2$

b. Substitute -4 for x.
$y = -\dfrac{1}{2}(-4) + 2 = 2 + 2 = 4$

16. $P = 2l + 2w$
$P - 2l = 2l + 2w - 2l$
$P - 2l = 2w$
$\dfrac{P - 2l}{2} = \dfrac{2w}{2}$
$\dfrac{P - 2l}{2} = w$

17. $\dfrac{50 \text{ miles}}{2 \text{ gallons}} = \dfrac{225 \text{ miles}}{x \text{ gallons}}$
$\dfrac{50}{2} = \dfrac{225}{x}$
$50x = 450$
$x = 9$
It will need 9 gallons.

18. Let x = number of minutes for the two plans to have the same cost.
Cost for Plan A = $19.95 + 0.35x$
Cost for Plan B = $29.95 + 0.10x$
The costs will be equal when the cost for Plan A = cost for Plan B.

$$19.95 + 0.35x = 29.95 + 0.10x$$
$$0.35x = 10 + 0.10x$$
$$0.25x = 10$$
$$x = 40$$

Lori would need to talk 40 minutes in a month for the plans to have the same cost.

19. Let x = smaller number.
Then $2x + 11$ = larger number.
smaller number + larger number = 29

$$x + (2x + 11) = 29$$
$$3x + 11 = 29$$
$$3x = 18$$
$$x = 6$$

The smaller number is 6 and the larger number is $2(6) + 11 = 23$.

20. Let x = smallest angle. Then $x + 5$ = second angle, $x + 50$ = third angle, and $4x + 25$ = fourth angle. Sum of the angle measures = $360°$

$$x + (x + 5) + (x + 50) + (4x + 25) = 360$$
$$7x + 80 = 360$$
$$7x = 280$$
$$x = 40$$

The angle measures are $40°$, $40 + 5 = 45°$, $40 + 50 = 90°$, and $4(40) + 25 = 185°$.

Chapter 4

Exercise Set 4.1

1. The *x*-coordinate is always listed first.

3. a. The horizontal axis is the *x*-axis.

   b. The vertical axis is the *y*-axis.

5. Axis is singular, while axes is plural.

7. The graph of a linear equation is an illustration of the set of points whose coordinates satisfy the equation.

9. a. Two points are needed to graph a linear equation.

   b. It is a good idea to use three or more points when graphing a linear equation to catch errors.

11. $ax + by = c$

13.

15. II

17. IV

19. I

21. III

23. III

25. II

27. $A(3, 1)$; $B(-3, 0)$; $C(1, -3)$; $D(-2, -3)$; $E(0, 3)$; $F\left(\dfrac{3}{2}, -1\right)$

29.

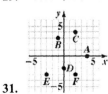

31.

33.

The points are collinear.

35.

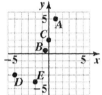

The points are not collinear since $(-5, -3)$ is not on the line.

37. a.
$$
\begin{array}{ll}
y = x + 2 & y = x + 2 \\
4 = 2 + 2 & 0 = -2 + 2 \\
4 = 4 \quad \text{True} & 0 = 0 \quad \text{True} \\
y = x + 2 & y = x + 2 \\
5 = -1 + 2 & 2 = 0 + 2 \\
5 = 1 \quad \text{False} & 2 = 2 \quad \text{True}
\end{array}
$$
Point c) does not satisfy the equation.

   b.

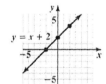

39. a.
$$
\begin{array}{ll}
3x - 2y = 6 & 3x - 2y = 6 \\
3(4) - 2(0) = 6 & 3(2) - 2(0) = 6 \\
12 = 6 \quad \text{False} & 6 = 6 \quad \text{True} \\[4pt]
3x - 2y = 6 & 3x - 2y = 6 \\
3\left(\dfrac{2}{3}\right) - 2(-2) = 6 & 3\left(\dfrac{4}{3}\right) - 2(-1) = 6 \\
2 + 4 = 6 & 4 + 2 = 6 \\
6 = 6 \quad \text{True} & 6 = 6 \quad \text{True}
\end{array}
$$
Point a) does not satisfy the equation.

   b.

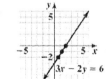

41. a.

$$\frac{1}{2}x + 4y = 4 \qquad\qquad \frac{1}{2}x + 4y = 4$$

$$\frac{1}{2}(-2) + 4(3) = 4 \qquad \frac{1}{2}(2) + 4\left(\frac{3}{4}\right) = 4$$

$$-1 + 12 = 4 \qquad\qquad 1 + 3 = 4$$

$$11 = 4 \text{ False} \qquad\qquad 4 = 4 \text{ True}$$

$$\frac{1}{2}x + 4y = 4 \qquad\qquad \frac{1}{2}x + 4y = 4$$

$$\frac{1}{2}(0) + 4(1) = 4 \qquad \frac{1}{2}(-4) + 4\left(\frac{3}{2}\right) = 4$$

$$0 + 4 = 4 \qquad\qquad -2 + 6 = 4$$

$$4 = 4 \text{ True} \qquad\qquad 4 = 4 \text{ True}$$

Point a) does not satisfy the equation.

b.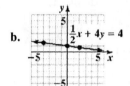

43. $y = 3x - 4$

$$y = 3(2) - 4$$

$$y = 6 - 4$$

$$y = 2$$

45. $y = 3x - 4$

$$y = 3(0) - 4$$

$$y = 0 - 4$$

$$y = -4$$

47. $2x + 3y = 12$

$$2x + 3(2) = 12$$

$$2x + 6 = 12$$

$$2x = 6$$

$$x = 3$$

49. $2x + 3y = 12$

$$2x + 3\left(\frac{11}{3}\right) = 12$$

$$2x + 11 = 12$$

$$2x = 1$$

$$x = \frac{1}{2}$$

51. The value of y is 0 when a straight line crosses the x-axis, because any point on the x-axis is neither above nor below the origin.

53. a. Latitude: 16°N   Longitude: 56°W
    b. Latitude: 29°N   Longitude: 90.5°W
    c. Latitude: 26°N   Longitude: 80.5°W
    d. Answers will vary.

57. The general form of a linear equation in one variable is $ax + b = c$.

58. It is a linear equation that has only one solution.

59.

$$-2(-3x + 5) + 6 = 4(x - 2)$$

$$-2(-3x) + (-2)(5) + 6 = 4(x) + 4(-2)$$

$$6x - 10 + 6 = 4x - 8$$

$$6x - 4 = 4x - 8$$

$$6x - 4x - 4 = 4x - 4x - 8$$

$$2x - 4 = -8$$

$$2x - 4 + 4 = -8 + 4$$

$$2x = -4$$

$$\frac{2x}{2} = \frac{-4}{2}$$

$$x = -2$$

60.

$$C = 2\pi r \qquad\qquad A = \pi r^2$$

$$= 2\pi(3) \qquad\qquad = \pi(3)^2$$

$$= 6\pi \qquad\qquad = 9\pi$$

$$\approx 18.84 \text{ inches} \qquad \approx 28.27 \text{ in.}^2$$

61.

$$2x - 5y = 6$$

$$2x - 2x - 5y = -2x + 6$$

$$-5y = -2x + 6$$

$$\frac{-5y}{-5} = \frac{-2x}{-5} + \frac{6}{-5}$$

$$y = \frac{2}{5}x - \frac{6}{5}$$

Exercise Set 4.2

1. To find the *x*-intercept, substitute 0 for *y* and find the corresponding value of *x*. To find the *y*-intercept, substitute 0 for *x* and find the corresponding value of *y*.

3. The graph of $y = b$ is a horizontal line.

5. You may not be able to read exact answers from a graph.

7. Yes. The equation goes through the origin because the point (0, 0) satisfies the equation.

9.
$$3x + y = 9$$
$$3(3) + y = 9$$
$$9 + y = 9$$
$$y = 0$$

11.
$$3x + y = 9$$
$$3x + (-6) = 9$$
$$3x - 6 = 9$$
$$3x = 15$$
$$x = 5$$

13.
$$3x + y = 9$$
$$3x + 0 = 9$$
$$3x = 9$$
$$x = 3$$

15.
$$3x - 2y = 8$$
$$3(4) - 2y = 8$$
$$12 - 2y = 8$$
$$-2y = -4$$
$$y = 2$$

17.
$$3x - 2y = 8$$
$$3x - 2(0) = 8$$
$$3x = 8$$
$$x = \frac{8}{3}$$

19.
$$3x - 2y = 8$$
$$3(-2) - 2y = 8$$
$$-6 - 2y = 8$$
$$-2y = 14$$
$$y = -7$$

21. $x = -3$ is a vertical line with *x*-intercept at $(-3, 0)$.

23. An equation of the form $y = 4$ is a horizontal line with *y*-intercept at $(0, 4)$.

25. Let $x = 0$, $y = 3(0) - 1 = -1$, $(0, -1)$

Let $x = 1$, $y = 3(1) - 1 = 2$, $(1, 2)$

Let $x = 2$, $y = 3(2) - 1 = 5$, $(2, 5)$

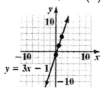

27. Let $x = 0$, $y = 4(0) - 2 = -2$, $(0, -2)$

Let $x = 1$, $y = 4(1) - 2 = 2$, $(1, 2)$

Let $x = 2$, $y = 4(2) - 2 = 6$, $(2, 6)$

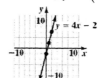

29. Let $x = 0$, then $0 + 2y = 6$

$$2y = 6$$

$$y = 3 \quad \rightarrow \quad (0,3)$$

Let $x = 2$, then $2 + 2y = 6$

$$2y = 4$$

$$y = 2 \quad \rightarrow \quad (2,2)$$

Let $x = 4$, then $4 + 2y = 6$

$$2y = 2$$

$$y = 1 \quad \rightarrow \quad (4,1)$$

31. $3x - 2y = 4$

$$-2y = -3x + 4$$

$$y = \frac{3}{2}x - 2$$

Let $x = 0$, $y = \frac{3}{2}(0) - 2 = -2$, $(0, -2)$

Let $x = 2$, $y = \frac{3}{2}(2) - 2 = 1$, $(2, 1)$

Let $x = 4$, $y = \frac{3}{2}(4) - 2 = 4$, $(4, 4)$

33. $4x + 3y = -9$

$$3y = -4x - 9$$

$$y = -\frac{4}{3}x - 3$$

Let $x = -3$, $y = -\frac{4}{3}(-3) - 3 = 1$, $(-3, 1)$

Let $x = 0$, $y = -\frac{4}{3}(0) - 3 = -3$, $(0, -3)$

Let $x = 3$, $y = -\frac{4}{3}(3) - 3 = -7$, $(3, -7)$

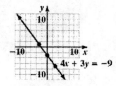

35. $6x + 5y = 30$

$$5y = -6x + 30$$

$$y = -\frac{6}{5}x + 6$$

Let $x = 0$, $y = -\frac{6}{5}(0) + 6 = 6$, $(0, 6)$

Let $x = 5$, $y = -\frac{6}{5}(5) + 6 = 0$, $(5, 0)$

Let $x = 10$, $y = -\frac{6}{5}(10) + 6 = -6$, $(10, -6)$

37. $-4x + 5y = 0$

$$5y = 4x$$

$$y = \frac{4}{5}x$$

Let $x = -5$, $y = \frac{4}{5}(-5) = -4$, $(-5, -4)$

Let $x = 0$, $y = \frac{4}{5}(0) = 0$, $(0, 0)$

Let $x = 5$, $y = \frac{4}{5}(5) = 4$, $(5, 4)$

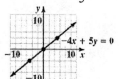

39. Let $x = 0$, $y = -20(0) + 60 = 60$, $(0, 60)$

Let $x = 2$, $y = -20(2) + 60 = 20$, $(2, 20)$

Let $x = 4$, $y = -20(4) + 60 = -20$, $(4, -20)$

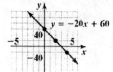

41. Let $x = -3$, $y = \frac{4}{3}(-3) = -4$, $(-3, -4)$

Let $x = 0$, $y = \frac{4}{3}(0) = 0$, $(0, 0)$

Let $x = 3$, $y = \frac{4}{3}(3) = 4$, $(3, 4)$

43. Let $x = 0$, $y = \frac{1}{2}(0) + 4 = 4$, $(0, 4)$

Let $x = 2$, $y = \frac{1}{2}(2) + 4 = 5$, $(2, 5)$

Let $x = 4$, $y = \frac{1}{2}(4) + 4 = 6$, $(4, 6)$

45. Let $x = 0$ Let $y = 0$

$y = 3x + 3$ $y = 3x + 3$

$y = 3(0) + 3$ $0 = 3x + 3$

$y = 3$ $-3x = 3$

         $x = -1$

47. Let $x = 0$ Let $y = 0$

$y = -4x + 2$ $y = -4x + 2$

$y = -4(0) + 2$ $0 = -4x + 2$

$y = 2$ $-2 = -4x$

         $x = \frac{1}{2}$

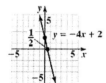

49. Let $x = 0$ Let $y = 0$

$y = 4x + 16$ $y = 4x + 16$

$y = 4(0) + 16$ $0 = 4x + 16$

$y = 16$ $-4x = 16$

         $x = -4$

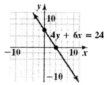

51. $4y + 6x = 24$

$4y = -6x + 24$

$y = -\frac{3}{2}x + 6$

Let $x = 0$ Let $y = 0$

$y = -\frac{3}{2}(0) + 6$ $0 = -\frac{3}{2}x + 6$

$y = 6$ $\frac{3}{2}x = 6$

         $x = 4$

53. Let $x = 0$ Let $y = 0$

$\frac{1}{2}x + 2y = 4$ $\frac{1}{2}x + 2y = 4$

$\frac{1}{2}(0) + 2y = 4$ $\frac{1}{2}x + 0 = 4$

$2y = 4$ $\frac{1}{2}x = 4$

$y = 2$ $x = 8$

55. Let $x = 0$ Let $y = 0$

$$12x - 24y = 48 \qquad 12x - 24y = 48$$
$$12(0) - 24y = 48 \quad 12x - 12(0) = 48$$
$$-24y = 48 \qquad 12x = 48$$
$$y = -2 \qquad x = 4$$

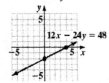

57. Let $x = 0$ Let $y = 0$

$$8y = 6x - 12 \qquad 8y = 6x - 12$$
$$8y = 6(0) - 12 \quad 8(0) = 6x - 12$$
$$8y = -12 \qquad 0 = 6x - 12$$
$$y = -\frac{3}{2} \qquad -6x = -12$$
$$x = 2$$

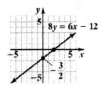

59. Let $x = 0$ Let $y = 0$

$$y = 15x + 45 \qquad y = 15x + 45$$
$$y = 15(0) + 45 \quad 0 = 15x + 45$$
$$y = 45 \qquad -15x = 45$$
$$x = -3$$

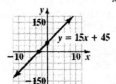

61. Let $x = 0$ Let $y = 0$

$$\frac{1}{3}x + \frac{1}{4}y = 12 \qquad \frac{1}{3}x + \frac{1}{4}y = 12$$
$$\frac{1}{3}(0) + \frac{1}{4}y = 12 \quad \frac{1}{3}x + \frac{1}{4}(0) = 12$$
$$\frac{1}{4}y = 12 \qquad \frac{1}{3}x = 12$$
$$y = 48 \qquad x = 36$$

63. Let $x = 0$ Let $y = 0$

$$\frac{1}{2}x = \frac{2}{5}y - 80 \qquad \frac{1}{2}x = \frac{2}{5}y - 80$$
$$\frac{1}{2}(0) = \frac{2}{5}y - 80 \quad \frac{1}{2}x = \frac{2}{5}(0) - 80$$
$$0 = \frac{2}{5}y - 80 \qquad \frac{1}{2}x = -80$$
$$-\frac{2}{5}y = -80 \qquad x = -160$$
$$y = 200$$

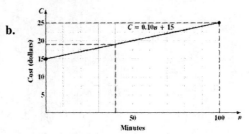

65. $x = -2$

67. $y = 6$

69.
$$ax + 3y = 10$$
$$a(2) + 3(0) = 10$$
$$2a + 0 = 10$$
$$2a = 10$$
$$a = 5$$

71.
$$3x + by = 14$$
$$3(0) + b(7) = 14$$
$$0 + 7b = 14$$
$$7b = 14$$
$$b = 2$$

73. Yes. For each 15 minutes of time, the number of calories burned increases by 200 calories.

75. a. $C = 0.10n + 15$

b.

c. $19

d. 100 minutes

77. a. $C = m + 40$

    b.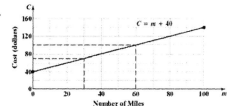

    c. $100

    d. 30 miles

79. a.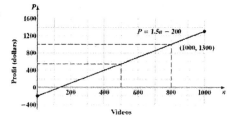

    b. $550

    c. 800 videos

81. Since each shaded area multiplied by the corresponding intercept must equal 6, the coefficients are 3 and 2 respectively.

83. Since the first shaded area multiplied by the x-intercept must equal -12, the coefficient of x is 6. Since the opposite of the second shaded area multiplied by the y-intercept must equal -12, the coefficient of y is 4.

85. a.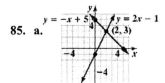

    b. $(2, 3)$

    c.
$$y = 2x - 1 \qquad y = -x + 5$$
$$3 = 2(2) - 1 \qquad 3 = -2 + 5$$
$$3 = 3 \qquad\qquad 3 = 3$$

    d. no

88.
$$2[6 - (4 - 5)] \div 2 - 8^2 = 2[6 - (-1)] \div 2 - 64$$
$$= 2[7] \div 2 - 64$$
$$= 14 \div 2 - 64$$
$$= 7 - 64$$
$$= -57$$

89.
$$\frac{-3^2 \cdot 4 \div 2}{\sqrt{9} - 2^2} = \frac{-9 \cdot 4 \div 2}{3 - 4}$$
$$= \frac{-36 \div 2}{-1}$$
$$= \frac{-18}{-1}$$
$$= 18$$

90.
$$\frac{8 \text{ ounces}}{3 \text{ gallons}} = \frac{x \text{ ounces}}{2.5 \text{ gallons}}$$
$$\frac{8}{3} = \frac{x}{2.5}$$
$$20 = 3x$$
$$6.67 \approx x$$
You should use 6.67 ounces of cleaner.

91. Let x = smaller integer. Then $3x + 1$ = larger integer.
$$x + (3x + 1) = 37$$
$$4x + 1 = 37$$
$$4x = 36$$
$$x = 9$$
The smaller integer is 9.
The larger is $3(9) + 1 = 27 + 1 = 28$.

Exercise Set 4.3

1. The slope of a line is the ratio of the vertical change to the horizontal change between any two points on the line.

3. A line with a positive slope rises from left to right.

5. Lines that rise from the left to right have a positive slope. Lines that fall from left to right have a negative slope.

7. No, since we cannot divide by 0, the slope is undefined.

9. Their slopes are the same.

11.
$$m = \frac{5 - 1}{6 - 4}$$
$$= \frac{4}{2}$$
$$= 2$$

13. $m = \dfrac{-2-0}{4-8}$

$= \dfrac{-2}{-4}$

$= \dfrac{1}{2}$

15. $m = \dfrac{\frac{1}{2}-\frac{1}{2}}{-3-9}$

$= \dfrac{0}{-12} = 0$

17. $m = \dfrac{-3-(-6)}{8-5}$

$= \dfrac{3}{3} = 1$

19. $m = \dfrac{2-4}{6-6}$

$= \dfrac{-2}{0}$

is undefined

21. $m = \dfrac{3-0}{-2-6}$

$= \dfrac{3}{-8}$

$= -\dfrac{3}{8}$

23. $m = \dfrac{2-\frac{5}{2}}{-\frac{3}{4}-0} = \dfrac{-\frac{1}{2}}{-\frac{3}{4}}$

$= \dfrac{-1}{2} \cdot \dfrac{4}{-3}$

$= \dfrac{-4}{-6}$

$= \dfrac{2}{3}$

25. $m = \dfrac{6}{3} = 2$

27. $m = \dfrac{-6}{3} = -2$

29. $m = \dfrac{4}{-7} = -\dfrac{4}{7}$

31. $m = \dfrac{7}{4}$

33. $m = \dfrac{0}{3} = 0$

35. $m = \dfrac{-2}{3} = -\dfrac{2}{3}$

37. Vertical line, slope is undefined.

39.

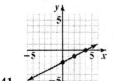

41.

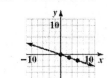

43.

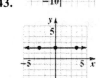

45.

47.

49. The lines are parallel because the slopes are the same.

51. The lines are perpendicular because the slopes are negative reciprocals.

53. The lines are perpendicular because the slopes are negative reciprocals.

55. The lines are neither parallel nor perpendicular.

57. The lines are neither parallel nor perpendicular.

59. The lines are parallel because the slopes are the same.

61. The lines are parallel because the lines are both vertical.

63. The lines are perpendicular because if the slope is 0, the line is horizontal and if the slope is undefined the line is vertical.

65. Its slope would be 3.

67. Its slope would be $\dfrac{1}{4}$.

69. The first graph appears to pass through the points $(-1, 0)$ and $(0, 6)$. Its slope

is $m = \dfrac{6-0}{0-(-1)} = \dfrac{6}{1} = 6$. The second graph

appears to pass through the points $(-4, 0)$ and

$(0, 6)$. Its slope is $m = \dfrac{6-0}{0-(-4)} = \dfrac{6}{4} = \dfrac{6}{4} = \dfrac{3}{2}$. The

first graph has the greater slope.

71. a. $m = \dfrac{22-45}{1961-1957} = \dfrac{-23}{4} = -\dfrac{23}{4}$

   b. $m = \dfrac{51-7}{1989-1985} = \dfrac{44}{4} = 11$

73. $m = \dfrac{-2-6}{4-2} = \dfrac{-8}{2} = -4$

Parallel lines have the same slope, so a line parallel to the given line would have a slope of -4.

75. $m = \dfrac{1-(-7)}{2-1} = \dfrac{1+7}{1} = 8$

Perpendicular lines have negative reciprocal slopes, so a line perpendicular to the given line

would have a slope of $-\dfrac{1}{8}$.

77. $m = \dfrac{-\dfrac{7}{2} - \left(-\dfrac{3}{8}\right)}{\dfrac{4}{9} - \dfrac{1}{2}}$

$= \dfrac{-\dfrac{28}{8} + \dfrac{3}{8}}{\dfrac{8}{18} - \dfrac{9}{18}} = \dfrac{-\dfrac{25}{8}}{-\dfrac{17}{18}}$

$= \left(-\dfrac{25}{8}\right)\left(-\dfrac{18^{9}}{17}\right)$

$= \dfrac{(25)(9)}{(4)(17)} = \dfrac{225}{68}$

79. a.

   b. $AC;\ m = \dfrac{4-1}{5-0} = \dfrac{3}{5}$

   $CB;\ m = \dfrac{4-2}{5-6} = \dfrac{2}{-1} = -2$

   $DB;\ m = \dfrac{-1-2}{1-6} = \dfrac{-3}{-5} = \dfrac{3}{5}$

   $AD;\ m = \dfrac{-1-1}{1-0} = \dfrac{-2}{1} = -2$

   c. Yes; opposite sides are parallel.

81. a. $AB;\ m = \dfrac{12-4}{2-0} = \dfrac{8}{2} = 4$

   $BC;\ m = \dfrac{8-12}{4-2} = \dfrac{-4}{2} = -2$

   $CD;\ m = \dfrac{16-8}{6-4} = \dfrac{8}{2} = 4$

   b. $\dfrac{4+(-2)+4}{3} = \dfrac{6}{3} = 2$

   c. $m = \dfrac{16-4}{6-0} = \dfrac{12}{6} = 2$

   d. yes

   e. Answers will vary.

83. $4x^2 + 9x + \dfrac{x}{3} = 4(0)^2 + 9(0) + \dfrac{0}{3}$

$$= 0 + 0 + 0$$
$$= 0$$

84. a. $\qquad -x = -\dfrac{5}{2}$

$$(-1)(-x) = (-1)\left(-\dfrac{5}{2}\right)$$

$$x = \dfrac{5}{2}$$

b. $\quad 8x = 0$

$$\dfrac{8x}{8} = \dfrac{0}{8}$$

$$x = 0$$

85. $\qquad \dfrac{2}{3}x + \dfrac{1}{7}x = 4$

$$21\left(\dfrac{2}{3}x + \dfrac{1}{7}x\right) = 21(4)$$

$$21\left(\dfrac{2}{3}x\right) + 21\left(\dfrac{1}{7}x\right) = 84$$

$$14x + 3x = 84$$

$$17x = 84$$

$$\dfrac{17x}{17} = \dfrac{84}{17}$$

$$x = \dfrac{84}{17}$$

86. $\qquad d = a + b + c$

$$d - a = a + b + c - a$$
$$d - a = b + c$$
$$d - a - b = b + c - b$$
$$d - a - b = c$$

87. $5x - 3y = 30$

y-intercept	x-intercept
$x = 0$	$y = 0$
$5(0) - 3y = 30$	$5x - 3(0) = 30$
$-3y = 30$	$5x = 30$
$y = -10$	$x = 6$
$(0, -10)$	$(6, 0)$

Mid-Chapter Test: 4.1-4.3

1. IV

2.

3. a. $\qquad \dfrac{1}{3}x + y = -2$

$$\dfrac{1}{3}(3) + (-3) = -2$$
$$1 - 3 = -2$$
$$-2 = -2 \text{ True}$$

b. $\qquad \dfrac{1}{3}x + y = -2$

$$\dfrac{1}{3}(0) + (2) = -2$$
$$0 + 2 = -2$$
$$2 = -2 \text{ False}$$

c. $\qquad \dfrac{1}{3}x + y = -2$

$$\dfrac{1}{3}(-6) + (0) = -2$$
$$-2 + 0 = -2$$
$$-2 = -2 \text{ True}$$

4. $y = 5x + 1$

$$y = 5(-1) + 1$$
$$y = -5 + 1$$
$$y = -4$$

5. $3x - 4y = 1$

$$3x - 4(2) = 1$$
$$3x - 8 = 1$$
$$3x = 9$$
$$x = 3$$

6. A graph of an equation in two variables is an illustration of a set of points whose coordinates satisfy the equation.

7.

8.

9. Let $x = -1$; $y = 3(-1) + 1 = -2$; $(-1, -2)$

    Let $x = 0$; $y = 3(0) + 1 = 1$; $(0, 1)$

    Let $x = 1$; $y = 3(1) + 1 = 4$; $(1, 4)$

10. Let $x = 2$; $y = -\dfrac{1}{2}(2) + 4 = 3$; $(2, 3)$

    Let $x = 4$; $y = -\dfrac{1}{2}(4) + 4 = 2$; $(4, 2)$

    Let $x = 6$; $y = -\dfrac{1}{2}(6) + 4 = 1$; $(6, 1)$

    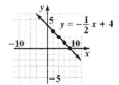

11. | y-intercept | x-intercept |
    |---|---|
    | Let $x = 0$ | Let $y = 0$ |
    | $3(0) - 4y = 12$ | $3x - 4(0) = 12$ |
    | $-4y = 12$ | $3x = 12$ |
    | $y = -3$ | $x = 4$ |

12. | y-intercept | x-intercept |
    |---|---|
    | Let $x = 0$ | Let $y = 0$ |
    | $\dfrac{1}{2}(0) + \dfrac{1}{5}y = 10$ | $\dfrac{1}{2}x + \dfrac{1}{5}(0) = 10$ |
    | $\dfrac{1}{5}y = 10$ | $\dfrac{1}{2}x = 10$ |
    | $y = 50$ | $x = 20$ |

    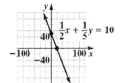

13. $m = \dfrac{3 - 5}{6 - (-1)} = -\dfrac{2}{7}$

14. $m = \dfrac{2 - 2}{7 - 4} = \dfrac{0}{3} = 0$

15. $m = \dfrac{5 - 0}{-3 - (-3)} = \dfrac{5}{0}$

    is undefined

16.   17.

18. The lines are neither parallel nor perpendicular.

19. The lines are perpendicular because the slopes are negative reciprocals.

20. Use the points $(0, 0)$ and $(10, 100)$.

    The slope is $m = \dfrac{100 - 0}{10 - 0} = \dfrac{100}{10} = 10$.

Exercise Set 4.4

1. $y = mx + b$

3. $y = 3x - 5$

5. Compare their slopes: If slopes are the same and their y-intercepts are different, the lines are parallel.

7. $y - y_1 = m(x - x_1)$

9. $m = 2$; y-intercept: $(0, -6)$

11. $m = \dfrac{4}{3}$; y-intercept: $(0, -7)$

13. $m = 1$; y-intercept: $(0, -3)$

15. $m = 3$; y-intercept: $(0, 2)$

17. $m = 2$; y-intercept: $(0, 0)$

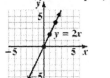

19. $-2x + y = -3$

$$y = 2x - 3$$

$m = 2$; y-intercept: $(0, -3)$

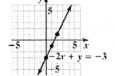

21. $5x - 2y = 10$

$$-2y = -5x + 10$$

$$y = \frac{5}{2}x - 5$$

$m = \frac{5}{2}$; y-intercept: $(0, -5)$

23. $6x + 12y = 18$

$$12y = -6x + 18$$

$$y = -\frac{1}{2}x + \frac{3}{2}$$

$m = -\frac{1}{2}$; y-intercept: $\left(0, \frac{3}{2}\right)$

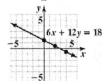

25. $-6x + 2y - 8 = 0$

$$2y = 6x + 8$$

$$y = 3x + 4$$

$m = 3$; y-intercept: $(0, 4)$

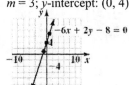

27. $3x = 2y - 4$

$$-2y = -3x - 4$$

$$y = \frac{3}{2}x + 2$$

$m = \frac{3}{2}$; y-intercept: $(0, 2)$

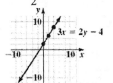

29. $m = \frac{4}{4} = 1$, $b = -2$

$$y = x - 2$$

31. $m = \frac{-2}{6} = -\frac{1}{3}$, $b = 2$

$$y = -\frac{1}{3}x + 2$$

33. $m = \frac{-3}{1} = -3$, $b = -5$

$$y = -3x - 5$$

35. $m = \frac{10}{30} = \frac{1}{3}$, $b = 5$

$$y = \frac{1}{3}x + 5$$

37. Since the slopes of the lines are the same and y-intercepts are different, the lines are parallel.

39.

$4x + 2y = 7$	$4x = 8y + 12$
$2y = -4x + 7$	$-8y = -4x + 12$
$y = -2x + \frac{7}{2}$	$y = \frac{1}{2}x - \frac{3}{2}$

Since the slopes of the lines are opposite reciprocals, the lines are perpendicular.

41.

$3x + 5y = 9$	$6x = -10y + 9$
$5y = -3x + 9$	$10y = -6x + 9$
$y = -\frac{3}{5}x + \frac{9}{5}$	$y = \frac{-6}{10}x + \frac{9}{10}$
	$y = -\frac{3}{5}x + \frac{9}{10}$

Since the slopes of the lines are the same and y-intercepts are different, the lines are parallel.

43. $y = \frac{1}{2}x - 2$ $2y = 6x + 9$

$\qquad\qquad\qquad y = 3x + \frac{9}{2}$

Since the slopes of the lines are not equal and are not opposite reciprocals, the lines are neither parallel nor perpendicular.

45. $5y = 2x + 9$ $-10x = 4y + 11$

$\quad y = \frac{2}{5}x + \frac{9}{5}$ $-4y = 10x + 11$

$\qquad\qquad\qquad y = -\frac{5}{2} - \frac{11}{4}$

Since the slopes of the lines are opposite reciprocals, the lines are perpendicular.

47. $3x + 7y = 21$ $7x + 3y = 21$

$\quad 7y = -3x + 21$ $3y = -7x + 21$

$\quad y = -\frac{3}{7}x + 3$ $y = -\frac{7}{3}x + 7$

Since the slopes of the lines are not equal and are not opposite reciprocals, the lines are neither parallel nor perpendicular.

49. $y - 2 = 3(x - 0)$

$\qquad y = 3x + 2$

51. $y - 5 = -3\left[x - (-4)\right]$

$\quad y - 5 = -3(x + 4)$

$\quad y - 5 = -3x - 12$

$\qquad y = -3x - 7$

53. $y - (-3) = \frac{1}{2}\left[x - (-1)\right]$

$\quad y + 3 = \frac{1}{2}(x + 1)$

$\quad y + 3 = \frac{1}{2}x + \frac{1}{2}$

$\qquad y = \frac{1}{2}x - \frac{5}{2}$

55. $y - 6 = \frac{2}{3}(x - 0)$

$\quad y - 6 = \frac{2}{3}x$

$\qquad y = \frac{2}{3}x + 6$

57. $m = \frac{4 - (-2)}{-2 - (-4)} = \frac{6}{2} = 3$

$\quad y - (-2) = 3\left[x - (-4)\right]$

$\quad y + 2 = 3(x + 4)$

$\quad y + 2 = 3x + 12$

$\qquad y = 3x + 10$

59. $m = \frac{-12 - 9}{8 - (-6)} = \frac{-21}{14} = -\frac{3}{2}$

$\quad y - 9 = -\frac{3}{2}\left(x - (-6)\right)$

$\quad y - 9 = -\frac{3}{2}(x + 6)$

$\quad y - 9 = -\frac{3}{2}x - 9$

$\qquad y = -\frac{3}{2}x$

61. $m = \frac{-2 - 3}{0 - 10} = \frac{-5}{-10} = \frac{1}{2}$

$\quad y - 3 = \frac{1}{2}(x - 10)$

$\quad y - 3 = \frac{1}{2}x - 5$

$\qquad y = \frac{1}{2}x - 2$

63. $y - (-4.5) = 7.4(x - 0)$

$\quad y + 4.5 = 7.4x$

$\qquad y = 7.4x - 4.5$

65. a. $y = 5x + 60$

  b. $y = 5(30) + 60 = 150 + 60 = \$210$

67. a. Use the slope-intercept form.

  b. Use the point-slope form.

  c. Use the point-slope form but first find the slope.

69. a. No. The equations will look different because different points are used.

  b. $y - (-4) = 2\left[x - (-5)\right]$

$\qquad\quad y + 4 = 2(x + 5)$

  c. $y - 12 = 2(x - 3)$

d. $y + 4 = 2(x + 5)$

$\qquad y + 4 = 2x + 10$

$\qquad\quad y = 2x + 6$

e. $y - 12 = 2(x - 3)$

$\qquad y - 12 = 2x - 6$

$\qquad\quad y = 2x + 6$

f. Yes

71. a. Use the points (0, 0) and (200, 293) and substitute into the slope formula.

$$m = \frac{293 - 0}{200 - 0} = \frac{293}{200} \approx 1.465$$

b. Using the point-slope formula, the equation of the line is:

$$y - y_1 = m(x - x_1)$$

$$f - 0 = 1.465(m - 0)$$

$$f = 1.465m$$

c. $f = 1.465m$

$\qquad f = 1.465(142.7)$

$\qquad f \approx 209.06$

The speed was 209.06 feet per second.

d. It is about 150 feet per second.

e. It is about 55 miles per hour.

73. First, find the slope of the line $2x + y = 6$.

$$2x + y = 6$$

$$\qquad y = -2x + 6$$

$m = -2$

Use $m = -2$ and $b = 5$ in the slope-intercept equation.

$y = -2x + 5$

75. $3x - 4y = 6$

$\qquad 4y = 3x - 6$

$$\qquad y = \frac{3}{4}x - \frac{3}{2}$$

$$\qquad m = \frac{3}{4}$$

$$y - (-1) = \frac{3}{4}\left[x - (-8)\right]$$

$$y + 1 = \frac{3}{4}(x + 8)$$

$$y + 1 = \frac{3}{4}x + 6$$

$$y = \frac{3}{4}x + 5$$

78. $|-4| < |-9|$ because $4 < 9$.

79. True; the product of two numbers with like signs is a positive number.

80. True; the sum of two negative numbers is a negative number.

81. False; the sum has the sign of the number with the larger absolute value.

82. False; the quotient of two numbers with like signs is a positive number.

83. $4^3 = 4 \cdot 4 \cdot 4 = 16 \cdot 4 = 64$

84. $i = prt$

$$\frac{i}{pt} = \frac{prt}{pt}$$

$$\frac{i}{pt} = r$$

Chapter 4 Review Exercises

1.

2.

The points are not collinear.

3. a. $\qquad 2x + 3y = 9$

$$2(5) + 3\left(-\frac{1}{3}\right) = 9$$

$$10 - 1 = 9$$

$$9 = 9 \quad \text{True}$$

b. $\qquad 2x + 3y = 9$

$$2(3) + 3(1) = 9$$

$$9 = 9 \quad \text{True}$$

c. $2x + 3y = 9$

$2(-2) + 3(4) = 9$

$8 = 9$ False

d. $2x + 3y = 9$

$2(2) + 3\left(\dfrac{5}{3}\right) = 9$

$9 = 9$ True

4. a. $3x - 2y = 8$ **b.** $3x - 2y = 8$

$3(2) - 2y = 8$ $3(0) - 2y = 8$

$-2y = 2$ $-2y = 8$

$y = -1$ $y = -4$

c. $3x - 2y = 8$ **d.** $3x - 2y = 8$

$3x - 2(5) = 8$ $3x - 2(0) = 8$

$3x - 10 = 8$ $3x = 8$

$3x = 18$ $x = \dfrac{8}{3}$

$x = 6$

5. $y = 4$ is a horizontal line with
   y-intercept = (0, 4).

6. $x = 2$ is a vertical line with x-intercept = (2, 0).

7. Let $x = -1$, $y = 3(-1) = -3$, $(-1, -3)$
   Let $x = 0$, $y = 3 \cdot 0 = 0$, $(0, 0)$
   Let $x = 1$, $y = 3 \cdot 1 = 3$, $(1, 3)$

8. Let $x = 0$, $y = 2 \cdot 0 - 1 = -1$, $(0, -1)$
   Let $x = 1$, $y = 2 \cdot 1 - 1 = 1$, $(1, 1)$
   Let $x = 2$, $y = 2 \cdot 2 - 1 = 3$, $(2, 3)$

9. Let $x = 0$, $y = -2 \cdot 0 + 5 = 5$, $(0, 5)$
   Let $x = 1$, $y = -2 \cdot 1 + 5 = 3$, $(1, 3)$
   Let $x = 2$, $y = -2 \cdot 2 + 5 = 1$, $(2, 1)$

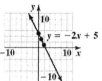

10. $2y + x = 8$

$2y = -x + 8$

$y = -\dfrac{1}{2}x + 4$

Let $x = 0$, $y = -\dfrac{1}{2}(0) + 4 = 4$, $(0, 4)$

Let $x = 2$, $y = -\dfrac{1}{2}(2) + 4 = 3$, $(2, 3)$

Let $x = 4$, $y = -\dfrac{1}{2}(4) + 4 = 2$, $(4, 2)$

11.   Let $x = 0$ Let $y = 0$

$-2x + 3y = 6$ $-2x + 3y = 6$

$-2 \cdot 0 + 3y = 6$ $-2x + 3 \cdot 0 = 6$

$3y = 6$ $-2x = 6$

$y = 2$ $x = -3$

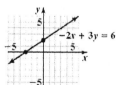

12. Let $x = 0$ Let $y = 0$

$5x + 2y = -10$ $5x + 2y = -10$

$5 \cdot 0 + 2y = -10$ $5x + 2 \cdot 0 = -10$

$2y = -10$ $5x = -10$

$y = -5$ $x = -2$

13. Let $x = 0$ Let $y = 0$

$5x + 10y = 20$ $5x + 10y = 20$

$5 \cdot 0 + 10y = 20$ $5x + 10 \cdot 0 = 20$

$10y = 20$ $5x = 20$

$y = 2$ $x = 4$

14. Let $x = 0$ Let $y = 0$

$\frac{2}{3}x = \frac{1}{4}y + 20$ $\frac{2}{3}x = \frac{1}{4}y + 20$

$\frac{2}{3} \cdot 0 = \frac{1}{4}y + 20$ $\frac{2}{3}x = \frac{1}{4} \cdot 0 + 20$

$0 = \frac{1}{4}y + 20$ $\frac{2}{3}x = 20$

$-\frac{1}{4}y = 20$ $x = 30$

$y = -80$

15. $m = \dfrac{5 - (-4)}{1 - 6}$

$= \dfrac{9}{-5}$

$= -\dfrac{9}{5}$

16. $m = \dfrac{-7 - (-6)}{8 - (-4)}$

$= \dfrac{-1}{12}$

$= -\dfrac{1}{12}$

17. $m = \dfrac{1 - (-3)}{-4 - (-2)}$

$= \dfrac{4}{-2}$

$= -2$

18. The slope of a horizontal line is 0.

19. The slope of a vertical line is undefined.

20. The slope of a straight line is the ratio of the vertical change to the horizontal change between any two points on the line.

21. $m = \dfrac{-2 - 4}{1 - (-2)} = \dfrac{-6}{3} = -2$

22. $m = \dfrac{2 - 0}{3 - (-5)} = \dfrac{2}{8} = \dfrac{1}{4}$

23. Neither. For the lines to be parallel the slopes have to be the same. For the lines to be perpendicular the slopes have to be opposite reciprocals.

24. Perpendicular, because the slopes are opposite reciprocals.

25. a. $m = \dfrac{415 - 201}{1996 - 1995} = \dfrac{214}{1} = 214$

    b. $m = \dfrac{415 - 268}{1996 - 1999} = \dfrac{147}{-3} = -49$

26. $6x + 7y = 21$

$7y = -6x + 14$

$y = -\dfrac{6}{7}x + \dfrac{21}{7}$

$y = -\dfrac{6}{7}x + 3$

$m = -\dfrac{6}{7}, b = 3$

The slope is $-\dfrac{6}{7}$; the y-intercept is $(0, 3)$.

27. $2x + 7 = 0$

$2x = -7$

$x = -\dfrac{7}{2}$

This is a vertical line, so the slope is undefined and there is no y-intercept.

28. $4y + 12 = 0$

$4y = -12$

$y = -3$

This is a horizontal line, so the slope is 0 and the y-intercept is $(0, -3)$.

29. $m = \dfrac{3}{1} = 3,\ b = -3$

$y = 3x - 3$

30. $m = \dfrac{-2}{4} = -\dfrac{1}{2},\ b = 2$

$y = -\dfrac{1}{2}x + 2$

31. $y = 2x - 7 \qquad 6y = 12x + 18$

$\qquad\qquad\qquad y = 2x + 3$

Since the slopes are the same and the y-intercepts are different, the lines are parallel.

32. $2x - 3y = 15 \qquad 3x + 2y = 12$

$-3y = -2x + 15 \qquad 2y = -3x + 12$

$y = \dfrac{2}{3}x - 5 \qquad y = -\dfrac{3}{2}x + 6$

Since the slopes are opposite reciprocals, the lines are perpendicular.

33. $y - 7 = 3(x - 2)$

$y - 7 = 3x - 6$

$y = 3x + 1$

34. $y - 2 = -\dfrac{2}{3}(x - 3)$

$y - 2 = -\dfrac{2}{3}x + 2$

$y = -\dfrac{2}{3}x + 4$

35. $y - 2 = 0(x - 6)$

$y - 2 = 0$

$y = 2$

36. Lines with undefined slopes are vertical and have the form $x = c$ where c is the value of x for any point on the line.

$x = 4$

37. $m = \dfrac{-3 - 4}{0 - (-2)} = \dfrac{-7}{2} = -\dfrac{7}{2}$

$y - (-3) = -\dfrac{7}{2}[x - 0]$

$y + 3 = -\dfrac{7}{2}x$

$y = -\dfrac{7}{2}x - 3$

38. $m = \dfrac{3 - (-2)}{-5 - (-5)} = \dfrac{5}{0}$ is undefined

Lines with undefined slopes are vertical and have the form $x = c$ where c is the value of x for any point on the line.

$x = -5$

Chapter 4 Practice Test

1. A graph is an illustration of the set of points that satisfy an equation.

2. a. IV   **b.** II

3. a. $ax + by = c$

   b. $y = mx + b$

   c. $y - y_1 = m(x - x_1)$

4. a. $\quad 3y = 5x - 9$

$\quad 3(2) = 5(4) - 9$

$\quad\quad 6 = 20 - 9$

$\quad\quad 6 = 11 \qquad$ False

   b. $\quad 3y = 5x - 9$

$\quad 3(0) = 5\left(\dfrac{9}{5}\right) - 9$

$\quad\quad 0 = 9 - 9$

$\quad\quad 0 = 0 \qquad$ True

   c. $\quad 3y = 5x - 9$

$\quad 3(-10) = 5(-1) - 9$

$\quad\quad -30 = -5 - 9$

$\quad\quad -30 = -14 \qquad$ False

d. $3y = 5x - 9$

$3(-3) = 5(0) - 9$

$-9 = -9$ True

$\left(\dfrac{9}{5}, 0\right)$ and $(0, -3)$ satisfy the equation.

5. $m = \dfrac{-3-5}{4-(-2)} = \dfrac{-8}{6} = -\dfrac{4}{3}$

6. $4x - 9y = 15$

$-9y = -4x + 15$

$y = \dfrac{4}{9}x - \dfrac{5}{3}$

$m = \dfrac{4}{9},\ b = -\dfrac{5}{3}$

The slope is $\dfrac{4}{9}$; the y-intercept is $\left(0, -\dfrac{5}{3}\right)$.

7. $m = \dfrac{-1}{1} = -1,\ b = -1$

$y = -x - 1$

8. $x = -4$ is a vertical line with x-intercept $= (-4, 0)$.

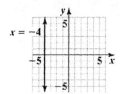

9. $y = 2$ is a horizontal line with y-intercept $= (0, 2)$.

10. Let $x = 0$, $y = 3 \cdot 0 - 2 = -2$, $(0, -2)$

Let $x = 1$, $y = 3 \cdot 1 - 2 = 1$, $(1, 1)$

Let $x = 2$, $y = 3 \cdot 2 - 2 = 4$, $(2, 4)$

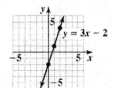

11. a. $3x - 6y = 12$

$-6y = -3x + 12$

$y = \dfrac{1}{2}x - 2$

b. Let $x = 0$, $y = \dfrac{1}{2}(0) - 2 = -2$, $(0, -2)$

Let $x = 2$, $y = \dfrac{1}{2}(2) - 2 = -1$, $(2, -1)$

Let $x = 4$, $y = \dfrac{1}{2}(4) - 2 = 0$, $(4, 0)$

12. $3x + 5y = 15$

Let $x = 0$	Let $y = 0$
$3(0) + 5y = 15$	$3x + 5(0) = 15$
$5y = 15$	$3x = 15$
$y = 3$	$x = 5$

13. $y - (-5) = 4(x - 2)$

$y + 5 = 4x - 8$

$y = 4x - 13$

14. $m = \dfrac{2-(-1)}{-4-3} = \dfrac{3}{-7} = -\dfrac{3}{7}$

$y - (-1) = -\dfrac{3}{7}(x - 3)$

$y + 1 = -\dfrac{3}{7}x + \dfrac{9}{7}$

$y = -\dfrac{3}{7}x + \dfrac{2}{7}$

15. $2y = 3x - 6$ $y - \dfrac{3}{2}x = -5$

$y = \dfrac{3}{2}x - 3$ $y = \dfrac{3}{2}x - 5$

The lines are parallel since they have the same slope but different y-intercepts.

119

16. slope = 3, y-intercept is $(0, -4)$.

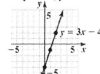

17. $4x - 2y = 6$

$$-2y = -4x + 6$$
$$y = 2x - 3$$

Slope = 2, y intercept is $(0, -3)$.

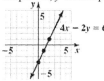

18. a.

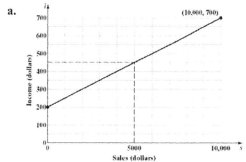

b. $450

Chapter 4 Cumulative Review Test

1. a. {1, 2, 3, ...}

b. {0, 1, 2, 3, ...}

2. $4 - 9 - (-10) + 13 = 4 - 9 + 10 + 13$
$$= -5 + 10 + 13$$
$$= 5 + 13$$
$$= 18$$

3. $(8 \div 4)^3 + 9^2 \div 3 = 2^3 + 9^2 \div 3$
$$= 8 + 81 \div 3$$
$$= 8 + 27$$
$$= 35$$

4. $(10 \div 5 \cdot 5 + 5 - 5)^2 = (2 \cdot 5 + 5 - 5)^2$
$$= (10 + 5 - 5)^2$$
$$= (15 - 5)^2$$
$$= 10^2$$
$$= 100$$

5. a. distributive property

b. commutative property of addition

6. $2x + 5 = 3(x - 5)$
$$2x + 5 = 3x - 15$$
$$-x + 5 = -15$$
$$-x = -20$$
$$x = 20$$

7. $3(x - 1) - (x + 4) = 2x - 7$
$$3x - 3 - x - 4 = 2x - 7$$
$$2x - 7 = 2x - 7$$
$$0 = 0$$

All real numbers are solutions

8. $\dfrac{2}{20} = \dfrac{x}{200}$
$$2(200) = 20x$$
$$400 = 20x$$
$$20 = x$$

9. $\dfrac{3 \text{ cans}}{\$1.50} = \dfrac{8 \text{ cans}}{x \text{ dollars}}$
$$\dfrac{3}{1.50} = \dfrac{8}{x}$$
$$3x = 12$$
$$x = \dfrac{12}{3} = 4$$

8 cans will cost $4.00.

10. $v = lwh$
$$\dfrac{v}{lh} = \dfrac{lwh}{lh}$$
$$\dfrac{v}{lh} = w$$

11. Let n = the number
$$11 + 2n = 19$$
$$11 - 11 + 2n = 19 - 11$$
$$2n = 8$$
$$\frac{2n}{2} = \frac{8}{2}$$
$$n = 4$$
The number is 4.

12. Let x = width of rectangle.
Then $2x + 3$ = length of rectangle.
$$P = 2l + 2w$$
$$36 = 2(2x + 3) + 2x$$
$$36 = 4x + 6 + 2x$$
$$36 = 6x + 6$$
$$30 = 6x$$
$$5 = x$$
The width is 5 feet and the length is
$2(5) + 3 = 13$ feet.

13. Let x = number of hours until the runners are 28 miles apart.

Runner	Rate	Time	Distance
First	6 mph	x	$6x$
Second	8 mph	x	$8x$

(Distance run by first runner) + (Distance run by second runner) = 28 miles
$$6x + 8x = 28$$
$$14x = 28$$
$$x = 2$$
It will take 2 hours.

14. Answers will vary. Some possible answers are $(0, 2), (4, 0), (2, 1)$.

15. Let $x = 0, y = f(0) = 3 \cdot 0 - 5 = -5$, $(0, -5)$

Let $x = 1, y = f(1) = 3 \cdot 1 - 5 = -2$, $(1, -2)$

Let $x = 2, y = f(2) = 3 \cdot 2 - 5 = 1$, $(2, 1)$

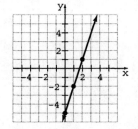

16. $6x - 3y = -12$

Let $x = 0$ $\qquad$ Let $y = 0$
$6(0) - 3y = -12$ $\quad$ $6x - 3(0) = -12$
$\qquad -3y = -12$ $\qquad$ $6x = -12$
$\qquad y = 4$ $\qquad\qquad$ $x = -2$

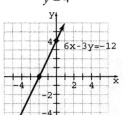

17. $m = \dfrac{-5 - (-6)}{6 - 5} = \dfrac{1}{1} = 1$

18. Since parallel lines have the same slope, the slope of the line will be -7.

19. Slope $= \dfrac{2}{3}$, y intercept is $(0, -3)$

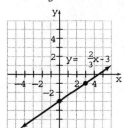

20. $y - 2 = 3(x - 5)$

Chapter 5

Exercise Set 5.1

1. In the expression t^p, t is the base, p is the exponent.

3. **a.** $\dfrac{x^m}{x^n} = x^{m-n}$, $x \neq 0$

    b. Answers will vary. One possible answer is that to divide terms with like bases, subtract the exponents.

5. **a.** $\left(x^m\right)^n = x^{m \cdot n}$

    b. Answers will vary. One possible answer is that to raise a power to a power, multiply the powers.

7. $a^2 b^5$

9. Answers will vary. See #2 and #5 above.

11. $x^5 \cdot x^4 = x^{5+4} = x^9$

13. $-z^4 \cdot z = -z^{4+1} = -z^5$

15. $y^3 \cdot y^2 = y^{3+2} = y^5$

17. $3^2 \cdot 3^3 = 3^{2+3} = 3^5 = 243$

19. $z^3 \cdot z^5 = z^{3+5} = z^8$

21. $\dfrac{6^2}{6} = \dfrac{6^2}{6^1} = 6^{2-1} = 6^1 = 6$

23. $\dfrac{x^{10}}{x^3} = x^{10-3} = x^7$

25. $\dfrac{3^6}{3^2} = 3^{6-2} = 3^4 = 81$

27. $\dfrac{y^4}{y^6} = \dfrac{y^4}{y^4 \cdot y^2} = \dfrac{1}{1 \cdot y^2} = \dfrac{1}{y^2}$

29. $\dfrac{c^4}{c^4} = c^{4-4} = c^0 = 1$

31. $\dfrac{a^3}{a^9} = \dfrac{a^3}{a^3 \cdot a^6} = \dfrac{1}{1 \cdot a^6} = \dfrac{1}{a^6}$

33. $x^0 = 1$

35. $3x^0 = 3 \cdot 1 = 3$

37. $4(5d)^0 = 4(5^0 d^0) = 4(1 \cdot 1) = 4 \cdot 1 = 4$

39. $-9(-4x)^0 = -9(-4)^0 \cdot x^0 = -9(1 \cdot 1) = -9(1) = -9$

41. $6x^3 y^2 z^0 = 6x^3 y^2 (1) = 6x^3 y^2$

43. $-8r(st)^0 = -8rs^0 t^0 = -8r \cdot 1 \cdot 1 = -8r$

45. $\left(x^4\right)^2 = x^{4 \cdot 2} = x^8$

47. $\left(x^5\right)^5 = x^{5 \cdot 5} = x^{25}$

49. $\left(x^3\right)^1 = x^{3 \cdot 1} = x^3$

51. $\left(x^4\right)^3 = x^{4 \cdot 3} = x^{12}$

53. $\left(n^6\right)^3 = n^{6 \cdot 3} = n^{18}$

55. $\left(-2w^2\right)^3 = (-2)^3 w^{2 \cdot 3} = -8w^6$

57. $\left(-3x^3\right)^3 = (-3)^3 x^{3 \cdot 3} = -27x^9$

59. $\left(4x^3 y^2\right)^3 = 4^3 \cdot x^{3 \cdot 3} y^{2 \cdot 3} = 64x^9 y^6$

61. $\left(\dfrac{x}{3}\right)^2 = \dfrac{x^2}{3^2} = \dfrac{x^2}{9}$

63. $\left(\dfrac{y}{x}\right)^4 = \dfrac{y^4}{x^4}$

65. $\left(\dfrac{-6}{x}\right)^3 = \dfrac{(-6)^3}{x^3} = \dfrac{-216}{x^3} = -\dfrac{216}{x^3}$

67. $\left(\dfrac{2x}{y}\right)^3 = \dfrac{2^3 x^3}{y^3} = \dfrac{8x^3}{y^3}$

69. $\left(\dfrac{4p}{5}\right)^2 = \dfrac{4^2 p^2}{5^2} = \dfrac{16p^2}{25}$

71. $\left(\dfrac{3x^4}{y}\right)^3 = \dfrac{3^3 x^{4\cdot3}}{y^3} = \dfrac{27x^{12}}{y^3}$

73. $\dfrac{a^8 b}{ab^4} = \dfrac{a\cdot a^7 \cdot b}{a\cdot b\cdot b^3} = \dfrac{a^7}{b^3}$

75. $\dfrac{5x^{12}y^2}{10xy^9} = \dfrac{5\cdot x\cdot x^{11}\cdot y^2}{2\cdot5\cdot x\cdot y^2\cdot y^7} = \dfrac{x^{11}}{2y^7}$

77. $\dfrac{30y^5 z^3}{5yz^6} = \dfrac{5\cdot6\cdot y\cdot y^4\cdot z^3}{5\cdot y\cdot z^3\cdot z^3} = \dfrac{6y^4}{z^3}$

79. $\dfrac{35x^4 y^9}{15x^9 y^{12}} = \dfrac{5\cdot7\cdot x^4\cdot y^9}{5\cdot3\cdot x^4\cdot x^5\cdot y^9\cdot y^3} = \dfrac{7}{3x^5 y^3}$

81. $\dfrac{-36xy^7 z}{12x^4 y^5 z} = -\dfrac{3\cdot12\cdot x\cdot y^5\cdot y^2\cdot z}{12\cdot x\cdot x^3\cdot y^5\cdot z} = -\dfrac{3y^2}{x^3}$

83. $-\dfrac{6x^2 y^7 z}{3x^5 y^9 z^6} = -\dfrac{2\cdot3\cdot x^2\cdot y^7\cdot z}{3\cdot x^2\cdot x^3\cdot y^7\cdot y^2\cdot z\cdot z^5} = -\dfrac{2}{x^3 y^2 z^5}$

85. $\left(\dfrac{10x^4}{5x^6}\right)^3 = \left(\dfrac{10}{5}\cdot\dfrac{x^4}{x^6}\right)^3 = \left(\dfrac{2}{x^2}\right)^3 = \dfrac{2^3}{x^{2\cdot3}} = \dfrac{8}{x^6}$

87. $\left(\dfrac{6y^6}{2y^3}\right)^3 = \left(\dfrac{6}{2}\cdot\dfrac{y^6}{y^3}\right)^3 = \left(3y^3\right)^3 = 3^3 y^{3\cdot3} = 27y^9$

89. $\left(\dfrac{6a^2 b^4}{3a^2 b^9}\right)^0 = 1$

91. $\left(\dfrac{x^4 y^3}{x^2 y^5}\right)^2 = \left(\dfrac{x^4}{x^2}\cdot\dfrac{y^3}{y^5}\right)^2 = \left(\dfrac{x^2}{y^2}\right)^2 = \dfrac{x^{2\cdot2}}{y^{2\cdot2}} = \dfrac{x^4}{y^4}$

93. $\left(\dfrac{9y^2 z^7}{18y^9 z}\right)^4 = \left(\dfrac{9}{18}\cdot\dfrac{y^2}{y^9}\cdot\dfrac{z^7}{z}\right)^4$
$= \left(\dfrac{z^6}{2y^7}\right)^4 = \dfrac{z^{6\cdot4}}{2^4 y^{7\cdot4}} = \dfrac{z^{24}}{16y^{28}}$

95. $\left(\dfrac{25s^4 t}{5s^6 t^4}\right)^3 = \left(\dfrac{25}{5}\cdot\dfrac{s^4}{s^6}\cdot\dfrac{t}{t^4}\right)^3$
$= \left(\dfrac{5}{s^2 t^3}\right)^3 = \dfrac{5^3}{s^{2\cdot3}\cdot t^{3\cdot3}} = \dfrac{125}{s^6 t^9}$

97. $\left(3xy^4\right)^2 = 3^2 x^2 y^{4\cdot2} = 9x^2 y^8$

99. $\left(5ab^3\right)(b) = 5a\left(b^{3+1}\right) = 5ab^4$

101. $(-2xy)(3xy) = (-2\cdot3)\left(x^{1+1}\right)\left(y^{1+1}\right) = -6x^2 y^2$

103. $\left(5x^2 y\right)\left(3xy^5\right) = (5\cdot3)\left(x^{2+1}\right)\left(y^{1+5}\right) = 15x^3 y^6$

105. $\left(-3p^2 q\right)^2\left(-p^2 q\right) = \left[(-3)^2 p^{2\cdot2} q^2\right]\left(-p^2 q\right)$
$= \left(9p^4 q^2\right)\left(-p^2 q\right)$
$= [9(-1)]p^{4+2}q^{2+1}$
$= -9p^6 q^3$

107. $\left(7r^3 s^2\right)^2\left(9r^3 s^4\right)^0 = (7^2\cdot r^{3\cdot2}s^{2\cdot2})(1) = 49r^6 s^4$

109. $(-x)^2 = (-x)(-x) = x^2$

111. $\left(\dfrac{x^5 y^5}{xy^5}\right)^3 = \left(\dfrac{x^5}{x}\cdot\dfrac{y^5}{y^5}\right)^3$
$= \left(x^4\cdot1\right)^3$
$= \left(x^4\right)^3$
$= x^{4\cdot3}$
$= x^{12}$

113. $\left(2.5x^3\right)^2 = 2.5^2\cdot x^{3\cdot2} = 6.25x^6$

115. $\dfrac{x^9 y^3}{x^2 y^7} = \dfrac{x^9}{x^2}\cdot\dfrac{y^3}{y^7} = \dfrac{x^7}{y^4}$

117. $\left(\dfrac{-m^4}{n^3}\right)^3 = \dfrac{(-1)^3 m^{4\cdot3}}{n^{3\cdot3}} = -\dfrac{m^{12}}{n^9}$

119. $\left(-6x^3 y^2\right)^3 = (-6)^3 x^{3\cdot3} y^{2\cdot3} = -216x^9 y^6$

123

121. $\left(-2x^4y^2z\right)^3 = (-2)^3x^{4\cdot3}y^{2\cdot3}z^{1\cdot3} = -8x^{12}y^6z^3$

123. $\left(9r^4s^5\right)^3 = 9^3\cdot r^{4\cdot3}\cdot s^{5\cdot3} = 729r^{12}s^{15}$

125. $\left(4x^2y\right)\left(3xy^2\right)^3 = \left(4x^2y\right)\left(3^3x^3y^{2\cdot3}\right)$
$$= \left(4x^2y\right)\left(27x^3y^6\right)$$
$$= 4\cdot27x^{2+3}y^{1+6}$$
$$= 108x^5y^7$$

127. $\left(7.3x^2y^4\right)^2 = 7.3^2x^{2\cdot2}y^{4\cdot2} = 53.29x^4y^8$

129. $\left(x^7y^5\right)\left(xy^2\right)^4 = \left(x^7y^5\right)\left(x^{1\cdot4}y^{2\cdot4}\right)$
$$= \left(x^7y^5\right)\left(x^4y^8\right)$$
$$= x^{7+4}y^{5+8}$$
$$= x^{11}y^{13}$$

131. $\left(\dfrac{-x^4z^7}{x^2z^5}\right)^4 = \left(-1\cdot\dfrac{x^4}{x^2}\cdot\dfrac{z^7}{z^5}\right)^4$
$$= (-x^2z^2)^4$$
$$= (-1)^4x^{2\cdot4}z^{2\cdot4}$$
$$= x^8z^8$$

133. $\dfrac{a+b}{b}$ cannot be simplified.

135. $\dfrac{y^2+3}{y}$ cannot be simplified.

137. $\dfrac{6yz^4}{yz^2} = 6y^{1-1}\cdot z^{4-2} = 6z^2$

139. $\dfrac{a^2+b^2}{a^2}$ cannot be simplified.

141. $a^3b = 2^3\cdot5 = 8\cdot5 = 40$

143. $(xy)^0 = (-5\cdot3)^0 = (-15)^0 = 1$

145. The sign will be positive because a negative number with an even number for an exponent will be positive. This is because $(-1)^m = 1$ when m is even.

147. Area $=$ Length $\cdot$ Width $= 8x\cdot x = 8x^2$

149. Area $=$ (Area of Top) $+$ (Area of Middle)
$+$ (Area of Bottom)
$$= (a\cdot b) + (a\cdot a) + (b\cdot b)$$
$$= ab + a^2 + b^2$$

151. $\left(3yz^2\right)^2\left(\dfrac{2y^3z^5}{10y^6z^4}\right)^0\left(4y^2z^3\right)^3$
$$= \left(3^2y^{1\cdot2}z^{2\cdot2}\right)\cdot1\cdot\left(4^3y^{2\cdot3}z^{3\cdot3}\right)$$
$$= \left(9y^2z^4\right)\left(64y^6z^9\right)$$
$$= 9\cdot64y^{2+6}z^{4+9}$$
$$= 576y^8z^{13}$$

154. $3^4\div3^3 - (5-8) + 7 = 81\div27 - (-3) + 7$
$$= 3 - (-3) + 7$$
$$= 6 + 7$$
$$= 13$$

155. $-4(x-3) + 5x - 2 = -4x + 12 + 5x - 2$
$$= -4x + 5x + 12 - 2$$
$$= x + 10$$

156. $2(x+4) - 3 = 5x + 4 - 3x + 1$
$$2x + 8 - 3 = 2x + 5$$
$$2x + 5 = 2x + 5 \text{ True}$$
All real numbers are solutions to this equation.

157. a.
$$P = 2l + 2w$$
$$26 = 2(x+5) + 2(x)$$
$$26 = 2x + 10 + 2x$$
$$26 = 4x + 10$$
$$16 = 4x$$
$$4 = x$$
$$x + 5 = 4 + 5 = 9$$
The sides are 4 and 9 inches long.

b.
$$P = 2l + 2w$$
$$P - 2l = 2l + 2w - 2l$$
$$P - 2l = 2w$$
$$\dfrac{P - 2l}{2} = w$$

Exercise Set 5.2

1. When a variable or number is raised to a negative exponent, the expression may be written as 1 divided by the variable or number raised to that positive exponent.

3. No, it is not simplified because of the negative exponent.

$$x^5 y^{-3} = \frac{x^5}{y^3}$$

5. The given simplification is not correct since
$$5^{-2} = \frac{1}{5^2} = \frac{1}{25}.$$

7. **a.** The numerator has one term, $x^5 y^2$.

   b. The factors of the numerator are x^5 and y^2.

9. The sign of the exponent changes when a factor is moved from the denominator to the numerator of a fraction.

11. $x^{-6} = \dfrac{1}{x^6}$

13. $5^{-1} = \dfrac{1}{5}$

15. $\dfrac{1}{x^{-3}} = x^3$

17. $\dfrac{1}{a^{-1}} = a^1 = a$

19. $\dfrac{1}{6^{-2}} = 6^2 = 36$

21. $\left(x^{-2}\right)^3 = x^{(-2)(3)} = x^{-6} = \dfrac{1}{x^6}$

23. $\left(y^{-5}\right)^4 = y^{(-5)(4)} = y^{-20} = \dfrac{1}{y^{20}}$

25. $\left(x^4\right)^{-2} = x^{4(-2)} = x^{-8} = \dfrac{1}{x^8}$

27. $\left(3^{-2}\right)^{-1} = 3^{(-2)(-1)} = 3^2 = 9$

29. $y^4 \cdot y^{-2} = y^{4+(-2)} = y^2$

31. $x^7 \cdot x^{-5} = x^{7+(-5)} = x^2$

33. $3^{-2} \cdot 3^4 = 3^{-2+4} = 3^2 = 9$

35. $\dfrac{r^5}{r^6} = r^{5-6} = r^{-1} = \dfrac{1}{r}$

37. $\dfrac{p^0}{p^{-3}} = p^{0-(-3)} = p^3$

39. $\dfrac{x^{-7}}{x^{-3}} = x^{-7-(-3)} = x^{-4} = \dfrac{1}{x^4}$

41. $\dfrac{3^2}{3^{-1}} = 3^{2-(-1)} = 3^3 = 27$

43. $5^{-3} = \dfrac{1}{5^3} = \dfrac{1}{125}$

45. $\dfrac{1}{z^{-9}} = z^9$

47. $\left(p^{-4}\right)^{-6} = p^{-4(-6)} = p^{24}$

49. $\left(y^{-2}\right)^{-3} = y^{(-2)(-3)} = y^6$

51. $x^3 \cdot x^{-7} = x^{3-7} = x^{-4} = \dfrac{1}{x^4}$

53. $x^{-8} \cdot x^{-7} = x^{-8-7} = x^{-15} = \dfrac{1}{x^{15}}$

55. $-4^{-2} = -\dfrac{1}{4^2} = -\dfrac{1}{16}$

57. $-(-4)^{-2} = -\dfrac{1}{(-4)^2} = -\dfrac{1}{16}$

59. $(-2)^{-3} = \dfrac{1}{(-2)^3} = -\dfrac{1}{8}$

61. $(-6)^{-2} = \dfrac{1}{(-6)^2} = \dfrac{1}{36}$

63. $\dfrac{x^{-5}}{x^5} = x^{-5-5} = x^{-10} = \dfrac{1}{x^{10}}$

65. $\dfrac{n^{-5}}{n^{-7}} = n^{-5-(-7)} = n^2$

67. $\dfrac{9^{-3}}{9^{-3}} = 9^{-3-(-3)} = 9^0 = 1$

69. $\left(2^{-1} + 3^{-1}\right)^0 = 1$

71. $\dfrac{2}{2^{-5}} = 2^{1-(-5)} = 2^6 = 64$

73. $\left(x^{-4}\right)^{-2} = x^{(-4)(-2)} = x^8$

75. $\left(x^0\right)^{-2} = (1)^{-2} = 1$

77. $2^{-3} \cdot 2 = 2^{-3+1} = 2^{-2} = \dfrac{1}{2^2} = \dfrac{1}{4}$

79. $7^{-5} \cdot 7^3 = 7^{-5+3} = 7^{-2} = \dfrac{1}{7^2} = \dfrac{1}{49}$

81. $\dfrac{x^{-1}}{x^{-4}} = x^{-1-(-4)} = x^3$

83. $\left(4^2\right)^{-1} = 4^{2(-1)} = 4^{-2} = \dfrac{1}{4^2} = \dfrac{1}{16}$

85. $\dfrac{5}{5^{-2}} = 5^{1-(-2)} = 5^3 = 125$

87. $\dfrac{3^{-4}}{3^{-2}} = 3^{-4-(-2)} = 3^{-2} = \dfrac{1}{3^2} = \dfrac{1}{9}$

89. $\dfrac{8^{-1}}{8^{-1}} = 8^{-1-(-1)} = 8^0 = 1$

91. $\left(-6x^2\right)^{-2} = (-6)^{-2} x^{2(-2)}$
$= (-6)^{-2} x^{-4} = \dfrac{1}{(-6)^2 x^4} = \dfrac{1}{36x^4}$

93. $3x^{-2}y^2 = 3 \cdot \dfrac{1}{x^2} \cdot y^2 = \dfrac{3y^2}{x^2}$

95. $\left(\dfrac{1}{2}\right)^{-2} = \left(\dfrac{2}{1}\right)^2 = 2^2 = 4$

97. $\left(\dfrac{5}{4}\right)^{-3} = \left(\dfrac{4}{5}\right)^3 = \dfrac{4^3}{5^3} = \dfrac{64}{125}$

99. $\left(\dfrac{c^4}{d^2}\right)^{-2} = \left(\dfrac{d^2}{c^4}\right)^2 = \dfrac{d^{2 \cdot 2}}{c^{4 \cdot 2}} = \dfrac{d^4}{c^8}$

101. $-\left(\dfrac{r^4}{s}\right)^{-4} = -\left(\dfrac{s}{r^4}\right)^4 = -\dfrac{s^4}{r^{4 \cdot 4}} = -\dfrac{s^4}{r^{16}}$

103. $-7a^{-3}b^{-4} = -7 \cdot \dfrac{1}{a^3} \cdot \dfrac{1}{b^4} = -\dfrac{7}{a^3 b^4}$

105. $\left(4x^5 y^{-3}\right)^{-3} = 4^{-3} x^{5(-3)} y^{(-3)(-3)}$
$= 4^{-3} x^{-15} y^9 = \dfrac{y^9}{4^3 x^{15}} = \dfrac{y^9}{64x^{15}}$

107. $\left(3z^{-4}\right)\left(6z^{-5}\right) = 3 \cdot 6 \cdot z^{-4} \cdot z^{-5} = 18z^{-9} = \dfrac{18}{z^9}$

109. $4x^4\left(-2x^{-4}\right) = 4 \cdot (-2) \cdot x^4 \cdot x^{-4} = -8x^0 = -8$

111. $\left(4x^2 y\right)\left(3x^3 y^{-1}\right) = 4 \cdot 3 \cdot x^2 \cdot x^3 \cdot y \cdot y^{-1} = 12x^5$

113. $\left(-5y^2\right)\left(4y^{-3}z^5\right) = -5 \cdot 4 \cdot y^2 \cdot y^{-3} \cdot z^5$
$= -20y^{-1}z^5$
$= -\dfrac{20z^5}{y}$

115. $\dfrac{24d^{12}}{3d^8} = \dfrac{24}{3} \cdot d^{12-8} = 8d^4$

117. $\dfrac{36x^{-4}}{9x^{-2}} = \dfrac{36}{4} \cdot \dfrac{x^{-4}}{x^{-2}} = 4 \cdot \dfrac{1}{x^2} = \dfrac{4}{x^2}$

119. $\dfrac{3x^4 y^{-2}}{6y^3} = \dfrac{3}{6} \cdot x^4 \cdot \dfrac{y^{-2}}{y^3} = \dfrac{1}{2} \cdot x^4 \cdot \dfrac{1}{y^5} = \dfrac{x^4}{2y^5}$

121. $\dfrac{32x^4 y^{-2}}{4x^{-2}y^0} = \left(\dfrac{32}{4}\right) x^{4-(-2)} y^{-2-0} = 8x^6 y^{-2} = \dfrac{8x^6}{y^2}$

123. $\left(\dfrac{5x^4y^{-7}}{z^3}\right)^{-2}=\left(\dfrac{z^3}{5x^4y^{-7}}\right)^2$

$\qquad = \dfrac{z^{3\cdot2}}{5^2x^{4\cdot2}y^{-7\cdot2}}$

$\qquad = \dfrac{z^6}{25x^8y^{-14}}$

$\qquad = \dfrac{y^{14}z^6}{25x^8}$

125. $\left(\dfrac{2r^{-5}s^9}{t^{12}}\right)^{-4}=\left(\dfrac{t^{12}}{2r^{-5}s^9}\right)^4$

$\qquad = \dfrac{t^{12\cdot4}}{2^4r^{-5\cdot4}s^{9\cdot4}}$

$\qquad = \dfrac{t^{48}}{16r^{-20}s^{36}}$

$\qquad = \dfrac{r^{20}t^{48}}{16s^{36}}$

127. $\left(\dfrac{x^3y^{-4}z}{y^{-2}}\right)^{-6}=\left(\dfrac{y^{-2}}{x^3y^{-4}z}\right)^6$

$\qquad = \dfrac{y^{-2\cdot6}}{x^{3\cdot6}y^{-4\cdot6}z^6}$

$\qquad = \dfrac{y^{-12}}{x^{18}y^{-24}z^6}$

$\qquad = \dfrac{y^{-12-(-24)}}{x^{18}z^6}$

$\qquad = \dfrac{y^{12}}{x^{18}z^6}$

129. $\left(\dfrac{p^6q^{-3}}{4p^8}\right)^2=\left(\dfrac{p^{6-8}q^{-3}}{4}\right)^2$

$\qquad = \left(\dfrac{p^{-2}q^{-3}}{4}\right)^2$

$\qquad = \dfrac{p^{-2\cdot2}q^{-3\cdot2}}{4^2}$

$\qquad = \dfrac{p^{-4}q^{-6}}{16}$

$\qquad = \dfrac{1}{16p^4q^6}$

131. **a.** Yes, $p^{-1}q^{-1}=\dfrac{1}{p}\cdot\dfrac{1}{q}=\dfrac{1}{pq}$.

b. No, $p^{-1}+q^{-1}=\dfrac{1}{p}+\dfrac{1}{q}\neq\dfrac{1}{p+q}$.

133. $4^2+4^{-2}=16+\dfrac{1}{4^2}=16+\dfrac{1}{16}=16\dfrac{1}{16}$

135. $5^3+5^{-3}=125+\dfrac{1}{5^3}=125+\dfrac{1}{125}=125\dfrac{1}{125}$

137. $5^0-3^{-1}=1-\dfrac{1}{3^1}=1-\dfrac{1}{3}=\dfrac{2}{3}$

139. $2^{-3}-2^3\cdot2^{-3}=2^{-3}-2^{3-3}=\dfrac{1}{2^3}-2^0=\dfrac{1}{8}-1=-\dfrac{7}{8}$

141. $2\cdot4^{-1}-4\cdot3^{-1}=2\cdot\dfrac{1}{4}-4\cdot\dfrac{1}{3}$

$\qquad = \dfrac{2}{4}-\dfrac{4}{3}$

$\qquad = \dfrac{6}{12}-\dfrac{16}{12}$

$\qquad = -\dfrac{10}{12}$

$\qquad = -\dfrac{5}{6}$

143. $3\cdot5^0-5\cdot3^{-2}=3\cdot1-5\cdot\dfrac{1}{3^2}$

$\qquad = 3-5\cdot\dfrac{1}{9}$

$\qquad = 3-\dfrac{5}{9}$

$\qquad = \dfrac{27}{9}-\dfrac{5}{9}$

$\qquad = \dfrac{22}{9}$

145. The missing number is -2 since $3^{-2}=\dfrac{1}{3^2}=\dfrac{1}{9}$.

147. The missing number is -3 since $\dfrac{1}{6^{-3}}=6^3=216$.

149. The missing number is -2 since
$(x^{-2})^{-2}=x^{(-2)(-2)}=x^4$.

127

151. The missing coefficient is 2 since $2^3 = 8$.
The missing exponent is -3 since

$$(x^{-3})^3 = x^{(-3)(3)} = x^{-9} = \frac{1}{x^9}.$$

153. The product rule is $(xy)^m = x^m y^m$, not

$$(x+y)^m = x^m + y^m.$$

155. Let x = the number of miles.

$$\frac{104 \text{ miles}}{52 \text{ minutes}} = \frac{x}{93 \text{ minutes}}$$

$$\frac{104}{52} = \frac{x}{93}$$

$$52x = 9672$$

$$x = 186$$

It will travel 186 miles.

156. Let x = the first integer.
Then $x + 2$ = the second integer.

$$x + (x+2) = 190$$

$$x + x + 2 = 190$$

$$2x + 2 = 190$$

$$2x = 188$$

$$x = 94$$

The numbers are 94 and $94 + 2 = 96$.

157. Let w = width of the shed.
Then $2w - 8$ = length of the shed.

$$P = 2l + 2w$$

$$56 = 2(2w - 8) + 2w$$

$$56 = 4w - 16 + 2w$$

$$56 = 6w - 16$$

$$72 = 6w$$

$$12 = w$$

The width is 12 ft and the length is $2(12) - 8 = 16$ ft.

158. Let x = the amount invested at 3%.
Then $9000 - x$ = the amount invested at 4%.

Principal	Rat	Time	Interest
x	3%	1	$0.03x$
$9000 - x$	4%	1	$0.04(9000 - x)$

$$0.03x - 0.04(9000 - x) = 32$$

$$0.03x - 360 + 0.04x = 32$$

$$0.07x - 360 = 32$$

$$0.07x = 392$$

$$x = 5600$$

$5600 was invested at 3% and $9000 - 5600 = 3400 invested at 4%.

159.
$$(6xy^5)(3x^2y^4) = 6 \cdot 3 \cdot x^1 \cdot x^2 \cdot y^5 \cdot y^4$$
$$= 18x^{1+2}y^{5+4}$$
$$= 18x^3y^9$$

Exercise Set 5.3

1. A number in scientific notation is written as a number greater than or equal to 1 and less than 10 that is multiplied by some power of 10.

3. a. Answers will vary.

b. $0.0000723 = 7.23 \times 10^{-5}$.

5. Move the decimal point 6 places to the right.

7. The exponent will be positive when the number is 10 or greater.

9. The exponent will be negative since $0.000937 < 1$.

11. $0.000001 = 1 \times 10^{-6}$

13. $350,000 = 3.5 \times 10^5$

15. $7950 = 7.95 \times 10^3$

17. $0.053 = 5.3 \times 10^{-2}$

19. $.000726 = 7.26 \times 10^{-4}$

21. $5,260,000,000 = 5.26 \times 10^9$

23. $0.00000914 = 9.14 \times 10^{-6}$

25. $220,300 = 2.203 \times 10^5$

27. $.005104 = 5.104 \times 10^{-3}$

29. $4.3 \times 10^4 = 43,000$

31. $9.32 \times 10^{-6} = 0.00000932$

33. $2.13 \times 10^{-5} = 0.0000213$

35. $6.25 \times 10^5 = 625,000$

37. $9.0 \times 10^6 = 9,000,000$

39. $5.35 \times 10^2 = 535$

41. $7.73 \times 10^{-7} = 0.000000773$

43. $1.0 \times 10^4 = 10,000$

45. 8 micrometers $= 8 \times 10^{-6} = 0.000008$ meters

47. 125 gigawatts $= 125 \times 10^9 = 125,000,000,000$ watts

49. 15.3 km $= 15.3 \times 10^3 = 15,300$ meters

51. 48.2 mm $= 48.2 \times 10^{-3} = 0.0482$ meters

53. $\left(2.0 \times 10^2\right)\left(3.0 \times 10^5\right) = (2.0 \times 3.0)\left(10^2 \times 10^5\right)$
$$= 6.0 \times 10^7$$
$$= 60,000,000$$

55. $\left(2.7 \times 10^{-6}\right)\left(9.0 \times 10^4\right) = (2.7 \times 9.0)\left(10^{-6} \times 10^4\right)$
$$= 24.3 \times 10^{-2}$$
$$= 0.243$$

57. $\left(1.6 \times 10^{-2}\right)\left(4.0 \times 10^{-3}\right) = (1.6 \times 4.0)\left(10^{-2} \times 10^{-3}\right)$
$$= 6.4 \times 10^{-5}$$
$$= 0.000064$$

59. $\dfrac{3.9 \times 10^{-5}}{3.0 \times 10^{-2}} = \left(\dfrac{3.9}{3.0}\right)\left(\dfrac{10^{-5}}{10^{-2}}\right) = 1.3 \times 10^{-3} = 0.0013$

61. $\dfrac{7.5 \times 10^6}{3.0 \times 10^3} = \left(\dfrac{7.5}{3.0}\right)\left(\dfrac{10^6}{10^3}\right) = 2.5 \times 10^3 = 2500$

63. $\dfrac{2.0 \times 10^4}{8.0 \times 10^{-2}} = \left(\dfrac{2.0}{8.0}\right)\left(\dfrac{10^4}{10^{-2}}\right) = 0.25 \times 10^6 = 250,000$

65. $(700,000)(6,000,000) = \left(7 \times 10^5\right)\left(6 \times 10^6\right)$
$$= (7 \times 6)\left(10^5 \times 10^6\right)$$
$$= 42 \times 10^{11}$$
$$= 4.2 \times 10^{12}$$

67. $(0.0004)(320) = \left(4 \times 10^{-4}\right)\left(4.3 \times 10^2\right)$
$$= (4 \times 3.2)\left(10^{-4} \times 10^2\right)$$
$$= 12.8 \times 10^{-2}$$
$$= 1.28 \times 10^{-1}$$

69. $\dfrac{5,600,000}{8000} = \dfrac{5.6 \times 10^6}{8.0 \times 10^3}$
$$= \left(\dfrac{5.6}{8.0}\right)\left(\dfrac{10^6}{10^3}\right)$$
$$= 0.7 \times 10^3$$
$$= 7.0 \times 10^2$$

71. $\dfrac{0.00035}{0.000002} = \dfrac{3.5 \times 10^{-4}}{2.0 \times 10^{-6}} = \left(\dfrac{3.5}{2.0}\right)\left(\dfrac{10^{-4}}{10^{-6}}\right) = 1.75 \times 10^2$

73. $3.3 \times 10^{-4}, \ 5.3, \ 7.3 \times 10^2, \ 1.75 \times 10^6$

75. a. $\left(6.55 \times 10^9\right) - \left(2.99 \times 10^8\right)$
$$= \left(65.5 \times 10^8\right) - \left(2.99 \times 10^8\right)$$
$$= (65.5 - 2.99) \times 10^8$$
$$= 62.51 \times 10^8$$
$$\approx 6,251,000,000$$
The people that live outside the U.S. total about 6,251,000,000.

b. $\dfrac{6.55 \times 10^9}{2.99 \times 10^8} = \left(\dfrac{6.55}{2.99}\right)\left(\dfrac{10^9}{10^8}\right) \approx 2.19 \times 10 \approx 21.9$

The world is about 21.9 times greater than the U.S. population.

77. Minimum volume
$$= \left(\dfrac{100,000 \ \text{ft}^3 \text{sec}}{\text{sec}}\right)\left(\dfrac{60 \ \text{sec}}{\text{min}}\right)\left(\dfrac{60 \ \text{min}}{\text{hr}}\right)(24 \ \text{hrs})$$
$$= \left(1.0 \times 10^5\right)\left(6.0 \times 10^1\right)\left(6.0 \times 10^1\right)\left(2.4 \times 10^1\right)\text{ft}^3$$
$$= (1 \times 6 \times 6 \times 2.4)\left(10^5 \times 10^1 \times 10^1 \times 10^1\right)\text{ft}^3$$
$$= 86.4 \times 10^8 \ \text{ft}^3$$
$$= 8,640,000,000 \ \text{ft}^3$$

79. $\left(2 \times 10^{-6}\right) \times \left(8 \times 10^{12}\right) = \left(2 \times 8\right)\left(10^{-6} \times 10^{12}\right)$

$$= 16 \times 10^{6}$$

$$= 1.6 \times 10^{7}$$

It would take 1.6×10^{7} seconds.

81. a. $\left(6.01 \times 10^{8}\right) - \left(4.33 \times 10^{8}\right)$

$$= \left(6.01 - 4.33\right) \times 10^{8}$$

$$= 1.68 \times 10^{8}$$

Titanic grossed $\$1.68 \times 10^{8}$ more than *E.T.*

b. $\dfrac{6.01 \times 10^{8}}{4.33 \times 10^{8}} = \left(\dfrac{6.01}{4.33}\right)\left(\dfrac{10^{8}}{10^{8}}\right)$

$$\approx 1.39(1)$$

$$\approx 1.39$$

The gross ticket sales of *Titanic* was about 1.39 times greater than *E.T.*

83. a. $\left(4.65 \times 10^{10}\right) - \left(9.9 \times 10^{9}\right)$

$$= \left(46.5 \times 10^{9}\right) - \left(9.9 \times 10^{9}\right)$$

$$= \left(46.5 - 9.9\right) \times 10^{9}$$

$$= 36.6 \times 10^{9}$$

$$= 36,600,000,000$$

Bill Gates was $\$36,600,000,000$ richer than Pierre Omidyar.

b. $\dfrac{4.65 \times 10^{10}}{9.9 \times 10^{9}} = \left(\dfrac{4.65}{9.9}\right)\left(\dfrac{10^{10}}{10^{9}}\right)$

$$\approx 0.47 \times 10$$

$$\approx 4.7$$

Bill Gates was worth 4.7 times greater than Pierre Omidyar.

85. a. Earth: 5.794×10^{24} metric tons

Moon: 7.34×10^{19} metric tons

Jupiter: 1.899×10^{27} metric tons

b. $\dfrac{5.794 \times 10^{24}}{7.340 \times 10^{19}} = \left(\dfrac{5.794}{7.340}\right)\left(\dfrac{10^{24}}{10^{19}}\right)$

$$\approx 0.789 \times 10^{5}$$

$$\approx 7.89 \times 10^{4}$$

c. $\dfrac{1.899 \times 10^{27}}{5.794 \times 10^{24}} = \left(\dfrac{1.899}{5.794}\right)\left(\dfrac{10^{27}}{10^{24}}\right)$

$$\approx 0.328 \times 10^{3}$$

$$\approx 3.28 \times 10^{2}$$

87. 47% of 2,800,000,000

$0.47\left(2,800,000,000\right) = \left(4.7 \times 10^{-1}\right)\left(2.8 \times 10^{9}\right)$

$$= \left(4.7 \times 2.8\right)\left(10^{-1} \times 10^{9}\right)$$

$$= 13.16 \times 10^{8}$$

$$= 1.316 \times 10^{9}$$

There was $\$1.316 \times 10^{9}$ revenue from digital products and services.

89. a. $55 - 59 : 200,000 = 2.0 \times 10^{5}$

$20 - 24 : 48,000 = 4.8 \times 10^{4}$

b. $\left(2.0 \times 10^{5}\right) + \left(4.8 \times 10^{4}\right)$

$$= \left(20 \times 10^{4}\right) + \left(4.8 \times 10^{4}\right)$$

$$= \left(20.0 + 4.8\right) \times 10^{4}$$

$$= 24.8 \times 10^{4}$$

$$= 2.48 \times 10^{5}$$

c. $\left(2.0 \times 10^{5}\right) - \left(4.8 \times 10^{4}\right)$

$$= \left(20 \times 10^{4}\right) - \left(4.8 \times 10^{4}\right)$$

$$= \left(20.0 - 4.8\right) \times 10^{4}$$

$$= 15.2 \times 10^{4}$$

$$= 1.52 \times 10^{5}$$

91. $10^{-18} = 0.000000000000000001$

Thus, there are 17 zeros after the decimal.

93. Answers will vary.

95. $\dfrac{10^{-12}}{10^{-18}} = 10^{-12-(-18)}$

$$= 10^{6}$$

$$= 1,000,000 \text{ times greater}$$

98. $4x^2 + 3x + \dfrac{x}{2} = 4 \cdot 0^2 + 3 \cdot 0 + \dfrac{0}{2}$

$\phantom{4x^2 + 3x + \dfrac{x}{2}} = 0 + 0 + 0$

$\phantom{4x^2 + 3x + \dfrac{x}{2}} = 0$

99. a. If $-x = -\dfrac{3}{2}$, then $x = \dfrac{3}{2}$.

    b. If $5x = 0$, then $x = 0$.

100. $2x - 3(x - 2) = x + 2$

$2x - 3x + 6 = x + 2$

$-x + 6 = x + 2$

$4 = 2x$

$2 = x$

101. $\left(\dfrac{-2x^5 y^7}{8x^8 y^3}\right)^3 = \left(\dfrac{-2}{8} \cdot \dfrac{x^5}{x^8} \cdot \dfrac{y^7}{y^3}\right)^3$

$\phantom{\left(\dfrac{-2x^5 y^7}{8x^8 y^3}\right)^3} = \left(\dfrac{-1}{4} \cdot \dfrac{1}{x^3} \cdot y^4\right)^3$

$\phantom{\left(\dfrac{-2x^5 y^7}{8x^8 y^3}\right)^3} = \left(-\dfrac{y^4}{4x^3}\right)^3$

$\phantom{\left(\dfrac{-2x^5 y^7}{8x^8 y^3}\right)^3} = \dfrac{(-1)^3 \, y^{4 \cdot 3}}{4^3 \, x^{3 \cdot 3}}$

$\phantom{\left(\dfrac{-2x^5 y^7}{8x^8 y^3}\right)^3} = -\dfrac{y^{12}}{64x^9}$

Mid-Chapter Test: 5.1-5.3

1. $y^{11} \cdot y^2 = y^{11+2} = y^{13}$

2. $\dfrac{x^{13}}{x^{10}} = x^{13-10} = x^3$

3. $(-3x^5 y^7)(-2xy^6) = (-3)(-2)x^{5+1} y^{7+6} = 6x^6 y^{13}$

4. $\dfrac{6a^{12} b^8}{9a^7 b^2} = \dfrac{6}{9} a^{12-7} b^{8-2} = \dfrac{2}{3} a^5 b^6 = \dfrac{2a^5 b^6}{3}$

5. $(-4x^2 y^4)^3 = (-4)^3 x^{2 \cdot 3} y^{4 \cdot 3} = -64x^6 y^{12}$

6. $\left(\dfrac{-5s^4 t^6}{10s^6 t^3}\right)^2 = \left(\dfrac{-5}{10} s^{4-6} t^{6-3}\right)^2$

$\phantom{\left(\dfrac{-5s^4 t^6}{10s^6 t^3}\right)^2} = \left(-\dfrac{1}{2} s^{-2} t^3\right)^2$

$\phantom{\left(\dfrac{-5s^4 t^6}{10s^6 t^3}\right)^2} = \left(-\dfrac{1}{2}\right)^2 s^{-2 \cdot 2} t^{3 \cdot 2}$

$\phantom{\left(\dfrac{-5s^4 t^6}{10s^6 t^3}\right)^2} = \dfrac{t^6}{4s^4}$

7. $(7x^9 y^5)(-3xy^4)^2 = (7x^9 y^5)\left[(-3)^2 x^2 y^{4 \cdot 2}\right]$

$ = (7x^9 y^5)(9x^2 y^8)$

$ = (7)(9)x^{9+2} y^{5+8}$

$ = 63x^{11} y^{13}$

8. $\dfrac{p^{-3}}{p^5} = p^{-3-5} = p^{-8} = \dfrac{1}{p^8}$

9. $x^{-4} \cdot x^{-6} = x^{-4+(-6)} = x^{-10} = \dfrac{1}{x^{10}}$

10. $(3^{-1} + 5^2)^0 = 1$

11. $\left(\dfrac{3}{7}\right)^{-2} = \left(\dfrac{7}{3}\right)^2 = \dfrac{7^2}{3^2} = \dfrac{49}{9}$

12. $(8x^{-2} y^5)(4x^3 y^{-6}) = (8)(4)x^{-2+3} y^{5-6}$

$\phantom{(8x^{-2} y^5)(4x^3 y^{-6})} = 32xy^{-1}$

$\phantom{(8x^{-2} y^5)(4x^3 y^{-6})} = \dfrac{32x}{y}$

13. $\dfrac{6m^{-4} n^{-7}}{2m^0 n^{-1}} = \dfrac{6}{2} m^{-4-0} n^{-7-(-1)} = 3m^{-4} n^{-6} = \dfrac{3}{m^4 n^6}$

14. $\left(\dfrac{2x^{-3} y^{-4}}{x^3 yz^{-2}}\right)^{-2} = \left(2x^{-3-3} y^{-4-1} z^2\right)^{-2}$

$\phantom{\left(\dfrac{2x^{-3} y^{-4}}{x^3 yz^{-2}}\right)^{-2}} = \left(2x^{-6} y^{-5} z^2\right)^{-2}$

$\phantom{\left(\dfrac{2x^{-3} y^{-4}}{x^3 yz^{-2}}\right)^{-2}} = 2^{-2} x^{-6 \cdot (-2)} y^{-5 \cdot (-2)} z^{2 \cdot (-2)}$

$\phantom{\left(\dfrac{2x^{-3} y^{-4}}{x^3 yz^{-2}}\right)^{-2}} = \dfrac{1}{4} x^{12} y^{10} z^{-4}$

$\phantom{\left(\dfrac{2x^{-3} y^{-4}}{x^3 yz^{-2}}\right)^{-2}} = \dfrac{x^{12} y^{10}}{4z^4}$

15. a. Answers will vary.

b. Answers will vary.

16. 6.54×10^9

17. 0.0000327

18. $18.9 \times 10^3 = 18,900$ meters

19. $(3.4 \times 10^{-6})(7.0 \times 10^3) = (3.4 \times 7.0)(10^{-6} \times 10^3)$
$$= 23.8 \times 10^{-6+3}$$
$$= 23.8 \times 10^{-3}$$
$$= 0.0238$$

20. $\dfrac{0.00006}{200} = \dfrac{6 \times 10^{-5}}{2 \times 10^2} = \dfrac{6}{2} \times 10^{-5-2} = 3 \times 10^{-7}$

Exercise Set 5.4

1. A polynomial is an expression containing the sum of a finite number of terms of the form ax^n where a is a real number and n is a whole number.

3. No, it contains a negative exponent.

5. a. A monomial is a one-termed polynomial.

b. A binomial is a two-termed polynomial.

c. A trinomial is a three-termed polynomial.

7. Answers will vary.

9. b) and c). If you add the exponents on the variables, the sum is 4.

11. Write with exponents on the variable decreasing from left to right.

13. To add polynomials, combine like terms.

15. To subtract polynomials, use the distributive property to remove parentheses and then combine like terms.

17. fifth

19. eighth

21. third

23. tenth

25. ninth

27. trinomial

29. monomial

31. binomial

33. monomial

35. not a polynomial

37. polynomial

39. trinomial

41. not a polynomial

43. Already in descending order, 0 degree

45. $x^2 - 2x - 4$, second

47. $3x^2 + x - 8$, second

49. Already in descending order, first

51. Already in descending order, second

53. $4x^3 - 3x^2 + x - 4$, third

55. $-2x^4 + 3x^2 + 5x - 6$, fourth

57. $(9x - 2) + (x - 7) = 9x - 2 + x - 7$
$$= 9x + x - 2 - 7$$
$$= 10x - 9$$

59. $(-3x + 8) + (2x + 3) = -3x + 8 + 2x + 3$
$$= -3x + 2x + 8 + 3$$
$$= -x + 11$$

61. $(t + 7) + (-3t - 8) = t + 7 - 3t - 8$
$$= t - 3t + 7 - 8$$
$$= -2t - 1$$

63. $(x^2 + 2.6x - 3) + (4x + 3.8)$
$$= x^2 + 2.6x - 3 + 4x + 3.8$$
$$= x^2 + 2.6x + 4x - 3 + 3.8$$
$$= x^2 + 6.6x + 0.8$$

65. $(4m-3)+(5m^2-4m+7)$

$=4m-3+5m^2-4m+7$

$=5m^2+4m-4m-3+7$

$=5m^2+4$

67. $(2x^2-3x+5)+(-x^2+6x-8)$

$=2x^2-3x+5-x^2+6x-8$

$=2x^2-x^2+6x-3x+5-8$

$=x^2+3x-3$

69. $(-x^2-4x+8)+\left(5x-2x^2+\dfrac{1}{2}\right)$

$=-x^2-4x+8+5x-2x^2+\dfrac{1}{2}$

$=-x^2-2x^2-4x+5x+8+\dfrac{1}{2}$

$=-3x^2+x+\dfrac{17}{2}$

71. $(5.2n^2-6n+1.7)+(3n^2+1.2n-2.3)$

$=5.2n^2-6n+1.7+3n^2+1.2n-2.3$

$=5.2n^2+3n^2-6n+1.2n+1.7-2.3$

$=8.2n^2-4.8n-0.6$

73. $(-7x^3-3x^2+4)+(4x+5x^3-7)$

$=-7x^3-3x^2+4+4x+5x^3-7$

$=-7x^3+5x^3-3x^2+4x+4-7$

$=-2x^3-3x^2+4x-3$

75. $(8x^2+2x-y)+(3x^2-9x+5)$

$=8x^2+2x-y+3x^2-9x+5$

$=8x^2+3x^2+2x-9x-y+5$

$=11x^2-7x-y+5$

77. $(2x^2y+2x-3)+(3x^2y-5x+5)$

$=2x^2y+2x-3+3x^2y-5x+5$

$=2x^2y+3x^2y+2x-5x-3+5$

$=5x^2y-3x+2$

79. $8x-7$

$\underline{3x+4}$

$11x-3$

81. $4y^2-2y+4$

$\underline{3y^2\qquad+1}$

$7y^2-2y+5$

83. $-x^2-3x+3$

$\underline{5x^2+5x-7}$

$4x^2+2x-4$

85. $2x^3+3x^2+6x\ -9$

$\underline{\quad-4x^2\qquad +7}$

$2x^3-\ x^2+6x\ -2$

87. $4n^3-5n^2+n-6$

$\underline{-n^3-6n^2-2n+8}$

$3n^3-11n^2-n+2$

89. $(4x-4)-(2x+2)=4x-4-2x-2$

$\qquad\qquad\qquad\qquad =4x-2x-4-2$

$\qquad\qquad\qquad\qquad =2x-6$

91. $(-2x-3)-(-5x-7)=-2x-3+5x+7$

$\qquad\qquad\qquad\qquad\quad =-2x+5x-3+7$

$\qquad\qquad\qquad\qquad\quad =3x+4$

93. $(-r+5)-(2r+5)=-r+5-2r-5$

$\qquad\qquad\qquad\qquad =-r-2r+5-5$

$\qquad\qquad\qquad\qquad =-3r$

95. $(-y^2+4y-5.2)-(5y^2+2.1y+7.5)$

$=-y^2+4y-5.2-5y^2-2.1y-7.5$

$=-y^2-5y^2+4y-2.1y-5.2-7.5$

$=-6y^2+1.9y-12.7$

97. $(5x^2-x-1)-(-3x^2-2x-5)$

$=5x^2-x-1+3x^2+2x+5$

$=5x^2+3x^2-x+2x-1+5$

$=8x^2+x+4$

99. $(-4.1n^2-3n)-(2.3n^2-9n+7.6)$

$=-4.1n^2-3n-2.3n^2+9n-7.6$

$=-4.1n^2-2.3n^2-3n+9n-7.6$

$=-6.4n^2+6n-7.6$

101. $\left(8x^3 - 2x^2 - 4x + 5\right) - \left(5x^2 + 8\right)$

$= 8x^3 - 2x^2 - 4x + 5 - 5x^2 - 8$

$= 8x^3 - 2x^2 - 5x^2 - 4x - 8 + 5$

$= 8x^3 - 7x^2 - 4x - 3$

103. $\left(2x^3 - 4x^2 + 5x - 7\right) - \left(3x + \dfrac{3}{5}x^2 - 5\right)$

$= 2x^3 - 4x^2 + 5x - 7 - 3x - \dfrac{3}{5}x^2 + 5$

$= 2x^3 - 4x^2 - \dfrac{3}{5}x^2 + 5x - 3x - 7 + 5$

$= 2x^3 - \dfrac{23}{5}x^2 + 2x - 2$

105. $\left(8x + 2\right) - \left(7x + 4\right) = 8x + 2 - 7x - 4$

$\qquad\qquad\qquad\qquad = 8x - 7x + 2 - 4$

$\qquad\qquad\qquad\qquad = x - 2$

107. $\left(2x^2 - 4x + 8\right) - \left(5x - 6\right)$

$= 2x^2 - 4x + 8 - 5x + 6$

$= 2x^2 - 4x - 5x + 8 + 6$

$= 2x^2 - 9x + 14$

109. $\left(-5c^3 - 6c^2 + 7\right) - \left(-2c^2 + 7c - 7\right)$

$= -5c^3 - 6c^2 + 7 + 2c^2 - 7c + 7$

$= -5c^3 - 6c^2 + 2c^2 - 7c + 7 + 7$

$= -5c^3 - 4c^2 - 7c + 14$

111.
$$
\begin{array}{ll}
6x + 5 \quad \text{or} & 6x + 5 \\
\underline{-(3x - 3)} & \underline{\;-3x + 3} \\
& 3x + 8
\end{array}
$$

113.
$$
\begin{array}{ll}
5a^2 - 13a + 19 \quad \text{or} & 5a^2 - 13a + 19 \\
\underline{-(2a^2 + 3a - 9)} & \underline{-2a^2 - 3a + 9} \\
& 3a^2 - 16a + 28
\end{array}
$$

115.
$$
\begin{array}{ll}
7x^2 - 3x - 4 \quad \text{or} & 7x^2 - 3x - 4 \\
\underline{-(6x^2 \quad -1)} & \underline{-6x^2 + 0x + 1} \\
& x^2 - 3x - 3
\end{array}
$$

117.
$$
\begin{array}{ll}
x^2 + 4 \quad \text{or} & x^2 + 4 \\
\underline{-(5x^2 + 4)} & \underline{-5x^2 - 4} \\
& -4x^2
\end{array}
$$

119.
$$
\begin{array}{ll}
4x^3 - 6x^2 + 7x - 9 \quad \text{or} & 4x^3 - 6x^2 + 7x - 9 \\
\underline{-(x^2 + 6x - 7)} & \underline{\;-x^2 - 6x + 7} \\
& 4x^3 - 7x^2 + x - 2
\end{array}
$$

121 – 123. Answers will vary.

125. Sometimes

127. Sometimes

129. Answers will vary.

One example is: $x^4 - 2x^3 + x$

131. No, all three terms must have degree 4 or 1.

133. $a^2 + 2ab + b^2$

135. $4x^2 + 3xy$

137. $\left(3x^2 - 6x + 3\right) - \left(2x^2 - x - 6\right) - \left(x^2 + 7x - 9\right)$

$= 3x^2 - 6x + 3 - 2x^2 + x + 6 - x^2 - 7x + 9$

$= \left(3x^2 - 2x^2 - x^2\right) + \left(-6x + x - 7x\right) + \left(3 + 6 + 9\right)$

$= -12x + 18$

139. $4\left(x^2 + 2x - 3\right) - 6\left(2 - 4x - x^2\right) - 2x\left(x + 2\right)$

$= 4x^2 + 8x - 12 - 12 + 24x + 6x^2 - 2x^2 - 4x$

$= \left(4x^2 + 6x^2 - 2x^2\right) + \left(8x + 24x - 4x\right) + \left(-12 - 12\right)$

$= 8x^2 + 28x - 24$

141. $\dfrac{5}{9} = \dfrac{2.5}{x}$

$5x = 2.5(9)$

$5x = 22.5$

$x = 4.5$

142. $n - 5$

143. $3y = 9$

$y = 3$

This equation represents a horizontal line with y-intercept $(0, 3)$.

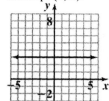

144. $y = -2x - 4$

This equation represents a line with y-intercept $(0, -4)$ and slope $m = \dfrac{-2}{1}$.

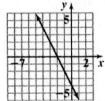

145. $\left(\dfrac{4x^3 y^5}{12x^7 y^4}\right)^3 = \left(\dfrac{x^{3-7} y^{5-4}}{3x^7}\right)^3$

$= \left(\dfrac{x^{-4} y^1}{3}\right)^3$

$= \left(\dfrac{y^1}{3x^4}\right)^3$

$= \dfrac{y^{1(3)}}{3^3 x^{4(3)}}$

$= \dfrac{y^3}{27x^{12}}$

146. $3.5 \times 10^{-2} = 0.035$

Exercise Set 5.5

1. To multiply two monomials, multiply their coefficients and use the product rule for exponents to determine the exponents on the variables.

3. First, Outer, Inner, Last

5. Yes, FOIL is simply a way to remember the procedure.

7. $(a+b)^2 = a^2 + 2ab + b^2$

$(a-b)^2 = a^2 - 2ab + b^2$

9. No, $(x-2)^2 = x^2 - 4x + 4$

11 – 13. Answers will vary.

15. $\left(3x^4\right)\left(-8x^2\right) = 3 \cdot (-8) x^{4+2} = -24x^6$

17. $5x^3 y^5 \left(4x^2 y\right) = (5 \cdot 4) x^{3+2} y^{5+1} = 20x^5 y^6$

19. $4xy^6 \left(-7x^2 y^9\right) = -28x^{1+2} y^{6+9} = -28x^3 y^{15}$

21. $9xy^6 \left(6x^5 y^8\right) = 9 \cdot 6x^{1+5} y^{6+8} = 54x^6 y^{14}$

23. $\left(6x^2 y\right)\left(\dfrac{1}{2}x^4\right) = 6 \cdot \dfrac{1}{2}x^{2+4} y = 3x^6 y$

25. $\left(3.3x^4\right)\left(1.8x^4 y^3\right) = (3.3 \cdot 1.8) x^{4+4} y^3 = 5.94x^8 y^3$

27. $9(x-5) = (9)(x) + (9)(-5) = 9x - 45$

29. $-3x(2x-2) = (-3x)(2x) + (-3x)(-2)$

$= -6x^2 + 6x$

31. $-2(8y+5) = (-2)(8y) + (-2)(5) = -16y - 10$

33. $-2x\left(x^2 - 2x + 5\right)$

$= (-2x)\left(x^2\right) + (-2x)(-2x) + (-2x)(5)$

$= -2x^3 + 4x^2 - 10x$

35. $5x\left(-4x^2 + 6x - 4\right)$

$= (5x)\left(-4x^2\right) + (5x)(6x) + (5x)(-4)$

$= -20x^3 + 30x^2 - 20x$

37. $0.5x^2 \left(x^3 - 6x^2 - 1\right)$

$= \left(0.5x^2\right)\left(x^3\right) + \left(0.5x^2\right)\left(-6x^2\right) + \left(0.5x^2\right)(-1)$

$= 0.5x^5 - 3x^4 - 0.5x^2$

39. $0.3x(2xy + 5x - 6y)$

$= (0.3x)(2xy) + (0.3x)(5x) + (0.3x)(-6y)$

$= 0.6x^2 y + 1.5x^2 - 1.8xy$

41. $\left(x^2 - 4y^3 - 3\right)y^4$

$= \left(x^2\right)\left(y^4\right) + \left(-4y^3\right)\left(y^4\right) + (-3)\left(y^4\right)$

$= x^2 y^4 - 4y^7 - 3y^4$

43. $(5x-2)(x+4)$

$\quad = (5x)(x)+(5x)(4)+(-2)(x)+(-2)(4)$

$\quad = 5x^2+20x-2x-8$

$\quad = 5x^2+18x-8$

45. $(2x+5)(3x-6)$

$\quad = (2x)(3x)+(2x)(-6)+(5)(3x)+(5)(-6)$

$\quad = 6x^2-12x+15x-30$

$\quad = 6x^2+3x-30$

47. $(2x-4)(2x+4)$

$\quad = (2x)(2x)+(2x)(4)+(-4)(2x)+(-4)(4)$

$\quad = 4x^2+8x-8x-16$

$\quad = 4x^2-16$

49. $(8-5x)(6+x)$

$\quad = (8)(6)+(8)(x)+(-5x)(6)+(-5x)(x)$

$\quad = 48+8x-30x-5x^2$

$\quad = 48-22x-5x^2$

$\quad = -5x^2-22x+48$

51. $(6x-1)(-2x+5)$

$\quad = (6x)(-2x)+(6x)(5)+(-1)(-2x)+(-1)(5)$

$\quad = -12x^2+30x+2x-5$

$\quad = -12x^2+32x-5$

53. $(x-2)(4x-2)$

$\quad = (x)(4x)+(x)(-2)+(-2)(4x)+(-2)(-2)$

$\quad = 4x^2-2x-8x+4$

$\quad = 4x^2-10x+4$

55. $(3k-6)(4k-2)$

$\quad = (3k)(4k)+(3k)(-2)+(-6)(4k)+(-6)(-2)$

$\quad = 12k^2-6k-24k+12$

$\quad = 12k^2-30k+12$

57. $(x-2)(x+2)$

$\quad = (x)(x)+(x)(2)+(-2)(x)+(-2)(2)$

$\quad = x^2+2x-2x-4$

$\quad = x^2-4$

59. $(2x-3)(2x-3)$

$\quad = (2x)(2x)+(2x)(-3)+(-3)(2x)+(-3)(-3)$

$\quad = 4x^2-6x-6x+9$

$\quad = 4x^2-12x+9$

61. $(6z-4)(7-z)$

$\quad = (6z)(7)+(6z)(-z)+(-4)(7)+(-4)(-z)$

$\quad = 42z-6z^2-28+4z$

$\quad = -6z^2+46z-28$

63. $(9-2x)(7-4x)$

$\quad = (9)(7)+(9)(-4x)+(-2x)(7)+(-2x)(-4x)$

$\quad = 63-36x-14x+8x^2$

$\quad = 8x^2-50x+63$

65. $(x+7)(y-3) = x(y)+(x)(-3)+7(y)+7(-3)$

$\qquad\qquad\qquad\quad = xy-3x+7y-21$

67. $(2x-3y)(3x+2y)$

$\quad = (2x)(3x)+(2x)(2y)+(-3y)(3x)+(-3y)(2y)$

$\quad = 6x^2+4xy-9xy-6y^2$

$\quad = 6x^2-5xy-6y^2$

69. $(9x+y)(4-3x)$

$\quad = (9x)(4)+(9x)(-3x)+y(4)+y(-3x)$

$\quad = 36x-27x^2+4y-3xy$

$\quad = -27x^2-3xy+36x+4y$

71. $(x+0.6)(x+0.3)$

$\quad = x(x)+x(0.3)+(0.6)x+(0.6)(0.3)$

$\quad = x^2+0.3x+0.6x+0.18$

$\quad = x^2+0.9x+0.18$

73. $(x+4)\left(x-\dfrac{1}{2}\right)$

$\quad = x(x)+x\left(-\dfrac{1}{2}\right)+4(x)+4\left(-\dfrac{1}{2}\right)$

$\quad = x^2-\dfrac{1}{2}x+4x-2$

$\quad = x^2+\dfrac{7}{2}x-2$

75. $(x+6)(x-6) = x^2 - 6^2 = x^2 - 36$

77. $(3x-8)(3x+8) = (3x)^2 - 8^2 = 9x^2 - 64$

79. $(x+y)^2 = (x)^2 + 2(x)(y) + (y)^2$
$= x^2 + 2xy + y^2$

81. $(x-0.2)^2 = (x)^2 - 2(x)(0.2) + (0.2)^2$
$= x^2 - 0.4x + 0.04$

83. $(4x+5)(4x+5) = (4x)^2 + 2(4x)(5) + (5)^2$
$= 16x^2 + 40x + 25$

85. $(0.4x+y)^2 = (0.4x)^2 + 2(0.4x)(y) + (y)^2$
$= 0.16x^2 + 0.8xy + y^2$

87. $(4c-5d)(4c+5d) = (4c)^2 - (5d)^2 = 16c^2 - 25d^2$

89. $(-2x+6)(-2x-6) = (-2x)^2 - 6^2 = 4x^2 - 36$

91. $(7s-3t)^2 = (7s)^2 - 2(7s)(3t) + (3t)^2$
$= 49s^2 - 42st + 9t^2$

93. $(4m+3)(4m^2 - 5m + 6)$
$= 4m(4m^2 - 5m + 6) + 3(4m^2 - 5m + 6)$
$= 16m^3 - 20m^2 + 24m + 12m^2 - 15m + 18$
$= 16m^3 - 8m^2 + 9m + 18$

95. $(3x+2)(4x^2 - x + 5)$
$= (3x)(4x^2 - x + 5) + 2(4x^2 - x + 5)$
$= 12x^3 - 3x^2 + 15x + 8x^2 - 2x + 10$
$= 12x^3 + 5x^2 + 13x + 10$

97. $(-2x^2 - 4x + 1)(7x - 3)$
$= -2x^2(7x-3) - 4x(7x-3) + 1(7x-3)$
$= -14x^3 + 6x^2 - 28x^2 + 12x + 7x - 3$
$= -14x^3 - 22x^2 + 19x - 3$

99. $(a+b)(a^2 - ab + b^2)$
$= a(a^2 - ab + b^2) + b(a^2 - ab + b^2)$
$= a^3 - a^2b + ab^2 + a^2b - ab^2 + b^3$
$= a^3 + b^3$

101. $(3x^2 - 2x + 4)(2x^2 + 3x + 1)$
$= 3x^2(2x^2 + 3x + 1) - 2x(2x^2 + 3x + 1)$
$\quad + 4(2x^2 + 3x + 1)$
$= 6x^4 + 9x^3 + 3x^2 - 4x^3 - 6x^2 - 2x$
$\quad + 8x^2 + 12x + 4$
$= 6x^4 + 5x^3 + 5x^2 + 10x + 4$

103. $(x^2 - x + 3)(x^2 - 2x)$
$= x^2(x^2 - 2x) - x(x^2 - 2x) + 3(x^2 - 2x)$
$= x^4 - 2x^3 - x^3 + 2x^2 + 3x^2 - 6x$
$= x^4 - 3x^3 + 5x^2 - 6x$

105. $(2x^3 - 6x^2 + x - 3)(x^2 + 4x)$
$= 2x^3(x^2 + 4x) - 6x^2(x^2 + 4x) + x(x^2 + 4x)$
$\quad - 3(x^2 + 4x)$
$= 2x^5 + 8x^4 - 6x^4 - 24x^3 + x^3 + 4x^2 - 3x^2 - 12x$
$= 2x^5 + 2x^4 - 23x^3 + x^2 - 12x$

107. $(b-1)^3 = (b-1)(b-1)^2$
$= (b-1)(b^2 - 2b + 1)$
$= b(b^2 - 2b + 1) - 1(b^2 - 2b + 1)$
$= b^3 - 2b^2 + b - b^2 + 2b - 1$
$= b^3 - 3b^2 + 3b - 1$

109. $(3a-5)^3$
$= (3a-5)(3a-5)^2$
$= (3a-5)(9a^2 - 30a + 25)$
$= 3a(9a^2 - 30a + 25) - 5(9a^2 - 30a + 25)$
$= 27a^3 - 90a^2 + 75a - 45a^2 + 150a - 125$
$= 27a^3 - 135a^2 + 225a - 125$

111. Yes, it will always be a monomial.

113. No, it could have 2 or 4 terms.

115. The missing exponents are 6, 3, and 1 since
$$3x^2(2x^6 - 5x^3 + 3x^1)$$
$$= 3x^2(2x^6) - 3x^2(5x^3) + 3x^2(3x^1)$$
$$= 6x^8 - 15x^5 + 9x^3$$

117. a. $A = (x+2)(2x+1)$
$$= x(2x) + x \cdot 1 + 2 \cdot 2x + 2 \cdot 1$$
$$= 2x^2 + x + 4x + 2$$
$$= 2x^2 + 5x + 2$$

   b. If $x = 4$, $A = 2 \cdot 4^2 + 5 \cdot 4 + 2 = 54$.
   The area is 54 square feet.

   c. For the rectangle to be a square, all sides must have the same length. Thus,
   $x + 2 = 2x + 1$.
   $$x + 2 = 2x + 1$$
   $$2 = x + 1$$
   $$1 = x$$
   The rectangle is a square when $x = 1$ foot.

119. a. Area of base $= (x+5)(3x+4)$
$$= 3x^2 + 4x + 15x + 20$$
$$= 3x^2 + 19x + 20$$

   b. Volume = height $\times$ Area
   $$= (2x-2)(3x^2 + 19x + 20)$$
   $$= 6x^3 + 38x^2 + 40x - 6x^2 - 38x - 40$$
   $$= 6x^3 + 32x^2 + 2x - 40$$

   c. $6x^3 + 32x^2 + 2x - 40$
   $$= 6(4)^3 + 32(4)^2 + 2 \cdot 4 - 40$$
   $$= 6 \cdot 64 + 32 \cdot 16 + 8 - 40$$
   $$= 384 + 512 - 32$$
   $$= 864$$
   The volume is 864 cubic feet.

   d. $(x+5)(3x+4)(2x-2)$
   $$= (4+5)(3 \cdot 4 + 4)(2 \cdot 4 - 2)$$
   $$= 9(12+4)(8-2)$$
   $$= 9 \cdot 16 \cdot 6$$
   $$= 864$$
   The volume is 864 cubic feet.

   e. The volumes are the same.

121. $\left(\dfrac{1}{2}x + \dfrac{2}{3}\right)\left(\dfrac{2}{3}x - \dfrac{2}{5}\right)$
$$= \left(\dfrac{1}{2}x\right)\left(\dfrac{2}{3}x\right) + \left(\dfrac{1}{2}x\right)\left(-\dfrac{2}{5}\right) + \left(\dfrac{2}{3}\right)\left(\dfrac{2}{3}x\right)$$
$$+ \left(\dfrac{2}{3}\right)\left(-\dfrac{2}{5}\right)$$
$$= \dfrac{1}{3}x^2 - \dfrac{1}{5}x + \dfrac{4}{9}x - \dfrac{4}{15}$$
$$= \dfrac{1}{3}x^2 - \dfrac{11}{45}x - \dfrac{4}{15}$$

123. $3(x+7) - 5 = 3x - 17$
$$3x + 21 - 5 = 3x - 17$$
$$3x + 16 = 3x - 17$$
$$16 = -17 \text{ False}$$
Since this is false, there is no solution.

124. Let x = measure of angle C
Then $x - 16$ = measure of angle D.
$$x + (x - 16) = 90$$
$$2x - 16 = 90$$
$$2x = 106$$
$$x = 53$$
Angle C measures $53°$ and angle D measures $53 - 16 = 37°$.

125. $\left(\dfrac{3xy^4}{6y^6}\right)^4 = \left(\dfrac{3}{6} \cdot x \cdot \dfrac{y^4}{y^6}\right)^4$
$$= \left(\dfrac{1}{2} \cdot x \cdot \dfrac{1}{y^2}\right)^4$$
$$= \left(\dfrac{x}{2y^2}\right)^4 = \dfrac{x^4}{2^4 y^{2 \cdot 4}} = \dfrac{x^4}{16y^8}$$

126. a. $-6^3 = -(6^3) = -216$

   b. $6^{-3} = \dfrac{1}{6^3} = \dfrac{1}{216}$

127. $(-x^2 - 6x + 5) - (4x^2 - 4x - 9)$
$$= -x^2 - 6x + 5 - 4x^2 + 4x + 9$$
$$= -x^2 - 4x^2 - 6x + 4x + 5 + 9$$
$$= -5x^2 - 2x + 14$$

Exercise Set 5.6

1. To divide a polynomial by a monomial, divide each term in the polynomial by the monomial.

3. $\dfrac{2x+8}{2} = \dfrac{2x}{2} + \dfrac{8}{2} = x+4$

5. Terms should be listed in descending order.

7. $\dfrac{x^2-7}{x-2} = \dfrac{x^2+0x-7}{x-2}$

9. $(x+5)(x-3)-2 = x^2+2x-15-2$
$$= x^2+2x-17$$

11. $\dfrac{x^2-x-42}{x-7} = x+6$ or $\dfrac{x^2-x-42}{x+6} = x-7$

13. $\dfrac{2x^2+5x+3}{2x+3} = x+1$ or $\dfrac{2x^2+5x+3}{x+1} = 2x+3$

15. $\dfrac{4x^2-9}{2x+3} = 2x-3$ or $\dfrac{4x^2-9}{2x-3} = 2x+3$

17. $\dfrac{3x+6}{3} = \dfrac{3x}{3} + \dfrac{6}{3} = x+2$

19. $\dfrac{4n+10}{2} = \dfrac{4n}{2} + \dfrac{10}{2} = 2n+5$

21. $\dfrac{7x+6}{3} = \dfrac{7x}{3} + \dfrac{6}{3} = \dfrac{7}{3}x+2$

23. $\dfrac{-6x+4}{2} = \dfrac{-6x}{2} + \dfrac{4}{2} = -3x+2$

25. $\dfrac{-9x-3}{-3} = \dfrac{-9x}{-3} + \dfrac{-3}{-3} = 3x+1$

27. $\dfrac{2x+16}{4} = \dfrac{2x}{4} + \dfrac{16}{4} = \dfrac{1}{2}x+4$

29. $\dfrac{4-10w}{-4} = \dfrac{(-1)(4-10w)}{(-1)(-4)}$
$$= \dfrac{-4+10w}{4}$$
$$= -\dfrac{4}{4} + \dfrac{10w}{4}$$
$$= -1 + \dfrac{5}{2}w$$

31. $(4x^2+8x-12) \div 4x^2 = \dfrac{4x^2+8x-12}{4x^2}$
$$= \dfrac{4x^2}{4x^2} + \dfrac{8x}{4x^2} + \dfrac{-12}{4x^2}$$
$$= 1 + \dfrac{2}{x} - \dfrac{3}{x^2}$$

33. $\dfrac{-4x^5+6x+8}{2x^2} = \dfrac{-4x^5}{2x^2} + \dfrac{6x}{2x^2} + \dfrac{8}{2x^2}$
$$= -2x^3 + \dfrac{3}{x} + \dfrac{4}{x^2}$$

35. $(x^5+3x^4-3) \div x^3 = \dfrac{x^5+3x^4-3}{x^3}$
$$= \dfrac{x^5}{x^3} + \dfrac{3x^4}{x^3} + \dfrac{-3}{x^3}$$
$$= x^2 + 3x - \dfrac{3}{x^3}$$

37. $\dfrac{6x^5-4x^4+12x^3-5x^2}{2x^3} = \dfrac{6x^5}{2x^3} - \dfrac{4x^4}{2x^3} + \dfrac{12x^3}{2x^3} - \dfrac{5x^2}{2x^3}$
$$= 3x^2 - 2x + 6 - \dfrac{5}{2x}$$

39. $\dfrac{8k^3+6k^2-8}{-4k} = \dfrac{(-1)(8k^3+6k^2-8)}{(-1)(-4k)}$
$$= \dfrac{-8k^3-6k^2+8}{4k}$$
$$= \dfrac{-8k^3}{4k} - \dfrac{6k^2}{4k} + \dfrac{8}{4k}$$
$$= -2k^2 - \dfrac{3}{2}k + \dfrac{2}{k}$$

41. $\dfrac{12x^5+3x^4-10x^2-9}{-3x^2}$
$$= \dfrac{12x^5}{-3x^2} + \dfrac{3x^4}{-3x^2} - \dfrac{10x^2}{-3x^2} - \dfrac{9}{-3x^2}$$
$$= -4x^3 - x^2 + \dfrac{10}{3} + \dfrac{3}{x^2}$$

43.
$$\begin{array}{r} x+3 \\ x+1\overline{\smash{)}x^2+4x+3} \\ \underline{x^2+x} \\ 3x+3 \\ \underline{3x+3} \\ 0 \end{array}$$

$$\frac{x^2+4x+3}{x+1}=x+3$$

45.
$$\begin{array}{r} 5y+1 \\ y-7\overline{\smash{)}5y^2-34y-7} \\ \underline{5y^2-35y} \\ y-7 \\ \underline{y-7} \\ 0 \end{array}$$

$$\frac{5y^2-34y-7}{y-7}=5y+1$$

47.
$$\begin{array}{r} 2x+4 \\ 3x+2\overline{\smash{)}6x^2+16x+8} \\ \underline{6x^2+4x} \\ 12x+8 \\ \underline{12x+8} \\ 0 \end{array}$$

$$\frac{6x^2+16x+8}{3x+2}=2x+4$$

49.
$$\frac{x^2-16}{-4+x}=\frac{x^2+0x-16}{x-4}$$

$$\begin{array}{r} x+4 \\ x-4\overline{\smash{)}x^2+0x-16} \\ \underline{x^2-4x} \\ 4x-16 \\ \underline{4x-16} \\ 0 \end{array}$$

$$\frac{x^2-16}{-4+x}=x+4$$

51.
$$\begin{array}{r} x+5 \\ 2x-3\overline{\smash{)}2x^2+7x-18} \\ \underline{2x^2-3x} \\ 10x-18 \\ \underline{10x-15} \\ -3 \end{array}$$

$$(2x^2+7x-18)\div(2x-3)=x+5-\frac{3}{2x-3}$$

53.
$$\frac{x^2-36}{x-6}=\frac{x^2+0x-36}{x-6}$$

$$\begin{array}{r} x+6 \\ x-6\overline{\smash{)}x^2+0x-36} \\ \underline{x^2-6x} \\ 6x-36 \\ \underline{6x-36} \\ 0 \end{array}$$

$$\frac{x^2-36}{x-6}=x+6$$

55.
$$\frac{-x+9x^3-16}{3x-4}=\frac{9x^3-x-16}{3x-4}$$

$$\begin{array}{r} 3x^2+4x+5 \\ 3x-4\overline{\smash{)}9x^3+0x^2-x-16} \\ \underline{9x^3-12x^2} \\ 12x^2-x-16 \\ \underline{12x^2-16x} \\ 15x-16 \\ \underline{15x-20} \\ 4 \end{array}$$

$$\frac{-x+9x^3-16}{3x-4}=3x^2+4x+5+\frac{4}{3x-4}$$

57.
$$\frac{6x+8x^2-12}{2x+3}=\frac{8x^2+6x-12}{2x+3}$$

$$\begin{array}{r} 4x-3 \\ 2x+3\overline{\smash{)}8x^2+6x-12} \\ \underline{8x^2+12x} \\ -6x-12 \\ \underline{-6x-9} \\ -3 \end{array}$$

$$\frac{6x+8x^2-12}{2x+3}=4x-3-\frac{3}{2x+3}$$

59.
$$x+4 \overline{) \begin{array}{r} 7x^2 \qquad -5 \\ 7x^3 + 28x^2 - 5x - 20 \end{array}}$$
$$\underline{7x^3 + 28x^2}$$
$$-5x - 20$$
$$\underline{-5x - 20}$$
$$0$$

$$\frac{7x^3 + 28x^2 - 5x - 20}{x+4} = 7x^2 - 5$$

61. $\dfrac{2x^3 - 4x^2 + 12}{x-2} = \dfrac{2x^3 - 4x^2 + 0x + 12}{x-2}$

$$x-2 \overline{) \begin{array}{r} 2x^2 \\ 2x^3 - 4x^2 + 0x + 12 \end{array}}$$
$$\underline{2x^3 - 4x^2}$$
$$0x + 12$$

$$\frac{2x^3 - 4x^2 + 12}{x-2} = 2x^2 + \frac{12}{x-2}$$

63. $\left(w^3 - 8\right) \div \left(w - 3\right) = \left(w^3 + 0w^2 + 0w - 8\right) \div \left(w - 3\right)$

$$w-3 \overline{) \begin{array}{r} w^2 + 3w + 9 \\ w^3 + 0w^2 + 0w - 8 \end{array}}$$
$$\underline{w^3 - 3w^2}$$
$$3w^2 + 0w$$
$$\underline{3w^2 - 9w}$$
$$9w - 8$$
$$\underline{9w - 27}$$
$$19$$

$$\left(w^3 - 8\right) \div \left(w - 3\right) = w^2 + 3w + 9 + \frac{19}{w-3}$$

65. $\dfrac{x^3 - 27}{x-3} = \dfrac{x^3 + 0x^2 + 0x - 27}{x-3}$

$$x-3 \overline{) \begin{array}{r} x^2 + 3x + 9 \\ x^3 + 0x^2 + 0x - 27 \end{array}}$$
$$\underline{x^3 - 3x^2}$$
$$3x^2 + 0x$$
$$\underline{3x^2 - 9x}$$
$$9x - 27$$
$$\underline{9x - 27}$$
$$0$$

$$\frac{x^3 - 27}{x-3} = x^2 + 3x + 9$$

67. $\dfrac{4x^3 - 5x}{2x-1} = \dfrac{4x^3 + 0x^2 - 5x + 0}{2x-1}$

$$2x-1 \overline{) \begin{array}{r} 2x^2 + x - 2 \\ 4x^3 + 0x^2 - 5x + 0 \end{array}}$$
$$\underline{4x^3 - 2x^2}$$
$$2x^2 - 5x$$
$$\underline{2x^2 - x}$$
$$-4x + 0$$
$$\underline{4x + 2}$$
$$-2$$

$$\frac{4x^3 - 5x}{2x-1} = 2x^2 + x - 2 - \frac{2}{2x-1}$$

69.
$$m-1 \overline{) \begin{array}{r} -m^2 - 7m - 5 \\ -m^3 - 6m^2 + 2m - 3 \end{array}}$$
$$\underline{-m^3 + m^2}$$
$$-7m^2 + 2m$$
$$\underline{-7m^2 + 7m}$$
$$-5m - 3$$
$$\underline{-5m + 5}$$
$$-8$$

$$\frac{-m^3 - 6m^2 + 2m - 3}{m-1} = -m^2 - 7m - 5 - \frac{8}{m-1}$$

71. $\dfrac{4t^3 - t + 4}{t+2} = \dfrac{4t^3 + 0t^2 - t + 4}{t+2}$

$$t+2 \overline{) \begin{array}{r} 4t^2 - 8t + 15 \\ 4t^3 + 0t^2 - t + 4 \end{array}}$$
$$\underline{4t^3 + 8t^2}$$
$$-8t^2 - t$$
$$\underline{-8t^2 - 16t}$$
$$15t + 4$$
$$\underline{15t + 30}$$
$$-26$$

$$\frac{4t^3 - t + 4}{t+2} = 4t^2 - 8t + 15 - \frac{26}{t+2}$$

73. No, $\dfrac{2x+1}{x^2} = \dfrac{2x}{x^2} + \dfrac{1}{x^2} = \dfrac{2}{x} + \dfrac{1}{x^2}$

75. $(x+4)(2x+3) + 4 = 2x^2 + 3x + 8x + 12 + 4$
$$= 2x^2 + 11x + 16$$

77. Third Degree since $\dfrac{x^4}{x} = x^3$

79. It has to be $4x$ since that is what must be multiplied with $4x^3$ in the quotient to get $16x^4$ in the dividend.

81. When dividing by $2x^2$, each exponent will decrease by two. So, the shaded areas must be 5, 3, 2, 1, respectively.

83. $\dfrac{3x^3 - 5}{3x - 2} = \dfrac{3x^3 + 0x^2 + 0x - 5}{3x - 2}$

$$
\begin{array}{r}
x^2 + \dfrac{2}{3}x + \dfrac{4}{9} \\
3x - 2 \overline{\smash{)}\,3x^3 + 0x^2 + 0x - 5} \\
\underline{3x^3 - 2x^2} \\
2x^2 + 0x \\
\underline{2x^2 - \dfrac{4}{3}x} \\
\dfrac{4}{3}x - 5 \\
\underline{\dfrac{4}{3}x - \dfrac{8}{9}} \\
-\dfrac{37}{9}
\end{array}
$$

$\dfrac{3x^3 - 5}{3x - 2} = x^2 + \dfrac{2}{3}x + \dfrac{4}{9} - \dfrac{37}{9(3x-2)}$

85.

$$
\begin{array}{r}
-3x + 3 \\
-x - 3 \overline{\smash{)}\,3x^2 + 6x - 10} \\
\underline{3x^2 + 9x} \\
-3x - 10 \\
\underline{-3x - 9} \\
-1
\end{array}
$$

$\dfrac{3x^2 + 6x - 10}{-x - 3} = -3x + 3 + \dfrac{1}{x+3}$

88. a. 2 is a natural number.

 b. 2 and 0 are whole numbers.

 c. $2, -5, 0, \dfrac{2}{5}, -6.3$, and $-\dfrac{23}{34}$ are rational numbers.

 d. $\sqrt{7}$ and $\sqrt{3}$ are irrational numbers.

 e. $2, -5, 0, \sqrt{7}, \dfrac{2}{5}, -6.3, \sqrt{3}$, and $-\dfrac{23}{34}$ are all real numbers.

89. a. $\dfrac{0}{1} = 0$

 b. $\dfrac{1}{0}$ is undefined

90. Evaluate expressions in parentheses first, then exponents, followed by multiplications and divisions from left to right, and finally additions and subtractions from left to right.

91. $2(x+3) + 2x = x + 4$

$2x + 6 + 2x = x + 4$

$4x + 6 = x + 4$

$4x = x - 2$

$3x = -2$

$x = -\dfrac{2}{3}$

92. Let p = the purchase price of the sweater, then $0.30p$ = the amount saved on the sweater.

$p - 0.30p = 27.65$

$0.7p = 27.65$

$p = \dfrac{27.65}{0.7}$

$p = 39.5$

The regular price of the sweater is $39.50.

93. $\dfrac{x^9}{x^{-4}} = x^{9-(-4)} = x^{9+4} = x^{13}$

Chapter 5 Review Exercises

1. $x^5 \cdot x^2 = x^{5+2} = x^7$

2. $x^2 \cdot x^4 = x^{2+4} = x^6$

3. $3^2 \cdot 3^3 = 3^{2+3} = 3^5 = 243$

4. $2^4 \cdot 2 = 2^{4+1} = 2^5 = 32$

5. $\dfrac{x^4}{x} = x^{4-1} = x^3$

6. $\dfrac{a^5}{a^5} = a^{5-5} = a^0 = 1$

7. $\dfrac{5^5}{5^3} = 5^{5-3} = 5^2 = 25$

8. $\dfrac{4^4}{4} = 4^{4-1} = 4^3 = 64$

9. $\dfrac{x^6}{x^8} = \dfrac{1}{x^{8-6}} = \dfrac{1}{x^2}$

10. $\dfrac{y^4}{y} = y^{4-1} = y^3$

11. $x^0 = 1$

12. $7y^0 = 7 \cdot 1 = 7$

13. $\left(-6z\right)^0 = 1$

14. $6^0 = 1$

15. $\left(5x\right)^2 = 5^2 x^2 = 25x^2$

16. $\left(3a\right)^3 = 3^3 a^3 = 27a^3$

17. $\left(-3x\right)^3 = \left(-3\right)^3 x^3 = -27x^3$

18. $\left(6s\right)^3 = 6^3 s^3 = 216s^3$

19. $\left(2x^2\right)^4 = 2^4 x^{2 \cdot 4} = 16x^8$

20. $\left(-x^4\right)^6 = \left(-1\right)^6 x^{4 \cdot 6} = x^{24}$

21. $\left(-p^8\right)^4 = \left(-1\right)^4 p^{8(4)} = p^{32}$

22. $\left(-\dfrac{2x^3}{y}\right)^2 = \dfrac{\left(-2\right)^2 x^{3(2)}}{y^2} = \dfrac{4x^6}{y^2}$

23. $\left(\dfrac{5y^2}{2b}\right)^2 = \dfrac{\left(-5\right)^2 y^{2(2)}}{2^2 b^2} = \dfrac{25y^4}{4b^2}$

24. $6x^2 \cdot 4x^3 = 6 \cdot 4 x^{2+3} = 24x^5$

25. $\dfrac{16x^2 y}{4xy^2} = \dfrac{16}{4} \cdot \dfrac{x^2}{x} \cdot \dfrac{y}{y^2} = 4x \cdot \dfrac{1}{y} = \dfrac{4x}{y}$

26.
$$2x(3xy^3)^3 = 2x(3^3 x^3 y^{3 \cdot 3})$$
$$= 2x(27x^3 y^9)$$
$$= 2 \cdot 27 x^{1+3} y^9$$
$$= 54x^4 y^9$$

27.
$$\left(\dfrac{9x^2 y}{3xy}\right)^2 = \left(\dfrac{9}{3} \cdot \dfrac{x^2}{x} \cdot \dfrac{y}{y}\right)^2$$
$$= \left(3x\right)^2$$
$$= 3^2 x^2$$
$$= 9x^2$$

28.
$$(2x^2 y)^3 (3xy^4) = (2^3 x^{2 \cdot 3} y^3)(3xy^4)$$
$$= (8x^6 y^3)(3xy^4)$$
$$= 8 \cdot 3 x^{6+1} y^{3+4}$$
$$= 24x^7 y^7$$

29.
$$4x^2 y^3 \left(2x^3 y^4\right)^2 = 4x^2 y^3 \left(2^2 x^{3(2)} y^{4(2)}\right)$$
$$= 4x^2 y^3 \left(4x^6 y^8\right)$$
$$= 4 \cdot 4 x^{2+6} y^{3+8}$$
$$= 16x^8 y^{11}$$

30. $3c^2 (2c^4 d^3) = 3 \cdot 2 \cdot c^{2+4} \cdot d^3 = 6c^6 d^3$

31.
$$\left(\dfrac{9a^3 b^2}{3ab^7}\right)^3 = \left(\dfrac{9}{3} \cdot \dfrac{a^3}{a} \cdot \dfrac{b^2}{b^7}\right)^3$$
$$= \left(3a^{3-1} \cdot \dfrac{1}{b^5}\right)^3$$
$$= \left(\dfrac{3a^2}{b^5}\right)^3$$
$$= \dfrac{3^3 a^{2 \cdot 3}}{b^{5 \cdot 3}}$$
$$= \dfrac{27a^6}{b^{15}}$$

32.
$$\left(\dfrac{21x^4 y^3}{7y^2}\right)^3 = \left(\dfrac{21}{7} \cdot x^4 \cdot \dfrac{y^3}{y^2}\right)^3$$
$$= \left(3x^4 y\right)^3$$
$$= 3^3 x^{4 \cdot 3} y^3$$
$$= 27x^{12} y^3$$

33. $b^{-9} = \dfrac{1}{b^9}$

34. $3^{-3} = \dfrac{1}{3^3} = \dfrac{1}{27}$

35. $5^{-2} = \dfrac{1}{5^2} = \dfrac{1}{25}$

36. $\dfrac{1}{z^{-2}} = z^2$

37. $\dfrac{1}{x^{-7}} = x^7$

38. $\dfrac{1}{4^{-2}} = 4^2 = 16$

39. $y^5 \cdot y^{-8} = y^{5-8} = y^{-3} = \dfrac{1}{y^3}$

40. $x^{-2} \cdot x^{-3} = x^{-2-3} = x^{-5} = \dfrac{1}{x^5}$

41. $p^{-6} \cdot p^4 = p^{-6+4} = p^{-2} = \dfrac{1}{p^2}$

42. $a^{-2} \cdot a^{-3} = a^{-2+(-3)} = a^{-5} = \dfrac{1}{a^5}$

43. $\dfrac{m^5}{m^{-5}} = m^{5-(-5)} = m^{10}$

44. $\dfrac{x^5}{x^{-2}} = x^{5-(-2)} = x^7$

45. $\dfrac{x^{-3}}{x^3} = \dfrac{1}{x^{3+3}} = \dfrac{1}{x^6}$

46. $\left(3x^4\right)^{-2} = 3^{-2} x^{4(-2)} = 3^{-2} x^{-8} = \dfrac{1}{3^2 x^8} = \dfrac{1}{9x^8}$

47. $\left(4x^{-3}y\right)^{-3} = 4^{-3} x^{(-3)(-3)} y^{-3}$

$\qquad = 4^{-3} x^9 y^{-3}$

$\qquad = \dfrac{x^9}{4^3 y^3}$

$\qquad = \dfrac{x^9}{64 y^3}$

48. $\left(-2m^{-3}n\right)^2 = (-2)^2 m^{-3(2)} n^{1 \cdot 2} = 4m^{-6}n^2 = \dfrac{4n^2}{m^6}$

49. $6y^{-2} \cdot 2y^4 = 6 \cdot 2 y^{-2+4} = 12y^2$

50. $\left(-5y^{-3}z\right)^3 = (-5)^3 y^{(-3)(3)} z^3$

$\qquad = -125 y^{-9} z^3$

$\qquad = -\dfrac{125z^3}{y^9}$

51. $\left(-4x^{-2}y^3\right)^{-2} = (-4)^{-2} x^{(-2)(-2)} y^{3(-2)}$

$\qquad = (-4)^{-2} x^4 y^{-6}$

$\qquad = \dfrac{x^4}{(-4)^2 y^6}$

$\qquad = \dfrac{x^4}{16y^6}$

52. $2x\left(3x^{-2}\right) = 2 \cdot 3 x^{1-2} = 6x^{-1} = \dfrac{6}{x}$

53. $\left(5x^{-2}y\right)\left(2x^4 y\right) = 5 \cdot 2 x^{-2+4} y^{1+1} = 10x^2 y^2$

54. $4y^{-2}\left(3x^2 y\right) = 4 \cdot 3 x^2 y^{-2+1} = 12x^2 y^{-1} = \dfrac{12x^2}{y}$

55. $4x^5\left(6x^{-7}y^2\right) = 4 \cdot 6 x^{5-7} y^2 = 24x^{-2} y^2 = \dfrac{24y^2}{x^2}$

56. $\dfrac{6xy^4}{2xy^{-1}} = \dfrac{6}{2} \cdot \dfrac{x}{x} \cdot \dfrac{y^4}{y^{-1}} = 3y^5$

57. $\dfrac{12x^{-2}y^3}{3xy^2} = \dfrac{12}{3} \cdot \dfrac{x^{-2}}{x} \cdot \dfrac{y^3}{y^2} = 4 \cdot \dfrac{1}{x^3} \cdot y = \dfrac{4y}{x^3}$

58. $\dfrac{49x^2 y^{-3}}{7x^{-3}y} = \dfrac{49}{7} \cdot \dfrac{x^2}{x^{-3}} \cdot \dfrac{y^{-3}}{y}$

$\qquad = \left(\dfrac{49}{7}\right) x^{2-(-3)} y^{-3-1}$

$\qquad = 7x^5 y^{-4}$

$\qquad = \dfrac{7x^5}{y^4}$

59. $\dfrac{4x^8 y^{-2}}{8x^7 y^3} = \dfrac{4}{8} \cdot \dfrac{x^8}{x^7} \cdot \dfrac{y^{-2}}{y^3} = \dfrac{1}{2} \cdot x \cdot \dfrac{1}{y^5} = \dfrac{x}{2y^5}$

60. $\dfrac{36x^4 y^7}{9x^5 y^{-3}} = \dfrac{36}{9} \cdot \dfrac{x^4}{x^5} \cdot \dfrac{y^7}{y^{-3}} = 4 \cdot \dfrac{1}{x} \cdot y^{10} = \dfrac{4y^{10}}{x}$

61. $1,720,000 = 1.72 \times 10^6$

62. $0.153 = 1.53 \times 10^{-1}$

63. $0.00763 = 7.63 \times 10^{-3}$

64. $47,000 = 4.7 \times 10^4$

65. $5,760 = 5.76 \times 10^3$

66. $0.000314 = 3.14 \times 10^{-4}$

67. $7.5 \times 10^{-3} = 0.0075$

68. $6.52 \times 10^{-4} = 0.000652$

69. $8.9 \times 10^6 = 8,900,000$

70. $5.12 \times 10^4 = 51,200$

71. $3.14 \times 10^{-5} = 0.0000314$

72. $1.103 \times 10^7 = 11,030,000$

73. 92 milliliters $= 92 \times 10^{-3} = 0.092$ liters

74. 6 gigameters $= 6 \times 10^9 = 6,000,000,000$ meters

75. 12.8 micrograms $= 12.8 \times 10^{-6}$
$= 0.0000128$ grams

76. 19.2 kilograms $= 19.2 \times 10^3 = 19,200$ grams

77. $\left(2.5 \times 10^2\right)\left(3.4 \times 10^{-4}\right) = \left(2.5 \times 3.4\right)\left(10^2 \times 10^{-4}\right)$
$= 8.5 \times 10^{-2}$
$= 0.085$

78. $\left(4.2 \times 10^{-3}\right)\left(3.0 \times 10^5\right) = \left(4.2 \times 3\right)\left(10^{-3} \times 10^5\right)$
$= 12.6 \times 10^2$
$= 1260$

79. $\left(3.5 \times 10^{-2}\right)\left(7.0 \times 10^3\right) = \left(3.5 \times 7.0\right)\left(10^{-2} \times 10^3\right)$
$= 24.5 \times 10^1$
$= 245$

80. $\dfrac{7.94 \times 10^6}{2.0 \times 10^{-2}} = \left(\dfrac{7.94}{2.0}\right)\left(\dfrac{10^6}{10^{-2}}\right)$
$= 3.97 \times 10^8$
$= 397,000,000$

81. $\dfrac{1.5 \times 10^{-2}}{5.0 \times 10^2} = \left(\dfrac{1.5}{5}\right)\left(\dfrac{10^{-2}}{10^2}\right) = 0.3 \times 10^{-4} = 0.00003$

82. $\dfrac{6.5 \times 10^4}{2.0 \times 10^6} = \left(\dfrac{6.5}{2.0}\right)\left(\dfrac{10^4}{10^6}\right) = 3.25 \times 10^{-2} = 0.0325$

83. $(14,000)(260,000) = \left(1.4 \times 10^4\right)\left(2.6 \times 10^5\right)$
$= (1.4 \times 2.6)\left(10^4 \times 10^5\right)$
$= 3.64 \times 10^9$

84. $(0.00053)(40,000) = \left(5.3 \times 10^{-4}\right)\left(4 \times 10^4\right)$
$= (5.3 \times 4)\left(10^{-4} \times 10^4\right)$
$= 21.2 \times 10^0$
$= 2.12 \times 10^1$

85. $(12,500)(400,000) = \left(1.25 \times 10^4\right)\left(4.0 \times 10^5\right)$
$= (1.25 \times 4.0)\left(10^4 \times 10^5\right)$
$= 5.0 \times 10^9$

86. $\dfrac{250}{500,000} = \dfrac{2.5 \times 10^2}{5.0 \times 10^5}$
$= \left(\dfrac{2.5}{5.0}\right)\left(\dfrac{10^2}{10^5}\right)$
$= 0.5 \times 10^{-3}$
$= 5.0 \times 10^{-4}$

87. $\dfrac{0.000068}{0.02} = \dfrac{6.8 \times 10^{-5}}{2 \times 10^{-2}} = \left(\dfrac{6.8}{2}\right)\left(\dfrac{10^{-5}}{10^{-2}}\right) = 3.4 \times 10^{-3}$

88. $\dfrac{850,000}{0.025} = \dfrac{8.5 \times 10^5}{2.5 \times 10^{-2}} = \left(\dfrac{8.50}{2.50}\right)\left(\dfrac{10^5}{10^{-2}}\right) = 3.4 \times 10^7$

89. $\dfrac{6.4 \times 10^6}{1.28 \times 10^2} = \left(\dfrac{6.4}{1.28}\right)\left(\dfrac{10^6}{10^2}\right) = 5 \times 10^4 = 50,000$

The milk tank holds 50,000 gallons.

90. a. $1,500,000,000,000

b. $0.075\left(1.5 \times 10^{12}\right) = \left(7.5 \times 10^{-2}\right)\left(1.5 \times 10^{12}\right)$
$= \left(7.5 \times 1.5\right)\left(10^{-2} \times 10^{12}\right)$
$= 11.25 \times 10^{10}$
$= \$1.125 \times 10^{11}$

91. Not a polynomial

92. monomial, zero degree

93. $x^2 + 3x - 4$, trinomial, second degree

94. $4x^2 - x - 3$, trinomial, second degree

95. Not a polynomial

96. $13x^3 - 4$, binomial, third degree

97. $-4x^2 + x$, binomial, second degree

98. Not a polynomial

99. $2x^3 + 4x^2 - 3x - 7$, polynomial, third degree

100. $(x + 8) + (4x - 11) = x + 8 + 4x - 11$
$= x + 4x + 8 - 11$
$= 5x - 3$

101. $(2d - 3) + (5d + 7) = 2d - 3 + 5d + 7$
$= 2d + 5d - 3 + 7$
$= 7d + 4$

102. $(-x - 10) + (-2x + 5) = -x - 10 - 2x + 5$
$= -x - 2x - 10 + 5$
$= -3x - 5$

103. $\left(-3x^2 + 9x + 5\right) + \left(-x^2 + 2x - 12\right)$
$= -3x^2 + 9x + 5 - x^2 + 2x - 12$
$= -3x^2 - x^2 + 9x + 2x + 5 - 12$
$= -4x^2 + 11x - 7$

104. $\left(-m^2 + 5m - 8\right) + \left(6m^2 - 5m - 2\right)$
$= -m^2 + 6m^2 + 5m - 5m - 8 - 2$
$= 5m^2 - 10$

105. $(6.2p - 4.3) + (1.9p + 7.1) = 6.2p + 1.9p - 4.3 + 7.1$
$= 8.1p + 2.8$

106. $(-6y - 7) - (-3y + 8) = -6y - 7 + 3y - 8$
$= -6y + 3y - 7 - 8$
$= -3y - 15$

107. $\left(4x^2 - 9x\right) - (3x + 15) = 4x^2 - 9x - 3x - 15$
$= 4x^2 - 12x - 15$

108. $\left(5a^2 - 6a - 9\right) - \left(2a^2 - a + 12\right)$
$= 5a^2 - 6a - 9 - 2a^2 + a - 12$
$= 5a^2 - 2a^2 - 6a + a - 9 - 12$
$= 3a^2 - 5a - 21$

109. $\left(x^2 + 7x - 3\right) - \left(x^2 + 3x - 5\right)$
$= x^2 + 7x - 3 - x^2 - 3x + 5$
$= x^2 - x^2 + 7x - 3x - 3 + 5$
$= 4x + 2$

110. $\left(-2x^2 + 8x - 7\right) - \left(3x^2 + 12\right)$
$= -2x^2 + 8x - 7 - 3x^2 - 12$
$= -2x^2 - 3x^2 + 8x - 7 - 12$
$= -5x^2 + 8x - 19$

111. $\dfrac{1}{7}x(21x + 21) = \dfrac{1}{7}x(21x) + \dfrac{1}{7}x(21)$
$= \dfrac{21}{7}x^2 + \dfrac{21}{7}x$
$= 3x^2 + 3x$

112. $-3x(5x + 4) = -3x \cdot 5x + (-3x)4$
$= -15x^2 - 12x$

113. $3x\left(2x^2 - 4x + 7\right) = 3x\left(2x^2\right) + 3x(-4x) + 3x(7)$
$= 6x^3 - 12x^2 + 21x$

114. $-c\left(2c^2-3c+5\right)$
$= (-c)\left(2c^2\right)+(-c)(-3c)+(-c)(5)$
$= -2c^3+3c^2-5c$

115. $-7b\left(-4b^2-3b-5\right)$
$= (-7b)(-4b^2)+(-7b)(-3b)+(-7b)(-5)$
$= 28b^3+21b^2+35b$

116. $(x+4)(x+5)=x\cdot x+x\cdot5+4\cdot x+4\cdot5$
$= x^2+5x+4x+20$
$= x^2+9x+20$

117. $(3x+6)(-4x+1)$
$= 3x(-4x)+3x(1)+6(-4x)+6(1)$
$= -12x^2+3x-24x+6$
$= -12x^2-21x+6$

118. $(-5x+3)^2=(-5x)^2+2(-5x)(3)+(3)^2$
$= 25x^2-30x+9$

119. $(6-2x)(2+3x)$
$= 6\cdot2+6\cdot3x+(-2x)(2)+(-2x)(3x)$
$= 12+18x-4x-6x^2$
$= 12+14x-6x^2$
$= -6x^2+14x+12$

120. $(r+5)(r-5)=(r)^2-(5)^2=r^2-25$

121. $(x-1)\left(3x^2+4x-6\right)$
$= x\left(3x^2+4x-6\right)-1\left(3x^2+4x-6\right)$
$= 3x^3+4x^2-6x-3x^2-4x+6$
$= 3x^3+x^2-10x+6$

122. $(3x+1)\left(x^2+2x+4\right)$
$= 3x\left(x^2+2x+4\right)+1\left(x^2+2x+4\right)$
$= 3x^3+6x^2+12x+x^2+2x+4$
$= 3x^3+7x^2+14x+4$

123. $(-4x+2)\left(3x^2-x+7\right)$
$= -4x\left(3x^2-x+7\right)+2\left(3x^2-x+7\right)$
$= -12x^3+4x^2-28x+6x^2-2x+14$
$= -12x^3+10x^2-30x+14$

124. $\dfrac{2x+4}{2}=\dfrac{2x}{2}+\dfrac{4}{2}=x+2$

125. $\dfrac{12y+18}{3}=\dfrac{12y}{3}+\dfrac{18}{3}=4y+6$

126. $\dfrac{8x^2+4x}{x}=\dfrac{8x^2}{x}+\dfrac{4x}{x}=8x+4$

127. $\dfrac{6x^2+9x-4}{3}=\dfrac{6x^2}{3}+\dfrac{9x}{3}-\dfrac{4}{3}=2x^2+3x-\dfrac{4}{3}$

128. $\dfrac{6w^2-5w+3}{3w}=\dfrac{6w^2}{3w}-\dfrac{5w}{3w}+\dfrac{3}{3w}=2w-\dfrac{5}{3}+\dfrac{1}{w}$

129. $\dfrac{16x^6-8x^5-3x^3+1}{4x}=\dfrac{16x^6}{4x}-\dfrac{8x^5}{4x}-\dfrac{3x^3}{4x}+\dfrac{1}{4x}$
$= 4x^5-2x^4-\dfrac{3}{4}x^2+\dfrac{1}{4x}$

130. $\dfrac{8m-4}{-2}=\dfrac{8m+(-4)}{-2}=\dfrac{8m}{-2}+\dfrac{-4}{-2}=-4m+2$

131. $\dfrac{5x^3+10x+2}{2x^2}=\dfrac{5x^3}{2x^2}+\dfrac{10x}{2x^2}+\dfrac{2}{2x^2}=\dfrac{5}{2}x+\dfrac{5}{x}+\dfrac{1}{x^2}$

132. $\dfrac{5x^2-6x+15}{3x}=\dfrac{5x^2}{3x}-\dfrac{6x}{3x}+\dfrac{15}{3x}=\dfrac{5}{3}x-2+\dfrac{5}{x}$

133.
$$\begin{array}{r}x+4\\x-3\overline{\smash{\big)}\,x^2+\ x-12}\\\underline{x^2-3x}\\4x-12\\\underline{4x-12}\\0\end{array}$$
$\dfrac{x^2+x-12}{x-3}=x+4$

134.
$$\begin{array}{r}5x-\ 2\\x+6\overline{\smash{\big)}\,5x^2+28x-10}\\\underline{5x^2+30x}\\-2x-10\\\underline{-2x-12}\\2\end{array}$$
$\dfrac{5x^2+28x-10}{x+6}=5x-2+\dfrac{2}{x+6}$

135.

$$\begin{array}{r} n+3 \\ 6n+1\overline{\smash{\big)}\ 6n^2+19n+3} \\ \underline{6n^2+n} \\ 18n+3 \\ \underline{18n+3} \\ 0 \end{array}$$

$$\frac{6n^2+19n+3}{6n+1}=n+3$$

136.

$$\begin{array}{r} 2x^2+3x-4 \\ 2x+3\overline{\smash{\big)}\ 4x^3+12x^2+x-12} \\ \underline{4x^3+6x^2} \\ 6x^2+x \\ \underline{6x^2+9x} \\ -8x-12 \\ \underline{-8x-12} \\ 0 \end{array}$$

$$\frac{4x^3+12x^2+x-12}{2x+3}=2x^2+3x-4$$

137.

$$\begin{array}{r} 2x-3 \\ 2x-3\overline{\smash{\big)}\ 4x^2-12x+9} \\ \underline{4x^2-6x} \\ -6x+9 \\ \underline{-6x+9} \\ 0 \end{array}$$

$$\frac{4x^2-12x+9}{2x-3}=2x-3$$

Chapter 5 Practice Test

1. $5x^4\cdot3x^2=5\cdot3x^{4+2}=15x^6$

2. $\left(3xy^2\right)^3=3^3x^3y^{2(3)}=27x^3y^6$

3. $\dfrac{24p^7}{3p^2}=\dfrac{24}{3}p^{7-2}=8p^5$

4. $\left(\dfrac{3x^2y}{6xy^3}\right)^3=\left(\dfrac{3}{6}\cdot\dfrac{x^2}{x}\cdot\dfrac{y}{y^3}\right)^3$

$$=\left(\frac{1}{2}\cdot x\cdot\frac{1}{y^2}\right)^3$$

$$=\left(\frac{x}{2y^2}\right)^3=\frac{x^3}{2^3y^{2\cdot3}}=\frac{x^3}{8y^6}$$

5. $\left(2x^3y^{-2}\right)^{-2}=2^{-2}x^{3(-2)}y^{(-2)(-2)}$

$$=2^{-2}x^{-6}y^4=\frac{y^4}{2^2x^6}=\frac{y^4}{4x^6}$$

6. $\left(4x^0\right)\left(3x^2\right)^0=(4\cdot1)\cdot1=4$

7. $\dfrac{30x^6y^2}{45x^{-1}y}=\dfrac{30}{45}x^{6-(-1)}y^{2-1}=\dfrac{2}{3}x^7y=\dfrac{2x^7y}{3}$

8. $(285,000)(50,000)=\left(2.85\times10^5\right)\left(5.0\times10^4\right)$

$$=(2.85\times5.0)\left(10^5\times10^4\right)$$

$$=14.25\times10^9$$

$$=1.425\times10^{10}$$

9. $\dfrac{0.0008}{4000}=\dfrac{8.0\times10^{-4}}{4.0\times10^3}$

$$=\left(\frac{8.0}{4.0}\right)\left(\frac{10^{-4}}{10^3}\right)$$

$$=2.0\times10^{-7}$$

10. $4x$ is a monomial

11. $-8c+5$, binomial

12. $x^{-2}+4$, not a polynomial

13. $-5+6x^3-2x^2+5x=6x^3-2x^2+5x-5$, third degree

14. $\left(6x-4\right)+\left(2x^2-5x-3\right)=6x-4+2x^2-5x-3$

$$=2x^2+6x-5x-4-3$$

$$=2x^2+x-7$$

15. $\left(y^2-7y+3\right)-\left(4y^2-5y-2\right)$

$$=y^2-7y+3-4y^2+5y+2$$

$$=y^2-4y^2-7y+5y+3+2$$

$$=-3y^2-2y+5$$

16. $\left(4x^2-5\right)-\left(x^2+x-8\right)=4x^2-5-x^2-x+8$

$$=4x^2-x^2-x-5+8$$

$$=3x^2-x+3$$

17. $-5d(-3d+8) = -5d(-3d) - 5d(8)$
$$= 15d^2 - 40d$$

18. $(5x+8)(3x-4)$
$$= (5x)(3x) + (5x)(-4) + 8(3x) + 8(-4)$$
$$= 15x^2 - 20x + 24x - 32$$
$$= 15x^2 + 4x - 32$$

19. $(9-4c)(5+3c)$
$$= 9 \cdot 5 + 9(3c) + (-4c) \cdot 5 + (-4c)(3c)$$
$$= 45 + 27c - 20c - 12c^2$$
$$= -12c^2 + 7c + 45$$

20. $(3x-5)(2x^2+4x-5)$
$$= 3x(2x^2+4x-5) - 5(2x^2+4x-5)$$
$$= 3x \cdot 2x^2 + 3x \cdot 4x - 3x \cdot 5 - 5 \cdot 2x^2 - 5 \cdot 4x - 5(-5)$$
$$= 6x^3 + 12x^2 - 15x - 10x^2 - 20x + 25$$
$$= 6x^3 + 2x^2 - 35x + 25$$

21. $\dfrac{16x^2 + 8x - 4}{4} = \dfrac{16x^2}{4} + \dfrac{8x}{4} - \dfrac{4}{4}$
$$= 4x^2 + 2x - 1$$

22. $\dfrac{-12x^2 - 6x + 5}{-3x} = \dfrac{(-1)(-12x^2 - 6x + 5)}{(-1)(-3x)}$
$$= \dfrac{12x^2 + 6x - 5}{3x}$$
$$= \dfrac{12x^2}{3x} + \dfrac{6x}{3x} - \dfrac{5}{3x}$$
$$= 4x + 2 - \dfrac{5}{3x}$$

23.
$$
\begin{array}{r}
4x+5 \\
2x-3\overline{\smash{)}\,8x^2 - 2x - 15} \\
\underline{8x^2 - 12x} \\
10x - 15 \\
\underline{10x - 15} \\
0
\end{array}
$$

$$\dfrac{8x^2 - 2x - 15}{2x - 3} = 4x + 5$$

24.
$$
\begin{array}{r}
3x-2 \\
4x+5\overline{\smash{)}\,12x^2 + 7x - 12} \\
\underline{12x^2 + 15x} \\
-8x - 12 \\
\underline{-8x - 10} \\
-2
\end{array}
$$

$$\dfrac{12x^2 + 7x - 12}{4x + 5} = 3x - 2 - \dfrac{2}{4x + 5}$$

25. a. $5730 = 5.73 \times 10^3$

b. $\dfrac{4.46 \times 10^9}{5.73 \times 10^3} = \left(\dfrac{4.46}{5.73}\right)\left(\dfrac{10^9}{10^3}\right)$
$$\approx 0.778 \times 10^6$$
$$\approx 7.78 \times 10^5$$

Chapter 5 Cumulative Review Test

1. $12 + 8 \div 2^2 + 3 = 12 + 8 \div 4 + 3$
$$= 12 + 2 + 3$$
$$= 17$$

2. $7 - (2x - 3) + 2x - 8(1 - x)$
$$= 7 - 2x + 3 + 2x - 8 + 8x$$
$$= 7 + 3 - 8 - 2x + 2x + 8x$$
$$= 2 + 8x$$
$$= 8x + 2$$

3. $-4x^2 + x - 7 = -4(-2)^2 + (-2) - 7$
$$= -4(4) - 2 - 7$$
$$= -16 - 2 - 7$$
$$= -18 - 7$$
$$= -25$$

4.
$$\dfrac{5}{8} = \dfrac{5t}{6} + 2$$
$$24\left(\dfrac{5}{8}\right) = 24\left(\dfrac{5t}{6} + 2\right)$$
$$15 = 20t + 48$$
$$15 - 48 = 20t + 48 - 48$$
$$-33 = 20t$$
$$\dfrac{-33}{20} = \dfrac{20t}{20}$$
$$-\dfrac{33}{20} = t$$

5.
$$3x+5=4(x-2)$$
$$3x+5=4x-8$$
$$3x-3x+5=4x-3x-8$$
$$5=x-8$$
$$5+8=x-8+8$$
$$13=x$$

6. $3(x+2)+3x-5=4x+1$
$$3x+6+3x-5=4x+1$$
$$6x+1=4x+1$$
$$2x=0$$
$$x=0$$

7. $3x-2=y-7$
$$y-7=3x-2$$
$$y=3x-2+7$$
$$y=3x+5$$

8. Let $(x_1,y_1)=(1,3)$ and $(x_2,y_2)=(5,1)$
$$m=\frac{y_2-y_1}{x_2-x_1}=\frac{1-3}{5-1}=\frac{-2}{4}=-\frac{1}{2}$$

9. Put each equation in slope intercept form.

$$3x-5y=7 \qquad\qquad 5y+3x=2$$
$$-5y=-3x+7 \qquad\qquad 5y=-3x+2$$
$$y=\frac{-3x+7}{-5} \qquad\qquad y=\frac{-3x+2}{5}$$
$$y=\frac{3}{5}x-\frac{7}{5} \qquad\qquad y=-\frac{3}{5}x+\frac{2}{5}$$

The slope of the first line is $\dfrac{3}{5}$ while the slope of the second lint is $-\dfrac{3}{5}$. Since the slopes are not equal, the two lines are not parallel.

10.
$$\left(\frac{5xy^{-3}}{x^{-2}y^5}\right)^2=\left(5x^{1-(-2)}y^{-3-5}\right)^2$$
$$=\left(5x^3y^{-8}\right)^2$$
$$=5^2x^{3\cdot2}y^{-8\cdot2}$$
$$=\frac{25x^6}{y^{16}}$$

11. $-5x+2-7x^2=-7x^2-5x+2$; second degree

12. $\left(x^2+4x-3\right)+\left(2x^2+5x+1\right)$
$$=x^2+4x-3+2x^2+5x+1$$
$$=x^2+2x^2+4x+5x-3+1$$
$$=3x^2+9x-2$$

13. $\left(6a^2+3a+2\right)-\left(a^2-3a-3\right)$
$$=6a^2+3a+2-a^2+3a+3$$
$$=6a^2-a^2+3a+3a+2+3$$
$$=5a^2+6a+5$$

14. $(5t-3)(2t-1)=5t\cdot2t+5t\cdot(-1)-3\cdot2t-3\cdot(-1)$
$$=10t^2-5t-6t+3$$
$$=10t^2-11t+3$$

15. $(2x-1)\left(3x^2-5x+2\right)$
$$=2x\left(3x^2-5x+2\right)-1\left(3x^2-5x+2\right)$$
$$=6x^3-10x^2+4x-3x^2+5x-2$$
$$=6x^3-13x^2+9x-2$$

16.
$$\frac{10d^2+12d-8}{4d}=\frac{10d^2}{4d}+\frac{12d}{4d}-\frac{8}{4d}$$
$$=\frac{5}{2}d+3-\frac{2}{d}$$

17.
$$3x-2\overline{)\begin{array}{l}2x+5\\6x^2+11x-10\end{array}}$$
$$\underline{6x^2-4x}$$
$$15x-10$$
$$\underline{15x-10}$$
$$0$$

$$\frac{6x^2+11x-10}{3x-2}=2x+5$$

18.
$$\frac{x}{8}=\frac{1.25}{3}$$
$$3x=8(1.25)$$
$$3x=10$$
$$x=\frac{10}{3}\approx3.33$$

Eight cans of soup cost $3.33.

19. Let b = Bob's average speed.
    Then $b + 7$ = Nick's average speed.
    $d = r \cdot t$. Both Bob and Nick drove for 0.5 hour
    and the total distance they covered was 60 miles.

    $0.5b + 0.5(b + 7) = 60$

    $0.5b + 0.5b + 3.5 = 60$

    $b + 3.5 = 60$

    $b = 56.5$

    Bob's average speed was 56.5 miles per hour,
    and Nick's average speed was $56.5 + 7 = 63.5$
    miles per hour.

20. Let x = the width of the rectangle.
    Then $3x - 2$ = the length of the rectangle.

    $P = 2l + 2w$

    $28 = 2(3x - 2) + 2x$

    $28 = 6x - 4 + 2x$

    $28 = 8x - 4$

    $32 = 8x$

    $4 = x$

    The width of the rectangle is 4 feet and the
    length is $3(4) - 2 = 10$ feet.

Chapter 6

Exercise Set 6.1

1. To factor an expression means to write the expression as the product of factors.

3. A composite number is a positive integer, other than 1, that is not a prime number.

5. The greatest common factor of two or more numbers is the greatest number that divides into all the numbers.

7. $1, 2, 4, x, 2x, 4x, x^2, 2x^2, 4x^2$ and the opposites of these factors.

9. $56 = 8 \cdot 7$
$ = 2 \cdot 4 \cdot 7$
$ = 2 \cdot 2 \cdot 2 \cdot 7$
$ = 2^3 \cdot 7$

11. $90 = 9 \cdot 10$
$ = 3 \cdot 3 \cdot 2 \cdot 5$
$ = 2 \cdot 3^2 \cdot 5$

13. $248 = 4 \cdot 62$
$ = 2 \cdot 2 \cdot 2 \cdot 31$
$ = 2^3 \cdot 31$

15. $20 = 2^2 \cdot 5$, $24 = 2^3 \cdot 3$, so the greatest common factor is 2^2 or 4.

17. $70 = 2 \cdot 5 \cdot 7$, $98 = 2 \cdot 7^2$, so the greatest common factor is $2 \cdot 7$ or 14.

19. $80 = 2^4 \cdot 5$, $126 = 2 \cdot 3^2 \cdot 7$, so the greatest common factor is 2.

21. The greatest common factor is x.

23. The greatest common factor is $3x$.

25. The greatest common factor is a.

27. The greatest common factor is qr.

29. The greatest common factor is $x^3 y^5$.

31. The greatest common factor is 1.

33. The greatest common factor is $x^2 y^2$.

35. The greatest common factor is x.

37. The greatest common factor is $x - 4$.

39. The greatest common factor is $2x - 3$.

41. The greatest common factor is $3w + 5$.

43. The greatest common factor is $x - 4$.

45. The greatest common factor is $x - 1$.

47. The greatest common factor is $x - 9$.

49. The greatest common factor is 4.
$4x - 8 = 4 \cdot x - 4 \cdot 2$
$ = 4(x - 2)$

51. The greatest common factor is 5.
$15x - 5 = 5 \cdot 3x - 5 \cdot 1$
$ = 5(3x - 1)$

53. The greatest common factor is 7.
$7q + 28 = 7 \cdot q + 7 \cdot 4$
$ = 7(q + 4)$

55. The greatest common factor is $3x$.
$9x^2 - 12x = 3x \cdot 3x - 3x \cdot 4$
$ = 3x(3x - 4)$

57. The greatest common factor is x^4.
$7x^5 - 9x^4 = x^4 \cdot 7x - x^4 \cdot 9$
$ = x^4(7x - 9)$

59. The greatest common factor is $3x^2$.
$3x^5 - 12x^2 = 3x^2 \cdot x^3 - 3x^2 \cdot 4$
$ = 3x^2(x^3 - 4)$

61. The greatest common factor is $12x^8$.
$36x^{12} + 24x^8 = 12x^8 \cdot 3x^4 + 12x^8 \cdot 2$
$\phantom{36x^{12} + 24x^8} = 12x^8(3x^4 + 2)$

63. The greatest common factor is $9y^3$.
$27y^{15} - 9y^3 = 9y^3 \cdot 3y^{12} - 9y^3 \cdot 1$
$\phantom{27y^{15} - 9y^3} = 9y^3(3y^{12} - 1)$

65. The greatest common factor is y.
$y + 6x^3 y = y \cdot 1 + y \cdot 6x^3$
$ = y(1 + 6x^3)$

67. The greatest common factor is a^2.
$$7a^4 + 3a^2 = a^2 \cdot 7a^2 + a^2 \cdot 3$$
$$= a^2\left(7a^2 + 3\right)$$

69. The greatest common factor is $4xy$.
$$16xy^2z + 4x^3y = 4xy \cdot 4yz + 4xy \cdot x^2$$
$$= 4xy\left(4yz + x^2\right)$$

71. The greatest common factor is $4x^2yz^3$.
$$80x^5y^3z^4 - 36x^2yz^3$$
$$= 4x^2yz^3 \cdot 20x^3y^2z - 4x^2yz^3 \cdot 9$$
$$= 4x^2yz^3\left(20x^3y^2z - 9\right)$$

73. The greatest common factor is $25x^2yz$.
$$25x^2yz^3 + 25x^3yz = 25x^2yz \cdot z^2 + 25x^2yz \cdot x$$
$$= 25x^2yz\left(z^2 + x\right)$$

75. The greatest common factor is $x^4y^3z^9$.
$$19x^4y^{12}z^{13} - 8x^5y^3z^9$$
$$= x^4y^3z^9 \cdot 19y^9z^4 - x^4y^3z^9 \cdot 8x$$
$$= x^4y^3z^9\left(19y^9z^4 - 8x\right)$$

77. The greatest common factor is 4.
$$8c^2 - 4c - 32 = 4 \cdot 2c^2 - 4 \cdot c - 4 \cdot 8$$
$$= 4\left(2c^2 - c - 8\right)$$

79. The greatest common factor is 3.
$$9x^2 + 18x + 3 = 3 \cdot 3x^2 + 3 \cdot 6x + 3 \cdot 1$$
$$= 3\left(3x^2 + 6x + 1\right)$$

81. The greatest common factor is $4x$.
$$4x^3 - 8x^2 + 12x = 4x \cdot x^2 - 4x \cdot 2x + 4x \cdot 3$$
$$= 4x\left(x^2 - 2x + 3\right)$$

83. The greatest common factor is 8.
$$40b^2 - 48c + 24 = 8 \cdot 5b^2 - 8 \cdot 6c + 8 \cdot 3$$
$$= 8\left(5b^2 - 6c + 3\right)$$

85. The greatest common factor is 3.
$$15p^2 - 6p + 9 = 3 \cdot 5p^2 - 3 \cdot 2p + 3 \cdot 3$$
$$= 3\left(5p^2 - 2p + 3\right)$$

87. The greatest common factor is $3a$.
$$9a^4 - 6a^3 + 3ab = 3a \cdot 3a^3 - 3a \cdot 2a^2 + 3a \cdot b$$
$$= 3a\left(3a^3 - 2a^2 + b\right)$$

89. The greatest common factor is xy.
$$8x^2y + 12xy^2 + 5xy = xy \cdot 8x + xy \cdot 12y + xy \cdot 5$$
$$= xy\left(8x + 12y + 5\right)$$

91. The greatest common factor is $x - 7$.
$$x(x - 7) + 6(x - 7) = (x - 7)(x + 6)$$

93. The greatest common factor is $a - 2$.
$$3b(a - 2) - 4(a - 2) = (a - 2)(3b - 4)$$

95. The greatest common factor is $2x + 1$.
$$4x(2x + 1) + 1(2x + 1) = (2x + 1)(4x + 1)$$

97. The greatest common factor is $2x + 1$.
$$5x(2x + 1) + 2x + 1 = 5x(2x + 1) + 1(2x + 1)$$
$$= (2x + 1)(5x + 1)$$

99. The greatest common factor is $6c + 7$.
$$3c(6c + 7) - 2(6c + 7) = (6c + 7)(3c - 2)$$

101. $12\nabla - 6\nabla^2 = 6\nabla \cdot 2 - 6\nabla \cdot \nabla$
$$= 6\nabla(2 - \nabla)$$

103. $12\square^3 - 4\square^2 + 4\square$
$$= 4\square \cdot 3\square^2 - 4\square \cdot \square + 4\square \cdot 1$$
$$= 4\square\left(3\square^2 - \square + 1\right)$$

105. The greatest common factor is $2x^2(2x + 7)$.
$$6x^5(2x + 7) + 4x^3(2x + 7) - 2x^2(2x + 7)$$
$$= 2x^2(2x + 7) \cdot 3x^3 + 2x^2(2x + 7) \cdot 2x$$
$$\quad + 2x^2(2x + 7) \cdot (-1)$$
$$= 2x^2(2x + 7)\left(3x^3 + 2x - 1\right)$$

107. $x^2 + 2x + 3x + 6$
$$= x \cdot x + x \cdot 2 + 3 \cdot x + 3 \cdot 2$$
$$= x(x + 2) + 3(x + 2)$$
$$= (x + 2)(x + 3)$$

108. $2x - (x - 5) + 4(3 - x) = 2x - x + 5 + 12 - 4x$
$$= x - 4x + 17$$
$$= -3x + 17$$

109. $4 + 3(x-8) = x - 4(x+2)$

$4 + 3x - 24 = x - 4x - 8$

$3x - 20 = -3x - 8$

$6x = 12$

$x = 2$

110. $4x - 5y = 20$

$-5y = -4x + 20$

$y = -\dfrac{-4x + 20}{-5}$

$y = \dfrac{4}{5}x - 4$

111. $V = \dfrac{1}{3}\pi r^2 h$

$= \dfrac{1}{3}\pi(4)^2(12)$

$= \dfrac{1}{3}\pi(16)(12)$

$= 64\pi$ in.^{3} or 201.06 in.^{3}

112. Let x = smaller number, then $2x - 1$ = the larger number.

$x + (2x - 1) = 41$

$x + 2x - 1 = 41$

$3x - 1 = 41$

$3x = 42$

$x = 14$

$2x - 1 = 2(14) - 1 = 28 - 1 = 27$

The numbers are 14 and 27.

113. $\left(\dfrac{3x^2 y^3}{2x^5 y^2}\right)^2 = \left(\dfrac{3y}{2x^3}\right)^2$

$= \dfrac{(3y)^2}{(2x^3)^2}$

$= \dfrac{3^2 y^2}{2^2 (x^3)^2}$

$= \dfrac{9y^2}{4x^6}$

Exercise Set 6.2

1. The first step in any factoring by grouping problem is to factor out a common factor, if one exists.

3. If you multiply $(x - 2y)(x - 3)$ using the FOIL method, you get the polynomial $x^2 - 3x - 2xy + 6y$

5. The number that changes the sign of each term in an expression when factored from each term is -1.

7. $x^2 + 3x + 2x + 6 = x(x+3) + 2(x+3)$
$= (x+3)(x+2)$

9. $x^2 + 5x + 4x + 20 = x(x+5) + 4(x+5)$
$= (x+5)(x+4)$

11. $x^2 + 2x + 5x + 10 = x(x+2) + 5(x+2)$
$= (x+2)(x+5)$

13. $c^2 - 4c + 7c - 28 = c(c-4) + 7(c-4)$
$= (c-4)(c+7)$

15. $4x^2 - 6x + 6x - 9 = 2x(2x-3) + 3(2x-3)$
$= (2x-3)(2x+3)$

17. $3x^2 + 9x + x + 3 = 3x(x+3) + 1(x+3)$
$= (x+3)(3x+1)$

19. $6x^2 + 3x - 2x - 1 = 3x(2x+1) - 1(2x+1)$
$= (2x+1)(3x-1)$

21. $8x^2 + 32x + x + 4 = 8x(x+4) + 1(x+4)$
$= (x+4)(8x+1)$

23. $12t^2 - 8t - 3t + 2$
$= 4t(3t-2) - 1(3t-2)$
$= (3t-2)(4t-1)$

25. $x^2 + 9x - x - 9 = x(x+9) - 1(x+9)$
$= (x+9)(x-1)$

27. $6p^2 + 15p - 4p - 10$
$= 3p(2p+5) - 2(2p+5)$
$= (2p+5)(3p-2)$

29. $x^2 + 2xy - 3xy - 6y^2$
$= x(x+2y) - 3y(x+2y)$
$= (x+2y)(x-3y)$

31. $3x^2 + 2xy - 9xy - 6y^2$
$= x(3x+2y) - 3y(3x+2y)$
$= (3x+2y)(x-3y)$

33. $10x^2 - 12xy - 25xy + 30y^2$
$= 2x(5x-6y) - 5y(5x-6y)$
$= (5x-6y)(2x-5y)$

35. $x^2 - bx - ax + ab$
$= x(x-b) - a(x-b)$
$= (x-b)(x-a)$

37. $xy + 9x - 5y - 45 = x(y+9) - 5(y+9)$
$= (y+9)(x-5)$

39. $a^2 + 3a + ab + 3b = a(a+3) + b(a+3)$
$= (a+3)(a+b)$

41. $xy - x + 5y - 5 = x(y-1) + 5(y-1)$
$= (y-1)(x+5)$

43. $12 + 8y - 3x - 2xy = 4(3+2y) - x(3+2y)$
$= (3+2y)(4-x)$

45. $z^3 + 5z^2 + z + 5 = z^2(z+5) + 1(z+5)$
$= (z+5)(z^2+1)$

47. $x^3 - 5x^2 + 8x - 40 = x^2(x-5) + 8(x-5)$
$= (x-5)(x^2+8)$

49. $2x^2 - 12x + 8x - 48$
$= 2 \cdot x^2 - 2 \cdot 6x + 2 \cdot 4x - 2 \cdot 24$
$= 2(x^2 - 6x + 4x - 24)$
$= 2[x(x-6) + 4(x-6)]$
$= 2(x-6)(x+4)$

51. $4x^2 + 8x + 8x + 16$
$= 4 \cdot x^2 + 4 \cdot 2x + 4 \cdot 2x + 4 \cdot 4$
$= 4(x^2 + 2x + 2x + 4)$
$= 4[x(x+2) + 2(x+2)]$
$= 4(x+2)(x+2)$
$= 4(x+2)^2$

53. $6x^3 + 9x^2 - 2x^2 - 3x$
$= x \cdot 6x^2 + x \cdot 9x - x \cdot 2x - x \cdot 3$
$= x(6x^2 + 9x - 2x - 3)$
$= x[3x(2x+3) - 1(2x+3)]$
$= x(2x+3)(3x-1)$

55. $p^3 - 6p^2q + 2p^2q - 12pq^2$
$= p \cdot p^2 - p \cdot 6pq + p \cdot 2pq - p \cdot 12q^2$
$= p(p^2 - 6pq + 2pq - 12q^2)$
$= p[p(p-6q) + 2q(p-6q)]$
$= p(p-6q)(p+2q)$

57. $5x + 3y + xy + 15 = xy + 5x + 3y + 15$
$= x(y+5) + 3(y+5)$
$= (y+5)(x+3)$

59. $6x + 5y + xy + 30 = 6x + xy + 5y + 30$
$= x(6+y) + 5(y+6)$
$= (x+5)(y+6)$

61. $ax + by + ay + bx = ax + ay + bx + by$
$= a(x+y) + b(x+y)$
$= (a+b)(x+y)$

63. $rs - 42 + 6s - 7r$
$= rs - 7r + 6s - 42$
$= r(s-7) + 6(s-7)$
$= (s-7)(r+6)$

65. $dc + 3c - ad - 3a = c(d+3) - a(d+3)$
$= (d+3)(c-a)$

67. Not *any* arrangement of the terms of a polynomial is factorable by grouping. $xy + 2x + 5y + 10$ is factorable but $xy + 10 + 2x + 5y$ is not factorable in this arrangement.

69. $\odot^2 + 3\odot - 5\odot - 15 = \odot(\odot+3) - 5(\odot+3)$
$$= (\odot+3)(\odot-5)$$

71. a. $2x^2 - 11x + 15 = 2x^2 - 5x - 6x + 15$

   b. $2x^2 - 5x - 6x + 15 = x(2x-5) - 3(2x-5)$
   $$= (x-3)(2x-5)$$

73. a. $2x^2 - 11x + 15 = 2x^2 - 6x - 5x + 15$

   b. $2x^2 - 6x - 5x + 15$
   $$= 2x(x-3) - 5(x-3)$$
   $$= (x-3)(2x-5)$$

75. a. $4x^2 - 17x - 15 = 4x^2 + 3x - 20x - 15$

   b. $4x^2 + 3x - 20x - 15 = x(4x+3) - 5(4x+3)$
   $$= (x-5)(4x+3)$$

77. $\star\odot + 3\star + 2\odot + 6 = \star(\odot+3) + 2(\odot+3)$
$$= (\odot+3)(\star+2)$$

79. $5 - 3(2x-7) = 4(x+5) - 6$
$$5 - 6x + 21 = 4x + 20 - 6$$
$$-6x + 26 = 4x + 14$$
$$12 = 10x$$
$$\frac{12}{10} = x$$
$$\frac{6}{5} = x$$

80. Let x = the number of pounds of jelly beans and $(50-x)$ = the number of pounds of gumdrops.

Item	Price per pound	Quantity	Cost
Jellybeans	6.25	x	$6.25x$
Gumdrops	2.50	$50-x$	$2.50(50-x)$
Mixture	4.75	50	$4.75(50)$

$$6.25x + 2.50(50-x) = 4.75(50)$$
$$6.25x + 125 - 2.50x = 237.5$$
$$3.75x + 125 = 237.5$$
$$3.75x = 112.5$$
$$x = 30$$

They should mix 30 pounds of jelly beans with 20 pounds of gumdrops.

81. $\dfrac{15x^3 - 6x^2 - 9x + 5}{3x} = \dfrac{15x^3}{3x} - \dfrac{6x^2}{3x} - \dfrac{9x}{3x} + \dfrac{5}{3x}$
$$= 5x^2 - 2x - 3 + \frac{5}{3x}$$

82.
$$
\begin{array}{r}
a - 4 \\
a+4 \overline{)\,a^2 \qquad\quad -16} \\
\underline{a^2 + 4a\qquad\quad} \\
-4a - 16 \\
\underline{-4a - 16} \\
0
\end{array}
$$

$$\frac{a^2 - 16}{a+4} = a - 4$$

Exercise Set 6.3

1. Since 960 is positive, both signs will be the same. Since 92 is positive, both signs will be positive.

3. Since −1500 is negative, one sign will be positive, the other will be negative.

5. Since 8000 is positive, both signs will be the same. Since −240 is negative, both signs will be negative.

7. The trinomial $x^2 - 11x + 24$ is obtained by multiplying the factors using the FOIL method.

9. The trinomial $2x^2 - 8xy - 10y^2$ is obtained by multiplying all the factors and combining like terms.

11. The answer is not fully factored. A 2 can be factored from $(2x - 4)$.

13. A trinomial factoring problem can be checked by multiplying the factors to see if the product is the same as the original expression.

15. $x^2 - 7x + 10 = (x - 5)(x - 2)$

17. $x^2 + 6x + 8 = (x + 4)(x + 2)$

19. $x^2 + 5x - 24 = (x + 8)(x - 3)$

21. $x^2 + 4x - 6$ is prime.

23. $y^2 - 13y + 12 = (y - 12)(y - 1)$

25. $a^2 - 2a - 8 = (a - 4)(a + 2)$

27. $r^2 - 2r - 15 = (r - 5)(r + 3)$

29. $b^2 - 11b + 18 = (b - 9)(b - 2)$

31. $x^2 - 8x - 15$ is prime.

33. $q^2 + 4q - 45 = (q + 9)(q - 5)$

35. $x^2 - 7x - 30 = (x - 10)(x + 3)$

37. $x^2 + 4x + 4 = (x + 2)(x + 2)$
$$= (x + 2)^2$$

39. $s^2 - 8s + 16 = (s - 4)(s - 4)$
$$= (s - 4)^2$$

41. $p^2 - 12p + 36 = (p - 6)(p - 6)$
$$= (p - 6)^2$$

43. $-18w + w^2 + 45$
$$= w^2 - 18w + 45$$
$$= (w - 15)(w - 3)$$

45. $10x - 39 + x^2$
$$= x^2 + 10x - 39$$
$$= (x + 13)(x - 3)$$

47. $x^2 - x - 20 = (x - 5)(x + 4)$

49. $y^2 + 13y + 40 = (y + 5)(y + 8)$

51. $x^2 + 12x - 64 = (x + 16)(x - 4)$

53. $s^2 + 14s - 24$ is prime.

55. $x^2 - 20x + 64 = (x - 16)(x - 4)$

57. $a^2 - 20a + 99 = (a - 11)(a - 9)$

59. $x^2 + 2 + 3x = x^2 + 3x + 2$
$$= (x + 1)(x + 2)$$

61. $7w - 18 + w^2 = w^2 + 7w - 18$
$$= (w + 9)(w - 2)$$

63. $x^2 - 8xy + 15y^2 = (x - 3y)(x - 5y)$

65. $m^2 - 6mn + 9n^2 = (m - 3n)(m - 3n)$
$$= (m - 3n)^2$$

67. $x^2 + 8xy + 12y^2 = (x + 6y)(x + 2y)$

69. $m^2 - 5mn - 24n^2 = (m - 8n)(m + 3n)$

71. $6x^2 - 30x + 24 = 6(x^2 - 5x + 4)$
$$= 6(x - 4)(x - 1)$$

73. $5x^2 + 20x + 15 = 5(x^2 + 4x + 3)$
$$= 5(x + 1)(x + 3)$$

75. $2x^2 - 18x + 40 = 2(x^2 - 9x + 20)$
$$= 2(x - 5)(x - 4)$$

77. $b^3 - 7b^2 + 10b = b(b^2 - 7b + 10)$
$$= b(b - 5)(b - 2)$$

79. $3z^3 - 21z^2 - 54z = 3z(z^2 - 7z - 18)$
$$= 3z(z - 9)(z + 2)$$

81. $x^3 + 8x^2 + 16x = x(x^2 + 8x + 16)$
$$= x(x + 4)(x + 4)$$
$$= x(x + 4)^2$$

83. $7a^2 - 35ab + 42b^2 = 7(a^2 - 5ab + 6b^2)$
$$= 7(a - 3b)(a - 2b)$$

85. $3r^3 + 6r^2t - 24rt^2 = 3r(r^2 + 2rt - 8t^2)$
$$= 3r(r + 4t)(r - 2t)$$

87. $x^4 - 4x^3 - 21x^2 = x^2(x^2 - 4x - 21)$
$$= x^2(x - 7)(x + 3)$$

89.

Sign of Coefficient of *x*-term	Sign of Constant of Trinomial	Signs of Constant Terms in the Binomial Factors
−	+	both negative
−	−	one positive and one negative
+	−	one positive and one negative
+	+	both positive

91. $x^2 - 12x + 32 = (x-8)(x-4)$

93. $x^2 - 2x - 35 = (x-7)(x+5)$

95. $x^2 + 0.6x + 0.08 = (x+0.4)(x+0.2)$

97. $x^2 + \dfrac{2}{5}x + \dfrac{1}{25} = \left(x+\dfrac{1}{5}\right)\left(x+\dfrac{1}{5}\right)$

$$= \left(x+\dfrac{1}{5}\right)^2$$

99. $x^2 - 24x - 256 = (x+8)(x-32)$

101.
$$4(2x-4) = 5x+11$$
$$8x-16 = 5x+11$$
$$8x-5x-16 = 5x-5x+11$$
$$3x-16 = 11$$
$$3x-16+16 = 11+16$$
$$3x = 27$$
$$x = 9$$

102. Let *x* be the percent of acid in the mixture.

Solution	Strength	Liters	Amount
18%	0.18	4	0.72
26%	0.26	1	0.26
Mixture	$\dfrac{x}{100}$	5	$\dfrac{5x}{100}$

$$0.72 + 0.26 = \dfrac{5x}{100}$$
$$0.98 = \dfrac{x}{20}$$
$$19.6 = x$$

The mixture is a 19.6% acid solution.

103.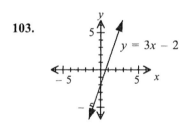

104. $\left(2x^2 + 5x - 6\right)(x-2)$

$$= 2x^2(x-2) + 5x(x-2) - 6(x-2)$$
$$= 2x^3 - 4x^2 + 5x^2 - 10x - 6x + 12$$
$$= 2x^3 + x^2 - 16x + 12$$

105.
$$\begin{array}{r} 3x+2 \\ x-4 \overline{\smash{\big)}\, 3x^2 - 10x - 10} \\ \underline{3x^2 - 12x} \\ 2x - 10 \\ \underline{2x - 8} \\ -2 \end{array}$$

$$\dfrac{3x^2 - 10x - 10}{x-4} = 3x + 2 - \dfrac{2}{x-4}$$

106. $20x^2 + 8x - 15x - 6 = 4x(5x+2) - 3(5x+2)$

$$= (5x+2)(4x-3)$$

Exercise Set 6.4

1. Factoring trinomials is the reverse process of multiplying binomials.

3. When factoring a trinomial of the form $ax^2 + bx + c$, the product of the first terms of the binomial factors must equal the first term of the trinomial, ax^2.

5. $2x^2 + 11x + 5 = (2x + 1)(x + 5)$

7. $3x^2 + 14x + 8 = (3x + 2)(x + 4)$

9. $5x^2 - 9x - 2 = (5x + 1)(x - 2)$

11. $3r^2 + 13r - 10 = (3r - 2)(r + 5)$

13. $4z^2 - 12z + 9 = (2z - 3)(2z - 3)$
$$= (2z - 3)^2$$

15. $6z^2 + z - 12 = (2z + 3)(3z - 4)$

17. $5a^2 - 12a + 6$ is prime.

19. $8x^2 + 19x + 6 = (8x + 3)(x + 2)$

21. $3x^2 + 11x + 4$ is prime.

23. $5y^2 - 16y + 3 = (5y - 1)(y - 3)$

25. $7x^2 + 43x + 6 = (7x + 1)(x + 6)$

27. $4x^2 + 4x - 15 = (2x + 5)(2x - 3)$

29. $49t^2 - 14t + 1 = (7t - 1)(7t - 1)$
$$= (7t - 1)^2$$

31. $5z^2 - 6z - 8 = (5z + 4)(z - 2)$

33. $4y^2 + 5y - 6 = (4y - 3)(y + 2)$

35. $10x^2 - 27x + 5 = (5x - 1)(2x - 5)$

37. $10d^2 - 7d - 12 = (5d + 4)(2d - 3)$

39. $8x^2 - 46x - 12 = 2 \cdot 4x^2 - 2 \cdot 23x - 2 \cdot 6$
$$= 2(4x^2 - 23x - 6)$$
$$= 2(4x + 1)(x - 6)$$

41. $10t + 3 + 7t^2 = 7t^2 + 10t + 3$
$$= (7t + 3)(t + 1)$$

43. $6x^2 + 16x + 10 = 2 \cdot 3x^2 + 2 \cdot 8x + 2 \cdot 5$
$$= 2(3x^2 + 8x + 5)$$
$$= 2(3x + 5)(x + 1)$$

45. $6x^3 - 5x^2 - 4x = x \cdot 6x^2 - x \cdot 5x - x \cdot 4$
$$= x(6x^2 - 5x - 4)$$
$$= x(2x + 1)(3x - 4)$$

47. $12x^3 + 28x^2 + 8x = 4x \cdot 3x^2 + 4x \cdot 7x + 4x \cdot 2$
$$= 4x(3x^2 + 7x + 2)$$
$$= 4x(3x + 1)(x + 2)$$

49. $4x^3 - 2x^2 - 12x = 2x \cdot 2x^2 - 2x \cdot x - 2x \cdot 6$
$$= 2x(2x^2 - x - 6)$$
$$= 2x(2x + 3)(x - 2)$$

51. $48c^2 + 8c - 16 = 8 \cdot 6c^2 + 8 \cdot c - 8 \cdot 2$
$$= 8(6c^2 + 8c - 2)$$
$$= 8(2c - 1)(3c + 2)$$

53. $4p - 12 + 8p^2 = 8p^2 + 4p - 12$
$$= 4 \cdot 2p^2 + 4 \cdot p - 4 \cdot 3$
$$= 4(2p^2 + p - 3)$$
$$= 4(2p + 3)(p - 1)$$

55. $8c^2 + 41cd + 5d^2 = (8c + d)(c + 5d)$

57. $15x^2 - xy - 6y^2 = (5x + 3y)(3x - 2y)$

59. $12x^2 + 10xy - 8y^2 = 2 \cdot 6x^2 + 2 \cdot 5xy - 2 \cdot 4y^2$
$$= 2(6x^2 + 5xy - 4y^2)$$
$$= 2(2x - y)(3x + 4y)$$

61. $7p^2 + 13pq + 6q^2 = (7p + 6q)(p + q)$

63. $6m^2 - mn - 2n^2 = (3m - 2n)(2m + n)$

65. $8x^3 + 10x^2 y + 3xy^2 = x \cdot 8x^2 + x \cdot 10xy + x \cdot 3y^2$
$$= x(8x^2 + 10xy + 3y^2)$$
$$= x(4x + 3y)(2x + y)$$

67. $4x^4 + 8x^3 y + 3x^2 y^2$
$$= x^2 \cdot 4x^2 + x^2 \cdot 8xy + x^2 \cdot 3y^2$$
$$= x^2 (4x^2 + 8xy + 3y^2)$$
$$= x^2 (2x + y)(2x + 3y)$$

69. $3x^2 - 20x - 7$. This polynomial was obtained by multiplying the factors.

71. $10x^2 + 35x + 15$. This polynomial was obtained by multiplying the factors.

73. $3t^4 + 11t^3 - 4t^2$. This polynomial was obtained by multiplying the factors.

75. a. The second factor can be found by dividing the trinomial by the binomial.

b.
$$\begin{array}{r} 6x + 11 \\ 3x + 10 \overline{\smash{\big)}\ 18x^2 + 93x + 110} \\ \underline{18x^2 + 60x} \\ 33x + 110 \\ \underline{33x + 110} \\ 0 \end{array}$$

The other factor is $6x + 11$.

77. $18x^2 + 9x - 20 = (6x - 5)(3x + 4)$

79. $15x^2 - 124x + 160 = (5x - 8)(3x - 20)$

81. $105a^2 - 220a - 160$
$$= 5 \cdot 21a^2 - 5 \cdot 44a - 5 \cdot 32$$
$$= 5(21a^2 - 44a - 32)$$
$$= 5(3a - 8)(7a + 4)$$

83. The other factor is $2x + 45$. The product of the three first terms must equal $6x^3$, and the product of the constants must equal $2250x$.

85. $-x^2 - 4(y + 3) + 2y^2$
$$= -(-3)^2 - 4(-5 + 3) + 2(-5)^2$$
$$= -9 - 4(-2) + 2(25)$$
$$= -9 + 8 + 50$$
$$= 49$$

86. $\dfrac{507.5}{3.56} \approx 142.56$

His average speed was about 142.56 miles per hour.

87. $36x^4 y^3 - 12xy^2 + 24x^5 y^6$
$$= 12xy^2 \cdot 3x^3 y - 12xy^2 \cdot 1 + 12xy^2 \cdot 2x^4 y^4$$
$$= 12xy^2 (3x^3 y - 1 + 2x^4 y^4)$$

88. $b^2 + 4b - 96 = (b + 12)(b - 8)$

Mid-Chapter Test: 6.1 – 6.4

1. Any factoring problem can be checked by multiplying the factors.

2. $18xy^2 : 2 \cdot 3^2 \cdot xy^2$
$27x^3 y^4 : 3^3 \cdot x^3 y^4$
$12x^2 y^3 : 2^2 \cdot 3 \cdot x^2 y^3$
GCF: $3 \cdot x \cdot y^2 = 3xy^2$

3. $4a^2 b^3 - 24a^3 b = 4a^2 b \cdot b^2 - 4a^2 b \cdot 6a$
$$= 4a^2 b (b^2 - 6a)$$

4. $5c(d - 6) - 3(d - 6) = (d - 6)(5c - 3)$

5. $7x(2x + 9) + 2x + 9$
$$= 7x(2x + 9) + 1(2x + 9)$$
$$= (2x + 9)(7x + 1)$$

6. $x^2 + 4x + 7x + 28$
$$= x(x + 4) + 7(x + 4)$$
$$= (x + 4)(x + 7)$$

7. $x^2 + 5x - 3x - 15$
$$= x(x + 5) - 3(x + 5)$$
$$= (x + 5)(x - 3)$$

8. $6a^2 + 15ab - 2ab - 5b^2$
$$= 3a(2a + 5b) - b(2a + 5b)$$
$$= (2a + 5b)(3a - b)$$

9. $5x^2 - 2xy - 45x + 18y$

$= x(5x - 2y) - 9(5x - 2y)$

$= (5x - 2y)(x - 9)$

10. $8x^3 + 4x^2 - 48x^2 - 24x$

$= 4x(2x^2 + x - 12x - 6)$

$= 4x[x(2x + 1) - 6(2x + 1)]$

$= 4x(2x + 1)(x - 6)$

11. $x^2 - 10x + 21 = (x - 3)(x - 7)$

12. $t^2 + 9t + 20 = (t + 5)(t + 4)$

13. $p^2 - 3p - 8$ is prime

14. $x^2 + 16x + 64 = (x + 8)(x + 8)$

$= (x + 8)^2$

15. $m^2 - 4mn - 45n^2 = (m + 5n)(m - 9n)$

16. $3x^2 + 17x + 10 = (3x + 2)(x + 5)$

17. $4z^2 - 11z + 6 = (4z - 3)(z - 2)$

18. $3y^2 + 13y + 6$ is prime

19. $9x^2 - 6x + 1 = (3x - 1)(3x - 1)$

$= (3x - 1)^2$

20. $6a^2 + 3ab - 3b^2 = 3(2a^2 + ab - b^2)$

$= 3(2a - b)(a + b)$

Exercise Set 6.5

1. a. $a^2 - b^2 = (a + b)(a - b)$

b. Answers will vary.

3. a. $a^3 + b^3 = (a + b)(a^2 - ab + b^2)$

b. Answers will vary.

5. No, there is no special formula for factoring the sum of two squares.

7. $x^2 + 9$ is prime

9. $3b^2 + 48 = 3(b^2 + 16)$

11. $16m^2 + 36n^2 = 4(4m^2 + 9n^2)$

13. $y^2 - 25 = y^2 - 5^2 = (y + 5)(y - 5)$

15. $81 - z^2 = 9^2 - z^2$

$= (9 + z)(9 - z)$

17. $x^2 - 49 = x^2 - 7^2$

$= (x + 7)(x - 7)$

19. $x^2 - y^2 = (x + y)(x - y)$

21. $9y^2 - 25z^2 = (3y)^2 - (5z)^2$

$= (3y + 5z)(3y - 5z)$

23. $64a^2 - 36b^2 = 4(16a^2 - 9b^2)$

$= 4[(4a)^2 - (3b)^2]$

$= 4(4a + 3b)(4a - 3b)$

25. $36 - 49x^2 = 6^2 - (7x)^2$

$= (6 + 7x)(6 - 7x)$

27. $z^4 - 81x^2 = (z^2)^2 - (9x)^2$

$= (z^2 + 9x)(z^2 - 9x)$

29. $25x^4 - 49y^4 = (5x^2)^2 - (7y^2)^2$

$= (5x^2 + 7y^2)(5x^2 - 7y^2)$

31. $36m^4 - 49n^2 = (6m^2)^2 - (7n)^2$

$= (6m^2 + 7n)(6m^2 - 7n)$

33. $2x^4 - 50y^2 = 2(x^4 - 25y^2)$

$= 2[(x^2)^2 - (5y)^2]$

$= 2(x^2 + 5y)(x^2 - 5y)$

161

35. $x^4 - 81 = \left(x^2\right)^2 - 9^2$

$\qquad = \left(x^2 + 9\right)\left(x^2 - 9\right)$

$\qquad = \left(x^2 + 9\right)\left(x^2 - 3^2\right)$

$\qquad = \left(x^2 + 9\right)(x+3)(x-3)$

37. $x^3 + y^3 = (x+y)\left(x^2 - xy + y^2\right)$

39. $x^3 - y^3 = (x-y)\left(x^2 + xy + y^2\right)$

41. $x^3 + 64 = x^3 + 4^3$

$\qquad = (x+4)\left(x^2 - 4x + 16\right)$

43. $x^3 - 27 = x^3 - 3^3$

$\qquad = (x-3)\left(x^2 + 3x + 9\right)$

45. $a^3 + 1 = a^3 + 1^3$

$\qquad = (a+1)\left(a^2 - a + 1\right)$

47. $27x^3 - 1 = (3x)^3 - 1^3$

$\qquad = (3x-1)\left(9x^2 + 3x + 1\right)$

49. $27a^3 - 125 = (3a)^3 - 5^3$

$\qquad = (3a-5)\left(9a^2 + 15a + 25\right)$

51. $27 - 8y^3 = 3^3 - (2y)^3$

$\qquad = (3-2y)\left(9 + 6y + 4y^2\right)$

53. $64m^3 + 27n^3 = (4m)^3 + (3n)^3$

$\qquad = (4m+3n)\left(16m^2 - 12mn + 9n^2\right)$

55. $8a^3 - 27b^3 = (2a)^3 - (3b)^3$

$\qquad = (2a-3b)(4a^2 + 6ab + 9b^2)$

57. $4x^2 - 24x + 36 = 4\left(x^2 - 6x + 9\right)$

$\qquad = 4(x-3)^2$

59. $50x^2 - 10x - 12 = 2\left(25x^2 - 5x - 6\right)$

$\qquad = 2(5x+2)(5x-3)$

61. $2d^2 + 16d + 32 = 2\left(d^2 + 8d + 16\right)$

$\qquad = 2(d+4)^2$

63. $5x^2 - 10x - 15 = 5\left(x^2 - 2x - 3\right)$

$\qquad = 5(x-3)(x+1)$

65. $5x^2 - 20 = 5\left(x^2 - 4\right)$

$\qquad = 5\left(x^2 - 2^2\right)$

$\qquad = 5(x+2)(x-2)$

67. $2x^2 - 50 = 2(x^2 - 25)$

$\qquad = 2(x^2 - 5^2)$

$\qquad = 2(x+5)(x-5)$

69. $2x^2 y - 18y = 2y\left(x^2 - 9\right)$

$\qquad = 2y\left(x^2 - 3^2\right)$

$\qquad = 2y(x+3)(x-3)$

71. $3x^3 y^2 + 3y^2 = 3y^2\left(x^3 + 1\right)$

$\qquad = 3y^2\left(x^3 + 1^3\right)$

$\qquad = 3y^2(x+1)\left(x^2 - x + 1\right)$

73. $2x^3 - 16 = 2\left(x^3 - 8\right)$

$\qquad = 2\left(x^3 - 2^3\right)$

$\qquad = 2(x-2)\left(x^2 + 2x + 4\right)$

75. $18x^2 - 50 = 2\left(9x^2 - 25\right)$

$\qquad = 2\left((3x)^2 - 5^2\right)$

$\qquad = 2(3x+5)(3x-5)$

77. $6t^2 r - 15tr + 21r = 3r\left(2t^2 - 5t + 7\right)$

79. $6x^2 - 4x + 24x - 16$

$\qquad = 2\left(3x^2 - 2x + 12x - 8\right)$

$\qquad = 2\left[x(3x-2) + 4(3x-2)\right]$

$\qquad = 2(3x-2)(x+4)$

81. $2rs^2 - 10rs - 48r = 2r\left(s^2 - 5s - 24\right)$

$\qquad = 2r(s+3)(s-8)$

83. $4x^2 + 5x - 6 = (x+2)(4x-3)$

85. $25b^2 - 100 = 25(b^2 - 4)$
$$= 25(b^2 - 2^2)$$
$$= 25(b+2)(b-2)$$

87. $a^5b^2 - 4a^3b^4 = a^3b^2(a^2 - 4b^2)$
$$= a^3b^2\left[a^2 - (2b)^2\right]$$
$$= a^3b^2(a+2b)(a-2b)$$

89. $5x^4 + 10x^3 + 5x^2 = 5x^2(x^2 + 2x + 1)$
$$= 5x^2(x+1)^2$$

91. $x^3 + 25x = x(x^2 + 25)$

93. $y^4 - 16 = (y^2)^2 - 4^2$
$$= (y^2 + 4)(y^2 - 4)$$
$$= (y^2 + 4)(y^2 - 2^2)$$
$$= (y^2 + 4)(y+2)(y-2)$$

95. $16m^3 + 250 = 2(8m^3 + 125)$
$$= 2\left[(2m)^3 + 5^3\right]$$
$$= 2(2m+5)(4m^2 - 10m + 25)$$

97. $ac + 2a + bc + 2b = a(c+2) + b(c+2)$
$$= (a+b)(c+2)$$

99. $9 - 9y^4 = 9(1 - y^4)$
$$= 9\left[1^2 - (y^2)^2\right]$$
$$= 9(1+y^2)(1-y^2)$$
$$= 9(1+y^2)(1^2 - y^2)$$
$$= 9(1+y^2)(1+y)(1-y)$$

101. You cannot divide both sides of the equation by $(a-b)$, because it equals 0.

103. $2\blacklozenge^6 + 4\blacklozenge^4\text{\ding{83}}^2$
$$= 2\blacklozenge^4\left(\blacklozenge^2 + 2\text{\ding{83}}^2\right)$$

105. $x^6 - 27y^9 = (x^2)^3 - (3y^3)^3$
$$= (x^2 - 3y^3)(x^4 + 3x^2y^3 + 9y^6)$$

107. $x^2 - 6x + 9 - 4y^2 = (x-3)^2 - (2y)^2$
$$= (x-3+2y)(x-3-2y)$$

109. $x^2 + 10x + 25 - y^2 + 4y - 4$
$$= (x+5)^2 - (y^2 - 4y + 4)$$
$$= (x+5)^2 - (y-2)^2$$
$$= \left[(x+5) + (y-2)\right]\left[(x+5) - (y-2)\right]$$
$$= (x+y+3)(x-y+7)$$

110. $7x - 2(x+6) = 2x - 5$
$$7x - 2x - 12 = 2x - 5$$
$$5x - 12 = 2x - 5$$
$$3x - 12 = -5$$
$$3x = 7$$
$$x = \frac{7}{3}$$

111. Substitute 36 for A, 6 for b, and 12 for d.
$$A = \frac{1}{2}h(b+d)$$
$$36 = \frac{1}{2}h(6+12)$$
$$36 = \frac{1}{2}h(18)$$
$$36 = 9h$$
$$4 = h$$
The height is 4 inches.

112. $-9(a^3b^2c^6)^0 = -9(1) = -9$

113. $\left(\dfrac{4x^4y}{6xy^5}\right)^3 = \left(\dfrac{4}{6} \cdot \dfrac{x^4}{x} \cdot \dfrac{y}{y^5}\right)^3$
$$= \left(\frac{2}{3} \cdot x^3 \cdot \frac{1}{y^4}\right)^3$$
$$= \left(\frac{2x^3}{3y^4}\right)^3$$
$$= \frac{2^3 x^{3\cdot3}}{3^3 y^{4\cdot3}} = \frac{8x^9}{27y^{12}}$$

114. $a^{-4}a^{-7} = a^{-4-7}$

$\qquad = a^{-11}$

$\qquad = \dfrac{1}{a^{11}}$

Exercise Set 6.6

1. Answers will vary.

3. The standard form of a quadratic equation is $ax^2 + bx + c = 0$, $a \neq 0$.

5. a. The zero-factor property may only be used when one side of the equation is equal to 0.

b. $(x+1)(x-2) = 4$

$\quad x^2 - 2x + x - 2 = 4$

$\qquad x^2 - x - 6 = 0$

$\qquad (x-3)(x+2) = 0$

$\quad x - 3 = 0 \quad$ or $\quad x + 2 = 0$

$\qquad x = 3 \qquad\qquad x = -2$

7. $(x+6)(x-7) = 0$

$\quad x + 6 = 0 \quad$ or $\quad x - 7 = 0$

$\qquad x = -6 \qquad\qquad x = 7$

9. $7x(x-8) = 0$

$\quad x = 0 \quad$ or $\quad x - 8 = 0$

$\qquad\qquad\qquad x = 8$

11. $(3x+7)(2x-11) = 0$

$\quad 3x + 7 = 0 \quad$ or $\quad 2x - 11 = 0$

$\qquad 3x = -7 \qquad\qquad 2x = 11$

$\qquad x = -\dfrac{7}{3} \qquad\qquad x = \dfrac{11}{2}$

13. $\qquad x^2 - 16 = 0$

$\quad (x+4)(x-4) = 0$

$\quad x + 4 = 0 \quad$ or $\quad x - 4 = 0$

$\qquad x = -4 \qquad\qquad x = 4$

15. $x^2 - 12x = 0$

$\quad x(x-12) = 0$

$\quad x = 0 \quad$ or $\quad x - 12 = 0$

$\qquad\qquad\qquad x = 12$

17. $\qquad x^2 + 7x = 0$

$\quad x(x+7) = 0$

$\quad x = 0 \quad$ or $\quad x + 7 = 0$

$\qquad\qquad\qquad x = -7$

19. $\qquad x^2 - 8x + 16 = 0$

$\quad (x-4)(x-4) = 0$

$\quad x - 4 = 0$

$\qquad x = 4$

21. $\qquad x^2 + 12x = -20$

$\quad x^2 + 12x + 20 = 0$

$\quad (x+10)(x+2) = 0$

$\quad x + 10 = 0 \qquad$ or $\quad x + 2 = 0$

$\qquad x = -10 \qquad\qquad\qquad x = -2$

23. $\qquad x^2 + 12x + 22 = 2$

$\quad x^2 + 12x + 20 = 0$

$\quad (x+2)(x+10) = 0$

$\quad x + 2 = 0 \qquad$ or $\quad x + 10 = 0$

$\qquad x = -2 \qquad\qquad\qquad x = -10$

25. $\quad 2x^2 - 5x - 24 = -3x$

$\quad 2x^2 - 2x - 24 = 0$

$\quad 2(x^2 - x - 12) = 0$

$\quad 2(x-4)(x+3) = 0$

$\quad x - 4 = 0 \quad$ or $\quad x + 3 = 0$

$\qquad x = 4 \qquad\qquad x = -3$

27. $\qquad 23p - 24 = -p^2$

$\quad p^2 + 23p - 24 = 0$

$\quad (p+24)(p-1) = 0$

$\quad p + 24 = 0 \qquad$ or $\quad p - 1 = 0$

$\qquad p = -24 \qquad\qquad\quad p = 1$

29. $\qquad 33w + 90 = -3w^2$

$\quad 3w^2 + 33w + 90 = 0$

$\quad 3(w^2 + 11w + 30) = 0$

$\quad 3(w+5)(w+6) = 0$

$\quad w + 5 = 0 \qquad$ or $\quad w + 6 = 0$

$\qquad w = -5 \qquad\qquad\quad w = -6$

31.
$$-2x-15=-x^2$$
$$x^2-2x-15=0$$
$$(x-5)(x+3)=0$$
$$x-5=0 \quad \text{or} \quad x+3=0$$
$$x=5 \qquad\qquad x=-3$$

33. $-x^2+29x+30=0$
$$x^2-29x-30=0$$
$$(x-30)(x+1)=0$$
$$x-30=0 \quad \text{or} \quad x+1=0$$
$$x=30 \qquad\qquad x=-1$$

35.
$$-15=4m^2+17m$$
$$4m^2+17m+15=0$$
$$(4m+5)(m+3)=0$$
$$4m+5=0 \quad \text{or} \quad m+3=0$$
$$4m=-5 \qquad\qquad m=-3$$
$$m=-\frac{5}{4}$$

37.
$$9p^2=-21p-6$$
$$9p^2+21p+6=0$$
$$3(3p^2+7p+2)=0$$
$$3(3p+1)(p+2)=0$$
$$3p+1=0 \quad \text{or} \quad p+2=0$$
$$3p=-1 \qquad\qquad p=-2$$
$$p=-\frac{1}{3}$$

39.
$$3r^2+13r=10$$
$$3r^2+13r-10=0$$
$$(3r-2)(r+5)=0$$
$$3r-2=0 \quad \text{or} \quad r+5=0$$
$$3r=2 \qquad\qquad r=-5$$
$$r=\frac{2}{3}$$

41.
$$4x^2+4x-48=0$$
$$4(x^2+x-12)=0$$
$$4(x+4)(x-3)=0$$
$$x+4=0 \quad \text{or} \quad x-3=0$$
$$x=-4 \qquad\qquad x=3$$

43.
$$8x^2+2x=3$$
$$8x^2+2x-3=0$$
$$(4x+3)(2x-1)=0$$
$$4x+3=0 \quad \text{or} \quad 2x-1=0$$
$$4x=-3 \qquad\qquad 2x=1$$
$$x=-\frac{3}{4} \qquad\qquad x=\frac{1}{2}$$

45.
$$c^2=64$$
$$c^2-64=0$$
$$c^2-8^2=0$$
$$(c+8)(c-8)=0$$
$$c+8=0 \quad \text{or} \quad c-8=0$$
$$c=-8 \qquad\qquad c=8$$

47.
$$2x^2=50x$$
$$2x^2-50x=0$$
$$2x(x-25)=0$$
$$2x=0 \quad \text{or} \quad x-25=0$$
$$x=0 \qquad\qquad x=25$$

49.
$$x^2=100$$
$$x^2-100=0$$
$$x^2-10^2=0$$
$$(x+10)(x-10)=0$$
$$x+10=0 \quad \text{or} \quad x-10=0$$
$$x=-10 \qquad\qquad x=10$$

51. $(x-2)(x-1)=12$
$$x^2-3x+2=12$$
$$x^2-3x-10=0$$
$$(x+2)(x-5)=0$$
$$x+2=0 \quad \text{or} \quad x-5=0$$
$$x=-2 \qquad\qquad x=5$$

53. $(3x+2)(x+1)=4$

$3x^2+5x+2=4$

$3x^2+5x-2=0$

$(3x-1)(x+2)=0$

$3x-1=0 \qquad x+2=0$

$x=\dfrac{1}{3} \qquad x=-2$

55. $2(a^2+9)=15a$

$2a^2+18=15a$

$2a^2-15a+18=0$

$(2a-3)(a-6)=0$

$2a-3=0 \quad$ or $\quad a-6=0$

$2a=3 \qquad a=6$

$a=\dfrac{3}{2}$

57. The solutions are 6 and –4, so the factors are $x-6$ and $x+4$.

$(x-6)(x+4)=x^2-2x-24$

The equation is $x^2-2x-24=0$.

59. The solutions are 6 and 0, so the factors are $x-0$ and $x-6$.

$(x-0)(x-6)=x(x-6)$

$=x^2-6x$

The equation is $x^2-6x=0$.

61. a. The solutions are

$x=\dfrac{1}{2} \qquad$ or $\qquad x=-\dfrac{1}{3}$

$2x=1 \qquad\qquad 3x=-1$

$2x-1=0 \qquad 3x+1=0$

Thus the factors are:

$(2x-1)$ and $(3x+1)$.

b. The equation is $(2x-1)(3x+1)$

$=6x^2-x-1=0$.

63. $(2x-3)(x-4)=(x-5)(x+3)+7$

$2x^2-11x+12=x^2-2x-15+7$

$2x^2-11x+12=x^2-2x-8$

$x^2-9x+20=0$

$(x-5)(x-4)=0$

$x-5=0 \text{ or } x-4=0$

$x=5 \qquad x=4$

65. $x(x-3)(x+2)=0$

$x=0 \text{ or } x-3=0 \text{ or } x+2=0$

$x=3 \qquad x=-2$

67. $\dfrac{3}{5}-\dfrac{2}{9}=\dfrac{27}{45}-\dfrac{10}{45}=\dfrac{27-10}{45}=\dfrac{17}{45}$

68. a. identity

b. contradiction

69. Let x be the number of people admitted in 60 minutes.

$\dfrac{160 \text{ people}}{13 \text{ minutes}}=\dfrac{x}{60 \text{ minutes}}$

$13x=160(60)$

$13x=9600$

$x\approx738$

About 738 people were admitted in 60 minutes.

70. $\left(\dfrac{3p^5q^7}{p^9q^8}\right)^2=\left(3\cdot p^{5-9}\cdot q^{7-8}\right)^2$

$=\left(3p^{-4}q^{-1}\right)^2$

$=\left(\dfrac{3}{p^4q}\right)^2$

$=\dfrac{3^2}{p^{4\cdot2}q^{1\cdot2}}=\dfrac{9}{p^8q^2}$

71. monomial

72. binomial

73. not a polynomial; can't have a variable in the denominator.

74. trinomial

Exercise Set 6.7

1. A right triangle is a triangle with a 90° angle.

3. The Pythagorean Theorem is $a^2 + b^2 = c^2$.

5.
$$a^2 + b^2 = c^2$$
$$a^2 + 4^2 = 5^2$$
$$a^2 + 16 = 25$$
$$a^2 = 9$$
$$a = 3$$

7.
$$a^2 + b^2 = c^2$$
$$12^2 + 5^2 = c^2$$
$$144 + 25 = c^2$$
$$169 = c^2$$
$$13 = c$$

9.
$$a^2 + b^2 = c^2$$
$$24^2 + b^2 = 30^2$$
$$576 + b^2 = 900$$
$$b^2 = 324$$
$$b = 18$$

11.
$$a^2 + b^2 = c^2$$
$$15^2 + 36^2 = c^2$$
$$225 + 1296 = c^2$$
$$1521 = c^2$$
$$39 = c$$

13. Let x be the smaller of the two positive numbers. Then $x + 4$ is the other positive number.
$$x(x+4) = 117$$
$$x^2 + 4x + 117 = 0$$
$$(x-9)(x+13) = 0$$
$$x - 9 = 0 \quad \text{or} \quad x + 13 = 0$$
$$x = 9 \qquad x = -13$$
Since x must be positive, the two numbers are 9 and $9 + 4 = 13$.

15. Let $x =$ first positive number. Then $2x + 2$ is the other number.
$$x(2x+2) = 84$$
$$2x^2 + 2x = 84$$
$$2x^2 + 2x - 84 = 0$$
$$2(x^2 + x - 42) = 0$$
$$2(x+7)(x-6) = 0$$
$$x + 7 = 0 \quad \text{or} \quad x - 6 = 0$$
$$x = -7 \qquad x = 6$$
The numbers have to be positive. Thus the numbers are 6 and $2(6) + 2 = 14$.

17. Let x be the first even integer. Then $x + 2$ is the next even integer.
$$x(x+2) = 288$$
$$x^2 + 2x = 288$$
$$x^2 + 2x - 288 = 0$$
$$(x+18)(x-16) = 0$$
$$x + 18 = 0 \quad \text{or} \quad x - 16 = 0$$
$$x = -18 \qquad x = 16$$
Since x must be positive, the numbers are 16 and 18.

19. Let $w =$ width. Then length $= 4w$.
$$A = lw$$
$$36 = (4w)(w)$$
$$36 = 4w^2$$
$$0 = 4w^2 - 36$$
$$0 = 4(w^2 - 9)$$
$$0 = 4(w+3)(w-3)$$
$$w + 3 = 0 \quad \text{or} \quad w - 3 = 0$$
$$w = -3 \qquad w = 3$$
Since dimensions must be positive, the width is 3 feet and the length is $4(3) = 12$ feet.

21. Let l = length of the garden and $\frac{2}{3}l$ = width of the garden.

$$lw = A$$
$$l\left(\frac{2}{3}l\right) = 150$$
$$2l^2 = 450$$
$$2l^2 - 450 = 0$$
$$2\left(l^2 - 225\right) = 0$$
$$2\left(l^2 - 15^2\right) = 0$$
$$2(l+15)(l-15) = 0$$
$$l+15 = 0 \quad \text{or} \quad l-15 = 0$$
$$l = -15 \qquad\qquad l = 15$$

Since dimensions must be positive, the length is

15 feet and the width is $\frac{2}{3}(15)$ = 10 feet.

23. Let a = length of a side of the original square. Then $a + 5$ is the length of the new side.

$$(\text{side}) \cdot (\text{side}) = \text{Area}$$
$$(a+5)(a+5) = 81$$
$$a^2 + 10a + 25 = 81$$
$$a^2 + 10a - 56 = 0$$
$$(a+14)(a-4) = 0$$
$$a+14 = 0 \quad \text{or} \quad a-4 = 0$$
$$a = -14 \quad \text{or} \qquad a = 4$$

Since length must be positive, the original square had sides of length 4 meters.

25. $d = 16t^2$
$$256 = 16t^2$$
$$\frac{256}{16} = t^2$$
$$16 = t^2$$
$$0 = t^2 - 16$$
$$0 = (t+4)(t-4)$$
$$t+4 = 0 \quad \text{or} \quad t-4 = 0$$
$$t = -4 \qquad\qquad t = 4$$

Since time must be positive, it would take the egg 4 seconds to hit the ground.

27. $a^2 + b^2 = c^2$
$$7^2 + 24^2 = 25^2$$
$$49 + 576 = 625$$
$$625 = 625 \quad \text{True}$$
Since these values are true for the Pythagorean Theorem, a right triangle can exist.

29. $a^2 + b^2 = c^2$
$$9^2 + 40^2 = 41^2$$
$$81 + 1600 = 1681$$
$$1681 = 1681 \quad \text{True}$$
Since these values are true for the Pythagorean Theorem, a right triangle can exist.

31. $a^2 + b^2 = c^2$
$$30^2 + b^2 = 34^2$$
$$900 + b^2 = 1156$$
$$b^2 = 256$$
$$b = 16 \text{ feet}$$

33. $a^2 + b^2 = c^2$
$$9^2 + 40^2 = c^2$$
$$81 + 1600 = c^2$$
$$1681 = c^2$$
$$41 = c$$
$$c = 41 \text{ feet}$$

35. Let x be the length of one leg of the triangle. Then $x + 2$ is the length of the other leg.

$$a^2 + b^2 = c^2$$
$$x^2 + (x+2)^2 = 10^2$$
$$x^2 + x^2 + 4x + 4 = 100$$
$$2x^2 + 4x + 4 = 100$$
$$2x^2 + 4x - 96 = 0$$
$$2\left(x^2 + 2x - 48\right) = 0$$
$$2(x+8)(x-6) = 0$$
$$x+8 = 0 \quad \text{or} \quad x-6 = 0$$
$$x = -8 \qquad\qquad x = 6$$

Since length is positive, the lengths are 6 ft, 8ft and 10ft.

37. Let w = width of the frame, then $w + 3$ = the length of the frame. The diagonal is 15.

$$a^2 + b^2 = c^2$$
$$(w+3)^2 + w^2 = 15^2$$
$$w^2 + 6w + 9 + w^2 = 225$$
$$2w^2 + 6w + 9 = 225$$
$$2w^2 + 6w - 216 = 0$$
$$2(w^2 + 3w - 108) = 0$$
$$2(w+12)(w-9) = 0$$
$$w + 12 = 0 \quad \text{or} \quad w - 9 = 0$$
$$w = -12 \qquad \qquad w = 9$$

The width is 9 in. and the length is 9 + 3 = 12 in.

39. Let w be the width of the rectangle. Then $3w + 3$ is the length and $3w + 4$ is the length of the diagonal.

$$a^2 + b^2 = c^2$$
$$w^2 + (3w+3)^2 = (3w+4)^2$$
$$w^2 + 9w^2 + 18w + 9 = 9w^2 + 24w + 16$$
$$10w^2 + 18w + 9 = 9w^2 + 24w + 16$$
$$w^2 - 6w - 7 = 0$$
$$(w-7)(w+1) = 0$$
$$w - 7 = 0 \quad \text{or} \quad w + 1 = 0$$
$$w = 7 \qquad \qquad w = -1$$

Since length is positive, the dimensions of the garden are 7 ft for the width and $3w + 3 = 3(7) + 3 = 24$ ft for the length.

41.
$$P = x^2 - 15x - 50$$
$$x^2 - 15x - 50 = 400$$
$$x^2 - 15 - 450 = 0$$
$$(x-30)(x+15) = 0$$
$$x - 30 = 0 \quad \text{or} \quad x + 15 = 0$$
$$x = 30 \qquad \qquad x = -15$$

Since x must be positive, she must sell 30 books for a profit of $400.

43. a.
$$n^2 + n = 20$$
$$n^2 + n - 20 = 0$$
$$(n-4)(n+5) = 0$$
$$n - 4 = 0 \quad \text{or} \quad n + 5 = 0$$
$$n = 4 \qquad \qquad n = -5$$

Since n must be positive, $n = 4$.

b.
$$n^2 + n = 90$$
$$n^2 + n - 90 = 0$$
$$(n-9)(n+10) = 0$$
$$n - 9 = 0 \quad \text{or} \quad n + 10 = 0$$
$$n = 9 \qquad \qquad n = -10$$

Since n must be positive, $n = 9$.

45. Before area can be determined, the length of the rectangle must be found first. Let x be the length of the rectangle.

$$a^2 + b^2 = c^2$$
$$x^2 + 18^2 = 30^2$$
$$x^2 + 324 = 900$$
$$x^2 = 576$$
$$x = 24$$

Area of a rectangle can be found by multiplying the length and width.

$$A = lw$$
$$= 24 \cdot 18$$
$$= 432$$

The area of the rectangle is 432 square feet.

47.
$$x^3 - 4x^2 - 32x = 0$$
$$x(x^2 - 4x - 32) = 0$$
$$x(x-8)(x+4) = 0$$
$$x = 0 \quad \text{or} \quad x - 8 = 0 \quad \text{or} \quad x + 4 = 0$$
$$x = 8 \qquad \qquad x = -4$$

49. Solutions are –2, 0, and 3, so the factors are $x + 2$, x, and $x - 3$.

$$x(x+2)(x-3) = x(x^2 - x - 6)$$
$$= x^3 - x^2 - 6x$$

The equation is $x^3 - x^2 - 6x = 0$.

51. $x + y = 9$ and $x^2 + y^2 = 45$

$y = 9 - x$

$$x^2 + (9-x)^2 = 45$$

$$x^2 + 81 - 18x + x^2 = 45$$

$$2x^2 - 18x + 81 - 45 = 0$$

$$2(x^2 - 9x + 18) = 0$$

$$2(x-3)(x-6) = 0$$

$x - 3 = 0$ or $x - 6 = 0$

$x = 3$ $\qquad$ $x = 6$

When $x = 3$, $y = 6$ and when $x = 6$, $y = 3$.

The numbers are 3 and 6.

55. "Seven less than three times a number" is $3x - 7$.

56. $(3x+2) - (x^2 - 4x + 6) = 3x + 2 - x^2 + 4x - 6$

$$= -x^2 + 3x + 4x + 2 - 6$$

$$= -x^2 + 7x - 4$$

57. $(3x^2 + 2x - 4)(2x - 1)$

$$= 3x^2(2x-1) + 2x(2x-1) - 4(2x-1)$$

$$= 6x^3 - 3x^2 + 4x^2 - 2x - 8x + 4$$

$$= 6x^3 + x^2 - 10x + 4$$

58.

$$\begin{array}{r} 2x - 3 \\ 3x-5 \overline{)\, 6x^2 - 19x + 15} \\ \underline{6x^2 - 10x} \\ -9x + 15 \\ \underline{-9x + 15} \\ 0 \end{array}$$

$$\frac{6x^2 - 19x + 15}{3x - 5} = 2x - 3$$

59. $\dfrac{6x^2 - 19x + 15}{3x - 5} = \dfrac{(3x-5)(2x-3)}{3x-5}$

$$= 2x - 3$$

Chapter 6 Review Exercises

1. The greatest common factor is y^3.

2. The greatest common factor is $3p$.

3. The greatest common factor is $6c^2$.

4. The greatest common factor is $5x^2 y^2$.

5. The greatest common factor is 1.

6. The greatest common factor is s.

7. The greatest common factor is $x - 3$.

8. The greatest common factor is $x + 5$.

9. $7x - 35 = 7(x - 5)$

10. $35x - 5 = 5(7x - 1)$

11. $24y^2 - 4y = 4y(6y - 1)$

12. $55p^3 - 20p^2 = 5p^2(11p - 4)$

13. $60a^2 b - 36ab^2 = 12ab(5a - 3b)$

14. $9xy - 36x^3 y^2 = 9xy(1 - 4x^2 y)$

15. $20x^3 y^2 + 8x^9 y^3 - 16x^5 y^2$

$$= 4x^3 y^2(5 + 2x^6 y - 4x^2)$$

16. $24x^2 - 13y^2 + 6xy$ is prime.

17. $14a^2 b - 7b - a^3$ is prime.

18. $x(5x+3) - 2(5x+3) = (5x+3)(x-2)$

19. $3x(x-1) + 4(x-1) = (x-1)(3x+4)$

20. $2x(4x-3) + 4x - 3 = 2x(4x-3) + 1(4x-3)$

$$= (4x-3)(2x+1)$$

21. $x^2 + 6x + 2x + 12 = x(x+6) + 2(x+6)$

$$= (x+6)(x+2)$$

22. $x^2 - 5x + 4x - 20 = x(x-5) + 4(x-5)$

$$= (x-5)(x+4)$$

23. $y^2 - 6y - 6y + 36 = y(y-6) - 6(y-6)$

$$= (y-6)(y-6)$$

$$= (y-6)^2$$

24. $3xy + 3x + 2y + 2 = 3x(y+1) + 2(y+1)$

$$= (y+1)(3x+2)$$

25. $4a^2 - 4ab - a + b = 4a(a-b) - 1(a-b)$

$$= (a-b)(4a-1)$$

26. $2x^2 + 12x - x - 6 = 2x(x+6) - 1(x+6)$

$$= (x+6)(2x-1)$$

27. $x^2 + 3x - 2xy - 6y = x(x+3) - 2y(x+3)$
$$= (x+3)(x-2y)$$

28. $5x^2 - xy + 20xy - 4y^2$
$$= x(5x-y) + 4y(5x-y)$$
$$= (5x-y)(x+4y)$$

29. $4x^2 + 12xy - 5xy - 15y^2$
$$= 4x(x+3y) - 5y(x+3y)$$
$$= (x+3y)(4x-5y)$$

30. $6a^2 - 10ab - 3ab + 5b^2$
$$= 2a(3a-5b) - b(3a-5b)$$
$$= (3a-5b)(2a-b)$$

31. $pq - 3q + 4p - 12 = q(p-3) + 4(p-3)$
$$= (p-3)(q+4)$$

32. $3x^2 - 9xy + 2xy - 6y^2 = 3x(x-3y) + 2y(x-3y)$
$$= (x-3y)(3x+2y)$$

33. $7a^2 + 14ab - ab - 2b^2 = 7a(a+2b) - b(a+2b)$
$$= (a+2b)(7a-b)$$

34. $8x^2 - 4x + 6x - 3 = 4x(2x-1) + 3(2x-1)$
$$= (2x-1)(4x+3)$$

35. $x^2 - x - 6 = (x+2)(x-3)$

36. $x^2 + 4x - 15$ is prime.

37. $x^2 + 11x + 18 = (x+9)(x+2)$

38. $n^2 + 3n - 40 = (n+8)(n-5)$

39. $b^2 + b - 20 = (b-4)(b+5)$

40. $x^2 - 15x + 56 = (x-8)(x-7)$

41. $c^2 - 10c - 20$ is prime.

42. $y^2 - 10y - 22$ is prime.

43. $x^3 - 17x^2 + 72x = x(x^2 - 17x + 72)$
$$= x(x-9)(x-8)$$

44. $t^3 - 5t^2 - 36t = t(t^2 - 5t - 36)$
$$= t(t-9)(t+4)$$

45. $x^2 - 2xy - 15y^2 = (x-5y)(x+3y)$

46. $4x^3 + 32x^2 y + 60xy^2 = 4x(x^2 + 8xy + 15y^2)$
$$= 4x(x+3y)(x+5y)$$

47. $2x^2 - x - 15 = (2x+5)(x-3)$

48. $6x^2 - 29x - 5 = (6x+1)(x-5)$

49. $4x^2 - 9x + 5 = (4x-5)(x-1)$

50. $5m^2 - 14m + 8 = (5m-4)(m-2)$

51. $16y^2 + 8y - 3 = (4y+3)(4y-1)$

52. $5x^2 - 32x + 12 = (5x-2)(x-6)$

53. $2t^2 + 14t + 9$ is prime.

54. $5x^2 + 37x - 24 = (5x-3)(x+8)$

55. $6s^2 + 13s + 5 = (2s+1)(3s+5)$

56. $6x^2 + 11x - 10 = (3x-2)(2x+5)$

57. $12x^2 + 2x - 4 = 2(6x^2 + x - 2)$
$$= 2(3x+2)(2x-1)$$

58. $25x^2 - 30x + 9 = (5x-3)(5x-3)$
$$= (5x-3)^2$$

59. $9x^3 - 12x^2 + 4x = x(9x^2 - 12x + 4)$
$$= x(3x-2)(3x-2)$$
$$= x(3x-2)^2$$

60. $18x^3 + 12x^2 - 16x = 2x(9x^2 + 6x - 8)$
$$= 2x(3x+4)(3x-2)$$

61. $4a^2 - 16ab + 15b^2 = (2a-3b)(2a-5b)$

62. $16a^2 - 22ab - 3b^2 = (8a+b)(2a-3b)$

63. $x^2 - 100 = x^2 - 10^2$
$$= (x+10)(x-10)$$

64. $x^2 - 36 = x^2 - 6^2$
$$= (x+6)(x-6)$$

65. $3x^2 - 48 = 3(x^2 - 16)$
$$= 3(x^2 - 4^2)$$
$$= 3(x+4)(x-4)$$

66. $81x^2 - 9y^2 = 9(9x^2 - y^2)$
$$= 9[(3x)^2 - y^2]$$
$$= 9(3x+y)(3x-y)$$

67. $81 - a^2 = 9^2 - a^2$
$$= (9+a)(9-a)$$

68. $64 - x^2 = 8^2 - x^2$
$$= (8+x)(8-x)$$

69. $16x^4 - 49y^2 = (4x^2)^2 - (7y)^2$
$$= (4x^2 + 7y)(4x^2 - 7y)$$

70. $64x^6 - 49y^6 = (8x^3)^2 - (7y^3)^2$
$$= (8x^3 + 7y^3)(8x^3 - 7y^3)$$

71. $a^3 + b^3 = (a+b)(a^2 - ab + b^2)$

72. $x^3 - y^3 = (x-y)(x^2 + xy + y^2)$

73. $x^3 - 1 = (x-1)(x^2 + x + 1)$

74. $x^3 + 8 = x^3 + 2^3$
$$= (x+2)(x^2 - 2x + 4)$$

75. $a^3 + 27 = a^3 + 3^3$
$$= (a+3)(a^2 - 3a + 9)$$

76. $b^3 - 64 = b^3 - 4^3$
$$= (b-4)(b^2 + 4b + 16)$$

77. $125a^3 + b^3 = (5a)^3 + b^3$
$$= (5a+b)(25a^2 - 5ab + b^2)$$

78. $27 - 8y^3 = 3^3 - (2y)^3$
$$= (3-2y)(9+6y+4y^2)$$

79. $3x^3 - 192y^3 = 3(x^3 - 64y^3)$
$$= 3[x^3 - (4y)^3]$$
$$= 3(x-4y)(x^2 + 4xy + 16y^2)$$

80. $27x^4 - 75y^2 = 3(9x^4 - 25y^2)$
$$= 3[(3x^2)^2 - (5y)^2]$$
$$= 3(3x^2 + 5y)(3x^2 - 5y)$$

81. $x^2 - 14x + 48 = (x-6)(x-8)$

82. $3x^2 - 18x + 27 = 3(x^2 - 6x + 9)$
$$= 3(x-3)^2$$

83. $5q^2 - 5 = 5(q^2 - 1)$
$$= 5(q^2 - 1^2)$$
$$= 5(q+1)(q-1)$$

84. $8x^2 + 16x - 24 = 8(x^2 + 2x - 3)$
$$= 8(x+3)(x-1)$$

85. $4y^2 - 36 = 4(y^2 - 9)$
$$= 4(y^2 - 3^2)$$
$$= 4(y+3)(y-3)$$

86. $x^2 - 6x - 27 = (x-9)(x+3)$

87. $9x^2 - 6x + 1 = (3x-1)(3x-1)$
$$= (3x-1)^2$$

88. $7x^2 + 25x - 12 = (7x-3)(x+4)$

89. $6b^3 - 6 = 6(b^3 - 1)$
$$= 6(b^3 - 1^3)$$
$$= 6(b-1)(b^2 + b + 1)$$

90. $x^3y - 27y = y(x^3 - 27)$
$$= y(x^3 - 3^3)$$
$$= y(x-3)(x^2 + 3x + 9)$$

91. $a^2b - 2ab - 15b = b(a^2 - 2a - 15)$
$$= b(a+3)(a-5)$$

92. $6x^3 + 30x^2 + 9x^2 + 45x$
$$= 3x(2x^2 + 10x + 3x + 15)$$
$$= 3x[2x(x+5) + 3(x+5)]$$
$$= 3x(2x+3)(x+5)$$

93. $x^2 - 4xy + 3y^2 = (x-3y)(x-y)$

94. $3m^2 + 2mn - 8n^2 = (3m-4n)(m+2n)$

95. $4x^2 + 12xy + 9y^2 = (2x+3y)(2x+3y)$
$$= (2x+3y)^2$$

96. $25a^2 - 49b^2 = (5a)^2 - (7b)^2$
$$= (5a+7b)(5a-7b)$$

97. $xy - 7x + 2y - 14 = x(y-7) + 2(y-7)$
$$= (x+2)(y-7)$$

98. $16y^5 - 25y^7 = y^5(16 - 25y^2)$
$$= y^5[4^2 - (5y)^2]$$
$$= y^5(4+5y)(4-5y)$$

99. $6x^2 + 5xy - 21y^2 = (2x-3y)(3x+7y)$

100. $4x^3 + 18x^2y + 20xy^2 = 2x(2x^2 + 9xy + 10y^2)$
$$= 2x(2x+5y)(x+2y)$$

101. $16x^4 - 8x^3 - 3x^2 = x^2(16x^2 - 8x - 3)$
$$= x^2(4x+1)(4x-3)$$

102. $d^4 - 16 = (d^2)^2 - 4^2$
$$= (d^2+4)(d^2-4)$$
$$= (d^2+4)(d+2)(d-2)$$

103. $x(x+9) = 0$
$x = 0$ or $x+9 = 0$
$$x = -9$$

104. $(a-2)(a+6) = 0$
$a-2 = 0$ or $a+6 = 0$
$a = 2$ $\qquad a = -6$

105. $(x+5)(4x-3) = 0$
$x+5 = 0$ or $4x-3 = 0$
$x = -5$ $\qquad 4x = 3$
$$x = \frac{3}{4}$$

106. $x^2 + 7x = 0$
$x(x+7) = 0$
$x = 0$ or $x+7 = 0$
$$x = -7$$

107. $6x^2 + 30x = 0$
$6x(x+5) = 0$
$6x = 0$ or $x+5 = 0$
$x = 0$ $\qquad x = -5$

108. $6x^2 + 18x = 0$
$6x(x+3) = 0$
$6x = 0$ or $x+3 = 0$
$x = 0$ $\qquad x = -3$

109. $r^2 + 9r + 18 = 0$
$(r+3)(r+6) = 0$
$r+3 = 0$ or $r+6 = 0$
$r = -3$ $\qquad r = -6$

110. $x^2 - 3x = -2$
$x^2 - 3x + 2 = 0$
$(x-1)(x-2) = 0$
$x-1 = 0$ or $x-2 = 0$
$x = 1$ $\qquad x = 2$

111. $x^2 - 12 = -x$
$x^2 + x - 12 = 0$
$(x+4)(x-3) = 0$
$x+4 = 0$ or $x-3 = 0$
$x = -4$ $\qquad x = 3$

112. $15x + 12 = -3x^2$

$3x^2 + 15x + 12 = 0$

$3\left(x^2 + 5x + 4\right) = 0$

$3(x+1)(x+4) = 0$

$x + 1 = 0 \quad \text{or} \quad x + 4 = 0$

$x = -1 \qquad x = -4$

113. $x^2 - 6x + 8 = 0$

$(x-4)(x-2) = 0$

$x - 4 = 0 \quad \text{or} \quad x - 2 = 0$

$x = 4 \qquad x = 2$

114. $3p^2 + 6p = 45$

$3p^2 + 6p - 45 = 0$

$3\left(p^2 + 2p - 15\right) = 0$

$3(x-3)(x+5) = 0$

$x - 3 = 0 \quad \text{or} \quad x + 5 = 0$

$x = 3 \qquad x = -5$

115. $8x^2 - 3 = -10x$

$8x^2 + 10x - 3 = 0$

$(4x-1)(2x+3) = 0$

$4x - 1 = 0 \quad \text{or} \quad 2x + 3 = 0$

$4x = 1 \qquad 2x = -3$

$x = \dfrac{1}{4} \qquad x = -\dfrac{3}{2}$

116. $3p^2 - 11p = 4$

$3p^2 - 11p - 4 = 0$

$(3p+1)(p-4) = 0$

$3p + 1 = 0 \quad \text{or} \quad p - 4 = 0$

$3p = -1 \qquad p = 4$

$p = -\dfrac{1}{3}$

117. $4x^2 - 16 = 0$

$4\left(x^2 - 4\right) = 0$

$4\left(x^2 - 2^2\right) = 0$

$4(x+2)(x-2) = 0$

$x + 2 = 0 \quad \text{or} \quad x - 2 = 0$

$x = -2 \qquad x = 2$

118. $49x^2 - 100 = 0$

$(7x)^2 - 10^2 = 0$

$(7x+10)(7x-10) = 0$

$7x + 10 = 0 \quad \text{or} \quad 7x - 10 = 0$

$7x = -10 \qquad 7x = 10$

$x = -\dfrac{10}{7} \qquad x = \dfrac{10}{7}$

119. $8x^2 - 14x + 3 = 0$

$(2x-3)(4x-1) = 0$

$2x - 3 = 0 \quad \text{or} \quad 4x - 1 = 0$

$2x = 3 \qquad 4x = 1$

$x = \dfrac{3}{2} \qquad x = \dfrac{1}{4}$

120. $-48x = -12x^2 - 45$

$12x^2 - 48x + 45 = 0$

$3\left(4x^2 - 16x + 15\right) = 0$

$3(2x-3)(2x-5) = 0$

$2x - 3 = 0 \quad \text{or} \quad 2x - 5 = 0$

$2x = 3 \qquad 2x = 5$

$x = \dfrac{3}{2} \qquad x = \dfrac{5}{2}$

121. $a^2 + b^2 = c^2$

122. hypotenuse

123. $a^2 + b^2 = c^2$

$6^2 + 8^2 = c^2$

$36 + 64 = c^2$

$100 = c^2$

$10 \text{ ft} = c$

124. $a^2 + b^2 = c^2$

$a^2 + 5^2 = 13^2$

$a^2 + 25 = 169$

$a^2 = 144$

$a = 12 \text{ m}$

125. Let x be the smaller integer. The larger is $x + 2$.

$$x(x+2) = 99$$
$$x^2 + 2x = 99$$
$$x^2 + 2x - 99 = 0$$
$$(x+11)(x-9) = 0$$
$$x + 11 = 0 \quad \text{or} \quad x - 9 = 0$$
$$x = -11 \qquad x = 9$$

The integers must be positive, so they are 9 and 11.

126. Let x be the smaller integer. Then the larger is $2x + 6$.

$$x(2x+6) = 56$$
$$2x^2 + 6x = 56$$
$$2x^2 + 6x - 56 = 0$$
$$2(x^2 + 3x - 28) = 0$$
$$(x+7)(x-4) = 0$$
$$x + 7 = 0 \quad \text{or} \quad x - 4 = 0$$
$$x = -7 \qquad x = 4$$

The integers must be positive, so they are 4 and 14.

127. Let w be the width of the rectangle. Then the length is $w + 3$.

$$w(w+3) = 180$$
$$w^2 + 3w = 180$$
$$w^2 + 3w - 180 = 0$$
$$(w+15)(w-12) = 0$$
$$w + 15 = 0 \quad \text{or} \quad w - 12 = 0$$
$$w = -15 \qquad w = 12$$

Since the width must be positive, it is 12 feet, and the length is 15 feet.

128. Let x be the length of one leg of the triangle. Then $x + 7$ is the length of the other leg and $x + 9$ is the length of the hypotenuse.

$$a^2 + b^2 = c^2$$
$$x^2 + (x+7)^2 = (x+9)^2$$
$$x^2 + x^2 + 14x + 49 = x^2 + 18x + 81$$
$$2x^2 + 14x + 49 = x^2 + 18x + 81$$
$$x^2 - 4x - 32 = 0$$
$$(x-8)(x+4) = 0$$
$$x - 8 = 0 \quad \text{or} \quad x + 4 = 0$$
$$x = 8 \qquad x = -4$$

Since lengths must be positive, the lengths of the three sides are 8 ft, 15 ft, and 17 ft.

129. Let x be the length of a side of the original square. Then $x - 4$ is the length of a side of the smaller square.

$$(x-4)^2 = 25$$
$$x^2 - 8x + 16 = 25$$
$$x^2 - 8x + 16 - 25 = 0$$
$$x^2 - 8x - 9 = 0$$
$$(x-9)(x+1) = 0$$
$$x - 9 = 0 \quad \text{or} \quad x + 1 = 0$$
$$x = 9 \qquad x = -1$$

Since lengths must be positive, the length of a side of the original square is 9 inches.

130. Let w be the width of the table. Then the length is $w + 2$ and the diagonal is $w + 4$.

$$a^2 + b^2 = c^2$$
$$w^2 + (w+2)^2 = (w+4)^2$$
$$w^2 + w^2 + 4x + 4 = w^2 + 8x + 16$$
$$2w^2 + 4x + 4 = w^2 + 8x + 16$$
$$w^2 - 4w - 12 = 0$$
$$(w-6)(w+2) = 0$$
$$w - 6 = 0 \quad \text{or} \quad w + 2 = 0$$
$$w = 6 \qquad w = -2$$

Since lengths must be positive, the diagonal is $w + 4 = 6 + 4 = 10$ ft.

131. $d = 16t^2$

$16 = 16t^2$

$1 = t^2$

$1 = t$

It will take 1 second for the pear to hit the ground.

132. $C = x^2 - 79x + 20$

$100 = x^2 - 79x + 20$

$0 = x^2 - 79x - 80$

$0 = (x - 80)(x + 1)$

$x - 80 = 0$ or $x + 1 = 0$

$x = 80$ $\qquad x = -1$

Since only a positive number of dozens of brownies can be made, the association can make 80 dozen brownies.

Chapter 6 Practice Test

1. The greatest common factor is $3y^3$.

2. The greatest common factor is $8p^2q^2$.

3. $5x^2y^3 - 15x^5y^2 = 5x^2y^2\left(y - 3x^3\right)$

4. $8a^3b - 12a^2b^2 + 28a^2b = 4a^2b\left(2a - 3b + 7\right)$

5. $4x^2 - 20x + x - 5 = 4x(x - 5) + 1(x - 5)$

$\qquad\qquad\qquad = (4x + 1)(x - 5)$

6. $a^2 - 4ab - 5ab + 20b^2 = a(a - 4b) - 5b(a - 4b)$

$\qquad\qquad\qquad\qquad = (a - 4b)(a - 5b)$

7. $r^2 + 5r - 24 = (r + 8)(r - 3)$

8. $25a^2 - 5ab - 6b^2 = (5a - 3b)(5a + 2b)$

9. $4x^2 - 16x - 48 = 4\left(x^2 - 4x - 12\right)$

$\qquad\qquad\qquad = 4(x + 2)(x - 6)$

10. $2y^3 - y^2 - 3y = y\left(2y^2 - y - 3\right)$

$\qquad\qquad\qquad = y(2y - 3)(y + 1)$

11. $12x^2 - xy - 6y^2 = (3x + 2y)(4x - 3y)$

12. $x^2 - 9y^2 = x^2 - (3y)^2$

$\qquad\qquad = (x + 3y)(x - 3y)$

13. $x^3 - 64 = x^3 - 4^3$

$\qquad\qquad = (x - 4)\left(x^2 + 4x + 16\right)$

14. $(6x - 5)(x + 3) = 0$

$6x - 5 = 0$ or $x + 3 = 0$

$6x = 5$ $\qquad\qquad x = -3$

$x = \dfrac{5}{6}$

15. $x^2 - 6x = 0$

$x(x - 6) = 0$

$x = 0$ or $x - 6 = 0$

$\qquad\qquad\qquad x = 6$

16. $x^2 = 64$

$x^2 - 64 = 0$

$x^2 - 8^2 = 0$

$(x + 8)(x - 8) = 0$

$x + 8 = 0$ or $x - 8 = 0$

$x = -8$ $\qquad\qquad x = 8$

17. $x^2 + 18x + 81 = 0$

$(x + 9)^2 = 0$

$x + 9 = 0$

$x = -9$

18. $x^2 - 7x + 12 = 0$

$(x - 3)(x - 4) = 0$

$x - 3 = 0$ or $x - 4 = 0$

$x = 3$ $\qquad\qquad x = 4$

19. $x^2 + 6 = -5x$

$x^2 + 5x + 6 = 0$

$(x + 2)(x + 3) = 0$

$x - 2 = 0$ or $x + 3 = 0$

$x = -2$ $\qquad\qquad x = -3$

20. Use Pythagorean Theorem.

$a^2 + b^2 = c^2$

$a^2 + 10^2 = 26^2$

$a^2 + 100 = 676$

$a^2 = 576$

$a = 24$ in.

21. Let x be the length of one leg. Then $2x - 2$ is the length of the other leg and $2x + 2$ is the length of the hypotenuse.

$$x^2 + (2x-2)^2 = (2x+2)^2$$
$$x^2 + 4x^2 - 8x + 4 = 4x^2 + 8x + 4$$
$$5x^2 - 8x + 4 = 4x^2 + 8x + 4$$
$$x^2 - 16x = 0$$
$$x(x-16) = 0$$
$$x = 0 \quad \text{or} \quad x - 16 = 0$$
$$x = 16$$

Since length has to be positive, the hypotenuse is $2x + 2 = 2(16) + 2 = 32 + 2 = 34$ ft.

22. Let x be the smaller of the two integers. Then $2x + 1$ is the larger.

$$x(2x+1) = 36$$
$$2x^2 + x - 36 = 0$$
$$(x-4)(2x+9) = 0$$
$$x - 4 = 0 \quad \text{or} \quad 2x + 9 = 0$$
$$x = 4 \qquad\qquad 2x = -9$$
$$x = -\frac{9}{2}$$

Since x must be positive and an integer, the smaller integer is 4 and the larger is $2 \cdot 4 + 1 = 9$.

23. Let x be the smaller of the two consecutive even integers. Then $x + 2$ is the larger.

$$x(x+2) = 168$$
$$x^2 + 2x - 168 = 0$$
$$(x+14)(x-12) = 0$$
$$x + 14 = 0 \quad \text{or} \quad x - 12 = 0$$
$$x = -14 \qquad\qquad x = 12$$

Since x must be positive, then the smaller integer is 12 and the larger is 14.

24. Let w be the width of the rectangle. Then the length is $w + 2$.

$$w(w+2) = 24$$
$$w^2 + 2w = 24$$
$$w^2 + 2w - 24 = 0$$
$$(w+6)(w-4) = 0$$
$$w + 6 = 0 \quad \text{or} \quad w - 4 = 0$$
$$w = -6 \qquad\qquad w = 4$$

Since the width is positive, it is 4 meters, and the length is 6 meters.

25. $d = 16t^2$

$$1600 = 16t^2$$
$$16t^2 - 1600 = 0$$
$$16(t^2 - 100) = 0$$
$$16(t^2 - 10^2) = 0$$
$$16(t+10)(t-10) = 0$$
$$t + 10 = 0 \quad \text{or} \quad t - 10 = 0$$
$$t = -10 \qquad\qquad t = 10$$

Since time must be positive, then it would take the object 10 seconds to fall 1600 feet to the ground.

Chapter 6 Cumulative Review Test

1. $4 - 5(2x + 4x^2 - 21)$

$$= 4 - 5[2(-4) + 4(-4)^2 - 21]$$
$$= 4 - 5[-8 + 4(16) - 21]$$
$$= 4 - 5(-8 + 64 - 21)$$
$$= 4 - 5(35)$$
$$= 4 - 175$$
$$= -171$$

2. $5x^2 - 3y + 7(2 + y^2 - 4x)$

$$= 5(3)^2 - 3(-2) + 7[2 + (-2)^2 - 4(3)]$$
$$= 5(9) + 6 + 7(2 + 4 - 12)$$
$$= 45 + 6 + 7(-6)$$
$$= 51 - 42$$
$$= 9$$

3. Let x = the cost of the room before tax.

$$x + (.15x) = 103.50$$
$$1.15x = 103.50$$
$$x = 90$$

The motel room costs $90 before taxes.

4. a. 7 is a natural number

b. $-6, -0.2, \dfrac{3}{5}, 7, 0, -\dfrac{5}{9}$, and 1.34 are rational numbers.

c. $\sqrt{7}$ and $-\sqrt{2}$ are irrational numbers.

d. All of the numbers are real numbers.

5. $\left|-8\right|$ is greater than $-\left|8\right|$ since $\left|-8\right| = 8$ and $-\left|8\right| = -(8) = -8$.

6. $4x - 2 = 4(x - 7) + 2x$

   $4x - 2 = 4x - 28 + 2x$

   $4x - 2 = 6x - 28$

   $\quad 26 = 2x$

   $\quad 13 = x$

7. $\dfrac{5}{12} = \dfrac{8}{x}$

   $5x = 8(12)$

   $5x = 96$

   $x = \dfrac{96}{5} = 19.2$

8. $\dfrac{1 \text{ gal}}{825 \text{ ft}^2} = \dfrac{x \text{ gal}}{5775 \text{ ft}^2}$

   $\dfrac{1}{825} = \dfrac{x}{5775}$

   $\dfrac{5775}{825} = x$

   $\quad 7 = x$

   It will take 7 gallons of paint to cover the house.

9. $4x + 3y = 7$

   $3y = -4x + 7$

   $y = \dfrac{-4x + 7}{3}$

   $y = -\dfrac{4}{3}x + \dfrac{7}{3}$

10. Let x be the amount of 10% acid solution needed.

    $0.10x + 0.04(3) = 0.08(x + 3)$

    $0.10x + 0.12 = 0.08x + 0.24$

    $0.02x + 0.12 = 0.24$

    $0.02x = 0.12$

    $x = 6$

    Six liters of the 10% solution is needed.

11. Let t = the number of hours that Brooke has been skiing. Then Bob has been skiing for $\left(t + \dfrac{1}{4}\right)$ hours.

	rate	time	distance
Brooke	8 kph	t	$8t$
Bob	4 kph	$t + \dfrac{1}{4}$	$4\left(t + \dfrac{1}{4}\right)$

    Brooke catches Bob when they have both gone the same distance.

    $8t = 4\left(t + \dfrac{1}{4}\right)$

    $8t = 4t + 1$

    $4t = 1$

    $t = \dfrac{1}{4}$

    It will take Brooke $\dfrac{1}{4}$ hour to catch Bob.

12. $y = -\dfrac{3}{5}x + 1$

    $m = -\dfrac{3}{5}$; y-intercept: $(0, 1)$

    Writing the slope as $m = \dfrac{-3}{5}$, we can use it and the y-intercept to obtain an additional point on the graph of the equation.

    $(0 + 5, 1 - 3) \rightarrow (5, -2)$

    The point $(5, -2)$ is also on the graph. Plot this point and the y-intercept, connect the points with a straight line and extend the line in both directions.

    Note: If we had written the slope as $m = \dfrac{3}{-5}$, we would have plotted the point $(-5, 4)$ with the y-intercept to obtain the graph.

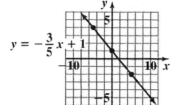

13. The slope must be found first.
Let $(3, 7)$ be (x_1, y_1) and $(-2, 4)$ be (x_2, y_2).

$$m = \frac{y_2 - y_1}{x_2 - x_1}$$

$$m = \frac{4-7}{-2-3} = \frac{-3}{-5} = \frac{3}{5}$$

Now choose one of the points and the point-slope formula.

$$y - y_1 = m(x - x_1)$$

$$y - 7 = \frac{3}{5}(x - 3)$$

$$y - 7 = \frac{3}{5}x - \frac{9}{5}$$

$$y - 7 + 7 = \frac{3}{5}x - \frac{9}{5} + 7$$

$$y = \frac{3}{5}x - \frac{9}{5} + \frac{35}{5}$$

$$y = \frac{3}{5}x + \frac{26}{5}$$

14.
$$(2x^{-3})^{-2}(4x^{-3}y^2)^3 = 2^{-2}x^{-3(-2)}4^3x^{-3(3)}y^{2(3)}$$
$$= 2^{-2}x^6 4^3 x^{-9}y^6$$
$$= 2^{-2}\cdot 4^3 x^{-3}y^6$$
$$= \frac{4^3 y^6}{2^2 x^3} = \frac{64y^6}{4x^3}$$
$$= \frac{16y^6}{x^3}$$

15.
$$(x^3 - x^2 + 6x - 5) - (4x^3 - 3x^2 + 7)$$
$$= x^3 - x^2 + 6x - 5 - 4x^3 + 3x^2 - 7$$
$$= x^3 - 4x^3 - x^2 + 3x^2 + 6x - 5 - 7$$
$$= -3x^3 + 2x^2 + 6x - 12$$

16.
$$(3x - 2)(x^2 + 5x - 6)$$
$$= 3x(x^2) - 2(x^2) + 3x(5x) - 2(5x) + 3x(-6) - 2(-6)$$
$$= 3x^3 - 2x^2 + 15x^2 - 10x - 18x + 12$$
$$= 3x^3 + 13x^2 - 28x + 12$$

17.
$$\begin{array}{r} x - 5 \\ x+3\overline{)x^2 - 2x + 6} \\ \underline{x^2 + 3x} \\ -5x + 6 \\ \underline{-5x - 15} \\ 21 \end{array}$$

$$\frac{x^2 - 2x + 6}{x + 3} = x - 5 + \frac{21}{x + 3}$$

18.
$$qr + 2q - 8r - 16 = q(r + 2) - 8(r + 2)$$
$$= (r + 2)(q - 8)$$

19. $5x^2 - 7x - 6 = (5x + 3)(x - 2)$

20.
$$7y^3 - 63y = 7y(y^2 - 9)$$
$$= 7y(y^2 - 3^2)$$
$$= 7y(y + 3)(y - 3)$$

Chapter 7

Exercise Set 7.1

1. Answers will vary.

3. The value of the variable does not make the denominator equal to 0.

5. There is no factor common to both the numerator and denominator of $\dfrac{4+3x}{9}$.

7. The denominator cannot be 0.

9. $x - 2 = 0$
$\quad\ x \neq 2$

11. $\dfrac{x+8}{8-x} = \dfrac{x+8}{-(-8+x)} = -\dfrac{x+8}{x-8} \neq -1$

No

13. The expression is defined for all real numbers except $x = 0$.

15. $4n - 16 = 0$
$\qquad 4n = 16$
$\qquad\ \ n = 4$
The expression is defined for all real numbers except $n = 4$.

17. $x^2 - 4 = 0$
$(x-2)(x+2) = 0$
$x - 2 = 0 \quad$ or $\quad x + 2 = 0$
$\quad\ x = 2 \qquad\qquad x = -2$
The expression is defined for all real numbers except $x = 2,\ x = -2$.

19. $\quad 2x^2 - 9x + 9 = 0$
$(2x-3)(x-3) = 0$
$2x - 3 = 0 \quad$ or $\quad x - 3 = 0$
$\qquad x = \dfrac{3}{2} \qquad\qquad x = 3$
The expression is defined for all real numbers except $x = \dfrac{3}{2},\ x = 3$.

21. All real numbers because $x^2 + 36 \neq 0$.

23. $\qquad 4p^2 - 25 = 0$
$(2p+5)(2p-5) = 0$
$2p + 5 = 0 \quad$ or $\quad 2p - 5 = 0$
$\qquad p = -\dfrac{5}{2} \qquad\qquad p = \dfrac{5}{2}$
The expression is defined for all real numbers except $p = \pm\dfrac{5}{2}$.

25. $\dfrac{8x^3 y}{24x^2 y^5} = \dfrac{8}{24} \cdot x^{3-2} \cdot y^{1-5}$
$\qquad = \dfrac{1}{3} xy^{-4}$
$\qquad = \dfrac{x}{3y^4}$

27. $\dfrac{\left(2a^4 b^5\right)^3}{2a^{12} b^{20}} = \dfrac{2^3 \cdot a^{4(3)} \cdot b^{5(3)}}{2a^{12} b^{20}}$
$\qquad\qquad = \dfrac{8a^{12} b^{15}}{2a^{12} b^{20}}$
$\qquad\qquad = 4a^{12-12} b^{15-20}$
$\qquad\qquad = 4a^0 b^{-5}$
$\qquad\qquad = \dfrac{4}{b^5}$

29. $\dfrac{2x}{x+xy} = \dfrac{2x}{x(1+y)}$
$\qquad\quad = \dfrac{2}{1+y}$

31. $\dfrac{5x+15}{x+3} = \dfrac{5(x+3)}{x+3}$
$\qquad\qquad = 5$

33. $\dfrac{x^3 + 6x^2 + 7x}{2x} = \dfrac{x\left(x^2 + 6x + 7\right)}{2x}$
$\qquad\qquad\qquad = \dfrac{x^2 + 6x + 7}{2}$

35. $\dfrac{r^2 - r - 2}{r - 2} = \dfrac{(r-2)(r+1)}{(r-2)}$

$\qquad = r + 1$

37. $\dfrac{x^2 + 2x}{x^2 + 4x + 4} = \dfrac{x(x+2)}{(x+2)^2}$

$\qquad = \dfrac{x}{x+2}$

39. $\dfrac{z^2 - 10z + 25}{z^2 - 25} = \dfrac{(z-5)^2}{(z-5)(z+5)}$

$\qquad = \dfrac{z-5}{z+5}$

41. $\dfrac{x^2 - 2x - 3}{x^2 - x - 6} = \dfrac{(x+1)(x-3)}{(x+2)(x-3)}$

$\qquad = \dfrac{x+1}{x+2}$

43. $\dfrac{4x - 3}{3 - 4x} = \dfrac{4x - 3}{-(4x - 3)}$

$\qquad = -1$

45. $\dfrac{x^2 - 2x - 8}{4 - x} = \dfrac{(x-4)(x+2)}{-(x-4)}$

$\qquad = -(x+2)$

47. $\dfrac{x^2 + 3x - 18}{-2x^2 + 6x} = \dfrac{(x+6)(x-3)}{-2x(x-3)}$

$\qquad = -\dfrac{x+6}{2x}$

49. $\dfrac{2x^2 + 5x - 3}{1 - 2x} = \dfrac{(2x-1)(x+3)}{-(2x-1)}$

$\qquad = -(x+3)$

51. $\dfrac{m-2}{4m^2 - 13m + 10} = \dfrac{m-2}{(4m-5)(m-2)} = \dfrac{1}{4m-5}$

53. $\dfrac{x^2 - 25}{(x+5)^2} = \dfrac{(x-5)(x+5)}{(x+5)^2}$

$\qquad = \dfrac{x-5}{x+5}$

55. $\dfrac{6x^2 - 13x + 6}{3x - 2} = \dfrac{(3x-2)(2x-3)}{3x - 2}$

$\qquad = 2x - 3$

57. $\dfrac{x^2 - 3x + 4x - 12}{x + 4} = \dfrac{x(x-3) + 4(x-3)}{x + 4}$

$\qquad = \dfrac{(x+4)(x-3)}{(x+4)}$

$\qquad = x - 3$

59. $\dfrac{2x^2 - 8x + 3x - 12}{2x^2 + 8x + 3x + 12} = \dfrac{2x(x-4) + 3(x-4)}{2x(x+4) + 3(x+4)}$

$\qquad = \dfrac{(x-4)(2x+3)}{(x+4)(2x+3)}$

$\qquad = \dfrac{x-4}{x+4}$

61. $\dfrac{a^3 - 8}{a - 2} = \dfrac{(a-2)(a^2 + 2a + 4)}{(a - 2)}$

$\qquad = a^2 + 2a + 4$

63. $\dfrac{9s^2 - 16t^2}{3s - 4t} = \dfrac{(3s + 4t)(3s - 4t)}{3s - 4t} = 3s + 4t$

65. $\dfrac{6x + 9y}{2x^2 + xy - 3y^2} = \dfrac{3(2x + 3y)}{(2x + 3y)(x - y)} = \dfrac{3}{x - y}$

67. $\dfrac{3\,☺}{15} = \dfrac{3\,☺}{3 \cdot 5} = \dfrac{☺}{5}$

69. $\dfrac{7\Delta}{14\Delta + 63} = \dfrac{7\Delta}{7(2\Delta + 9)} = \dfrac{\Delta}{2\Delta + 9}$

71. $\dfrac{3\Delta - 4}{4 - 3\Delta} = \dfrac{-(2 - 4\Delta)}{(2 - 4\Delta)} = -1$

73. $x^2 - x - 6 = (x - 3)(x + 2)$

Denominator $= x + 2$

75. $(x + 4)(x + 5) = x^2 + 9x + 20$

Numerator $= x^2 + 9x + 20$

77. a. $\dfrac{x - 2}{x^2 - 2x + 3x - 6} = \dfrac{x - 2}{x(x-2) + 3(x-2)}$

$\qquad = \dfrac{(x-2)}{(x+3)(x-2)}$

$\qquad x \neq -3, x \neq 2$

b. $\dfrac{(x-2)}{(x+3)(x-2)} = \dfrac{1}{x+3}$

79. a. $\dfrac{x+5}{2x^3+7x^2-15x} = \dfrac{x+5}{x\left(2x^2+7x-15\right)}$

$= \dfrac{x+5}{x(2x-3)(x+5)}$

$x \neq 0,\ x \neq \dfrac{3}{2},\ x \neq -5$

b. $\dfrac{(x+5)}{x(2x-3)(x+5)} = \dfrac{1}{x(2x-3)}$

81. $\dfrac{\left(\dfrac{1}{5}x^5 - \dfrac{2}{3}x^4\right)}{\left(\dfrac{1}{5}x^5 - \dfrac{2}{3}x^4\right)} = 1$

84. $z = \dfrac{x-y}{4}$

$4z = x - y$

$4z - x = -y$

$y = x - 4z$

85. Let x = measure of the smallest angle. Then the second angle $= x + 30$ and third angle $= 3x + 10$.
angle 1 + angle 2 + angle 3 = 180°
$x + (x+30) + (3x+10) = 180$

$5x + 40 = 180$

$5x = 140$

$x = 28$

$x + 30 = 28 + 30 = 58$

$3x + 10 = 3(28) + 10 = 84 + 10 = 94$.

The three angles are 28°, 58°, and 94°.

86. $\left(\dfrac{5x^2 y^2}{9x^4 y^3}\right)^2 = \left(\dfrac{5}{9x^2 y}\right)^2$

$= \dfrac{25}{81x^4 y^2}$

87. $3x^2 - 4x - 8 - \left(-5x^2 + 6x + 11\right)$

$= 3x^2 - 4x - 8 + 5x^2 - 6x - 11$

$= 8x^2 - 10x - 19$

88. $3a^2 - 6a - 72 = 3(a^2 - 2a - 24)$

$= 3(a-6)(a+4)$

89. $a^2 + b^2 = c^2$

$5^2 + 12^2 = c^2$

$25 + 144 = c^2$

$169 = c^2$

$\sqrt{169} = \sqrt{c^2}$

$13 = c$

The hypotenuse is 13 inches long.

Exercise Set 7.2

1. Answers will vary.

3. $\dfrac{x+3}{x-4} \cdot \dfrac{\square}{x+3} = x+5$

Numerator must be $(x+5)(x-4) = x^2 + x - 20$

5. $\dfrac{x-5}{x+5} \cdot \dfrac{x+5}{\square} = \dfrac{1}{x+7}$

Denominator must be
$(x+7)(x-5) = x^2 + 2x - 35$

7. $\left(\dfrac{2}{5}\right)\left(\dfrac{15}{19}\right) = \dfrac{2}{{}_1\cancel{5}} \cdot \dfrac{\cancel{15}^3}{19} = \dfrac{2 \cdot 3}{1 \cdot 19} = \dfrac{6}{19}$

9. $\left(\dfrac{6}{8}\right)\left(-\dfrac{10}{14}\right) = \dfrac{6}{8} \cdot \dfrac{10}{14} = \dfrac{3}{4} \cdot \dfrac{-5}{7} = \dfrac{3 \cdot (-5)}{4 \cdot 7} = -\dfrac{15}{28}$

11. $\left(-\dfrac{4}{11}\right)\left(-\dfrac{55}{64}\right) = \dfrac{-{}^1\cancel{4}}{{}_1\cancel{11}} \cdot \dfrac{-\cancel{55}^5}{\cancel{64}_{16}} = \dfrac{-1 \cdot (-5)}{1 \cdot 16} = \dfrac{5}{16}$

13. $\dfrac{3}{7} \div \dfrac{5}{7} = \dfrac{3}{\cancel{7}} \cdot \dfrac{\cancel{7}}{5} = \dfrac{3}{5}$

15. $-\dfrac{2}{9} \div \dfrac{32}{39} = \dfrac{-{}^1\cancel{2}}{{}_3\cancel{9}} \cdot \dfrac{\cancel{39}^{13}}{\cancel{32}_{16}} = \dfrac{-1 \cdot 13}{3 \cdot 16} = -\dfrac{13}{48}$

17. $\dfrac{6x}{4y} \cdot \dfrac{y^2}{12} = \dfrac{6x}{4y} \cdot \dfrac{y^2}{6 \cdot 2}$

$= \dfrac{xy}{8}$

19. $\dfrac{14x^2}{y^4} \cdot \dfrac{5x^2}{y^2} = \dfrac{70x^4}{y^6}$

21. $\dfrac{6x^5 y^3}{5z^3} \cdot \dfrac{6x^4}{5yz^4} = \dfrac{36x^9 y^2}{25z^7}$

23. $\dfrac{3x-2}{3x+2}\cdot\dfrac{x-1}{1-x}=\dfrac{3x-2}{3x+2}\cdot\dfrac{\cancel{(x-1)}}{-\cancel{(x-1)}}$

$=\dfrac{-3x+2}{3x+2}$

25. $\dfrac{x^2+7x+6}{x+6}\cdot\dfrac{1}{x+1}=\dfrac{\cancel{(x+6)}\,\cancel{(x+1)}}{\cancel{(x+6)}\,\cancel{(x+1)}}$

$=1$

27. $\dfrac{a}{a^2-b^2}\cdot\dfrac{a+b}{a^2+ab}=\dfrac{\cancel{a}}{(a+b)(a-b)}\cdot\dfrac{\cancel{(a+b)}}{\cancel{a}\,\cancel{(a+b)}}$

$=\dfrac{1}{(a-b)(a+b)}$

$=\dfrac{1}{a^2-b^2}$

29. $\dfrac{6x^2-14x-12}{6x+4}\cdot\dfrac{2x+4}{2x^2-2x-12}$

$=\dfrac{2\left(3x^2-7x-6\right)}{2(3x+2)}\cdot\dfrac{2(x+2)}{2\left(x^2-x-6\right)}$

$=\dfrac{\cancel{2}\,\cancel{(3x+2)}\,\cancel{(x-3)}}{\cancel{2}\,\cancel{(3x+2)}}\cdot\dfrac{\cancel{2}\,(x+2)}{\cancel{2}\,\cancel{(x-3)}\,\cancel{(x+2)}}$

$=1$

31. $\dfrac{3x^2-13x-10}{x^2-2x-15}\cdot\dfrac{x^2+x-2}{3x^2-x-2}$

$=\dfrac{\cancel{(3x+2)}\,\cancel{(x-5)}}{(x+3)\,\cancel{(x-5)}}\cdot\dfrac{(x+2)\,\cancel{(x-1)}}{\cancel{(3x+2)}\,\cancel{(x-1)}}$

$=\dfrac{x+2}{x+3}$

33. $\dfrac{x+9}{x-3}\cdot\dfrac{x^3-27}{x^2+3x+9}$

$=\dfrac{(x+9)}{\cancel{(x-3)}}\cdot\dfrac{\cancel{(x-3)}\,\cancel{(x^2+3x+9)}}{\cancel{(x^2+3x+9)}}$

$=x+9$

35. $\dfrac{12x^3}{y^2}\div\dfrac{3x}{y^3}=\dfrac{12x^3}{y^2}\cdot\dfrac{y^3}{3x}$

$=\dfrac{3\cdot4x^3}{y^2}\cdot\dfrac{y^3}{3x}$

$=4x^2y$

37. $\dfrac{15xy^2}{4z}\div\dfrac{5x^2y^2}{12z^2}=\dfrac{3\cdot5xy^2}{4z}\cdot\dfrac{12z^2}{5x^2y^2}$

$=\dfrac{3\cdot5xy^2}{4z}\cdot\dfrac{3\cdot4z^2}{5x^2y^2}$

$=\dfrac{9z}{x}$

39. $\dfrac{11xy}{7ab^2}\div\dfrac{6xy}{7}=\dfrac{11xy}{7ab^2}\cdot\dfrac{7}{6xy}$

$=\dfrac{11}{6ab^2}$

41. $\dfrac{12r+6}{r}\div\dfrac{2r+1}{r^3}=\dfrac{6\,\cancel{(2r+1)}}{\cancel{r}}\cdot\dfrac{\cancel{r}\cdot r^2}{\cancel{(2r+1)}}$

$=6r^2$

43. $\dfrac{x^2+11x+18}{x}\div\dfrac{x+2}{x}=\dfrac{(x+9)\,\cancel{(x+2)}}{\cancel{x}}\cdot\dfrac{\cancel{x}}{\cancel{(x+2)}}$

$=x+9$

45. $\dfrac{x^2-12x+32}{x^2-6x-16}\div\dfrac{x^2-x-12}{x^2-5x-24}$

$=\dfrac{x^2-12x+32}{x^2-6x-16}\cdot\dfrac{x^2-5x-24}{x^2-x-12}$

$=\dfrac{(x-8)\,\cancel{(x-4)}}{\cancel{(x-8)}\,(x+2)}\cdot\dfrac{\cancel{(x-8)}\,\cancel{(x+3)}}{\cancel{(x-4)}\,\cancel{(x+3)}}$

$=\dfrac{x-8}{x+2}$

47. $\dfrac{2x^2+9x+4}{x^2+7x+12}\div\dfrac{2x^2-x-1}{(x+3)^2}$

$=\dfrac{\cancel{(2x+1)}\,\cancel{(x+4)}}{\cancel{(x+3)}\,\cancel{(x+4)}}\cdot\dfrac{\cancel{(x+3)}\,(x+3)}{\cancel{(2x+1)}\,(x-1)}$

$=\dfrac{x+3}{x-1}$

49. $\dfrac{x^2 - y^2}{x^2 - 2xy + y^2} \div \dfrac{x + y}{y - x}$

$= \dfrac{\cancel{(x-y)}\,\cancel{(x+y)}}{\cancel{(x-y)}\,(x-y)} \cdot \dfrac{-1\cancel{(x-y)}}{\cancel{(x+y)}}$

$= -1$

51. $\dfrac{5x^2 - 4x - 1}{5x^2 + 6x + 1} \div \dfrac{x^2 - 5x + 4}{x^2 + 2x + 1}$

$= \dfrac{\cancel{(5x+1)}\,\cancel{(x-1)}}{\cancel{(5x+1)}\,\cancel{(x+1)}} \cdot \dfrac{\cancel{(x+1)}\,(x+1)}{(x-4)\,\cancel{(x-1)}}$

$= \dfrac{x+1}{x-4}$

53. $\dfrac{11z}{6y^2} \cdot \dfrac{24x^2 y^4}{11z} = \dfrac{\cancel{11z}}{\cancel{6}\,y^2} \cdot \dfrac{\cancel{6} \cdot 4 \cdot x^2 \cdot y^2 \cdot \cancel{y^2}}{\cancel{11z}}$

$= 4x^2 y^2$

55. $\dfrac{63a^2 b^3}{20c^3} \cdot \dfrac{4c^4}{9a^3 b^5} = \dfrac{7 \cdot \cancel{9} \cdot \cancel{4} \cdot \cancel{a^2}\,\cancel{b^3}\,\cancel{c^3}\,c}{5 \cdot \cancel{4} \cdot \cancel{9}\,\cancel{a^2}\,a\cancel{b^3}\,b^2\,\cancel{c^3}} = \dfrac{7c}{5ab^2}$

57. $\dfrac{-xy}{a} \div \dfrac{-2ax}{6y} = \dfrac{-xy}{a} \cdot \dfrac{6y}{-2ax}$

$= \dfrac{\cancel{-}\,\cancel{x}y}{a} \cdot \dfrac{\cancel{2} \cdot 3y}{\cancel{-}\,\cancel{2}a\cancel{x}}$

$= \dfrac{3y^2}{a^2}$

59. $\dfrac{64m^6}{21x^5 y^7} \cdot \dfrac{14x^{12} y^5}{16m^5}$

$= \dfrac{\cancel{16} \cdot 4\cancel{m^5}\,m}{3 \cdot \cancel{7}\,\cancel{x^5}\,\cancel{y^5}\,y^2} \cdot \dfrac{2 \cdot \cancel{7}\,\cancel{x^5}\,x^7\,\cancel{y^5}}{\cancel{16}\,\cancel{m^5}}$

$= \dfrac{8mx^7}{3y^2}$

61. $\dfrac{(x+3)^2}{5x^2} \cdot \dfrac{10x}{x^2 - 9}$

$= \dfrac{(x+3)\,\cancel{(x+3)}}{\cancel{5}\,\cancel{x}\,x} \cdot \dfrac{\cancel{5} \cdot 2\cancel{x}}{\cancel{(x+3)}\,(x-3)}$

$= \dfrac{2(x+3)}{x(x-3)}$

63. $\dfrac{1}{5x^2 y^2} \div \dfrac{1}{35x^3 y} = \dfrac{1}{\cancel{5}\,\cancel{x^2}\,\cancel{y}\,y} \cdot \dfrac{\cancel{5} \cdot 7\,\cancel{x^2}\,x\,\cancel{y}}{1}$

$= \dfrac{7x}{y}$

65. $\dfrac{(4m)^2}{8n^3} \div \dfrac{m^6 n^8}{2} = \dfrac{16m^2}{8n^3} \cdot \dfrac{2}{m^6 n^8}$

$= \dfrac{\cancel{8} \cdot 2\,\cancel{m^2}}{\cancel{8}n^3} \cdot \dfrac{2}{m^4\,\cancel{m^2}\,n^8}$

$= \dfrac{4}{m^4 n^{11}}$

67. $\dfrac{r^2 + 5r + 6}{r^2 + 9r + 18} \cdot \dfrac{r^2 + 4r - 12}{r^2 - 5r + 6}$

$= \dfrac{(r+2)\,\cancel{(r+3)}}{\cancel{(r+6)}\,\cancel{(r+3)}} \cdot \dfrac{\cancel{(r+6)}\,\cancel{(r-2)}}{(r-3)\,\cancel{(r-2)}}$

$= \dfrac{r+2}{r-3}$

69. $\dfrac{x^2 - 12x + 36}{x^2 - 8x + 12} \div \dfrac{x^2 - 7x + 12}{x^2 - 6x + 8}$

$= \dfrac{x^2 - 12x + 36}{x^2 - 8x + 12} \cdot \dfrac{x^2 - 6x + 8}{x^2 - 7x + 12}$

$= \dfrac{(x-6)\,(x-6)}{\cancel{(x-6)}\,\cancel{(x-2)}} \cdot \dfrac{\cancel{(x-4)}\,\cancel{(x-2)}}{\cancel{(x-4)}\,(x-3)}$

$= \dfrac{x-6}{x-3}$

71. $\dfrac{2w^2 + 3w - 35}{w^2 - 7w - 8} \cdot \dfrac{w^2 - 5w - 24}{w^2 + 8w + 15}$

$= \dfrac{(2w-7)\,\cancel{(w+5)}}{\cancel{(w-8)}\,(w+1)} \cdot \dfrac{\cancel{(w-8)}\,\cancel{(w+3)}}{\cancel{(w+5)}\,\cancel{(w+3)}}$

$= \dfrac{2w-7}{w+1}$

73. $\dfrac{q^2-11q+30}{2q^2-7q-15} \div \dfrac{q^2-2q-24}{q^2-q-20}$

$= \dfrac{q^2-11q+30}{2q^2-7q-15} \cdot \dfrac{q^2-q-20}{q^2-2q-24}$

$= \dfrac{(q-6)(q-5)}{(2q+3)(q-5)} \cdot \dfrac{(q-5)(q+4)}{(q-6)(q+4)}$

$= \dfrac{q-5}{2q+3}$

75. $\dfrac{4n^2-9}{9n^2-1} \cdot \dfrac{3n^2-2n-1}{2n^2-5n+3}$

$= \dfrac{(2n+3)(2n-3)}{(3n+1)(3n-1)} \cdot \dfrac{(3n+1)(n-1)}{(2n-3)(n-1)}$

$= \dfrac{2n+3}{3n-1}$

77. $\dfrac{6\Delta^2}{13} \cdot \dfrac{13}{36\Delta^5} = \dfrac{6\Delta^2}{13} \cdot \dfrac{13}{6 \cdot 6\Delta^2 \Delta^3} = \dfrac{1}{6\Delta^3}$

79. $\dfrac{\Delta-\odot}{9\Delta-9\odot} \div \dfrac{\Delta^2-\odot^2}{\Delta^2+2\Delta\odot+\odot^2}$

$= \dfrac{\Delta-\odot}{9\Delta-9\odot} \cdot \dfrac{\Delta^2+2\Delta\odot+\odot^2}{\Delta^2-\odot^2}$

$= \dfrac{\Delta-\odot}{9(\Delta-\odot)} \cdot \dfrac{(\Delta+\odot)(\Delta+\odot)}{(\Delta+\odot)(\Delta-\odot)}$

$= \dfrac{\Delta+\odot}{9(\Delta-\odot)}$

81. $(x+2)(x+3) = x^2+3x+2x+6$

$\qquad\qquad\qquad = x^2+5x+6$

Numerator is x^2+5x+6.

83. $(x-6)(x+2) = x^2+2x-6x-12$

$\qquad\qquad\qquad = x^2-4x-12$

Numerator is $x^2-4x-12$.

85. $(x^2-4)(x-1) \div (x+2) =$

$(x-2)(x+2)(x-1) \cdot \dfrac{1}{x+2} = x^2-x-2x+2$

$\qquad\qquad\qquad\qquad\qquad = x^2-3x+2$

Numerator is x^2-3x+2.

87. $\left(\dfrac{x+2}{x^2-4x-12} \cdot \dfrac{x^2-9x+18}{x-2}\right) \div \dfrac{x^2+5x+6}{x^2-4} = \dfrac{x+2}{x^2-4x-12} \cdot \dfrac{x^2-9x+18}{x-2} \cdot \dfrac{x^2-4}{x^2+5x+6}$

$= \dfrac{(x+2)}{(x-6)(x+2)} \cdot \dfrac{(x-3)(x-6)}{(x-2)} \cdot \dfrac{(x+2)(x-2)}{(x+2)(x+3)}$

$= \dfrac{x-3}{x+3}$

89. $\left(\dfrac{x^2-x-6}{2x^2-9x+9} \div \dfrac{x^2+x-12}{x^2+3x-4}\right) \cdot \dfrac{2x^2-5x+3}{x^2+x-2} = \dfrac{x^2-x-6}{2x^2-9x+9} \cdot \dfrac{x^2+3x-4}{x^2+x-12} \cdot \dfrac{2x^2-5x+3}{x^2+x-2}$

$= \dfrac{(x-3)(x+2)}{(2x-3)(x-3)} \cdot \dfrac{(x+4)(x-1)}{(x+4)(x-3)} \cdot \dfrac{(2x-3)(x-1)}{(x+2)(x-1)}$

$= \dfrac{x-1}{x-3}$

91. $\dfrac{(x-3)(x-2)}{(x+4)(x-5)} \cdot \dfrac{(x+4)(x-1)}{(x-3)(x-1)} = \dfrac{x-2}{x-5}$

The numerator is $(x-3)(x-2) = x^2 - 5x + 6$.

The denominator is $(x+4)(x-5) = x^2 - x - 20$.

94. Let x = the time it takes the tug boat to reach the barge.

Then $x + 2$ = the time it takes the tug boat to return to the dock.

	Rate	Time	Distance
Trip Out	15	x	15(x)
Return Trip	5	$x+2$	5($x+2$)

Distance to barge = Distance back to dock
$15(x) = 5(x+2)$
$15x = 5x + 10$
$10x = 10$
$x = 1$
It took 1 hour for the tug boat to reach the barge.

95. $\left(4x^3y^2z^4\right)\left(3xy^3z^7\right) = 4 \cdot 3 \cdot x^3xy^2y^3z^4z^7$

$$= 12x^4y^5z^{11}$$

96.

$$
\require{enclose}
\begin{array}{r}
2x^2 + x - 2 \\
2x-1 \enclose{longdiv}{4x^3 + 0x^2 - 5x + 0} \\
\underline{4x^3 - 2x^2} \\
2x^2 - 5x \\
\underline{2x^2 - x} \\
-4x + 0 \\
\underline{-4x + 2} \\
-2
\end{array}
$$

$\dfrac{4x^3 - 5x}{2x-1} = 2x^2 + x - 2 - \dfrac{2}{2x-1}$

97. $6x^2 - 18x - 60 = 6\left(x^2 - 3x - 10\right)$

$$= 6(x-5)(x+2)$$

98. $3x^2 - 9x - 30 = 0$

$3\left(x^2 - 3x - 10\right) = 0$

$3(x-5)(x+2) = 0$

$x - 5 = 0 \quad$ or $\quad x + 2 = 0$

$x = 5 \qquad\qquad x = -2$

Exercise Set 7.3

1. Answers will vary.

3. Answers will vary.

5. $\dfrac{9}{x+6} - \dfrac{2}{x}$

The only factor (other than 1) of the first denominator is $x + 6$. The only factor (other than 1) of the second denominator is x. The LCD is therefore $x(x+6)$.

7. $\dfrac{2}{x+3} + \dfrac{1}{x} + \dfrac{1}{4}$

The only factor (other than 1) of the first denominator is $x + 3$. The only factor (other than 1) of the second denominator is x. The only factor (other than 1) of the third denominator is 4. The LCD is therefore $4x(x+3)$.

9. a. The negative sign in $-(2x - 9)$ was not distributed.

   b. $\dfrac{4x-3}{5x+4} - \dfrac{2x-9}{5x+4} = \dfrac{4x-3-(2x-9)}{5x+4}$

   $$= \dfrac{4x-3-2x+9}{5x+4}$$

   $$\neq \dfrac{4x-3-2x-9}{5x+4}$$

11. a. The negative sign in $-\left(3x^2 - 4x + 5\right)$ was not distributed.

   b. $\dfrac{8x-2}{x^2-4x+3} - \dfrac{3x^2-4x+5}{x^2-4x+3}$

   $$= \dfrac{8x-2-\left(3x^2-4x+5\right)}{x^2-4x+3}$$

   $$= \dfrac{8x-2-3x^2+4x-5}{x^2-4x+3}$$

   $$\neq \dfrac{8x-2-3x^2-4x+5}{x^2-4x+3}$$

13. $\dfrac{4}{7} + \dfrac{2}{7} = \dfrac{4+2}{7} = \dfrac{6}{7}$

15. $\dfrac{5r+2}{4} - \dfrac{3}{4} = \dfrac{5r+2-3}{4} = \dfrac{5r-1}{4}$

17. $\dfrac{2}{x} + \dfrac{x+4}{x} = \dfrac{2+x+4}{x}$

$\qquad = \dfrac{x+6}{x}$

19. $\dfrac{6}{n+1} + \dfrac{n+2}{n+1} = \dfrac{6+n+2}{n+1}$

$\qquad = \dfrac{n+8}{n+1}$

21. $\dfrac{x}{x-3} + \dfrac{4x+9}{x-3} = \dfrac{x+4x+9}{x-3} = \dfrac{5x+9}{x-3}$

23. $\dfrac{4t+7}{5t^2} - \dfrac{3t+4}{5t^2} = \dfrac{4t+7-(3t+4)}{5t^2}$

$\qquad = \dfrac{4t+7-3t-4}{5t^2}$

$\qquad = \dfrac{t+3}{5t^2}$

25. $\dfrac{5x+4}{x^2-x-12} + \dfrac{-4x-1}{x^2-x-12} = \dfrac{5x+4-4x-1}{x^2-x-12}$

$\qquad = \dfrac{x+3}{(x+3)(x-4)}$

$\qquad = \dfrac{1}{x-4}$

27. $\dfrac{2m+5}{(m+4)(m-3)} - \dfrac{m+1}{(m+4)(m-3)}$

$\qquad = \dfrac{2m+5-(m+1)}{(m+4)(m-3)} = \dfrac{2m+5-m-1}{(m+4)(m-3)}$

$\qquad = \dfrac{m+4}{(m+4)(m-3)}$

$\qquad = \dfrac{1}{m-3}$

29. $\dfrac{2p-6}{p-5} - \dfrac{p+6}{p-5} = \dfrac{2p-6-(p+6)}{p-5}$

$\qquad = \dfrac{2p-6-p-6}{p-5}$

$\qquad = \dfrac{p-12}{p-5}$

31. $\dfrac{x^2+4x+1}{x+2} - \dfrac{5x+7}{x+2} = \dfrac{x^2+4x+1-(5x+7)}{x+2}$

$\qquad = \dfrac{x^2+4x+1-5x-7}{x+2}$

$\qquad = \dfrac{x^2-x-6}{x+2}$

$\qquad = \dfrac{(x-3)(x+2)}{x+2}$

$\qquad = x-3$

33. $\dfrac{3x+13}{2x+10} - \dfrac{2(x+4)}{2x+10} = \dfrac{3x+13-2(x+4)}{2x+10}$

$\qquad = \dfrac{3x+13-2x-8}{2x+10}$

$\qquad = \dfrac{x+5}{2(x+5)}$

$\qquad = \dfrac{1}{2}$

35. $\dfrac{b^2-2b-2}{b^2-b-6} + \dfrac{b-4}{b^2-b-6} = \dfrac{b^2-2b-2+b-4}{b^2-b-6}$

$\qquad = \dfrac{b^2-b-6}{b^2-b-6}$

$\qquad = 1$

37. $\dfrac{t-3}{t+3} - \dfrac{-3t-15}{t+3} = \dfrac{t-3-(-3t-15)}{t+3}$

$\qquad = \dfrac{t-3+3t+15}{t+3}$

$\qquad = \dfrac{4t+12}{t+3}$

$\qquad = \dfrac{4(t+3)}{t+3}$

$\qquad = 4$

39. $\dfrac{3x^2+15x}{x^3+2x^2-8x} + \dfrac{2x^2+5x}{x^3+2x^2-8x}$

$\qquad = \dfrac{3x^2+15x+2x^2+5x}{x^3+2x^2-8x}$

$\qquad = \dfrac{5x^2+20x}{x(x^2+2x-8)}$

$\qquad = \dfrac{5x(x+4)}{x(x-2)(x+4)}$

$\qquad = \dfrac{5}{x-2}$

41. $\dfrac{3x^2-9x}{4x^2-8x}+\dfrac{3x}{4x^2-8x}=\dfrac{3x^2-9x+3x}{4x^2-8x}$

$\qquad\qquad =\dfrac{3x^2-6x}{4x^2-8x}$

$\qquad\qquad =\dfrac{3x(x-2)}{4x(x-2)}$

$\qquad\qquad =\dfrac{3}{4}$

43. $\dfrac{3x^2-4x+6}{3x^2+7x+2}-\dfrac{10x+11}{3x^2+7x+2}$

$\qquad =\dfrac{3x^2-4x+6-(10x+11)}{3x^2+7x+2}$

$\qquad =\dfrac{3x^2-4x+6-10x-11}{3x^2+7x+2}$

$\qquad =\dfrac{3x^2-14x-5}{3x^2+7x+2}$

$\qquad =\dfrac{(3x+1)(x-5)}{(3x+1)(x+2)}$

$\qquad =\dfrac{x-5}{x+2}$

45. $\dfrac{x^2+3x-6}{x^2-5x+4}-\dfrac{-2x^2+4x-4}{x^2-5x+4}$

$\qquad =\dfrac{x^2+3x-6-(-2x^2+4x-4)}{x^2-5x+4}$

$\qquad =\dfrac{x^2+3x-6+2x^2-4x+4}{x^2-5x+4}$

$\qquad =\dfrac{3x^2-x-2}{x^2-5x+4}$

$\qquad =\dfrac{(3x+2)(x-1)}{(x-4)(x-1)}$

$\qquad =\dfrac{3x+2}{x-4}$

47. $\dfrac{5x^2+30x+8}{x^2-64}+\dfrac{x^2+19x}{x^2-64}$

$\qquad =\dfrac{5x^2+30x+8+x^2+19x}{x^2-64}$

$\qquad =\dfrac{6x^2+49x+8}{x^2-64}$

$\qquad =\dfrac{(6x+1)(x+8)}{(x-8)(x+8)}$

$\qquad =\dfrac{6x+1}{x-8}$

49. $\dfrac{x}{5}+\dfrac{x+4}{5}$

Least common denominator = 5

51. $\dfrac{3}{n}+\dfrac{1}{9n}$

Least common denominator = $9n$

53. $\dfrac{3}{5x}+\dfrac{7}{3}$

Least common denominator $=5x\cdot 3=15x$

55. $\dfrac{6}{p}+\dfrac{9}{p^3}$

Least common denominator = p^3

57. $\dfrac{m+3}{3m-4}+m=\dfrac{m+3}{3m-4}+\dfrac{m}{1}$

Least common denominator = $3m-4$

59. $\dfrac{x}{6x}+\dfrac{4}{x^2}$

Least common denominator = $6x^2$

61. $\dfrac{x+1}{12x^2y}-\dfrac{7}{9x^3}=\dfrac{x+1}{3\cdot 4x^2y}-\dfrac{7}{3^2x^3}$

Least common denominator
$=4\cdot 3^2\cdot x^3\cdot y=36x^3y$

63. $\dfrac{4}{2r^4s^5}-\dfrac{5}{9r^3s^7}=\dfrac{4}{2r^4s^5}-\dfrac{5}{3\cdot 3r^3s^7}$

Least common denominator
$=2\cdot 3\cdot 3r^4s^7=18r^4s^7$

65. $\dfrac{3}{m}-\dfrac{17m}{m+2}$

Least common denominator $=m(m+2)$

67. $\dfrac{5x-2}{x^2+x}-\dfrac{13}{x}=\dfrac{5x-2}{x(x+1)}-\dfrac{13}{x}$

Least common denominator $=x(x+1)$

69. $\dfrac{n}{4n-1}+\dfrac{n-8}{1-4n}=\dfrac{n}{4n-1}+\dfrac{(-1)(n-8)}{(-1)(1-4n)}$

$\qquad =\dfrac{n}{4n-1}+\dfrac{-n+8}{4n-1}$

Least common denominator $=4n-1$ or $1-4n$

71. $\dfrac{3}{4k-5r}-\dfrac{10}{-4k+5r}$

$=\dfrac{3}{4k-5r}-\dfrac{(-1)10}{(-1)(-4k+5r)}$

$=\dfrac{3}{4k-5r}-\dfrac{-10}{4k-5r}$

Least common denominator $=4k-5r$ or $-4k+5r$

73. $\dfrac{4}{2q^2+2q}-\dfrac{5}{9q}=\dfrac{4}{2q(q+1)}-\dfrac{5}{3\cdot3q}$

Least common denominator $=2\cdot3\cdot3q(q+1)$

$\qquad\qquad\qquad\qquad\qquad =18q(q+1)$

75. $\dfrac{21}{24x^2y}+\dfrac{x+4}{15xy^3}=\dfrac{21}{3\cdot8x^2y}+\dfrac{x+4}{3\cdot5xy^3}$

Least common denominator

$=8\cdot3\cdot5x^2y^3=120x^2y^3$

77. $\dfrac{11}{3x+12}+\dfrac{3x+1}{2x+4}=\dfrac{11}{3(x+4)}+\dfrac{3x+1}{2(x+2)}$

Least common denominator

$=3\cdot2(x+4)(x+2)=6(x+4)(x+2)$

79. $\dfrac{9x+4}{x+1}-\dfrac{2x-6}{x+8}$

Least common denominator $=(x+1)(x+8)$

81. $\dfrac{x-2}{x^2-5x-24}+\dfrac{3}{x^2+11x+24}$

$=\dfrac{x-2}{(x-8)(x+3)}+\dfrac{3}{(x+8)(x+3)}$

Least common denominator

$=(x-8)(x+3)(x+8)$

83. $\dfrac{5}{(a-4)^2}-\dfrac{a+2}{a^2-7a+12}$

$=\dfrac{5}{(a-4)^2}-\dfrac{a+2}{(a-4)(a-3)}$

Least common denominator

$=(a-4)^2(a-3)$

85. $\dfrac{9x}{x^2+6x+5}-\dfrac{5x^2}{x^2+4x+3}$

$=\dfrac{9x}{(x+5)(x+1)}-\dfrac{5x^2}{(x+3)(x+1)}$

Least common denominator

$=(x+5)(x+1)(x+3)$

87. $\dfrac{3x-5}{x^2-6x+9}+\dfrac{3}{x-3}$

$=\dfrac{3x+5}{(x-3)^2}+\dfrac{3}{x-3}$

Least common denominator $=(x-3)^2$

89. $\dfrac{8x^2}{x^2-7x+6}+x-9$

$=\dfrac{8x^2}{(x-6)(x-1)}+\dfrac{x-9}{1}$

Least common denominator $=(x-6)(x-1)$

91. $\dfrac{t-1}{3t^2+10t-8}-\dfrac{11}{3t^2+11t-4}$

$=\dfrac{t-1}{(3t-2)(t+4)}-\dfrac{11}{(3t-1)(t+4)}$

Least common denominator

$=(3t-2)(t+4)(3t-1)$

93. $\dfrac{3x-1}{4x^2+4x+1}+\dfrac{x^2+x-9}{8x^2+10x+3}$

$=\dfrac{3x-1}{(2x+1)^2}+\dfrac{x^2+x-9}{(2x+1)(4x+3)}$

Least common denominator

$=(2x+1)^2(4x+3)$

95. $\dfrac{1}{7}+\dfrac{2}{5}=\dfrac{5}{5}\cdot\dfrac{1}{7}+\dfrac{2}{5}\cdot\dfrac{7}{7}=\dfrac{5}{35}+\dfrac{14}{35}=\dfrac{19}{35}$

97. $\dfrac{2}{9}+\dfrac{3}{4}=\dfrac{4}{4}\cdot\dfrac{2}{9}+\dfrac{3}{4}\cdot\dfrac{9}{9}=\dfrac{8}{36}+\dfrac{27}{36}=\dfrac{35}{36}$

99. $\dfrac{5}{9} - \dfrac{1}{2} = \dfrac{2}{2} \cdot \dfrac{5}{9} - \dfrac{1}{9} \cdot \dfrac{9}{2} = \dfrac{10}{18} - \dfrac{9}{18} = \dfrac{1}{18}$

101. $x^2 - 6x + 3 + \boxed{} = 2x^2 - 5x - 6$

$\boxed{} = 2x^2 - 5x - 6 - (x^2 - 6x + 3)$

$= 2x^2 - 5x - 6 - x^2 + 6x - 3$

$= x^2 + x - 9$

Sum of numerators must be $2x^2 - 5x - 6$

103. $-x^2 - 4x + 3 + \boxed{} = 5x - 7$

$\boxed{} = 5x - 7 - (-x^2 - 4x + 3)$

$= 5x - 7 + x^2 + 4x - 3$

$= x^2 + 9x - 10$

Sum of numerator must be $5x - 7$

105. $\dfrac{3}{\copyright} + \dfrac{4}{5\copyright}$

Least common denominator $= 5\copyright$

107. $\dfrac{8}{\Delta^2 - 9} - \dfrac{2}{\Delta + 3} = \dfrac{8}{(\Delta + 3)(\Delta - 3)} - \dfrac{2}{\Delta + 3}$

Least common denominator $= (\Delta + 3)(\Delta - 3)$

109. $\dfrac{4x - 1}{x^2 - 25} - \dfrac{3x^2 - 8}{x^2 - 25} + \dfrac{8x - 7}{x^2 - 25}$

$= \dfrac{4x - 1 - (3x^2 - 8) + (8x - 7)}{x^2 - 25}$

$= \dfrac{4x - 1 - 3x^2 + 8 + 8x - 7}{x^2 - 25}$

$= \dfrac{-3x^2 + 12x}{x^2 - 25}$

111. $\dfrac{7}{6x^5 y^9} - \dfrac{9}{2x^3 y} + \dfrac{6}{5x^{12} y^2}$

$= \dfrac{7}{2 \cdot 3x^5 y^9} - \dfrac{9}{2x^3 y} + \dfrac{6}{5x^{12} y^2}$

Least common denominator
$= 2 \cdot 3 \cdot 5x^{12} y^9 = 30x^{12} y^9$

113. $\dfrac{3x}{x^2 - x - 12} + \dfrac{2}{x^2 - 6x + 8} + \dfrac{3}{x^2 + x - 6}$

$= \dfrac{3x}{(x - 4)(x + 3)} + \dfrac{2}{(x - 4)(x - 2)} + \dfrac{3}{(x + 3)(x - 2)}$

Least common denominator
$= (x - 4)(x + 3)(x - 2)$

115. $4\dfrac{3}{5} - 2\dfrac{5}{9} = \dfrac{23}{5} - \dfrac{23}{9}$

$= \dfrac{207}{45} - \dfrac{115}{45}$

$= \dfrac{92}{45}$

$= 2\dfrac{2}{45}$

116. $6x + 4 = -(x + 2) - 3x + 4$

$6x + 4 = -x - 2 - 3x + 4$

$6x + 4 = -4x + 2$

$6x + 4x = 2 - 4$

$10x = -2$

$x = \dfrac{-2}{10} = -\dfrac{1}{5}$

117. $\dfrac{6}{128} = \dfrac{x}{48}$

$128x = 6 \cdot 48$

$128x = 288$

$x = \dfrac{288}{128}$

$x = \dfrac{9}{4} = 2\dfrac{1}{4}$

You should use 2.25 ounces of concentrate.

118. Let h = the number of hours played.
The cost under Plan 1 is $C = 250 + 5h$ while the cost under Plan 2 is $C = 600$.

Set the two costs equal
$250 + 5h = 600$

$5h = 350$

$h = 70$

If Malcolm plays 70 hours in a year, the cost of the two plans is equal.

119. Use the two points given and find the slope between the two points. This is the slope of the line.
$$m = \frac{y_2 - y_1}{x_2 - x_1} = \frac{5 - (-2)}{4 - (-3)} = \frac{7}{7} = 1$$
The slope of the line is 1.

120. $\dfrac{8.4 \times 10^8}{2.1 \times 10^{-3}} = 4.0 \times 10^{8-(-3)} = 4.0 \times 10^{11}$

121.
$$2x^2 - 3 = x$$
$$2x^2 - x - 3 = 0$$
$$(2x - 3)(x + 1) = 0$$
$$2x - 3 = 0 \quad \text{or} \quad x + 1 = 0$$
$$2x = 3 \qquad\qquad x = -1$$
$$x = \frac{3}{2}$$

Exercise Set 7.4

1. For each fraction, divide the LCD by the denominator.

3. a. Answers will vary.

b. $\dfrac{x}{x^2 - x - 6} + \dfrac{3}{x^2 - 4}$

$= \dfrac{x}{(x-3)(x+2)} + \dfrac{3}{(x-2)(x+2)}$

$= \dfrac{x(x-2)}{(x-3)(x+2)(x-2)} + \dfrac{3(x-3)}{(x-2)(x+2)(x-3)}$

$= \dfrac{x^2 - 2x + 3x - 9}{(x-3)(x+2)(x-2)}$

$= \dfrac{x^2 + x - 9}{(x-3)(x+2)(x-2)}$

5. a. $\dfrac{y}{4z} + \dfrac{5}{6z^2}$

$4z = 2 \cdot 2 \cdot z$

$6z^2 = 2 \cdot 3 \cdot z^2$

Least common denominator

$2 \cdot 2 \cdot 3 \cdot z^2 = 12z^2$

b. $\dfrac{y}{4z} + \dfrac{5}{6z^2} = \dfrac{y}{4z} \cdot \dfrac{3z}{3z} + \dfrac{5}{6z^2} \cdot \dfrac{2}{2}$

$= \dfrac{3yz}{12z^2} + \dfrac{10}{12z^2}$

$= \dfrac{3yz + 10}{12z^2}$

c. Yes. After factoring out the common factors, the reduced form would be the same.

7. $\dfrac{2}{x} + \dfrac{3}{y} = \dfrac{y}{y} \cdot \dfrac{2}{x} + \dfrac{3}{y} \cdot \dfrac{x}{x}$

$= \dfrac{2y}{xy} + \dfrac{3x}{xy}$

$= \dfrac{2y + 3x}{xy}$

9. $\dfrac{5}{x^2} + \dfrac{1}{2x} = \dfrac{2}{2} \cdot \dfrac{5}{x^2} + \dfrac{1}{2x} \cdot \dfrac{x}{x}$

$= \dfrac{10}{2x^2} + \dfrac{x}{2x^2}$

$= \dfrac{x + 10}{2x^2}$

11. $3 + \dfrac{8}{x} = \dfrac{x}{x} \cdot 3 + \dfrac{8}{x} = \dfrac{3x}{x} + \dfrac{8}{x} = \dfrac{3x + 8}{x}$

13. $\dfrac{2}{x^2} + \dfrac{3}{5x} = \dfrac{5}{5} \cdot \dfrac{2}{x^2} + \dfrac{3}{5x} \cdot \dfrac{x}{x}$

$= \dfrac{10}{5x^2} + \dfrac{3x}{5x^2}$

$= \dfrac{3x + 10}{5x^2}$

15. $\dfrac{9}{4x^2 y} + \dfrac{3}{5xy^2} = \dfrac{5y}{5y} \cdot \dfrac{9}{4x^2 y} + \dfrac{3}{5xy^2} \cdot \dfrac{4x}{4x}$

$= \dfrac{45y}{20x^2 y^2} + \dfrac{12x}{20x^2 y^2}$

$= \dfrac{45y + 12x}{20x^2 y^2}$

17. $4y + \dfrac{x}{y} = \dfrac{y}{y} \cdot \dfrac{4y}{1} + \dfrac{x}{y} = \dfrac{4y^2}{y} + \dfrac{x}{y} = \dfrac{4y^2 + x}{y}$

19. $\dfrac{3a-1}{2a}+\dfrac{2}{3a}=\dfrac{3}{3}\cdot\dfrac{3a-1}{2a}+\dfrac{2}{3a}\cdot\dfrac{2}{2}$

$\qquad =\dfrac{3(3a-1)}{3\cdot 2a}+\dfrac{2\cdot 2}{3a\cdot 2}$

$\qquad =\dfrac{9a-3}{6a}+\dfrac{4}{6a}$

$\qquad =\dfrac{9a-3+4}{6a}$

$\qquad =\dfrac{9a+1}{6a}$

21. $\dfrac{6x}{y}+\dfrac{2y}{xy}=\dfrac{x}{x}\cdot\dfrac{6x}{y}+\dfrac{2y}{xy}$

$\qquad =\dfrac{6x^2}{xy}+\dfrac{2y}{xy}$

$\qquad =\dfrac{6x^2+2y}{xy}$

23. $\dfrac{9}{b}-\dfrac{4}{5a^2}=\dfrac{5a^2}{5a^2}\cdot\dfrac{9}{b}-\dfrac{4}{5a^2}\cdot\dfrac{b}{b}$

$\qquad =\dfrac{45a^2}{5a^2b}-\dfrac{4b}{5a^2b}$

$\qquad =\dfrac{45a^2-4b}{5a^2b}$

25. $\dfrac{4}{x}+\dfrac{9}{x-3}=\dfrac{x-3}{x-3}\cdot\dfrac{4}{x}+\dfrac{9}{(x-3)}\cdot\dfrac{x}{x}$

$\qquad =\dfrac{4(x-3)}{x(x-3)}+\dfrac{9x}{x(x-3)}$

$\qquad =\dfrac{4x-12}{x(x-3)}+\dfrac{9x}{x(x-3)}$

$\qquad =\dfrac{4x-12+9x}{x(x-3)}$

$\qquad =\dfrac{13x-12}{x(x-3)}$

27. $\dfrac{9}{p+3}+\dfrac{2}{p}=\dfrac{p}{p}\cdot\dfrac{9}{p+3}+\dfrac{2}{p}\cdot\dfrac{p+3}{p+3}$

$\qquad =\dfrac{9p}{p(p+3)}+\dfrac{2(p+3)}{p(p+3)}$

$\qquad =\dfrac{9p}{p(p+3)}+\dfrac{2p+6}{p(p+3)}$

$\qquad =\dfrac{9p+2p+6}{p(p+3)}$

$\qquad =\dfrac{11p+6}{p(p+3)}$

29. $\dfrac{5}{d+1}-\dfrac{d}{3d+5}$

$\qquad =\dfrac{3d+5}{3d+5}\cdot\dfrac{5}{d+1}-\dfrac{d}{3d+5}\cdot\dfrac{d+1}{d+1}$

$\qquad =\dfrac{5(3d+5)}{(3d+5)(d+1)}-\dfrac{d(d+1)}{(3d+5)(d+1)}$

$\qquad =\dfrac{15d+25}{(3d+5)(d+1)}-\dfrac{d^2+d}{(3d+5)(d+1)}$

$\qquad =\dfrac{15d+25-d^2-d}{(3d+5)(d+1)}$

$\qquad =\dfrac{-d^2+14d+25}{(3d+5)(d+1)}$

31. $\dfrac{8}{p-3}+\dfrac{2}{3-p}=\dfrac{8}{p-3}+\dfrac{2}{3-p}\cdot\dfrac{-1}{-1}$

$\qquad\qquad =\dfrac{8}{p-3}+\dfrac{-2}{p-3}$

$\qquad\qquad =\dfrac{6}{p-3}$

33. $\dfrac{9}{x+7}-\dfrac{5}{-x-7}=\dfrac{9}{x+7}-\dfrac{5}{-x-7}\cdot\dfrac{-1}{-1}$

$\qquad\qquad =\dfrac{9}{x+7}-\dfrac{-5}{x+7}$

$\qquad\qquad =\dfrac{14}{x+7}$

35. $\dfrac{8}{a-2}+\dfrac{a}{2a-4}=\dfrac{2}{2}\cdot\dfrac{8}{a-2}+\dfrac{a}{2(a-2)}$

$\qquad\qquad =\dfrac{16}{2(a-2)}+\dfrac{a}{2(a-2)}$

$\qquad\qquad =\dfrac{a+16}{2(a-2)}$

37. $\dfrac{x+5}{x-5}-\dfrac{x-5}{x+5}=\dfrac{x+5}{x+5}\cdot\dfrac{x+5}{x-5}-\dfrac{x-5}{x+5}\cdot\dfrac{x-5}{x-5}$

$\quad=\dfrac{(x+5)^2}{(x+5)(x-5)}-\dfrac{(x-5)^2}{(x+5)(x-5)}$

$\quad=\dfrac{x^2+10x+25-\left(x^2-10x+25\right)}{(x-5)(x+5)}$

$\quad=\dfrac{x^2+10x+25-x^2+10x-25}{(x-5)(x+5)}$

$\quad=\dfrac{20x}{(x-5)(x+5)}$

39. $\dfrac{5}{6n+3}-\dfrac{2}{n}=\dfrac{n}{n}\cdot\dfrac{5}{3(2n+1)}-\dfrac{2}{n}\cdot\dfrac{3(2n+1)}{3(2n+1)}$

$\quad=\dfrac{5n}{3n(2n+1)}-\dfrac{6(2n+1)}{3n(2n+1)}$

$\quad=\dfrac{5n}{3n(2n+1)}-\dfrac{12n+6}{3n(2n+1)}$

$\quad=\dfrac{5n-12n-6}{3n(2n+1)}$

$\quad=\dfrac{-7n-6}{3n(2n+1)}$

41. $\dfrac{3}{2w+10}+\dfrac{6}{w+2}$

$\quad=\dfrac{w+2}{w+2}\cdot\dfrac{3}{2(w+5)}+\dfrac{6}{w+2}\cdot\dfrac{2(w+5)}{2(w+5)}$

$\quad=\dfrac{3(w+2)}{2(w+2)(w+5)}+\dfrac{12(w+5)}{2(w+2)(w+5)}$

$\quad=\dfrac{3w+6+12w+60}{2(w+2)(w+5)}$

$\quad=\dfrac{15w+66}{2(w+2)(w+5)}$

43. $\dfrac{z}{z^2-16}+\dfrac{4}{z+4}=\dfrac{z}{(z+4)(z-4)}+\dfrac{4}{(z+4)}\cdot\dfrac{z-4}{z-4}$

$\quad=\dfrac{z+4z-16}{(z+4)(z-4)}$

$\quad=\dfrac{5z-16}{(z+4)(z-4)}$

45. $\dfrac{x+2}{x^2-4}-\dfrac{2}{x+2}=\dfrac{x+2}{(x+2)(x-2)}-\dfrac{2}{x+2}\cdot\dfrac{x-2}{x-2}$

$\quad=\dfrac{x+2}{(x+2)(x-2)}-\dfrac{2(x-2)}{(x+2)(x-2)}$

$\quad=\dfrac{x+2-(2x-4)}{(x+2)(x-2)}$

$\quad=\dfrac{x+2-2x+4}{(x+2)(x-2)}$

$\quad=\dfrac{-x+6}{(x+2)(x-2)}$

47. $\dfrac{3r+4}{r^2-10r+24}-\dfrac{2}{r-6}$

$\quad=\dfrac{3r+4}{(r-6)(r-4)}-\dfrac{2}{r-6}\cdot\dfrac{r-4}{r-4}$

$\quad=\dfrac{3r+4}{(r-6)(r-4)}-\dfrac{2(r-4)}{(r-6)(r-4)}$

$\quad=\dfrac{3r+4-(2r-4)}{(r-6)(r-4)}$

$\quad=\dfrac{3r+4-2r+8}{(r-6)(r-4)}$

$\quad=\dfrac{r+12}{(r-6)(r-4)}$

49. $\dfrac{x^2-3}{x^2+2x-8}-\dfrac{x-4}{x+4}$

$\quad=\dfrac{x^2-3}{(x+4)(x-2)}-\dfrac{x-4}{x+4}\cdot\dfrac{x-2}{x-2}$

$\quad=\dfrac{x^2-3}{(x+4)(x-2)}-\dfrac{(x-4)(x-2)}{(x+4)(x-2)}$

$\quad=\dfrac{x^2-3-\left(x^2-6x+8\right)}{(x+4)(x-2)}$

$\quad=\dfrac{x^2-3-x^2+6x-8}{(x+4)(x-2)}$

$\quad=\dfrac{6x-11}{(x+4)(x-2)}$

51. $\dfrac{x-6}{x^2+10x+25}+\dfrac{x-3}{x+5}$

$=\dfrac{x-6}{(x+5)^2}+\dfrac{x-3}{x+5}\cdot\dfrac{x+5}{x+5}$

$=\dfrac{x-6}{(x+5)^2}+\dfrac{(x-3)(x+5)}{(x+5)^2}$

$=\dfrac{x-6+x^2+2x-15}{(x+5)^2}$

$=\dfrac{x^2+3x-21}{(x+5)^2}$

53. $\dfrac{5}{a^2-9a+8}-\dfrac{6}{a^2-6a-16}$

$=\dfrac{a+2}{a+2}\cdot\dfrac{5}{(a-8)(a-1)}-\dfrac{6}{(a-8)(a+2)}\cdot\dfrac{a-1}{a-1}$

$=\dfrac{5(a+2)}{(a-8)(a-1)(a+2)}-\dfrac{6(a-1)}{(a-8)(a-1)(a+2)}$

$=\dfrac{5a+10-(6a-6)}{(a-8)(a-1)(a+2)}$

$=\dfrac{5a+10-6a+6}{(a-8)(a-1)(a+2)}$

$=\dfrac{-a+16}{(a-8)(a-1)(a+2)}$

55. $\dfrac{2}{x^2+6x+9}+\dfrac{7}{x^2+x-6}$

$=\dfrac{x-2}{x-2}\cdot\dfrac{2}{(x+3)(x+3)}+\dfrac{7}{(x+3)(x-2)}\cdot\dfrac{x+3}{x+3}$

$=\dfrac{2(x-2)}{(x+3)(x+3)(x-2)}+\dfrac{7(x+3)}{(x+3)(x+3)(x-2)}$

$=\dfrac{2x-4+7x+21}{(x+3)(x+3)(x-2)}$

$=\dfrac{9x+17}{(x+3)^2(x-2)}$

57. $\dfrac{x}{2x^2+7x+3}-\dfrac{5}{3x^2+7x-6}$

$=\dfrac{3x-2}{3x-2}\cdot\dfrac{x}{(2x+1)(x+3)}-\dfrac{5}{(3x-2)(x+3)}\cdot\dfrac{2x+1}{2x+1}$

$=\dfrac{x(3x-2)}{(2x+1)(3x-2)(x+3)}-\dfrac{5(2x+1)}{(2x+1)(3x-2)(x+3)}$

$=\dfrac{3x^2-2x-(10x+5)}{(2x+1)(3x-2)(x+3)}$

$=\dfrac{3x^2-2x-10x-5}{(2x+1)(3x-2)(x+3)}$

$=\dfrac{3x^2-12x-5}{(2x+1)(3x-2)(x+3)}$

59. $\dfrac{x}{4x^2+11x+6}-\dfrac{2}{8x^2+2x-3}$

$=\dfrac{2x-1}{2x-1}\cdot\dfrac{x}{(4x+3)(x+2)}-\dfrac{2}{(4x+3)(2x-1)}\cdot\dfrac{x+2}{x+2}$

$=\dfrac{x(2x-1)}{(4x+3)(x+2)(2x-1)}-\dfrac{2(x+2)}{(4x+3)(x+2)(2x-1)}$

$=\dfrac{2x^2-x-(2x+4)}{(4x+3)(x+2)(2x-1)}$

$=\dfrac{2x^2-x-2x-4}{(4x+3)(x+2)(2x-1)}$

$=\dfrac{2x^2-3x-4}{(4x+3)(x+2)(2x-1)}$

61. $\dfrac{3w+12}{w^2+w-12}-\dfrac{2}{w-3}=\dfrac{3(w+4)}{(w-3)(w+4)}-\dfrac{2}{w-3}$

$=\dfrac{3\cancel{(w+4)}}{(w-3)\cancel{(w+4)}}-\dfrac{2}{w-3}$

$=\dfrac{3}{w-3}-\dfrac{2}{w-3}$

$=\dfrac{1}{w-3}$

63. $\dfrac{4r}{2r^2-10r+12}+\dfrac{4}{r-2}$

$=\dfrac{4r}{2\left(r^2-5r+6\right)}+\dfrac{4}{r-2}$

$=\dfrac{\cancel{2}\cdot 2r}{\cancel{2}(r-3)(r-2)}+\dfrac{4}{r-2}\cdot\dfrac{r-3}{r-3}$

$=\dfrac{2r}{(r-3)(r-2)}+\dfrac{4r-12}{(r-3)(r-2)}$

$=\dfrac{6r-12}{(r-3)(r-2)}$

$=\dfrac{6\cancel{(r-2)}}{(r-3)\cancel{(r-2)}}$

$=\dfrac{6}{(r-3)}$

65. $\dfrac{4}{x^2-4x}-\dfrac{x}{4x-16}$

$=\dfrac{4}{x(x-4)}-\dfrac{x}{4(x-4)}$

$=\dfrac{4}{4}\cdot\dfrac{4}{x(x-4)}-\dfrac{x}{4(x-4)}\cdot\dfrac{x}{x}$

$=\dfrac{16}{4x(x-4)}-\dfrac{x^2}{4x(x-4)}$

$=\dfrac{16-x^2}{4x(x-4)}$

$=\dfrac{-1(x+4)\cancel{(x-4)}}{4x\cancel{(x-4)}}$

$=-\dfrac{x+4}{4x}$

67. $\dfrac{8}{x}+6$ is defined for all real numbers

except $x=0$

69. $\dfrac{3}{x-4}+\dfrac{7}{x+6}$

$x-4=0$ when $x=4$ and $x+6=0$ when $x=-6$
The expression is defined for all real

numbers except $x=4$ and $x=-6$.

71. $\dfrac{3}{\Delta-2}-\dfrac{4}{2-\Delta}=\dfrac{3}{\Delta-2}+\dfrac{4}{\Delta-2}$

$=\dfrac{3+4}{\Delta-2}$

$=\dfrac{7}{\Delta-2}$

73. $\dfrac{5}{a+b}+\dfrac{4}{a}$

$a+b=0$ when $a=-b$. The expression is defined
for all real numbers except $a=0$ and $a=-b$.

75. $\dfrac{x}{x^2-9}+\dfrac{2x}{x+3}+\dfrac{2x^2-5x}{9-x^2}$

$=\dfrac{x}{x^2-9}+\dfrac{2x}{x+3}+\dfrac{2x^2-5x}{-1\left(x^2-9\right)}$

$=\dfrac{x}{(x+3)(x-3)}+\dfrac{2x}{(x+3)}\cdot\dfrac{x-3}{x-3}-\dfrac{2x^2-5x}{(x+3)(x-3)}$

$=\dfrac{x+2x^2-6x-(2x^2-5x)}{(x+3)(x-3)}$

$=\dfrac{x+2x^2-6x-2x^2+5x}{(x+3)(x-3)}=\dfrac{0}{(x+3)(x-3)}$

$=0$

77. $\dfrac{x+6}{4-x^2}-\dfrac{x+3}{x+2}+\dfrac{x-3}{2-x}$

$=\dfrac{x+6}{(2-x)(2+x)}-\dfrac{x+3}{x+2}+\dfrac{x-3}{2-x}$

$=\dfrac{x+6}{(2-x)(2+x)}-\dfrac{x+3}{x+2}\cdot\dfrac{2-x}{2-x}+\dfrac{x-3}{2-x}\cdot\dfrac{x+2}{x+2}$

$=\dfrac{x+6-(2x-x^2+6-3x)+(x^2-x-6)}{(2-x)(2+x)}$

$=\dfrac{x+6+x^2+x-6+x^2-x-6}{(2-x)(2+x)}$

$=\dfrac{2x^2+x-6}{(2-x)(2+x)}=\dfrac{(2x-3)\cancel{(x+2)}}{(2-x)\cancel{(2+x)}}$

$=\dfrac{2x-3}{2-x}$

79. $\dfrac{2}{x^2-x-6}+\dfrac{3}{x^2-2x-3}+\dfrac{1}{x^2+3x+2}=\dfrac{2}{(x+2)(x-3)}+\dfrac{3}{(x-3)(x+1)}+\dfrac{1}{(x+2)(x+1)}$

$$=\dfrac{2(x+1)}{(x+2)(x-3)(x+1)}+\dfrac{3(x+2)}{(x+2)(x-3)(x+1)}+\dfrac{x-3}{(x+2)(x-3)(x+1)}$$

$$=\dfrac{2x+2+3x+6+x-3}{(x+2)(x-3)(x+1)}$$

$$=\dfrac{6x+5}{(x+2)(x-3)(x+1)}$$

82. Let x = the number of hours it takes the train to travel 42 miles.

$$\dfrac{22\text{ miles}}{0.8\text{ hours}}=\dfrac{42\text{ miles}}{x\text{ hours}}$$

$$22x=0.8(42)$$

$$22x=33.6$$

$$x\approx 1.53$$

It takes about 1.53 hours for the train to travel 42 miles.

83. $m=\dfrac{-2-6}{3-(-4)}=\dfrac{-8}{7}=-\dfrac{8}{7}$

84.

$$\begin{array}{r} 4x-3 \\ 2x+3\overline{\smash{\big)}\,8x^2+6x-15} \\ \underline{8x^2+12x} \\ -6x-15 \\ \underline{-6x-9} \\ -6 \end{array}$$

$$(8x^2+6x-15)\div(2x+3)=4x-3-\dfrac{6}{2x+3}$$

85. $\dfrac{x^2+xy-6y^2}{x^2-xy-2y^2}\cdot\dfrac{y^2-x^2}{x^2+2xy-3y^2}$

$$=\dfrac{(x+3y)(x-2y)}{(x+y)(x-2y)}\cdot\dfrac{-1(x-y)(x+y)}{(x+3y)(x-y)}$$

$$=-1$$

Mid-Chapter Test: 7.1-7.4

1. $\dfrac{9}{3x-2}$

$3x-2=0$ when $x=\dfrac{2}{3}$

The expression is defined for all real numbers except $x=\dfrac{2}{3}$.

2. $\dfrac{2x+1}{x^2-5x-14}$

$$x^2-5x-14=0$$

$$(x-7)(x+2)=0$$

$$x-7=0 \quad\text{or}\quad x+2=0$$

$$x=7 \qquad\qquad x=-2$$

The expression is defined for all real numbers except $x=7$, $x=-2$.

3. $\dfrac{9x+18}{x+2}=\dfrac{9\cancel{(x+2)}}{\cancel{(x+2)}}=9$

4. $\dfrac{2x^2+13x+15}{3x^2+14x-5}=\dfrac{(2x+3)\cancel{(x+5)}}{(3x-1)\cancel{(x+5)}}=\dfrac{2x+3}{3x-1}$

5. $\dfrac{25r^2-36t^2}{5r-6t}=\dfrac{(5r+6t)\cancel{(5r-6t)}}{\cancel{(5r-6t)}}=5r+6t$

6. $\dfrac{15x^2}{2y}\cdot\dfrac{4y^4}{5x^5}=\dfrac{3\cdot\cancel{5}\cdot\cancel{x^2}}{\cancel{2}\cancel{y}}\cdot\dfrac{2\cdot\cancel{2}\cdot y^3\cdot\cancel{y}}{\cancel{5}\cdot x^3\cdot\cancel{x^2}}=\dfrac{6y^3}{x^3}$

7. $\dfrac{m-3}{m+4} \cdot \dfrac{m^2+8m+16}{3-m}$

$= \dfrac{\cancel{m-3}}{\cancel{m+4}} \cdot \dfrac{\cancel{(m+4)}\,(m+4)}{-1\cancel{(m-3)}}$

$= -(m+4)$ or $-m-4$

8. $\dfrac{x^3+27}{x^2-2x-15} \cdot \dfrac{x^2-7x+10}{x^2-3x+9}$

$= \dfrac{(x+3)\cancel{(x^2-3x+9)}}{\cancel{(x-5)}\cancel{(x+3)}} \cdot \dfrac{\cancel{(x-5)}\,(x-2)}{\cancel{(x^2-3x+9)}}$

$= x-2$

9. $\dfrac{5x-1}{x^2+11x+10} \div \dfrac{10x-2}{x^2+17x+70}$

$= \dfrac{5x-1}{x^2+11x+10} \cdot \dfrac{x^2+17x+70}{10x-2}$

$= \dfrac{\cancel{5x-1}}{\cancel{(x+10)}\,(x+1)} \cdot \dfrac{\cancel{(x+10)}\,(x+7)}{2\cancel{(5x-1)}}$

$= \dfrac{x+7}{2(x+1)}$

10. $\dfrac{5x^2+7x+2}{x^2+6x+5} \div \dfrac{7x^2-39x-18}{x^2-x-30}$

$= \dfrac{5x^2+7x+2}{x^2+6x+5} \cdot \dfrac{x^2-x-30}{7x^2-39x-18}$

$= \dfrac{(5x+2)\cancel{(x+1)}}{\cancel{(x+5)}\cancel{(x+1)}} \cdot \dfrac{\cancel{(x-6)}\cancel{(x+5)}}{(7x+3)\cancel{(x-6)}}$

$= \dfrac{5x+2}{7x+3}$

11. $\dfrac{x^2}{x+6} - \dfrac{36}{x+6} = \dfrac{x^2-36}{x+6}$

$= \dfrac{\cancel{(x+6)}\,(x-6)}{\cancel{(x+6)}}$

$= x-6$

12. $\dfrac{2x^2-2x}{2x+5} + \dfrac{x-15}{2x+5} = \dfrac{2x^2-2x+x-15}{2x+5}$

$= \dfrac{2x^2-x-15}{2x+5}$

$= \dfrac{\cancel{(2x+5)}\,(x-3)}{\cancel{(2x+5)}}$

$= x-3$

13. $\dfrac{3x^2-x}{4x^2-9x+2} - \dfrac{3x+4}{4x^2-9x+2} = \dfrac{3x^2-x-3x-4}{4x^2-9x+2}$

$= \dfrac{3x^2-4x-4}{4x^2-9x+2}$

$= \dfrac{(3x+2)\cancel{(x-2)}}{(4x-1)\cancel{(x-2)}}$

$= \dfrac{3x+2}{4x-1}$

14. $\dfrac{2m}{6m^2+3m} + \dfrac{m+7}{2m+1} = \dfrac{2m}{3m(2m+1)} + \dfrac{m+7}{2m+1}$

The LCD is $3m(2m+1)$.

15. $\dfrac{9x+8}{2x^2-5x-12} + \dfrac{2x+3}{x^2-9x+20}$

$= \dfrac{9x+8}{(2x+3)(x-4)} + \dfrac{2x+3}{(x-5)(x-4)}$

The LCD is $(2x+3)(x-4)(x-5)$.

16. $\dfrac{x+1}{2x} + \dfrac{4x-3}{5x} = \dfrac{5}{5} \cdot \dfrac{x+1}{2x} + \dfrac{4x-3}{5x} \cdot \dfrac{2}{2}$

$= \dfrac{5x+5}{10x} + \dfrac{8x-6}{10x}$

$= \dfrac{5x+5+8x-6}{10x}$

$= \dfrac{13x-1}{10x}$

17. $\dfrac{2a+5}{a+3} - \dfrac{3a+1}{a-4}$

$= \dfrac{a-4}{a-4} \cdot \dfrac{2a+5}{a+3} - \dfrac{3a+1}{a-4} \cdot \dfrac{a+3}{a+3}$

$= \dfrac{(a-4)(2a+5)}{(a-4)(a+3)} - \dfrac{(3a+1)(a+3)}{(a-4)(a+3)}$

$= \dfrac{2a^2-3a-20}{(a-4)(a+3)} - \dfrac{3a^2+10a+3}{(a-4)(a+3)}$

$= \dfrac{2a^2-3a-20-3a^2-10a-3}{(a-4)(a+3)}$

$= \dfrac{-a^2-13a-23}{(a-4)(a+3)} = \dfrac{-1(a^2+13a+23)}{a^2-a-12}$

$= -\dfrac{a^2+13a+23}{a^2-a-12}$

18. $\dfrac{x^2+5}{2x^2+13x+6} + \dfrac{3x-1}{2x+1}$

$= \dfrac{x^2+5}{(2x+1)(x+6)} + \dfrac{3x-1}{2x+1}$

$= \dfrac{x^2+5}{(2x+1)(x+6)} + \dfrac{3x-1}{2x+1} \cdot \dfrac{x+6}{x+6}$

$= \dfrac{x^2+5}{(2x+1)(x+6)} + \dfrac{3x^2+17x-6}{(2x+1)(x+6)}$

$= \dfrac{x^2+5+3x^2+17x-6}{(2x+1)(x+6)}$

$= \dfrac{4x^2+17x-1}{2x^2+13x+6}$

19. $\dfrac{x}{x^2+3x+2} - \dfrac{4}{x^2-x-6}$

$= \dfrac{x}{(x+2)(x+1)} - \dfrac{4}{(x-3)(x+2)}$

$= \dfrac{x-3}{x-3} \cdot \dfrac{x}{(x+2)(x+1)} - \dfrac{4}{(x-3)(x+2)} \cdot \dfrac{x+1}{x+1}$

$= \dfrac{x^2-3x}{(x-3)(x+2)(x+1)} - \dfrac{4x+4}{(x-3)(x+2)(x+1)}$

$= \dfrac{x^2-3x-4x-4}{(x-3)(x+2)(x+1)}$

$= \dfrac{x^2-7x-4}{(x-3)(x+2)(x+1)}$

20. To add these fractions a common denominator of $x(x+1)$ is needed.

$\dfrac{7}{x+1} + \dfrac{8}{x} = \dfrac{x}{x} \cdot \dfrac{7}{x+1} + \dfrac{8}{x} \cdot \dfrac{x+1}{x+1}$

$= \dfrac{7x}{x(x+1)} + \dfrac{8x+8}{x(x+1)}$

$= \dfrac{7x+8x+8}{x(x+1)}$

$= \dfrac{15x+8}{x(x+1)}$

Exercise Set 7.5

1. A complex fraction is a fraction whose numerator or denominator (or both) contains a fraction.

3. a. $\dfrac{\dfrac{x+9}{4}}{\dfrac{7}{x^2+5x+6}}$

Numerator: $\dfrac{x+9}{4}$

Denominator: $\dfrac{7}{x^2+5x+6}$

b. $\dfrac{\dfrac{1}{2y}+x}{\dfrac{3}{y}+x^2}$

Numerator, $\dfrac{1}{2y}+x$

Denominator, $\dfrac{3}{y}+x^2$

5. $\dfrac{4+\dfrac{2}{3}}{5+\dfrac{1}{3}} = \dfrac{\left(4+\dfrac{2}{3}\right)3}{\left(5+\dfrac{1}{3}\right)3}$

$= \dfrac{12+2}{15+1}$

$= \dfrac{14}{16}$

$= \dfrac{7}{8}$

7. $\dfrac{2+\frac{3}{8}}{1+\frac{1}{3}} = \dfrac{\left(2+\frac{3}{8}\right)24}{\left(1+\frac{1}{3}\right)24}$

$\qquad = \dfrac{48+9}{24+8}$

$\qquad = \dfrac{57}{32}$

9. $\dfrac{\frac{2}{3}+\frac{1}{4}}{\frac{5}{6}-\frac{1}{3}} = \dfrac{\left(\frac{2}{3}+\frac{1}{4}\right)12}{\left(\frac{5}{6}-\frac{1}{3}\right)12}$

$\qquad = \dfrac{8+3}{10-4}$

$\qquad = \dfrac{11}{6}$

11. $\dfrac{\frac{xy^2}{7}}{\frac{3}{x^2}} = \dfrac{xy^2}{7} \cdot \dfrac{x^2}{3} = \dfrac{x^3 y^2}{21}$

13. $\dfrac{\frac{6a^2 b}{7}}{\frac{9ac^2}{b^2}} = \dfrac{6a^2 b}{7} \cdot \dfrac{b^2}{9ac^2}$

$\qquad = \dfrac{6a^2 b^3}{63ac^2}$

$\qquad = \dfrac{2ab^3}{21c^2}$

15. $\dfrac{a-\frac{a}{b}}{\frac{3+a}{b}} = \dfrac{\left(a-\frac{a}{b}\right)b}{\left(\frac{3+a}{b}\right)b} = \dfrac{ab-a}{3+a}$

17. $\dfrac{\frac{9}{x}+\frac{3}{x^2}}{3+\frac{1}{x}} = \dfrac{\left(\frac{9}{x}+\frac{3}{x^2}\right)x^2}{\left(3+\frac{1}{x}\right)x^2}$

$\qquad = \dfrac{9x+3}{3x^2 + x}$

$\qquad = \dfrac{3(3x+1)}{x(3x+1)}$

$\qquad = \dfrac{3}{x}$

19. $\dfrac{5-\frac{1}{x}}{4-\frac{1}{x}} = \dfrac{\left(5-\frac{1}{x}\right)x}{\left(4-\frac{1}{x}\right)x} = \dfrac{5x-1}{4x-1}$

21. $\dfrac{\frac{m}{n}-\frac{n}{m}}{\frac{m+n}{n}} = \dfrac{\left(\frac{m}{n}-\frac{n}{m}\right)mn}{\left(\frac{m+n}{n}\right)mn}$

$\qquad = \dfrac{m^2 - n^2}{m(m+n)}$

$\qquad = \dfrac{(m+n)(m-n)}{m(m+n)}$

$\qquad = \dfrac{m-n}{m}$

23. $\dfrac{\frac{a^2}{b}-b}{\frac{b^2}{a}-a} = \dfrac{\left(\frac{a^2}{b}-b\right)ab}{\left(\frac{b^2}{a}-a\right)ab}$

$\qquad = \dfrac{\left(a^2 - b^2\right)a}{\left(b^2 - a^2\right)b}$

$\qquad = \dfrac{a\left(a^2 - b^2\right)}{-b\left(a^2 - b^2\right)}$

$\qquad = -\dfrac{a}{b}$

25. $\dfrac{2-\frac{a}{b}}{\frac{a}{b}-2} = \dfrac{\left(2-\frac{a}{b}\right)b}{\left(\frac{a}{b}-2\right)b}$

$\qquad = \dfrac{2b-a}{a-2b}$

$\qquad = \dfrac{-1(-2b+a)}{a-2b}$

$\qquad = \dfrac{-1(a-2b)}{(a-2b)}$

$\qquad = -1$

27. $\dfrac{\frac{4}{x^2}+\frac{4}{x}}{\frac{4}{x}+\frac{4}{x^2}} = \dfrac{\frac{4}{x^2}+\frac{4}{x}}{\frac{4}{x^2}+\frac{4}{x}} = 1$

29. $\dfrac{\dfrac{1}{a}-\dfrac{1}{b}}{\dfrac{1}{ab}}=\dfrac{\left(\dfrac{1}{a}-\dfrac{1}{b}\right)ab}{\left(\dfrac{1}{ab}\right)ab}=\dfrac{b-a}{1}=b-a$

31. $\dfrac{\dfrac{a}{b}+\dfrac{1}{a}}{\dfrac{b}{a}+\dfrac{1}{a}}=\dfrac{\left(\dfrac{a}{b}+\dfrac{1}{a}\right)ab}{\left(\dfrac{b}{a}+\dfrac{1}{a}\right)ab}=\dfrac{a^2+b}{b^2+b}=\dfrac{a^2+b}{b(b+1)}$

33. $\dfrac{x}{\dfrac{1}{x}-\dfrac{1}{y}}=\dfrac{(x)xy}{\left(\dfrac{1}{x}-\dfrac{1}{y}\right)xy}=\dfrac{x^2y}{y-x}$

35. $\dfrac{\dfrac{5}{a}+\dfrac{5}{a^2}}{\dfrac{5}{b}+\dfrac{5}{b^2}}=\dfrac{\left(\dfrac{5}{a}+\dfrac{5}{a^2}\right)a^2b^2}{\left(\dfrac{5}{b}+\dfrac{5}{b^2}\right)a^2b^2}$

$=\dfrac{5ab^2+5b^2}{5a^2b+5a^2}$

$=\dfrac{5b^2(a+1)}{5a^2(b+1)}$

$=\dfrac{ab^2+b^2}{a^2(b+1)}$

37. a. Answers will vary.

b. $\dfrac{5+\dfrac{3}{5}}{\dfrac{1}{8}-4}=\dfrac{\dfrac{25}{5}+\dfrac{3}{5}}{\dfrac{1}{8}-\dfrac{32}{8}}$

$=\dfrac{\dfrac{28}{5}}{-\dfrac{31}{8}}$

$=\dfrac{28}{5}\cdot\left(-\dfrac{8}{31}\right)$

$=-\dfrac{224}{155}$

c. $\dfrac{5+\dfrac{3}{5}}{\dfrac{1}{8}-4}=\dfrac{\left(5+\dfrac{3}{5}\right)40}{\left(\dfrac{1}{8}-4\right)40}=\dfrac{200+24}{5-160}=-\dfrac{224}{155}$

39. a. Answers will vary.

b. $\dfrac{\dfrac{x-y}{x+y}+\dfrac{6}{x+y}}{2-\dfrac{7}{x+y}}=\dfrac{\dfrac{x-y+6}{x+y}}{\dfrac{2(x+y)}{x+y}-\dfrac{7}{x+y}}$

$=\dfrac{\dfrac{x-y+6}{x+y}}{\dfrac{2x+2y-7}{x+y}}$

$=\dfrac{x-y+6}{\cancel{x+y}}\cdot\dfrac{\cancel{x+y}}{2x+2y-7}$

$=\dfrac{x-y+6}{2x+2y-7}$

c. $\dfrac{\dfrac{x-y}{x+y}+\dfrac{6}{x+y}}{2-\dfrac{7}{x+y}}=\dfrac{\dfrac{x-y+6}{x+y}}{2-\dfrac{7}{x+y}}$

$=\dfrac{\left(\dfrac{x-y+6}{x+y}\right)(x+y)}{\left(2-\dfrac{7}{x+y}\right)(x+y)}$

$=\dfrac{x-y+6}{2(x+y)-7}$

$=\dfrac{x-y+6}{2x+2y-7}$

41. a. $\dfrac{\dfrac{5}{12x}}{\dfrac{8}{x^2}-\dfrac{4}{3x}}$

b. $\dfrac{\dfrac{5}{12x}}{\dfrac{8}{x^2}-\dfrac{4}{3x}}=\dfrac{\dfrac{5}{12x}}{\dfrac{24-4x}{3x^2}}$

$=\dfrac{5}{12x}\cdot\dfrac{3x^2}{(24-4x)}$

$=\dfrac{5x}{4(24-4x)}$

$=\dfrac{5x}{96-16x}$

43. $\dfrac{x^{-1}+y^{-1}}{3} = \dfrac{\dfrac{1}{x}+\dfrac{1}{y}}{3}$

$\qquad = \dfrac{\left(\dfrac{1}{x}+\dfrac{1}{y}\right)xy}{3xy}$

$\qquad = \dfrac{y+x}{3xy}$

45. $\dfrac{x^{-1}+y^{-1}}{x^{-1}y^{-1}} = \dfrac{\dfrac{1}{x}+\dfrac{1}{y}}{\dfrac{1}{xy}}$

$\qquad = \dfrac{\left(\dfrac{1}{x}+\dfrac{1}{y}\right)xy}{\left(\dfrac{1}{xy}\right)xy}$

$\qquad = \dfrac{y+x}{1}$

$\qquad = x+y$

47. a. $E = \dfrac{\dfrac{1}{2}\left(\dfrac{2}{3}\right)}{\dfrac{2}{3}+\dfrac{1}{2}} = \dfrac{\dfrac{2}{6}}{\dfrac{4+3}{6}} = \dfrac{2}{\cancel{6}}\cdot\dfrac{\cancel{6}}{7} = \dfrac{2}{7}$

b. $E = \dfrac{\dfrac{1}{2}\left(\dfrac{4}{5}\right)}{\dfrac{4}{5}+\dfrac{1}{2}} = \dfrac{\dfrac{4}{10}}{\dfrac{8+5}{10}} = \dfrac{4}{\cancel{10}}\cdot\dfrac{\cancel{10}}{13} = \dfrac{4}{13}$

49. $\dfrac{\dfrac{a}{b}+b-\dfrac{1}{a}}{\dfrac{a}{b^2}-\dfrac{b}{a}+\dfrac{3}{a^2}} = \dfrac{\dfrac{a^2+b^2a-b}{ba}}{\dfrac{a^3-ab^3+3b^2}{a^2b^2}}$

$\qquad = \dfrac{a^2+b^2a-b}{\cancel{a}\,\cancel{b}}\cdot\dfrac{\overset{a}{\cancel{a^2}}\,\overset{b}{\cancel{b^2}}}{a^3-ab^3+3b^2}$

$\qquad = \dfrac{(a^2+b^2a-b)ab}{a^3-ab^3+3b^2}$

$\qquad = \dfrac{a^3b+a^2b^3-ab^2}{a^3-ab^3+3b^2}$

51. $2x-8(5-x) = 9x-3(x+2)$

$\qquad 2x-40+8x = 9x-3x-6$

$\qquad\quad 10x-40 = 6x-6$

$\qquad\qquad\quad 4x = 34$

$\qquad\qquad\quad\ x = \dfrac{34}{4} = \dfrac{17}{2}$

52. A polynomial is an expression containing a finite number of terms of the form ax^n where a is a real number and n is a whole number.

53. $x^2-13x+40 = (x-8)(x-5)$

54. $\dfrac{x}{3x^2+17x-6} - \dfrac{2}{x^2+3x-18}$

$= \dfrac{x}{(3x-1)(x+6)} - \dfrac{2}{(x+6)(x-3)}$

$= \dfrac{x}{(3x-1)(x+6)}\cdot\dfrac{x-3}{x-3} - \dfrac{2}{(x+6)(x-3)}\cdot\dfrac{3x-1}{3x-1}$

$= \dfrac{x^2-3x-(6x-2)}{(3x-1)(x+6)(x-3)}$

$= \dfrac{x^2-3x-6x+2}{(3x-1)(x+6)(x-3)}$

$= \dfrac{x^2-9x+2}{(3x-1)(x+6)(x-3)}$

Exercise Set 7.6

1. a. Answers will vary.

b. $\dfrac{1}{x-1} - \dfrac{1}{x+1} = \dfrac{3x}{x^2-1}$

$\dfrac{1}{x-1} - \dfrac{1}{x+1} = \dfrac{3x}{(x-1)(x+1)}$

Multiply both sides of the equation by the least common denominator, $(x-1)(x+1)$.

$$(x-1)(x+1)\left(\dfrac{1}{x-1} - \dfrac{1}{x+1}\right) = \left(\dfrac{3x}{(x-1)(x+1)}\right)(x-1)(x+1)$$

$$(x-1)(x+1)\left(\dfrac{1}{x-1}\right) - (x-1)(x+1)\left(\dfrac{1}{x+1}\right) = 3x$$

$$x+1-(x-1) = 3x$$

$$x+1-x+1 = 3x$$

$$2 = 3x$$

$$\dfrac{2}{3} = x$$

3. a. The problem on the left is an expression to be simplified while the problem on the right is an equation to be solved.

b. Left: Write the fractions with the LCD, $12(x-1)$, then combine numerators.
Right: Multiply both sides of the equation by the LCD, $12(x-1)$, then solve.

c. Left: $\dfrac{x}{3} - \dfrac{x}{4} + \dfrac{1}{x-1} = \dfrac{x \cdot 4(x-1)}{3 \cdot 4(x-1)} - \dfrac{x \cdot 3(x-1)}{4 \cdot 3(x-1)} + \dfrac{1 \cdot 3 \cdot 4}{(x-1)3 \cdot 4}$

$= \dfrac{4x(x-1) - 3x(x-1) + 12}{3 \cdot 4(x-1)}$

$= \dfrac{4x^2 - 4x - 3x^2 + 3x + 12}{12(x-1)}$

$= \dfrac{x^2 - x + 12}{12(x-1)}$

Right: $\dfrac{x}{3} - \dfrac{x}{4} = \dfrac{1}{x-1}$

$$12(x-1)\left(\dfrac{x}{3} - \dfrac{x}{4}\right) = \left(\dfrac{1}{x-1}\right)12(x-1)$$

$$12(x-1)\left(\dfrac{x}{3}\right) - 12(x-1)\left(\dfrac{x}{4}\right) = 12$$

$$4x(x-1) - 3x(x-1) = 12$$

$$4x^2 - 4x - 3x^2 + 3x = 12$$

$$x^2 - x - 12 = 0$$

$$(x-4)(x+3) = 0$$

$$x - 4 = 0 \quad \text{or} \quad x + 3 = 0$$

$$x = 4 \qquad\qquad x = -3$$

5. You must check for extraneous solutions when there is a variable in the denominator.

7. 2 cannot be a solution because it makes the denominator zero in the first term.

9. No, because there are no variables in the denominator.

11. Yes, because there is a variable in the denominator.

13.
$$\frac{x}{3} - \frac{x}{4} = 1$$
$$12\left(\frac{x}{3} - \frac{x}{4}\right) = 12(1)$$
$$4x - 3x = 12$$
$$x = 12$$

15.
$$\frac{r}{6} = \frac{r}{4} + \frac{1}{3}$$
$$12\left(\frac{r}{6}\right) = 12\left(\frac{r}{4} + \frac{1}{3}\right)$$
$$2r = 3r + 4$$
$$-r = 4$$
$$r = -4$$

17.
$$\frac{z}{2} + 6 = \frac{z}{5}$$
$$10\left(\frac{z}{2} + 6\right) = 10\left(\frac{z}{5}\right)$$
$$5z + 60 = 2z$$
$$3z = -60$$
$$z = -20$$

19.
$$\frac{z}{6} + \frac{2}{3} = \frac{z}{5} - \frac{1}{3}$$
$$30\left(\frac{z}{6} + \frac{2}{3}\right) = 30\left(\frac{z}{5} - \frac{1}{3}\right)$$
$$5z + 20 = 6z - 10$$
$$-z = -30$$
$$z = 30$$

21.
$$d + 7 = \frac{3}{2}d + 5$$
$$2(d + 7) = 2\left(\frac{3}{2}d + 5\right)$$
$$2d + 14 = 3d + 10$$
$$-d = -4$$
$$d = 4$$

23.
$$3k + \frac{1}{6} = 4k - 4$$
$$6\left(3k + \frac{1}{6}\right) = 6(4k - 4)$$
$$18k + 1 = 24k - 24$$
$$-6k = -25$$
$$k = \frac{25}{6}$$

25.
$$\frac{(n+6)}{3} = \frac{5(n-8)}{10}$$
$$10(n+6) = 3 \cdot 5(n-8)$$
$$10n + 60 = 15n - 120$$
$$180 = 5n$$
$$36 = n$$

27.
$$\frac{x-5}{15} = \frac{3}{5} - \frac{x-4}{10}$$
$$30\left(\frac{x-5}{15}\right) = 30\left(\frac{3}{5} - \frac{x-4}{10}\right)$$
$$2(x-5) = 6(3) - 3(x-4)$$
$$2x - 10 = 18 - 3x + 12$$
$$2x - 10 = 30 - 3x$$
$$5x = 40$$
$$x = 8$$

29.
$$\frac{-p+1}{4} + \frac{13}{20} = \frac{p}{5} - \frac{p-1}{2}$$
$$20\left(\frac{-p+1}{4} + \frac{13}{20}\right) = 20\left(\frac{p}{5} - \frac{p-1}{2}\right)$$
$$5(-p+1) + 13 = 4p - 10(p-1)$$
$$-5p + 5 + 13 = 4p - 10p + 10$$
$$-5p + 18 = -6p + 10$$
$$p = -8$$

31.
$$\frac{d-3}{4} + \frac{1}{15} = \frac{2d+1}{3} - \frac{34}{15}$$
$$60\left(\frac{d-3}{4} + \frac{1}{15}\right) = 60\left(\frac{2d+1}{3} - \frac{34}{15}\right)$$
$$15(d-3) + 4 = 20(2d+1) - 4(34)$$
$$15d - 45 + 4 = 40d + 20 - 136$$
$$15d - 41 = 40d - 116$$
$$-25d = -75$$
$$d = 3$$

33.
$$2 + \frac{3}{x} = \frac{11}{4}$$
$$4x\left(2 + \frac{3}{x}\right) = 4x\left(\frac{11}{4}\right)$$
$$8x + 12 = 44$$
$$8x = 32$$
$$x = 4$$

Check: $2 + \dfrac{3}{4} = \dfrac{11}{4}$
$$\frac{8}{4} + \frac{3}{4} = \frac{11}{4}$$
$$\frac{11}{4} = \frac{11}{4} \quad \text{True}$$

35.
$$7 - \frac{5}{x} = \frac{9}{2}$$
$$2x\left(7 - \frac{5}{x}\right) = 2x\left(\frac{9}{2}\right)$$
$$14x - 10 = 9x$$
$$5x = 10$$
$$x = 2$$

Check: $7 - \dfrac{5}{2} = \dfrac{9}{2}$
$$\frac{14}{2} - \frac{5}{2} = \frac{9}{2}$$
$$\frac{9}{2} = \frac{9}{2} \quad \text{True}$$

37.
$$\frac{4}{n} - \frac{3}{2n} = \frac{1}{2}$$
$$2n\left(\frac{4}{n} - \frac{3}{2n}\right) = 2n\left(\frac{1}{2}\right)$$
$$8 - 3 = n$$
$$5 = n$$

Check: $\dfrac{4}{5} - \dfrac{3}{10} = \dfrac{1}{2}$
$$\frac{8}{10} - \frac{3}{10} = \frac{1}{2}$$
$$\frac{5}{10} = \frac{1}{2}$$
$$\frac{1}{2} = \frac{1}{2} \quad \text{True}$$

39.
$$\frac{x-1}{x-5} = \frac{4}{x-5}$$
$$(x-5)\left(\frac{x-1}{x-5}\right) = \left(\frac{4}{x-5}\right)(x-5)$$
$$x - 1 = 4$$
$$x = 5$$

Check: $\dfrac{x-1}{x-5} = \dfrac{4}{x-5}$
$$\frac{5-1}{5-5} = \frac{4}{5-5}$$
$$\frac{4}{0} = \frac{4}{0}$$

Since $\dfrac{4}{0}$ is not a real number, 5 is an extraneous solution. This equation has no solution.

41.
$$\frac{5}{a+3} = \frac{4}{a+1}$$
$$5(a+1) = 4(a+3)$$
$$5a + 5 = 4a + 12$$
$$a = 7$$

Check: $\dfrac{5}{a+3} = \dfrac{4}{a+1}$
$$\frac{5}{7+3} = \frac{4}{7+1}$$
$$\frac{5}{10} = \frac{4}{8}$$
$$\frac{1}{2} = \frac{1}{2} \quad \text{True}$$

43.
$$\frac{y+3}{y-3} = \frac{6}{4}$$
$$4(y+3) = 6(y-3)$$
$$4y + 12 = 6y - 18$$
$$30 = 2y$$
$$15 = y$$

Check: $\dfrac{y+3}{y-3} = \dfrac{6}{4}$
$$\frac{15+3}{15-3} = \frac{6}{4}$$
$$\frac{18}{12} = \frac{6}{4}$$
$$\frac{3}{2} = \frac{3}{2} \quad \text{True}$$

45.

$$\frac{2x-3}{x-4} = \frac{5}{x-4}$$

$$(x-4)\left(\frac{2x-3}{x-4}\right) = \left(\frac{5}{x-4}\right)(x-4)$$

$$2x-3 = 5$$

$$2x = 8$$

$$x = 4$$

Check: $\dfrac{2x-3}{x-4} = \dfrac{5}{x-4}$

$$\frac{2(4)-3}{4-4} = \frac{5}{4-4}$$

$$\frac{5}{0} = \frac{5}{0}$$

Since $\dfrac{5}{0}$ is not a real number, 4 is an extraneous solution. This equation has no solution.

47.

$$\frac{x^2}{x-3} = \frac{9}{x-3}$$

$$(x-3)\left(\frac{x^2}{x-3}\right) = (x-3)\left(\frac{9}{x-3}\right)$$

$$x^2 = 9$$

$$x^2 - 9 = 0$$

$$(x+3)(x-3) = 0$$

$$x+3 = 0 \quad \text{or} \quad x-3 = 0$$

$$x = -3 \qquad\quad x = 3$$

Check: $\quad x = -3 \qquad\qquad x = 3$

$$\frac{(-3)^2}{-3-3} = \frac{9}{-3-3} \qquad \frac{(3)^2}{3-3} = \frac{9}{3-3}$$

$$\frac{9}{-6} = \frac{9}{-6}\text{ True} \qquad \frac{9}{0} = \frac{9}{0}$$

Since $\dfrac{9}{0}$ is not a real number, 3 is an extraneous solution. The solution is $x = -3$.

49.

$$\frac{n-3}{n+2} = \frac{n+4}{n+10}$$

$$(n-3)(n+10) = (n+2)(n+4)$$

$$n^2 + 7n - 30 = n^2 + 6n + 8$$

$$n = 38$$

Check: $\dfrac{38-3}{38+2} = \dfrac{38+4}{38+10}$

$$\frac{35}{40} = \frac{42}{48}$$

$$\frac{7}{8} = \frac{7}{8} \quad \text{True}$$

51.

$$\frac{1}{r} = \frac{3r}{8r+3}$$

$$1(8r+3) = 3r^2$$

$$0 = 3r^2 - 8r - 3$$

$$0 = (3r+1)(r-3)$$

$$3r+1 = 0 \quad \text{or} \quad r-3 = 0$$

$$3r = -1 \qquad\qquad r = 3$$

$$r = -\frac{1}{3}$$

Check: $\dfrac{1}{\frac{-1}{3}} = \dfrac{3\left(-\frac{1}{3}\right)}{8\left(\frac{-1}{3}\right)+3} \qquad \dfrac{1}{3} = \dfrac{3(3)}{8(3)+3}$

$$\frac{1}{\frac{-1}{3}} = \frac{-1}{\frac{-8}{3}+\frac{9}{3}} \qquad \frac{1}{3} = \frac{9}{27}$$

$$-3 = \frac{-1}{\frac{1}{3}} \qquad\qquad \frac{1}{3} = \frac{1}{3} \quad \text{True}$$

$$-3 = -3 \quad \text{True}$$

53.
$$\frac{k}{k+2} = \frac{3}{k-2}$$
$$k(k-2) = 3(k+2)$$
$$k^2 - 2k = 3k + 6$$
$$k^2 - 5k - 6 = 0$$
$$(k-6)(k+1) = 0$$
$$k - 6 = 0 \quad \text{or} \quad k + 1 = 0$$
$$k = 6 \qquad\qquad k = -1$$

Check: $\dfrac{6}{6+2} = \dfrac{3}{6-2}$ $\qquad\qquad \dfrac{-1}{-1+2} = \dfrac{3}{-1-2}$

$\qquad\qquad \dfrac{6}{8} = \dfrac{3}{4}$ $\qquad\qquad\qquad \dfrac{-1}{1} = \dfrac{3}{-3}$

$\qquad\qquad \dfrac{3}{4} = \dfrac{3}{4}$ True $\qquad\qquad -1 = -1$ True

55.
$$\frac{4}{r} + r = \frac{20}{r}$$
$$r\left(\frac{4}{r} + r\right) = r\left(\frac{20}{r}\right)$$
$$4 + r^2 = 20$$
$$r^2 - 16 = 0$$
$$(r+4)(r-4) = 0$$
$$r + 4 = 0 \quad \text{or} \quad r - 4 = 0$$
$$r = -4 \qquad\qquad r = 4$$

Check: $\dfrac{4}{4} + 4 = \dfrac{20}{4}$ $\qquad\qquad \dfrac{4}{-4} - 4 = \dfrac{20}{-4}$

$\qquad\qquad \dfrac{4}{4} + \dfrac{16}{4} = \dfrac{20}{4}$ $\qquad\qquad \dfrac{4}{-4} - \dfrac{16}{4} = \dfrac{20}{-4}$

$\qquad\qquad \dfrac{20}{4} = \dfrac{20}{4}$ True $\qquad\qquad -\dfrac{20}{4} = -\dfrac{20}{4}$ True

57.
$$x + \frac{20}{x} = -9$$
$$x\left(x + \frac{20}{x}\right) = -9x$$
$$x^2 + 20 = -9x$$
$$x^2 + 9x + 20 = 0$$
$$(x+4)(x+5) = 0$$
$$x + 4 = 0 \quad \text{or} \quad x + 5 = 0$$
$$x = -4 \qquad\qquad x = -5$$

Check $x = -4$: $\quad x + \dfrac{20}{x} = -9$

$$-4 + \frac{20}{-4} = -9$$
$$-4 + (-5) = -9 \quad \text{True}$$

Check $x = -5$: $\quad x + \dfrac{20}{x} = -9$

$$-5 + \frac{20}{-5} - -9$$
$$5 + (-4) = -9 \quad \text{True}$$

59.
$$\frac{3y-2}{y+1} = 4 - \frac{y+2}{y-1}$$
$$(y+1)(y-1)\left(\frac{3y-2}{y+1}\right) = \left(4 - \frac{y+2}{y-1}\right)(y+1)(y-1)$$
$$(y-1)(3y-2) = 4(y+1)(y-1) - \left(\frac{y+2}{y-1}\right)(y+1)(y-1)$$
$$3y^2 - 5y + 2 = 4\left(y^2 - 1\right) - (y+2)(y+1)$$
$$3y^2 - 5y + 2 = 4y^2 - 4 - (y^2 + 3y + 2)$$
$$3y^2 - 5y + 2 = 3y^2 - 3y - 6$$
$$-5y + 2 = -3y - 6$$
$$8 = 2y$$
$$4 = y$$

Check: $\dfrac{3y-2}{y+1} = 4 - \dfrac{y+2}{y-1}$

$$\frac{3(4)-2}{4+1} = 4 - \frac{4+2}{4-1}$$
$$\frac{12-2}{5} = 4 - \frac{6}{3}$$
$$\frac{10}{5} = 4 - 2$$
$$2 = 2 \quad \text{True}$$

61.

$$\frac{1}{x+3}+\frac{1}{x-3}=\frac{-5}{x^2-9}$$

$$\frac{1}{x+3}+\frac{1}{x-3}=\frac{-5}{(x-3)(x+3)}$$

$$(x-3)(x+3)\left[\frac{1}{x+3}+\frac{1}{x-3}\right]=\left[\frac{-5}{(x-3)(x+3)}\right](x-3)(x+3)$$

$$(x-3)(x+3)\left(\frac{1}{x+3}\right)+(x-3)(x+3)\left(\frac{1}{x-3}\right)=-5$$

$$x-3+x+3=-5$$

$$2x=-5$$

$$x=-\frac{5}{2}$$

Check: $\dfrac{1}{x+3}+\dfrac{1}{x-3}=\dfrac{-5}{x^2-9}$

$$\frac{1}{-\frac{5}{2}+3}+\frac{1}{-\frac{5}{2}-3}=\frac{-5}{\left(-\frac{5}{2}\right)^2-9}$$

$$\frac{1}{\frac{1}{2}}-\frac{1}{\left(-\frac{11}{2}\right)}=\frac{-5}{\frac{25}{4}-9}$$

$$2-\frac{2}{11}=\frac{-5}{-\frac{11}{4}}$$

$$\frac{20}{11}=\frac{20}{11}\ \text{True}$$

63.

$$\frac{x}{x-3}+\frac{3}{2}=\frac{3}{x-3}$$

$$2(x-3)\left(\frac{x}{x-3}+\frac{3}{2}\right)=2(x-3)\left(\frac{3}{x-3}\right)$$

$$2x+3(x-3)=2(3)$$

$$2x+3x-9=6$$

$$5x=15$$

$$x=3$$

Check: $\dfrac{x}{x-3}+\dfrac{3}{2}=\dfrac{3}{x-3}$

$$\frac{3}{3-3}+\frac{3}{2}=\frac{3}{3-3}$$

$$\frac{3}{0}+\frac{3}{2}=\frac{3}{0}$$

Since $\dfrac{3}{0}$ is not a real number, 3 is an extraneous solution. This equation has no solution.

65.

$$\frac{3}{x-5}-\frac{4}{x+5}=\frac{11}{x^2-25}$$

$$\frac{3}{x-5}-\frac{4}{x+5}=\frac{11}{(x-5)(x+5)}$$

$$(x-5)(x+5)\left[\frac{3}{x-5}-\frac{4}{x+5}\right]=\left[\frac{11}{(x-5)(x+5)}\right](x-5)(x+5)$$

$$(x-5)(x+5)\left(\frac{3}{x-5}\right)-(x-5)(x+5)\left(\frac{4}{x+5}\right)=11$$

$$3(x+5)-4(x-5)=11$$

$$3x+15-4x+20=11$$

$$-x+35=11$$

$$24=x$$

Check: $\dfrac{3}{x-5}-\dfrac{4}{x+5}=\dfrac{11}{x^2-25}$

$$\frac{3}{24-5}-\frac{4}{24+5}=\frac{11}{24^2-25}$$

$$\frac{3}{19}-\frac{4}{29}=\frac{11}{576-25}$$

$$\frac{87}{551}-\frac{76}{551}=\frac{11}{551}$$

$$\frac{11}{551}=\frac{11}{551}\ \text{True}$$

67.

$$\frac{3x}{x^2-9}+\frac{1}{x-3}=\frac{3}{x+3}$$

$$\frac{3x}{(x-3)(x+3)}+\frac{1}{x-3}=\frac{3}{x+3}$$

$$(x-3)(x+3)\left[\frac{3x}{(x-3)(x+3)}+\frac{1}{x-3}\right]=\left[\frac{3}{x+3}\right](x-3)(x+3)$$

$$(x-3)(x+3)\left(\frac{3x}{(x-3)(x+3)}\right)+(x-3)(x+3)\left(\frac{1}{x-3}\right)=3(x-3)$$

$$3x+x+3=3x-9$$
$$4x+3=3x-9$$
$$x+3=-9$$
$$x=-12$$

Check: $\dfrac{3x}{x^2-9}+\dfrac{1}{x-3}=\dfrac{3}{x+3}$

$$\frac{3(-12)}{(-12)^2-9}+\frac{1}{-12-3}=\frac{3}{-12+3}$$

$$\frac{-36}{144-9}+\frac{1}{-15}=\frac{3}{-9}$$

$$-\frac{36}{135}-\frac{1}{15}=-\frac{1}{3}$$

$$-\frac{4}{15}-\frac{1}{15}=-\frac{1}{3}$$

$$-\frac{5}{15}=-\frac{1}{3}$$

$$-\frac{1}{3}=-\frac{1}{3}\ \ \text{True}$$

69.

$$\frac{1}{y-1}+\frac{1}{2}=\frac{2}{y^2-1}$$

$$\frac{1}{y-1}+\frac{1}{2}=\frac{2}{(y+1)(y-1)}$$

$$2(y+1)(y-1)\left[\frac{1}{y-1}+\frac{1}{2}\right]=2(y+1)(y-1)\left[\frac{2}{(y+1)(y-1)}\right]$$

$$2(y+1)(y-1)\left(\frac{1}{y-1}\right)+2(y+1)(y-1)\left(\frac{1}{2}\right)=2(2)$$

$$2(y+1)+(y+1)(y-1)=4$$
$$2y+2+y^2-1=4$$
$$y^2+2y-3=0$$
$$(y+3)(y-1)=0$$
$$y+3=0\ \ \text{or}\ \ y-1=0$$
$$y=-3\qquad y=1$$

Check: $y=-3$

$$\frac{1}{y-1}+\frac{1}{2}=\frac{2}{y^2-1}$$

$$\frac{1}{-3-1}+\frac{1}{2}=\frac{2}{(-3)^2-1}$$

$$\frac{1}{-4}+\frac{1}{2}=\frac{2}{8}$$

$$\frac{1}{4}=\frac{1}{4}\qquad\text{True}$$

Check: $y=1$

$$\frac{1}{y-1}+\frac{1}{2}=\frac{2}{y^2-1}$$

$$\frac{1}{1-1}+\frac{1}{2}=\frac{2}{1^2-1}$$

$$\frac{1}{0}+\frac{1}{2}=\frac{2}{0}$$

The solution to the equation is –3. Since $\dfrac{1}{0}$ and $\dfrac{2}{0}$ are not real numbers, 1 is an extraneous solution.

71.
$$\frac{3t}{6t+6}+\frac{t}{2t+2}=\frac{2t-3}{t+1}$$

$$\frac{3t}{6(t+1)}+\frac{t}{2(t+1)}=\frac{2t-3}{t+1}$$

$$6(t+1)\left[\frac{3t}{6(t+1)}+\frac{t}{2(t+1)}\right]=6(t+1)\left(\frac{2t-3}{t+1}\right)$$

$$3t+3t=6(2t-3)$$

$$6t=12t-18$$

$$-6t=-18$$

$$t=3$$

Check:
$$\frac{3(3)}{6(3)+6}+\frac{3}{2(3)+2}=\frac{2(3)-3}{3+1}$$

$$\frac{9}{18+6}+\frac{3}{6+2}=\frac{6-3}{4}$$

$$\frac{9}{24}+\frac{3}{8}=\frac{3}{4}$$

$$\frac{3}{8}+\frac{3}{8}=\frac{3}{4}$$

$$\frac{6}{8}=\frac{3}{4}$$

$$\frac{3}{4}=\frac{3}{4} \quad \text{True}$$

73. The solution is 5. Since $3=x-2$, $x=5$.

75. The solution is 0.
Since $x+x=0$, $x=0$.

77. x can be any real number.
$x-2+x-2=2x-4$.

79.
$$\frac{1}{p}+\frac{1}{q}=\frac{1}{f}$$

$$\frac{1}{30}+\frac{1}{q}=\frac{1}{10}$$

$$\frac{1}{q}=\frac{1}{10}-\frac{1}{30}$$

$$\frac{1}{q}=\frac{2}{30}$$

$$2q=30$$

$$q=15$$

The image will appear 15 cm from the mirror.

81.
$$\frac{x-4}{x^2-2x}=\frac{-4}{x^2-4}$$

$$\frac{x-4}{x(x-2)}=\frac{-4}{(x+2)(x-2)}$$

$$x(x+2)(x-2)\left(\frac{x-4}{x(x-2)}\right)=x(x+2)(x-2)\left(\frac{-4}{(x+2)(x-2)}\right)$$

$$(x+2)(x-4)=-4x$$

$$x^2-2x-8=-4x$$

$$x^2+2x-8=0$$

$$(x+4)(x-2)=0$$

$$x+4=0 \text{ or } x-2=0$$

$$x=-4 \qquad x=2$$

Since $\dfrac{-2}{0}$ and $\dfrac{-4}{0}$ are not a real numbers, 2 is an extraneous solution.

The solution to the equation is -4.

83. No, it is impossible for both sides of the equation to be equal.

85. Let x be the number of minutes of internet access over 5 hours.

Plan 1: $7.95 + 0.15x$

Plan 2: 19.95

$7.95 + 0.15x = 19.95$

$0.15x = 12$

$x = 80$

$80 \text{ minutes} = \dfrac{80}{60} \text{ hours} = 1\dfrac{1}{3} \text{ hours}$

Jake would have to use the internet more than

$5 + 1\dfrac{1}{3} = 6\dfrac{1}{3}$ hours.

86. $\dfrac{600 \text{ gallons}}{4 \text{ gallons/minute}} = 150 \text{ minutes}$

87. Let x = measure of larger angle

$\dfrac{1}{2}x - 30$ = measure of smaller angle

$x + \left(\dfrac{1}{2}x - 30\right) = 180$

$\dfrac{3}{2}x - 30 = 180$

$\dfrac{3}{2}x = 210$

$\dfrac{2}{3} \cdot \dfrac{3}{2}x = \dfrac{2}{3} \cdot 210$

$x = 140$

The angles measure 40° and 140°.

88. $(3.4 \times 10^{-5})(2 \times 10^{13})$

$= (3.4 \cdot 2) \times (10^{-5} \cdot 10^{13})$

$= 6.8 \times 10^{-5+13}$

$= 6.8 \times 10^{8}$

89. A linear equation is an equation that can be written in the form $ax + b = c$ where a, b, and c are real numbers and $a \neq 0$. An example would be $5x + 8 = 19$. A quadratic equation has the form $ax^2 + bx + c = 0$ where a, b, and c are real numbers and $a \neq 0$. An example would be $4x^2 + 7x + 2 = 0$.

Exercise Set 7.7

1. Some examples are:

$A = \dfrac{1}{2}bh$, $A = \dfrac{1}{2}h(b_1 + b_2)$, $V = \dfrac{1}{3}\pi r^2 h$, and

$V = \dfrac{4}{3}\pi r^3$

3. It represents 1 complete task.

5. Let w = width, then $\dfrac{2}{3}w + 5$ = length

area = width · length

$99 = w\left(\dfrac{2}{3}w + 5\right)$

$99 = \dfrac{2w^2}{3} + 5w$

$3(99) = 3\left(\dfrac{2w^2}{3} + 5w\right)$

$297 = 2w^2 + 15w$

$2w^2 + 15w - 297 = 0$

$2w^2 + 15w - 297 = 0$

$(2w + 33)(w - 9) = 0$

$2w + 33 = 0 \qquad \text{or} \quad w - 9 = 0$

$w - -\dfrac{33}{2} \qquad\qquad w = 9$

Since the width cannot be negative, $w = 9$

$l = \dfrac{2}{3}w + 5$

$l = \dfrac{2}{3}(9) + 5$

$l = 6 + 5 = 11$

The length is 11 inches and the width is 9 inches.

7. Let x = height, then $x + 5$ = base

area $= \dfrac{1}{2} \cdot$ height $\cdot$ base

$$42 = \frac{1}{2}x(x+5)$$
$$2(42) = 2\left[\frac{1}{2}x(x+5)\right]$$
$$84 = x(x+5)$$
$$84 = x^2 + 5x$$
$$0 = x^2 + 5x - 84$$
$$0 = (x-7)(x+12)$$

$x - 7 = 0$ or $x + 12 = 0$
$x = 7 \qquad\qquad x = -12$

Since the height cannot be negative, $x = 7$.
base $= x + 5 = 7 + 15 = 12$
The base is 12 cm and the height is 7 cm.

9. Let b = the base of the triangle, then $\left(\dfrac{1}{2}b - 1\right)$ is the height.

$$A = \frac{1}{2}bh$$
$$12 = \frac{1}{2}b\left(\frac{1}{2}b - 1\right)$$
$$12 = \frac{1}{4}b^2 - \frac{1}{2}b$$
$$4(12) = 4\left(\frac{1}{4}b^2 - \frac{1}{2}b\right)$$
$$48 = b^2 - 2b$$
$$0 = b^2 - 2b - 48$$
$$0 = (b-8)(b+6)$$

$b - 8 = 0$ or $b + 6 = 0$
$b = 8$ or $b = -6$

Since a length cannot be negative, the base is 8 feet.

11. Let one number be x, then the other number is $9x$.

$$\frac{1}{x} - \frac{1}{9x} = 1$$
$$9x\left(\frac{1}{x} - \frac{1}{9x}\right) = 9x(1)$$
$$9 - 1 = 9x$$
$$8 = 9x$$
$$\frac{8}{9} = x$$

The numbers are $\dfrac{8}{9}$ and $9\left(\dfrac{8}{9}\right) = 8$.

13. Let x = amount by which the numerator was increased.

$$\frac{3+x}{4} = \frac{5}{2}$$
$$4\left(\frac{3+x}{4}\right) = \left(\frac{5}{2}\right)4$$
$$3 + x = 10$$
$$x = 7$$

The numerator was increased by 7.

15. Let r = speed of Creole Queen paddle boat.

$$t = \frac{d}{r}$$

Time upstream = time downstream
$$\frac{6}{r-3} = \frac{12}{r+3}$$
$$6(r+3) = 12(r-3)$$
$$6r + 18 = 12r - 36$$
$$54 = 6r$$
$$9 = r$$

The boat's speed in still water is 9 mph.

17. Let d = distance and $t = \dfrac{d}{r}$.

Time going + Time returning $= \dfrac{5}{2}$.

$$\frac{d}{12} + \frac{d}{12} = \frac{5}{2}$$
$$\frac{2d}{12} = \frac{5}{2}$$
$$4d = 60$$
$$d = 15$$

The trolley traveled 15 miles in one direction.

19. Let r be the speed of the propeller plane, then $4r$ is the speed of the jet.

$$\frac{d}{r} = t$$

time by jet + time by propeller plane = 6 hr

$$\frac{1600}{4r} + \frac{500}{r} = 6$$

$$\frac{400}{r} + \frac{300}{r} = 6$$

$$\frac{900}{r} = 6$$

$$r = \frac{900}{6} = 150$$

The speed of the propeller plane is 150 mph and the speed of the jet is 600 mph.

21. Let $d =$ distance from the dock to the no wake zone, then $36.6 - d =$ distance from the no wake zone to Paradise Island.

$$t = \frac{d}{r}$$

time traveled to no wake zone + time traveled from no wake zone = total time

$$\frac{d}{4} + \frac{36.6 - d}{28} = 1.7$$

$$28\left(\frac{d}{4} + \frac{36.6 - d}{28}\right) = 1.7(28)$$

$$7d + 36.6 - d = 47.6$$

$$6d = 11$$

$$d = \frac{11}{6} \approx 1.83$$

$$36.6 - d = 36.6 - 1.83 \approx 34.77$$

The dock is about 1.83 miles from the no wake zone and about 34.77 miles from the no wake zone to the island.

23. Let $d =$ the distance flown with the wind, then $2900 - d =$ the time flown against the wind.

time with wind $= \dfrac{d}{600}$

time against wind $= \dfrac{2900 - d}{550}$

$$\frac{d}{600} + \frac{2900 - d}{550} = 5$$

$$3300\left(\frac{d}{600} + \frac{2900 - d}{550}\right) = 5(3300)$$

$$5.5d + 17,400 - 6d = 16,500$$

$$-0.5d = -900$$

$$d = 1800$$

time with wind $= \dfrac{1800}{600} = 3$

time against wind $= \dfrac{2900 - 800}{550} = 2$

It flew 3 hours at 600 mph and 2 hours at 550 mph.

25. time at 30 ft/s $= \dfrac{d}{30}$

time at 25 ft/s $= \dfrac{d}{25}$

$$\frac{d}{25} = \frac{d}{30} + 8$$

$$150\left(\frac{d}{25}\right) = \left(\frac{d}{30} + 8\right)150$$

$$6d = 5d + 1200$$

$$d = 1200$$

The boat traveled 1200 feet in one direction.

27. Felicia's rate $= \dfrac{1}{6}$

Reynaldo's rate $= \dfrac{1}{8}$

$$\frac{t}{6} + \frac{t}{8} = 1$$

$$48\left(\frac{t}{6} + \frac{t}{8}\right) = 48(1)$$

$$8t + 6t = 48$$

$$14t = 48$$

$$t = 3\frac{3}{7}$$

It will take them $3\dfrac{3}{7}$ hours.

29. Gary's rate $= \dfrac{1}{6}$

Alex's rate $= \dfrac{1}{12}$

$$\frac{t}{6} + \frac{t}{12} = 1$$

$$12\left(\frac{t}{6} + \frac{t}{12}\right) = 12(1)$$

$$2t + t = 12$$

$$3t = 12$$

$$t = 4$$

It will take them 4 hours.

31. Eric's rate $= \dfrac{1}{60}$

Jessup's rate $= \dfrac{1}{40}$

$$\dfrac{t}{60} + \dfrac{t}{40} = 1$$

$$120\left(\dfrac{t}{60} + \dfrac{t}{40}\right) = 120(1)$$

$$2t + 3t = 120$$

$$5t = 120$$

$$t = 24$$

It will take them 24 minutes.

33. input rate $= \dfrac{1}{40}$

output rate $= \dfrac{1}{60}$

$$\dfrac{t}{40} - \dfrac{t}{60} = 1$$

$$120\left(\dfrac{t}{40} - \dfrac{t}{60}\right) = 120(1)$$

$$3t - 2t = 120$$

$$t = 120$$

It will take 120 minutes or 2 hours for the tub to fill.

35. Rate for first $= \dfrac{1}{40}$

Rate for second $= \dfrac{1}{t}$

In 24 minutes, the first computer completes

$\dfrac{24}{40} = \dfrac{3}{5}$ of the checks and the second computer

completes $\dfrac{24}{t}$.

$$\dfrac{3}{5} + \dfrac{24}{t} = 1$$

$$5t\left(\dfrac{3}{5} + \dfrac{24}{t}\right) = 5t(1)$$

$$3t + 120 = 5t$$

$$120 = 2t$$

$$60 = t$$

It would take the second computer 60 minutes or 1 hour.

37. Rate for first backhoe $= \dfrac{1}{12}$

Rate for second backhoe $= \dfrac{1}{15}$

Work done by first $= \dfrac{1}{12} \cdot 5 = \dfrac{5}{12}$

Work done by second $= \dfrac{1}{15} \cdot t = \dfrac{t}{15}$

$$\dfrac{5}{12} + \dfrac{t}{15} = 1$$

$$60\left(\dfrac{5}{12} + \dfrac{t}{15}\right) = 1 \cdot 60$$

$$25 + 4t = 60$$

$$4t = 35$$

$$t = \dfrac{35}{4} = 8\dfrac{3}{4}$$

It takes the smaller backhoe $8\dfrac{3}{4}$ days to finish the trench.

39. Ken's rate $= \dfrac{1}{4}$

Bettina's rate $= \dfrac{1}{6}$

$$\dfrac{t+3}{6} + \dfrac{t}{4} = 1$$

$$12\left(\dfrac{t+3}{6} + \dfrac{t}{4}\right) = 12 \cdot 1$$

$$2(t+3) + 3t = 12$$

$$2t + 6 + 3t = 12$$

$$5t = 6$$

$$t = \dfrac{6}{5}$$

It will take them $\dfrac{6}{5}$ hour or 1 hr, 12 min longer.

41. Rate of first skimmer $= \dfrac{1}{60}$

Rate of second skimmer $= \dfrac{1}{50}$

Rate of transfer $= \dfrac{1}{30}$

$$\frac{t}{60} + \frac{t}{50} - \frac{t}{30} = 1$$

$$300\left(\frac{t}{60} + \frac{t}{50} - \frac{t}{30}\right) = 300$$

$$5t + 6t - 10t = 300$$

$$t = 300$$

It will take 300 hours to fill the tank.

43. Let x be the number.

$$\frac{3}{x} + 2x = 7$$

$$x\left(\frac{3}{x} + 2x\right) = 7x$$

$$3 + 2x^2 = 7x$$

$$2x^2 - 7x + 3 = 0$$

$$(2x - 1)(x - 3) = 0$$

$$2x - 1 = 0 \quad \text{or} \quad x - 3 = 0$$

$$x = \frac{1}{2} \qquad\qquad x = 3$$

The numbers are 3 or $\frac{1}{2}$.

45. Ed's rate $= \dfrac{1}{8}$

Samantha's rate $= \dfrac{1}{4}$

$$\frac{p}{4} - 1 = \frac{p}{8}$$

$$8\left(\frac{p}{4} - 1\right) = 8\left(\frac{p}{8}\right)$$

$$2p - 8 = p$$

$$p = 8$$

Each must pick 8 pints.

47. $\dfrac{1}{2}(x + 3) - (2x + 5) = \dfrac{1}{2}x + \dfrac{3}{2} - 2x - 5$

$$= \frac{x}{2} - \frac{4x}{2} + \frac{3}{2} - \frac{10}{2}$$

$$= -\frac{3x}{2} - \frac{7}{2}$$

48.

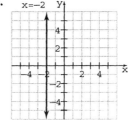

49. $\dfrac{x^2 - 14x + 48}{x^2 - 5x - 24} \div \dfrac{2x^2 - 13x + 6}{2x^2 + 5x - 3} = \dfrac{x^2 - 14x + 48}{x^2 - 5x - 24} \cdot \dfrac{2x^2 + 5x - 3}{2x^2 - 13x + 6}$

$$= \frac{(x - 6)(x - 8)}{(x + 3)(x - 8)} \cdot \frac{(2x - 1)(x + 3)}{(2x - 1)(x - 6)}$$

$$= 1$$

50. $\dfrac{x}{6x^2 - x - 15} - \dfrac{5}{9x^2 - 12x - 5} = \dfrac{x}{(2x + 3)(3x - 5)} - \dfrac{5}{(3x + 1)(3x - 5)}$

$$= \frac{x(3x + 1)}{(2x + 3)(3x - 5)(3x + 1)} - \frac{5(2x + 3)}{(2x + 3)(3x - 5)(3x + 1)}$$

$$= \frac{3x^2 + x - (10x + 15)}{(2x + 3)(3x - 5)(3x + 1)}$$

$$= \frac{3x^2 + x - 10x - 15}{(2x + 3)(3x - 5)(3x + 1)}$$

$$= \frac{3x^2 - 9x - 15}{(2x + 3)(3x - 5)(3x + 1)}$$

Exercise Set 7.8

1. **a.**   As one quantity increases, the other increases.

    b.   Answers will vary.

    c.   Answers will vary.

3. One quantity varies as a product of two or more quantities.

5. **a.** y decreases; as x increases, the denominator increases which makes the value of the whole fraction smaller.

    b.   inverse variation; as one quantity increases, the other decreases.

7. Direct   9. Inverse

11. Direct   13. Direct

15. Direct   17. Inverse

19. Direct   21. Inverse

23. Inverse

25. **a.** $x = ky$

    b.   Substitute 12 for y and 6 for k.
    $x = 6(12)$
    $x = 72$

27. **a.** $y = kR$

    b.   Substitute 180 for R and 1.7 for k.
    $y = 1.7(180)$
    $y = 306$

29. **a.** $R = \dfrac{k}{W}$

    b.   Substitute 160 for W and 8 for k.
    $$R = \frac{8}{160}$$
    $$R = \frac{1}{20} = 0.05$$

31. **a.** $A = \dfrac{kB}{C}$

    b.   Substitute 12 for B, 4 for C, and 3 for k.
    $$A = \frac{3(12)}{4} = \frac{36}{4} = 9$$

33. **a.** $x = ky$

    b.   To find k, substitute 12 for x and 3 for y.
    $12 = k(3)$
    $$\frac{12}{3} = k$$
    $4 = k$
    Thus, $x = 4y$.
    Now substitute 5 for y.
    $x = 4(5) = 20$

35. **a.** $y = kR^2$

    b.   To find k, substitute 5 for y and 5 for R.
    $5 = k(5)^2$
    $5 = k(25)$
    $$\frac{5}{25} = k$$
    $$\frac{1}{5} = k$$
    Thus $y = \dfrac{1}{5}R^2$.
    Now substitute 10 for R.
    $$y = \frac{1}{5}(10)^2 = \frac{1}{5}(100) = 20$$

37. **a.** $S = \dfrac{k}{G}$

    b.   To find k, substitute 12 for S and 0.4 for G.
    $$12 = \frac{k}{0.4}$$
    $12(0.4) = k$
    $4.8 = k$
    Thus $S = \dfrac{4.8}{G}$.
    Now substitute 5 for G.
    $$S = \frac{4.8}{5} = 0.96$$

39. a. $x = \dfrac{k}{P^2}$

b. To find k, substitute 5 for P and 4 for x.

$$4 = \frac{k}{5^2}$$

$$4 = \frac{k}{25}$$

$$100 = k$$

Thus $x = \dfrac{100}{P^2}$.

Now substitute 2 for P.

$$x = \frac{100}{(2)^2} = \frac{100}{4} = 25$$

41. a. $F = \dfrac{kM_1M_2}{d}$

b. To find k, substitute 5 for M_1, 10 for M_2, 0.2 for d, and 20 for F.

$$20 = \frac{k(5)(10)}{0.2}$$

$$20 = k(250)$$

$$k = \frac{20}{250} = 0.08$$

Thus $F = \dfrac{0.08M_1M_2}{d}$.

Now substitute 10 for M_1, 20 for M_2, and 0.4 for d.

$$F = \frac{0.08(10)(20)}{0.4} = \frac{16}{0.4} = 40$$

43. $a = kb$
$k(2b) = 2(kb) = 2a$
If b is doubled, a is doubled.

45. $y = \dfrac{k}{x}$

$$\frac{k}{2x} = \frac{1}{2}\left(\frac{k}{x}\right) = \frac{1}{2}y$$

If x is doubled, y is halved.

47. $F = \dfrac{km_1m_2}{d^2}$

$$\frac{k(2m_1)m_2}{d^2} = \frac{2km_1m_2}{d^2} = 2 \cdot \frac{km_1m_2}{d^2} = 2F$$

If m_1 is doubled, F is doubled.

49. $F = \dfrac{km_1m_2}{d^2}$

$$\frac{k(2m_1)\left(\frac{1}{2}m_2\right)}{d^2} = \frac{2 \cdot \frac{1}{2}km_1m_2}{d^2} = \frac{1 \cdot km_1m_2}{d^2} = F$$

If m_1 is doubled and m_2 is halved, F is unchanged.

51. $F = \dfrac{km_1m_2}{d^2}$

$$\frac{k\left(\frac{1}{2}m_1\right)(4m_2)}{d^2} = \frac{\frac{1}{2} \cdot 4km_1m_2}{d^2} =$$

$$\frac{2 \cdot km_1m_2}{d^2} = 2 \cdot \frac{km_1m_2}{d^2} = 2F$$

If m_1 is halved and m_2 is quadrupled, F is doubled.

53. Notice that as x gets bigger, y gets smaller. This suggests that the variation is inverse rather than direct. Therefore use the equation $y = \dfrac{k}{x}$. To determine the value of k, choose one of the ordered pairs and substitute the values into the equation $y = \dfrac{k}{x}$ and solve for k. We'll use the ordered pair (5, 1).

$$y = \frac{k}{x}$$

$$1 = \frac{k}{5} \Rightarrow k = 5$$

55. The equation is $p = kl$ To find k substitute 150 for l and 2542.50 for p.

$$2542.50 = k(150)$$

$$k = \frac{2542.50}{150}$$

$$k = 16.95$$

Thus $p = 16.95l$.

Now substitute 520 for l.

$$p = 16.95(520)$$

$$p = 8814$$

The profit would be $8814.

57. The equation is $d = kw$. To find k, substitute 2376 for d and 132 for w.

$2376 = k132$

$k = \frac{2376}{132}$

$k = 18$

Thus $d = 18w$. Now substitute 172 for w.

$d = 18(172)$

$d = 3096$

The recommended dosage for Nathan is 3096 mg.

59. The equation is $S = kF$. To find k, substitute 1.4 for S and 20 for F.

$1.4 = k(20)$

$\frac{1.4}{20} = k$

$0.07 = k$

Thus, $S = 0.07F$.

Now substitute 15 for F.

$S = 0.07(15) = 1.05$

The spring will stretch 1.05 inches.

61. The equation is $V = \frac{k}{P}$. To find k, substitute 800 for V and 200 for P.

$800 = \frac{k}{200}$

$800(200) = k$

$160,000 = k$

Thus $V = \frac{160,000}{P}$.

Now substitute 25 for P.

$V = \frac{160,000}{25} = 6400$

The volume is 6400 cc.

63. The equation is $t = \frac{k}{s}$. To find k, substitute 6 for s and 2.6 for t.

$2.6 = \frac{k}{6}$

$k = 6(2.6)$

$k = 15.6$

Thus $t = \frac{15.6}{s}$.

Now substitute 5 for s.

$t = \frac{15.6}{5}$

$t = 3.12$

Jackie will take 3.12 hours.

65. The equation is $I = \frac{k}{d^2}$. To find k, substitute 20 for I and 15 for d.

$20 = \frac{k}{(15)^2}$

$20 = \frac{k}{225}$

$k = 20(225) = 4500$

Thus $I = \frac{4500}{d^2}$.

Now substitute 10 for d.

$I = \frac{4500}{(10)^2}$

$I = \frac{4500}{100}$

$I = 45$

The intensity is 45 foot-candles.

67. The equation is $d = ks^2$. To find k, substitute 40 for s and 60 for d.

$d = ks^2$

$60 = k(40)^2$

$60 = 1600k$

$k = \frac{60}{1600} = 0.0375$

Thus $d = 0.0375s^2$. Now substitute 56 for s.

$d = 0.0375s^2$

$d = 0.0375(56)^2$

$d = 117.6$

The stopping distance is 117.6 feet.

69. The equation is $V = kBh$. To find k, substitute 160 for V, 48 for B, and 10 for h.

$160 = k(48)(10)$

$160 = 480k$

$\frac{160}{480} = k$

$\frac{1}{3} = k$

Thus, $V = \frac{1}{3}Bh$. Now substitute 42 for B and 9 for h.

$V = \frac{1}{3}(42)(9) = 14(9) = 126$

The volume of the pyramid would be 126 m^3.

71. The equation is $R = \dfrac{kA}{P}$. To find k, substitute 400 for A, 2 for P, and 4600 for R.

$$4600 = \frac{k(400)}{2}$$

$$4600 = 200k$$

$$k = \frac{4600}{200} = 23$$

Thus $R = \dfrac{23A}{P}$. Now substitute 500 for A and 2.50 for P.

$$R = \frac{23(500)}{2.50}$$

$$R = \frac{11,500}{2.50}$$

$$R = 4600 \text{ DVDs}$$

They would still rent 4600 DVDs per week.

73. The equation is $w = \dfrac{k}{d^2}$. To find k, substitute 140 for w and 4000 for d.

$$140 = \frac{k}{(4000)^2}$$

$$140 = \frac{k}{16,000,000}$$

$$140(16,000,000) = k$$

$$2,240,000,000 = k$$

Thus $w = \dfrac{2,240,000,000}{d^2}$. Now substitute 4100 for d.

$$w = \frac{2,240,000,000}{(4100)^2}$$

$$w = \frac{2,240,000,000}{16,810,000}$$

$$w \approx 133.25$$

The weight is about 133.25 pounds.

75. The equation is $N = \dfrac{kp_1 p_2}{d}$. To find k, substitute 100,000 for N, 300 for d, 60,000 for p_1, and 200,000 for p_2.

$$100,000 = \frac{k(60,000)(200,000)}{300}$$

$$100,000 = 40,000,000k$$

$$\frac{100,000}{40,000,000} = k$$

$$0.0025 = k$$

Thus $N = \dfrac{0.0025 p_1 p_2}{d}$.

Now substitute 450 for d, 125,000 for p_1, and 175,000 for p_2.

$$N = \frac{0.0025(125,000)(175,000)}{450}$$

$$N \approx 121,528$$

About 121,528 calls are made.

77. Let I be the intensity of the illumination and d be the distance the subject is from the flash.

The equation is $I = \dfrac{k}{d^2}$. To find k, substitute $\dfrac{1}{16}$ for I and 4 for d.

$$\frac{1}{16} = \frac{k}{4^2}$$

$$\frac{1}{16} = \frac{k}{16}$$

$$1 = k$$

Thus $I = \dfrac{1}{d^2}$.

Now, substitute 7 for d.

$$I = \frac{1}{7^2} = \frac{1}{49}$$

The illumination is $\dfrac{1}{49}$ of the light of the flash.

79. a. The equation is $P = 14.7 + kx$

b. The find k, substitute 40.5 for P and 60 for x.

$$40.5 = 14.7 + 60k$$

$$25.8 = 60k$$

$$\frac{25.8}{60} = k$$

$$0.43 = k$$

c. The equation is $P = 14.7 + 0.43x$.
Substitute 160 for P.
$$160 = 14.7 + 0.43x$$
$$145.3 = 0.43x$$
$$\frac{145.3}{0.43} = x$$
$$337.9 \approx x$$
The submarine can go about 337.9 feet deep.

80.
$$\begin{array}{r} 2x - 3 \\ 4x+9\overline{)8x^2 + 6x - 21} \\ \underline{8x^2 + 18x} \\ -12x - 21 \\ \underline{-12x - 27} \\ 6 \end{array}$$
$$\frac{8x^2 + 6x - 21}{4x+9} = 2x - 3 + \frac{6}{4x+9}$$

81. $y(z-2) + 8(z-2) = (z-2)(y+8)$

82.
$$3x^2 - 24 = -6x$$
$$3x^2 + 6x - 24 = 0$$
$$x^2 + 2x - 8 = 0$$
$$(x+4)(x-2) = 0$$
$$x+4=0 \quad \text{or} \quad x-2=0$$
$$x=-4 \qquad x=2$$

83. $\frac{x+8}{x-3} \cdot \frac{x^3-27}{x^2+3x+9} = \frac{(x+8)(x-3)(x^2+3x+9)}{(x-3)(x^2+3x+9)}$
$$= x+8$$

Chapter 7 Review Exercises

1. $\frac{5}{2x-18}$
$$2x - 18 = 0$$
$$2(x-9) = 0$$
$$x - 9 = 0$$
$$x = 9$$
The expression is defined for all real numbers except $x = 9$.

2. $\frac{2x+1}{x^2 - 8x + 15}$
$$x^2 - 8x + 15 = 0$$
$$(x-3)(x-5) = 0$$
$$x-3=0 \quad \text{or} \quad x-5=0$$
$$x=3 \qquad x=5$$
The expression is defined for all real numbers except $x = 3$, $x = 5$.

3. $\frac{7x-1}{5x^2 + 4x - 1}$
$$5x^2 + 4x - 1 = 0$$
$$(5x-1)(x+1) = 0$$
$$5x-1=0 \quad \text{or} \quad x+1=0$$
$$x=\frac{1}{5} \qquad x=-1$$
The expression is defined for all real numbers except $x = \frac{1}{5}$, $x = -1$.

4. $\frac{y}{xy-3y} = \frac{y}{y(x-3)} = \frac{1}{x-3}$

5. $\frac{x^3 + 5x^2 + 12x}{x} = \frac{x(x^2 + 5x + 12)}{x}$
$$= x^2 + 5x + 12$$

6. $\frac{9x^2 + 3xy}{3x} = \frac{3x(3x+y)}{3x} = 3x + y$

7. $\frac{x^2 + 2x - 8}{x-2} = \frac{(x-2)(x+4)}{x-2} = x+4$

8. $\frac{a^2 - 81}{a-9} = \frac{(a-9)(a+9)}{a-9} = a+9$

9. $\frac{-2x^2 + 7x + 4}{x-4} = \frac{-1(2x^2 + 7x + 4)}{(x-4)}$
$$= \frac{-1(2x+1)(x-4)}{(x-4)}$$
$$= -(2x+1)$$

10. $\frac{b^2 - 2b + 10}{b^2 - 3b - 10} = \frac{(b-5)(b-2)}{(b-5)(b+2)} = \frac{b-2}{b+2}$

11. $\dfrac{4x^2-11x-3}{4x^2-7x-2} = \dfrac{\cancel{(4x+1)}(x-3)}{\cancel{(4x+1)}(x-2)}$

$= \dfrac{x-3}{x-2}$

12. $\dfrac{2x^2-21x+40}{4x^2-4x-15} = \dfrac{(x-8)\cancel{(2x-5)}}{(2x+3)\cancel{(2x-5)}}$

$= \dfrac{x-8}{2x+3}$

13. $\dfrac{5a^2}{6b} \cdot \dfrac{2}{4a^2b} = \dfrac{5\cdot2}{6\cdot4} \cdot \dfrac{a^2}{a^2} \cdot \dfrac{1}{b\cdot b}$

$= \dfrac{10}{24b^2}$

$= \dfrac{5}{12b^2}$

14. $\dfrac{30x^2y^3}{3z} \cdot \dfrac{6z^3}{5xy^3} = \dfrac{15\cdot2\cdot6}{15}\left(\dfrac{x^2}{x}\right)\left(\dfrac{y^3}{y^3}\right)\left(\dfrac{z^3}{z}\right)$

$= 12xz^2$

15. $\dfrac{20a^3b^4}{7c^3} \cdot \dfrac{14c^7}{5a^5b} = \dfrac{20}{5} \cdot \dfrac{14}{7} \cdot \dfrac{a^3}{a^5} \cdot \dfrac{b^4}{b} \cdot \dfrac{c^7}{c^3}$

$= 8\dfrac{1}{a^2}b^3c^4$

$= \dfrac{8b^3c^4}{a^2}$

16. $\dfrac{1}{x-4} \cdot \dfrac{4-x}{9} = \dfrac{1}{\cancel{x-4}} \cdot \dfrac{-1\cancel{(x-4)}}{9}$

$= -\dfrac{1}{9}$

17. $\dfrac{-m+4}{15m} \cdot \dfrac{10m}{m-4} = \dfrac{-1\cancel{(m-4)}}{3\cdot\cancel{5m}} \cdot \dfrac{2\cdot\cancel{5m}}{\cancel{m-4}}$

$= \dfrac{-2}{3}$

18. $\dfrac{a-2}{a+3} \cdot \dfrac{a^2+4a+3}{a^2-a-2} = \dfrac{\cancel{(a-2)}\cancel{(a+3)}\cancel{(a+1)}}{\cancel{(a+3)}\cancel{(a-2)}\cancel{(a+1)}}$

$= 1$

19. $\dfrac{9x^6}{y^2} \div \dfrac{x^4}{4y} = \dfrac{9x^6}{y^2} \cdot \dfrac{4y}{x^4} = \dfrac{36x^6y}{x^4y^2} = \dfrac{36x^2}{y}$

20. $\dfrac{5xy^2}{z} \div \dfrac{x^4y^2}{4z^2} = \dfrac{5xy^2}{z} \cdot \dfrac{4z^2}{x^4y^2} = \dfrac{20z}{x^3}$

21. $\dfrac{6a+6b}{a^2} \div \dfrac{a^2-b^2}{a^2} = \dfrac{6(a+b)}{\cancel{a^2}} \cdot \dfrac{\cancel{a^2}}{(a+b)(a-b)}$

$= \dfrac{6}{a-b}$

22. $\dfrac{1}{a^2+8a+15} \div \dfrac{8}{a+5} = \dfrac{1}{\cancel{(a+5)}(a+3)} \cdot \dfrac{\cancel{a+5}}{8}$

$= \dfrac{1}{8(a+3)}$

23. $(t+8) \div \dfrac{t^2+5t-24}{t-3} = \dfrac{\cancel{(t+8)}}{1} \cdot \dfrac{\cancel{t-3}}{\cancel{(t-3)}\cancel{(t+8)}}$

$= 1$

24. $\dfrac{x^2+xy-2y^2}{2y} \div \dfrac{x+2y}{12y^2} = \dfrac{(x-y)\cancel{(x+2y)}}{2y} \cdot \dfrac{12y^2}{\cancel{x+2y}}$

$= 6y(x-y)$

25. $\dfrac{n}{n+5} - \dfrac{2}{n+5} = \dfrac{n-2}{n+5}$

26. $\dfrac{4x}{x+7} + \dfrac{28}{x+7} = \dfrac{4x+28}{x+7} = \dfrac{4\cancel{(x+7)}}{\cancel{x+7}} = 4$

27. $\dfrac{5x-4}{x+8} + \dfrac{44}{x+8} = \dfrac{5x-4+44}{x+8}$

$= \dfrac{5x+40}{x+8} = \dfrac{5\cancel{(x+8)}}{\cancel{x+8}}$

$= 5$

28. $\dfrac{7x-3}{x^2+7x-30} - \dfrac{3x+9}{x^2+7x-30} = \dfrac{7x-3-(3x+9)}{x^2+7x-30}$

$= \dfrac{7x-3-3x-9}{x^2+7x-30}$

$= \dfrac{4x-12}{x^2+7x-30}$

$= \dfrac{4\cancel{(x-3)}}{(x+10)\cancel{(x-3)}}$

$= \dfrac{4}{x+10}$

29. $\dfrac{5h^2+12h-1}{h+5}-\dfrac{h^2-5h+14}{h+5}$

$=\dfrac{5h^2+12h-1-\left(h^2-5h+14\right)}{h+5}$

$=\dfrac{5h^2+12h-1-h^2+5h-14}{h+5}$

$=\dfrac{4h^2+17h-15}{h+5}$

$\dfrac{(4h-3)\cancel{(h+5)}}{\cancel{h+5}}$

$=4h-3$

30. $\dfrac{6x^2-4x}{2x-3}-\dfrac{-3x+12}{2x-3}$

$=\dfrac{6x^2+4x-(-3x+12)}{2x-3}$

$=\dfrac{6x^2-4x+3x-12}{2x-3}$

$=\dfrac{6x^2-x-12}{2x-3}$

$=\dfrac{(3x+4)\cancel{(2x-3)}}{\cancel{2x-3}}$

$=3x+4$

31. $\dfrac{a}{8}+\dfrac{5a}{3}$

Least common denominator $=2(2)(2)(3)=24$

32. $\dfrac{10}{x+3}+\dfrac{2x}{x+3}$

Least common denominator $=x+3$

33. $\dfrac{10}{4xy^3}-\dfrac{11}{10x^2y}$

Least common denominator $=20x^2y^3$

34. $\dfrac{6}{x-3}-\dfrac{2}{x}$

Least common denominator $=x(x-3)$

35. $\dfrac{8}{n+5}+\dfrac{2n-3}{n-4}$

Least common denominator $=(n+5)(n-4)$

36. $\dfrac{5x-12}{x^2+2x}-\dfrac{4}{x+2}=\dfrac{5x-12}{x(x+2)}-\dfrac{4}{x+2}$

Least common denominator $=x(x+2)$

37. $\dfrac{2r+1}{r-s}-\dfrac{6}{r^2-s^2}=\dfrac{2r+1}{r-s}-\dfrac{6}{(r+s)(r-s)}$

Least common denominator $=(r+s)(r-s)$

38. $\dfrac{3x^2}{x-9}+10x^3=\dfrac{4x^2}{x-9}+\dfrac{10x^3}{1}$

Least common denominator $=x-9$

39. $\dfrac{19x-5}{x^2+2x-35}+\dfrac{-10x+1}{x^2+9x+14}$

$=\dfrac{19x-5}{(x+7)(x-5)}+\dfrac{-10x+1}{(x+7)(x+2)}$

Least common denominator

$=(x+7)(x-5)(x+2)$

40. $\dfrac{5}{3y^2}+\dfrac{y}{2y}=\dfrac{5}{3y^2}+\dfrac{1}{2}$

$=\dfrac{5}{3y^2}\cdot\dfrac{2}{2}+\dfrac{1}{2}\cdot\dfrac{3y^2}{3y^2}$

$=\dfrac{10}{6y^2}+\dfrac{3y^2}{6y^2}$

$=\dfrac{10+3y^2}{6y^2}$

$=\dfrac{3y^2+10}{6y^2}$

41. $\dfrac{3x}{xy}+\dfrac{1}{4x}=\dfrac{3x}{xy}\cdot\dfrac{4}{4}+\dfrac{1}{4x}\cdot\dfrac{y}{y}$

$=\dfrac{12x}{4xy}+\dfrac{y}{4xy}$

$=\dfrac{12x+y}{4xy}$

42. $\dfrac{5x}{3xy}-\dfrac{6}{x^2}=\dfrac{5x}{3xy}\cdot\dfrac{x}{x}-\dfrac{6}{x^2}\cdot\dfrac{3y}{3y}$

$=\dfrac{5x^2}{3x^2y}-\dfrac{18y}{3x^2y}$

$=\dfrac{5x^2-18y}{3x^2y}$

43. $7 - \dfrac{2}{x+2} = 7\left(\dfrac{x+2}{x+2}\right) - \dfrac{2}{x+2}$

$= \dfrac{7x+14-2}{x+2}$

$= \dfrac{7x+12}{x+2}$

44. $\dfrac{x-y}{y} - \dfrac{x+y}{x} = \dfrac{x-y}{y} \cdot \dfrac{x}{x} - \dfrac{x+y}{x} \cdot \dfrac{y}{y}$

$= \dfrac{x(x-y)}{xy} - \dfrac{y(x+y)}{xy}$

$= \dfrac{x^2 - xy - xy - y^2}{xy}$

$= \dfrac{x^2 - 2xy - y^2}{xy}$

45. $\dfrac{7}{x+4} + \dfrac{2}{x} = \dfrac{7}{x+4} \cdot \dfrac{x}{x} + \dfrac{2}{x} \cdot \dfrac{x+4}{x+4}$

$= \dfrac{7x}{x(x+4)} + \dfrac{2(x+4)}{x(x+4)}$

$= \dfrac{7x + 2x + 8}{x(x+4)}$

$= \dfrac{9x+8}{x(x+4)}$

46. $\dfrac{2}{3x} - \dfrac{3}{3x-6} = \dfrac{2}{3x} - \dfrac{3}{3(x-2)}$

$= \dfrac{2}{3x} \cdot \dfrac{x-2}{x-2} - \dfrac{3}{3(x-2)} \cdot \dfrac{x}{x}$

$= \dfrac{2(x-2)}{3x(x-2)} - \dfrac{3x}{3x(x-2)}$

$= \dfrac{2x - 4 - 3x}{3x(x-2)}$

$= \dfrac{-x-4}{3x(x-2)}$

47. $\dfrac{1}{z+5} + \dfrac{9}{(z+5)^2} = \dfrac{1}{z+5} \cdot \dfrac{z+5}{z+5} + \dfrac{9}{(z+5)^2}$

$= \dfrac{1(z+5)}{(z+5)^2} + \dfrac{9}{(z+5)^2}$

$= \dfrac{z+5+9}{(z+5)^2}$

$= \dfrac{z+14}{(z+5)^2}$

48. $\dfrac{x+2}{x^2-x-6} + \dfrac{x-3}{x^2-8x+15}$

$= \dfrac{x+2}{(x-3)(x+2)} + \dfrac{x-3}{(x-3)(x-5)}$

$= \dfrac{1}{x-3} + \dfrac{1}{x-5}$

$= \dfrac{1}{x-3} \cdot \dfrac{x-5}{x-5} + \dfrac{1}{x-5} \cdot \dfrac{x-3}{x-3}$

$= \dfrac{x-5}{(x-3)(x-5)} + \dfrac{x-3}{(x-5)(x-3)}$

$= \dfrac{x-5+x-3}{(x-5)(x-3)}$

$= \dfrac{2x-8}{(x-5)(x-3)}$

49. $\dfrac{x+4}{x+6} - \dfrac{x-5}{x+2} = \dfrac{(x+4)(x+2)}{(x+6)(x+2)} - \dfrac{(x-5)(x+6)}{(x+2)(x+6)}$

$= \dfrac{x^2 + 6x + 8 - (x^2 + x - 30)}{(x+6)(x+2)}$

$= \dfrac{x^2 + 6x + 8 - x^2 + x + 30}{(x+6)(x+2)}$

$= \dfrac{5x+38}{(x+6)(x+2)}$

50. $2 + \dfrac{x}{x-4} = \dfrac{2(x-4)}{x-4} + \dfrac{x}{x-4}$

$= \dfrac{2x - 8 + x}{x-4}$

$= \dfrac{3x-8}{x-4}$

51. $\dfrac{a+2}{b} \div \dfrac{a-2}{5b^2} = \dfrac{a+2}{b} \cdot \dfrac{5b^2}{a-2}$

$\qquad = \dfrac{5b(a+2)}{a-2}$

$\qquad = \dfrac{5ab+10b}{a-2}$

52. $\dfrac{x+5}{x^2-9} + \dfrac{2}{x+3} = \dfrac{x+5}{(x-3)(x+3)} + \dfrac{2}{x+3}$

$\qquad = \dfrac{x+5}{(x-3)(x+3)} + \dfrac{2(x-3)}{(x-3)(x+3)}$

$\qquad = \dfrac{x+5+2x-6}{(x-3)(x+3)}$

$\qquad = \dfrac{3x-1}{(x-3)(x+3)}$

53. $\dfrac{6p+12q}{p^2q} \cdot \dfrac{p^4}{p+2q} = \dfrac{6\cancel{(p+2q)}p^4}{\cancel{(p+2q)}p^2q} = \dfrac{6p^2}{q}$

54. $\dfrac{8}{(x+2)(x-3)} - \dfrac{6}{(x-2)(x+2)}$

$\qquad = \dfrac{8(x-2)}{(x+2)(x-3)(x-2)} - \dfrac{6(x-3)}{(x+2)(x-3)(x-2)}$

$\qquad = \dfrac{8(x-2)-6(x-3)}{(x+2)(x-3)(x-2)}$

$\qquad = \dfrac{8x-16-6x+18}{(x+2)(x-3)(x-2)}$

$\qquad = \dfrac{2x+2}{(x+2)(x-3)(x-2)}$

55. $\dfrac{x+7}{x^2+9x+14} - \dfrac{x-10}{x^2-49}$

$\qquad = \dfrac{x+7}{(x+7)(x+2)} - \dfrac{x-10}{(x+7)(x-7)}$

$\qquad = \dfrac{x+7}{(x+7)(x+2)} \cdot \dfrac{(x-7)}{(x-7)} - \dfrac{x-10}{(x+7)(x-7)} \cdot \dfrac{(x+2)}{(x+2)}$

$\qquad = \dfrac{x^2-49-\left(x^2-8x-20\right)}{(x+7)(x-7)(x+2)}$

$\qquad = \dfrac{8x-29}{(x+7)(x-7)(x+2)}$

56. $\dfrac{x-y}{x+y} \cdot \dfrac{xy+x^2}{x^2-y^2} = \dfrac{\cancel{x-y}}{x+y} \cdot \dfrac{x(y+x)}{(x+y)\cancel{(x-y)}}$

$\qquad = \dfrac{x}{x+y}$

57. $\dfrac{3x^2-27y^2}{30} \div \dfrac{(x-3y)^2}{6}$

$\qquad = \dfrac{3\left(x^2-9y^2\right)}{30} \cdot \dfrac{6}{(x-3y)^2}$

$\qquad = \dfrac{3\cancel{(x-3y)}(x+3y)}{\cancel{6}\cdot 5} \cdot \dfrac{\cancel{6}}{\cancel{(x-3y)}(x-3y)}$

$\qquad = \dfrac{3(x+3y)}{5(x-3y)}$

58. $\dfrac{a^2-11a+30}{a-6} \cdot \dfrac{a^2-8a+15}{a^2-10a+25}$

$\qquad = \dfrac{\cancel{(a-6)}\cancel{(a-5)}}{\cancel{a-6}} \cdot \dfrac{\cancel{(a-5)}(a-3)}{\cancel{(a-5)^2}}$

$\qquad = a-3$

59. $\dfrac{a}{a^2-1} - \dfrac{3}{3a^2-2a-5}$

$\qquad = \dfrac{a}{(a-1)(a+1)} - \dfrac{3}{(3a-5)(a+1)}$

$\qquad = \dfrac{a(3a-5)}{(a-1)(a+1)(3a-5)} - \dfrac{3(a-1)}{(a-1)(a+1)(3a-5)}$

$\qquad = \dfrac{a(3a-5)-3(a-1)}{(a-1)(a+1)(3a-5)}$

$\qquad = \dfrac{3a^2-5a-3a+3}{(a-1)(a+1)(3a-5)}$

$\qquad = \dfrac{3a^2-8a+3}{(a-1)(a+1)(3a-5)}$

60. $\dfrac{2x^2+6x-20}{x^2-2x} \div \dfrac{x^2+7x+10}{2x^2-8}$

$= \dfrac{2\left(x^2+3x-10\right)}{x\left(x-2\right)} \cdot \dfrac{2\left(x^2-4\right)}{\left(x+5\right)\left(x+2\right)}$

$= \dfrac{2\left(x+5\right)\left(x-2\right)}{x\left(x-2\right)} \cdot \dfrac{2\left(x+2\right)\left(x-2\right)}{\left(x+5\right)\left(x+2\right)}$

$= \dfrac{4\left(x-2\right)}{x}$

$= \dfrac{4x-8}{x}$

61. $\dfrac{5+\dfrac{2}{3}}{\dfrac{3}{4}} = \dfrac{12\left(5+\dfrac{2}{3}\right)}{12\left(\dfrac{3}{4}\right)} = \dfrac{60+8}{9} = \dfrac{68}{9}$

62. $\dfrac{1+\dfrac{5}{8}}{3-\dfrac{9}{16}} = \dfrac{16\left(1+\dfrac{5}{8}\right)}{16\left(3-\dfrac{9}{16}\right)}$

$= \dfrac{16+10}{48-9}$

$= \dfrac{26}{39}$

$= \dfrac{2}{3}$

63. $\dfrac{\dfrac{12ab}{9c}}{\dfrac{4a}{c^2}} = \dfrac{12ab}{9c} \cdot \dfrac{c^2}{4a} = \dfrac{bc}{3}$

64. $\dfrac{\dfrac{18x^4y^2}{9xy^5}}{4z^2} = \dfrac{18x^4y^2 \cdot 4z^2}{\dfrac{9xy^5}{4z^2} \cdot 4z^2} = \dfrac{72x^4y^2z^2}{9xy^5} = \dfrac{8x^3z^2}{y^3}$

65. $\dfrac{a-\dfrac{a}{b}}{\dfrac{1+a}{b}} = \dfrac{\left(a-\dfrac{a}{b}\right)b}{\left(\dfrac{1+a}{b}\right)b} = \dfrac{ab-a}{1+a}$

66. $\dfrac{r^2+\dfrac{7}{s}}{s^2} = \dfrac{\left(r^2+\dfrac{7}{s}\right)s}{\left(s^2\right)s} = \dfrac{r^2s+7}{s^3}$

67. $\dfrac{\dfrac{3}{x}+\dfrac{2}{x^2}}{5-\dfrac{1}{x}} = \dfrac{\left(\dfrac{3}{x}+\dfrac{2}{x^2}\right)x^2}{\left(5-\dfrac{1}{x}\right)x^2} = \dfrac{3x+2}{5x^2-x} = \dfrac{3x+2}{x\left(5x-1\right)}$

68. $\dfrac{\dfrac{x}{x+y}}{\dfrac{x^2}{4x+4y}} = \dfrac{x}{x+y} \cdot \dfrac{4x+4y}{x^2}$

$= \dfrac{1}{x+y} \cdot \dfrac{4\left(x+y\right)}{x}$

$= \dfrac{4}{x}$

69. $\dfrac{\dfrac{9}{x}}{\dfrac{9}{x^2}} = \dfrac{9}{x} \cdot \dfrac{x^2}{9} = x$

70. $\dfrac{\dfrac{1}{a}+2}{\dfrac{1}{a}+\dfrac{3}{a}} = \dfrac{\dfrac{1}{a}+2}{\dfrac{4}{a}} = \dfrac{\left(\dfrac{1}{a}+2\right)a}{\left(\dfrac{4}{a}\right)a} = \dfrac{1+2a}{4}$

71. $\dfrac{\dfrac{1}{x^2}-\dfrac{1}{x}}{\dfrac{1}{x^2}+\dfrac{1}{x}} = \dfrac{x^2\left(\dfrac{1}{x^2}-\dfrac{1}{x}\right)}{x^2\left(\dfrac{1}{x^2}+\dfrac{1}{x}\right)}$

$= \dfrac{1-x}{1+x}$

$= \dfrac{-x+1}{x+1}$

72. $\dfrac{\dfrac{8x}{y}-x}{\dfrac{y}{x}-1} = \dfrac{\left(\dfrac{8x}{y}-x\right)xy}{\left(\dfrac{y}{x}-1\right)xy}$

$= \dfrac{8x^2-x^2y}{y^2-xy}$

$= \dfrac{8x^2-x^2y}{y\left(y-x\right)}$

73. $\dfrac{5}{9} = \dfrac{10}{x+3}$

$5\left(x+3\right) = 10\left(9\right)$

$5x+15 = 90$

$5x = 75$

$x = 15$

74. $\dfrac{x}{4} = \dfrac{x-3}{2}$

$2x = 4(x-3)$

$2x = 4x - 12$

$12 = 2x$

$6 = x$

75. $\dfrac{12}{n} + 2 = \dfrac{n}{4}$

$4n\left(\dfrac{12}{n} + 2\right) = 4n\left(\dfrac{n}{4}\right)$

$48 + 8n = n^2$

$0 = n^2 - 8n - 48$

$0 = (n-12)(n+4)$

$n - 12 = 0 \quad$ or $\quad n + 4 = 0$

$n = 12 \qquad\qquad n = -4$

Since –4 is an extraneous root, $n = 12$ is the solution.

76. $\dfrac{10}{m} + \dfrac{3}{2} = \dfrac{m}{10}$

$10m\left(\dfrac{10}{m} + \dfrac{3}{2}\right) = 10m\left(\dfrac{m}{10}\right)$

$100 + 15m = m^2$

$0 = m^2 - 15m - 100$

$0 = (m-20)(m+5)$

$m - 20 = 0 \quad$ or $\quad m + 5 = 0$

$m = 20 \qquad\qquad m = -5$

Since –5 is an extraneous root, $m = 20$ is the solution.

77. $\dfrac{-4}{d} = \dfrac{3}{2} + \dfrac{4-d}{d}$

$2d\left(\dfrac{-4}{d}\right) = 2d\left(\dfrac{3}{2} + \dfrac{4-d}{d}\right)$

$-8 = 3d + 8 - 2d$

$-8 = d + 8$

$-16 = d$

78. $\dfrac{1}{x-7} + \dfrac{1}{x+7} = \dfrac{1}{x^2 - 49}$

$\dfrac{1}{x-7} + \dfrac{1}{x+7} = \dfrac{1}{(x-7)(x+7)}$

$(x-7)(x+7)\left[\dfrac{1}{x-7} + \dfrac{1}{x+7}\right] = \left[\dfrac{1}{(x-7)(x+7)}\right](x-7)(x+7)$

$x + 7 + x - 7 = 1$

$2x = 1$

$x = \dfrac{1}{2}$

79. $\dfrac{x-3}{x-2} + \dfrac{x+1}{x+3} = \dfrac{2x^2 + x + 1}{x^2 + x - 6}$

$\dfrac{x-3}{x-2} + \dfrac{x+1}{x+3} = \dfrac{2x^2 + x + 1}{(x-2)(x+3)}$

$(x-2)(x+3)\left(\dfrac{x-3}{x-2} + \dfrac{x+1}{x+3}\right) = (x-2)(x+3)\left[\dfrac{2x^2 + x + 1}{(x-2)(x+3)}\right]$

$(x+3)(x-3) + (x-2)(x+1) = 2x^2 + x + 1$

$x^2 - 9 + x^2 - x - 2 = 2x^2 + x + 1$

$2x^2 - x - 11 = 2x^2 + x + 1$

$-12 = 2x$

$-6 = x$

80.

$$\frac{a}{a^2-64}+\frac{4}{a+8}=\frac{3}{a-8}$$

$$\frac{a}{(a+8)(a-8)}+\frac{4}{a+8}=\frac{3}{a-8}$$

$$(a+8)(a-8)\left[\frac{a}{(a+8)(a-8)}+\frac{4}{a+8}\right]=(a+8)(a-8)\left(\frac{3}{a-8}\right)$$

$$a+4(a-8)=3(a+8)$$

$$a+4a-32=3a+24$$

$$5a-32=3a+24$$

$$2a=56$$

$$a=28$$

81.

$$\frac{d}{d-4}-4=\frac{4}{d-4}$$

$$(d-4)\left(\frac{d}{d-4}-4\right)=(d-4)\left(\frac{4}{d-4}\right)$$

$$d-4(d-4)=4$$

$$d-4d+16=4$$

$$-3d+16=4$$

$$-3d=-12$$

$$d=4$$

Since $\dfrac{4}{0}$ is not a real number, there is no solution.

82. John and Amy rate $=\dfrac{1}{6}$

Paul and Cindy rate $=\dfrac{1}{4}$

$$\frac{t}{6}+\frac{t}{4}=1$$

$$24\left(\frac{t}{6}+\frac{t}{4}\right)=24(1)$$

$$4t+6t=24$$

$$10t=24$$

$$t=\frac{24}{10}=2.4$$

It will take the 4 people 2.4 hours.

83. $\dfrac{3}{4}$-inch hose's rate $=\dfrac{1}{7}$

$\dfrac{5}{16}$-inch hose's rate $=\dfrac{1}{12}$

$$\frac{t}{7}-\frac{t}{12}=1$$

$$\frac{12t-7t}{84}=1$$

$$5t=84$$

$$t=\frac{84}{5}=16\frac{4}{5}$$

It will take $16\dfrac{4}{5}$ hours to fill the pool.

84. Let x be one number then, $6x$ is the other number.

$$\frac{1}{x}+\frac{1}{6x}=7$$

$$6x\left(\frac{1}{x}+\frac{1}{6x}\right)=6x(7)$$

$$6+1=42x$$

$$7=42x$$

$$\frac{7}{42}=x$$

$$\frac{1}{6}=x$$

$$6x=6\left(\frac{1}{6}\right)=1$$

The numbers are $\dfrac{1}{6}$ and 1.

85. Let x = Robert's speed, then
$3.5 + x$ = Tran's speed

$$t = \frac{d}{r}$$

Robert's time = Tran's time

$$\frac{3}{x} = \frac{8}{3.5 + x}$$

$$3(3.5 + x) = 8x$$

$$10.5 + 3x = 8x$$

$$10.5 = 5x$$

$$2.1 = x$$

$$x + 3.5 = 2.1 + 3.5 = 5.6$$

Robert's speed is 2.1 mph and Tran's speed is 5.6 mph.

86. The equation is $W = \dfrac{kL^2}{A}$. To find k, substitute 4 for W, 2 for L, and 10 for A.

$$4 = \frac{k(2)^2}{10}$$

$$4 = \frac{4k}{10}$$

$$40 = 4k$$

$$10 = k$$

Thus $W = \dfrac{10L^2}{A}$. Now substitute 5 for L and 20 for A.

$$W = \frac{10 \cdot (5)^2}{20} = \frac{250}{20} = \frac{25}{2}$$

87. The equation is $z = \dfrac{kxy}{r^2}$. To find k, substitute 12 for z, 20 for x, 8 for y, and 8 for r.

$$12 = \frac{k(20)(8)}{(8)^2}$$

$$12 = \frac{160k}{64}$$

$$12(64) = 160k$$

$$768 = 160k$$

$$\frac{768}{160} = k$$

$$k = 4.8$$

Thus $z = \dfrac{4.8xy}{r^2}$. Now substitute 10 for x, 80 for y, and 3 for r.

$$z = \frac{4.8(10)(80)}{3^2} \approx 426.7$$

88.
$$d = kw$$

$$182 = k \cdot 132$$

$$182 = 132k$$

$$1.379 \approx k$$

Now we have to substitute 198 for w and 1.379 for k to find the recommended dosage for Bill.

$$d = kw$$

$$d = 1.379 \cdot 198$$

$$d \approx 273$$

Thus, 273 mg is needed for Bill.

89.
$$t = \frac{k}{s}$$

$$1.4 = \frac{k}{6}$$

$$8.4 = k$$

Now we have to substitute 5 for s and 8.4 for k to find Leif's time.

$$t = \frac{k}{s}$$

$$t = \frac{8.4}{5}$$

$$t = 1.68$$

Thus, it will take Leif 1.68 hours.

Chapter 7 Practice Test

1. $\dfrac{-8 + x}{x - 8} = \dfrac{x - 8}{x - 8} = 1$

2. $\dfrac{x^3 - 1}{x^2 - 1} = \dfrac{(x - 1)(x^2 + x + 1)}{(x - 1)(x + 1)} = \dfrac{x^2 + x + 1}{x + 1}$

3. $\dfrac{20x^2 y^3}{4z^2} \cdot \dfrac{8xz^3}{5xy^4} = \dfrac{4 \cdot 5x^2 y^3}{4z^2} \cdot \dfrac{2 \cdot 4xz^3}{5xy^4}$

$$= \frac{8x^3 y^3 z^3}{xy^4 z^2}$$

$$= \frac{8x^2 z}{y}$$

4. $\dfrac{a^2 - 9a + 14}{a - 2} \cdot \dfrac{a^2 - 4a - 21}{(a - 7)^2}$

$$= \frac{(a - 7)(a - 2)}{a - 2} \cdot \frac{(a - 7)(a + 3)}{(a - 7)^2}$$

$$= a + 3$$

5. $\dfrac{x^2-x-6}{x^2-9}\cdot\dfrac{x^2-6x+9}{x^2+4x+4}$

$=\dfrac{\cancel{(x-3)}\,\cancel{(x+2)}}{\cancel{(x-3)}\,(x+3)}\cdot\dfrac{(x-3)(x-3)}{\cancel{(x+2)}\,(x+2)}$

$=\dfrac{(x-3)^2}{(x+3)(x+2)}$

$=\dfrac{x^2-6x+9}{(x+3)(x+2)}$

6. $\dfrac{x^2-1}{x+2}\cdot\dfrac{x+2}{1-x^2}$

$=\dfrac{\cancel{(x-1)}\,\cancel{(x+1)}}{\cancel{x+2}}\cdot\dfrac{\cancel{x+2}}{-1\cancel{(x-1)}\,\cancel{(x+1)}}$

$=-1$

7. $\dfrac{x^2-4y^2}{5x+20y}\div\dfrac{x+2y}{x+4y}=\dfrac{(x-2y)\,\cancel{(x+2y)}}{5\cancel{(x+4y)}}\cdot\dfrac{\cancel{x+4y}}{\cancel{x+2y}}$

$=\dfrac{x-2y}{5}$

8. $\dfrac{15}{y^2+2y-15}\div\dfrac{5}{y-3}=\dfrac{\overset{3}{\cancel{15}}}{\cancel{(y-3)}\,(y+5)}\cdot\dfrac{\cancel{y-3}}{\underset{1}{\cancel{5}}}$

$=\dfrac{3}{y+5}$

9. $\dfrac{m^2+3m-18}{m-3}\div\dfrac{m^2-8m+15}{3-m}$

$=\dfrac{(m+6)(m-3)}{m-3}\cdot\dfrac{-1(m-3)}{(m-5)(m-3)}$

$=\dfrac{-(m+6)}{m-5}$

$=-\dfrac{m+6}{m-5}$

10. $\dfrac{4x+3}{8y}+\dfrac{2x-5}{8y}=\dfrac{4x+3+2x-5}{8y}$

$=\dfrac{6x-2}{8y}$

$=\dfrac{2(3x-1)}{8y}$

$=\dfrac{3x-1}{4y}$

11. $\dfrac{7x^2-4}{x+3}-\dfrac{6x+9}{x+3}=\dfrac{7x^2-4-(6x+9)}{x+3}$

$=\dfrac{7x^2-4-6x-9}{x+3}$

$=\dfrac{7x^2-6x-13}{x+3}$

12. $\dfrac{2}{xy}-\dfrac{8}{xy^3}=\dfrac{2}{xy}\cdot\dfrac{y^2}{y^2}-\dfrac{8}{xy^3}$

$=\dfrac{2y^2}{xy^3}-\dfrac{8}{xy^3}$

$=\dfrac{2y^2-8}{xy^3}$

13. $3-\dfrac{5z}{z-5}=3\left(\dfrac{z-5}{z-5}\right)-\dfrac{5z}{z-5}$

$=\dfrac{3z-15}{z-5}-\dfrac{5z}{z-5}$

$=\dfrac{3z-15-5z}{z-5}$

$=\dfrac{-2z-15}{z-5}$

$=-\dfrac{2z+15}{z-5}$

14. $\dfrac{x-5}{x^2-16}-\dfrac{x-2}{x^2+2x-8}$

$=\dfrac{x-5}{(x-4)(x+4)}-\dfrac{x-2}{(x-2)(x+4)}$

$=\dfrac{x-5}{(x-4)(x+4)}-\dfrac{1}{x+4}$

$=\dfrac{x-5}{(x-4)(x+4)}-\dfrac{1}{x+4}\cdot\dfrac{x-4}{x-4}$

$=\dfrac{x-5-(x-4)}{(x-4)(x+4)}$

$=\dfrac{x-5-x+4}{(x-4)(x+4)}$

$=\dfrac{-1}{(x-4)(x+4)}$

15. $\dfrac{2+\dfrac{1}{2}}{3-\dfrac{1}{5}}=\dfrac{10\left(2+\dfrac{1}{2}\right)}{10\left(3-\dfrac{1}{5}\right)}=\dfrac{20+5}{30-2}=\dfrac{25}{28}$

16. $\dfrac{x+\dfrac{x}{y}}{\dfrac{7}{x}} = \left(x+\dfrac{x}{y}\right)\dfrac{x}{7}$

$\qquad = \left(\dfrac{xy+x}{y}\right)\left(\dfrac{x}{7}\right)$

$\qquad = \dfrac{yx^2+x^2}{7y}$

17. $\dfrac{4+\dfrac{3}{x}}{\dfrac{9}{x}-5} = \dfrac{x\left(4+\dfrac{3}{x}\right)}{x\left(\dfrac{9}{x}-5\right)} = \dfrac{4x+3}{9-5x}$

18. $2+\dfrac{8}{x} = 6$

$x\left(2+\dfrac{8}{x}\right) = 6x$

$2x+8 = 6x$

$8 = 4x$

$2 = x$

19. $\dfrac{2x}{3} - \dfrac{x}{4} = x+1$

$12\left(\dfrac{2x}{3}-\dfrac{x}{4}\right) = 12(x+1)$

$8x-3x = 12x+12$

$5x = 12x+12$

$-7x = 12$

$x = -\dfrac{12}{7}$

20. $\dfrac{x}{x-8} + \dfrac{6}{x-2} = \dfrac{x^2}{x^2-10x+16}$

$\dfrac{x}{x-8} + \dfrac{6}{x-2} = \dfrac{x^2}{(x-8)(x-2)}$

$(x-8)(x-2)\left(\dfrac{x}{x-8}+\dfrac{6}{x-2}\right) = (x-8)(x-2)\left[\dfrac{x^2}{(x-8)(x-2)}\right]$

$x(x-2)+6(x-8) = x^2$

$x^2-2x+6x-48 = x^2$

$4x-48 = 0$

$4x = 48$

$x = 12$

21. $\dfrac{t}{10} + \dfrac{t}{15} = 1$

$30\left(\dfrac{t}{10}+\dfrac{t}{15}\right) = 30(1)$

$3t+2t = 30$

$5t = 30$

$t = 6$

It will take them 6 hours to level one acre together.

22. Let x be the number.

$$x + \frac{1}{x} = 2$$

$$x\left(x + \frac{1}{x}\right) = x(2)$$

$$x^2 + 1 = 2x$$

$$x^2 - 2x + 1 = 0$$

$$(x-1)(x-1) = 0$$

$$x - 1 = 0$$

$$x = 1$$

The number is 1.

23. Let $x =$ base, then $2x - 2 =$ height.

$$\text{area} = \frac{1}{2} \cdot \text{base} \cdot \text{height}$$

$$30 = \frac{1}{2}x(2x - 2)$$

$$2(30) = 2\left[\frac{1}{2}x(2x - 2)\right]$$

$$60 = x(2x - 2)$$

$$60 = 2x^2 - 2x$$

$$0 = 2x^2 - 2x - 60$$

$$0 = 2\left(x^2 - x - 30\right)$$

$$0 = 2(x - 6)(x + 5)$$

$$x - 6 = 0 \quad \text{or} \quad x + 5 = 0$$

$$x = 6 \qquad\qquad x = -5$$

Since the base cannot be negative, the base is 6 inches and the height is $2(6) - 2 = 10$ inches.

24. Let $d =$ the distance she rollerblades, then $12 - d$ is the distance she bicycles.

$$t = \frac{d}{r}$$

$$\frac{d}{4} + \frac{12 - d}{10} = 1.5$$

$$20\left(\frac{d}{4} + \frac{12 - d}{10}\right) = 20(1.5)$$

$$5d + 2(12 - d) = 30$$

$$5d + 24 - 2d = 30$$

$$3d = 6$$

$$d = 2$$

She rollerblades for 2 miles.

25.

$$w = \frac{k}{f}$$

$$4.3 = \frac{k}{263}$$

$$1130.9 = k$$

Now substitute 1000 for f and 1130.9 for k.

$$w = \frac{k}{f}$$

$$w = \frac{1130.9}{1000}$$

$$w \approx 1.1309$$

The wavelength would be about 1.13 feet.

Chapter 7 Cumulative Review Test

1.
$$3x^2 - 5xy^2 - 7 = 3(-4)^2 - 5(-4)(-2)^2 - 7$$
$$= 3(16) - 5(-4)(4) - 7$$
$$= 48 + 80 - 7$$
$$= 121$$

2.
$$[6 - [3(8 \div 4)]^2 + 9 \cdot 4]^2$$
$$= [6 - [3(2)]^2 + 9 \cdot 4]^2$$
$$= [6 - (6)^2 + 9 \cdot 4]^2$$
$$= [6 - 36 + 9 \cdot 4]^2$$
$$= [6 - 36 + 36]^2$$
$$= [6]^2$$
$$= 36$$

3.
$$5z + 4 = -3(z - 7)$$
$$5z + 4 = -3z + 21$$
$$5z + 3z = 21 - 4$$
$$8z = 17$$
$$z = \frac{17}{8}$$

4. $c + (c + 0.012c)$

230

5. Let $(-7, 8)$ be (x_1, y_1) and $(3, 8)$ be (x_2, y_2).

$$m = \frac{y_2 - y_1}{x_2 - x_1}$$

$$m = \frac{8-8}{3-(-7)}$$

$$m = \frac{0}{10} = 0$$

The slope is 0.

6. To put in slope-intercept form, solve for y.

$$3x - 4y = 12$$

$$3x - 3x - 4y = -3x + 12$$

$$-4y = -3x + 12$$

$$\frac{-4y}{-4} = \frac{-3x}{-4} + \frac{12}{-4}$$

$$y = \frac{3}{4}x - 3$$

7. $\left(6x^2 - 3x - 5\right) - \left(-2x^2 - 8x - 19\right)$

$$= 6x^2 - 3x - 5 + 2x^2 + 8x + 19$$

$$= 6x^2 + 2x^2 - 3x + 8x - 5 + 19$$

$$= 8x^2 + 5x + 14$$

8. $\left(3n^2 - 4n + 3\right)\left(2n - 5\right)$

$$= 3n^2\left(2n - 5\right) - 4n\left(2n - 5\right) + 3\left(2n - 5\right)$$

$$= 6n^3 - 15n^2 - 8n^2 + 20n + 6n - 15$$

$$= 6n^3 - 23n^2 + 26n - 15$$

9. $\dfrac{4x - 38}{8} = \dfrac{4x}{8} - \dfrac{38}{8} = \dfrac{1}{2}x - \dfrac{19}{4}$

10. $8a^2 - 8a - 5a + 5 = 8a(a - 1) - 5(a - 1)$

$$= (a - 1)(8a - 5)$$

11. $13x^2 + 26x - 39 = 13\left(x^2 + 2x - 3\right)$

$$= 13(x + 3)(x - 1)$$

12.

$$2x^2 = 11x - 12$$

$$2x^2 - 11x + 12 = 0$$

$$(x - 4)(2x - 3) = 0$$

$$x - 4 = 0 \quad \text{or} \quad 2x - 3 = 0$$

$$x = 4 \qquad\qquad x = \frac{3}{2}$$

13. $\dfrac{x^2 + x - 12}{x^2 - x - 6} \cdot \dfrac{x^2 - 2x - 8}{2x^2 - 7x - 4}$

$$= \frac{(x+4)\cancel{(x-3)}}{\cancel{(x-3)}\cancel{(x+2)}} \cdot \frac{\cancel{(x-4)}\cancel{(x+2)}}{(2x+1)\cancel{(x-4)}}$$

$$= \frac{x+4}{2x+1}$$

14. $\dfrac{r}{r+2} - \dfrac{3}{r-5} = \dfrac{r}{r+2} \cdot \dfrac{r-5}{r-5} - \dfrac{3}{r-5} \cdot \dfrac{r+2}{r+2}$

$$= \frac{r(r-5)}{(r+2)(r-5)} - \frac{3(r+2)}{(r+2)(r-5)}$$

$$= \frac{r^2 - 5r - (3r + 6)}{(r+2)(r-5)}$$

$$= \frac{r^2 - 5r - 3r - 6}{(r+2)(r-5)}$$

$$= \frac{r^2 - 8r - 6}{(r+2)(r-5)}$$

15. $\dfrac{4}{x^2 - 3x - 10} + \dfrac{6}{x^2 + 5x + 6}$

$$= \frac{4}{(x-5)(x+2)} + \frac{6}{(x+2)(x+3)}$$

$$= \frac{4(x+3)}{(x-5)(x+2)(x+3)} + \frac{6(x-5)}{(x-5)(x+2)(x+3)}$$

$$= \frac{4(x+3) + 6(x-5)}{(x-5)(x+2)(x+3)}$$

$$= \frac{4x + 12 + 6x - 30}{(x-5)(x+2)(x+3)}$$

$$= \frac{10x - 18}{(x-5)(x+2)(x+3)}$$

16.

$$\frac{x}{9} - \frac{x}{6} = \frac{1}{12}$$

$$36\left(\frac{x}{9} - \frac{x}{6}\right) = 36\left(\frac{1}{12}\right)$$

$$4x - 6x = 3$$

$$-2x = 3$$

$$x = -\frac{3}{2}$$

17.

$$\frac{7}{x+3} + \frac{5}{x+2} = \frac{5}{x^2+5x+6}$$

$$\frac{7}{x+3} + \frac{5}{x+2} = \frac{5}{(x+3)(x+2)}$$

$$(x+3)(x+2)\left(\frac{7}{x+3} + \frac{5}{x+2}\right) = (x+3)(x+2)\left[\frac{5}{(x+3)(x+2)}\right]$$

$$7(x+2) + 5(x+3) = 5$$

$$7x + 14 + 5x + 15 = 5$$

$$12x + 29 = 5$$

$$12x = -24$$

$$x = -2$$

Check:

$$\frac{7}{x+3} + \frac{5}{x+2} = \frac{5}{x^2+5x+6}$$

$$\frac{7}{-2+3} + \frac{5}{-2+2} = \frac{5}{(-2)^2+5(-2)+6}$$

$$\frac{7}{1} + \frac{5}{0} = \frac{5}{0}$$

Since $\dfrac{5}{0}$ is not a real number, there is no solution.

18. Let x = the total medical bills. The cost under plan 1 is $0.10x$, while the cost under plan 2 is $100 + 0.05x$.

$$0.10x = 150 + 0.05x$$

$$0.05x = 150$$

$$x = 3000$$

The cost under both plans is the same for $3000 in total medical bills.

19. Let x = pounds of sunflower seed and
y = pounds of premixed assorted seed mix
$$x + y = 50$$
$$0.50x + 0.20y = 16.00$$
Solve the first equation for y.
$$y = 50 - x$$
Substitute $50 - x$ for y in the second equation.
$$0.50x + 0.20(50 - x) = 16$$
$$0.50x + 10 - 0.20x = 16$$
$$0.30x = 6$$
$$x = 20$$
$$y = 50 - x = 50 - 20 = 30$$
He will have to use 20 pounds of sunflower seed and 30 pounds of premixed assorted seed mix.

20. Let d = distance on first leg, then the distance on the second leg is $12.75 - d$.

$$t = \frac{d}{r}$$

Time for first leg + time for second leg = total time

$$\frac{d}{6.5} + \frac{12.75 - d}{9.5} = 1.5$$

$$(9.5)(6.5)\left[\frac{d}{6.5} + \frac{12.75 - d}{9.5}\right] = (9.5)(6.5)(1.5)$$

$$9.5d + 6.5(12.75 - d) = 92.625$$

$$9.5d + 82.875 - 6.5d = 92.625$$

$$3d = 9.75$$

$$d = 3.25$$

$$12.75 - d = 12.75 - 3.25 = 9.5$$

The distance traveled during the first leg of the race was 3.25 miles and the distance traveled in the second leg of the race was 9.5 miles.

Chapter 8

Exercise Set 8.1

1. No, a nonlinear equation will not always have a y-intercept because the graph does not have to intersect the y-axis. For example, the graph of $y = \dfrac{1}{x}$ does not have a y-intercept.

3. When graphing $y = \dfrac{1}{x}$, we cannot substitute 0 for x because $\dfrac{1}{0}$ is undefined.

5.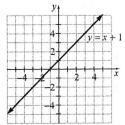

7.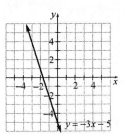

9.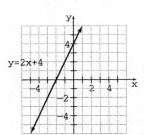

11.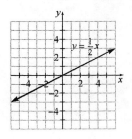

13.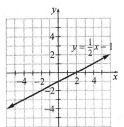

15.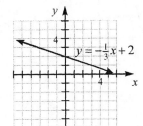

17.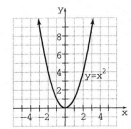

19.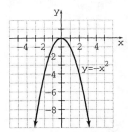

21.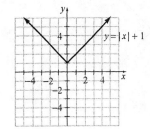

23.

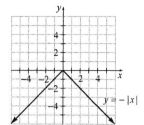

25.

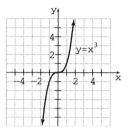

27.

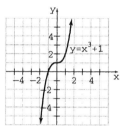

29.

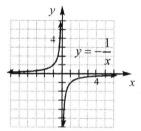

31.

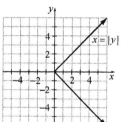

33.

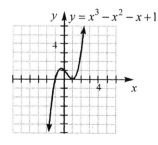

35.

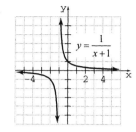

37.

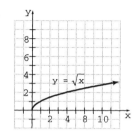

39.

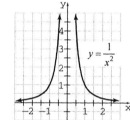

41. $y = \dfrac{x^2}{x+1}$

$$\left(\frac{1}{12}\right) \overset{?}{=} \frac{\left(\frac{1}{3}\right)^2}{\left(\frac{1}{3}\right)+1}$$

$$\frac{1}{12} \overset{?}{=} \frac{\frac{1}{9}}{\frac{4}{3}}$$

$$\frac{1}{12} \overset{?}{=} \frac{1}{9} \cdot \frac{3}{4}$$

$$\frac{1}{12} = \frac{1}{12} \quad \text{True}$$

Yes, $\left(\dfrac{1}{3}, \dfrac{1}{12}\right)$ is on the graph of the equation

$y = \dfrac{x^2}{x+1}$.

43. a.

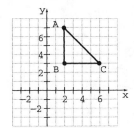

b. $\text{Area} = \dfrac{1}{2}bh = \dfrac{1}{2}(4)(4) = 8$

The area is **8** square units.

45. a. The average length of the golf course at the majors in 1980 was about 6875 yards.

b. The average length of the golf course at the majors in 2005 was about 7300 yards.

c. The average length of the golf course at the majors was greater than 7000 yards for the years 1990, 2000, and 2005.

d. No, the average length of the golf courses in the majors from 1995 to 2005 does not appear to be linear. The increase in length is not the same from 1995 to 2000 as it is from 2000 to 2005. That is, the increase in the length is not constant.

47.

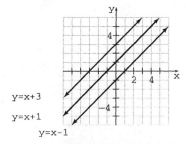

a. Each graph crosses the y-axis at the point corresponding to the constant term in the graph's equation.

b. Yes, all the equations seem to have the same slant or slope.

49.

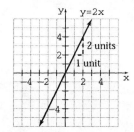

For each unit change in x, y changes 2 units.
Thus, the rate of change of y with respect to x is 2.

51.

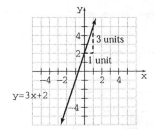

For each unit change in x, y changes 3 units.
Thus, the rate of change of y with respect to x is 3.

53. Answers may vary. One possible answer is the points $(4, -3)$ and $(5, 1)$.

Starting at $(3, -7)$: For a unit change, x changes from 3 to $3 + 1 = 4$. At the same time, y changes from -7 to $-7 + 4 = -3$. So, $(4, -3)$ is a solution to the equation.

Starting at $(4, -3)$: For a unit change, x changes from 4 to $4 + 1 = 5$. At the same time, y changes from -3 to $-3 + 4 = 1$. So, $(5, 1)$ is a solution to the equation.

55. c

57. a

59. d

61. b

63. b

65. d

67. b

69. d

71.

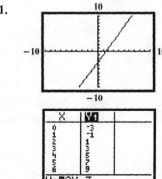

73.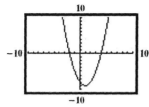

75.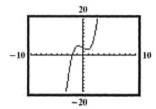

77.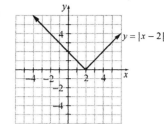

81. $\begin{aligned} x^2 - 4x + 5 &= (-3)^2 - 4(-3) + 5 \\ &= 9 + 12 + 5 \\ &= 26 \end{aligned}$

82. $m = \dfrac{y_2 - y_1}{x_2 - x_1} = \dfrac{7-3}{6-(-5)} = \dfrac{7-3}{6+5} = \dfrac{4}{11}$

83.

$$\frac{2x}{x^2-4} + \frac{1}{x-2} = \frac{2}{x+2}$$

$$\frac{2x}{(x-2)(x+2)} + \frac{1}{x-2} = \frac{2}{x+2}$$

$$(x-2)(x+2)\left(\frac{2x}{(x-2)(x+2)} + \frac{1}{x-2}\right) = (x-2)(x+2)\left(\frac{2}{x+2}\right)$$

$$2x + (x+2) = 2(x-2)$$

$$3x + 2 = 2x - 4$$

$$x + 2 = -4$$

$$x = -6$$

Note that -6 does not cause a denominator to be zero, so it is not extraneous.
In other words -6 checks, so the solution to the equation is $x = -6$.

84.

$$\frac{4x}{x^2+6x+9}-\frac{2x}{x+3}=\frac{x+1}{x+3}$$

$$\frac{4x}{(x+3)^2}-\frac{2x}{x+3}=\frac{x+1}{x+3}$$

$$(x+3)^2\left(\frac{4x}{(x+3)^2}-\frac{2x}{x+3}\right)=(x+3)^2\left(\frac{x+1}{x+3}\right)$$

$$4x-2x(x+3)=(x+3)(x+1)$$

$$4x-2x^2-6x=x^2+x+3x+3$$

$$4x-2x^2-6x=x^2+x+3x+3$$

$$-2x^2-2x=x^2+4x+3$$

$$0=3x^2+6x+3$$

$$0=3(x^2+2x+1)$$

$$0=3(x+1)^2$$

$$x+1=0$$

$$x=-1$$

Note that -1 does not cause a denominator to be zero, so it is not extraneous. In other words, -1 checks. The solution to the equation is $x=-1$.

Exercise Set 8.2

1. A function is a correspondence between a first set of elements, the domain, and a second set of elements, the range, such that each element of the domain corresponds to exactly one element in the range.

3. Yes, all functions are also relations. A function is a set of ordered pairs so it is a relation.

5. $\{5, 6, 7, 8\}$

7. If each vertical line drawn through any part of the graph intersects the graph in at most one point, the graph represents a function. If any vertical line intersects the graph at more than one point, then the graph is not a function.

9. The range is the set of values for the dependent variable.

11. Domain: $\mathbb{R}$ or $(-\infty, \infty)$
There are no restrictions on values of x that can be used.
Range: $\mathbb{R}$ or $(-\infty, \infty)$
All values of y are represented in the function.

13. If y depends on x, then y is the dependent variable.

15. "$f(x)$" is read "f of x."

17. a. Yes, the relation is a function.

   b. Domain: $\{3, 5, 11\}$,
   Range: $\{6, 10, 22\}$

19. a. Yes, the relation is a function.

   b. Domain: {Cameron, Tyrone, Vishnu},
   Range: $\{3, 6\}$

21. a. No, the relation is not a function.

   b. Domain: $\{1990, 2001, 2002\}$,
   Range: $\{20, 34, 37\}$

23. a. The relation is a function.

   b. Domain: $\{1, 2, 3, 4, 5\}$,
   Range: $\{1, 2, 3, 4, 5\}$

25. a. The relation is a function.

   b. Domain: $\{1, 2, 3, 4, 5, 7\}$,
   Range: $\{-1, 0, 2, 4, 9\}$

27. a. The relation is not a function.

   b. Domain: $\{1, 2, 3\}$,
   Range: $\{1, 2, 4, 5, 6\}$

29. a. The relation is not a function.

   b. Domain: $\{0, 1, 2\}$, Range: $\{-7, -1, 2, 3\}$

31. $A = \{0\}$

33. $C = \{18, 20\}$

35. $E = \{0, 1, 2\}$

37. $H = \{0, 7, 14, 21, 28, ...\}$

39. $J = \{1, 2, 3, 4, ...\}$ or $J = N$

41. a. Set A is the set of all x such that x is a natural number less than 7.

   b. $A = \{1, 2, 3, 4, 5, 6\}$

43. $\{x \mid x \geq 0\}$

45. $\{z \mid z \leq 2\}$

47. $\{p|-6 \leq p < 3\}$

49. $\{q|q > -3 \text{ and } q \in N\}$

51. $\{r|r \leq \pi \text{ and } r \in W\}$

53. $\{x|x \geq 1\}$

55. $\{x|x \leq 4 \text{ and } x \in I\}$ or $\{x|x < 5 \text{ and } x \in I\}$

57. $\{x|-3 < x \leq 5\}$

59. $\{x|-2.5 \leq x < 4.2\}$

61. $\{x|-3 \leq x \leq 1 \text{ and } x \in I\}$

63. a. The graph represents a function.

    b. Domain: $\mathbb{R}$, Range: $\mathbb{R}$

    c. $x = 2$

65. a. The graph does not represent a function.

    b. Domain: $\{x| \ 0 \leq x \leq 2\}$,

        Range: $\{y|-3 \leq y \leq 3\}$

    c. $x \approx 1.5$

67. a. The graph represents a function.

    b. Domain: $\mathbb{R}$, Range: $\{y|y \geq 0\}$

    c. $x = -3$ or $x = -1$

69. a. The graph represents a function.

    b. Domain: $\{-1, 0, 1, 2, 3\}$,
        Range: $\{-1, 0, 1, 2, 3\}$

    c. $x = 2$

71. a. The graph does not represent a function.

    b. Domain: $\{x|x \geq 2\}$, Range: $\mathbb{R}$

    c. $x = 3$

73. a. The graph represents a function.

    b. Domain: $\{x|-2 \leq x \leq 2\}$,

        Range : $\{y|-1 \leq y \leq 2\}$

    c. $x = -2$ or $x = 2$

75. a. $f(2) = -2(2) + 7 = -4 + 7 = 3$

    b. $f(-3) = -2(-3) + 7 = 6 + 7 = 13$

77. a. $h(0) = (0)^2 - (0) - 6 = -6$

    b. $h(-1) = (-1)^2 - (-1) - 6 = 1 + 1 - 6 = -4$

79. a. $r(1) = -(1)^3 - 2(1)^2 + (1) + 4$
           $= -(1) - 2(1) + (1) + 4$
           $= -1 - 2 + 1 + 4$
           $= 2$

    b. $r(-2) = -(-2)^3 - 2(-2)^2 + (-2) + 4$
             $= -(-8) - 2(4) + (-2) + 4$
             $= 8 - 8 - 2 + 4$
             $= 2$

81. a. $h(6) = |5 - 2(6)| = |5 - 12| = |-7| = 7$

    b. $h\left(\dfrac{5}{2}\right) = \left|5 - 2\left(\dfrac{5}{2}\right)\right| = |5 - 5| = 0$

83. a. $s(-3) = \sqrt{(-3) + 3} = \sqrt{0} = 0$

    b. $s(6) = \sqrt{(6) + 3} = \sqrt{9} = 3$

85. a. $g(0) = \dfrac{(0)^3 - 2}{(0) - 2} = \dfrac{-2}{-2} = 1$

    b. $g(0) = \dfrac{(2)^3 - 2}{2 - 2} = \dfrac{8 - 2}{0} = \dfrac{6}{0}$ undefined

87. a. $A(4) = 6(4) = 24$
       The area is 24 square feet.

    b. $A(6.5) = 6(6.5) = 39$
       The area is 39 square feet.

89. a. $A(r) = \pi r^2$

    b. $A(12) = \pi(12)^2 = 144\pi \approx 452.4$
       The area is about 452.4 square yards.

91. a. $C(F) = \dfrac{5}{9}(F - 32)$

b. $C(-31) = \dfrac{5}{9}(-31 - 32) = \dfrac{5}{9}(-63) = -35$

The Celsius temperature that corresponds to $-31°F$ is $-35°C$.

93. a. $T(3) = -0.03(3)^2 + 1.5(3) + 14$
$= -0.27 + 4.5 + 14$
$= 18.23$
The temperature is $18.23°C$.

b. $T(12) = -0.03(12)^2 + 1.5(12) + 14$
$= -4.32 + 18 + 14$
$= 27.68$
The temperature is $27.68°C$.

95. a. $T(4) = -0.02(4)^2 - 0.34(4) + 80$
$= -0.32 - 1.36 + 80$
$= 78.32$
The temperature is $78.32°$.

b. $T(12) = -0.02(12)^2 - 0.34(12) + 80$
$= -2.88 - 4.08 + 80$
$= 73.04$
The temperature is $73.04°$.

97. a. $T(6) = \dfrac{1}{3}(6)^3 + \dfrac{1}{2}(6)^2 + \dfrac{1}{6}(6)$
$= 72 + 18 + 1$
$= 91$
If the base is 6 by 6 oranges, then there will be 91 oranges in the pyramid.

b. $T(8) = \dfrac{1}{3}(8)^3 + \dfrac{1}{2}(8)^2 + \dfrac{1}{6}(8)$
$= \dfrac{512}{3} + 32 + \dfrac{4}{3}$
$= 204$
If the base is 8 by 8 oranges, then there will be 204 oranges in the pyramid.

99. Answers will vary. One possible interpretation: The person warms up slowly, possibly by walking for 5 minutes, then begins jogging slowly over a period of 5 minutes. For the next 15 minutes, the person jogs at a steady pace. For the next 5 minutes, he walks slowly and his heart rate decreases to his normal resting heart rate. The rate stays the same for the next 5 minutes.

101. Answers will vary. One possible interpretation: The man walks on level ground, about 30 feet above sea level, for 5 minutes. For the next 5 minutes he walks uphill to 45 feet above sea level. For 5 minutes he walks on level ground then walks quickly downhill for 3 minutes to an elevation of 20 feet above sea level. For 7 minutes he walks on level ground. Then he walks quickly uphill.

103. Answers may vary. One possible interpretation: A woman drives in stop-and-go traffic for five minutes. Then she drives on the highway for fifteen minutes, gets off onto a country road for a few minutes, stops for a couple of minutes, and returns to stop-and-go traffic.

105. a. Yes, both graphs pass the vertical line test.

b. The independent variable is the year.

c. $f(2005) = \$218{,}600$.

d. $g(2005) = \$865{,}000$.

e. percent increase $= \dfrac{218{,}600 - 89{,}300}{89{,}300}$
≈ 1.448
$\approx 144.8\%$
The percent increase from 1988 through 2005 was about 144.8%.

107. a. Yes, both graphs pass the vertical line test.

b. $f(1998) \approx 6.0$ million viewers.

c. $g(1998) \approx 4.4$ million viewers.

d. Yes, both graphs are approximately linear from 1998 to 2005. The number of Today Show viewers is remained relatively constant, while the number of Good Morning America viewers increased at a relatively constant rate.

e. The two shows will have the same number of viewers around the year 2006 or 2007.

109. a.

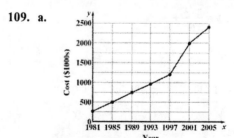

b. No, the points don't lie on a straight line.

c. The cost of a 30-second commercial in 2004 was about $2,300,000.

111. a.

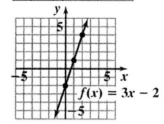

b. $f(40,000) = -0.00004(40,000) + 4.25$

$= -1.6 + 4.25$

$= 2.65$

The cost of a bushel of soybeans if 40,000 bushels are produced is approximately $2.65 per bushel.

114. $3x - 2 = \frac{1}{3}(3x - 3)$

$3x - 2 = \frac{1}{3}(3x) - \frac{1}{3}(3)$

$3x - 2 = x - 1$

$3x - x = -1 + 2$

$2x = 1$

$x = \frac{1}{2}$

The solution set is $\left\{\frac{1}{2}\right\}$.

115. $E = a_1 p_1 + a_2 p_2 + a_3 p_3$

$E - a_1 p_1 - a_3 p_3 = a_2 p_2$

$p_2 = \frac{E - a_1 p_1 - a_3 p_3}{a_2}$

116. $3x + 6y = 9$

$6y = -3x + 9$

$y = \frac{-3x + 9}{6}$

$y = \frac{-3x}{6} + \frac{9}{6}$

$y = -\frac{1}{2}x + \frac{3}{2}$

The slope is $m = -\frac{1}{2}$ and the y-intercept is $\left(0, \frac{3}{2}\right)$.

117. $\frac{3x^2 - 16x - 12}{3x^2 - 10x - 8} \div \frac{x^2 - 7x + 6}{3x^2 - 11x - 4}$

$= \frac{3x^2 - 16x - 12}{3x^2 - 10x - 8} \cdot \frac{3x^2 - 11x - 4}{x^2 - 7x + 6}$

$= \frac{(3x + 2)(x - 6)}{(3x + 2)(x - 4)} \cdot \frac{(3x + 1)(x - 4)}{(x - 1)(x - 6)}$

$= \frac{\cancel{(3x+2)}\ \cancel{(x-6)}}{\cancel{(3x+2)}\ \cancel{(x-4)}} \cdot \frac{(3x + 1)\cancel{(x-4)}}{(x - 1)\cancel{(x-6)}}$

$= \frac{3x + 1}{x - 1}$

Exercise Set 8.3

1. The graph of a linear function will be a straight line.

3. To find the x-intercept, set $y = 0$ and solve for x. To find the y-intercept, set $x = 0$ and solve for y.

5. To solve an equation in one variable graphically, graph both sides of the equation. The solution is the x-coordinate of the point of intersection.

7. $f(x) = 3x - 2$

x	calculation	$f(x)$
0	$3(0) - 2$	-2
1	$3(1) - 2$	1
2	$3(2) - 2$	4

9. $g(x) = -\frac{2}{3}x + 4$

x	calculation	$g(x)$
-3	$-\frac{2}{3}(-3) + 4$	6
0	$-\frac{2}{3}(0) + 4$	4
3	$-\frac{2}{3}(0) + 4$	2

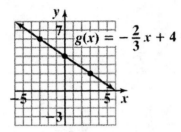

11. $f(x) = \dfrac{3}{4}x + 1$

x	calculation	$f(x)$
-4	$\dfrac{3}{4}(-4)+1$	-2
0	$\dfrac{3}{4}(0)+1$	1
4	$\dfrac{3}{4}(4)+1$	4

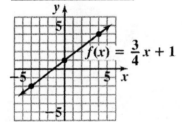

13. $f(x) = 3x - 6$

For the y-intercept, set $x = 0$:
$f(0) = 3(0) - 6 = 0 - 6 = -6$
The y-intercept is $(0, -6)$.
For the x-intercept, set $f(x) = 0$ and solve for x:
$0 = 3x - 6$
$6 = 3x$
$2 = x$
The x-intercept is $(2, 0)$.

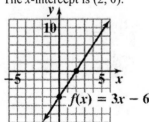

15. $f(x) = 2x + 3$

For the y-intercept, set $x = 0$:
$f(0) = 2(0) + 3 = 0 + 3 = 3$
The y-intercept is $(0, 3)$.
For the x-intercept, set $f(x) = 0$ and solve for x:
$0 = 2x + 3$
$-3 = 2x$
$-\dfrac{3}{2} = x$

The x-intercept is $\left(-\dfrac{3}{2}, 0\right)$.

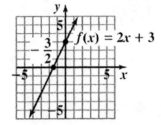

17. $g(x) = 4x - 8$

For the y-intercept, set $x = 0$:
$g(0) = 4(0) - 8 = 0 - 8 = -8$
The y-intercept is $(0, -8)$.
For the x-intercept, set $g(x) = 0$ and solve for x:
$0 = 4x - 8$
$8 = 4x$
$2 = x$
The x-intercept is $(2, 0)$.

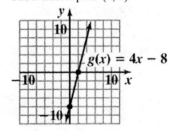

241

19. $s(x) = \dfrac{4}{3}x + 3$

For the y-intercept, set $x = 0$:

$$s(0) = \dfrac{4}{3}(0) + 3 = 0 + 3 = 3$$

The y-intercept is $(0, 3)$.

For the x-intercept, set $s(x) = 0$ and solve for x:

$$0 = \dfrac{4}{3}x + 3$$

$$-\dfrac{4}{3}x = 3$$

$$-\dfrac{3}{4}\left(-\dfrac{4}{3}x\right) = -\dfrac{3}{4}(3)$$

$$x = -\dfrac{9}{4}$$

The x-intercept is $\left(-\dfrac{9}{4}, 0\right)$.

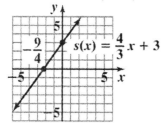

21. $h(x) = -\dfrac{6}{5}x + 2$

For the y-intercept, set $x = 0$:

$$h(0) = -\dfrac{6}{5}(0) + 2 = 0 + 2 = 2$$

The y-intercept is $(0, 2)$.

For the x-intercept, set $h(x) = 0$ and solve for x:

$$0 = -\dfrac{6}{5}x + 2$$

$$\dfrac{6}{5}x = 2$$

$$\dfrac{5}{6}\left(\dfrac{6}{5}x\right) = \dfrac{5}{6}(2)$$

$$x = \dfrac{5}{3}$$

The x-intercept is $\left(\dfrac{5}{3}, 0\right)$.

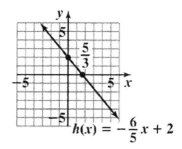

23. $g(x) = -\dfrac{1}{2}x + 2$

For the y-intercept, set $x = 0$:

$$g(0) = -\dfrac{1}{2}(0) + 2 = 0 + 2 = 2$$

The y-intercept is $(0, 2)$.

For the x-intercept, set $g(x) = 0$ and solve for x:

$$0 = -\dfrac{1}{2}x + 2$$

$$\dfrac{1}{2}x = 2$$

$$2\left(\dfrac{1}{2}x\right) = 2(2)$$

$$x = 4$$

The x-intercept is $(4, 0)$.

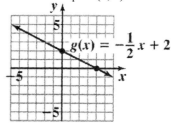

25. $w(x) = \dfrac{1}{3}x - 2$

For the y-intercept, set $x = 0$:

$$w(0) = \dfrac{1}{3}(0) - 2 = 0 - 2 = -2$$

The y-intercept is $(0, -2)$.

For the x-intercept, set $w(x) = 0$ and solve for x:

$$0 = \dfrac{1}{3}x - 2$$

$$2 = \dfrac{1}{3}x$$

$$3(2) = 3\left(\dfrac{1}{3}x\right)$$

$$6 = x$$

The x-intercept is $(6, 0)$.

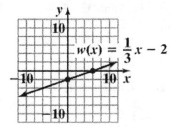

27. $s(x) = -\dfrac{4}{3}x + 48$

For the y-intercept, set $x = 0$:

$s(0) = -\dfrac{4}{3}(0) + 48 = 0 + 48 = 48$

The y-intercept is $(0, 48)$.
For the x-intercept, set $s(x) = 0$ and solve for x:

$0 = -\dfrac{4}{3}x + 48$

$\dfrac{4}{3}x = 48$

$\dfrac{3}{4}\left(\dfrac{4}{3}x\right) = \dfrac{3}{4}(48)$

$x = 36$

The x-intercept is $(36, 0)$.

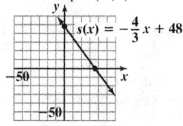

29. The equation is $d = 30t$. To graph, plot a few points.

t	Calculation	d
0	30(0)	0
1	30(1)	30
4	30(4)	120

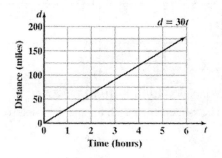

31. **a.** $p = 60x - 80{,}000$. To graph, plot a few points.

x	Calculation	p
0	60(0) − 80,000	−80,000
2500	60(2500) − 80,000	70,000
5000	60(5000) − 80,000	220,000

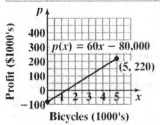

b. To break even, the profit would be zero. That is, set $p = 0$ and solve for x:

$0 = 60x - 80{,}000$

$80{,}000 = 60x$

$x = \dfrac{-80{,}000}{-60} \approx 1333$

The company must sell about 1,300 bicycles to break even.

c. To earn a profit of \$150,000, set $p = 150{,}000$ and solve for x:

$150{,}000 = 60x - 80{,}000$

$230{,}000 = 60x$

$x = \dfrac{230{,}000}{60} \approx 3833$

The company must sell about 3,800 bicycles to make a \$150,000 profit.

33. **a.** $s(x) = 500 + 0.15x$

b. To graph, plot a few points.

x	Calculation	s
0	500 + 0.15(0)	500
1000	500 + 0.15(1000)	650
5000	500 + 0.15(5000)	1250

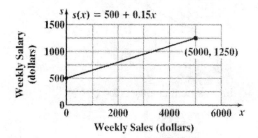

c. For weekly sales of $3000,
$$s(3000) = 500 + 0.15(3000)$$
$$= 500 + 450$$
$$= 950$$
Her salary is $950.

d. For a salary of $1100, set $s = 1100$ and solve for x.
$$1100 = 500 + 0.15x$$
$$600 = 0.15x$$
$$4000 = x$$
Her weekly sales are $4000.

35. a. There is only one y-value for each x-value. (It passes the vertical line test.)

b. The independent variable is length. The dependent variable is weight.

c. Yes, the graph of weight versus length is approximately linear.

d. The weight of the average girl who is 85 centimeters long is 11.5 kilograms.

e. The average length of a girl with a weight of 7 kilograms is 65 centimeters.

f. For a girl 95 centimeters long, the weights 12.0 – 15.5 kilograms are considered normal.

g. As the lengths increase, the normal range of weights increases. Yes, this is expected: as the girls grow, it is reasonable that their weights will vary more.

37. The x- and y-intercepts of a graph will be the same when the graph goes through the origin.

39. Answers may vary. One possible answer is, $f(x) = 4$ is a function whose graph has no x-intercept but has a y-intercept of $(0, 4)$.

41. The x- and y-intercepts will both be $(0, 0)$.

43. a.

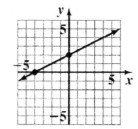

b. vertical change $= 2 - 0 = 2$

c. horizontal change $= 0 - (-4) = 4$

d. $\dfrac{\text{vertical change}}{\text{horizontal change}} = \dfrac{2}{4} = \dfrac{1}{2}$

The ratio represents the slope of the line.

45. Graph $Y_1 = 2x + 5$ and $Y_2 = 8x - 1$, and find the intersection.

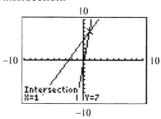

The solution is $x = 1$

47. Graph $Y_1 = 0.3(x + 5)$ and $Y_2 = -0.6(x + 2)$, and find the intersection.

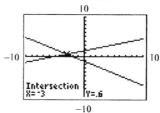

The solution is $x = -3$.

49. Let $Y_1 = 2(x + 3.2)$.

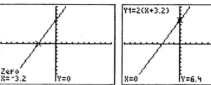

Window: $-10, 10, 1, -10, 10, 1$

The x-intercept is $(-3.2, 0)$, and the y-intercept is $(0, 6.4)$.

51. To use the graphing calculator, we must rewrite the equation in the form $y = f(x)$.
$$-4x - 3.2y = 8$$
$$-3.2y = 4x + 8$$
$$y = \frac{4x + 8}{-3.2}$$
$$y = -1.25x - 2.5$$
Let $Y_1 = -1.25x - 2.5$.

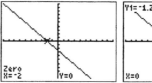 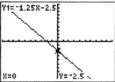

Window: $-10, 10, 1, -10, 10, 1$

The x-intercept is $(-2, 0)$, and the y-intercept is $(0, -2.5)$.

53. $(7x^2 - 3x - 100) + (-4x^2 - 9x + 12)$

$= 7x^2 - 3x - 100 - 4x^2 - 9x + 12$

$= 7x^2 - 4x^2 - 3x - 9x - 100 + 12$

$= 3x^2 - 12x - 88$

54. $3x^2 - 12x - 96 = 3(x^2 - 4x - 32)$

$= 3(x + 4)(x - 8)$

55. $\dfrac{x-2}{x^2-4} = \dfrac{x-2}{(x+2)(x-2)} = \dfrac{1}{x+2}$

56. $x + \dfrac{24}{x} = 10$

$x\left(x + \dfrac{24}{x}\right) = x(10)$

$x^2 + 24 = 10x$

$x^2 - 10x + 24 = 0$

$(x - 4)(x - 6) = 0$

$x - 4 = 0$ or $x - 6 = 0$

$x = 4$ $\qquad x = 6$

Note that neither 4 nor 6 is extraneous. In other words, both 4 and 6 check. The solutions to the equation are $x = 4$ and $x = 6$.

Mid-Chapter Test: 8.1-8.3

1.
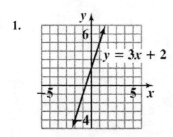

2.
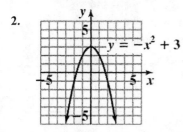

3.
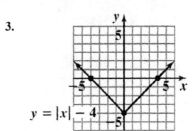

4.
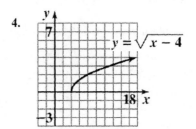

5. a. A relation is any set of ordered pairs.

b. A function is a correspondence between a first set of elements, the domain, and a second set of elements, the range, such that each element of the domain corresponds to exactly one element in the range.

c. No, every relation is not a function. A relation can have two ordered pairs with the same first element but a function cannot.

d. Yes, every function is also a relation. A function is a set of ordered pairs so it is a relation.

6. The relation is a function.
Domain: $\{1, 2, 7, -5\}$
Range: $\{5, -3, -1, 6\}$

7. The relation is not a function.
Domain: $\{x \mid -2 \le x \le 2\}$
Range: $\{y \mid -4 \le y \le 4\}$

8. The relation is a function.
Domain: $\{x \mid -5 \le x \le 3\}$
Range: $\{y \mid -1 \le y \le 3\}$

9. $E = \{0, 9, 18, 27, 36, \ldots\}$

10. $\{z \mid z \le -5\}$

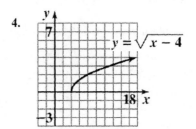

11. $g(-2) = 2(-2)^2 + 8(-2) - 13$
$= 2(4) + 8(-2) - 13$
$= 8 - 16 - 13$
$= -21$

12. $h(3) = -6(3)^2 + 3(3) + 150$
$= -6(9) + 3(3) + 150$
$= -54 + 9 + 150$
$= 105$
At $t = 3$ seconds, the apple is 105 feet above the ground.

13. $f(x) = 2x + 3$
For the *y*-intercept, set $x = 0$:
$f(0) = 2(0) + 3 = 0 + 3 = 3$
The *y*-intercept is $(0, 3)$.
For the *x*-intercept, set $f(x) = 0$ and solve for *x*:
$0 = 2x + 3$
$-3 = 2x$
$-\dfrac{3}{2} = x$
The *x*-intercept is $\left(-\dfrac{3}{2}, 0\right)$.

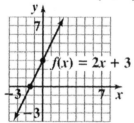

14. $g(x) = \dfrac{1}{2}x - 4$
For the *y*-intercept, set $x = 0$:
$g(0) = \dfrac{1}{2}(0) - 4 = 0 - 4 = -4$
The *y*-intercept is $(0, -4)$.
For the *x*-intercept, set $g(x) = 0$ and solve for *x*:
$0 = \dfrac{1}{2}x - 4$
$4 = \dfrac{1}{2}x$
$2(4) = 2\left(\dfrac{1}{2}\right)x$
$8 = x$
The *x*-intercept is $(8, 0)$.

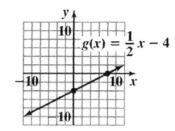

15. $m(x) = -\dfrac{1}{3}x + 2$
For the *y*-intercept, set $x = 0$:
$m(0) = -\dfrac{1}{3}(0) + 2 = 0 + 2 = 2$
The *y*-intercept is $(0, 2)$.
For the *x*-intercept, set $m(x) = 0$ and solve for *x*:
$0 = -\dfrac{1}{3}x + 2$
$\dfrac{1}{3}x = 2$
$3\left(\dfrac{1}{3}\right)x = 3(2)$
$x = 6$
The *x*-intercept is $(6, 0)$.

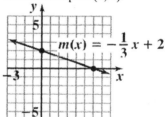

16. a. Let $x = 0$: $p(0) = 30(0) - 660$
$= 0 - 660$
$= -660$
Let $x = 40$: $p(40) = 30(40) - 660$
$= 1200 - 660$
$= 540$
Plots the points $(0, -660)$ and $(40, 540)$ and connect to obtain the graph.

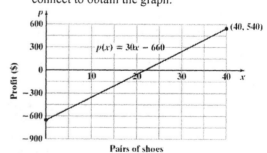

b. $p(x) = 0$

$$30x - 660 = 0$$
$$30x = 660$$
$$x = 22$$

The company will break even if 22 pairs of shoes are produced and sold.

c. $p(x) = 360$

$$30x - 660 = 360$$
$$30x = 1020$$
$$x = 34$$

The company will make a \$360 daily profit if 34 pairs of shoes are produced and sold.

Exercise Set 8.4

1. When the slope is given as a rate of change it means the change in y for a unit change in x.

3. **a.**   If a graph is translated up 3 units, it is lifted or moved up 3 units.

    b.   If the y-intercept is $(0, -4)$ and the graph is translated up 5 units, the new y-intercept will be at $y = -4 + 5 = 1$. The new y-intercept is $(0, 1)$.

5. Two lines are perpendicular if their slopes are negative reciprocals, or if one line is horizontal and the other is vertical.

7. The slope is negative and y decreases 6 units when x increases 2 units. Thus, $m = -\dfrac{6}{2} = -3$.
    The line crosses the y-axis at 0 so $b = 0$. Hence, $m = -3$ and $b = 0$, and the equation of the line is $f(x) = -3x + 0$ or $f(x) = -3x$.

9. The slope is undefined since the change in x is 0. The equation of this vertical line is $x = -2$.

11. The slope is negative and y decreases 1 unit when x increases 3 units. Thus, $m = -\dfrac{1}{3}$. The line crosses the y-axis at 2 so $b = 2$. Hence, $m = -\dfrac{1}{3}$ and $b = 2$, and the equation of the line is $f(x) = -\dfrac{1}{3}x + 2$.

13. The slope is negative and y decreases 15 units when x increases 10 units. Thus, $m = -\dfrac{15}{10} = -\dfrac{3}{2}$. The line crosses the y-axis at 15 so $b = 15$. Hence, $m = -\dfrac{3}{2}$ and $b = 15$, and the equation of the line is $f(x) = -\dfrac{3}{2}x + 15$.

15. The graph of $f(x) = -5$ is a horizontal line with y-intercept $(0, -5)$.

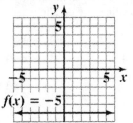

17. The graph of $g(x) = -15$ is a horizontal line with y-intercept $(0, -15)$.

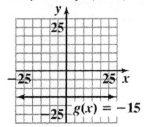

19. $m_1 = \dfrac{2-0}{0-2} = \dfrac{2}{-2} = -1$; $m_2 = \dfrac{3-0}{0-3} = \dfrac{3}{-3} = -1$
    Since their slopes are equal, l_1 and l_2 are parallel.

21. $m_1 = \dfrac{7-6}{5-4} = \dfrac{1}{1} = 1$; $m_2 = \dfrac{4-(-1)}{1-(-1)} = \dfrac{5}{2}$
    Since their slopes are different and since the product of their slopes is not -1, l_1 and l_2 are neither parallel nor perpendicular.

23. $m_1 = \dfrac{-2-2}{-1-3} = \dfrac{-4}{-4} = 1$; $m_2 = \dfrac{-1-0}{3-2} = \dfrac{-1}{1} = -1$
    Since the product of their slopes is -1, the lines are perpendicular.

25. $y = \dfrac{1}{5}x + 9$, so $m_1 = \dfrac{1}{5}$

$y = -5x + 2$, so $m_2 = -5$

Since the product of their slopes is -1, the lines are perpendicular.

27. $4x + 2y = 8$ $8x = 4 - 4y$

$2y = -4x + 8$ $4y = -8x + 4$

$y = -2x + 4$ $y = -2x + 1$

$m_1 = -2$ $m_2 = -2$

Since their slopes are equal and their y-intercepts are different, the lines are parallel.

29. $2x - y = 4$ $-x + 4y = 4$

$-y = -2x + 4$ $4y = x + 4$

$y = 2x - 4$ $y = \dfrac{1}{4}x + 1$

$m_1 = 2$ $m_2 = \dfrac{1}{4}$

Since their slopes are different and since the product of their slopes is not -1, the lines are neither parallel nor perpendicular.

31. $y = \dfrac{1}{2}x - 6$ $-4y = 8x + 15$

$m_1 = \dfrac{1}{2}$ $y = \dfrac{8x + 15}{-4}$

               $y = -2x - \dfrac{15}{4}$

               $m_2 = -2$

Since the product of their slopes is -1, the lines are perpendicular.

33. $y = \dfrac{1}{2}x + 6$ $-2x + 4y = 8$

$m_1 = \dfrac{1}{2}$ $4y = 2x + 8$

              $y = \dfrac{1}{2}x + 2$

              $m_2 = \dfrac{1}{2}$

Since their slopes are equal and their y-intercepts are different, the lines are parallel.

35. $x - 2y = -9$ $y = x + 6$

$-2y = -x - 9$ $m_2 = 1$

$y = \dfrac{-x - 9}{-2}$

$y = \dfrac{1}{2}x + \dfrac{9}{2}$

$m_1 = \dfrac{1}{2}$

Since their slopes are different and since the product of their slopes is not -1, the lines are neither parallel nor perpendicular.

37. The slope of the given line, $y = 2x + 4$, is $m_1 = 2$. So $m_2 = 2$. Now use the point-slope form with $m = 2$ and $(x_1, y_1) = (2, 5)$ to obtain the slope-intercept form.

$y - y_1 = m(x - x_1)$

$y - 5 = 2(x - 2)$

$y - 5 = 2x - 4$

$y = 2x + 1$

39. Find the slope of the given line.

$2x - 5y = 7$

$-5y = -2x + 7$

$y = \dfrac{-2x + 7}{-5}$

$y = \dfrac{2}{5}x - \dfrac{7}{5}$

$m_1 = \dfrac{2}{5}$, so $m_2 = \dfrac{2}{5}$. Now use the point-slope form with $m = \dfrac{2}{5}$ and $(x_1, y_1) = (-3, -5)$ to obtain the standard form.

$y - y_1 = m(x - x_1)$

$y - (-5) = \dfrac{2}{5}(x - (-3))$

$y + 5 = \dfrac{2}{5}(x + 3)$

$y + 5 = \dfrac{2}{5}x + \dfrac{6}{5}$

$5(y + 5) = 5\left(\dfrac{2}{5}x + \dfrac{6}{5}\right)$

$5y + 25 = 2x + 6$

$-2x + 5y = -19$

$2x - 5y = 19$

41. Find the slope of the line with the given intercepts:

$m = \dfrac{5-0}{0-3} = -\dfrac{5}{3}$. The *y*-intercept (0, 5), so *b* = 5.

Thus, the slope-intercept form of the equation is

$y = -\dfrac{5}{3}x + 5$.

43. The slope of the given line $y = \dfrac{1}{3}x + 1$ is

$m_1 = \dfrac{1}{3}$. So m_2 is the negative reciprocal, or

$m_2 = -\dfrac{1}{m_1} = -\dfrac{1}{\frac{1}{3}} = -1 \cdot 3 = -3$. Use the point-

slope form with $m = -3$ and $(x_1, y_1) = (5, -2)$ to obtain the function notation.

$$y - y_1 = m(x - x_1)$$
$$y - (-2) = -3(x - 5)$$
$$y + 2 = -3x + 15$$
$$y = -3x + 13$$
$$f(x) = -3x + 13$$

45. Find the slope of the line with the given intercepts:

$m_1 = \dfrac{-3-0}{0-2} = \dfrac{-3}{-2} = \dfrac{3}{2}$. So m_2 is the negative

reciprocal, or $m_2 = -\dfrac{1}{m_1} = -\dfrac{1}{\frac{3}{2}} = -\dfrac{2}{3}$. Now use the

point-slope form with $m = -\dfrac{2}{3}$ and

$(x_1, y_1) = (6, 2)$ and obtain the slope-intercept form.

$$y - y_1 = m(x - x_1)$$
$$y - 2 = -\dfrac{2}{3}(x - 6)$$
$$y - 2 = -\dfrac{2}{3}x + 4$$
$$y = -\dfrac{2}{3}x + 6$$

47. $m = \dfrac{2-4}{-4-6} = \dfrac{-2}{-10} = \dfrac{1}{5}$

For a 1-unit change in *x*, *y* changes $\dfrac{1}{5}$ or 0.2 unit.

49. a. Let $(x_1, y_1) = (2005, 19.7)$ and

$(x_2, y_2) = (2006, 31.0)$. Then

$m = \dfrac{31.0 - 19.7}{2006 - 2005} = \dfrac{11.3}{1} = 11.3$.

b. positive

c. Let $(x_1, y_1) = (2004, 7.3)$ and

$(x_2, y_2) = (2008, 35.6)$. Then

$m = \dfrac{35.6 - 7.3}{2008 - 2004} = \dfrac{28.3}{4} = 7.075$.

The average rate of change in digital TV sales from 2004 to 2008 is 7.075 million sales per year.

51. a, b.

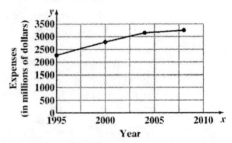

c. From 1995 to 2000,

$m = \dfrac{2876 - 2257}{2000 - 1995} = \dfrac{619}{5} = 123.8$

From 2000 to 2004,

$m = \dfrac{3133 - 2876}{2004 - 2000} = \dfrac{257}{4} = 64.25$

From 2004 to 2008,

$m = \dfrac{3260 - 3133}{2008 - 2004} = \dfrac{127}{4} = 31.75$

d. The greatest average rate of change occurred during the period 1995 to 2000, because the largest slope corresponds to these years.

53. a. If *x* is the number of years after age 20, two points on the graph are (0, 200) and (50, 150).

The slope is $m = \dfrac{150 - 200}{50 - 0} = \dfrac{-50}{50} = -1$ and

the *y*-intercept is (0, 200), so *b* = 200. Thus, the equation for the line is $h(x) = -1x + 200$, or $h(x) = -x + 200$.

b. Note that a 34-year-old man is represented by $x = 14$. Now, $h(14) = -(14) + 200 = 186$. Thus, the maximum recommended heart rate for a 34-year-old man is 186 beats per minute.

55. a. Note that $t = 0$ represents 1997 and $t = 7$ represents 2004. Thus, two ordered pairs of the function are (0, 159.5) and (7, 294.9). The slope of the line is

$$m = \frac{294.9 - 159.50}{2004 - 1997} = \frac{135.4}{7} \approx 19.34$$

The y-intercept is (0, 159.5), so $b = 159.5$. The linear function is $M(t) \approx 19.34t + 159.5$.

b. Note that 2003 is represented by $t = 6$:
$$M(6) = 19.34(6) + 159.5$$
$$= 116.04 + 159.5$$
$$= 275.54$$

The function indicates that Medicaid spending for 2003 was \$275.54 billion. This estimate is similar to the spending shown in the graph.

c. Note that 2010 is represented by $t = 13$:
$$M(13) = 19.34(13) + 159.5$$
$$= 251.42 + 159.5$$
$$= 410.92$$

The function indicates that Medicaid spending for 2010 will be \$410.92 billion.

d.
$$M(t) = 340$$
$$19.34t + 159.5 = 340$$
$$19.34t = 180.5$$
$$t = \frac{180.5}{19.34} \approx 9.3$$

Note that $t = 9$ represents the year 2006. Thus, according to the function, Medicaid spending reached \$340 billion during the year 2006.

57. a. Note that $t = 0$ represents 2001 and $t = 3$ represents 2004. Thus, two ordered pairs of the function are (0, 19.4) and (3, 16.1). The slope of the line is

$$m = \frac{16.1 - 19.4}{3 - 0} = \frac{-3.3}{3} = -1.1$$

The y-intercept is (0, 19.4), so, $b = 19.4$. The linear function is $P(t) = -1.1t + 19.4$.

b. Negative. The negative slope occurs because the percent of teenagers who used illicit drugs decreased from 2001 through 2004.

c. Note that 2003 is represented by $t = 2$:
$$P(2) = -1.1(2) + 19.4 = -2.2 + 19.4 = 17.2$$

According to the function, 17.2% of teenagers used illicit drugs in 2003. This estimate is close to the percent shown in the graph.

d. Note that 2010 is represented by $t = 9$:
$$P(9) = -1.1(9) + 19.4 = -9.9 + 19.4 = 9.5$$

According to the function, 9.5% of teenagers will be using illicit drugs in 2010.

59. a. Note that $t = 0$ represents 1995 and that $t = 9$ represents 2004. Thus, two ordered pairs of the function are (0, 110500) and (9, 185200). The slope of the line is

$$m = \frac{185,200 - 110,500}{9 - 0} = \frac{74,700}{9} = 8300$$

The y-intercept is (0, 110500), so $b = 110,500$. The linear function is $P(t) = 8300t + 110,500$.

b. Note that $t = 5$ represents 2000:
$$P(5) = 8300(5) + 110,500$$
$$= 41,500 + 110,500$$
$$= 152,000$$

The median home sale price in the year 2000 was about \$152,000.

c. Note that $t = 15$ represents 2010:
$$P(15) = 8300(15) + 110,500$$
$$= 124,500 + 110,500$$
$$= 235,000$$

The median home sale price in the year 2010 will be about \$235,000.

d.
$$8300t + 110,500 = 200,000$$
$$8300t = 89,500$$
$$t = \frac{89,500}{8300} \approx 10.8$$

Now $t \approx 10.8$ corresponds to the year 2005. Thus, assuming the trend continues, the median home sale price will reach \$200,000 near the end of the year 2005.

61. a. To find the function, use the points (2.5, 210) and (6, 370) to determine the slope.

$$m = \frac{370 - 210}{6 - 2.5} = \frac{160}{3.5} \approx 45.7$$

Now use the point-slope form with $m = 45.7$ and $(s_1, C_1) = (2.5, 210)$
$$C - C_1 = m(s - s_1)$$
$$C - 210 = 45.7(s - 2.5)$$
$$C - 210 = 45.7s - 114.25$$
$$C = 45.7s + 95.75$$
$$C(s) \approx 45.7s + 95.8$$

b. $C(5) \approx 45.7(5) + 95.8 = 228.5 + 95.8 = 324.3$
The average person will burn about 324.3 calories.

63. a. To find the function, use the points (200, 50) and (300, 30) to determine the slope:

$$m = \frac{30 - 50}{300 - 200} = \frac{-20}{100} = -0.20$$

Now use the point-slope form with $m = -0.20$ and $(p_1, d_1) = (200, 50)$

$$d - d_1 = m(p - p_1)$$
$$d - 50 = -0.20(p - 200)$$
$$d - 50 = -0.20p + 40$$
$$d = -0.20p + 90$$
$$d(p) = -0.20p + 90$$

b. $d(260) = -0.20(260) + 90 = -52 + 90 = 38$

When the price is \$260, the demand will be 38 DVD players.

c.
$$d(p) = 45$$
$$-0.20p + 90 = 45$$
$$-0.20p = -45$$
$$p = 225$$

In order to have a demand of 45 DVD players, the price should be \$225.

65. a. To find the function, use the points (2, 130) and (4, 320) to determine the slope:

$$m = \frac{320 - 130}{4 - 2} = \frac{190}{2} = 95.$$

Now use the point-slope form with $m = 95$ and $(p_1, s_1) = (2, 130)$.

$$s - s_1 = m(p - p_1)$$
$$s - 130 = 95(p - 2.00)$$
$$s - 130 = 95p - 190$$
$$s = 95p - 60$$
$$s(p) = 95p - 60$$

b. $s(2.80) = 95(2.80) - 60 = 266 - 60 = 206$

If the price is \$2.80, the supply will be 206 kites.

c.
$$s(p) = 255$$
$$95p - 60 = 225$$
$$95p = 285$$
$$p = 3$$

For the supply to be 225 kites, the price should be \$3.00.

67. a. To find the function, use the points (80, 1000) and (200, 2500) to determine the slope:

$$m = \frac{2500 - 1000}{200 - 80} = \frac{1500}{120} = 12.5.$$

Now use the point-slope form with $m = 12.5$ and $(t_1, i_1) = (80, 1000)$.

$$i - i_1 = m(t - t_1)$$
$$i - 1000 = 12.5(t - 80)$$
$$i - 1000 = 12.5t - 1000$$
$$i = 12.5t$$
$$i(t) = 12.5t$$

b. $i(120) = 12.5(120) = 1500$

If 120 tickets are sold, the income will be \$1500.

c.
$$i(t) = 2200$$
$$12.5t = 2200$$
$$t = 176$$

If the income is \$2200, then 176 tickets were sold.

69. a. To find the function, use the points (2000, 30) and (4000, 50) to determine the slope:

$$m = \frac{50 - 30}{4000 - 2000} = \frac{20}{2000} = 0.01.$$

Now use the point-slope form with $m = 0.01$ and $(w_1, r_1) = (2000, 30)$.

$$r - r_1 = m(w - w_1)$$
$$r - 30 = 0.01(w - 2000)$$
$$r - 30 = 0.01w - 20$$
$$r = 0.01w + 10$$
$$r(w) = 0.01w + 10$$

b.
$$r(3613) = 0.01(3613) + 10$$
$$= 36.13 + 10$$
$$= 46.13$$

If the weight of the care is 3613 pounds, then the registration fee will be \$46.13.

c.
$$r(w) = 60$$
$$0.01w + 10 = 60$$
$$0.01w = 50$$
$$w = 5000$$

If the registration fee is \$60, then the weight of the car is 5000 pounds.

71. a. To find the function, use the points (50, 36.0) and (70, 18.7) to determine the slope:

$$m = \frac{18.7 - 36.0}{70 - 50} = \frac{-17.3}{20} = -0.865 .$$

Now use the point-slope form with $m = -0.865$ and $(a_1, y_1) = (50, 36.0)$:

$$y - y_1 = m(a - a_1)$$
$$y - 36.0 = -0.865(a - 50)$$
$$y - 36.0 = -0.865a + 43.25$$
$$y = -0.865a + 79.25$$
$$y(a) = -0.865a + 79.25$$

b. $y(37) = -0.865(37) + 79.25$
$$= -32.005 + 79.25$$
$$= 47.245$$

The additional life expectancy will be about 47.2 years.

c. $y(a) = 25$
$$-0.865a + 79.25 = 25$$
$$-0.865a = -54.25$$
$$a \approx 62.7$$

In order to have an additional life expectancy of 25 years, one would need to be currently about 62.7 years old.

73. a. To find the function, use the points (18, 14) and (36, 17.4) to determine the slope:

$$m = \frac{17.4 - 14}{36 - 18} = \frac{3.4}{18} \approx 0.189 .$$

Now use the point-slope form with $m = 0.189$ and $(a_1, w_1) = (18, 14)$:

$$w - w_1 = m(a - a_1)$$
$$w - 14 = 0.189(a - 18)$$
$$w - 14 = 0.189a - 3.402$$
$$w = 0.189a + 10.598$$
$$w(a) \approx 0.189a + 10.6$$

b. $w(22) = 0.189(22) + 10.6$
$$= 4.158 + 10.6$$
$$= 14.758$$

A 22-month-old boy who is in the 95th percentile will weigh about 14.758 kg. This is close to the weight shown in the graph.

75. If the *y*-intercept is $(0, -13)$ and the line is translated up 17 units, the new *y*-intercept will be at $y = -13 + 17 = 4$. The *y*-intercept of the translated graph is (0, 4).

77. If the *y*-intercept is $(0, 6)$ and the line is translated down 9 units, the new *y*-intercept will be at $y = 6 - 9 = -3$. The *y*-intercept of the translated graph is $(0, -3)$.

79. a. The slope is 3 and the *y*-intercept is (0, 1), so the equation is $y = 3x + 1$.

b. The slope is 3 and the *y*-intercept is (0, –5), so the equation is $y = 3x - 5$.

81. a. The slope of the translated graph is 1.

b. If the *y*-intercept $b = -1$ is translated up 3 units, the *y*-intercept of the translated graph is at $y = -1 + 3 = 2$. The new *y*-intercept is (0, 2).

c. Using $m = 1$ and $b = 2$, the equation of the translated graph is $y = x + 2$.

83. First, rewrite the equation in the slope-intercept form by solving for *y* in terms of *x*.
$$3x - 2y = 6$$
$$-2y = -3x + 6$$
$$y = \frac{-3x + 6}{-2}$$
$$y = \frac{3}{2}x - 3$$

Thus, $m = \frac{3}{2}$ and $b = -3$. If the graph is translated down 4 units, then the *y*-intercept of the translated graph is at $y = -3 - 4 = -7$. Hence, the equation of the translated graph is $y = \frac{3}{2}x - 7$.

85. The *y*-intercept of $y = 3x + 6$ is 6. On the screen, the *y*-intercept is not 6. The *y*-intercept is wrong.

87. The slope of $y = \frac{1}{2}x + 4$ is $\frac{1}{2}$. On the screen, the slope is not $\frac{1}{2}$. The slope is wrong.

89. There are 91 steps and the total vertical distance is 1292.2 in. Thus, the average height of a step is $\frac{1292.2}{91} = 14.2$ inches. If the slope is 2.21875 and the average height, or "rise", is 14.2 inches, the average width, or "run", is found as follows:

$$\text{slope} = \frac{\text{rise}}{\text{run}} = \frac{\text{height}}{\text{width}}$$

$$2.21875 = \frac{14.2}{\text{width}}$$

$$\text{width} = \frac{14.2}{2.21875} = 6.4$$

The average width is 6.4 inches.

93.
$$\frac{-6^2 - 16 \div 2 \div |-4|}{5 - 3 \cdot 2 - 4 \div 2^2} = \frac{-36 - 16 \div 2 \div 4}{5 - 6 - 4 \div 4}$$
$$= \frac{-36 - 8 \div 4}{5 - 6 - 1}$$
$$= \frac{-36 - 2}{-1 - 1}$$
$$= \frac{-38}{-2}$$
$$= 19$$

94.
$$2.6x - (-1.4x + 3.4) = 6.2$$
$$2.6x + 1.4x - 3.4 = 6.2$$
$$4.0x - 3.4 = 6.2$$
$$4.0x = 9.6$$
$$x = 2.4$$

95.
$$\frac{3}{4}x + \frac{1}{5} = \frac{2}{3}(x - 2)$$
$$60\left[\frac{3}{4}x + \frac{1}{5}\right] = 60\left[\frac{2}{3}(x - 2)\right]$$
$$45x + 12 = 40(x - 2)$$
$$45x + 12 = 40x - 80$$
$$5x + 12 = -80$$
$$5x = -92$$
$$x = -\frac{92}{5}$$

96. a. A relation is any set of ordered pairs.

b. A function is a correspondence between a first set of elements, the domain, and a second set of elements, the range, such that each element of the domain corresponds to exactly one element in the range.

c. Answers will vary.

97. Domain: {3, 4, 5, 6}; Range: {−2, −1, 2, 3}

Exercise Set 8.5

1. Yes, $f(x) + g(x) = (f + g)(x)$ for all values of x. This is how addition of functions is defined.

3. $f(x)/g(x) = (f/g)(x)$ provided $g(x) \neq 0$. This is because division by zero is undefined.

5. No, $(f - g)(x) \neq (g - f)(x)$ for all values of x since subtraction is not commutative. For example, if $f(x) = x^2 + 1$ and $g(x) = x$, then
$$(f - g)(x) = f(x) - g(x)$$
$$= (x^2 + 1) - (x)$$
$$= x^2 - x + 1$$
$$(g - f)(x) = g(x) - f(x)$$
$$= (x) - (x^2 + 1)$$
$$= -x^2 + x - 1$$
So, $(f - g)(x) \neq (g - f)(x)$.

7. a. $(f + g)(-2) = f(-2) + g(-2) = (-3) + 5 = 2$

b. $(f - g)(-2) = f(-2) - g(-2) = (-3) - 5 = -8$

c. $(f \cdot g)(-2) = f(-2) \cdot g(-2) = (-3) \cdot 5 = -15$

d. $(f/g)(-2) = f(-2)/g(-2) = (-3)/5 = -\frac{3}{5}$

9. a. $(f + g)(x) = f(x) + g(x)$
$$= (x + 5) + (x^2 + x)$$
$$= x^2 + 2x + 5$$

b. $(f + g)(a) = a^2 + 2a + 5$

c. $(f + g)(2) = (2)^2 + 2(2) + 5$
$$= 4 + 2(2) + 5$$
$$= 4 + 4 + 5$$
$$= 13$$

11. a. $(f + g)(x) = f(x) + g(x)$
$$= (-3x^2 + x - 4) + (x^3 + 3x^2)$$
$$= x^3 + x - 4$$

b. $(f + g)(a) = a^3 + a - 4$

c. $(f + g)(2) = (2)^3 + (2) - 4$
$$= 8 + 2 - 4$$
$$= 6$$

13. a. $(f+g)(x) = f(x)+g(x)$
$$= (4x^3 - 3x^2 - x) + (3x^2 + 4)$$
$$= 4x^3 - x + 4$$

b. $(f+g)(a) = 4a^3 - a + 4$

c. $(f+g)(2) = 4(2)^3 - (2) + 4$
$$= 4(8) - 2 + 4$$
$$= 32 - 2 + 4$$
$$= 30 + 4$$
$$= 34$$

15. $f(2) = (2)^2 - 4 = 4 - 4 = 0$
$g(2) = -5(2) + 3 = -10 + 3 = -7$
$f(2) + g(2) = 0 + (-7) = -7$

17. $f(4) = (4)^2 - 4 = 16 - 4 = 12$
$g(4) = -5(4) + 3 = -20 + 3 = -17$
$f(4) - g(4) = 12 - (-17) = 29$

19. $f(3) = (3)^2 - 4 = 9 - 4 = 5$
$g(3) = -5(3) + 3 = -15 + 3 = -12$
$f(3) \cdot g(3) = 5(-12) = -60$

21. $f\left(\dfrac{3}{5}\right) = \left(\dfrac{3}{5}\right)^2 - 4 = \dfrac{9}{25} - \dfrac{100}{25} = -\dfrac{91}{25}$
$g\left(\dfrac{3}{5}\right) = -5\left(\dfrac{3}{5}\right) + 3 = -3 + 3 = 0$
$\dfrac{f\left(\dfrac{3}{5}\right)}{g\left(\dfrac{3}{5}\right)} = \dfrac{-\dfrac{91}{25}}{0}$ which is undefined.

23. $f(-3) = (-3)^2 - 4 = 9 - 4 = 5$
$g(-3) = -5(-3) + 3 = 15 + 3 = 18$
$g(-3) - f(-3) = 18 - 5 = 13$

25. $f(0) = (0)^2 - 4 = 0 - 4 = -4$
$g(0) = -5(0) + 3 = 0 + 3 = 3$
$g(0)/f(0) = 3/(-4) = -\dfrac{3}{4}$

27. $(f+g)(x) = f(x) + g(x)$
$$= (2x^2 - x) + (x - 6)$$
$$= 2x^2 - 6$$

29. $(f+g)(x) = 2x^2 - 6$
$(f+g)(2) = 2(2)^2 - 6 = 2(4) - 6 = 8 - 6 = 2$

31. $(f-g)(-2) = f(-2) - g(-2)$
$$= \left(2 \cdot (-2)^2 - (-2)\right) - \left((-2) - 6\right)$$
$$= (8+2) - (-8)$$
$$= 10 + 8$$
$$= 18$$

33. $(f \cdot g)(0) = f(0) \cdot g(0)$
$$= \left(2 \cdot 0^2 - 0\right) \cdot (0 - 6)$$
$$= (0)(-6)$$
$$= 0$$

35. $(f/g)(-1) = f(-1)/g(-1)$
$$= \left(2 \cdot (-1)^2 - (-1)\right) \Big/ \left((-1) - 6\right)$$
$$= (2+1)/(-7)$$
$$= 3/(-7)$$
$$= -\dfrac{3}{7}$$

37. $(g/f)(5) = g(5)/f(5)$
$$= (5-6) \Big/ \left(2 \cdot 5^2 - 5\right)$$
$$= (-1)/(50-5)$$
$$= (-1)/45$$
$$= -\dfrac{1}{45}$$

39. $(g-f)(x) = g(x) - f(x)$
$$= (x-6) - (2x^2 - x)$$
$$= x - 6 - 2x^2 + x$$
$$= -2x^2 + 2x - 6$$

41. $(f+g)(0) = f(0) + g(0) = 2 + 1 = 3$

43. $(f \cdot g)(2) = f(2) \cdot g(2) = 4 \cdot (-1) = -4$

45. $(g-f)(-1) = g(-1) - f(-1) = 2 - 1 = 1$

47. $(g/f)(4) = g(4)/f(4) = 1/0$, which is undefined

49. $(f+g)(-2) = f(-2) + g(-2) = 1 + (-1) = 0$

51. $(f \cdot g)(1) = f(1) \cdot g(1) = 0 \cdot 1 = 0$

53. $(f/g)(4) = f(4)/g(4) = -3/1 = -3$

55. $(g/f)(2) = g(2)/f(2) = 2/(-1) = -2$

57. a. Frank contributed $1000 in 2004.

 b. $1400 - 600 = 800$
 In 2006, Sharon contributed about $800 more than Frank.

 c. $1600 + 2000 + 1400 + 900 + 2000 = 7900$
 Frank and Sharon contributed about $7900 over the five-year period.

 d. $(F + S)(2005) = \$900$

59. a. The import of crude oil to China was the greatest in 2003. That year approximately 1.8 million barrels per day were imported.

 b. In 1998 and 2001, the import of crude oil decrease from the year before.

 c. $I(2002) \approx 1.4$ million barrels

 d. $5.8 - 1.8 = 4.0$
 In 2003, about 4.0 million barrels of crude oil were produced in China per day.

61. a. About 20 houses were sold in Fuller in the summer of 2006.

 b. $28 - 20 = 8$
 About 8 houses were sold in Fuller at other times in 2006.

 c. $Y(2005) \approx 30 - 18 = 12$ houses

 d. $(S + Y)(2003) \approx 23$ houses

63. a.

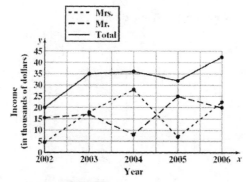

 b.

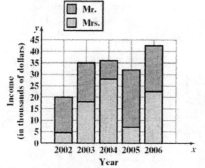

 c.

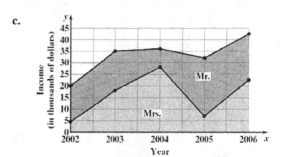

65. a.

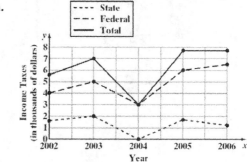

 b.

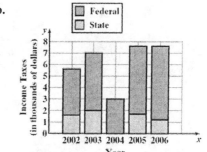

 c.

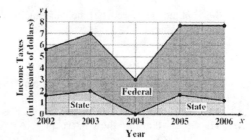

67. If $(f + g)(a) = 0$, then, $f(a)$ and $g(a)$ must either be opposites or both be equal to 0.

69. If $(f - g)(a) = 0$, then $f(a) = g(a)$.

71. If $(f / g)(a) < 0$, then $f(a)$ and $g(a)$ must have opposite signs.

73.

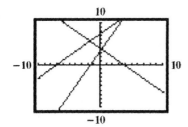

75.

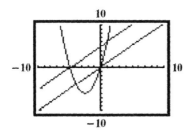

78. $(-4)^{-3} = \dfrac{1}{(-4)^3} = \dfrac{1}{-64} = -\dfrac{1}{64}$

79.
$$A = \frac{1}{2}bh$$
$$2 \cdot A = 2 \cdot \frac{1}{2}bh$$
$$2A = bh$$
$$\frac{2A}{b} = \frac{bh}{b}$$
$$\frac{2A}{b} = h \text{ or } h = \frac{2A}{b}$$

80. Let the pre-tax cost of the washing machine be x.
$$x + 0.06x = 477$$
$$1.06x = 477$$
$$x = 450$$
The pre-tax cost of the washing machine was $450.

81. Set $y = 0$ to find the x-intercept.
$$3x - 4(0) = 12$$
$$3x = 12$$
$$x = 4$$
The x-intercept is (4, 0).
Set $x = 0$ to find the y-intercept.
$$3(0) - 4y = 12$$
$$-4y = 12$$
$$y = -3$$
The y-intercept is (0, −3).

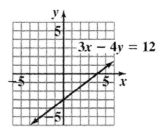

82. $2,960,000 = 2.96 \times 10^6$

83.

x	y
−3	1
−2	0
−1	−1
0	−2
1	−1
2	0
3	1

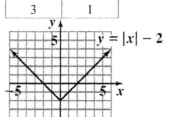

Chapter 8 Review Exercises

1. $y = \dfrac{1}{2}x$

x	y
−2	$y = \dfrac{1}{2}(-2) = -1$
0	$y = \dfrac{1}{2}(0) = 0$
2	$y = \dfrac{1}{2}(2) = 1$

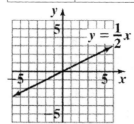

2. $y = -2x - 1$

x	y
0	$y = -2(0) - 1 = -1$
1	$y = -2(1) - 1 = -3$
2	$y = -2(2) - 1 = -4$

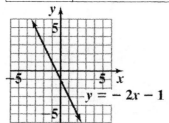

3. $y = \frac{1}{2}x + 3$

x	y
0	$y = \frac{1}{2}(0) + 3 = 3$
-2	$y = \frac{1}{2}(-2) + 3 = 2$
-4	$y = \frac{1}{2}(-4) + 3 = 1$

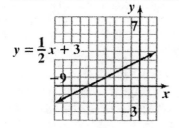

4. $y = -\frac{3}{2}x + 1$

x	y
-2	$y = -\frac{3}{2}(-2) + 1 = 4$
0	$y = -\frac{3}{2}(0) + 1 = 1$
2	$y = -\frac{3}{2}(2) + 1 = -2$

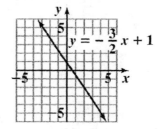

5. $y = x^2$

x	y
-3	$y = (-3)^2 = 9$
-1	$y = (-1)^2 = 1$
0	$y = 0^2 = 0$
2	$y = 2^2 = 4$

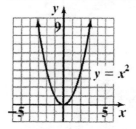

6. $y = x^2 - 1$

x	y
-3	$y = (-3)^2 - 1 = 8$
-1	$y = (-1)^2 - 1 = 0$
0	$y = 0^2 - 1 = 0$
1	$y = 1^2 - 1 = 0$
2	$y = 2^2 - 1 = 3$

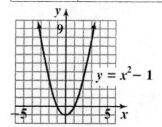

7. $y = |x|$

x	y
-4	$y = |-4| = 4$
-1	$y = |-1| = 1$
0	$y = |0| = 0$
2	$y = |2| = 2$

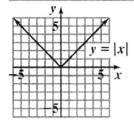

8. $y = |x| - 1$

x	y
-4	$y = |-4| - 1 = 3$
-1	$y = |-1| - 1 = 0$
0	$y = |0| - 1 = -1$
2	$y - |2| - 1 = 1$

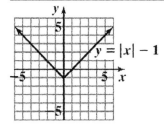

9. $y = x^3$

x	y
-2	$y = (-2)^3 = -8$
-1	$y = (-1)^3 = -1$
0	$y = 0^3 = 0$
1	$y = 1^3 = 1$
2	$y = 2^3 = 8$

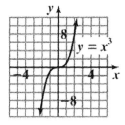

10. $y = x^3 + 4$

x	y
-2	$y = (-2)^3 + 4 = -4$
-1	$y = (-1)^3 + 4 = 3$
0	$y = 0^3 + 4 = 4$
1	$y = 1^3 + 4 = 5$

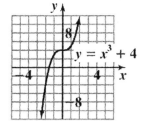

11. A function is a correspondence between a first set of elements, the domain, and a second set of elements, the range, such that each element of the domain corresponds to exactly one element in the range.

12. No, every relation is not a function. For example, $\{(4, 2), (4, -2)\}$ is a relation but not a function.

    Yes, every function is a relation because it is a set of ordered pairs.

13. Yes, each member of the domain corresponds to exactly one member of the range.

14. No, the domain element 2 corresponds to more than one member of the range (5 and –2).

15. $A = \{3, 4, 5, 6\}$

16. $B = \{6, 7, 8, ...\}$

17. $(4, \infty)$

18. $(-\infty, -2]$

19. $(-1.5, 2.7]$

20. a. Yes, the relation is a function.

    b. Domain: $\mathbb{R}$; Range: $\mathbb{R}$

21. a. Yes, the relation is a function.

    b. Domain: $\mathbb{R}$; Range: $\{y \mid y \le 0\}$

22. a. No, the relation is not a function.

    b. Domain: $\{x \mid -3 \le x \le 3\}$;

       Range: $\{y \mid -3 \le y \le 3\}$

23. a. No, the relation is not a function.

    b. Domain: $\{x \mid -2 \le x \le 2\}$;

       Range: $\{y \mid -1 \le y \le 1\}$

24. $f(x) = -x^2 + 3x - 4$

    a. $f(2) = -(2)^2 + 3(2) - 4 = -4 + 6 - 4 = -2$

    b. $f(h) = -h^2 + 3h - 4$

25. $g(t) = 2t^3 - 3t^2 + 6$

    a. $g(-1) = 2(-1)^3 - 3(-1)^2 + 6 = -2 - 3 + 6 = 1$

    b. $g(a) = 2a^3 - 3a^2 + 6$

26. Answers will vary. One possible interpretation: Car speeds up to 50 mph and stays at 50 mph for about 11 minutes. It then speeds up to about 68 mph and stays at that speed for 5 minutes. It then stops quickly and stays stopped for 5 minutes. It then travels through stop and go traffic for 5 minutes.

27. $N(x) = 40x - 0.2x^2$

    a. $N(30) = 40(30) - 0.2(30)^2$

             $= 1200 - 180$

             $= 1020$

       1020 baskets of apples are produced by 20 trees.

    b. $N(50) = 40(50) - 0.2(50)^2$

             $= 2000 - 500$

             $= 1500$

       1500 baskets of apples are produced by 50 trees.

28. $h(t) = -16t^2 + 196$

    a. $h(1) = -16(1)^2 + 196 = -16 + 196 = 180$

       After 1 second, the height of the ball is 180 feet.

    b. $h(3) = -16(3)^2 + 196 = -144 + 196 = 52$

       After 3 seconds, the height of the ball is 52 feet.

29. $f(x) = \dfrac{1}{2}x - 4$

For the *y*-intercept, set $x = 0$:

$$f(0) = \frac{1}{2}(0) - 4 = 0 - 4 = -4$$

The *y*-intercept is $(0, -4)$.

For the *x*-intercept, set $f(x) = 0$ and solve for x:

$$0 = \frac{1}{2}x - 4$$

$$4 = \frac{1}{2}x$$

$$2(4) = 2\left(\frac{1}{2}\right)x$$

$$8 = x$$

The *x*-intercept is $(8, 0)$.

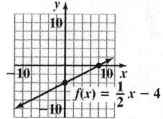

30. $f(x) = \dfrac{8}{3}x - 80$

For the *y*-intercept, set $x = 0$:

$$f(0) = \frac{8}{3}(0) - 80 = 0 - 80 = -80$$

The *y*-intercept is $(0, -80)$.

For the *x*-intercept, set $f(x) = 0$ and solve for x:

$$0 = \frac{8}{3}x - 80$$

$$80 = \frac{8}{3}x$$

$$\frac{3}{8}(80) = \frac{3}{8}\left(\frac{8}{3}x\right)$$

$$30 = x$$

The *x*-intercept is $(4, 0)$.

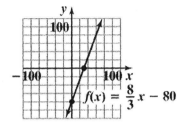

31. $f(x) = 4$ is a horizontal line 4 units above the x-axis.

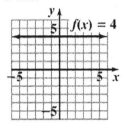

32. a. $p(x) = 0.1x - 5000$

x	p
0	$p = 0.1(0) - 5000 = -5000$
50,000	$p = 0.1(50,000) - 5000 = 0$
100,000	$p = 0.1(100,000) - 5000 = 5000$

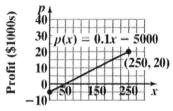

b. Approximately 50,000 bagels are sold when the company breaks even.

c. If the company has $22,000 profit, then approximately 270,000 bagels were sold.

33. The principle is $12,000 and the time is one year. Use the decimal form of the interest rate.
$I = 12,000r$

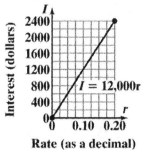

34. a. Yes, the graph represents a function. It passes the vertical line test.

b. This is a horizontal line (constant function), so the slope is 0. The y-intercept is (0, 3), so the equation is $f(x) = 3$.

35. a. No, the graph does not represent a function. It fails the vertical line test.

36. a. Yes, the graph represents a function. It passes the vertical line test.

b. The x-intercept is (4, 0) and the y-intercept is (0, 2), so $m = \dfrac{2-0}{0-4} = \dfrac{2}{-4} = -\dfrac{1}{2}$ and $b = 2$.

Thus, the equation is $f(x) = -\dfrac{1}{2}x + 2$.

37. a. The slope of the translated graph is the same as the slope of the original graph: $m = -2$.

b. The y-intercept of the translated graph is the 4 less than the y-intercept of the original graph: $5 - 4 = 1$. The y-intercept of the translated graph is (0, 1).

c. Since $m = -2$ and $b = 1$, the equation of the translated graph is $y = -2x + 1$.

38. Use the point-slope form.
$$y - y_1 = m(x - x_1)$$
$$y - (-4) = \frac{2}{3}(x - (-6))$$
$$y + 4 = \frac{2}{3}(x + 6)$$
$$y + 4 = \frac{2}{3}x + 4$$
$$y = \frac{2}{3}x$$
Since $b = 0$, the y-intercept is (0, 0).

39. a.

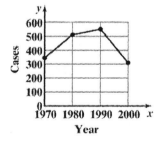

260

b. 1970 to 1980:

$$m_1 = \frac{510 - 346}{1980 - 1970} = \frac{164}{10} = 16.4$$

1980 to 1990:

$$m_2 = \frac{552 - 510}{1990 - 1980} = \frac{42}{10} = 4.2$$

1990 to 2000:

$$m_3 = \frac{317 - 552}{2000 - 1990} = \frac{-235}{10} = -23.5$$

c. The number of reported cases of typhoid fever increased the most during the 10-year period of 1970 – 1980.

40. Let t represent the number of years after 1980. Then 1980 is represented by $t = 0$ and 2070 is represented by $t = 90$. Therefore, the points $(0, 35.6)$ and $(90, 98.2)$ are on the line. Find the slope: $m = \dfrac{98.2 - 35.6}{2070 - 1980} = \dfrac{62.6}{90} \approx 0.7$.

The t-intercept is $(0, 35,6)$, so $b = 35.6$. Thus, the equation of the line is: $n = 0.7t + 35.6$. With function notation, we have $n(t) = 0.7t + 35.6$.

41.

$$2x - 3y = 10 \qquad\qquad y = \frac{2}{3}x - 5$$
$$-3y = -2x + 10 \qquad\qquad m_2 = \frac{2}{3}$$
$$\frac{-3y}{-3} = \frac{-2x + 10}{-3}$$
$$y = \frac{2}{3}x - \frac{10}{3}$$
$$m_1 = \frac{2}{3}$$

Since their slopes are equal and their y-intercepts are different, the lines are parallel.

42.

$$2x - 3y = 7 \qquad\qquad -3x - 2y = 8$$
$$-3y = -2x + 7 \qquad\qquad -2y = 3x + 8$$
$$y = \frac{-2x + 7}{-3} \qquad\qquad y = \frac{3x + 8}{-2}$$
$$y = \frac{2}{3}x - \frac{7}{3} \qquad\qquad y = -\frac{3}{2}x - 4$$
$$m_1 = \frac{2}{3} \qquad\qquad m_2 = -\frac{3}{2}$$

Since the slopes are negative reciprocals, the lines are perpendicular.

43.

$$4x - 2y = 13 \qquad\qquad -2x + 4y = -9$$
$$-2y = -4x + 13 \qquad\qquad 4y = 2x - 9$$
$$y = \frac{-4x + 13}{-2} \qquad\qquad y = \frac{2x - 9}{4}$$
$$y = 2x - \frac{13}{2} \qquad\qquad y = \frac{1}{2}x - \frac{9}{4}$$
$$m_1 = 2 \qquad\qquad m_2 = \frac{1}{2}$$

Since the slopes not equal and are not negative reciprocals, the lines are neither parallel nor perpendicular.

44. Use the point-slope form with $m = \dfrac{1}{2}$ and $(x_1, y_1) = (4, 9)$.

$$y - 9 = \frac{1}{2}(x - 4)$$
$$y - 9 = \frac{1}{2}x - 2$$
$$y = \frac{1}{2}x + 7$$

45. First, find the slope: $m = \dfrac{-6 - 1}{4 - (-3)} = \dfrac{-7}{7} = -1$.

Now, use the point-slope form with $m = -1$ and $(x_1, y_1) = (-3, 1)$.

$$y - 1 = -1\left(x - (-3)\right)$$
$$y - 1 = -1(x + 3)$$
$$y - 1 = -x - 3$$
$$y = -x - 2$$

46. The slope of the line $y = -\dfrac{2}{3}x + 1$ is $-\dfrac{2}{3}$. Since the new line is parallel to this line, its slope is also $-\dfrac{2}{3}$. Use the slope-intercept form with $m = -\dfrac{2}{3}$ and y-intercept of $(0, 6)$.

$$y = -\frac{2}{3}x + 6$$

47. $-2y = -5x + 7$

$$y = \frac{-5x + 7}{-2}$$

$$y = \frac{5}{2}x - \frac{7}{2}$$

The slope of this line is $\frac{5}{2}$. The new line is

parallel to this line, so its slope is also $\frac{5}{2}$. Use

the point-slope form with $m = \frac{5}{2}$ and

$(x_1, y_1) = (2, 8)$.

$$y - 8 = \frac{5}{2}(x - 2)$$

$$y - 8 = \frac{5}{2}x - 5$$

$$y = \frac{5}{2}x + 3$$

48. The slope of the line $y = \frac{3}{5}x + 5$ is $\frac{3}{5}$. Since the

new line is perpendicular to this line, its slope is

$-\frac{5}{3}$. Use the point-slope form with $m = -\frac{5}{3}$ and

$(x_1, y_1) = (-3, 1)$.

$$y - 1 = -\frac{5}{3}\left(x - (-3)\right)$$

$$y - 1 = -\frac{5}{3}(x + 3)$$

$$y - 1 = -\frac{5}{3}x - 5$$

$$y = -\frac{5}{3}x - 4$$

49. $4x - 2y = 8$

$$-2y = -4x + 8$$

$$y = \frac{-4x + 8}{-2}$$

$$y = 2x - 4$$

The slope of this line is 2. Since the new line is

perpendicular to this line, its slope is $-\frac{1}{2}$. Use

the point-slope form with $m = -\frac{1}{2}$ and

$(x_1, y_1) = (4, 5)$.

$$y - 5 = -\frac{1}{2}(x - 4)$$

$$y - 5 = -\frac{1}{2}x + 2$$

$$y = -\frac{1}{2}x + 7$$

50. $m_1 = \frac{-3 - 3}{0 - 5} = \frac{-6}{-5} = \frac{6}{5}$

$m_2 = \frac{-2 - (-1)}{2 - 1} = \frac{-2 + 1}{2 - 1} = \frac{-1}{1} = -1$

Since the slopes are not equal and are not
negative reciprocals, the lines are neither parallel
nor perpendicular.

51. $m_1 = \frac{3 - 2}{2 - 3} = \frac{1}{-1} = -1$

$m_2 = \frac{4 - 1}{1 - 4} = \frac{3}{-3} = -1$

Since the slopes are equal, the lines are parallel.

52. $m_1 = \frac{6 - 3}{4 - 7} = \frac{3}{-3} = -1$

$m_2 = \frac{3 - 2}{6 - 5} = \frac{1}{1} = 1$

Since the slopes are negative reciprocals, the
lines are perpendicular.

53. $m_1 = \frac{3 - 5}{2 - (-3)} = \frac{3 - 5}{2 + 3} = \frac{-2}{5} = -\frac{2}{5}$

$m_2 = \frac{2 - (-2)}{-1 - (-4)} = \frac{2 + 2}{-1 + 4} = \frac{4}{3}$

Since the slopes are not equal and are not
negative reciprocals, the lines are neither parallel
nor perpendicular.

54. a. First, find the slope of the linear function
using the points (35, 10.76) and (50, 19.91):

$$m = \frac{19.91 - 10.76}{50 - 35} = \frac{9.15}{15} = 0.61.$$

Use the point-slope form with $m = 0.61$
and $(a_1, r_1) = (35, 10.76)$:

$$r - r_1 = m(a - a_1)$$

$$r - 10.76 = 0.61(a - 35)$$

$$r - 10.76 = 0.61a - 21.35$$

$$r = 0.61a - 10.59$$

$$r(a) = 0.61a - 10.59$$

b. $r(40) = 0.61(40) - 10.59$

$\qquad = 24.40 - 10.59$

$\qquad = 13.81$

The monthly rate for a 40-year-old man is about $13.81.

55. a. First, find the slope of the linear function using the points (30, 489) and (50, 525):

$m = \dfrac{525 - 489}{50 - 30} = \dfrac{36}{20} = 1.8$.

Use the point-slope form with $m = 1.8$ and $(r_1, C_1) = (30, 489)$:

$C - C_1 = m(r - r_1)$

$C - 489 = 1.8(r - 30)$

$C - 489 = 1.8r - 54$

$\qquad C = 1.8r + 435$

$\quad C(r) = 1.8r + 435$

b. $C(40) = 1.8(40) + 435 = 72 + 435 = 507$

When a person swims at 40 yards per minute for one hour, he or she will burn 507 calories.

c. $\qquad C(r) = 600$

$\quad 1.8r + 435 = 600$

$\qquad\quad 1.8r = 165$

$\qquad\qquad r \approx 91.7$

To burn 600 calories in 1 hour, the person needs to swim at a speed of about 91.7 yards per minute.

56. $(f + g)(x) = f(x) + g(x)$

$\qquad\qquad = (x^2 - 3x + 4) + (2x - 5)$

$\qquad\qquad = x^2 - x - 1$

57. $(f + g)(x) = x^2 - x - 1$

$(f + g)(4) = (4)^2 - 4 - 1 = 16 - 4 - 1 = 11$

58. $(g - f)(x) = (2x - 5) - (x^2 - 3x + 4)$

$\qquad\qquad = 2x - 5 - x^2 + 3x - 4$

$\qquad\qquad = -x^2 + 5x - 9$

59. $(g - f)(x) = -x^2 + 5x - 9$

$(g - f)(-1) = -(-1)^2 + 5(-1) - 9$

$\qquad\qquad = -1 - 5 - 9$

$\qquad\qquad = -15$

60. $(f \cdot g)(-1) = f(-1) \cdot g(-1)$

$\qquad = \left((-1)^2 - 3(-1) + 4\right) \cdot \left(2(-1) - 5\right)$

$\qquad = (1 + 3 + 4) \cdot (-2 - 5)$

$\qquad = 8(-7)$

$\qquad = -56$

61. $(f \cdot g)(3) = f(3) \cdot g(3)$

$\qquad = \left(3^2 - 3(3) + 4\right) \cdot \left(2(3) - 5\right)$

$\qquad = (9 - 9 + 4) \cdot (6 - 5)$

$\qquad = 4(1)$

$\qquad = 4$

62. $(f / g)(1) = \dfrac{f(1)}{g(1)}$

$\qquad = \dfrac{1^2 - 3(1) + 4}{2(1) - 5}$

$\qquad = \dfrac{1 - 3 + 4}{2 - 5}$

$\qquad = \dfrac{2}{-3}$

$\qquad = -\dfrac{2}{3}$

63. $(f / g)(2) = \dfrac{f(2)}{g(2)}$

$\qquad = \dfrac{2^2 - 3(2) + 4}{2(2) - 5}$

$\qquad = \dfrac{4 - 6 + 4}{4 - 5}$

$\qquad = \dfrac{2}{-1}$

$\qquad = -2$

64. a. The projected female population worldwide in 2050 is about 4.6 billion.

b. 2.9 billion $-$ 0.8 billion $=$ 2.1 billion

The projected number of women 15–49 years of age in 2050 about 2.1 billion.

c. 3.5 billion $-$ 2.7 billion $=$ 0.8 billion

The number of women who are projected to be in the 50 years and older age group in 2010 is about 0.8 billion.

d. 2002: 3.1 billion $-$ 2.5 billion $=$ 0.6 billion

2010: 3.5 billion $-$ 2.7 billion $=$ 0.8 billion

$\dfrac{0.8\ \text{billion} - 0.6\ \text{billion}}{0.6\ \text{billion}} = \dfrac{0.2\ \text{billion}}{0.6\ \text{billion}} \approx 0.33$

There is approximately a 33% increase in the projected number of women 50 years and older from 2002 to 2010.

65. **a.** Ginny's total retirement income in 2006 was about $47,000.

   b. Ginny's pension income in 2005 was about $28,000.

   c. $25,000 - 22,000 = 3000$
   Ginny's interest and dividend income in 2003 was about $3000.

Chapter 8 Practice Test

1. $y = -2x + 1$

x	y
-1	$y = -2(-1) + 1 = 3$
0	$y = -2(0) + 1 = 1$
1	$y = -2(1) + 1 = -1$

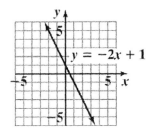

2. $y = \sqrt{x}$

x	y
-1	$y = \sqrt{-1}$ undefined
0	$y = \sqrt{0} = 0$
1	$y = \sqrt{1} = 1$
4	$y = \sqrt{4} = 2$

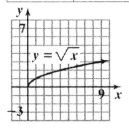

3. $y = x^2 - 4$

x	y
-2	$y = (-2)^2 - 4 = 0$
-1	$y = (-1)^2 - 4 = -3$
0	$y = (0)^2 - 4 = -4$
1	$y = (1)^2 - 4 = -3$
2	$y = (2)^2 - 4 = 0$

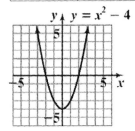

4. $y = |x|$

x	y
-3	$y = |-3| = 3$
0	$y = |0| = 0$
4	$y = |4| = 4$

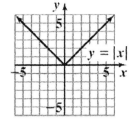

5. A function is a correspondence between a first set of elements, the domain, and a second set of elements, the range, such that each element of the domain corresponds to exactly one element in the range.

6. Yes, the set of ordered pairs represents a function because each member in the domain corresponds with exactly one member in the range.

7. Yes, the graph represents a function because it passes the vertical line test.

Domain: $\mathbb{R}$

Range: $\{y | y \le 4\}$

8. No, the graph does not represent a function because it fails the vertical line test.

Domain: $\{x | -3 \le x \le 3\}$

Range: $\{y | -2 \le y \le 2\}$

9. $f(x) = 3x^2 - 6x + 5$

$f(-2) = 3(-2)^2 - 6(-2) + 5 = 12 + 12 + 5 = 29$

10. The graph of $f(x) = -3$ is a horizontal line 3 units below the x-axis.

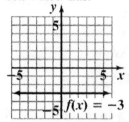

11. a. $p(x) = 10.2x - 50,000$

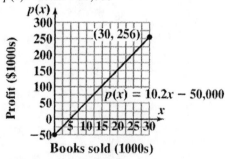

b. The company breaks even when $p(x) = 0$.

$10.2x - 50,000 = 0$

$10.2x = 50,000$

$x = \dfrac{50,000}{10.2} = 4900$

The company breaks even when it sells 4900 books.

c. $10.2x - 50,000 = 100,000$

$10.2x = 150,000$

$x = \dfrac{150,000}{10.2} \approx 14,700$

The company needs to sell about 14,700 books to break even.

12. The year 2000 correspond to $t = 0$, and the year 2050 corresponds to $t = 50$. Thus, the points $(0, 274.634)$ and $(50, 419.854)$ lie on the graph.

$m = \dfrac{419.854 - 274.634}{50 - 0} = \dfrac{145.220}{50} = 2.9044$

The y-intercept is $(0, 274.634)$, so $b = 274.634$. Thus, the equation is $p(t) = 2.9044t + 274.634$.

13. Write each equation in slope-intercept form by solving for y.

$2x - 3y = 12$

$-3y = -2x + 12$

$y = \dfrac{-2x + 12}{-3}$

$y = \dfrac{2}{3}x - 4$

$4x + 10 = 6y$

$\dfrac{4x + 10}{6} = y$

$\dfrac{4}{6}x + \dfrac{10}{6} = y$

$y = \dfrac{2}{3}x + \dfrac{5}{3}$

Since the slopes are the same and the y-intercepts are different, the lines are parallel.

14. The slope of $y = \dfrac{1}{2}x + 1$ is $\dfrac{1}{2}$. Any line perpendicular to this line must have slope -2. Use the point-slope form with $m = -2$ and $(x_1, y_1) = (6, -5)$.

$y - y_1 = m(x - x_1)$

$y - (-5) = -2(x - 6)$

$y + 5 = -2x + 12$

$y = -2x + 7$

15. a. Let t be the number of years since 2000. Then 2000 is represented by $t = 0$ and 2010 is represented by $t = 10$. Find the slope of the linear function using the points $(0, 266)$ and $(10, 236)$.

$m = \dfrac{236 - 266}{10 - 0} = \dfrac{-30}{10} = -3$

The y-intercept is $(0, 266)$, so $b = 266$. Thus, the equation is $r(t) = -3t + 266$.

b. Note that 2006 is represented by $t = 6$:

$r(6) = -3(6) + 266 = -18 + 266 = 248$

From the function, the death rate due to heart disease in 2006 was 248 per 100,000.

c. Note that 2020 is represented by $t = 20$:

$r(20) = -3(20) + 266 = -60 + 266 = 206$

Assuming the trend continues, the death rate due to heart disease in 2020 will be 206 per 100,000.

16. $(f+g)(3) = f(3)+g(3)$

$$= \left(2(3)^2 - (3)\right) + \left((3) - 6\right)$$
$$= (2(9) - 3) + (3 - 6)$$
$$= (18 - 3) + (3 - 6)$$
$$= 15 + (-3)$$
$$= 12$$

17. $(f/g)(-1) = \dfrac{f(-1)}{g(-1)}$

$$= \dfrac{2(-1)^2 - (-1)}{(-1) - 6}$$
$$= \dfrac{2+1}{-7}$$
$$= -\dfrac{3}{7}$$

18. $f(a) = 2a^2 - a$

19. a. The total number of tons of paper to be used in 2010 will be about 44 million tons.

   b. The number of tons of paper to be used by businesses in 2010 will be about 18 million tons.

   c. The number of tons of paper to be used for reference, print media, and household use in 2010 will be about $44 - 18 = 26$, or 26 million tons.

Chapter 8 Cumulative Review Test

1. Substitute -3 for x and -2 for y and simplify:
$$4x^2 - 7xy^2 + 6 = 4(-3)^2 - 7(-3)(-2)^2 + 6$$
$$= 4(9) - 7(-3)(4) + 6$$
$$= 36 + 84 + 6$$
$$= 126$$

2. $2 - \{3[6 - 4(6^2 \div 4)]\} = 2 - \{3[6 - 4(36 \div 4)]\}$
$$= 2 - \{3[6 - 4(9)]\}$$
$$= 2 - \{3[6 - 36]\}$$
$$= 2 - \{3[-30]\}$$
$$= 2 - \{-90\}$$
$$= 2 + 90$$
$$= 92$$

3. $2(x+4) - 5 = -3[x - (2x+1)]$
$$2x + 8 - 5 = -3[x - 2x - 1]$$
$$2x + 3 = -3[-x - 1]$$
$$2x + 3 = 3x + 3$$
$$3 = x + 3$$
$$0 = x$$

4. $\qquad P = 2E + 3R$
$$P - 2E = 2E + 3R - 2E$$
$$P - 2E = 3R$$
$$\dfrac{P - 2E}{3} = \dfrac{3R}{3}$$
$$\dfrac{P - 2E}{3} = R \quad \text{or} \quad R = \dfrac{P - 2E}{3}$$

5. $\qquad 5x - 2y = 12$
$$-5x + 5x - 2y = -5x + 12$$
$$-2y = -5x + 12$$
$$\dfrac{-2y}{-2} = \dfrac{-5x + 12}{-2}$$
$$y = \dfrac{5}{2}x - 6$$

6. $\left(\dfrac{6x^2 y^3}{2x^5 y}\right)^3 = \left(\dfrac{6}{2} \cdot \dfrac{y^{3-1}}{x^{5-2}}\right)^3$
$$= \left(\dfrac{3y^2}{x^3}\right)^3$$
$$= \dfrac{3^3 (y^2)^3}{(x^3)^3}$$
$$= \dfrac{27 y^{2(3)}}{x^{3(3)}}$$
$$= \dfrac{27 y^6}{x^9}$$

7. $(6x^2 - 3x - 2) - (-2x^2 - 8x - 1)$
$$= 6x^2 - 3x - 2 + 2x^2 + 8x + 1$$
$$= 6x^2 + 2x^2 - 3x + 8x - 2 + 1$$
$$= 8x^2 + 5x - 1$$

8. $(4x^2 - 6x + 3)(3x - 5)$
$$= 4x^2(3x - 5) - 6x(3x - 5) + 3(3x - 5)$$
$$= 12^3 - 20x^2 - 18x^2 + 30x + 9x - 15$$
$$= 12x^3 - 38x^2 + 39x - 15$$

9. $6a^2 - 6a - 5a + 5 = 6a(a-1) - 5(a-1)$
$$= (a-1)(6a-5)$$

10.
$$2x^2 = x + 15$$
$$2x^2 - x - 15 = 0$$
$$(x-3)(2x+5) = 0$$
$$x - 3 = 0 \quad \text{or} \quad 2x + 5 = 0$$
$$x = 3 \qquad x = -\frac{5}{2}$$

11. $\dfrac{x^2-9}{x^2-x-6} \cdot \dfrac{x^2-2x-8}{2x^2-7x-4}$

$= \dfrac{(x+3)\cancel{(x-3)}}{\cancel{(x-3)}(x+2)} \cdot \dfrac{\cancel{(x-4)}(x+2)}{(2x+1)\cancel{(x-4)}}$

$= \dfrac{x+3}{2x+1}$

12. $\dfrac{x}{x+4} - \dfrac{3}{x-5} = \dfrac{x}{x+4} \cdot \dfrac{x-5}{x-5} - \dfrac{3}{x-5} \cdot \dfrac{x+4}{x+4}$

$= \dfrac{x(x-5)}{(x+4)(x-5)} - \dfrac{3(x+4)}{(x+4)(x-5)}$

$= \dfrac{x^2 - 5x - (3x+12)}{(x+4)(x-5)}$

$= \dfrac{x^2 - 5x - 3x - 12}{(x+4)(x-5)}$

$= \dfrac{x^2 - 8x - 12}{(x+4)(x-5)}$

13. $\dfrac{4}{x^2-3x-10} + \dfrac{2}{x^2+5x+6}$

$= \dfrac{4}{(x-5)(x+2)} + \dfrac{2}{(x+2)(x+3)}$

$= \dfrac{4(x+3)}{(x-5)(x+2)(x+3)} + \dfrac{2(x-5)}{(x-5)(x+2)(x+3)}$

$= \dfrac{4(x+3) + 2(x-5)}{(x-5)(x+2)(x+3)}$

$= \dfrac{4x + 12 + 2x - 10}{(x-5)(x+2)(x+3)}$

$= \dfrac{6x+2}{(x-5)(x+2)(x+3)}$

14.
$$\frac{x}{9} - \frac{x}{6} = \frac{1}{12}$$
$$36\left(\frac{x}{9} - \frac{x}{6}\right) = 36\left(\frac{1}{12}\right)$$
$$4x - 6x = 3$$
$$-2x = 3$$
$$x = -\frac{3}{2}$$

15.
$$\frac{7}{x+3} + \frac{5}{x+2} = \frac{5}{x^2+5x+6}$$
$$\frac{7}{x+3} + \frac{5}{x+2} = \frac{5}{(x+3)(x+2)}$$
$$(x+3)(x+2)\left(\frac{7}{x+3} + \frac{5}{x+2}\right) = (x+3)(x+2)\left[\frac{5}{(x+3)(x+2)}\right]$$
$$7(x+2) + 5(x+3) = 5$$
$$7x + 14 + 5x + 15 = 5$$
$$12x + 29 = 5$$
$$12x = -24$$
$$x = -2$$

Check:
$$\frac{7}{x+3} + \frac{5}{x+2} = \frac{5}{x^2+5x+6}$$
$$\frac{7}{-2+3} + \frac{5}{-2+2} = \frac{5}{(-2)^2+5(-2)+6}$$
$$\frac{7}{1} + \frac{5}{0} = \frac{5}{0}$$

Since $\dfrac{5}{0}$ is not a real number, there is no solution.

16. a. The graph is not a function because it fails the vertical line test.

b. Domain: $\{x \mid x \le 2\}$
Range: $\mathbb{R}$

17. Write each equation in slope-intercept form by solving for y.

$2x - 5y = 6$ $\qquad\qquad$ $5x - 2y = 9$

$\quad -5y = -2x + 6$ $\qquad$ $\quad -2y = -5x + 9$

$\qquad y = \dfrac{-2x+6}{-5}$ $\qquad\quad$ $y = \dfrac{-5x+9}{-2}$

$\qquad y = \dfrac{2}{5}x - \dfrac{6}{5}$ $\qquad\quad$ $y = \dfrac{5}{2}x - \dfrac{9}{2}$

Since the slopes are $\dfrac{2}{5}$ and $\dfrac{5}{2}$ which are neither equal nor negative reciprocals, the lines are neither parallel nor perpendicular.

18. $(f+g)(x) = f(x) + g(x)$
$$= (x^2 + 3x - 2) + (4x - 6)$$
$$= x^2 + 7x - 8$$

19. $(f \cdot g)(4) = f(4) \cdot g(4)$
$$= [(4)^2 + 3(4) - 2] \cdot [4(4) - 6]$$
$$= (16 + 12 - 2)(16 - 6)$$
$$= (26)(10)$$
$$= 260$$

20. a. Commercial Sector $= 14\%$ of 2.18×10^{13}
$$= 0.14 \times 2.18 \times 10^{13}$$
$$= 0.3052 \times 10^{13}$$
$$= 3.052 \times 10^{12}$$
Commercial sector consumption of natural gas in 2003 was 3.052×10^{12} cubic feet.

b. Industrial Sector $= 37\%$ of 2.18×10^{13}
$$= 0.37 \times 2.18 \times 10^{13}$$
$$= 0.8066 \times 10^{13}$$
$$= 8.066 \times 10^{12}$$
Transportation Sector $= 3\%$ of 2.18×10^{13}
$$= 0.03 \times 2.18 \times 10^{13}$$
$$= 0.0654 \times 10^{13}$$
$$= 6.54 \times 10^{11}$$
$$8.066 \times 10^{12} - 6.54 \times 10^{11}$$
$$= 8.066 \times 10^{12} - 0.654 \times 10^{12}$$
$$= (8.066 - 0.654) \times 10^{12}$$
$$= 7.412 \times 10^{12}$$
In 2003, the industrial sector consumed 7.412×10^{12} cubic feet more of natural gas than the transportation sector.

c. $2.18 \times 10^{13} + 10\%$ of 2.18×10^{13}
$$= 110\% \text{ of } 2.18 \times 10^{13}$$
$$= 1.1 \times 2.18 \times 10^{13}$$
$$= 2.398 \times 10^{13}$$
Consumption of natural gas in 2006 will be 2.398×10^{13} cubic feet.

268

Chapter 9

Exercise Set 9.1

1. The solution to a system of equations represents the ordered pairs that satisfy all the equations in the system.

3. Write the equation in slope-intercept form and compare their slopes and y-intercepts. If the slopes are different, the system has exactly one solution. If the slopes and y-intercepts are the same, the system has an infinite number of solutions. If the slopes are the same and the y-intercepts are different, the system has no solution.

5. The point of intersection can only be estimated.

7. **a.**

$y = 3x - 6$	$y = -3x$
$0 = 3(2) - 6$	$0 = -3(2)$
$0 = 6 - 6$	$0 = -6$ False
$0 = 0$ True	

   Since (2, 0) does not satisfy both equations, it is not a solution to the system of equations.

   b.
   $y = 3x - 6$
   $0 = 3(0) - 6$
   $0 = -6$ False

   Since (0, 0) does not satisfy the first equation, it is not a solution to the system of equations.

   c.

$y = 3x - 6$	$y = -3x$
$-3 = 3(1) - 6$	$-3 = -3(1)$
$-3 = 3 - 6$	$-3 = -3$ True
$-3 = -3$ True	

   Since (1, −3) satisfies both equations, it is a solution to the system of equations.

9. **a.**

$y = 2x - 3$	$y = x + 5$
$13 = 2(8) - 3$	$13 = 8 + 5$
$13 = 13$ True	$13 = 13$ True

   Since (8, 13) satisfies both equations, it is a solution to the system.

   b.

$y = 2x - 3$	$y = x + 5$
$5 = 2(4) - 3$	$5 = 4 + 5$
$5 = 5$ True	$5 = 9$ False

   Since (4, 5) does not satisfy both equations, it is not a solution to the system.

   c.

$y = 2x - 3$	$y = x + 5$
$7 = 2(5) - 3$	$7 = 5 + 5$
$7 = 7$ True	$7 = 10$ False

   Since (5, 7) does not satisfy both equations, it is not a solution to the system.

11. **a.**

$4x + y = 15$	$5x + y = 10$
$4(3) + 3 = 15$	$5(3) + 3 = 10$
$15 = 15$ True	$18 = 10$ False

    Since (3, 3) does not satisfy both equations, it is not a solution to the system.

    b.
    $4x + y = 15$
    $4(2) + (0) = 15$
    $8 = 15$ False

    Since (2, 0) does not satisfy the first equation, it is not a solution to the system.

    c.

$4x + y = 15$	$5x + y = 10$
$4(-1) + 19 = 15$	$5(-1) + 19 = 10$
$15 = 15$ True	$14 = 10$ False

    Since (−1, 19) does not satisfy both equations, it is not a solution to the system.

13. Solve the first equation for y.
    $$4x - 6y = 12$$
    $$-6y = -4x + 12$$
    $$y = \frac{2}{3}x - 2$$

    Notice that it is the same as the second equation. If the ordered pair satisfies the first equation, then it also satisfies the second equation.

    a.
    $4x - 6y = 12$
    $4(3) - 6(0) = 12$
    $12 = 12$ True

    Since (3, 0) satisfies both equations, it is a solution to the system.

    b.
    $4x - 6y = 12$
    $4(9) - 6(4) = 12$
    $12 = 12$ True

    Since (9, 4) satisfies both equations, it is a solution to the system.

    c.
    $4x - 6y = 12$
    $4(6) - 6(1) = 12$
    $18 = 12$ False

    Since (6, 1) does not satisfy the first equation, it is not a solution to the system.

15. a.
$$3x - 4y = 8 \qquad 2y = \frac{2}{3}x - 4$$
$$3(0) - 4(-2) = 8$$
$$8 = 8 \text{ True} \qquad 2(-2) = \frac{2}{3}(0) - 4$$
$$-4 = -4 \text{ True}$$

Since $(0, -2)$ satisfies both equations, it is a solution to the system.

b.
$$3x - 4y = 8$$
$$3(1) - 4(-6) = 8$$
$$27 = 8 \text{ False}$$

Since $(1, -6)$ does not satisfy the first equation, it is not a solution to the system.

c.
$$3x - 4y = 8 \qquad 2y = \frac{2}{3}x - 4$$
$$3\left(-\frac{1}{3}\right) - 4\left(-\frac{9}{4}\right) = 8 \qquad 2\left(-\frac{9}{4}\right) = \frac{2}{3}\left(-\frac{1}{3}\right) - 4$$
$$8 = 8 \qquad\qquad -\frac{9}{2} = -\frac{38}{9} \text{ False}$$
$$\text{True}$$

Since $\left(-\frac{1}{3}, -\frac{9}{4}\right)$ does not satisfy both equations, it is not a solution to the system.

17. consistent—one solution

19. dependent—infinite number of solutions

21. consistent—one solution

23. inconsistent—no solution

25. Write each equation in slope-intercept form.
$$y = 2x - 1 \qquad 3y = 5x - 6$$
$$y = \frac{5}{3}x - 2$$

Since the slopes of the lines are not the same, the lines intersect to produce one solution. This is a consistent system.

27. Write each equation in slope-intercept form.
$$2y = 3x + 3 \qquad y = \frac{3}{2}x - 2$$
$$y = \frac{3}{2}x + \frac{3}{2}$$

Since the lines have the same slope, $\frac{3}{2}$, and different y-intercepts, the lines are parallel. There is no solution. This is an inconsistent system.

29. Write each equation in slope-intercept form.
$$2x = y - 6 \qquad 3x = 3y + 5$$
$$y = 2x + 6 \qquad 3y = 3x - 5$$
$$y = x - \frac{5}{3}$$

Since the slopes of the lines are not the same, the lines intersect to produce one solution. This is a consistent system.

31. Write each equation in slope-intercept form.
$$3x + 5y = -7 \qquad -3x - 5y = -10$$
$$5y = -3x - 7 \qquad -5y = 3x - 10$$
$$y = -\frac{3}{5}x - \frac{7}{5} \qquad y = -\frac{3}{5}x + 2$$

Since the lines have the same slope and different y-intercepts, the lines are parallel. There is no solution. This is an inconsistent system.

33. Write each equation in slope-intercept form.
$$x = 3y + 5 \qquad 2x - 6y = 10$$
$$x - 5 = 3y \qquad -6y = -2x + 10$$
$$\frac{1}{3}x - \frac{5}{3} = y \qquad y = \frac{1}{3}x - \frac{5}{3}$$

Since both equations are identical, the line is the same for both of them. There are an infinite number of solutions. This is a dependent system.

35. Write each equation in slope-intercept form.
$$y = \frac{3}{2}x + \frac{1}{2} \qquad 3x - 2y = \frac{5}{2}$$
$$-2y = -3x - \frac{5}{2}$$
$$y = \frac{3}{2}x + \frac{5}{4}$$

Since the lines have the same slope and different y-intercepts, the lines are parallel. There is no solution. This is an inconsistent system.

37. Graph the equations $y = x + 3$ and $y = -x + 3$.

The lines intersect and the point of intersection is $(0, 3)$. This is a consistent system.

39. Graph the equations $y = 3x - 6$ and $y = -x + 6$.

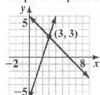

The lines intersect and the point of intersection is (3, 3). This is a consistent system.

41. Graph the equations $4x = 8$ or $x = 2$ and $y = -3$.

The lines intersect and the point of intersection is (2, –3). This is a consistent system.

43. Graph the equations $x + y = 5$ or $y = -x + 5$ and $-x + y = 1$ or $y = x + 1$.

The lines intersect and the point of intersection is (2, 3). This is a consistent system.

45. Graph the equations $y = -\dfrac{1}{2}x + 4$ and

$x + 2y = 6$ or $y = -\dfrac{1}{2}x + 3$.

The lines are parallel. The system is inconsistent and there is no solution.

47. Graph the equations $x + 2y = 8$ or $y = -\dfrac{1}{2}x + 4$

and $5x + 2y = 0$ or $y = -\dfrac{5}{2}x$.

The lines intersect and the point of intersection is (–2, 5). This is a consistent system.

49. Graph the equations $2x + 3y = 6$ or

$y = -\dfrac{2}{3}x + 2$ and $4x = -6y + 12$

or $y = -\dfrac{2}{3}x + 2$.

The lines are identical. There are an infinite number of solutions. This is a dependent system.

51. Graph the equations $y = 3$ and $y = 2x - 3$.

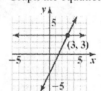

The lines intersect and the point of intersection is (3, 3). This is a consistent system.

53. Graph the equations $x - 2y = 4$ or $y = \dfrac{1}{2}x - 2$

and $2x - 4y = 8$ or $y = \dfrac{1}{2}x - 2$.

The lines are identical. There are an infinite number of solutions. This is a dependent system.

55. Graph the equations $2x + y = -2$ or $y = -2x - 2$ and $6x + 3y = 6$ or $y = -2x + 2$.

The lines are parallel. The system is inconsistent and there is no solution.

57. Graph the equations $4x - 3y = 6$ or $y = \frac{4}{3}x - 2$

and $2x + 4y = 14$ or $y = -\frac{1}{2}x + \frac{7}{2}$.

The lines intersect and the point of intersection is (3, 2). This is a consistent system.

59. Graph the equations $2x - 3y = 0$ or $y = \frac{2}{3}x$ and

$x + 2y = 0$ or $y = -\frac{1}{2}x$.

The lines intersect and the point of intersection is (0, 0). This is a consistent system.

61. Write each equation in slope-intercept form.

$$6x - 4y = 12 \qquad\qquad 12y = 18x - 24$$
$$-4y = -6x + 12$$
$$y = \frac{3}{2}x - 2$$
$$y = \frac{3}{2}x - 3$$

The lines are parallel because they have the same slope, $\frac{3}{2}$, and different y-intercepts.

63. The system has an infinite number of solutions. If the two lines have two points in common then they must be the same line.

65. The system has no solutions. Distinct parallel lines do not intersect.

67. $x = 5, y = 3$ has one solution, (5, 3).

69. (repair) $c = 600 + 650n$
(replacement) $c = 1800 + 450n$
Graph the equations and determine the intersection.

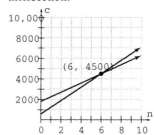

The solution is (6, 4500). Therefore, the total cost of repair equals the total cost of replacement at 6 years.

71. $c = 25h$

$c = 21h + 28$

Graph the equations and determine the intersection.

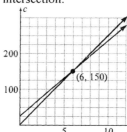

The solution (6, 150). Therefore, the boats must be rented for 6 hours for the cost to be the same.

79. $3x - (x - 6) + 4(3 - x) = 3x - x + 6 + 12 - 4x$
$$= 3x - x - 4x + 6 + 12$$
$$= -2x + 18$$

80. $2(x + 3) - x = 5x + 2$
$$2x + 6 - x = 5x + 2$$
$$x + 6 = 5x + 2$$
$$-4x = -4$$
$$x = 1$$

81.
$$A = p(1+rt)$$
$$1000 = 500(1 + r \cdot 20)$$
$$1000 = 500(1 + 20r)$$
$$1000 = 500(1) + 500(20r)$$
$$1000 = 500 + 10,000r$$
$$1000 - 500 = 500 + 10,000r - 500$$
$$500 = 10,000r$$
$$\frac{500}{10,000} = \frac{10,000r}{10,000}$$
$$0.05 = r$$

82. a. For the x-intercept, set $y = 0$.
$$2x + 3y = 12$$
$$2x + 3(0) = 12$$
$$2x = 12$$
$$x = 6$$
The x-intercept is (6, 0).

For the y-intercept, set $x = 0$.
$$2x + 3y = 12$$
$$2(0) + 3y = 12$$
$$3y = 12$$
$$y = 4$$
The y-intercept is (0, 4).

b.
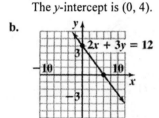

83. $\dfrac{x^2 - 9x + 14}{2 - x} = \dfrac{(x-7)(x-2)}{-(x-2)} = -(x-7)$

84.
$$\frac{4}{b} + 2b = \frac{38}{3}$$
$$3b\left(\frac{4}{b} + 2b\right) = 3b\left(\frac{38}{3}\right)$$
$$12 + 6b^2 = 38b$$
$$6b^2 - 38b + 12 = 0$$
$$2(3b^2 - 19b + 6) = 0$$
$$2(3b - 1)(b - 6) = 0$$
$$3b - 1 = 0 \quad \text{or} \quad b - 6 = 0$$
$$b = \frac{1}{3} \qquad b = 6$$

Exercise Set 9.2

1. The x in the first equation, since both 6 and 12 are divisible by 3.

3. You will obtain a false statement, such as $3 = 0$.

5. $x + 2y = 6$
$2x - 3y = 5$
Solve the first equation for x, $x = 6 - 2y$.
Substitute $6 - 2y$ for x in the second equation.
$$2(6 - 2y) - 3y = 5$$
$$12 - 4y - 3y = 5$$
$$-7y = -7$$
$$y = 1$$
Substitute 1 for y in the equation $x = 6 - 2y$.
$$x = 6 - 2(1) = 4$$
The solution is (4, 1).

7. $x + y = -2$
$x - y = 0$
Solve the first equation for y, $y = -2 - x$.
Substitute $-2 - x$ for y in the second equation.
$$x - (-2 - x) = 0$$
$$x + 2 + x = 0$$
$$2x = -2$$
$$x = -1$$
Substitute -1 for x in the equation $y = -2 - x$.
$$y = -2 - (-1) = -2 + 1 = -1$$
The solution is $(-1, -1)$.

9. $3x + y = 3$
$3x + y + 5 = 0$
Solve the first equation for y, $y = -3x + 3$.
Substitute $-3x + 3$ for y in the second equation.
$$3x + y + 5 = 0$$
$$3x - 3x + 3 + 5 = 0$$
$$8 = 0 \text{ False}$$
There is no solution.

11. $x = 3$
$x + y + 5 = 0$
Substitute 3 for x in the second equation.
$$x + y + 5 = 0$$
$$3 + y + 5 = 0$$
$$y = -8$$
The solution is (3, –8).

13. $x = y + 1$

    $4x + 2y = -14$

    Substitute $y + 1$ for x in the second equation.
    $$4x + 2y = -14$$
    $$4(y+1) + 2y = -14$$
    $$4y + 4 + 2y = -14$$
    $$6y + 4 = -14$$
    $$6y = -18$$
    $$y = \frac{-18}{6} = -3$$

    Now, substitute –3 for y in the first equation.
    $$x = y + 1$$
    $$x = -3 + 1 = -2$$
    The solution is (–2, –3).

15. $2x + y = 11$

    $y = 3x - 4$

    Substitute $3x - 4$ for y in the first equation.
    $$2x + y = 11$$
    $$2x + 3x - 4 = 11$$
    $$5x - 4 = 11$$
    $$5x = 15$$
    $$x = 3$$

    Now substitute 3 for x in the second equation.
    $$y = 3x - 4$$
    $$y = 3(3) - 4$$
    $$y = 9 - 4$$
    $$y = 5$$
    The solution is (3, 5).

17. $y = \frac{1}{3}x - 2$

    $x - 3y = 6$

    Substitute $\frac{1}{3}x - 2$ for y in the second equation.
    $$x - 3y = 6$$
    $$x - 3\left(\frac{1}{3}x - 2\right) = 6$$
    $$x - x + 6 = 6$$
    $$6 = 6$$

    Since this is a true statement, there are an infinite number of solutions. This is a dependent system.

19. $2x + 5y = 9$

    $6x - 2y = 10$

    First solve the second equation for y.
    $$\frac{1}{2}(6x - 2y) = \frac{1}{2}(10)$$
    $$3x - y = 5$$
    $$y = 3x - 5$$

    Now substitute $3x - 5$ for y in the first equation.
    $$2x + 5y = 9$$
    $$2x + 5(3x - 5) = 9$$
    $$2x + 15x - 25 = 9$$
    $$17x = 34$$
    $$x = 2$$

    Finally substitute 2 for x in the equation $y = 3x - 5$.
    $$y = 3(2) - 5$$
    $$y = 6 - 5$$
    $$y = 1$$
    The solution is (2, 1).

21. $y = 2x - 13$

    $-4x - 7 = 9y$

    Substitute $2x - 13$ for y in the second equation.
    $$-4x - 7 = 9y$$
    $$-4x - 7 = 9(2x - 13)$$
    $$-4x - 7 = 18x - 117$$
    $$-22x - 7 = -117$$
    $$-22x = -110$$
    $$x = \frac{-110}{-22} = 5$$

    Now substitute 5 for x in the first equation.
    $$y = 2x - 13$$
    $$y = 2(5) - 13$$
    $$y = 10 - 13$$
    $$y = -3$$
    The solution is (5, –3).

23. $4x - 5y = -4$

$3x = 2y - 3$

First solve the second equation for x.

$\dfrac{1}{3}(3x) = \dfrac{1}{3}(2y - 3)$

$x = \dfrac{2}{3}y - 1$

Now substitute $\dfrac{2}{3}y - 1$ for x in the first equation.

$4x - 5y = -4$

$4\left(\dfrac{2}{3}y - 1\right) - 5y = -4$

$\dfrac{8}{3}y - 4 - 5y = -4$

$\dfrac{8}{3}y - \dfrac{15}{3}y = 0$

$-\dfrac{7}{3}y = 0$

$y = 0$

Finally substitute 0 for y in the equation $x = \dfrac{2}{3}y - 1$.

$x = \dfrac{2}{3}(0) - 1 = -1$

The solution is $(-1, 0)$.

25. $4x + 5y = -6$

$2x - \dfrac{10}{3}y = -4$

First solve the second equation for x,

$2x - \dfrac{10}{3}y = -4$

$3\left(2x - \dfrac{10}{3}y\right) = 3(-4)$

$6x - 10y = -12$

$6x = 10y - 12$

$x = \dfrac{5}{3}y - 2$

Now substitute $\dfrac{5}{3}y - 2$ for x in the first equation.

$4x + 5y = -6$

$4\left(\dfrac{5}{3}y - 2\right) + 5y = -6$

$\dfrac{20}{3}y - 8 + 5y = -6$

$\dfrac{20}{3}y + \dfrac{15}{3}y = 8 - 6$

$\dfrac{35}{3}y = 2$

$y = \dfrac{6}{35}$

Finally, substitute $\dfrac{9}{37}$ for y in the equation $x = \dfrac{5}{3}y - 2$.

$x = \dfrac{5}{3}\left(\dfrac{6}{35}\right) - 2$

$x = \dfrac{2}{7} - \dfrac{14}{7}$

$x = -\dfrac{12}{7}$

The solution is $\left(-\dfrac{12}{7}, \dfrac{6}{35}\right)$.

27. $3x - 4y = 15$

$-6x + 8y = -14$

First solve the second equation for y,

$-6x + 8y = -14$

$8y = 6x - 14$

$y = \dfrac{3}{4}x - \dfrac{7}{4}$

Now substitute $\dfrac{3}{4}x - \dfrac{7}{4}$ for y in the first equation.

$3x - 4\left(\dfrac{3}{4}x - \dfrac{7}{4}\right) = 15$

$3x - 3x + 7 = 15$

$7 = 15$ False

Since this is a false statement, there is no solution to this system.

29.
$$4x - y = 1$$
$$10x + \frac{1}{2}y = 1$$

Solve the first equation for y.
$$y = 4x - 1$$
Now substitute $4x - 1$ for y in the second equation.
$$10x + \frac{1}{2}(4x - 1) = 1$$
$$10x + 2x - \frac{1}{2} = 1$$
$$12x - \frac{1}{2} = 1$$
$$12x = \frac{3}{2}$$
$$\frac{1}{12}(12x) = \frac{1}{12}\left(\frac{3}{2}\right)$$
$$x = \frac{1}{8}$$

Now substitute $\frac{1}{8}$ for x in the first equation;
$$y = 4x - 1$$
$$y = 4\left(\frac{1}{8}\right) - 1 = \frac{1}{2} - 1 = -\frac{1}{2}$$
The solution to the system is $\left(\frac{1}{8}, -\frac{1}{2}\right)$.

31. Let $x =$ the smaller number,
then $y =$ the larger number.
$$x + y = 80$$
$$x + 8 = y$$
Substitute $x + 8$ for y in the first equation.
$$x + y = 80$$
$$x + (x + 8) = 80$$
$$2x + 8 = 80$$
$$2x = 72$$
$$x = 36$$
Now substitute 36 for x in the second equation:
$$x + 8 = y$$
$$36 + 8 = y$$
$$44 = y$$
The two integers are 36 and 44.

33. Let $l =$ the length of the rectangle,
then $w =$ the width.
$$2l + 2w = 50$$
$$l = 9 + w$$
Substitute $9 + w$ for l in the first equation and solve for w.
$$2l + 2w = 50$$
$$2(9 + w) + 2w = 50$$
$$18 + 2w + 2w = 50$$
$$4w = 32$$
$$w = 8$$
Now substitute 8 for w in the second equation to find l.
$$l = 9 + w$$
$$l = 9 + 8 = 17$$
The length of the rectangle is 17 feet and the width is 8 feet.

35. Let $f =$ the number of attendees who paid the full fee and $d =$ the number of attendees who received the discount.
$$f + d = 2500$$
$$d = f - 622$$
Substitute $f - 622$ for d in the first equation.
$$f + d = 2500$$
$$f + (f - 622) = 2500$$
$$2f - 622 = 2500$$
$$2f = 3122$$
$$f = 1561$$
Now substitute 1561 for f in the second equation and solve for d.
$$d = f - 622 = 1561 - 622 = 939$$
The number of attendees who paid the full amount was 1561 and then 939 got the discount fee.

37. Let $c =$ the client's portion of the award and $a =$ the attorneys portion.
$$c + a = 40{,}000$$
$$c = 3a$$
Now solve the first equation for a and substitute it for a in the second equation.
$$c + a = 40000$$
$$a = 40000 - c$$

$$c = 3a$$
$$c = 3(40000 - c)$$
$$c = 120{,}000 - 3c$$
$$4c = 120{,}000$$
$$c = 30{,}000$$
The client received \$30,000.

39. $c = 1280 + 794n$

$c = 874n$

   a. Substitute $874n$ for c in the first equation.

$874n = 1280 + 794n$

$80n = 1280$

$n = 16$

The mortgage plans will have the same total cost at 16 months.

   b. $12 \text{ years} \cdot \left(\dfrac{12 \text{ months}}{1 \text{ year}} \right) = 144 \text{ months}$

$c = 1280 + 794(144) = 115{,}616$

$c = 874(144) = 125{,}856$

Yes, since \$115,616 is less than \$125,856, she should refinance.

41. a. Substitute $65 + 72t$ for m in the first equation and solve for t.

$65 + 72t = 80 + 60t$

$12t = 15$

$t = \dfrac{15}{12} = 1.25$

It will take Roberta 1.25 hours to catch up to Jean.

   b. Now substitute 1.25 for t in one of the original equtions.

$m = 80 + 60(1.25)$

$m = 80 + 75$

$m = 155$

They will be at mile marker 155 when they meet.

43. a. $T = 180 - 10t$

   b. $T = 20 + 6t$

   c. $T = 180 - 10t$

$T = 20 + 6t$

Substitute $20 + 6t$ for T in the first equation.

$20 + 6t = 180 - 10t$

$16t = 160$

$t = 10$

It will take 10 minutes for the ball and oil to reach the same temperature.

   d. Substitute 10 for t in the original equation $T = 180 - 10t$.

$T = 180 - 10(10) = 180 - 100 = 80$

The temperature will be 80°F.

45. $\dfrac{27.6}{1.2} = 23$

The willow tree is about 23 years old.

46. $4x + 5y = 22$

Substitute 3 for x and solve for y.

$4(3) + 5y = 22$

$12 + 5y = 22$

$5y = 10$

$y = 2$

The ordered pair is (3, 2).

47. $4x - 8y = 16$

Let $x = 0$ and solve for y.

$4(0) - 8y = 16$

$-8y = 16$

$y = -2$

Let $y = 0$ and solve for x.

$4x - 8(0) = 16$

$4x = 16$

$x = 4$

The intercepts are (0, –2) and (4, 0).

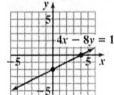

48. $3x - 5y = 25$

Write in slope-intercept form.

$-5y = -3x + 25$

$y = \dfrac{3}{5}x - 5$

The slope is $\dfrac{3}{5}$ and the y intercept is $(0, -5)$.

49. Use the points (–2, 0) and (0, 4) to find the slope of the line: $m = \dfrac{4 - 0}{0 - (-2)} = \dfrac{4}{2} = 2$

The y-intercept is 4, so $b = 4$.

Thus, the equation of the line is:

$y = mx + b$

$y = 2x + 4$

50. $(6x + 7)(3x - 2) = 18x^2 - 12x + 21x - 14$

$= 18x^2 + 9x - 14$

Exercise Set 9.3

1. Multiply the top equation by 2.
   Now the top equation contains a $-2x$ and the bottom equation contains a $2x$. When the equations are added together, the variable x will be eliminated.

3. You will obtain a false statement, such as $0 = 6$

5. $\qquad x + y = 6$
   $\qquad \underline{x - y = 4}$
   Add: $2x \quad = 10$
   $\qquad\quad x = 5$
   Substitute 5 for x in the first equation.
   $x + y = 6$
   $5 + y = 6$
   $\quad\ y = 1$
   The solution is (5, 1).

7. Add: $\quad x + y = 9$
   $\qquad\quad \underline{-x + y = 1}$
   $\qquad\qquad 2y = 10$
   $\qquad\qquad\ \ y = 5$
   Substitute 5 for y in the first equation.
   $x + y = 9$
   $x + 5 = 9$
   $\quad\ x = 4$
   The solution is (4, 5).

9. Add: $\quad x + 2y = 21$
   $\qquad\quad \underline{2x - 2y = -6}$
   $\qquad\quad 3x \qquad = 15$
   $\qquad\qquad\ x = 5$
   Substitute 5 for x in the first equation.
   $x + 2y = 21$
   $5 + 2y = 21$
   $\quad\ 2y = 16$
   $\qquad y = 8$
   The solution is (5, 8).

11. $\quad 4x + y = 6 \quad (eq.\ 1)$
    $-8x - 2y = 20 \quad (eq.\ 2)$
    Multiply the first equation by 2, then add to the second equation.
    $2[4x + y = 6]$

gives
$\quad 8x + 2y = 12$
$\underline{-8x - 2y = 20}$
$\qquad\qquad 0 = 32 \quad$ False
There is no solution.

13. $-5x + y = 14 \quad (eq.\ 1)$
    $-3x + y = -2 \quad (eq.\ 2)$
    To eliminate y, multiply the first equation by -1 and then add.
    $-1[-5x + y = 14]$
    gives
    $\quad 5x - y = -14$
    $\underline{-3x + y = -2}$
    $\quad 2x \qquad = -16$
    $\qquad\quad x = -8$
    Substitute -8 for x in the first equation.
    $\qquad -5x + y = 14$
    $-5(-8) + y = 14$
    $\qquad 40 + y = 14$
    $\qquad\qquad\ y = -26$
    The solution is (–8, –26).

15. $2x + \ y = -6$
    $2x - 2y = \ 3$
    To eliminate y, multiply the first equation by 2 and then add.
    $2[2x + y = -6]$
    gives
    $4x + 2y = -12$
    $\underline{2x - 2y = \ 3}$
    $6x \qquad = -9$
    $\qquad x = -\dfrac{3}{2}$
    Substitute $-\dfrac{3}{2}$ for x in the first equation.
    $\qquad 2x + y = -6$
    $2\left(-\dfrac{3}{2}\right) + y = -6$
    $\qquad -3 + y = -6$
    $\qquad\qquad y = -3$
    The solution is $\left(-\dfrac{3}{2}, -3\right)$.

17. $2y = 6x + 16 \quad (eq.\ 1)$

$y = -3x - 4 \quad (eq.\ 2)$

Rewrite the equations to align the variables on the left hand side of the equal sign.

$-6x + 2y = 16$

$3x + y = -4$

To eliminate x, multiply the second equation by 2, then add.

$2[3x + y = -4]$

gives

$-6x + 2y = 16$

$\underline{6x + 2y = -8}$

$4y = 8$

$y = 2$

Substitute 2 for y in the second equation.

$3x + y = -4$

$3x + 2 = -4$

$3x = -6$

$x = -2$

The solution is $(-2, 2)$.

19. $5x + 3y = 12 \quad (eq.\ 1)$

$3x - 6y = 15 \quad (eq.\ 2)$

To eliminate y, multiply the first equation by 2 and then add.

$2[5x + 3y = 12]$

gives

$10x + 6y = 24$

$\underline{3x - 6y = 15}$

$13x \quad = 39$

$x = 3$

Substitute 3 for x in the first equation.

$5x + 3y = 12$

$5(3) + 3y = 12$

$15 + 3y = 12$

$3y = -3$

$y = -1$

The solution is $(3, -1)$.

21. $-2y = -4x + 12$

$y = 2x - 6$

Align x- and y-terms on the left side.

$4x - 2y = 12 \quad (eq.\ 1)$

$-2x + y = -6 \quad (eq.\ 2)$

To eliminate x, multiply the second equation by 2 and then add.

$2[-2x + y = -6]$

gives

$4x - 2y = 12$

$\underline{-4x + 2y = -12}$

$0 = 0$

Since this is a true statement, there are an infinite number of solutions. This is a dependent system.

23. $5x - 4y = -3$

$7y = 2x + 12$

Align x- and y-terms on the left side.

$5x - 4y = -3 \quad (eq.\ 1)$

$-2x + 7y = 12 \quad (eq.\ 2)$

To eliminate x, multiply the first equation by 2 and the second equation by 5 and then add.

$2[5x - 4y = -3]$

$5[-2x + 7y = 12]$

gives

$10x - \ 8y = -6$

$\underline{-10x + 35y = 60}$

$27y = 54$

$y = \ 2$

Substitute 2 for y in the first equation.

$5x - 4y = -3$

$5x - 4(2) = -3$

$5x - 8 = -3$

$5x = 5$

$x = 1$

The solution is $(1, 2)$.

25. $5x - 4y = 1 \quad (eq.\ 1)$

$-10x + 8y = -3 \quad (eq.\ 2)$

Multiply the first equation by 2, then add.

$2[5x - 4y = 1]$

gives

$10x - 8y = \ 1$

$\underline{-10x + 8y = -3}$

$0 = -2 \ \text{False}$

There is no solution.

27. $5x - 6y = 0$ $(eq. 1)$

$3x + 4y = 0$ $(eq. 2)$

To eliminate y, multiply the first equation by 3 and the second equation by -5 and then add.

$3[5x - 6y = 0]$

$-5[3x + 4y = 0]$

gives

$15x - 18y = 0$

$\underline{-15x - 20y = 0}$

$-38y = 0$

$y = 0$

Substitute 0 for y in the second equation.

$3x + 4y = 0$

$3x + 4(0) = 0$

$3x = 0$

$x = 0$

The solution is $(0, 0)$.

29. $-5x + 4y = -20$ $(eq. 1)$

$3x - 2y = 15$ $(eq. 2)$

To eliminate y, multiply the second equation by 2, then add.

$2[3x - 2y = 15]$

gives

$-5x + 4y = -20$

$\underline{6x - 4y = \ \ 30}$

$x \qquad = \ 10$

Substitute 10 for x in the first equation.

$-5x + 4y = -20$

$-5(10) + 4y = -20$

$-50 + 4y = -20$

$4y = 30$

$y = \dfrac{15}{2}$

The solution is $\left(10, \dfrac{15}{2}\right)$.

31. $6x = 4y + 12$

$3y - 5x = -6$

Align the x- and y-terms on the left side.

$6x - 4y = 12$

$5x - 3y = 6$

To eliminate x, multiply the first equation by 5 and the second equation by -6 and then add.

$5(6x - 4y = 12)$

$-6(5x - 3y = 6)$

gives

$30x - 20y = 60$

$\underline{-30x + 18y = -36}$

$-2y = 24$

$y = -12$

Substitute -12 for y in the first equation.

$6x - 4y = 12$

$6x - 4(-12) = 12$

$6x + 48 = 12$

$6x = -36$

$x = -6$

The solution is $(-6, -12)$.

33. $4x + 5y = 0$

$3x = 6y + 4$

Align the x- and y-terms on the left side.

$4x + 5y = 0$

$3x - 6y = 4$

To eliminate y, multiply the first equation by 6 and the second equation by 5 and then add.

$6[4x + 5y = 0]$

$5[3x - 6y = 4]$

gives

$24x + 30y = 0$

$\underline{15x - 30y = 20}$

$39x \qquad = 20$

$x = \dfrac{20}{39}$

Substitute $\dfrac{20}{39}$ for x in the first equation.

$4x + 5y = 0$

$4\left(\dfrac{20}{39}\right) + 5y = 0$

$5y = -\dfrac{80}{39}$

$y = -\dfrac{16}{39}$

The solution is $\left(\dfrac{20}{39}, -\dfrac{16}{39}\right)$.

35. $x - \dfrac{1}{2}y = 4$

$3x + y = 6$

To eliminate y, multiply the first equation by 2 and then add.

$2\left[x - \dfrac{1}{2}y = 4\right]$

gives

$2x - y = 8$

$\dfrac{3x + y = 6}{5x \quad = 14}$

$x = \dfrac{14}{5}$

Substitute $\dfrac{14}{5}$ for x in the second equation.

$3x + y = 6$

$3\left(\dfrac{14}{5}\right) + y = 6$

$\dfrac{42}{5} + y = 6$

$y = 6 - \dfrac{42}{5}$

$y = \dfrac{30}{5} - \dfrac{42}{5}$

$y = -\dfrac{12}{5}$

The solution is $\left(\dfrac{14}{5}, -\dfrac{12}{5}\right)$.

37. $3x - y = 4$

$2x - \dfrac{2}{3}y = 8$

To eliminate y, multiply the first equation by $-\dfrac{2}{3}$ and then add.

$-\dfrac{2}{3}\left[3x - y = 4\right]$

gives

$-2x + \dfrac{2}{3}y = \dfrac{8}{3}$

$\dfrac{2x - \dfrac{2}{3}y = 8}{0 = \dfrac{32}{3} \quad \text{False}}$

There is no solution.

39. Add: $x + y = 20$

$\dfrac{x - y = 8}{2x = 28}$

$x = 14$

Substitute 14 for x in the first equation to find the second number.

$x + y = 20$

$14 + y = 20$

$y = 6$

The numbers are 14 and 6.

41. $x + 2y = 14$

$x - \ y = \ 2$

To eliminate y, multiply the second equation by 2 and then add.

$2\left[x - y = 2\right]$

gives

$x + 2y = 14$

$\dfrac{2x - 2y = 4}{3x = 18}$

$x = 6$

Substitute 6 for x in the second equation to find the other number.

$x - y = 2$

$6 - y = 2$

$y = 4$

The numbers are 6 and 4.

43. Let x be the length of the rectangle and y be the width.

$2x + 2y = 18$

$x = 2y$

Align the x and y variables on the left hand side of the equal sign.

$2x + 2y = 18$

$\dfrac{x - 2y = 0}{}$

Now add the equations to eliminate the y.

$3x = 18$

$x = 6$

Substitute 6 for x in the second equation and solve for y.

$x = 2y$

$6 = 2y$

$3 = y$

The width is 3 inches and the length is 6 inches.

45. Let l = the length and w = the width of the photograph.

$2l + 2w = 36$

$l - w = 2$

Multiply the second equation by 2

$2[l - w = 2]$

gives

$2l + 2w = 36$

$\underline{2l - 2w = 4}$

$4l = 40$

$l = 10$

$10 - w = 2$

$w = 8$

The length is 10 inches and the width is 8 inches.

47. Answers will vary.

49. a. $4x + 2y = 1000$

$2x + 4y = 800$

To eliminate x, multiply the first equation by -2.

$-2[4x + 2y = 1000]$

gives

$-8x - 4y = -2000$

$\underline{2x + 4y = 800}$

$-6x = -1200$

$x = 200$

Substitute 200 for x in the first equation.

$4x + 2y = 1000$

$4(200) + 2y = 1000$

$800 + 2y = 1000$

$2y = 200$

$y = 100$

The solution is (200, 100).

b. They will have the same solution. Dividing an equation by a nonzero number does not change the solutions.

$2x + y = 500$

$2x + 4y = 800$

To eliminate x, multiply the first equation by -1 and then add.

$-1[2x + y = 500]$

gives

$-2x - y = -500$

$\underline{2x + 4y = 800}$

$3y = 300$

$y = 100$

Substitute 100 for y in the first equation.

$2x + y = 500$

$2x + 100 = 500$

$2x = 400$

$x = 200$

The solution is (200, 100).

51. $\dfrac{x+2}{2} - \dfrac{y+4}{3} = 4$

$\dfrac{x+y}{2} = \dfrac{1}{2} + \dfrac{x-y}{3}$

Start by writing each equation in standard form after clearing fractions.

For the first equation:

$6\left[\dfrac{x+2}{2} - \dfrac{y+4}{3} = 4\right]$

$3(x+2) - 2(y+4) = 24$

$3x + 6 - 2y - 8 = 24$

$3x - 2y - 2 = 24$

$3x - 2y = 26$

For the second equation:

$6\left[\dfrac{x+y}{2} = \dfrac{1}{2} + \dfrac{x-y}{3}\right]$

$3(x+y) = 3 + 2(x-y)$

$3x + 3y = 3 + 2x - 2y$

$x + 5y = 3$

The new system is:

$3x - 2y = 26$

$x + 5y = 3$

To eliminate x, multiply the second equation by -3 and then add.

$-3[x + 5y = 3]$

gives

$3x - 2y = 26$

$\underline{-3x - 15y = -9}$

$-17y = 17$

$y = -1$

Now, substitute -1 for y in the equation $x + 5y = 3$.

$x + 5(-1) = 3$

$x - 5 = 3$

$x = 8$

The solution is $(8, -1)$.

53. $x + 2y - z = 2$

$2x - y + z = 3$

$3x + y + z = 8$

Add the second and third equations to eliminate y.

$2x - y + z = 3$

$\underline{3x + y + z = 8}$

$5x + 2z = 11$

Multiply the second equation by 2 and then add the first equation to eliminate y.

$2[2x - y + z = 3]$

gives

$4x - 2y + 2z = 6$

$\underline{x + 2y - z = 2}$

$5x + z = 8$

Now we have two equations with two unknowns.

$5x + 2z = 11$

$5x + z = 8$

To eliminate z, multiply the second equation by -2 and then add.

$-2[5x + z = 8]$

gives

$5x + 2z = 11$

$\underline{-10x - 2z = -16}$

$-5x = -5$

$x = 1$

Substitute 1 for x in the second equation and solve for z

$5x + z = 8$

$5(1) + z = 8$

$5 + z = 8$

$z = 3$

Now go back to one of the original equations and substitute 1 for x and 3 for z and then find y. The third equation is used below.

$3x + y + z = 8$

$3(1) + y + 3 = 8$

$3 + y + 3 = 8$

$y + 6 = 8$

$y = 2$

The solution to the system is $(1, 2, 3)$.

55. $5^3 = 5 \cdot 5 \cdot 5 = 125$

56. $2(2x - 3) = 2x + 8$

$4x - 6 = 2x + 8$

$4x - 2x = 6 + 8$

$2x = 14$

$x = 7$

57. $(4x^2y - 3xy + y) - (2x^2y + 6xy - 3y)$

$= 4x^2y - 3xy + y - 2x^2y - 6xy + 3y$

$= 4x^2y - 2x^2y - 3xy - 6xy + y + 3y$

$= 2x^2y - 9xy + 4y$

58. $\left(8a^4b^2c\right)\left(4a^2b^7c^4\right) = 8 \cdot 4 \cdot a^4 \cdot a^2 \cdot b^2 \cdot b^7 \cdot c \cdot c^4$

$= 32a^{4+2}b^{2+7}c^{1+4}$

$= 32a^6b^9c^5$

59. $xy + xc - ay - ac = x(y + c) - a(y + c)$

$= (y + c)(x - a)$

60. $f(x) = 2x^2 - 4$

$f(-3) = 2(-3)^2 - 4$

$= 2(9) - 4$

$= 18 - 4$

$= 14$

Exercise Set 9.4

1. The graph will be a plane.

3.

$x = 1$

$2x - y = 4$

$-3x + 2y - 2z = 1$

Substitute 1 for x in the second equation.

$2(1) - y = 4$

$2 - y = 4$

$-y = 2$

$y = -2$

Substitute 1 for x and -2 for y in the third equation.

$-3(1) + 2(-2) - 2z = 1$

$-3 - 4 - 2z = 1$

$-7 - 2z = 1$

$-2z = 8$

$z = -4$

The solution is $(1, -2, -4)$.

5.
$$5x - 6z = -17$$
$$3x - 4y + 5z = -1$$
$$2z = -6$$

Solve the third equation for z.
$$2z = -6$$
$$z = -3$$

Substitute –3 for z in the first equation.
$$5x - 6z = -17$$
$$5x - 6(-3) = -17$$
$$5x + 18 = -17$$
$$5x = -35$$
$$x = -7$$

Substitute –7 for x and –3 for z in the second equation.
$$3x - 4y + 5z = -1$$
$$3(-7) - 4y + 5(-3) = -1$$
$$-21 - 4y - 15 = -1$$
$$-4y - 36 = -1$$
$$-4y = 35$$
$$y = \frac{35}{-4} = -\frac{35}{4}$$

The solution is $\left(-7, -\dfrac{35}{4}, -3\right)$.

7.
$$x + 2y = 6$$
$$3y = 9$$
$$x + 2z = 12$$

Solve the second equation for y.
$$3y = 9$$
$$y = 3$$

Substitute 3 for y in the first equation.
$$x + 2y = 6$$
$$x + 2(3) = 6$$
$$x + 6 = 6$$
$$x = 0$$

Substitute 0 for x in the third equation.
$$x + 2z = 12$$
$$0 + 2z = 12$$
$$2z = 12$$
$$z = 6$$

The solution is $(0, 3, 6)$.

9.
$$x - 2y = -3 \quad (1)$$
$$3x + 2y = 7 \quad (2)$$
$$2x - 4y + z = -6 \quad (3)$$

To eliminate y between equations (1) and (2), add equations (1) and (2).

$$x - 2y = -3$$
$$\underline{3x + 2y = 7}$$
Add: $\quad 4x = 4$
$$x = 1$$

Substitute 1 for x in equation (1).
$$(1) - 2y = -3$$
$$-2y = -4$$
$$y = 2$$

Substitute 1 for x, and 2 for y in equation (3).
$$2(1) - 4(2) + z = -6$$
$$2 - 8 + z = -6$$
$$-6 + z = -6$$
$$z = 0$$

The solution is $(1, 2, 0)$.

11.
$$2y + 4z = 2 \quad (1)$$
$$x + y + 2z = -2 \quad (2)$$
$$2x + y + z = 2 \quad (3)$$

To eliminate x between equations (2) and (3), multiply equation (2) by –2 and then add.
$$-2[x + y + 2z = -2]$$
$$2x + y + z = 2$$
gives
$$-2x - 2y - 4z = 4$$
$$\underline{2x + y + z = 2}$$
Add: $\quad -y - 3z = 6 \quad (4)$

To eliminate y between equations (1) and (4), multiply equation (4) by 2 and then add.
$$2y + 4z = 2$$
$$2\left[-y - 3z = 6\right]$$
gives
$$2y + 4z = 2$$
$$\underline{-2y - 6z = 12}$$
Add: $\quad -2z = 14$
$$z = -7$$

Substitute –7 for z in equation (1).

$$2y + 4(-7) = 2$$
$$2y - 28 = 2$$
$$2y = 30$$
$$y = 15$$

Substitute 15 for y, and -7 for z in equation (3).

$$2x + (15) + (-7) = 2$$
$$2x + 8 = 2$$
$$2x = -6$$
$$x = -3$$

The solution is $(-3, 15, -7)$.

13.
$$3p + 2q = 11 \quad (1)$$
$$4q - r = 6 \quad (2)$$
$$6p + 7r = 4 \quad (3)$$

To eliminate r between equations (2) and (3), multiply equation (2) by 7 and add to equation (3).

$$7[4q - r = 6]$$
$$6p + 7r = 4$$

gives

$$28q - 7r = 42$$
$$\underline{6p + 7r = 4}$$

Add: $6p + 28q = 46 \quad (4)$

Equations (1) and (4) are two equations in two unknowns. To eliminate p, multiply equation (1) by -2 and add to equation (4).

$$-2[3p + 2q = 11]$$
$$6p + 28q = 46$$

gives

$$-6p - 4q = -22$$
$$\underline{6p + 28q = 46}$$

Add: $\quad 24q = 24$
$$q = 1$$

Substitute 1 for q in equation (1).

$$3p + 2q = 11$$
$$3p + 2(1) = 11$$
$$3p + 2 = 11$$
$$3p = 9$$
$$p = 3$$

Substitute 1 for q in equation (2).

$$4q - r = 6$$
$$4(1) - r = 6$$
$$4 - r = 6$$
$$-r = 2$$
$$r = -2$$

The solution is $(3, 1, -2)$.

15.
$$p + q + r = 4 \quad (1)$$
$$p - 2q - r = 1 \quad (2)$$
$$2p - q - 2r = -1 \quad (3)$$

To eliminate q between equations (1) and (3), simply add.

$$p + q + r = 4$$
$$\underline{2p - q - 2r = -1}$$

Add: $3p \quad - r = 3 \quad (4)$

To eliminate q between equations (1) and (2), multiply equation (1) by 2 and then add.

$$2[p + q + r = 4]$$
$$p - 2q - r = 1$$

gives

$$2p + 2q + 2r = 8$$
$$\underline{p - 2q - r = 1}$$

Add: $3p \quad + r = 9 \quad (5)$

Equations (4) and (5) are two equations in two unknowns.

$$3p - r = 3$$
$$3p + r = 9$$

To eliminate r, simply add these two equations.

$$3p - r = 3$$
$$\underline{3p + r = 9}$$

Add: $6p \quad = 12$
$$p = 2$$

Substitute 2 for p in equation (5).

$$3p + r = 9$$
$$3(2) + r = 9$$
$$6 + r = 9$$
$$r = 3$$

Substitute 2 for p and 3 for r in equation (1).

$$p + q + r = 4$$
$$2 + q + 3 = 4$$
$$q + 5 = 4$$
$$q = -1$$

The solution is $(2, -1, 3)$.

17.
$$2x - 2y + 3z = 5 \quad (1)$$
$$2x + y - 2z = -1 \quad (2)$$
$$4x - y - 3z = 0 \quad (3)$$

To eliminate y between equations (2) and (3), simply add.

$$2x + y - 2z = -1$$
$$\underline{4x - y - 3z = 0}$$

Add: $6x \quad - 5z = -1 \quad (4)$

To eliminate y between equations (1) and (2), multiply equation (2) by 2 and then add.
$$2x - 2y + 3z = 5$$
$$2[2x + y - 2z = -1]$$
gives
$$2x - 2y + 3z = 5$$
$$\underline{4x + 2y - 4z = -2}$$
Add: $6x \quad - z = 3$ (5)

Equations (4) and (5) are two equations in two unknowns.
$$6x - 5z = -1$$
$$6x - z = 3$$
To eliminate x, multiply equation (5) by -1 and then add.
$$6x - 5z = -1$$
$$-1[6x - z = 3]$$
gives
$$6x - 5z = -1$$
$$\underline{-6x + z = -3}$$
Add: $-4z = -4$
$$z = 1$$

Substitute 1 for z in equation (5).
$$6x - z = 3$$
$$6x - 1 = 3$$
$$6x = 4$$
$$x = \frac{4}{6} = \frac{2}{3}$$

Substitute $\frac{2}{3}$ for x and 1 for z in equation (2).
$$2x + y - 2z = -1$$
$$2\left(\frac{2}{3}\right) + y - 2(1) = -1$$
$$\frac{4}{3} + y - 2 = -1$$
$$y - \frac{2}{3} = -1$$
$$y = -1 + \frac{2}{3} = -\frac{1}{3}$$

The solution is $\left(\frac{2}{3}, -\frac{1}{3}, 1\right)$.

19. $\quad r - 2s + t = 2$ (1)
$$2r + 3s - t = -3 \text{ (2)}$$
$$2r - s - 2t = 1 \quad \text{(3)}$$
To eliminate t between equations (1) and (2), add the two equations.
$$r - 2s + t = 2$$
$$\underline{2r + 3s - t = -3}$$
Add: $\quad 3r + s = -1$ (4)

To eliminate t between equations (2) and (3), multiply equation (2) by -2 and then add.
$$-2[2r + 3s - t = -3]$$
$$2r - s - 2t = -1$$
gives
$$-4r - 6s + 2t = 6$$
$$\underline{2r - s - 2t = 1}$$
Add: $\quad -2r - 7s = 7$ (5)

Equations (4) and (5) make a system of two equations in two variables. To eliminate r, multiply equation (4) by 2 and equation (5) by 3, then add the equations.
$$2\left[3r + s = -1\right]$$
$$3\left[-2r - 7s = 7\right]$$
gives:
$$6r + 2s = -2$$
$$\underline{-6r - 21s = 21}$$
Add: $\quad -19s = 19$
$$s = -1$$

Substitute -1 for s into equation (4) and solve for r.
$$3r + s = -1$$
$$3r - 1 = -1$$
$$3r = 0$$
$$r = 0$$

Finally, substitute 0 for r and -1 for s into equation (1)
$$r - 2s + t = 2$$
$$(0) - 2(-1) + t = 2$$
$$2 + t = 2$$
$$t = 0$$
The solution is $(0, -1, 0)$.

21. $2a+2b-c=2$ (1)

$3a+4b+c=-4$ (2)

$5a-2b-3c=5$ (3)

To eliminate c between equations (1) and (2), simply add.

$2a+2b-c\ \ =2$

$\underline{3a+4b+c=-4}$

Add: $5a+6b\ \ \ \ =-2$ (4)

To eliminate c between equations (2) and (3), multiply equation (2) by 3 and then add.

$3[3a+4b+c=-4]$

$5a-2b-3c=5$

gives

$9a+12b+3c=-12$

$\underline{5a-\ 2b-3c=\ \ \ 5}$

Add: $14a+10b\ \ \ \ \ =-7$ (5)

Equations (4) and (5) are two equations in two unknowns.

$5a+6b=-2$

$14a+10b=-7$

To eliminate b, multiply equation (4) by -5 and multiply equation (5) by 3 and then add.

$-5[5a+6b=-2]$

$3[14a+10b=-7]$

gives

$-25a-30b=\ \ 10$

$\underline{42a+30b=-21}$

Add: $17a\ \ \ \ \ \ \ =-11$

$a=-\dfrac{11}{17}$

Substitute $-\dfrac{11}{17}$ for a in equation (4).

$5a+6b=-2$

$5\left(-\dfrac{11}{17}\right)+6b=-2$

$-\dfrac{55}{17}+6b=-2$

$6b=-2+\dfrac{55}{17}$

$b=\dfrac{1}{6}\cdot\dfrac{21}{17}=\dfrac{7}{34}$

Substitute $-\dfrac{11}{17}$ for a and $\dfrac{7}{34}$ for b in equation (2).

$3a+4b+c=-4$

$3\left(-\dfrac{11}{17}\right)+4\left(\dfrac{7}{34}\right)+c=-4$

$-\dfrac{33}{17}+\dfrac{14}{17}+c=-4$

$-\dfrac{19}{17}+c=-4$

$c=-4+\dfrac{19}{17}$

$c=-\dfrac{49}{17}$

The solution is $\left(-\dfrac{11}{17},\dfrac{7}{34},-\dfrac{49}{17}\right)$.

23. $-x+3y+z=0$ (1)

$-2x+4y-z=0$ (2)

$3x-y+2z=0$ (3)

To eliminate z between equations (1) and (2), simply add.

$-x+3y+z=0$

$\underline{-2x+4y-z=0}$

Add: $-3x+7y\ \ \ \ =0$ (4)

To eliminate z between equations (2) and (3), multiply equation (2) by 2 and then add.

$2[-2x+4y-z]=0$

$3x-y+2z=0$

gives

$-4x+8y-2z=0$

$\underline{3x-\ y+2z=0}$

Add: $-x+7y\ \ \ \ =0$ (5)

Equations (4) and (5) are two equations in two unknowns.

$-3x+7y=0$

$-x+7y=0$

To eliminate y, multiply equation (4) by -1 and then add.

$-1[-3x+7y=0]$

$-x+7y=0$

gives

$3x-7y=0$

$\underline{-x+7y=0}$

Add: $2x\ \ \ \ =0$

$x=0$

Substitute 0 for x in equation (5).

$$-x + 7y = 0$$
$$-0 + 7y = 0$$
$$7y = 0$$
$$y = 0$$

Finally, substitute 0 for x and 0 for y into equation (1).
$$-x + 3y + z = 0$$
$$-0 + 3(0) + z = 0$$
$$0 + z = 0$$
$$z = 0$$

The solution is (0, 0, 0).

25. $-\dfrac{1}{4}x + \dfrac{1}{2}y - \dfrac{1}{2}z = -2$ (1)

   $\dfrac{1}{2}x + \dfrac{1}{3}y - \dfrac{1}{4}z = 2$ (2)

   $\dfrac{1}{2}x - \dfrac{1}{2}y + \dfrac{1}{4}z = 1$ (3)

To clear fractions, multiply equation (1) by 4, equation (2) by 12, and equation (3) by 4.

$$4\left(-\dfrac{1}{4}x + \dfrac{1}{2}y - \dfrac{1}{2}z = -2\right)$$

$$12\left(\dfrac{1}{2}x + \dfrac{1}{3}y - \dfrac{1}{4}z = 2\right)$$

$$4\left(\dfrac{1}{2}x - \dfrac{1}{2}y + \dfrac{1}{4}z = 1\right)$$

gives
$$-x + 2y - 2z = -8 \quad (4)$$
$$6x + 4y - 3z = 24 \quad (5)$$
$$2x - 2y + z = 4 \quad (6)$$

To eliminate y between equations (4) and (6), simply add.
$$-x + 2y - 2z = -8$$
$$\underline{2x - 2y + z = 4}$$
Add: $x - z = -4$ (7)

To eliminate y between equations (5) and (6), multiply equation (6) by 2 and then add to equation (5).
$$6x + 4y - 3z = 24$$
$$2[2x - 2y + z = 4]$$
gives
$$6x + 4y - 3z = 24$$
$$\underline{4x - 4y + 2z = 8}$$
Add: $10x - z = 32$ (8)

Equations (7) and (8) are two equations in two unknowns.

$$x - z = -4$$
$$10x - z = 32$$

To eliminate z, multiply equation (7) by -1 and then add.
$$-1[x - z = -4]$$
$$10x - z = 32$$
gives
$$-x + z = 4$$
$$\underline{10x - z = 32}$$
Add: $9x = 36$
$$x = \frac{36}{9} = 4$$

Substitute 4 for x in equation (7).
$$x - z = -4$$
$$4 - z = -4$$
$$-z = -8$$
$$z = 8$$

Finally, substitute 4 for x and 8 for z in equation (4).
$$-x + 2y - 2z = -8$$
$$-4 + 2y - 2(8) = -8$$
$$-4 + 2y - 16 = -8$$
$$2y - 20 = -8$$
$$2y = 12$$
$$y = \frac{12}{2} = 6$$

The solution is (4, 6, 8).

27. $x - \dfrac{2}{3}y - \dfrac{2}{3}z = -2$ (1)

   $\dfrac{2}{3}x + y - \dfrac{2}{3}z = \dfrac{1}{3}$ (2)

   $-\dfrac{1}{4}x + y - \dfrac{1}{4}z = \dfrac{3}{4}$ (3)

To clear fractions, multiply equation (1) by 3, equation (2) by 3, and equation (3) by 4. The resulting system is

$$3\left(x - \dfrac{2}{3}y - \dfrac{2}{3}z = -2\right)$$

$$3\left(\dfrac{2}{3}x + y - \dfrac{2}{3}z = \dfrac{1}{3}\right)$$

$$4\left(-\dfrac{1}{4}x + y - \dfrac{1}{4}z = \dfrac{3}{4}\right)$$

gives

$3x-2y-2z=-6\,(4)$

$2x+3y-2z=1\quad(5)$

$-x+4y-z=3\quad(6)$

To eliminate x between equations (4) and (6), multiply equation (6) by 3 and then add.

$3x-2y-2z=-6$

$3[-x+4y-z=3]$

gives

$\quad 3x-2y-2z=-6$

$\underline{-3x+12y-3z=\;9}$

Add: $\quad 10y-5z=\;3\quad(7)$

To eliminate x between equations (5) and (6), multiply equation (6) by 2 and then add.

$2x+3y-2z=1$

$2[-x+4y-z=3]$

gives

$\quad 2x+3y-2z=1$

$\underline{-2x+8y-2z=6}$

Add: $\quad 11y-4z=7\quad(8)$

Equations (7) and (8) are two equations in two unknowns.

$10y-5z=3$

$11y-4z=7$

To eliminate z, multiply equation (7) by -4 and equation (8) by 5 and then add.

$-4[10y-5z=3]$

$5[11y-4z=7]$

gives

$\quad -40y+20z=-12$

$\underline{\quad 55y-20z=\;\;35}$

Add: $15y\qquad=\;23$

$\qquad\qquad y=\dfrac{23}{15}$

Substitute $\dfrac{23}{15}$ for y into equation (7).

$10y-5z=3$

$10\left(\dfrac{23}{15}\right)-5z=3$

$\dfrac{46}{3}-5z=3$

$-5z=-\dfrac{37}{3}$

$z=\left(-\dfrac{1}{5}\right)\left(-\dfrac{37}{3}\right)=\dfrac{37}{15}$

Substitute $\dfrac{23}{15}$ for y and $\dfrac{37}{15}$ for z in equation (6).

$-x+4y-z=3$

$-x+4\left(\dfrac{23}{15}\right)-\dfrac{37}{15}=3$

$-x+\dfrac{92}{15}-\dfrac{37}{15}=3$

$-x+\dfrac{11}{3}=3$

$-x=-\dfrac{2}{3}$

$x=\dfrac{2}{3}$

The solution is $\left(\dfrac{2}{3},\dfrac{23}{15},\dfrac{37}{15}\right)$.

29. $0.2x+0.3y+0.3z=1.1$

$0.4x-0.2y+0.1z=0.4$

$-0.1x-0.1y+0.3z=0.4$

Multiply each equation by 10.

$10(0.2x+0.3y+0.3z=1.1)$

$10(0.4x-0.2y+0.1z=0.4)$

$10(-0.1x-0.1y+0.3z=0.4)$

gives

$2x+3y+3z=11\quad(1)$

$4x-2y+z=4\quad(2)$

$-x-y+3z=4\quad(3)$

To eliminate z between equations (1) and (2), multiply equation (2) by -3 and then add.

$2x+3y+3z=11$

$-3[4x-2y+z=4]$

gives

$\quad 2x+3y+3z=\;11$

$\underline{-12x+6y-3z=-12}$

Add: $-10x+9y\quad=-1\quad(4)$

To eliminate z between equations (1) and (3) multiply equation (1) by -1 and then add.

$-1[2x+3y+3z=11]$

$\quad -x-y+3z=4$

gives

$\quad -2x-3y-3z=-11$

$\underline{\;-x-\;y+3z=\;\;4}$

Add: $-3x-4y\quad=-7\quad(5)$

Equations (4) and (5) are two equations in two

unknowns.

$$-10x + 9y = -1$$
$$-3x - 4y = -7$$

To eliminate x, multiply equation (4) by -3 and equation (5) by 10.

$$-3[-10x + 9y = -1]$$
$$10[-3x - 4y = -7]$$

gives

$$30x - 27y = 3$$
$$\underline{-30x - 40y = -70}$$

Add: $\quad -67y = -67$

$$y = 1$$

Substitute 1 for y in equation (4).

$$-10x + 9y = -1$$
$$-10x + 9(1) = -1$$
$$-10x = -10$$
$$x = 1$$

Substitute 1 for x and 1 for y in equation (1).

$$2x + 3y + 3z = 11$$
$$2(1) + 3(1) + 3z = 11$$
$$5 + 3z = 11$$
$$3z = 6$$
$$z = 2$$

The solution is (1, 1, 2).

31. $\quad 2x + y + 2z = 1 \quad$ (1)
$$x - 2y - z = 0 \quad (2)$$
$$3x - y + z = 2 \quad (3)$$

To eliminate z between equations (2) and (3), simply add.

$$x - 2y - z = 0$$
$$\underline{3x - y + z = 2}$$

Add: $4x - 3y \quad = 2 \quad$ (4)

To eliminate z between equations (1) and (2), multiply equation (2) by 2 and then add.

$$2x + y + 2z = 1$$
$$2[x - 2y - z = 0]$$

gives

$$2x + y + 2z = 1$$
$$\underline{2x - 4y - 2z = 0}$$

Add: $4x - 3y \quad = 1 \quad$ (5)

Equations (4) and (5) are two equations in two unknowns.

$$4x - 3y = 2$$
$$4x - 3y = 1$$

To eliminate x, multiply equation (4) by -1 and

then add.

$$-1[4x - 3y = 2]$$
$$4x - 3y = 1$$

gives

$$-4x + 3y = -2$$
$$\underline{4x - 3y = 1}$$

Add: $\quad 0 = -1 \quad$ False

Since this is a false statement, there is no solution and the system is inconsistent.

33. $\quad x - 4y - 3z = -1 \quad$ (1)
$$-3x + 12y + 9z = 3 \quad (2)$$
$$2x - 10y - 7z = 5 \quad (3)$$

To eliminate x between equations (1) and (2), multiply equation (1) by 3 and then add.

$$3[x - 4y - 3z = -1]$$
$$-3x + 12y + 9z = 3$$

gives

$$3x - 12y - 9z = -3$$
$$\underline{-3x + 12y + 9z = 3}$$

Add: $\quad 0 = 0$

Since $0 = 0$ is a true statement, we suspect that the system is dependent and therefore has infinitely many solutions. However, we could still encounter a contradiction if we used a different pair of equations.

To eliminate x from equations (1) and (3), multiply equation (1) by -2 and add.

$$-2[x - 4y - 3z = -1]$$
$$2x - 10y - 7z = 5$$

gives

$$-2x + 8y + 6z = 2$$
$$\underline{2x - 10y - 7z = 5}$$

Add: $\quad -2y - z = 7$

To eliminate x from equations (2) and (3), multiply equation (2) by 2 and equation (3) by 3, then add.

$$2[-3x + 12y + 9z = 3]$$
$$3[2x - 10y - 7z = 5]$$

gives

$$-6x + 24y + 18z = 6$$
$$\underline{6x - 30y - 21z = 15}$$

Add: $\quad -6y - 3z = 21$

Since neither of the remaining pairs yields a contradiction, the system is indeed dependent and has an infinite number of solutions.

35.
$$x + 3y + 2z = 6 \quad (1)$$
$$x - 2y - z = 8 \quad (2)$$
$$-3x - 9y - 6z = -7 \quad (3)$$

To eliminate x between equations (1) and (3), multiply equation (1) by 3 and then add.
$$3[x + 3y + 2z = 6]$$
$$-3x - 9y - 6z = -7$$
gives
$$\begin{aligned} 3x + 9y + 6z &= 18 \\ -3x - 9y - 6z &= -7 \end{aligned}$$
Add: $\qquad\qquad 0 = 11$

Since $0 = 11$ is a false statement, the system is inconsistent.

37. No point is common to all three planes. Therefore, the system is inconsistent.

39. One point is common to all three planes. There is one solution and the system is consistent.

41. a. Yes, if two or more of the planes are parallel, there will be no solution.

b. Yes, three planes may intersect at a single point.

c. No, the possibilities are no solution, one solution, or infinitely many solutions.

43. $Ax + By + Cz = 1$

Substitute $(-1, 2, -1)$, $(-1, 1, 2)$, and $(1, -2, 2)$ into the equation forming three equations in the three unknowns A, B, and C.
$$A(-1) + B(2) + C(-1) = 1$$
$$A(-1) + B(1) + C(2) = 1$$
$$A(1) + B(-2) + C(2) = 1$$
gives
$$-A + 2B - C = 1 \quad (1)$$
$$-A + B + 2C = 1 \quad (2)$$
$$A - 2B + 2C = 1 \quad (3)$$

To eliminate A between equations (1) and (2), multiply equation (2) by -1 and then add.
$$-A + 2B - C = 1$$
$$-1[-A + B + 2C = 1]$$
gives
$$\begin{aligned} -A + 2B - C &= 1 \\ A - B - 2C &= -1 \end{aligned}$$
Add: $\qquad B - 3C = 0 \quad (4)$

To eliminate A between equations (1) and (3), simply add.

$$-A + 2B \ - C = 1$$
Add: $\dfrac{A - 2B + 2C = 1}{C = 2}$

Substitute 2 for C in equation (4).
$$B - 3(2) = 0$$
$$B - 6 = 0$$
$$B = 6$$

Substitute 6 for B and 2 for C in equation (1).
$$-A + 2B - C = 1$$
$$-A + 2(6) - (2) = 1$$
$$-A + 12 - 2 = 1$$
$$-A + 10 = 1$$
$$-A = -9$$
$$A = 9$$

The equation is $9x + 6y + 2z = 1$.

45. Answers will vary. One example is
$$x + y + z = 10$$
$$x + 2y + z = 11$$
$$x + y + 2z = 16$$

Choose coefficients for x, y, and z, then use the given coordinates to find the constants.

47. a. $y = ax^2 + bx + c$

For the point $(1, -1)$, let $y = -1$ and $x = 1$.
$$-1 = a(1)^2 + b(1) + c$$
$$-1 = a + b + c \qquad\qquad (1)$$

For the point $(-1, -5)$, let $y = -5$ and $x = -1$.
$$-5 = a(-1)^2 + b(-1) + c$$
$$-5 = a - b + c \qquad\qquad (2)$$

For the point $(3, 11)$, let $y = 11$ and $x = 3$.
$$11 = a(3)^2 + b(3) + c$$
$$11 = 9a + 3b + c \qquad (3)$$

Equations (1), (2), and (3) give us a system of three equations.
$$a + b + c = -1 \,(1)$$
$$a - b + c = -5 \,(2)$$
$$9a + 3b + c = 11 \,(3)$$

To eliminate a and c between equations (1) and (2) multiply equation (2) by -1 and then add.
$$a + b + c = -1$$
$$-1[a - b + c = -5]$$
gives

$$a+b+c=-1$$
$$\underline{-a+b-c=\ 5}$$
Add: $2b\quad\ \ =4$
$$b=2$$
Substitute 2 for b in equations (1) and (3).
Equation (1) becomes
$$a+b+c=-1$$
$$a+2+c=-1$$
$$a+c=-3\qquad(4)$$
Equation (3) becomes
$$9a+3b+c=11$$
$$9a+3(2)+c=11$$
$$9a+c=5\qquad(5)$$
Equations (4) and (5) are two equations in two unknowns. To eliminate c, multiply equation (4) by -1 and then add.
$$-1[a+c=-3]$$
$$9a+c=5$$
gives
$$-a-c=\ 3$$
$$\underline{9a+c=5}$$
Add: $8a\quad\ =8$
$$a=1$$
Finally, substitute 1 for a in equation (4).
$$a+c=-3$$
$$1+c=-3$$
$$c=-4$$
Thus, $a=1$, $b=2$, and $c=-4$.

b. The quadratic equation is $y=x^2+2x-4$.
This is the equation determined by the values found in part (a).

49.
$$3p+4q=11\qquad(1)$$
$$2p+r+s=9\qquad(2)$$
$$q-s=-2\qquad(3)$$
$$p+2q-r=2\qquad(4)$$
To eliminate r between equations (2) and (4), simply add.
$$2p\quad\ \ +r+s=\ 9$$
$$\underline{p+2q-r\quad\ =\ 2}$$
Add: $3p+2q\quad\ +s=11\qquad(5)$
To eliminate s between equations (3) and (5), simply add.

$$q-s=-2$$
$$\underline{3p+2q+s=11}$$
Add: $3p+3q\quad\ =9\qquad(6)$
Equations (1) and (6) give us a system of two equations in two unknowns.
To eliminate p, multiply equation (6) by -1 and then add.
$$3p+4q=11$$
$$-1[3p+3q=9]$$
gives
$$3p+4q=\ 11$$
$$\underline{-3p-3q=-9}$$
Add: $q=\ 2$
Substitute 2 for q in equation (3).
$$q-s=-2$$
$$2-s=-2$$
$$-s=-4$$
$$s=4$$
Substitute 2 for q in equation (1).
$$3p+4q=11$$
$$3p+4(2)=11$$
$$3p+8=11$$
$$3p=3$$
$$p=1$$
Finally, substitute 1 for p and 4 for s in equation (2).
$$2p+r+s=9$$
$$2(1)+r+4=9$$
$$r+6=9$$
$$r=3$$
The solution is (1, 2, 3, 4).

51. a. commutative property of addition

b. associative property of multiplication

c. distributive property

52.
$$3x+4=-(x-6)$$
$$3x+4=-x+6$$
$$3x+x=6-4$$
$$4x=2$$
$$x=\frac{2}{4}=\frac{1}{2}$$

53. Let t be the time for Margie.

Then, $t - \dfrac{1}{6}$ is the time for David.

	rate	time	distance
David	5	t	$5t$
Margie	3	$t + \dfrac{1}{6}$	$3\left(t + \dfrac{1}{6}\right)$

a. The distances traveled are the same.

$$3\left(t + \frac{1}{6}\right) = 5t$$
$$3t + \frac{1}{2} = 5t$$
$$\frac{1}{2} = 2t$$
$$t = \frac{1}{4} \text{ hour (or 15 minutes)}$$

b. The distance is

$$5t = 5\left(\frac{1}{4}\right) = \frac{5}{4} = 1\frac{1}{4} \text{ or 1.25 miles.}$$

54. Let w = the width of the rectangle. Then $2w + 2$ = the length of the rectangle.

$$2(2w + 2) + 2w = 22$$
$$4w + 4 + 2w = 22$$
$$6w + 4 = 22$$
$$6w = 18$$
$$w = 3$$

The width is 3 ft, and the length is $2(3) + 2 = 8$ ft.

Mid-Chapter Test: 9.1-9.4

1. a. $(-1, 1)$

$$4(-1) + 3(1) = -1$$
$$-4 + 3 = -1$$
$$-1 = -1 \quad \text{True}$$
$$-1 - 2(1) = 8$$
$$-1 - 2 = 8$$
$$-3 = 8 \quad \text{False}$$

$(-1, 1)$ is not a solution.

b. $(2, -3)$

$$4(2) + 3(-3) = -1$$
$$8 - 9 = -1$$
$$-1 = -1 \quad \text{True}$$
$$2 - 2(-3) = 8$$
$$2 + 6 = 8$$
$$8 = 8 \quad \text{True}$$

$(2, -3)$ is a solution.

2. a. $\left(\dfrac{1}{2}, 5\right)$

$$6\left(\frac{1}{2}\right) - 5 = -2$$
$$3 - 5 = -2$$
$$-2 = -2 \quad \text{True}$$
$$7\left(\frac{1}{2}\right) + \frac{1}{2}(5) = 6$$
$$\frac{7}{2} + \frac{5}{2} = 6$$
$$6 = 6 \quad \text{True}$$

$\left(\dfrac{1}{2}, 5\right)$ is a solution.

b. $\left(\dfrac{1}{3}, 4\right)$

$$6\left(\frac{1}{3}\right) - 4 = -2$$
$$2 - 4 = -2$$
$$-2 = -2 \quad \text{True}$$
$$7\left(\frac{1}{3}\right) + \frac{1}{2}(4) = 6$$
$$\frac{7}{3} + 2 = 6$$
$$\frac{13}{3} = 6 \quad \text{False}$$

$\left(\dfrac{1}{3}, 4\right)$ is not a solution.

3. Solve each equation for y.

$$2x + y = 8 \qquad\qquad 3x - 4y = 1$$
$$y = -2x + 8 \qquad\quad -4y = -3x + 1$$
$$y = \frac{3}{4}x - \frac{1}{4}$$

Since the equations have different slopes the system will have one solution.

4. Solve each equation for y.

$$\frac{1}{2}x - 3y = 5 \qquad\qquad -2x + 12y = -20$$

$$-3y = -\frac{1}{2}x + 5 \qquad 12y = 2x - 20$$

$$y = \frac{1}{6}x - \frac{5}{3} \qquad\qquad y = \frac{1}{6}x - \frac{5}{3}$$

Since both equations are the same, the system has infinite solutions.

5. Solve each equation for y.

$$y = \frac{3}{2}x + \frac{5}{2} \qquad 3x - 2y = 7$$

$$-2y = -3x + 7$$

$$y = \frac{3}{2}x - \frac{7}{2}$$

Since the two equations have the same slope but different y-intercepts, the lines are parallel which means there is no solution.

6. $y = 2x + 1$ has slope of 2 and y-intercept of 1.

$y = -x + 4$ has slope of -1 and y-intercept of 4.

The solution is (1, 3).

7. $x = 5$ is a vertical line with x-intercept (5, 0).

$y = -3$ is a horizontal line with y-intercept (0, -3).

The solution is (5, -3).

8. $3x + y = -2$

$2x - 3y = -16$

Solve the first equation for y, $y = -3x - 2$.

Substitute $-3x - 2$ for y in the second equation.

$$2x - 3(-3x - 2) = -16$$

$$2x + 9x + 6 = -16$$

$$11x + 6 = -16$$

$$11x = -22$$

$$x = -2$$

To find y, substitute -2 for x in the equation $y = -3x - 2$.

$$y = -3(-2) - 2 = 6 - 2 = 4$$

The solution is (-2, 4).

9. $x - 3y = 2$

$4x + 9y = 1$

Solve the first equation for x, $x = 3y + 2$.

Substitute $-3x - 2$ for x in the second equation.

$$4(3y + 2) + 9y = 1$$

$$12y + 8 + 9y = 1$$

$$21y + 8 = 1$$

$$21y = -7$$

$$y = -\frac{1}{3}$$

To find x, substitute $-\frac{1}{3}$ for y in the equation

$$x = 3\left(-\frac{1}{3}\right) + 2 = -1 + 2 = 1.$$

The solution is $\left(1, -\frac{1}{3}\right)$.

10. $3x - y = 5$

$x - \frac{1}{3}y = 2$

Solve the second equation for x, $x = \frac{1}{3}y + 2$.

Substitute $\frac{1}{3}y + 2$ for x in the first equation.

$$3\left(\frac{1}{3}y + 2\right) - y = 5$$

$$y + 2 - y = 5$$

$$2 = 5 \text{ False}$$

Since this is a false statement, the system has no solution.

11. Let l be the length and w be the width of the rectangle.

$2l + 2w = 44$

$l = w + 8$

Substitute $w + 8$ for l in the first equation and solve for w.

$$2(w + 8) + 2w = 44$$

$$2w + 16 + 2w = 44$$

$$4w + 16 = 44$$

$$4w = 28$$

$$w = 7$$

Now substitute 7 for w in the equation $l = w + 8$.

$l = 7 + 8 = 15$.

The length of the rectangle is 15 feet and the width is 7 feet.

12.
$$x + 3y = 1$$
$$\underline{2x - 3y = 11}$$
$$3x \qquad = 12$$
$$x = 4$$

Substitute 4 for x in the first equation and then find y.
$$4 + 3y = 1$$
$$3y = -3$$
$$y = -1$$
The solution is $(4, -1)$.

13. $4x + 3y = 4$ (1)
   $-8x + 5y = 14$ (2)

Multiply equation 1 by 2 giving $8x + 6y = 8$.

Now add the two equations to eliminate x and solve for y.
$$8x + 6y = 8$$
$$\underline{-8x + 5y = 14}$$
$$11y = 22$$
$$y = 2$$

Substitute 2 for y in equation 1 and solve for x.
$$4x + 3(2) = 4$$
$$4x + 6 = 4$$
$$4x = -2$$
$$x = -\frac{1}{2}$$

The solution is $\left(-\frac{1}{2}, 2\right)$.

14. $5x - 2y = 1$ (1)
   $-10x + 4y = -2$ (2)

Multiply equation 1 by 2 giving $10x - 4y = 2$.
Now add the two equations.
$$10x - 4y = 2$$
$$\underline{-10x + 4y = -2}$$
$$0 = 0$$

The system has an infinite number of solutions.

15. A system's solution has to have two values, one for x and one for y.
$$3x - 5y = -16 \quad (1)$$
$$2x + 3y = 21 \quad (2)$$

Multiply equation 1 by 3 and equation 2 by 5, then add to eliminate y and solve for x.
$$3(3x - 5y = -16)$$
$$\underline{5(2x + 3y = 21)}$$

$$9x - 15y = -48$$
$$\underline{10x + 15y = 105}$$
$$19x = 57$$
$$x = 3$$

Substitute 3 for x in equation 2 and solve for y.
$$2(3) + 3y = 21$$
$$6 + 3y = 21$$
$$3y = 15$$
$$y = 5$$
The solution is $(3, 5)$.

16. $x + y + z = 2$ (1)
   $2x - y + 2z = -2$ (2)
   $3x + 2y + 6z = 1$ (3)

To eliminate y between equations (1) and (2), simply add.
$$x + y + z = 2$$
Add: $\underline{2x - y + 2z = -2}$
$$3x + 3z = 0 \quad (4)$$

To eliminate y between equations (2) and (3), multiply equation (2) by 2 and then add.
$$4x - 2y + 4z = -4$$
$$\underline{3x + 2y + 6z = 1}$$
Add: $7x + 10z = -3$ (5)

Equations (4) and (5) are two equations in two unknowns.
$$3x + 3z = 0 \quad (4)$$
$$7x + 10z = -3 \quad (5)$$

To eliminate x, multiply equation (4) by -7, multiply equation (5) by 3 and then add.
$$-21x - 21z = 0$$
$$\underline{21x + 30z = -9}$$
Add: $9z = -9$
$$z = -1$$

Substitute -1 for z in equation (4).
$$3x + 3(-1) = 0$$
$$3x - 3 = 0$$
$$3x = 3$$
$$x = 1$$

Substitute 1 for x and -1 for z in equation (1).
$$x + y + z = 2$$
$$1 + y + (-1) = 2$$
$$y = 2$$

The solution is $(1, 2, -1)$.

17.

$$2x - y - z = 1 \quad (1)$$
$$3x + 5y + 2z = 12 \quad (2)$$
$$-6x - 4y + 5z = 3 \quad (3)$$

To eliminate x between equations (1) and (2), multiply equation (1) by -3 and equation (2) by 2, then add.

$$-6x + 3y + 3z = -3$$
$$\underline{6x + 10y + 4z = 24}$$

Add: $\quad 13y + 7z = 21 \quad (4)$

To eliminate x between equations (2) and (3), multiply equation (2) by 2 and then add.

$$6x + 10y + 4z = 24$$
$$\underline{-6x - 4y + 5z = 3}$$

Add: $\quad 6y + 9z = 27 \quad (5)$

Equations (4) and (5) are two equations in two unknowns.

$$13y + 7z = 21 \quad (4)$$
$$6y + 9z = 27 \quad (5)$$

To eliminate z, multiply equation (4) by -9, multiply equation (5) by 7 and then add.

$$-117y - 63z = -189$$
$$\underline{42y + 63z = 189}$$

Add: $\quad -75y = 0$
$$y = 0$$

Substitute 0 for y in equation (4).

$$13(0) + 7z = 21$$
$$7z = 21$$
$$z = 3$$

Substitute 0 for y and 3 for z in equation (1).

$$2x - y - z = 1$$
$$2x - 0 - 3 = 1$$
$$2x - 3 = 1$$
$$2x = 4$$
$$x = 2$$

The solution is $(2, 0, 3)$.

Exercise Set 9.5

1. Let x = the land area of Georgia and y = the land area of Ireland.

$$x + y = 139,973$$
$$y = x + 573$$

Substitute $x + 573$ for y in the first equation.

$$x + (x + 573) = 139,973$$
$$2x + 573 = 139,973$$
$$2x = 139,400$$
$$x = 69,700$$

Substitute 69,700 for x in the second equation.

$$y = 69,700 + 573 = 70,273$$

The land area of Georgia is 69,700 km^2 and the land area of Ireland is 70,273 km^2.

3. Let F = grams of fat in fries and H = grams of fat in hamburger

$$F = 3H + 4$$
$$F - H = 46$$

Substitute $3H + 4$ for F in the second equation.

$$F - H = 46$$
$$3H + 4 - H = 46$$
$$2H = 42$$
$$H = 21$$

Substitute 21 for H in the first equation.

$$F = 3H + 4$$
$$F = 3(21) + 4 = 63 + 4 = 67$$

The hamburger has 21 grams of fat and the fries have 67 grams of fat.

5. Let h = the cost of a hot dog and s = the cost of a soda.

$$2h + 3s = 7$$
$$4h + 2s = 10$$

Multiply the first equation by -2 and then add.

$$-2[2h + 3s = 7]$$
$$4h + 2s = 10$$

gives

$$-4h - 6s = -14$$
$$\underline{4h + 2s = 10}$$

Add: $\quad -4s = -4$
$$s = 1$$

Substitute 1 for s in the first equation.

$$2h + 3(1) = 7$$
$$2h + 3 = 7$$
$$2h = 4$$
$$h = 2$$

Each hot dog costs \$2 and each soda costs \$1.

7. Let x = the number of photos on the 128MB card and y = the number on the 516MB card.
$$y = 4x$$
$$x + y = 360$$
Substitute $4x$ for y in the second equation.
$$x + 4x = 360$$
$$5x = 360$$
$$x = 72$$
Substitute 72 for x in the first equation.
$$y = 4(72) = 288$$
The 128-megabyte memory card can store 72 photos and the 516-megabyte card can store 288 photos.

9. Let x = the measure of the larger angle and y = the measure of the smaller angle.
$$x + y = 90$$
$$x = 2y + 15$$
Substitute $2y + 15$ for x in the first equation.
$$x + y = 90$$
$$2y + 15 + y = 90$$
$$3y + 15 = 90$$
$$3y = 75$$
$$y = 25$$
Now, substitute 25 for y in the second equation.
$$x = 2y + 15$$
$$x = 2(25) + 15 = 50 + 15 = 65$$
The two angles measure 25° and 65°.

11. Let A and B be the measures of the two angles.
$$A + B = 180$$
$$A = 3B - 28$$
Substitute $3B - 28$ for A in the first equation.
$$A + B = 180$$
$$3B - 28 + B = 180$$
$$4B - 28 = 180$$
$$4B = 208$$
$$B = 52$$
Now substitute 52 for B in the second equation.
$$A = 3B - 28$$
$$A = 3(52) - 28 = 128$$
The two angles measure 52° and 128°.

13. Let t = team's rowing speed in still water and c = speed of current
$$t + c = 15.6$$
$$t - c = 8.8$$
Add the equations to eliminate variable c.

$$t + c = 15.6$$
$$\underline{t - c = 8.8}$$
$$2t = 24.4$$
$$t = 12.2$$
Substitute 12.2 for t in the first equation.
$$(12.2) + c = 15.6$$
$$c = 3.4$$
The team's speed in still water is 12.2 mph and the speed of the current is 3.4 mph.

15. Let x = the weekly salary and y = the commission rate.
$$x + 4000y = 660$$
$$x + 6000y = 740$$
Multiply the first equation by -1 and then add.
$$-1[x + 4000y = 660]$$
$$x + 6000y = 740$$
gives
$$-x - 4000y = -660$$
$$\underline{x + 6000y = 740}$$
Add: $2000y = 80$

$$y = \frac{80}{2000} = 0.04$$
Substitute 0.04 for y in the first equation.
$$x + 4000y = 660$$
$$x + 4000(0.04) = 660$$
$$x + 160 = 660$$
$$x = 500 \text{ dollars}$$
His weekly salary is $500 and the commission rate is 4%.

17. Let x = the amount of 5% solution and y = the amount of 30% solution.
$$x + y = 3$$
$$0.05x + 0.30y = 0.20(3)$$
Solve the first equation for x.
$$x = 3 - y$$
Substitute $3 - y$ for x in the second equation.
$$0.05x + 0.30y = 0.20(3)$$
$$0.05(3 - y) + 0.30y = 0.6$$
$$0.15 - 0.05y + 0.30y = 0.6$$
$$0.25y = 0.45$$
$$y = 1.8$$
Substitute 1.8 for y in the first equation.
$$x + y = 3$$
$$x + 1.8 = 3$$
$$x = 1.2$$
Pola should mix 1.2 ounces of the 5% solution with 1.8 ounces of the 30% solution.

19. Let x = gallons of concentrate (18% solution) and y = gallons of water (0% solution).
$$x + y = 200$$
$$0.18x + 0y = 0.009(200)$$
Solve the second equation for x.
$$0.18x + 0y = 0.009(200)$$
$$0.18x = 1.8$$
$$x = 10$$
Substitute 10 for x in the first equation.
$$x + y = 200$$
$$10 + y = 200$$
$$y = 190$$
The mixture should contain 10 gallons of concentrate and 190 gallons of water.

21. Let x = pounds of birdseed and y = pounds of sunflower seeds
$$0.59x + 0.89y = 0.76(40)$$
$$x + y = 40$$
Solve the second equation for y.
$$y = 40 - x$$
Substitute $40 - x$ for y in the first equation.
$$0.59x + 0.89y = 0.76(40)$$
$$0.59x + 0.89(40 - x) = 30.4$$
$$0.59x + 35.6 - 0.89x = 30.4$$
$$-0.3x = -5.2$$
$$x = 17\frac{1}{3}$$
Substitute $17\frac{1}{3}$ for x in the second equation.
$$x + y = 40$$
$$17\frac{1}{3} + y = 40$$
$$y = 22\frac{2}{3}$$
Angela Leinenbach should mix $17\frac{1}{3}$ pounds of birdseed at \$0.59 per pound with $22\frac{2}{3}$ pounds of sunflower seeds at \$0.89 per pound.

23. Let x = the price of an adult ticket and y = the price of a child ticket.
$$3x + 4y = 159$$
$$2x + 3y = 112$$
Multiply the first equation by 2 and the second equation by -3, then add.

$$2[3x + 4y = 159]$$
$$-3[2x + 3y = 112]$$
gives
$$6x + 8y = 318$$
$$-6x - 9y = -336$$
Add: $\quad -y = -18$
$$y = 18$$
Substitute 18 for y in the second equation.
$$2x + 3(18) = 112$$
$$2x + 54 = 112$$
$$2x = 58$$
$$x = 29$$
Adult tickets cost \$29 and child tickets cost \$18.

25. Let x = amount invested at 5% and y = amount invested at 6%
$$x + y = 10,000$$
$$0.05x + 0.06y = 540$$
Solve the first equation for y.
$$y = 10,000 - x$$
Substitute $10,000 - x$ for y in the second equation.
$$0.05x + 0.06y = 540$$
$$0.05x + 0.06(10,000 - x) = 540$$
$$0.05x + 600 - 0.06x = 540$$
$$-0.01x = -60$$
$$x = 6000$$
Substitute 6000 for x in the first equation.
$$x + y = 10,000$$
$$6000 + y = 10,000$$
$$y = 4000$$
Mr. and Mrs. Gamton invested \$6000 at 5% and \$4000 at 6%.

27. Let x = the amount of the whole milk (3.25% fat) and y = the amount of the skim milk (0% fat).
$$x + y = 260$$
$$0.0325x + 0y = 0.02(260)$$
Solve the second equation for x.
$$0.0325x + 0y = 0.02(260)$$
$$0.0325x = 5.2$$
$$x = 160$$
Now substitute 160 for x in the first equation.
$$(160) + y = 260$$
$$y = 100$$
Becky needs to mix 160 gallons of the whole milk with 100 gallons of skim milk to produce 260 gallons of 2% fat milk.

29. Let x = pounds of *Season's Choice* birdseed at $1.79/lb and y = pounds of *Garden Mix* birdseed at $1.19/lb
$$1.79x + 1.19y = 28.00$$
$$x + y = 20$$
Solve the second equation for y.
$$y = 20 - x$$
Substitute $20 - x$ for y in the first equation.
$$1.79x + 1.19(20 - x) = 28.00$$
$$1.79x + 23.80 - 1.19x = 28.00$$
$$0.60x = 4.20$$
$$x = 7$$
Substitute 7 for x in the second equation.
$$(7) + y = 20$$
$$y = 13$$
The Carters should buy 7 pounds *of Season's Choice* and 13 pounds of *Garden Mix*.

31. Let x = the rate of the slower car and y = the rate of the faster car.
$$4x + 4y = 420$$
$$y = x + 5$$
Substitute $x + 5$ for y in the first equation.
$$4x + 4y = 420$$
$$4x + 4(x + 5) = 420$$
$$4x + 4x + 20 = 420$$
$$8x + 20 = 420$$
$$8x = 400$$
$$x = 50$$
Now substitute 50 for x in the second equation.
$$y = x + 5$$
$$y = 50 + 5 = 55$$
The rate of the slower car is 50 mph and the rate of the faster car is 55 mph.

33. Let x = the amount of time traveled at 65 mph and y = the amount of time traveled at 50 mph.
$$x + y = 11.4$$
$$65x + 50y = 690$$
Solve the first equation for x: $x = 11.4 - y$
Now substitute $11.4 - y$ into x in the second equation.
$$65(11.4 - y) + 50y = 690$$
$$741 - 65y + 50y = 690$$
$$-15y = -51$$
$$y = 3.4$$
Substitute 3.4 for y in the first equation.
$$x + (3.4) = 11.4$$
$$x = 8$$
Cabrina traveled for 8 hours at 65 mph and Dabney traveled for 3.4 hours at 50 mph.

35. Let x = the number of grams of Mix A and y = the number of grams of Mix B.
$$0.1x + 0.2y = 20$$
$$0.06x + 0.02y = 6$$
To solve, multiply the second equation by -10 and then add.
$$0.1x + 0.2y = 20$$
$$-10[0.06x + 0.02y = 6]$$
gives
$$0.1x + 0.2y = 20$$
$$\underline{-0.6x - 0.2y = -60}$$
Add: $-0.5x \qquad = -40$
$$x = \frac{-40}{-0.5} = 80$$
Now substitute 80 for x in the first equation.
$$0.1x + 0.2y = 20$$
$$0.1(80) + 0.2y = 20$$
$$8 + 0.2y = 20$$
$$0.2y = 12$$
$$y = \frac{12}{0.2} = 60$$
The scientist should feed each animal 80 grams of Mix A and 60 grams of Mix B.

37. Let x = the amount of the first alloy and y = the amount of the second alloy.
$$0.7x + 0.4y = 0.6(300)$$
$$0.3x + 0.6y = 0.4(300)$$
To solve, multiply the first equation by 3 and the second equation by -2 and then add.
$$3[0.7x + 0.4y = 0.6(300)]$$
$$-2[0.3x + 0.6y = 0.4(300)]$$
gives
$$2.1x + 1.2y = 540$$
$$\underline{-0.6x - 1.2y = -240}$$
Add: $1.5x \qquad = 300$
$$x = \frac{300}{1.5} = 200$$
Now substitute 200 for x in the first equation.
$$0.7x + 0.4y = 0.6(300)$$
$$0.7(200) + 0.4y = 0.6(300)$$
$$140 + 0.4y = 180$$
$$0.4y = 40$$
$$y = \frac{40}{0.4} = 100$$
200 grams of the first alloy should be combined with 100 grams of the second alloy to produce the desired mixture.

39. Let t = the number of years since 2002 and y = the number of returns filed. We can use the given functions to form a system of equations.

Paper $\quad y = -2.73t + 58.37$

Online $\quad y = 1.95t + 10.58$

Subsitute $-2.73t + 58.37$ for y in the second equation.

$$-2.73t + 58.37 = 1.95t + 10.58$$
$$58.37 = 4.68t + 10.58$$
$$47.79 = 4.68t$$
$$\frac{47.79}{4.68} = t \quad \text{or} \quad t \approx 10.2$$

Thus, if the trends continue, the number of paper Form 1040 returns will be about the same as the number of online Form 1040, 1040A, 1040EZ returns in 2012 (2002 + 10 = 2012).

41. Let x = speed of Melissa's car and y = speed of Tom's car

$$x = y + 15$$
$$\frac{150}{x} = \frac{120}{y} \quad \text{(travel times are equal)}$$

Substitute $y + 15$ for x in the second equation.

$$\frac{150}{y+15} = \frac{120}{y}$$
$$150y = 120y + 1800$$
$$30y = 1800$$
$$y = 60$$

Substitute 60 for y in the first equation.

$$x = y + 15$$
$$x = 60 + 15$$
$$x = 75$$

Tom traveled at 60 mph and Melissa traveled at 75 mph.

43. Let x = pieces of personal mail

y = number of bills and statements

z = number of advertisements

$$x + y + z = 24$$
$$y = 2x - 2$$
$$z = 5x + 2$$

Substitute $2x - 2$ for y and $5x + 2$ for z in the first equation.

$$x + y + z = 24$$
$$x + 2x - 2 + 5x + 2 = 24$$
$$8x = 24$$
$$x = 3$$

Substitute 3 for x in the second and third equations.

$$y = 2x - 2 \qquad\qquad z = 5x + 2$$
$$y = 2(3) - 2 \qquad\quad z = 5(3) + 2$$
$$y = 4 \qquad\qquad\quad z = 17$$

An average American household receives 3 pieces of personal mail, 4 bills and statements, and 17 advertisements per week.

45. Let x = the number of bowl game appearances by Alabama, y = the number of bowl game appearnces by Tennessee, and z = the number of bowl game appearances by Texas.

$$x + y + z = 141$$
$$x = z + 8$$
$$y + z = x + 37$$

Solve the third equation for y.

$$y + z = x + 37$$
$$y = x + 37 - z$$

Substitute $x + 37 - z$ for y in the first equation.

$$x + (x + 37 - z) + z = 141$$
$$2x + 37 = 141$$
$$2x = 104$$
$$x = 52$$

Substitute 52 for x in the second equation.

$$52 = z + 8$$
$$44 = z$$

Substitute 44 for z and 52 for x in the third equation.

$$y + 44 = 52 + 37$$
$$y + 44 = 89$$
$$y = 45$$

Through 2004, Alabama has appeared in 52 bowl games, Tennessee has appeared in 45 bowl games, and Texas has appeared in 44 bowl games.

47. Let x = the number of top-ten finishes for Vijay Singh, y = the number of top-ten finishes for Tiger Woods, and z = the number of top-ten finishes for Phil Mickelson.

$$x + y + z = 191$$
$$y = z + 8$$
$$x = z + 12$$

Substitute $z + 8$ for y and $z + 12$ for x in the first equation.

$$(z + 12) + (z + 8) + z = 191$$
$$3z + 20 = 191$$
$$3z = 171$$
$$z = 57$$

Substitute 57 for z in the second equation.
$$y = 57 + 8 = 65$$
Substitute 57 for z in the third equation.
$$x = 57 + 12 = 69$$
Vijay Singh had 69 top-ten finishes, Tiger Woods had 65 top-ten finishes, and Phil Mickelson had 57 top-ten finishes.

49. Let x = the amount of snowfall (inches) in Haverhill, y = the amount of snowfall in Plymouth, and z = the amount of snowfall in Salem.
$$x + y + z = 112.5$$
$$y = z$$
$$y = x + 1.5$$
Substitute $x + 1.5$ for both y and z in the first equation.
$$x + (x + 1.5) + (x + 1.5) = 112.5$$
$$3x + 3 = 112.5$$
$$3x = 109.5$$
$$x = 36.5$$
Substitute 36.5 for x in the third equation.
$$y = 36.5 + 1.5$$
$$y = 38$$
Haverhill received 36.5 inches of snow while Plymouth and Salem each received 38 inches.

51. Let F = the number of Super Bowls held in Florida, L = the number held in Louisiana, and C = the number held in California.
$$F + L + C = 32 \qquad (1)$$
$$F = L + 3 \qquad (2)$$
$$F + L = 2C - 1 \qquad (3)$$
Substitute $L + 3$ for F in the first and third equations.

$(L+3) + L + C = 32 \qquad (L+3) + L = 2C - 1$
$\quad 2L + C + 3 = 32 \qquad\quad 2L + 3 = 2C - 1$
$\quad\quad 2L + C = 29 \qquad\qquad 2L - 2C = -4$

We now have two equations in two unknowns.
$$2L + C = 29 \quad (4)$$
$$2L - 2C = -4 \quad (5)$$
Eliminate L by multiplying equation (5) by -1 and adding to equation (4).
$$2L + C = 29$$
$$-1\left[2L - 2C = -4\right]$$
gives

$$2L + C = 29$$
$$\underline{-2L + 2C = 4}$$
Add: $\qquad 3C = 33$
$$C = 11$$
Substitute 11 for C in equation (4).
$$2L + 11 = 29$$
$$2L = 18$$
$$L = 9$$
Substitute 9 for L in equation (2).
$$F = 9 + 3 = 12$$
Florida hosted 12 Super Bowls, Louisiana hosted 9 Super Bowls, and California hosted 11 Super Bowls.

53. Let x be the smallest angle measure, y the second smallest, and z the largest.
$$x + y + z = 180$$
$$x = \frac{2}{3}y$$
$$z = 3y - 30$$
Substitute $\frac{2}{3}y$ for x and $3y - 30$ for z in the first equation.
$$x + y + z = 180$$
$$\frac{2}{3}y + y + 3y - 30 = 180$$
$$\frac{14}{3}y - 30 = 180$$
$$\frac{14}{3}y = 210$$
$$y = 45$$
Substitute 45 for y in the second equation.
$$x = \frac{2}{3}y$$
$$x = \frac{2}{3}(45) = 30$$
Substitute 45 for y in the third equation.
$$z = 3y - 30$$
$$z = 3(45) - 30 = 135 - 30 = 105$$
The three angles are 30°, 45°, and 105°.

55. Let x = the amount invested at 3%, y = the amount invested at 5%, and z = the amount invested at 6%.

$$y = 2x \quad (1)$$
$$x + y + z = 10,000 \, (2)$$
$$0.03x + 0.05y + 0.06z = 525 \quad (3)$$

Substitute $2x$ for y in equation (2).
$$x + y + z = 10,000$$
$$x + 2x + z = 10,000$$
$$3x + z = 10,000 \quad (4)$$

Substitute $2x$ for y in equation (3).
$$0.03x + 0.05y + 0.06z = 525$$
$$0.03x + 0.05(2x) + 0.06z = 525$$
$$0.03x + 0.10x + 0.06z = 525$$
$$0.13x + 0.06z = 525 \, (5)$$

Equations (4) and (5) are a system of two equations in two unknowns.
$$3x + z = 10,000$$
$$0.13x + 0.06z = 525$$

To eliminate z, multiply equation (4) by -3 and equation (5) by 50 and add.
$$-3[3x + z = 10,000]$$
$$50[0.13x + 0.06x = 525]$$
gives
$$-9x - 3z = -30,000$$
$$6.5x + 3z = 26,250$$

Add: $\quad \overline{-2.5x = -3750}$

$$x = \frac{-3750}{-2.5} = 1500$$

Substitute 1500 for x in equation (4).
$$3x + z = 10,000$$
$$3(1500) + z = 10,000$$
$$4500 + z = 10,000$$
$$z = 5500$$

Substitute 1500 for x and 5500 for z in equation (2).
$$x + y + z = 10,000$$
$$1500 + y + 5500 = 10,000$$
$$y + 7000 = 10,000$$
$$y = 3000$$

Tam invested $1500 at 3%, $3000 at 5%, and $5500 at 6%.

57. Let x = the amount of the 10% solution, y = the amount of the 12% solution, and z = the amount of the 20% solution.

$$x + y + z = 8 \quad (1)$$
$$0.10x + 0.12y + 0.20z = (0.13)8 \, (2)$$
$$z = x - 2 \quad (3)$$

Substitute $x - 2$ for z in equation (1).
$$x + y + z = 8$$
$$x + y + (x - 2) = 8$$
$$2x + y - 2 = 8$$
$$2x + y = 10 \quad (4)$$

Substitute $x - 2$ for z in equation (2).
$$0.10x + 0.12y + 0.20z = (0.13)8$$
$$0.10x + 0.12y + 0.20(x - 2) = (0.13)8$$
$$0.10x + 0.12y + 0.20x - 0.40 = 1.04$$
$$0.30x + 0.12y = 1.44 \quad (5)$$

Equations (4) and (5) are a system of two equations in two unknowns.
$$2x + y = 10$$
$$0.30x + 0.12y = 1.44$$

To solve, multiply equation (5) by 100 and equation (4) by -12 and then add.
$$-12[2x + y = 10]$$
$$100[0.30x + 0.12y = 1.44]$$
gives
$$-24x - 12y = -120$$
$$30x + 12y = 144$$

Add: $6x \quad\;\; = 24$

$$x = 4$$

Substitute 4 for x in equation (4).
$$2x + y = 10$$
$$2(4) + y = 10$$
$$8 + y = 10$$
$$y = 2$$

Finally, substitute 4 for x in equation (3).
$$z = x - 2$$
$$z = 4 - 2$$
$$z = 2$$

The mixture consists of 4 liters of the 10% solution, 2 liters of the 12% solution, and 2 liters of the 20% solution.

59. Let x = the number of children's chairs, y = the number of standard chairs, and z = the number of executive chairs.

$$5x+4y+7z=154 \ (1)$$
$$3x+2y+5z=94 \ (2)$$
$$2x+2y+4z=76 \ (3)$$

To eliminate y between equations (1) and (2), multiply equation (2) by -2 and add.

$$5x+4y+7z=154$$
$$-2[3x+2y+5z=94]$$

gives

$$5x+4y+7z=154$$
$$-6x-4y-10z=-188$$

Add: $-x \quad -3z=-34 \quad$ (4)

To eliminate y between equations (2) and (3), multiply equation (3) by -1 and add.

$$3x+2y+5z=94$$
$$-1[2x+2y+4z=76]$$

gives

$$3x+2y+5z=94$$
$$-2x-2y-4z=-76$$

Add: $x \quad +z=18 \quad$ (5)

Equations (4) and (5) are a system of two equations in two unknowns. To eliminate x, simply add.

$$-x-3z=-34$$
$$x+z=18$$

Add: $-2z=-16$

$$z=\frac{-16}{-2}=8$$

Substitute 8 for z in equation (5).
$$x+z=18$$
$$x+8=18$$
$$x=10$$

Substitute 10 for x and 8 for z in equation (3).
$$2x+2y+4z=76$$
$$2(10)+2y+4(8)=76$$
$$20+2y+32=76$$
$$2y+52=76$$
$$2y=24$$
$$y=12$$

The Donaldson Furniture Company should produce 10 children's chairs, 12 standard chairs, and 8 executive chairs.

61. $I_A+I_B+I_C=0 \ (1)$
$$-8I_B+10I_C=0 \ (2)$$
$$4I_A-8I_B=6 \ (3)$$

To eliminate I_A between equations (1) and (3), multiply equation (1) by -4 and add.
$$-4[I_A+I_B+C=0]$$
$$4I_A-8I_B=6$$

gives

$$-4I_A-4I_B-4I_C=0$$
$$4I_A-8I_B \quad =6$$

Add: $\quad -12I_B-4I_C=6$
or $\quad -6I_B-2I_C=3 \quad$ (4)

Equations (4) and (2) are a system of two equations in two unknowns.
$$-8I_B+10I_C=0$$
$$-6I_B-2I_C=3$$

Multiply equation (4) by 5 and add this result to equation (2).
$$-8I_B+10I_C=0$$
$$5[-6I_B-2I_C=3]$$

gives

$$-8I_B+10I_C=0$$
$$-30I_B-10I_C=15$$

Add: $-38I_B \quad =15$

$$I_B=\frac{15}{-38}=-\frac{15}{38}$$

Substitute $-\frac{15}{38}$ for I_B in equation (2).
$$-8I_B+10I_C=0$$
$$-8\left(-\frac{15}{38}\right)+10I_C=0$$
$$\frac{120}{38}+10I_C=0$$
$$10I_C=-\frac{120}{38}$$
$$\frac{1}{10}(10I_C)=\frac{1}{10}\left(-\frac{120}{38}\right)$$
$$I_C=-\frac{12}{38}=-\frac{6}{19}$$

Finally, substitute $-\frac{15}{38}$ for I_B in equation (3).

$$4I_A - 8I_B = 6$$

$$4I_A - 8\left(-\frac{15}{38}\right) = 6$$

$$4I_A + \frac{120}{38} = 6$$

$$4I_A = 6 - \frac{120}{38}$$

$$4I_A = 6 - \frac{60}{19}$$

$$4I_A = \frac{114}{19} - \frac{60}{19}$$

$$4I_A = \frac{54}{19}$$

$$\frac{1}{4}(4I_A) = \frac{1}{4}\left(\frac{54}{19}\right)$$

$$I_A = \frac{27}{38}$$

The current in branch A is $\frac{27}{38}$, the current in branch B is $-\frac{15}{38}$ and the current in branch C is $-\frac{6}{19}$.

64. Substitute -2 for x and 5 for y.

$$\frac{1}{2}x + \frac{2}{5}xy + \frac{1}{8}y = \frac{1}{2}(-2) + \frac{2}{5}(-2)(5) + \frac{1}{8}(5)$$

$$= -1 - 4 + \frac{5}{8}$$

$$= -5 + \frac{5}{8}$$

$$= -\frac{35}{8}$$

65. $4 - 2\left[(x-5) + 2x\right] = -(x+6)$

$$4 - 2(x - 5 + 2x) = -x - 6$$

$$4 - 2(3x - 5) = -x - 6$$

$$4 - 6x + 10 = -x - 6$$

$$-6x + 14 = -x - 6$$

$$-6x + x = -6 - 14$$

$$-5x = -20$$

$$x = 4$$

66. The slope is $m = \dfrac{-8 - (-4)}{2 - 6} = \dfrac{-8 + 4}{2 - 6} = \dfrac{-4}{-4} = 1$.

Use the point-slope form with $m = 1$ and $(x_1, y_1) = (6, -4)$.

$$y - y_1 = m(x - x_1)$$

$$y - (-4) = 1(x - 6)$$

$$y + 4 = x - 6$$

$$y = x - 10$$

67. Use the vertical line test. If a vertical line cannot be drawn to intersect the graph in more than one point, the graph represents a function.

Exercise Set 9.6

1. A square matrix has the same number of rows and columns.

3. $-\dfrac{1}{2}R_2$. The next step is to change the -2 in the second row to 1 by multiplying the second row of numbers by $-\dfrac{1}{2}$.

5. Switch R_1 and R_2 in order to continue placing ones along the diagonal.

7. Dependent (assuming no contradictions exist)

9. $\begin{bmatrix} 5 & -10 & | & -25 \\ 3 & -7 & | & -4 \end{bmatrix} \Rightarrow \begin{bmatrix} 1 & -2 & | & -5 \\ 3 & -7 & | & -4 \end{bmatrix} \frac{1}{5}R_1$

11. $\begin{bmatrix} 4 & 7 & 2 & | & -1 \\ 3 & 2 & 1 & | & -5 \\ 1 & 1 & 3 & | & -8 \end{bmatrix} \Rightarrow \begin{bmatrix} 1 & 1 & 3 & | & -8 \\ 3 & 2 & 1 & | & -5 \\ 4 & 7 & 2 & | & -1 \end{bmatrix} \begin{smallmatrix} \text{switch} \\ \\ R_1 \text{ and } R_3 \end{smallmatrix}$

13. $\begin{bmatrix} 1 & 3 & | & 12 \\ -4 & 11 & | & -6 \end{bmatrix} \Rightarrow \begin{bmatrix} 1 & 3 & | & 12 \\ 0 & 23 & | & 42 \end{bmatrix} 4R_1 + R_2$

15. $\begin{bmatrix} 1 & 0 & 8 & | & \frac{1}{4} \\ 5 & 2 & 2 & | & -2 \\ 6 & -3 & 1 & | & 0 \end{bmatrix} \Rightarrow$

$\begin{bmatrix} 1 & 0 & 8 & | & \frac{1}{4} \\ 0 & 2 & -38 & | & -\frac{13}{4} \\ 6 & -3 & 1 & | & 0 \end{bmatrix} -5R_1 + R_2$

17. $x + 3y = 3$
$-x + y = -3$

$$\begin{bmatrix} 1 & 3 & | & 3 \\ -1 & 1 & | & -3 \end{bmatrix}$$

$$\begin{bmatrix} 1 & 3 & | & 3 \\ 0 & 4 & | & 0 \end{bmatrix} R_1 + R_2$$

$$\begin{bmatrix} 1 & 3 & | & 3 \\ 0 & 1 & | & 0 \end{bmatrix} \tfrac{1}{4} R_2$$

The system is
$x + 3y = 3$
$y = 0$
Substitute 0 for y in the first equation.
$x + 3y = 3$
$x + 3(0) = 3$
$x + 0 = 3$
$x = 3$
The solution is (3, 0).

19. $x + 3y = -2$
$-2x - 7y = 3$

$$\begin{bmatrix} 1 & 3 & | & -2 \\ -2 & -7 & | & 3 \end{bmatrix}$$

$$\begin{bmatrix} 1 & 3 & | & -2 \\ 0 & -1 & | & -1 \end{bmatrix} 2R_1 + R_2$$

$$\begin{bmatrix} 1 & 3 & | & -2 \\ 0 & 1 & | & 1 \end{bmatrix} -R_2$$

The system is
$x + 3y = -2$
$y = 1$
Substitute 1 for y in the first equation.
$x + 3(1) = -2$
$x + 3 = -2$
$x = -5$
The solution is $(-5, 1)$.

21. $5a - 10b = -10$
$2a + b = 1$

$$\begin{bmatrix} 5 & -10 & | & -10 \\ 2 & 1 & | & 1 \end{bmatrix}$$

$$\begin{bmatrix} 1 & -2 & | & -2 \\ 2 & 1 & | & 1 \end{bmatrix} \tfrac{1}{5} R_1$$

$$\begin{bmatrix} 1 & -2 & | & -2 \\ 0 & 5 & | & 5 \end{bmatrix} -2R_1 + R_2$$

$$\begin{bmatrix} 1 & -2 & | & -2 \\ 0 & 1 & | & 1 \end{bmatrix} \tfrac{1}{5} R_2$$

The system is
$a - 2b = -2$
$b = 1$
Substitute 1 for b in the first equation.
$a - 2(1) = -2$
$a - 2 = -2$
$a = 0$
The solution is (0, 1).

23. $2x - 5y = -6$
$-4x + 10y = 12$

$$\begin{bmatrix} 2 & -5 & | & -6 \\ -4 & 10 & | & 12 \end{bmatrix}$$

$$\begin{bmatrix} 1 & -\tfrac{5}{2} & | & -3 \\ -4 & 10 & | & 12 \end{bmatrix} \tfrac{1}{2} R_1$$

$$\begin{bmatrix} 1 & -\tfrac{5}{2} & | & -3 \\ 0 & 0 & | & 0 \end{bmatrix} 4R_1 + R_2$$

Since the last row contains all 0's, this is a dependent system of equations.

25. $12x + 2y = 2$
$6x - 3y = -11$

$$\begin{bmatrix} 12 & 2 & | & 2 \\ 6 & -3 & | & -11 \end{bmatrix}$$

$$\begin{bmatrix} 1 & \tfrac{1}{6} & | & \tfrac{1}{6} \\ 6 & -3 & | & -11 \end{bmatrix} \tfrac{1}{12} R_1$$

$$\begin{bmatrix} 1 & \tfrac{1}{6} & | & \tfrac{1}{6} \\ 0 & -4 & | & -12 \end{bmatrix} -6R_1 + R_2$$

$$\begin{bmatrix} 1 & \tfrac{1}{6} & | & \tfrac{1}{6} \\ 0 & 1 & | & 3 \end{bmatrix} -\tfrac{1}{4} R_2$$

The system is
$$x + \frac{1}{6}y = \frac{1}{6}$$
$$y = 3$$
Substitute 3 for y in the first equation.
$$x + \frac{1}{6}y = \frac{1}{6}$$
$$x + \frac{1}{6}(3) = \frac{1}{6}$$
$$x + \frac{3}{6} = \frac{1}{6}$$
$$x = -\frac{2}{6} = -\frac{1}{3}$$
The solution is $\left(-\frac{1}{3}, 3 \right)$.

27. $-3x+6y=5$
$2x-4y=7$

$$\begin{bmatrix} -3 & 6 & | & 5 \\ 2 & -4 & | & 7 \end{bmatrix}$$

$$\begin{bmatrix} 1 & -2 & | & -\frac{5}{3} \\ 2 & -4 & | & 7 \end{bmatrix} -\frac{1}{3}R_1$$

$$\begin{bmatrix} 1 & -2 & | & -\frac{5}{3} \\ 0 & 0 & | & \frac{31}{3} \end{bmatrix} -2R_1+R_2$$

Since the last row contains zeros on the left and a nonzero number on the right, this is an inconsistent system and there is no solution.

29. $12x-8y=6$
$-3x+4y=-1$

$$\begin{bmatrix} 12 & -8 & | & 6 \\ -3 & 4 & | & -1 \end{bmatrix}$$

$$\begin{bmatrix} 1 & -\frac{2}{3} & | & \frac{1}{2} \\ -3 & 4 & | & -1 \end{bmatrix} \frac{1}{12}R_1$$

$$\begin{bmatrix} 1 & -\frac{2}{3} & | & \frac{1}{2} \\ 0 & 2 & | & \frac{1}{2} \end{bmatrix} 3R_1+R_2$$

$$\begin{bmatrix} 1 & -\frac{2}{3} & | & \frac{1}{2} \\ 0 & 1 & | & \frac{1}{4} \end{bmatrix} \frac{1}{2}R_2$$

The system is
$$x-\frac{2}{3}y=\frac{1}{2}$$
$$y=\frac{1}{4}$$

Substitute $\frac{1}{4}$ for y in the first equation.

$$x-\frac{2}{3}y=\frac{1}{2}$$
$$x-\frac{2}{3}\left(\frac{1}{4}\right)=\frac{1}{2}$$
$$x-\frac{1}{6}=\frac{1}{2}$$
$$x=\frac{1}{2}+\frac{1}{6}$$
$$x=\frac{4}{6}=\frac{2}{3}$$

The solution is $\left(\frac{2}{3},\frac{1}{4}\right)$.

31. $10m=8n+15$
$16n=-15m-2$
Write the system in standard form.
$10m-8n=15$
$15m+16n=-2$

$$\begin{bmatrix} 10 & -8 & | & 15 \\ 15 & 16 & | & -2 \end{bmatrix}$$

$$\begin{bmatrix} 1 & -\frac{4}{5} & | & \frac{3}{2} \\ 15 & 16 & | & -2 \end{bmatrix} \frac{1}{10}R_1$$

$$\begin{bmatrix} 1 & -\frac{4}{5} & | & \frac{3}{2} \\ 0 & 28 & | & -\frac{49}{2} \end{bmatrix} -15R_1+R_2$$

$$\begin{bmatrix} 1 & -\frac{4}{5} & | & \frac{3}{2} \\ 0 & 1 & | & -\frac{7}{8} \end{bmatrix} \frac{1}{28}R_2$$

The system is
$$m-\frac{4}{5}n=\frac{3}{2}$$
$$n=-\frac{7}{8}$$

Substitute $-\frac{7}{8}$ for n in the first equation.

$$m-\frac{4}{5}n=\frac{3}{2}$$
$$m-\frac{4}{5}\left(-\frac{7}{8}\right)=\frac{3}{2}$$
$$m+\frac{7}{10}=\frac{3}{2}$$
$$m=\frac{3}{2}-\frac{7}{10}=\frac{15}{10}-\frac{7}{10}=\frac{8}{10}=\frac{4}{5}$$

The solution is $\left(\frac{4}{5},-\frac{7}{8}\right)$.

33. $x-3y+2z=5$
$2x+5y-4z=-3$
$-3x+y-2z=-11$

$$\begin{bmatrix} 1 & -3 & 2 & | & 5 \\ 2 & 5 & -4 & | & -3 \\ -3 & 1 & -2 & | & -11 \end{bmatrix}$$

$$\begin{bmatrix} 1 & -3 & 2 & | & 5 \\ 0 & 11 & -8 & | & -13 \\ -3 & 1 & -2 & | & -11 \end{bmatrix} -2R_1+R_2$$

$$\begin{bmatrix} 1 & -3 & 2 & | & 5 \\ 0 & 11 & -8 & | & -13 \\ 0 & -8 & 4 & | & 4 \end{bmatrix} 3R_1+R_3$$

$$\begin{bmatrix} 1 & -3 & 2 & \bigm| & 5 \\ 0 & 1 & -\dfrac{8}{11} & \bigm| & -\dfrac{13}{11} \\ 0 & -8 & 4 & \bigm| & 4 \end{bmatrix} \dfrac{1}{11}R_2$$

$$\begin{bmatrix} 1 & -3 & 2 & \bigm| & 5 \\ 0 & 1 & -\dfrac{8}{11} & \bigm| & -\dfrac{13}{11} \\ 0 & 0 & -\dfrac{20}{11} & \bigm| & -\dfrac{60}{11} \end{bmatrix} 8R_2 + R_3$$

$$\begin{bmatrix} 1 & -3 & 2 & \bigm| & 5 \\ 0 & 1 & -\dfrac{8}{11} & \bigm| & -\dfrac{13}{11} \\ 0 & 0 & 1 & \bigm| & 3 \end{bmatrix} -\dfrac{11}{20}R_3$$

The system is

$$x - 3y + 2z = 5$$
$$y - \dfrac{8}{11}z = -\dfrac{13}{11}$$
$$z = 3$$

Substitute 3 for z in the second equation.

$$y - \dfrac{8}{11}(3) = -\dfrac{13}{11}$$
$$y - \dfrac{24}{11} = -\dfrac{13}{11}$$
$$y = \dfrac{11}{11}$$
$$y = 1$$

Substitute 1 for y and 3 for z in the first equation.

$$x - 3(1) + 2(3) = 5$$
$$x - 3 + 6 = 5$$
$$x + 3 = 5$$
$$x = 2$$

The solution is (2, 1, 3).

35. $x + 2y = 5$
$y - z = -1$
$2x - 3z = 0$

Write the system in standard form.

$$x + 2y + 0z = 5$$
$$0x + y - z = -1$$
$$2x + 0y - 3z = 0$$

$$\begin{bmatrix} 1 & 2 & 0 & \bigm| & 5 \\ 0 & 1 & -1 & \bigm| & -1 \\ 2 & 0 & -3 & \bigm| & 0 \end{bmatrix}$$

$$\begin{bmatrix} 1 & 2 & 0 & \bigm| & 5 \\ 0 & 1 & -1 & \bigm| & -1 \\ 0 & -4 & -3 & \bigm| & -10 \end{bmatrix} -2R_1 + R_3$$

$$\begin{bmatrix} 1 & 2 & 0 & \bigm| & 5 \\ 0 & 1 & -1 & \bigm| & -1 \\ 0 & 0 & -7 & \bigm| & -14 \end{bmatrix} 4R_2 + R_3$$

$$\begin{bmatrix} 1 & 2 & 0 & \bigm| & 5 \\ 0 & 1 & -1 & \bigm| & -1 \\ 0 & 0 & 1 & \bigm| & 2 \end{bmatrix} -\dfrac{1}{7}R_3$$

The system is

$$x + 2y = 5$$
$$y - z = -1$$
$$z = 2$$

Substitute 2 for z in the second equation.

$$y - z = -1$$
$$y - 2 = -1$$
$$y = 1$$

Substitute 1 for y in the first equation.

$$x + 2y = 5$$
$$x + 2(1) = 5$$
$$x = 3$$

The solution is (3, 1, 2).

37. $x - 2y + 4z = 5$
$-3x + 4y - 2z = -8$
$4x + 5y - 4z = -3$

$$\begin{bmatrix} 1 & -2 & 4 & \bigm| & 5 \\ -3 & 4 & -2 & \bigm| & -8 \\ 4 & 5 & -4 & \bigm| & -3 \end{bmatrix}$$

$$\begin{bmatrix} 1 & -2 & 4 & \bigm| & 5 \\ 0 & -2 & 10 & \bigm| & 7 \\ 4 & 5 & -4 & \bigm| & -3 \end{bmatrix} 3R_1 + R_2$$

$$\begin{bmatrix} 1 & -2 & 4 & \bigm| & 5 \\ 0 & -2 & 10 & \bigm| & 7 \\ 0 & 13 & -20 & \bigm| & -23 \end{bmatrix} -4R_1 + R_3$$

$$\begin{bmatrix} 1 & -2 & 4 & \bigm| & 5 \\ 0 & 1 & -5 & \bigm| & -\dfrac{7}{2} \\ 0 & 13 & -20 & \bigm| & -23 \end{bmatrix} -\dfrac{1}{2}R_2$$

$$\begin{bmatrix} 1 & -2 & 4 & \bigm| & 5 \\ 0 & 1 & -5 & \bigm| & -\dfrac{7}{2} \\ 0 & 0 & 45 & \bigm| & \dfrac{45}{2} \end{bmatrix} -13R_2 + R_3$$

$$\begin{bmatrix} 1 & -2 & 4 & \bigm| & 5 \\ 0 & 1 & -5 & \bigm| & -\dfrac{7}{2} \\ 0 & 0 & 1 & \bigm| & \dfrac{1}{2} \end{bmatrix} \dfrac{1}{45}R_3$$

The system is

$$x - 2y + 4z = 5$$
$$y - 5z = -\frac{7}{2}$$
$$z = \frac{1}{2}$$

Substitute $\frac{1}{2}$ for z in the second equation.

$$y - 5\left(\frac{1}{2}\right) = -\frac{7}{2}$$
$$y - \frac{5}{2} = -\frac{7}{2}$$
$$y = -\frac{7}{2} + \frac{5}{2} = -\frac{2}{2} = -1$$

Substitute -1 for y and $\frac{1}{2}$ for z in the first equation.

$$x - 2(-1) + 4\left(\frac{1}{2}\right) = 5$$
$$x + 2 + 2 = 5$$
$$x + 4 = 5$$
$$x = 1$$

The solution is $\left(1, -1, \frac{1}{2}\right)$.

39.
$$2x - 5y + z = 1$$
$$3x - 5y + z = 3$$
$$-4x + 10y - 2z = -2$$

$$\begin{bmatrix} 2 & -5 & 1 & | & 1 \\ 3 & -5 & 1 & | & 3 \\ -4 & 10 & -2 & | & -2 \end{bmatrix}$$

$$\begin{bmatrix} 1 & -\frac{5}{2} & \frac{1}{2} & | & \frac{1}{2} \\ 3 & -5 & 1 & | & 3 \\ -4 & 10 & -2 & | & -2 \end{bmatrix} \frac{1}{2}R_1$$

$$\begin{bmatrix} 1 & -\frac{5}{2} & \frac{1}{2} & | & \frac{1}{2} \\ 0 & \frac{5}{2} & -\frac{1}{2} & | & \frac{3}{2} \\ -4 & 10 & -2 & | & -2 \end{bmatrix} -3R_1 + R_2$$

$$\begin{bmatrix} 1 & -\frac{5}{2} & \frac{1}{2} & | & \frac{1}{2} \\ 0 & \frac{5}{2} & -\frac{1}{2} & | & \frac{3}{2} \\ 0 & 0 & 0 & | & 0 \end{bmatrix} 4R_1 + R_3$$

Since there is a row of all zeros, the system is dependent.

41.
$$4p - q + r = 4$$
$$-6p + 3q - 2r = -5$$
$$2p + 5q - r = 7$$

$$\begin{bmatrix} 4 & -1 & 1 & | & 4 \\ -6 & 3 & -2 & | & -5 \\ 2 & 5 & -1 & | & 7 \end{bmatrix}$$

$$\begin{bmatrix} 1 & -\frac{1}{4} & \frac{1}{4} & | & 1 \\ -6 & 3 & -2 & | & -5 \\ 2 & 5 & -1 & | & 7 \end{bmatrix} \frac{1}{4}R_1$$

$$\begin{bmatrix} 1 & -\frac{1}{4} & \frac{1}{4} & | & 1 \\ 0 & \frac{3}{2} & -\frac{1}{2} & | & 1 \\ 2 & 5 & -1 & | & 7 \end{bmatrix} 6R_1 + R_2$$

$$\begin{bmatrix} 1 & -\frac{1}{4} & \frac{1}{4} & | & 1 \\ 0 & \frac{3}{2} & -\frac{1}{2} & | & 1 \\ 0 & \frac{11}{2} & -\frac{3}{2} & | & 5 \end{bmatrix} -2R_1 + R_3$$

$$\begin{bmatrix} 1 & -\frac{1}{4} & \frac{1}{4} & | & 1 \\ 0 & 1 & -\frac{1}{3} & | & \frac{2}{3} \\ 0 & \frac{11}{2} & -\frac{3}{2} & | & 5 \end{bmatrix} \frac{2}{3}R_2$$

$$\begin{bmatrix} 1 & -\frac{1}{4} & \frac{1}{4} & | & 1 \\ 0 & 1 & -\frac{1}{3} & | & \frac{2}{3} \\ 0 & 0 & \frac{1}{3} & | & \frac{4}{3} \end{bmatrix} -\frac{11}{2}R_2 + R_3$$

$$\begin{bmatrix} 1 & -\frac{1}{4} & \frac{1}{4} & | & 1 \\ 0 & 1 & -\frac{1}{3} & | & \frac{2}{3} \\ 0 & 0 & 1 & | & 4 \end{bmatrix} 3R_3$$

The system is
$$x - \frac{1}{4}y + \frac{1}{4}z = 1$$
$$y - \frac{1}{3}z = \frac{2}{3}$$
$$z = 4$$

Substitute 4 for z in the second equation.

$$y - \frac{1}{3}z = \frac{2}{3}$$
$$y - \frac{1}{3}(4) = \frac{2}{3}$$
$$y - \frac{4}{3} = \frac{2}{3}$$
$$y = \frac{6}{3} = 2$$

Substitute 2 for y and 4 for z in the first equation.

$$x - \frac{1}{4}y + \frac{1}{4}z = 1$$
$$x - \frac{1}{4}(2) + \frac{1}{4}(4) = 1$$
$$x - \frac{1}{2} + 1 = 1$$
$$x + \frac{1}{2} = 1$$
$$x = \frac{1}{2}$$

The solution is $\left(\frac{1}{2}, 2, 4\right)$.

43.
$$2x - 4y + 3z = -12$$
$$3x - y + 2z = -3$$
$$-4x + 8y - 6z = 10$$

$$\begin{bmatrix} 2 & -4 & 3 & | & -12 \\ 3 & -1 & 2 & | & -3 \\ -4 & 8 & -6 & | & 10 \end{bmatrix}$$

$$\begin{bmatrix} 1 & -2 & \frac{3}{2} & | & -6 \\ 3 & -1 & 2 & | & -3 \\ -4 & 8 & -6 & | & 10 \end{bmatrix} \frac{1}{2}R_1$$

$$\begin{bmatrix} 1 & -2 & \frac{3}{2} & | & -6 \\ 0 & 5 & -\frac{5}{2} & | & 15 \\ -4 & 8 & -6 & | & 10 \end{bmatrix} -3R_1 + R_2$$

$$\begin{bmatrix} 1 & -2 & \frac{3}{2} & | & -6 \\ 0 & 5 & -\frac{5}{2} & | & 15 \\ 0 & 0 & 0 & | & -14 \end{bmatrix} 4R_1 + R_3$$

Since the last row contains zeros on the left and a nonzero number on the right, the system is inconsistent and there is no solution.

45.
$$5x - 3y + 4z = 22$$
$$-x - 15y + 10z = -15$$
$$-3x + 9y - 12z = -6$$

$$\begin{bmatrix} 5 & -3 & 4 & | & 22 \\ -1 & -15 & 10 & | & -15 \\ -3 & 9 & -12 & | & -6 \end{bmatrix}$$

$$\begin{bmatrix} 1 & -\frac{3}{5} & \frac{4}{5} & | & \frac{22}{5} \\ -1 & -15 & 10 & | & -15 \\ -3 & 9 & -12 & | & -6 \end{bmatrix} \frac{1}{5}R_1$$

$$\begin{bmatrix} 1 & -\frac{3}{5} & \frac{4}{5} & | & \frac{22}{5} \\ 0 & -\frac{78}{5} & \frac{54}{5} & | & -\frac{53}{5} \\ -3 & 9 & -12 & | & -6 \end{bmatrix} R_1 + R_2$$

$$\begin{bmatrix} 1 & -\frac{3}{5} & \frac{4}{5} & | & \frac{22}{5} \\ 0 & -\frac{78}{5} & \frac{54}{5} & | & -\frac{53}{5} \\ 0 & \frac{36}{5} & -\frac{48}{5} & | & \frac{36}{5} \end{bmatrix} 3R_1 + R_3$$

$$\begin{bmatrix} 1 & -\frac{3}{5} & \frac{4}{5} & | & \frac{22}{5} \\ 0 & 1 & -\frac{9}{13} & | & \frac{53}{78} \\ 0 & \frac{36}{5} & -\frac{48}{5} & | & \frac{36}{5} \end{bmatrix} -\frac{5}{78}R_2$$

$$\begin{bmatrix} 1 & -\frac{3}{5} & \frac{4}{5} & | & \frac{22}{5} \\ 0 & 1 & -\frac{9}{13} & | & \frac{53}{78} \\ 0 & 0 & -\frac{60}{13} & | & \frac{30}{13} \end{bmatrix} -\frac{36}{5}R_2 + R_3$$

$$\begin{bmatrix} 1 & -\frac{3}{5} & \frac{4}{5} & | & \frac{22}{5} \\ 0 & 1 & -\frac{9}{13} & | & \frac{53}{78} \\ 0 & 0 & 1 & | & -\frac{1}{2} \end{bmatrix}$$

The system is
$$x - \frac{3}{5}y + \frac{4}{5}z = \frac{22}{5}$$
$$y - \frac{9}{13}z = \frac{53}{78}$$
$$z = -\frac{1}{2}$$

Substitute $-\frac{1}{2}$ for z in the second equation.
$$y - \frac{9}{13}z = \frac{53}{78}$$
$$y - \frac{9}{13}\left(-\frac{1}{2}\right) = \frac{53}{78}$$
$$y + \frac{9}{26} = \frac{53}{78}$$
$$y = \frac{53}{78} - \frac{27}{78} = \frac{26}{78} = \frac{1}{3}$$

Substitute $\frac{1}{3}$ for y and $-\frac{1}{2}$ for z in the first equation.
$$x - \frac{3}{5}y + \frac{4}{5}z = \frac{22}{5}$$
$$x - \frac{3}{5}\left(\frac{1}{3}\right) + \frac{4}{5}\left(-\frac{1}{2}\right) = \frac{22}{5}$$
$$x - \frac{1}{5} - \frac{2}{5} = \frac{22}{5}$$
$$x = \frac{22}{5} + \frac{1}{5} + \frac{2}{5} = \frac{25}{5} = 5$$

The solution is $\left(5, \frac{1}{3}, -\frac{1}{2}\right)$.

47. No, this is the same as switching the order of the equations.

49. Let x = smallest angle
  y = remaining angle
  z = largest angle
  $z = x + 55$

  $z = y + 20$

$x + y + z = 180$

Write the system in standard form:

  $x - z = -55$

  $y - z = -20$

$x + y + z = 180$

$$\begin{bmatrix} 1 & 0 & -1 & | & -55 \\ 0 & 1 & -1 & | & -20 \\ 1 & 1 & 1 & | & 180 \end{bmatrix}$$

$$\begin{bmatrix} 1 & 0 & -1 & | & -55 \\ 0 & 1 & -1 & | & -20 \\ 0 & 1 & 2 & | & 235 \end{bmatrix} -1R_1 + R_3$$

$$\begin{bmatrix} 1 & 0 & -1 & | & -55 \\ 0 & 1 & -1 & | & -20 \\ 0 & 0 & 3 & | & 255 \end{bmatrix} -1R_2 + R_3$$

$$\begin{bmatrix} 1 & 0 & -1 & | & -55 \\ 0 & 1 & -1 & | & -20 \\ 0 & 0 & 1 & | & 85 \end{bmatrix} \frac{1}{3}R_3$$

The system is

$x - z = -55$

$y - z = -20$

  $z = 85$

Substitute 85 for z in the second equation.

  $y - z = -20$

$y - 85 = -20$

   $y = 65$

Substitute 85 for z in the first equation.

  $x - z = -55$

$x - 85 = -55$

   $x = 30$

The angles are 30°, 65°, and 85°.

51. Let x = amount Chiquita controls,
  y = amount Dole controls,
  z = amount Del Monte controls
   $x = z + 12$
   $y = 2z - 3$
$x + y + z = 65$

Write the system in standard form.

  $x - z = 12$

  $y - 2z = -3$

$x + y + z = 65$

$$\begin{bmatrix} 1 & 0 & -1 & | & 12 \\ 0 & 1 & -2 & | & -3 \\ 1 & 1 & 1 & | & 65 \end{bmatrix}$$

$$\begin{bmatrix} 1 & 0 & -1 & | & 12 \\ 0 & 1 & -2 & | & -3 \\ 0 & 1 & 2 & | & 53 \end{bmatrix} -1R_1 + R_3$$

$$\begin{bmatrix} 1 & 0 & -1 & | & 12 \\ 0 & 1 & -2 & | & -3 \\ 0 & 0 & 4 & | & 56 \end{bmatrix} -1R_2 + R_3$$

$$\begin{bmatrix} 1 & 0 & -1 & | & 12 \\ 0 & 1 & -2 & | & -3 \\ 0 & 0 & 1 & | & 14 \end{bmatrix} \frac{1}{4}R_3$$

The system is

  $x - z = 12$

  $y - 2z = -3$

    $z = 14$

Substitute 14 for z in the second equation.

   $y - 2z = -3$

$y - 2(14) = -3$

  $y - 28 = -3$

     $y = 25$

Substitute 14 for z in the first equation.

  $x - z = 12$

$x - 14 = 12$

   $x = 26$

Del Monte controls 14% of the bananas, Dole controls 25%, Chiquita controls 26%, and others controlling the remaining $100\% - 65\% = 35\%$.

53. $\dfrac{3x^2 + 4x - 23}{x + 4}$

$$\begin{array}{r} 3x - 8 \\ x + 4 \overline{\smash{)}\ 3x^2 + 4x - 23} \\ \underline{3x^2 + 12x} \\ -8x - 23 \\ \underline{-8x - 32} \\ 9 \end{array}$$

$\dfrac{3x^2 + 4x - 23}{x + 4} = 3x - 8 + \dfrac{9}{x + 4}$

54. $2x^2 - x - 36 = 0$

$(2x-9)(x+4) = 0$

$2x-9 = 0$ or $x+4 = 0$

$x = \dfrac{9}{2}$ $x = -4$

55. $\dfrac{1}{x^2-4} - \dfrac{2}{x-2} = \dfrac{1}{(x+2)(x-2)} - \dfrac{2}{x-2}$

$= \dfrac{1}{(x+2)(x-2)} - \dfrac{2(x+2)}{(x+2)(x-2)}$

$= \dfrac{1-2(x+2)}{(x+2)(x-2)}$

$= \dfrac{1-2x-4}{(x+2)(x-2)}$

$= \dfrac{-2x-3}{(x+2)(x-2)}$

56. Let b be the time (in minutes) it takes Terry to stach the wood by herself.

$\dfrac{t}{a} + \dfrac{t}{b} = 1$

$\dfrac{12}{20} + \dfrac{12}{b} = 1$

$20b\left(\dfrac{12}{20} + \dfrac{12}{b}\right) = 20b(1)$

$12b + 240 = 20b$

$240 = 8b$

$30 = b$

It will take Terry 30 minutes to stach the wood by herself.

Exercise Set 9.7

1. Answers will vary.

3. If $D = 0$ and D_x, D_y, or D_z is not equal to 0, the system is inconsistent.

5. $x = \dfrac{D_x}{D} = \dfrac{12}{4} = 3$

$y = \dfrac{D_y}{D} = \dfrac{-2}{4} = -\dfrac{1}{2}$

The solution is $\left(3, -\dfrac{1}{2}\right)$.

7. $\begin{vmatrix} 2 & 4 \\ 1 & 5 \end{vmatrix} = (2)(5) - (1)(4) = 10 - 4 = 6$

9. $\begin{vmatrix} \frac{1}{2} & 3 \\ 2 & -4 \end{vmatrix} = \dfrac{1}{2}(-4) - (2)(3) = -2 - 6 = -8$

11. $\begin{vmatrix} 3 & 2 & 0 \\ 0 & 5 & 3 \\ -1 & 4 & 2 \end{vmatrix} = 3\begin{vmatrix} 5 & 3 \\ 4 & 2 \end{vmatrix} - 0\begin{vmatrix} 2 & 0 \\ 4 & 2 \end{vmatrix} + (-1)\begin{vmatrix} 2 & 0 \\ 5 & 3 \end{vmatrix}$

$= 3(10-12) - 0(4-0) - 1(6-0)$

$= 3(-2) - 0(4) - 1(6)$

$= -6 - 0 - 6$

$= -12$

13. $\begin{vmatrix} 2 & 3 & 1 \\ 1 & -3 & -6 \\ -4 & 5 & 9 \end{vmatrix}$

$= 2\begin{vmatrix} -3 & -6 \\ 5 & 9 \end{vmatrix} - 1\begin{vmatrix} 3 & 1 \\ 5 & 9 \end{vmatrix} + (-4)\begin{vmatrix} 3 & 1 \\ -3 & -6 \end{vmatrix}$

$= 2[-27-(-30)] - 1(27-5) - 4[-18-(-3)]$

$= 2(3) - 1(22) - 4(-15)$

$= 6 - 22 + 60$

$= 44$

15. $x + 3y = 1$

$-2x - 3y = 4$

To solve, first calculate D, D_x, and D_y.

$D = \begin{vmatrix} 1 & 3 \\ -2 & -3 \end{vmatrix} = (1)(-3) - (-2)(3) = -3 - (-6) = 3$

$D_x = \begin{vmatrix} 1 & 3 \\ 4 & -3 \end{vmatrix} = (1)(-3) - (4)(3) = -15$

$D_y = \begin{vmatrix} 1 & 1 \\ -2 & 4 \end{vmatrix} = (1)(4) - (-2)(1) = 4 - (-2) = 6$

$x = \dfrac{D_x}{D} = \dfrac{-15}{3} = -5$ and $y = \dfrac{D_y}{D} = \dfrac{6}{3} = 2$

The solution is $(-5, 2)$.

17. $-x - 2y = 2$

    $x + 3y = -6$

To solve, first calculate D, D_x, and D_y.

$$D = \begin{vmatrix} -1 & -2 \\ 1 & 3 \end{vmatrix} = (-1)(3) - (1)(-2) = -3 + 2 = -1$$

$$D_x = \begin{vmatrix} 2 & -2 \\ -6 & 3 \end{vmatrix}$$

$$= (2)(3) - (-6)(-2)$$

$$= 6 - 12$$

$$= -6$$

$$D_y = \begin{vmatrix} -1 & 2 \\ 1 & -6 \end{vmatrix} = (-1)(-6) - (1)(2) = 6 - 2 = 4$$

$$x = \frac{D_x}{D} = \frac{-6}{-1} = 6 \text{ and } y = \frac{D_y}{D} = \frac{4}{-1} = -4$$

The solution is $(6, -4)$.

19. $6x = 4y + 7$

    $8x - 1 = -3y$

Rewrite the system in standard form:

$6x - 4y = 7$

$8x + 3y = 1$

Now calculate D, D_x, and D_y.

$$D = \begin{vmatrix} 6 & -4 \\ 8 & 3 \end{vmatrix} = (6)(3) - (8)(-4) = 18 + 32 = 50$$

$$D_x = \begin{vmatrix} 7 & -4 \\ 1 & 3 \end{vmatrix} = (7)(3) - (1)(-4) = 21 + 4 = 25$$

$$D_y = \begin{vmatrix} 6 & 7 \\ 8 & 1 \end{vmatrix} = (6)(1) - (8)(7) = 6 - 56 = -50$$

$$x = \frac{D_x}{D} = \frac{25}{50} = \frac{1}{2} \text{ and } y = \frac{D_y}{D} = \frac{-50}{50} = -1$$

The solution is $\left(\frac{1}{2}, -1 \right)$.

21. $5p - 7q = -21$

    $-4p + 3q = 22$

To solve, first calculate D, D_p, and D_q.

$$D = \begin{vmatrix} 5 & -7 \\ -4 & 3 \end{vmatrix} = (5)(3) - (-4)(-7)$$

$$= 15 - 28$$

$$= -13$$

$$D_p = \begin{vmatrix} -21 & -7 \\ 22 & 3 \end{vmatrix} = (-21)(3) - (22)(-7)$$

$$= -63 - (-154)$$

$$= 91$$

$$D_q = \begin{vmatrix} 5 & -21 \\ -4 & 22 \end{vmatrix} = (5)(22) - (-4)(-21)$$

$$= 110 - 84$$

$$= 26$$

$$p = \frac{D_p}{D} = \frac{91}{-13} = -7 \text{ and } q = \frac{D_q}{D} = \frac{26}{-13} = -2$$

The solution is $(-7, -2)$.

23. $x + 5y = 3$

    $2x - 6 = -10y$

Rewrite the system in standard form:

    $x + 5y = 3$

    $2x + 10y = 6$

To solve, first calculate D, D_x, and D_y.

$$D = \begin{vmatrix} 1 & 5 \\ 2 & 10 \end{vmatrix} = (1)(10) - (2)(5) = 10 - 10 = 0$$

$$D_x = \begin{vmatrix} 3 & 5 \\ 6 & 10 \end{vmatrix} = (3)(10) - (6)(5) = 30 - 30 = 0$$

$$D_y = \begin{vmatrix} 1 & 3 \\ 2 & 6 \end{vmatrix} = (1)(6) - (2)(3) = 6 - 6 = 0$$

Since $D = 0$, $D_x = 0$, and $D_y = 0$, the system is dependent so there are an infinite number of solutions.

25. $3r = -4s - 6$

    $3s = -5r + 1$

Rewrite the system in standard form.

    $3r + 4s = -6$

    $5r + 3s = 1$

Now calculate D, D_r, and D_s.

$$D = \begin{vmatrix} 3 & 4 \\ 5 & 3 \end{vmatrix} = (3)(3) - (5)(4) = 9 - 20 = -11$$

$$D_r = \begin{vmatrix} -6 & 4 \\ 1 & 3 \end{vmatrix} = (-6)(3) - (1)(4) = -18 - 4 = -22$$

$$D_s = \begin{vmatrix} 3 & -6 \\ 5 & 1 \end{vmatrix} = (3)(1) - (5)(-6) = 3 + 30 = 33$$

$$r = \frac{D_r}{D} = \frac{-22}{-11} = 2 \text{ and } s = \frac{D_s}{D} = \frac{33}{-11} = -3$$

The solution is $(2, -3)$.

27. $5x - 5y = 3$

$-x + y = -4$

To solve, first calculate D, D_x, and D_y.

$$D = \begin{vmatrix} 5 & -5 \\ -1 & 1 \end{vmatrix} = (5)(1) - (-1)(-5) = 5 - 5 = 0$$

$$D_x = \begin{vmatrix} 3 & -5 \\ -4 & 1 \end{vmatrix} = (3)(1) - (-4)(-5) = 3 - 20 = -17$$

Since $D = 0$ and $D_x \neq 0$, the system is inconsistent, so there is no solution.

29. $6.3x - 4.5y = -9.9$

$-9.1x + 3.2y = -2.2$

Here, you can work with decimals in the determinants. If you do not want to use decimals, then you need to multiply each equation by 10 to clear the decimals.

First, calculate D, D_x, and D_y.

$$D = \begin{vmatrix} 6.3 & -4.5 \\ -9.1 & 3.2 \end{vmatrix} = (6.3)(3.2) - (-9.1)(-4.5)$$

$$= 20.16 - 40.95 = -20.79$$

$$D_x = \begin{vmatrix} -9.9 & -4.5 \\ -2.2 & 3.2 \end{vmatrix} = (-9.9)(3.2) - (-2.2)(-4.5)$$

$$= -31.68 - 9.90 = -41.58$$

$$D_y = \begin{vmatrix} 6.3 & -9.9 \\ -9.1 & -2.2 \end{vmatrix} = (6.3)(-2.2) - (-9.1)(-9.9)$$

$$= -13.86 - 90.09 = -103.95$$

$$x = \frac{D_x}{D} = \frac{-41.58}{-20.79} = 2 \text{ and}$$

$$y = \frac{D_y}{D} = \frac{-103.95}{-20.79} = 5$$

The solution is (2, 5).

31.
$$x + y + z = 3$$
$$0x - 3y + 4z = 15$$
$$-3x + 4y - 2z = -13$$

To solve, first calculate D, D_x, D_y, and D_z.

$$D = \begin{vmatrix} 1 & 1 & 1 \\ 0 & -3 & 4 \\ -3 & 4 & -2 \end{vmatrix} \text{ (using first column)}$$

$$= 1\begin{vmatrix} -3 & 4 \\ 4 & -2 \end{vmatrix} - 0\begin{vmatrix} 1 & 1 \\ 4 & -2 \end{vmatrix} + (-3)\begin{vmatrix} 1 & 1 \\ -3 & 4 \end{vmatrix}$$

$$= 1(6 - 16) - 0(-2 - 4) - 3(4 + 3)$$

$$= 1(-10) - 0(-6) - 3(7)$$

$$= -10 - 0 - 21$$

$$= -31$$

$$D_x = \begin{vmatrix} 3 & 1 & 1 \\ 15 & -3 & 4 \\ -13 & 4 & -2 \end{vmatrix} \text{ (using first row)}$$

$$= 3\begin{vmatrix} -3 & 4 \\ 4 & -2 \end{vmatrix} - 1\begin{vmatrix} 15 & 4 \\ -13 & -2 \end{vmatrix} + 1\begin{vmatrix} 15 & -3 \\ -13 & 4 \end{vmatrix}$$

$$= 3(6 - 16) - 1(-30 + 52) + 1(60 - 39)$$

$$= 3(-10) - 1(22) + 1(21)$$

$$= -30 - 22 + 21$$

$$= -31$$

$$D_y = \begin{vmatrix} 1 & 3 & 1 \\ 0 & 15 & 4 \\ -3 & -13 & -2 \end{vmatrix} \text{ (using first column)}$$

$$= 1\begin{vmatrix} 15 & 4 \\ -13 & -2 \end{vmatrix} - 0\begin{vmatrix} 3 & 1 \\ -13 & -2 \end{vmatrix} + (-3)\begin{vmatrix} 3 & 1 \\ 15 & 4 \end{vmatrix}$$

$$= 1(-30 + 52) - 0(-6 + 13) - 3(12 - 15)$$

$$= 1(22) - 0(7) - 3(-3)$$

$$= 22 - 0 + 9$$

$$= 31$$

$$D_z = \begin{vmatrix} 1 & 1 & 3 \\ 0 & -3 & 15 \\ -3 & 4 & -13 \end{vmatrix} \text{ (using first column)}$$

$$= 1\begin{vmatrix} -3 & 15 \\ 4 & -13 \end{vmatrix} - 0\begin{vmatrix} 1 & 3 \\ 4 & -13 \end{vmatrix} + (-3)\begin{vmatrix} 1 & 3 \\ -3 & 15 \end{vmatrix}$$

$$= 1(39 - 60) - 0(-13 - 12) - 3(15 + 9)$$

$$= 1(-21) - 0(-25) - 3(24)$$

$$= -21 - 0 - 72$$

$$= -93$$

$$x = \frac{D_x}{D} = \frac{-31}{-31} = 1, \ y = \frac{D_y}{D} = \frac{31}{-31} = -1, \text{ and}$$

$$z = \frac{D_z}{D} = \frac{-93}{-31} = 3$$

The solution is $(1, -1, 3)$.

33. $3x - 5y - 4z = -4$

$4x + 2y + 0z = 1$

$0x + 6y - 4z = -11$

To solve, first calculate D, D_x, D_y, and D_z.

$D = \begin{vmatrix} 3 & -5 & -4 \\ 4 & 2 & 0 \\ 0 & 6 & -4 \end{vmatrix}$ (using the first row)

$= 3\begin{vmatrix} 2 & 0 \\ 6 & -4 \end{vmatrix} - (-5)\begin{vmatrix} 4 & 0 \\ 0 & -4 \end{vmatrix} + (-4)\begin{vmatrix} 4 & 2 \\ 0 & 6 \end{vmatrix}$

$= 3(-8-0) + 5(-16-0) - 4(24-0)$

$= 3(-8) + 5(-16) - 4(24)$

$= -24 - 80 - 96$

$= -200$

$D_x = \begin{vmatrix} -4 & -5 & -4 \\ 1 & 2 & 0 \\ -11 & 6 & -4 \end{vmatrix}$ (using the first row)

$= -4\begin{vmatrix} 2 & 0 \\ 6 & -4 \end{vmatrix} - (-5)\begin{vmatrix} 1 & 0 \\ -11 & -4 \end{vmatrix} + (-4)\begin{vmatrix} 1 & 2 \\ -11 & 6 \end{vmatrix}$

$= -4(-8-0) + 5(-4+0) - 4(6+22)$

$= -4(-8) + 5(-4) - 4(28)$

$= 32 - 20 - 112$

$= -100$

$D_y = \begin{vmatrix} 3 & -4 & -4 \\ 4 & 1 & 0 \\ 0 & -11 & -4 \end{vmatrix}$ (using the first row)

$= 3\begin{vmatrix} 1 & 0 \\ -11 & -4 \end{vmatrix} - (-4)\begin{vmatrix} 4 & 0 \\ 0 & -4 \end{vmatrix} + (-4)\begin{vmatrix} 4 & 1 \\ 0 & -11 \end{vmatrix}$

$= 3(-4+0) + 4(-16-0) - 4(-44-0)$

$= 3(-4) + 4(-16) - 4(-44)$

$= -12 - 64 + 176$

$= 100$

$D_z = \begin{vmatrix} 3 & -5 & -4 \\ 4 & 2 & 1 \\ 0 & 6 & -11 \end{vmatrix}$ (using the first row)

$= 3\begin{vmatrix} 2 & 1 \\ 6 & -11 \end{vmatrix} - (-5)\begin{vmatrix} 4 & 1 \\ 0 & -11 \end{vmatrix} + (-4)\begin{vmatrix} 4 & 2 \\ 0 & 6 \end{vmatrix}$

$= 3(-22-6) + 5(-44-0) - 4(24-0)$

$= 3(-28) + 5(-44) - 4(24)$

$= -84 - 220 - 96$

$= -400$

$x = \dfrac{D_x}{D} = \dfrac{-100}{-200} = \dfrac{1}{2}$, $y = \dfrac{D_y}{D} = \dfrac{100}{-200} = -\dfrac{1}{2}$, and

$z = \dfrac{D_z}{D} = \dfrac{-400}{-200} = 2$

The solution is $\left(\dfrac{1}{2}, -\dfrac{1}{2}, 2\right)$.

35. $x + 4y - 3z = -6$

$2x - 8y + 5z = 12$

$3x + 4y - 2z = -3$

To solve, first calculate D, D_x, D_y, and D_z

$D = \begin{vmatrix} 1 & 4 & -3 \\ 2 & -8 & 5 \\ 3 & 4 & -2 \end{vmatrix}$ (using the first row)

$= 1\begin{vmatrix} -8 & 5 \\ 4 & -2 \end{vmatrix} - 4\begin{vmatrix} 2 & 5 \\ 3 & -2 \end{vmatrix} + (-3)\begin{vmatrix} 2 & -8 \\ 3 & 4 \end{vmatrix}$

$= 1(16-20) - 4(-4-15) - 3(8+24)$

$= 1(-4) - 4(-19) - 3(32)$

$= -4 + 76 - 96$

$= -24$

$D_x = \begin{vmatrix} -6 & 4 & -3 \\ 12 & -8 & 5 \\ -3 & 4 & -2 \end{vmatrix}$ (using the first row)

$= -6\begin{vmatrix} -8 & 5 \\ 4 & -2 \end{vmatrix} - 4\begin{vmatrix} 12 & 5 \\ -3 & -2 \end{vmatrix} + (-3)\begin{vmatrix} 12 & -8 \\ -3 & 4 \end{vmatrix}$

$= -6(16-20) - 4(-24+15) - 3(48-24)$

$= -6(-4) - 4(-9) - 3(24)$

$= 24 + 36 - 72$

$= -12$

$D_y = \begin{vmatrix} 1 & -6 & -3 \\ 2 & 12 & 5 \\ 3 & -3 & -2 \end{vmatrix}$ (using the first row)

$= 1\begin{vmatrix} 12 & 5 \\ -3 & -2 \end{vmatrix} - (-6)\begin{vmatrix} 2 & 5 \\ 3 & -2 \end{vmatrix} + (-3)\begin{vmatrix} 2 & 12 \\ 3 & -3 \end{vmatrix}$

$= 1(-24+15) + 6(-4-15) - 3(-6-36)$

$= 1(-9) + 6(-19) - 3(-42)$

$= -9 - 114 + 126$

$= 3$

$D_z = \begin{vmatrix} 1 & 4 & -6 \\ 2 & -8 & 12 \\ 3 & 4 & -3 \end{vmatrix}$ (using the first row)

$= 1\begin{vmatrix} -8 & 12 \\ 4 & -3 \end{vmatrix} - 4\begin{vmatrix} 2 & 12 \\ 3 & -3 \end{vmatrix} + (-6)\begin{vmatrix} 2 & -8 \\ 3 & 4 \end{vmatrix}$

$= 1(24-48) - 4(-6-36) - 6(8+24)$

$= 1(-24) - 4(-42) - 6(32)$

$= -24 + 168 - 192$

$= -48$

$x = \dfrac{D_x}{D} = \dfrac{-12}{-24} = \dfrac{1}{2}$, $y = \dfrac{D_y}{D} = \dfrac{3}{-24} = -\dfrac{1}{8}$, and

$z = \dfrac{D_z}{D} = \dfrac{-48}{-24} = 2$.

The solution is $\left(\dfrac{1}{2}, -\dfrac{1}{8}, 2\right)$.

37. $a - b + 2c = 3$
   $a - b + c = 1$
   $2a + b + 2c = 2$
   To solve, first calculate D, D_a, D_b, and D_c.

$D = \begin{vmatrix} 1 & -1 & 2 \\ 1 & -1 & 1 \\ 2 & 1 & 2 \end{vmatrix}$ (using first column)

$\quad = 1\begin{vmatrix} -1 & 1 \\ 1 & 2 \end{vmatrix} - 1\begin{vmatrix} -1 & 2 \\ 1 & 2 \end{vmatrix} + 2\begin{vmatrix} -1 & 2 \\ -1 & 1 \end{vmatrix}$

$\quad = 1(-2-1) - 1(-2-2) + 2(-1+2)$

$\quad = 1(-3) - 1(-4) + 2(1)$

$\quad = -3 + 4 + 2$

$\quad = 3$

$D_a = \begin{vmatrix} 3 & -1 & 2 \\ 1 & -1 & 1 \\ 2 & 1 & 2 \end{vmatrix}$ (using first column)

$\quad = 3\begin{vmatrix} -1 & 1 \\ 1 & 2 \end{vmatrix} - 1\begin{vmatrix} -1 & 2 \\ 1 & 2 \end{vmatrix} + 2\begin{vmatrix} -1 & 2 \\ -1 & 1 \end{vmatrix}$

$\quad = 3(-2-1) - 1(-2-2) + 2(-1+2)$

$\quad = 3(-3) - 1(-4) + 2(1)$

$\quad = -9 + 4 + 2$

$\quad = -3$

$D_b = \begin{vmatrix} 1 & 3 & 2 \\ 1 & 1 & 1 \\ 2 & 2 & 2 \end{vmatrix}$ (using first column)

$\quad = 1\begin{vmatrix} 1 & 1 \\ 2 & 2 \end{vmatrix} - 1\begin{vmatrix} 3 & 2 \\ 2 & 2 \end{vmatrix} + 2\begin{vmatrix} 3 & 2 \\ 1 & 1 \end{vmatrix}$

$\quad = 1(2-2) - 1(6-4) + 2(3-2)$

$\quad = 1(0) - 1(2) + 2(1)$

$\quad = 0 - 2 + 2$

$\quad = 0$

$D_c = \begin{vmatrix} 1 & -1 & 3 \\ 1 & -1 & 1 \\ 2 & 1 & 2 \end{vmatrix}$ (using first column)

$\quad = 1\begin{vmatrix} -1 & 1 \\ 1 & 2 \end{vmatrix} - 1\begin{vmatrix} -1 & 3 \\ 1 & 2 \end{vmatrix} + 2\begin{vmatrix} -1 & 3 \\ -1 & 1 \end{vmatrix}$

$\quad = 1(-2-1) - 1(-2-3) + 2(-1+3)$

$\quad = 1(-3) - 1(-5) + 2(2)$

$\quad = -3 + 5 + 4$

$\quad = 6$

$a = \dfrac{D_a}{D} = \dfrac{-3}{3} = -1$, $b = \dfrac{D_b}{D} = \dfrac{0}{3} = 0$, and

$c = \dfrac{D_c}{D} = \dfrac{6}{3} = 2$

The solution is $(-1, 0, 2)$.

39. $a + 2b + c = 1$
   $a - b + 3c = 2$
   $2a + b + 4c = 3$
   To solve, first calculate D, D_a, D_b, and D_c.

$D = \begin{vmatrix} 1 & 2 & 1 \\ 1 & -1 & 3 \\ 2 & 1 & 4 \end{vmatrix}$ (using first column)

$\quad = 1\begin{vmatrix} -1 & 3 \\ 1 & 4 \end{vmatrix} - 1\begin{vmatrix} 2 & 1 \\ 1 & 4 \end{vmatrix} + 2\begin{vmatrix} 2 & 1 \\ -1 & 3 \end{vmatrix}$

$\quad = 1(-4-3) - 1(8-1) + 2(6+1)$

$\quad = 1(-7) - 1(7) + 2(7)$

$\quad = -7 - 7 + 14 = 0$

$D_a = \begin{vmatrix} 1 & 2 & 1 \\ 2 & -1 & 3 \\ 3 & 1 & 4 \end{vmatrix}$ (using first column)

$\quad = 1\begin{vmatrix} -1 & 3 \\ 1 & 4 \end{vmatrix} - 2\begin{vmatrix} 2 & 1 \\ 1 & 4 \end{vmatrix} + 3\begin{vmatrix} 2 & 1 \\ -1 & 3 \end{vmatrix}$

$\quad = 1(-4-3) - 2(8-1) + 3(6+1)$

$\quad = 1(-7) - 2(7) + 3(7)$

$\quad = -7 - 14 + 21 = 0$

$D_b = \begin{vmatrix} 1 & 1 & 1 \\ 1 & 2 & 3 \\ 2 & 3 & 4 \end{vmatrix}$ (using first row)

$\quad = 1\begin{vmatrix} 2 & 3 \\ 3 & 4 \end{vmatrix} - 1\begin{vmatrix} 1 & 3 \\ 2 & 4 \end{vmatrix} + 1\begin{vmatrix} 1 & 2 \\ 2 & 3 \end{vmatrix}$

$\quad = 1(8-9) - 1(4-6) + 1(3-4)$

$\quad = 1(-1) - 1(-2) + 1(-1)$

$\quad = -1 + 2 - 1 = 0$

$D_c = \begin{vmatrix} 1 & 2 & 1 \\ 1 & -1 & 2 \\ 2 & 1 & 3 \end{vmatrix}$ (using first column)

$\quad = 1\begin{vmatrix} -1 & 2 \\ 1 & 3 \end{vmatrix} - 1\begin{vmatrix} 2 & 1 \\ 1 & 3 \end{vmatrix} + 2\begin{vmatrix} 2 & 1 \\ -1 & 2 \end{vmatrix}$

$\quad = 1(-3-2) - 1(6-1) + 2(4+1)$

$\quad = 1(-5) - 1(5) + 2(5)$

$\quad = -5 - 5 + 10 = 0$

Since $D = 0$, $D_a = 0$, $D_b = 0$, and $D_c = 0$, there are an infinite number of solutions to the system and it is a dependent system.

41. $1.1x + 2.3y - 4.0z = -9.2$

$-2.3x + 0y + 4.6z = 6.9$

$0x - 8.2y - 7.5z = -6.8$

Here, you can work with decimals in the determinants. If you do not want to use decimals, then you need to multiply each equation by 10 to clear the decimals. To solve, first calculate D, D_x, D_y, and D_z.

$$D = \begin{vmatrix} 1.1 & 2.3 & -4.0 \\ -2.3 & 0 & 4.6 \\ 0 & -8.2 & -7.5 \end{vmatrix} = 1.1\begin{vmatrix} 0 & 4.6 \\ -8.2 & -7.5 \end{vmatrix} - (-2.3)\begin{vmatrix} 2.3 & -4.0 \\ -8.2 & -7.5 \end{vmatrix} + 0\begin{vmatrix} 2.3 & -4.0 \\ 0 & 4.6 \end{vmatrix} \quad \text{(using first column)}$$

$$= 1.1(0 + 37.72) + 2.3(-17.25 - 32.8) + 0(10.58 - 0)$$

$$= 1.1(37.72) + 2.3(-50.05) + 0(10.58)$$

$$= 41.492 - 115.115 + 0$$

$$= -73.623$$

$$D_x = \begin{vmatrix} -9.2 & 2.3 & -4.0 \\ 6.9 & 0 & 4.6 \\ -6.8 & -8.2 & -7.5 \end{vmatrix} = -9.2\begin{vmatrix} 0 & 4.6 \\ -8.2 & -7.5 \end{vmatrix} - 6.9\begin{vmatrix} 2.3 & -4.0 \\ -8.2 & -7.5 \end{vmatrix} + (-6.8)\begin{vmatrix} 2.3 & -4.0 \\ 0 & 4.6 \end{vmatrix} \quad \text{(using first column)}$$

$$= -9.2(0 + 37.72) - 6.9(-17.25 - 32.8) - 6.8(10.58 - 0)$$

$$= -9.2(37.72) - 6.9(-50.05) - 6.8(10.58)$$

$$= -347.024 + 345.345 - 71.944$$

$$= -73.623$$

$$D_y = \begin{vmatrix} 1.1 & -9.2 & -4.0 \\ -2.3 & 6.9 & 4.6 \\ 0 & -6.8 & -7.5 \end{vmatrix} = 1.1\begin{vmatrix} 6.9 & 4.6 \\ -6.8 & -7.5 \end{vmatrix} - (-2.3)\begin{vmatrix} -9.2 & -4.0 \\ -6.8 & -7.5 \end{vmatrix} + 0\begin{vmatrix} -9.2 & -4.0 \\ 6.9 & 4.6 \end{vmatrix} \quad \text{(using first column)}$$

$$= 1.1(-51.75 + 31.28) + 2.3(69 - 27.2) + 0(-42.32 + 27.6)$$

$$= 1.1(-20.47) + 2.3(41.8) + 0(-14.72)$$

$$= -22.517 + 96.14 + 0$$

$$= 73.623$$

$$D_z = \begin{vmatrix} 1.1 & 2.3 & -9.2 \\ -2.3 & 0 & 6.9 \\ 0 & -8.2 & -6.8 \end{vmatrix} = 1.1\begin{vmatrix} 0 & 6.9 \\ -8.2 & -6.8 \end{vmatrix} - (-2.3)\begin{vmatrix} 2.3 & -9.2 \\ -8.2 & -6.8 \end{vmatrix} + 0\begin{vmatrix} 2.3 & -9.2 \\ 0 & 6.9 \end{vmatrix} \quad \text{(using first column)}$$

$$= 1.1(0 + 56.58) + 2.3(-15.64 - 75.44) + 0(15.87 - 0)$$

$$= 1.1(56.58) + 2.3(-91.08) + 0(15.87)$$

$$= 62.238 - 209.484 + 0$$

$$= -147.246$$

$$x = \frac{D_x}{D} = \frac{-73.623}{-73.623} = 1, \quad y = \frac{D_y}{D} = \frac{73.623}{-73.623} = -1, \text{ and } z = \frac{D_z}{D} = \frac{-147.246}{-73.623} = 2$$

The solution is $(1, -1, 2)$.

43. $-6x + 3y - 12z = -13$

$\qquad 5x + 2y - 3z = 1$

$\qquad 2x - y + 4z = -5$

To solve, first calculate D, D_x, D_y, and D_z.

$$D = \begin{vmatrix} -6 & 3 & -12 \\ 5 & 2 & -3 \\ 2 & -1 & 4 \end{vmatrix} \text{ (using the first row)}$$

$$= -6\begin{vmatrix} 2 & -3 \\ -1 & 4 \end{vmatrix} - 3\begin{vmatrix} 5 & -3 \\ 2 & 4 \end{vmatrix} + (-12)\begin{vmatrix} 5 & 2 \\ 2 & -1 \end{vmatrix}$$

$$= -6(8-3) - 3(20+6) - 12(-5-4)$$

$$= -6(5) - 3(26) - 12(-9)$$

$$= -30 - 78 + 108$$

$$= 0$$

$$D_x = \begin{vmatrix} -13 & 3 & -12 \\ 1 & 2 & -3 \\ -5 & -1 & 4 \end{vmatrix} \text{ (using the first row)}$$

$$= -13\begin{vmatrix} 2 & -3 \\ -1 & 4 \end{vmatrix} - 3\begin{vmatrix} 1 & -3 \\ -5 & 4 \end{vmatrix} + (-12)\begin{vmatrix} 1 & 2 \\ -5 & -1 \end{vmatrix}$$

$$= -13(8-3) - 3(4-15) - 12(-1+10)$$

$$= -13(5) - 3(-11) - 12(9)$$

$$= -65 + 33 - 108$$

$$= -140$$

Since $D = 0$ and $D_x = -140 \neq 0$, there is no solution to the system and it is an inconsistent system.

45. $2x + \dfrac{1}{2}y - 3z = 5$

$\qquad -3x + 2y + 2z = 1$

$\qquad 4x - \dfrac{1}{4}y - 7z = 4$

To clear the system of fractions, multiply the first equation by 2 and the third equation by 4.

$\qquad 4x + y - 6z = 10$

$\qquad -3x + 2y + 2z = 1$

$\qquad 16x - y - 28z = 16$

To solve, first calculate D, D_x, D_y, and D_z.

$$D = \begin{vmatrix} 4 & 1 & -6 \\ -3 & 2 & 2 \\ 16 & -1 & -28 \end{vmatrix} \text{ (use the first row)}$$

$$= 4\begin{vmatrix} 2 & 2 \\ -1 & -28 \end{vmatrix} - 1\begin{vmatrix} -3 & 2 \\ 16 & -28 \end{vmatrix} + (-6)\begin{vmatrix} -3 & 2 \\ 16 & -1 \end{vmatrix}$$

$$= 4(-56+2) - 1(84-32) - 6(3-32)$$

$$= 4(-54) - 1(52) - 6(-29)$$

$$= -216 - 52 + 174$$

$$= -94$$

$$D_x = \begin{vmatrix} 10 & 1 & -6 \\ 1 & 2 & 2 \\ 16 & -1 & -28 \end{vmatrix} \text{ (use the first row)}$$

$$= 10\begin{vmatrix} 2 & 2 \\ -1 & -28 \end{vmatrix} - 1\begin{vmatrix} 1 & 2 \\ 16 & -28 \end{vmatrix} + (-6)\begin{vmatrix} 1 & 2 \\ 16 & -1 \end{vmatrix}$$

$$= 10(-56+2) - 1(-28-32) - 6(-1-32)$$

$$= 10(-54) - 1(-60) - 6(-33)$$

$$= -540 + 60 + 198$$

$$= -282$$

$$D_y = \begin{vmatrix} 4 & 10 & -6 \\ -3 & 1 & 2 \\ 16 & 16 & -28 \end{vmatrix} \text{ (use the first row)}$$

$$= 4\begin{vmatrix} 1 & 2 \\ 16 & -28 \end{vmatrix} - 10\begin{vmatrix} -3 & 2 \\ 16 & -28 \end{vmatrix} + (-6)\begin{vmatrix} -3 & 1 \\ 16 & 16 \end{vmatrix}$$

$$= 4(-28-32) - 10(84-32) - 6(-48-16)$$

$$= 4(-60) - 10(52) - 6(-64)$$

$$= -240 - 520 + 384$$

$$= -376$$

$$D_z = \begin{vmatrix} 4 & 1 & 10 \\ -3 & 2 & 1 \\ 16 & -1 & 16 \end{vmatrix} \text{ (use the first row)}$$

$$= 4\begin{vmatrix} 2 & 1 \\ -1 & 16 \end{vmatrix} - 1\begin{vmatrix} -3 & 1 \\ 16 & 16 \end{vmatrix} + 10\begin{vmatrix} -3 & 2 \\ 16 & -1 \end{vmatrix}$$

$$= 4(32+1) - 1(-48-16) + 10(3-32)$$

$$= 4(33) - 1(-64) + 10(-29)$$

$$= 132 + 64 - 290$$

$$= -94$$

$$x = \frac{D_x}{D} = \frac{-282}{-94} = 3, \quad y = \frac{D_y}{D} = \frac{-376}{-94} = 4,$$

$$z = \frac{D_z}{D} = \frac{-94}{-94} = 1$$

The solution is $(3, 4, 1)$.

47. $0.3x - 0.1y - 0.3z = -0.2$

$0.2x - 0.1y + 0.1z = -0.9$

$0.1x + 0.2y - 0.4z = 1.7$

To clear decimals multiply each equation by 10.

$3x - y - 3x = -2$

$2x - y + z = -9$

$x + 2y - 4z = 17$

To solve, first calculate D, D_x, D_y, and D_z.

$D = \begin{vmatrix} 3 & -1 & -3 \\ 2 & -1 & 1 \\ 1 & 2 & -4 \end{vmatrix}$ (use first column)

$= 3\begin{vmatrix} -1 & 1 \\ 2 & -4 \end{vmatrix} - 2\begin{vmatrix} -1 & -3 \\ 2 & -4 \end{vmatrix} + 1\begin{vmatrix} -1 & -3 \\ -1 & 1 \end{vmatrix}$

$= 3(4-2) - 2(4+6) + 1(-1-3)$

$= 3(2) - 2(10) + 1(-4)$

$= 6 - 20 - 4$

$= -18$

$D_x = \begin{vmatrix} -2 & -1 & -3 \\ -9 & -1 & 1 \\ 17 & 2 & -4 \end{vmatrix}$ (use first column)

$= -2\begin{vmatrix} -1 & 1 \\ 2 & -4 \end{vmatrix} - (-9)\begin{vmatrix} -1 & -3 \\ 2 & -4 \end{vmatrix} + 17\begin{vmatrix} -1 & -3 \\ -1 & 1 \end{vmatrix}$

$= -2(4-2) + 9(4+6) + 17(-1-3)$

$= -2(2) + 9(10) + 17(-4)$

$= -4 + 90 - 68$

$= 18$

$D_y = \begin{vmatrix} 3 & -2 & -3 \\ 2 & -9 & 1 \\ 1 & 17 & -4 \end{vmatrix}$ (use first column)

$= 3\begin{vmatrix} -9 & 1 \\ 17 & -4 \end{vmatrix} - 2\begin{vmatrix} -2 & -3 \\ 17 & -4 \end{vmatrix} + 1\begin{vmatrix} -2 & -3 \\ -9 & 1 \end{vmatrix}$

$= 3(36-17) - 2(8+51) + 1(-2-27)$

$= 3(19) - 2(59) + 1(-29)$

$= 57 - 118 - 29$

$= -90$

$D_z = \begin{vmatrix} 3 & -1 & -2 \\ 2 & -1 & -9 \\ 1 & 2 & 17 \end{vmatrix}$ (use first column)

$= 3\begin{vmatrix} -1 & -9 \\ 2 & 17 \end{vmatrix} - 2\begin{vmatrix} -1 & -2 \\ 2 & 17 \end{vmatrix} + 1\begin{vmatrix} -1 & -2 \\ -1 & -9 \end{vmatrix}$

$= 3(-17+18) - 2(-17+4) + 1(9-2)$

$= 3(1) - 2(-13) + 1(7)$

$= 3 + 26 + 7$

$= 36$

$x = \dfrac{D_x}{D} = \dfrac{18}{-18} = -1$, $y = \dfrac{D_y}{D} = \dfrac{-90}{-18} = 5$, and

$z = \dfrac{D_z}{D} = \dfrac{36}{-18} = -2$

The solution is $(-1, 5, -2)$.

49. $\begin{vmatrix} a_1 & b_1 \\ a_2 & b_2 \end{vmatrix} = a_1 b_2 - a_2 b_1$

$\begin{vmatrix} a_2 & b_2 \\ a_1 & b_1 \end{vmatrix} = a_2 b_1 - a_1 b_2$

The second result is the negative of the first result. Thus, the second determinant has the opposite sign.

51. 0; $\begin{vmatrix} a & b \\ a & b \end{vmatrix} = (a)(b) - (a)(b) = ab - ab = 0$

53. 0; for example

$\begin{vmatrix} 0 & 0 & 0 \\ a & b & c \\ d & e & f \end{vmatrix} = 0\begin{vmatrix} b & c \\ e & f \end{vmatrix} - 0\begin{vmatrix} a & c \\ d & f \end{vmatrix} + 0\begin{vmatrix} a & b \\ d & e \end{vmatrix}$

$= 0 - 0 + 0 = 0$

55. Yes, the determinant will become the opposite of the original value.

57. No; the value of the new determinant will be the same as the original value. Each time a row (or column) is multiplied by -1, the value of the determinant changes sign. Since two rows were multiplied by -1, the value of the determinant changes sign twice, which yields the original value.

59. Yes; the value of the new determinant will be double the original value. That is, we would multiply the original determinant by 2 as well.

61. $\begin{vmatrix} 4 & 6 \\ -2 & y \end{vmatrix} = 32$

$(4)(y) - (-2)(6) = 32$

$4y + 12 = 32$

$4y = 20$

$y = \dfrac{20}{4} = 5$

63. $\begin{vmatrix} 4 & 7 & y \\ 3 & -1 & 2 \\ 4 & 1 & 5 \end{vmatrix} = -35$

$4\begin{vmatrix} -1 & 2 \\ 1 & 5 \end{vmatrix} - 3\begin{vmatrix} 7 & y \\ 1 & 5 \end{vmatrix} + 4\begin{vmatrix} 7 & y \\ -1 & 2 \end{vmatrix} = -35$

$4(-5-2) - 3(35-y) + 4(14+y) = -35$

$4(-7) - 3(35-y) + 4(14+y) = -35$

$-28 - 105 + 3y + 56 + 4y = -35$

$-77 + 7y = -35$

$7y = 42$

$y = 6$

65. a. To eliminate y, multiply the first equation by b_2 and the second equation by $-b_1$ and then add.

$b_2[a_1 x + b_1 y = c_1]$

$-b_1[a_2 x + b_2 y = c_2]$

gives

$a_1 b_2 x + b_1 b_2 y = c_1 b_2$

$\underline{-a_2 b_1 x - b_1 b_2 y = -c_2 b_1}$

Add: $(a_1 b_2 - a_2 b_1)x \qquad = c_1 b_2 - c_2 b_1$

$x = \dfrac{c_1 b_2 - c_2 b_1}{a_1 b_2 - a_2 b_1}$

b. To eliminate x, multiply the first equation by $-a_2$ and the second equation by a_1 and then add.

$-a_2[a_1 x + b_1 y = c_1]$

$a_1[a_2 x + b_2 y = c_2]$

gives

$-a_1 a_2 x - a_2 b_1 y = -a_2 c_1$

$\underline{a_1 a_2 x + a_1 b_2 y = a_1 c_2}$

Add: $(a_1 b_2 - a_2 b_1)y = a_1 c_2 - a_2 c_1$

$y = \dfrac{a_1 c_2 - a_2 c_1}{a_1 b_2 - a_2 b_1}$

66. $2x - 5y = 6$

$-5y = -2x + 6$

$y = \dfrac{-2x + 6}{-5}$

$y = \dfrac{2}{5}x - \dfrac{6}{5}$

67. $3x + 4y = 8$

Solve for y.

$4y = -3x + 8$

$y = -\dfrac{3}{4}x + 2$

x	y
-4	$y = -\frac{3}{4}(-4) + 2 = 3 + 2 = 5$
0	$y = -\frac{3}{4}(0) + 2 = 0 + 2 = 2$
4	$y = -\frac{3}{4}(4) + 2 = -3 + 2 = -1$

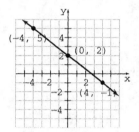

68. $3x + 4y = 8$

For the x-intercept, let $y = 0$.

$3x + 4(0) = 8$

$3x + 0 = 8$

$3x = 8$

$x = \dfrac{8}{3} = 2\dfrac{2}{3}$

For the y-intercept, let $x = 0$.

$3(0) + 4y = 8$

$0 + 4y = 8$

$4y = 8$

$y = 2$

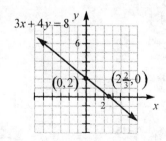

69. $3x + 4y = 8$

Solve for y.

$4y = -3x + 8$

$y = -\dfrac{3}{4}x + 2$

The slope is $-\dfrac{3}{4}$ and the y-intercept is 2.

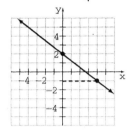

Chapter 9 Review Exercises

1. a. $\quad y = 4x - 2 \qquad 2x + 3y = 8$

$\qquad 6 = 4(2) - 2 \qquad 2(2) + 3(6) = 8$

$\qquad\quad 6 = 6\ \text{True} \qquad\qquad 22 = 8\ \text{False}$

Since (2, 6) does not satisfy both equations, it is not a solution to the system.

b. $\quad y = 4x - 2$

$\qquad 0 = 4(4) - 2$

$\qquad 0 = 14\ \text{False}$

Since (4, 0) does not satisfy the first equation, it is not a solution to the system.

c. $\quad y = 4x - 2 \qquad 2x + 3y = 8$

$\qquad 2 = 4(1) - 2 \qquad 2(1) + 3(2) = 8$

$\qquad 2 = 2\ \text{True} \qquad\qquad 8 = 8\ \text{True}$

Since (1, 2) satisfies both equations, it is a solution to the system.

2. a. $\quad y = -x + 4$

$\qquad \dfrac{3}{2} = -\dfrac{5}{2} + 4$

$\qquad \dfrac{3}{2} = \dfrac{3}{2}\ \text{True}$

$\qquad\qquad 3x + 5y = 15$

$\qquad 3\left(\dfrac{5}{2}\right) + 5\left(\dfrac{3}{2}\right) = 15$

$\qquad\qquad\quad 15 = 15\ \text{True}$

Since $\left(\dfrac{5}{2}, \dfrac{3}{2}\right)$ satisfies both equations, it is a solution to the system.

b. $\quad y = -x + 4$

$\qquad 5 = -(-1) + 4$

$\qquad 5 = 5\ \text{True}$

$\qquad\qquad 3x + 5y = 15$

$\qquad 3(-1) + 5(5) = 15$

$\qquad\qquad 22 = 15\ \text{False}$

Since (−1, 5) does not satisfy both equations, it is not a solution to the system.

c. $\quad y = -x + 4$

$\qquad \dfrac{3}{5} = -\dfrac{1}{2} + 4$

$\qquad \dfrac{3}{5} = \dfrac{7}{2}\ \text{False}$

Since $\left(\dfrac{1}{2}, \dfrac{3}{5}\right)$ does not satisfy the first equation, it is not a solution to the system.

3. consistent, one solution

4. inconsistent, no solutions

5. dependent, infinite number of solutions

6. consistent, one solution

7. Write each equation in slope-intercept form.

$x + 2y = 10 \qquad\qquad 4x = -8y + 16$

$\quad 2y = -x + 10 \qquad\quad 8y = -4x + 16$

$\qquad y = -\dfrac{1}{2}x + 5 \qquad\quad y = -\dfrac{1}{2}x + 2$

Since the slope of each line is $-\dfrac{1}{2}$ but the y-intercepts are different, the two lines are parallel. There is no solution. This is an inconsistent system.

8. Write each equation in slope-intercept form.

$y = -2x + 5$ is already in this form.

$2x + 5y = 10$

$\quad 5y = -2x + 10$

$\qquad y = -\dfrac{2}{5}x + 2$

Since the slopes of the lines are different, the lines intersect to produce one solution. This is a consistent system.

9. Write each equation in slope-intercept form.

$y = \frac{1}{3}x + \frac{2}{3}$ is already in this form.

$6y - 2x = 4$

$\quad 6y = 2x + 4$

$\quad\quad y = \frac{1}{3}x + \frac{2}{3}$

Since both equations are identical, the line is the same for both of them. There are an infinite number of solutions. This is a dependent system.

10. Write each equation in slope-intercept form.

$\quad 6x = 4y - 20 \quad\quad\quad 4x = 6y + 20$

$6x + 20 = 4y \quad\quad\quad 4x - 20 = 6y$

$\frac{3}{2}x + 5 = y \quad\quad\quad \frac{2}{3}x - \frac{10}{5} = y$

Since the slopes of the lines are different, the lines intersect to produce one solution. This is a consistent system.

11. Graph $y = x - 4$ and $y = 2x - 7$.

The lines intersect and the point of intersection is $(3, -1)$. This is a consistent system.

12. Graph $x = -2$ and $y = 5$.

The lines intersect and the point of intersection is $(-2, 5)$. This is a consistent system.

13. Graph the equations $x + 2y = 8$ and $2x - y = -4$.

The lines intersect and the point of intersection is $(0, 4)$. This is a consistent system.

14. Graph $x + 4y = 8$ and $y = 2$.

The lines intersect and the point of intersection is $(0, 2)$. This is a consistent system.

15. Graph $y = 3$ and $y = -2x + 5$.

The lines intersect and the point of intersection is $(1, 3)$. This is a consistent system.

16. Graph the equations $y = x - 3$ and $3x - 3y = 9$.

Both equations produce the same line. This is a dependent system. There are an infinite number of solutions.

17. Graph $3x + y = 0$ and $3x - 3y = 12$.

The lines intersect and the point of intersection is $(1, -3)$. This is a consistent system.

18. Graph $x + 5y = 10$ and $\frac{1}{5}x + y = -1$.

The lines are parallel and do not intersect. The system is inconsistent and there is no solution.

19.
$$y = 4x - 18$$
$$2x - 5y = 0$$
Substitute $4x - 18$ for y in the second equation.
$$2x - 5y = 0$$
$$2x - 5(4x - 18) = 0$$
$$2x - 20x + 90 = 0$$
$$-18x + 90 = 0$$
$$-18x = -90$$
$$x = 5$$
Now, substitute 5 for x in the first equation.
$$y = 4x - 18$$
$$y = 4(5) - 18$$
$$y = 20 - 18$$
$$y = 2$$
The solution is (5, 2).

20.
$$x = 3y - 9$$
$$x + 2y = 1$$
Substitute $3y - 9$ for x in the second equation.
$$x + 2y = 1$$
$$3y - 9 + 2y = 1$$
$$5y - 9 = 1$$
$$5y = 10$$
$$y = 2$$
Now substitute 2 for y in the first equation.
$$x = 3y - 9$$
$$x = 3(2) - 9$$
$$x = 6 - 9$$
$$x = -3$$
The solution is (–3, 2).

21. $2x - y = 7$
$$x + 2y = 6$$
Solve the second equation for x, $x = 6 - 2y$.
Substitute $6 - 2y$ for x in the first equation.
$$2x - y = 7$$
$$2(6 - 2y) - y = 7$$
$$12 - 4y - y = 7$$
$$-5y = -5$$
$$y = 1$$
Substitute 1 for y in the equation $x = 6 - 2y$.
$$x = 6 - 2(1)$$
$$x = 6 - 2$$
$$x = 4$$
The solution is (4, 1).

22.
$$x = -3y$$
$$x + 4y = 5$$
Substitute $-3y$ for x in the second equation.
$$x + 4y = 5$$
$$-3y + 4y = 5$$
$$y = 5$$
Substitute 5 for y in the first equation.
$$x = -3y$$
$$x = -3(5)$$
$$x = -15$$
The solution is (–15, 5).

23. $4x - 2y = 7$
$$y = 2x + 3$$
Substitute $2x + 3$ for y in the first equation.
$$4x - 2y = 7$$
$$4x - 2(2x + 3) = 7$$
$$4x - 4x - 6 = 7$$
$$-6 = 7 \text{ False}$$
There is no solution.

24. $2x - 4y = 7$
$$-4x + 8y = -14$$
Solve the first equation for x.
$$2x - 4y = 7$$
$$2x = 4y + 7$$
$$x = 2y + \frac{7}{2}$$
Substitute $2y + \frac{7}{2}$ for x in the second equation.
$$-4x + 8y = -14$$
$$-4\left(2y + \frac{7}{2}\right) + 8y = -14$$
$$-8y - 14 + 8y = -14$$
$$-14 = -14 \text{ True}$$
There are an infinite number of solutions.

25. $2x - 3y = 8$
$$6x - 5y = 20$$
Solve the first equation for x.
$$2x - 3y = 8$$
$$2x = 3y + 8$$
$$x = \frac{3}{2}y + 4$$
Substitute $\frac{3}{2}y + 4$ for x in the second equation.

$$6x - 5y = 20$$

$$6\left(\frac{3}{2}y + 4\right) - 5y = 20$$

$$9y + 24 - 5y = 20$$

$$4y + 24 = 20$$

$$4y = -4$$

$$y = -1$$

Substitute -1 for y in the equation $x = \frac{3}{2}y + 4$.

$$x = \frac{3}{2}(-1) + 4 = -\frac{3}{2} + \frac{8}{2} = \frac{5}{2}$$

The solution is $\left(\frac{5}{2}, -1\right)$.

26. $3x - y = -5$

$x + 2y = 8$

Solve the second equation for x, $x = 8 - 2y$.

Substitute $8 - 2y$ for x in the first equation.

$$3x - y = -5$$

$$3(8 - 2y) - y = -5$$

$$24 - 6y - y = -5$$

$$-7y = -29$$

$$y = \frac{29}{7}$$

Substitute $\frac{29}{7}$ for y in the equation $x = 8 - 2y$.

$$x = 8 - 2\left(\frac{29}{7}\right)$$

$$x = \frac{56}{7} - \frac{58}{7}$$

$$x = -\frac{2}{7}$$

The solution is $\left(-\frac{2}{7}, \frac{29}{7}\right)$.

27.
$$\begin{aligned} x - y &= -4 \\ -x + 6y &= -6 \\ \hline 5y &= -10 \\ y &= -2 \end{aligned}$$

Substitute -2 for y in the first equation.

$$x - y = -4$$

$$x - (-2) = -4$$

$$x + 2 = -4$$

$$x = -6$$

The solution is $(-6, -2)$.

28.
$$\begin{aligned} x + 2y &= -3 \\ 5x - 2y &= 9 \\ \hline 6x &= 6 \\ x &= 1 \end{aligned}$$

Substitute 1 for x in the first equation.

$$x + 2y = -3$$

$$1 + 2y = -3$$

$$2y = -4$$

$$y = -2$$

The solution is $(1, -2)$.

29. $x + y = 12$

$2x + y = 5$

To eliminate y, multiply the first equation by -1 and then add.

$-1[x + y = 12]$

gives

$$\begin{aligned} -x - y &= -12 \\ 2x + y &= 5 \\ \hline x &= -7 \end{aligned}$$

Substitute -7 for x in the first equation.

$$x + y = 12$$

$$-7 + y = 12$$

$$y = 19$$

The solution is $(-7, 19)$.

30. $4x - 3y = 8$

$2x + 5y = 8$

To eliminate x, multiply the second equation by -2 and then add.

$-2[2x + 5y = 8]$

gives

$$\begin{aligned} 4x - 3y &= 8 \\ -4x - 10y &= -16 \\ \hline -13y &= -8 \\ y &= \frac{8}{13} \end{aligned}$$

Substitute $\frac{8}{13}$ for y in the first equation.

$$4x - 3y = 8$$

$$4x - 3\left(\frac{8}{13}\right) = 8$$

$$4x - \frac{24}{13} = 8$$

$$4x = \frac{104}{13} + \frac{24}{13} = \frac{128}{13}$$

$$x = \frac{1}{4} \cdot \frac{128}{13} = \frac{32}{13}$$

The solution is $\left(\frac{32}{13}, \frac{8}{13}\right)$.

31. $-2x + 3y = 15$

$7x + 3y = 6$

To eliminate y, multiply the second equation by -1 and then add.

$-1[7x + 3y = 6]$

gives

$-2x + 3y = 15$

$\underline{-7x - 3y = -6}$

$-9x \quad\quad = \quad 9$

$x = -1$

Substitute -1 for x in the second equation.

$7x + 3y = 6$

$7(-1) + 3y = 6$

$-7 + 3y = 6$

$3y = 13$

$y = \dfrac{13}{3}$

The solution is $\left(-1, \dfrac{13}{3}\right)$.

32. $\quad 2x + y = 3$

$-4x - 2y = 5$

Multiply the first equation by 2, and then add.

$2[2x + y = 3]$

gives

$4x + 2y = 6$

$\underline{-4x - 2y = 3}$

$0 = 9 \quad$ False

There is no solution.

33. $3x = -4y + 15$

$8y = -6x + 30$

Align the x and y terms.

$3x + 4y = 15$

$6x + 8y = 30$

To eliminate x, multiply the first equation by -2, and then add.

$-2[3x + 4y = 15]$

gives

$-6x - 8y = -30$

$\underline{6x + 8y = \quad 30}$

$0 = 0 \quad$ True

There are an infinite number of solutions.

34. $2x - 5y = 12$

$3x - 4y = -6$

To eliminate x, multiply the first equation by -3 and the second equation by 2 and then add.

$-3[2x - 5y = 12]$

$2[3x - 4y = -6]$

gives

$-6x + 15y = -36$

$\underline{6x - 8y = -12}$

$7y = -48$

$y = -\dfrac{48}{7}$

Now, substitute $-\dfrac{48}{7}$ for y in the first equation.

$2x - 5y = 12$

$2x - 5\left(-\dfrac{48}{7}\right) = 12$

$2x + \dfrac{240}{7} = 12$

$2x = \dfrac{84}{7} - \dfrac{240}{7}$

$2x = -\dfrac{156}{7}$

$x = -\dfrac{78}{7}$

The solution is $\left(-\dfrac{78}{7}, -\dfrac{48}{7}\right)$.

35. $x - 2y - 4z = 13 \quad (1)$

$3y + 2z = -2 \quad (2)$

$5z = -20 \quad (3)$

Solve equation (3) for z.

$5z = -20$

$z = -4$

Substitute -4 for z in equation (2).

$3y + 2(-4) = -2$

$3y - 8 = -2$

$3y = 6$

$y = 2$

Substitute -4 for z and 2 for y in equation (1).

$x - 2(2) - 4(-4) = 13$

$x - 4 + 16 = 13$

$x + 12 = 13$

$x = 1$

The solution is $(1, 2, -4)$.

36. $2a+b-2c=5$
$$3b+4c=1$$
$$3c=-6$$
Solve the third equation for c.
$$3c=-6$$
$$c=-2$$
Substitute -2 for c in the second equation.
$$3b+4(-2)=1$$
$$3b-8=1$$
$$3b=9$$
$$b=3$$
Substitute 3 for b and -2 for c in the first equation.
$$2a+(3)-2(-2)=5$$
$$2a+3+4=5$$
$$2a+7=5$$
$$2a=-2$$
$$a=-1$$
The solution is $(-1, 3, -2)$.

37. $x+2y+3z=3$ (1)
$$-2x-3y-z=5 \quad (2)$$
$$3x+3y+7z=2 \quad (3)$$
To eliminate x between equations (1) and (2), multiply equation (1) by 2 and then add.
$$2[x+2y+3z=3]$$
$$-2x-3y-z=5$$
gives
$$2x+4y+6z=6$$
$$\underline{-2x-3y\ -z=5}$$
Add: $y+5z=11$ (4)
To eliminate x between equations (2) and (3), multiply equation (2) by 3 and equation (3) by 2, then add.
$$3[-2x-3y-z=5]$$
$$2[3x+3y+7z=2]$$
gives
$$-6x-9y-3z=15$$
$$\underline{6x+6y+14z=\ 4}$$
Add: $-3y+11z=19$ (5)
Equations (4) and (5) are two equations in two unknowns.
$$y+5z=11 \quad (4)$$
$$-3y+11z=19 \quad (5)$$
To eliminate y, multiply equation (4) by 3 and then add.

$$3[y+5z=11]$$
$$-3y+11z=19$$
gives
$$3y+15z=33$$
$$\underline{-3y+11z=19}$$
Add: $26z=52$
$$z=2$$
Substitute 2 for z in equation (4).
$$y+5(2)=11$$
$$y+10=11$$
$$y=1$$
Finally, substitute 1 for y and 2 for z in equation (1).
$$x+2(1)+3(2)=3$$
$$x+2+6=3$$
$$x+8=3$$
$$x=-5$$
The solution is $(-5, 1, 2)$.

38. $-x-4y+2z=1$ (1)
$$2x+2y+z=0 \quad (2)$$
$$-3x-2y-5z=5 \quad (3)$$
To eliminate y between equations (1) and (2), multiply equation (2) by 2 and add.
$$-x-4y+2z=1$$
$$2[2x+2y+z=0]$$
gives
$$-x-4y+2z=1$$
$$\underline{4x+4y+2z=0}$$
Add: $3x\quad+4z=1$ (4)
To eliminate y between equations (2) and (3), simply add.
$$2x+2y+z=0$$
$$\underline{-3x-2y-5z=5}$$
Add: $-x\quad-4z=5$ (5)
Equations (4) and (5) are two equations in two unknowns.
$$3x+4z=1 \quad (4)$$
$$-x-4z=5 \quad (5)$$
To eliminate z, simply add equations (4) and (5).
$$3x+4z=1$$
$$\underline{-x-4z=5}$$
Add: $2x\quad=6$
$$x=3$$

Substitute 3 for x in equation (4).

$$3(3) + 4z = 1$$
$$9 + 4z = 1$$
$$4z = -8$$
$$z = -2$$

Finally, substitute 3 for x and -2 for z in equation (1).

$$-(3) - 4y + 2(-2) = 1$$
$$-3 - 4y - 4 = 1$$
$$-7 - 4y = 1$$
$$-4y = 8$$
$$y = -2$$

The solution is $(3, -2, -2)$.

39.
$$3y - 2z = -4 \ (1)$$
$$3x - 5z = -7 \ (2)$$
$$2x + y = 6 \ (3)$$

To eliminate y between equations (1) and (3), multiply equation (3) by -3 and then add.

$$3y - 2z = -4$$
$$-3[2x + y = 6]$$

gives

$$3y - 2z = -4$$
$$\underline{-6x - 3y \quad = -18}$$

Add: $\quad -6x \quad - 2z = -22$

or

$$-3x - z = -11 \qquad (4)$$

Equations (4) and (2) are two equations in two unknowns. To eliminate x, simply add.

$$3x - 5z = -7$$
$$\underline{-3x - \ z = -11}$$

Add: $\quad -6z = -18$

$$z = 3$$

Substitute 3 for z in equation (2).

$$3x - 5z = -7$$
$$3x - 5(3) = -7$$
$$3x - 15 = -7$$
$$3x = 8$$
$$x = \frac{8}{3}$$

Substitute $\dfrac{8}{3}$ for x in equation (3).

$$2x + y = 6$$
$$2\left(\frac{8}{3}\right) + y = 6$$
$$\frac{16}{3} + y = 6$$
$$y = 6 - \frac{16}{3}$$
$$y = \frac{18}{3} - \frac{16}{3}$$
$$y = \frac{2}{3}$$

The solution is $\left(\dfrac{8}{3}, \dfrac{2}{3}, 3\right)$.

40.
$$a + 2b - 5c = 19 \quad (1)$$
$$2a - 3b + 3c = -15 \ (2)$$
$$5a - 4b - 2c = -2 \quad (3)$$

To eliminate a between equations (1) and (2), multiply equation (1) by -2 and then add.

$$-2[a + 2b - 5c = 19]$$
$$2a - 3b + 3c = -15$$

gives

$$-2a - 4b + 10c = -38$$
$$\underline{2a - 3b + \ 3c = -15}$$

Add: $\quad -7b + 13c = -53 \quad (4)$

To eliminate a between equations (1) and (3), multiply equation (1) by -5 and then add.

$$-5[a + 2b - 5c = 19]$$
$$5a - 4b - 2c = -2$$

gives

$$-5a - 10b + 25c = -95$$
$$\underline{5a - \ 4b - \ 2c = -2}$$

Add: $\quad -14b + 23c = -97 \quad (5)$

Equations (4) and (5) are two equations in two unknowns.

$$-7b + 13c = -53$$
$$-14b + 23c = -97$$

To eliminate b, multiply equation (4) by -2 and then add.

$$-2[-7b + 13c = -53]$$
$$-14b + 23c = -97$$

gives

$$14b - 26c = \ 106$$
$$\underline{-14b + 23c = -97}$$

Add: $\quad -3c = 9$

$$c = -3$$

Substitute -3 for c in equation (4).

$$-7b + 13c = -53$$
$$-7b + 13(-3) = -53$$
$$-7b - 39 = -53$$
$$-7b = -14$$
$$b = 2$$

Finally, substitute 2 for b and -3 for c in equation (1).

$$a + 2b - 5c = 19$$
$$a + 2(2) - 5(-3) = 19$$
$$a + 4 + 15 = 19$$
$$a + 19 = 19$$
$$a = 0$$

The solution is $(0, 2, -3)$.

41.
$$x - y + 3z = 1 \quad (1)$$
$$-x + 2y - 2z = 1 \quad (2)$$
$$x - 3y + z = 2 \quad (3)$$

To eliminate x between equations (1) and (2), simply add.

$$x - y + 3z = 1$$
$$\underline{-x + 2y - 2z = 1}$$

Add: $\quad y + z = 2 \quad\quad (4)$

To eliminate x between equations (2) and (3), simply add.

$$-x + 2y - 2z = 1$$
$$\underline{x - 3y + z = 2}$$

Add: $-y - z = 3 \quad (5)$

Equations (4) and (5) are two equations in two unknowns. To eliminate y and z simply add.

$$y + z = 2$$
$$\underline{-y - z = 3}$$

Add: $\quad 0 = 5 \quad$ False

Since this is a false statement, there is no solution to the system. This is an inconsistent system.

42.
$$-2x + 2y - 3z = 6 \quad (1)$$
$$4x - y + 2z = -2 \quad (2)$$
$$2x + y - z = 4 \quad (3)$$

To eliminate x between equations (1) and (2), multiply equation (1) by 2 and then add.

$$2[-2x + 2y - 3z = 6]$$
$$4x - y + 2z = -2$$

gives

$$-4x + 4y - 6z = 12$$
$$\underline{4x - y + 2z = -2}$$

Add: $\quad 3y - 4z = 10 \quad (4)$

To eliminate x between equations (1) and (3), simply add.

$$-2x + 2y - 3z = 6$$
$$\underline{2x + y - z = 4}$$

Add: $\quad 3y - 4z = 10 \quad (5)$

Equations (4) and (5) are two equations in two unknowns.

$$3y - 4z = 10$$
$$3y - 4z = 10$$

Since they are identical, there are an infinite number of solutions. This is a dependent system.

43. Let $x =$ Jennifer's age and $y =$ Luan's age.

$$y = x + 10$$
$$x + y = 66$$

Substitute $x + 10$ for y in the second equation.

$$x + (x + 10) = 66$$
$$2x + 10 = 66$$
$$2x = 56$$
$$x = 28$$

Now substitute 28 for x in the first equation.

$$y = (28) + 10$$
$$y = 38$$

Jennifer is 28 years old and Luan is 38 years old.

44. Let x = the speed of the plane in still air and y = the speed of the wind.

$$x + y = 560$$
$$x - y = 480$$

To eliminate y, simply add.

$$x + y = 560$$
$$\underline{x - y = 480}$$

Add: $2x \quad = 1040$

$x \quad = 520$

Substitute 520 for x in the first equation.

$$x + y = 560$$
$$(520) + y = 560$$
$$y = 40$$

The speed of the plane in still air is 520 mph and the speed of the wind is 40 mph.

45. Let x = the amount of 20% acid solution and y = the amount of 50% acid solution.

$$x + y = 6$$
$$0.2x + 0.5y = 0.4(6)$$

To clear decimals, multiply the second equation by 10.

$$x + y = 6$$
$$2x + 5y = 24$$

Solve the first equation for y.

$$x + y = 6$$
$$y = -x + 6$$

Substitute $-x + 6$ for y in the second equation.

$$2x + 5y = 24$$
$$2x + 5(-x + 6) = 24$$
$$2x - 5x + 30 = -24$$
$$-3x + 30 = 24$$
$$-3x = -6$$
$$x = \frac{-6}{-3} = 2$$

Finally, substitute 2 for x in the equation $y = -x + 6$.

$$y = -x + 6$$
$$y = -2 + 6$$
$$y = 4$$

Sally should combine 2 liters of solution A with 4 liters of solution B.

46. Let x = the number of adult tickets and y = the number of children's tickets.

$$x + y = 650$$
$$15x + 11y = 8790$$

To solve, multiply the first equation by -11 and then add.

$$-11[x + y = 650]$$
$$15x + 11y = 8790$$

gives

$$-11x - 11y = -7150$$
$$\underline{15x + 11y = 8790}$$

Add: $4x \quad = 1640$

$$x = \frac{1640}{4} = 410$$

Substitute 410 for x in the first equation.

$$x + y = 650$$
$$410 + y = 650$$
$$y = 240$$

Thus, 410 adult tickets and 240 children's tickets were sold.

47. Let x = age at first time and y = age at second time.

$$y = 2x - 5$$
$$x + y = 118$$

Substitute $2x - 5$ for y in the second equation.

$$x + y = 118$$
$$x + 2x - 5 = 118$$
$$3x - 5 = 118$$
$$3x = 123$$
$$x = 41$$

Substitute 41 for x in the first equation.

$$y = 2x - 5$$
$$y = 2(41) - 5$$
$$y = 82 - 5$$
$$y = 77$$

His ages were 41 years and 77 years.

48. Let x = the amount invested at 7%, y = the amount invested at 5%, and z = the amount invested at 3%.

$$x + y + z = 40{,}000 \quad (1)$$
$$y = x - 5000 \quad (2)$$
$$0.07x + 0.05y + 0.03z = 2300 \quad (3)$$

Substitute $x - 5000$ for y in equations (1) and (3).
Equation (1) becomes
$$x + y + z = 40{,}000$$
$$x + x - 5000 + z = 40{,}000$$
$$2x + z = 45{,}000 \, (4)$$

Equation (3) becomes
$$0.07x + 0.05y + 0.03z = 2300$$
$$0.07x + 0.05(x - 5000) + 0.03z = 2300$$
$$0.07x + 0.05x - 250 + 0.03z = 2300$$
$$0.12x + 0.03z = 2550 \, (5)$$

Equation (4) and (5) are a system of two equations in two unknowns. Solve equation (4) for z.
$$2x + z = 45{,}000$$
$$z = -2x + 45{,}000$$

Substitute $-2x + 45{,}000$ for z in equation (5).
$$0.12x + 0.03z = 2550$$
$$0.12x + 0.03(-2x + 45{,}000) = 2550$$
$$0.12x - 0.06x + 1350 = 2550$$
$$0.06x = 1200$$
$$x = 20{,}000$$

Now substitute 20,000 for x in equation (2).
$$y = x - 5000$$
$$y = 20{,}000 - 5000 = 15{,}000$$

Finally, substitute 20,000 for x and 15,000 for y in equation (1).
$$x + y + z = 40{,}000$$
$$20{,}000 + 15{,}000 + z = 40{,}000$$
$$35{,}000 + z = 40{,}000$$
$$z = 5000$$

Thus, $20,000 was invested at 7%, $15,000 at 5%, and $5000 at 3%.

49.
$$x + 5y = 1$$
$$-2x - 8y = -6$$

$$\begin{bmatrix} 1 & 5 & | & 1 \\ -2 & -8 & | & -6 \end{bmatrix}$$

$$\begin{bmatrix} 1 & 5 & | & 1 \\ 0 & 2 & | & -4 \end{bmatrix} 2R_1 + R_2$$

$$\begin{bmatrix} 1 & 5 & | & 1 \\ 0 & 1 & | & -2 \end{bmatrix} \tfrac{1}{2}R_2$$

The system is
$$x + 5y = 1$$
$$y = -2$$

Substitute –2 for y in the first equation.
$$x + 5(-2) = 1$$
$$x - 10 = 1$$
$$x = 11$$
The solution is $(11, -2)$.

50.
$$2x - 5y = 1$$
$$2x + 4y = 10$$

$$\begin{bmatrix} 2 & -5 & | & 1 \\ 2 & 4 & | & 10 \end{bmatrix}$$

$$\begin{bmatrix} 1 & -\tfrac{5}{2} & | & \tfrac{1}{2} \\ 2 & 4 & | & 10 \end{bmatrix} \tfrac{1}{2}R_1$$

$$\begin{bmatrix} 1 & -\tfrac{5}{2} & | & \tfrac{1}{2} \\ 0 & 9 & | & 9 \end{bmatrix} -2R_1 + R_2$$

$$\begin{bmatrix} 1 & -\tfrac{5}{2} & | & \tfrac{1}{2} \\ 0 & 1 & | & 1 \end{bmatrix} \tfrac{1}{9}R_2$$

The system is
$$x - \frac{5}{2}y = \frac{1}{2}$$
$$y = 1$$

Substitute 1 for y in the first equation.
$$x - \frac{5}{2}(1) = \frac{1}{2}$$
$$x - \frac{5}{2} = \frac{1}{2}$$
$$x = \frac{6}{2}$$
$$x = 3$$
The solution is $(3, 1)$.

51. $3y = 6x - 12$

$4x = 2y + 8$

Write the system in standard form.

$-6x + 3y = -12$

$4x - 2y = 8$

$$\begin{bmatrix} -6 & 3 & | & -12 \\ 4 & -2 & | & 8 \end{bmatrix}$$

$$\begin{bmatrix} 1 & -\frac{1}{2} & | & 2 \\ 4 & -2 & | & 8 \end{bmatrix} -\frac{1}{6}R_1$$

$$\begin{bmatrix} 1 & -\frac{1}{2} & | & 2 \\ 0 & 0 & | & 0 \end{bmatrix} -4R_1 + R_2$$

Since the last row is all zeros, the system is dependent. There are an infinite number of solutions.

52. $2x - y - z = 5$

$x + 2y + 3z = -2$

$3x - 2y + z = 2$

$$\begin{bmatrix} 2 & -1 & -1 & | & 5 \\ 1 & 2 & 3 & | & -2 \\ 3 & -2 & 1 & | & 2 \end{bmatrix}$$

$$\begin{bmatrix} 1 & -\frac{1}{2} & -\frac{1}{2} & | & \frac{5}{2} \\ 1 & 2 & 3 & | & -2 \\ 3 & -2 & 1 & | & 2 \end{bmatrix} \frac{1}{2}R_1$$

$$\begin{bmatrix} 1 & -\frac{1}{2} & -\frac{1}{2} & | & \frac{5}{2} \\ 0 & \frac{5}{2} & \frac{7}{2} & | & -\frac{9}{2} \\ 3 & -2 & 1 & | & 2 \end{bmatrix} -1R_1 + R_2$$

$$\begin{bmatrix} 1 & -\frac{1}{2} & -\frac{1}{2} & | & \frac{5}{2} \\ 0 & \frac{5}{2} & \frac{7}{2} & | & -\frac{9}{2} \\ 0 & -\frac{1}{2} & \frac{5}{2} & | & -\frac{11}{2} \end{bmatrix} -3R_1 + R_3$$

$$\begin{bmatrix} 1 & -\frac{1}{2} & -\frac{1}{2} & | & \frac{5}{2} \\ 0 & 1 & \frac{7}{5} & | & -\frac{9}{5} \\ 0 & -\frac{1}{2} & \frac{5}{2} & | & -\frac{11}{2} \end{bmatrix} \frac{2}{5}R_2$$

$$\begin{bmatrix} 1 & -\frac{1}{2} & -\frac{1}{2} & | & \frac{5}{2} \\ 0 & 1 & \frac{7}{5} & | & -\frac{9}{5} \\ 0 & 0 & \frac{16}{5} & | & -\frac{32}{5} \end{bmatrix} \frac{1}{2}R_2 + R_3$$

$$\begin{bmatrix} 1 & -\frac{1}{2} & -\frac{1}{2} & | & \frac{5}{2} \\ 0 & 1 & \frac{7}{5} & | & -\frac{9}{5} \\ 0 & 0 & 1 & | & -2 \end{bmatrix} \frac{5}{16}R_3$$

The system is

$$x - \frac{1}{2}y - \frac{1}{2}z = \frac{5}{2}$$

$$y + \frac{7}{5}z = -\frac{9}{5}$$

$$z = -2$$

Substitute –2 for z in the second equation.

$$y + \frac{7}{5}z = -\frac{9}{5}$$

$$y + \frac{7}{5}(-2) = -\frac{9}{5}$$

$$y - \frac{14}{5} = -\frac{9}{5}$$

$$y = \frac{5}{5} = 1$$

Substitute 1 for y and –2 for z in the first equation.

$$x - \frac{1}{2}y - \frac{1}{2}z = \frac{5}{2}$$

$$x - \frac{1}{2}(1) - \frac{1}{2}(-2) = \frac{5}{2}$$

$$x - \frac{1}{2} + 1 = \frac{5}{2}$$

$$x + \frac{1}{2} = \frac{5}{2}$$

$$x = \frac{4}{2} = 2$$

The solution is (2, 1, –2).

53. $3a - b + c = 2$

$2a - 3b + 4c = 4$

$a + 2b - 3c = -6$

$$\begin{bmatrix} 3 & -1 & 1 & | & 2 \\ 2 & -3 & 4 & | & 4 \\ 1 & 2 & -3 & | & -6 \end{bmatrix}$$

$$\begin{bmatrix} 1 & -\frac{1}{3} & \frac{1}{3} & | & \frac{2}{3} \\ 2 & -3 & 4 & | & 4 \\ 1 & 2 & -3 & | & -6 \end{bmatrix} \frac{1}{3}R_1$$

$$\begin{bmatrix} 1 & -\frac{1}{3} & \frac{1}{3} & | & \frac{2}{3} \\ 0 & -\frac{7}{3} & \frac{10}{3} & | & \frac{8}{3} \\ 1 & 2 & -3 & | & -6 \end{bmatrix} -2R_1 + R_2$$

$$\begin{bmatrix} 1 & -\frac{1}{3} & \frac{1}{3} & | & \frac{2}{3} \\ 0 & -\frac{7}{3} & \frac{10}{3} & | & \frac{8}{3} \\ 0 & \frac{7}{3} & -\frac{10}{3} & | & -\frac{20}{3} \end{bmatrix} -1R_1 + R_3$$

$$\begin{bmatrix} 1 & -\frac{1}{3} & \frac{1}{3} & \Big| & \frac{2}{3} \\ 0 & 1 & -\frac{10}{7} & \Big| & -\frac{8}{7} \\ 0 & \frac{7}{3} & -\frac{10}{3} & \Big| & -\frac{20}{3} \end{bmatrix} -\frac{3}{7}R_2$$

$$\begin{bmatrix} 1 & -\frac{1}{3} & \frac{1}{3} & \Big| & \frac{2}{3} \\ 0 & 1 & -\frac{10}{7} & \Big| & -\frac{8}{7} \\ 0 & 0 & 0 & \Big| & -4 \end{bmatrix} -\frac{7}{3}R_1 + R_3$$

Since the last row has all zeros on the left side and a nonzero number on the right side, the system is inconsistent and has no solution.

54. $x + y + z = 3$

$\quad\ 3x + 4y = -1$

$\quad\quad\ \ y - 3z = -10$

$$\begin{bmatrix} 1 & 1 & 1 & \Big| & 3 \\ 3 & 4 & 0 & \Big| & -1 \\ 0 & 1 & -3 & \Big| & -10 \end{bmatrix}$$

$$\begin{bmatrix} 1 & 1 & 1 & \Big| & 3 \\ 0 & 1 & -3 & \Big| & -10 \\ 0 & 1 & -3 & \Big| & -10 \end{bmatrix} -3R_1 + R_2$$

$$\begin{bmatrix} 1 & 1 & 1 & \Big| & 3 \\ 0 & 1 & -3 & \Big| & -10 \\ 0 & 0 & 0 & \Big| & 0 \end{bmatrix} -1R_2 + R_3$$

Since the last row is all zeros and there are no contradictions, the system is dependent. There are an infinite number of solutions.

55. $\quad 7x - 8y = -10$

$\quad\ -5x + 4y = 2$

To solve, first calculate D, D_x, and D_y.

$$D = \begin{vmatrix} 7 & -8 \\ -5 & 4 \end{vmatrix} = (7)(4) - (-5)(-8)$$

$$= 28 - 40 = -12$$

$$D_x = \begin{vmatrix} -10 & -8 \\ 2 & 4 \end{vmatrix} = (-10)(4) - (2)(-8)$$

$$= -40 + 16 = -24$$

$$D_y = \begin{vmatrix} 7 & -10 \\ -5 & 2 \end{vmatrix} = (7)(2) - (-5)(-10)$$

$$= 14 - 50 = -36$$

$$x = \frac{D_x}{D} = \frac{-24}{-12} = 2 \text{ and } y = \frac{D_y}{D} = \frac{-36}{-12} = 3$$

The solution is $(2, 3)$.

56. $\quad x + 4y = 5$

$\quad\ 5x + 3y = -9$

To solve, first calculate D, D_x, and D_y.

$$D = \begin{vmatrix} 1 & 4 \\ 5 & 3 \end{vmatrix}$$

$$= (1)(3) - (5)(4)$$

$$= 3 - 20$$

$$= -17$$

$$D_x = \begin{vmatrix} 5 & 4 \\ -9 & 3 \end{vmatrix}$$

$$= (5)(3) - (-9)(4)$$

$$= 15 + 36$$

$$= 51$$

$$D_y = \begin{vmatrix} 1 & 5 \\ 5 & -9 \end{vmatrix}$$

$$= (1)(-9) - (5)(5)$$

$$= -9 - 25$$

$$= -34$$

$$x = \frac{D_x}{D} = \frac{51}{-17} = -3 \text{ and } y = \frac{D_y}{D} = \frac{-34}{-17} = 2$$

The solution is $(-3, 2)$.

57. $\quad 9m + 4n = -1$

$\quad\ 7m - 2n = -11$

To solve, first calculate D, D_m, and D_n.

$$D = \begin{vmatrix} 9 & 4 \\ 7 & -2 \end{vmatrix}$$

$$= (9)(-2) - (7)(4)$$

$$= -18 - 28$$

$$= -46$$

$$D_m = \begin{vmatrix} -1 & 4 \\ -11 & -2 \end{vmatrix}$$

$$= (-1)(-2) - (-11)(4)$$

$$= 2 + 44$$

$$= 46$$

$$D_n = \begin{vmatrix} 9 & -1 \\ 7 & -11 \end{vmatrix}$$

$$= (9)(-11) - (7)(-1)$$

$$= -99 + 7$$

$$= -92$$

$$m = \frac{D_m}{D} = \frac{46}{-46} = -1 \text{ and } n = \frac{D_n}{D} = \frac{-92}{-46} = 2.$$

The solution is $(-1, 2)$.

58.
$$p+q+r=5$$
$$2p+q-r=-5$$
$$3p+2q-3r=-12$$

To solve, calculate D, D_p, D_q, and D_r.

$$D = \begin{vmatrix} 1 & 1 & 1 \\ 2 & 1 & -1 \\ 3 & 2 & -3 \end{vmatrix}$$

$$= 1\begin{vmatrix} 1 & -1 \\ 2 & -3 \end{vmatrix} - 1\begin{vmatrix} 2 & -1 \\ 3 & -3 \end{vmatrix} + 1\begin{vmatrix} 2 & 1 \\ 3 & 2 \end{vmatrix}$$

$$= 1(-3+2) - 1(-6+3) + 1(4-3)$$

$$= 1(-1) - 1(-3) + 1(1)$$

$$= -1+3+1 = 3$$

$$D_p = \begin{vmatrix} 5 & 1 & 1 \\ -5 & 1 & -1 \\ -12 & 2 & -3 \end{vmatrix}$$

$$= 5\begin{vmatrix} 1 & -1 \\ 2 & -3 \end{vmatrix} - 1\begin{vmatrix} -5 & -1 \\ -12 & -3 \end{vmatrix} + 1\begin{vmatrix} -5 & 1 \\ -12 & 2 \end{vmatrix}$$

$$= 5(-3+2) - 1(15-12) + 1(-10+12)$$

$$= 5(-1) - 1(3) + 1(2)$$

$$= -5-3+2 = -6$$

$$D_q = \begin{vmatrix} 1 & 5 & 1 \\ 2 & -5 & -1 \\ 3 & -12 & -3 \end{vmatrix}$$

$$= 1\begin{vmatrix} -5 & -1 \\ -12 & -3 \end{vmatrix} - 5\begin{vmatrix} 2 & -1 \\ 3 & -3 \end{vmatrix} + 1\begin{vmatrix} 2 & -5 \\ 3 & -12 \end{vmatrix}$$

$$= 1(15-12) - 5(-6+3) + 1(-24+15)$$

$$= 1(3) - 5(-3) + 1(-9)$$

$$= 3+15-9 = 9$$

$$D_r = \begin{vmatrix} 1 & 1 & 5 \\ 2 & 1 & -5 \\ 3 & 2 & -12 \end{vmatrix}$$

$$= 1\begin{vmatrix} 1 & -5 \\ 2 & -12 \end{vmatrix} - 1\begin{vmatrix} 2 & -5 \\ 3 & -12 \end{vmatrix} + 5\begin{vmatrix} 2 & 1 \\ 3 & 2 \end{vmatrix}$$

$$= 1(-12+10) - 1(-24+15) + 5(4-3)$$

$$= 1(-2) - 1(-9) + 5(1)$$

$$= -2+9+5 = 12$$

$$p = \frac{D_p}{D} = \frac{-6}{3} = -2, \quad q = \frac{D_q}{D} = \frac{9}{3} = 3, \text{ and}$$

$$r = \frac{D_r}{D} = \frac{12}{3} = 4$$

The solution is (–2, 3, 4).

59.
$$-2a+3b-4c=-7$$
$$2a+b+c=5$$
$$-2a-3b+4c=3$$

To solve, calculate D, D_a, D_b, and D_c.

$$D = \begin{vmatrix} -2 & 3 & -4 \\ 2 & 1 & 1 \\ -2 & -3 & 4 \end{vmatrix}$$

$$= -2\begin{vmatrix} 1 & 1 \\ -3 & 4 \end{vmatrix} - 3\begin{vmatrix} 2 & 1 \\ -2 & 4 \end{vmatrix} + (-4)\begin{vmatrix} 2 & 1 \\ -2 & -3 \end{vmatrix}$$

$$= -2(4+3) - 3(8+2) - 4(-6+2)$$

$$= -2(7) - 3(10) - 4(-4)$$

$$= -14-30+16$$

$$= -28$$

$$D_a = \begin{vmatrix} -7 & 3 & -4 \\ 5 & 1 & 1 \\ 3 & -3 & 4 \end{vmatrix}$$

$$= -7\begin{vmatrix} 1 & 1 \\ -3 & 4 \end{vmatrix} - 3\begin{vmatrix} 5 & 1 \\ 3 & 4 \end{vmatrix} + (-4)\begin{vmatrix} 5 & 1 \\ 3 & -3 \end{vmatrix}$$

$$= -7(4+3) - 3(20-3) - 4(-15-3)$$

$$= -7(7) - 3(17) - 4(-18)$$

$$= -49-51+72$$

$$= -28$$

$$D_b = \begin{vmatrix} -2 & -7 & -4 \\ 2 & 5 & 1 \\ -2 & 3 & 4 \end{vmatrix}$$

$$= -2\begin{vmatrix} 5 & 1 \\ 3 & 4 \end{vmatrix} - (-7)\begin{vmatrix} 2 & 1 \\ -2 & 4 \end{vmatrix} + (-4)\begin{vmatrix} 2 & 5 \\ -2 & 3 \end{vmatrix}$$

$$= -2(20-3) + 7(8+2) - 4(6+10)$$

$$= -2(17) + 7(10) - 4(16)$$

$$= -34+70-64$$

$$= -28$$

$$D_c = \begin{vmatrix} -2 & 3 & -7 \\ 2 & 1 & 5 \\ -2 & -3 & 3 \end{vmatrix}$$

$$= -2\begin{vmatrix} 1 & 5 \\ -3 & 3 \end{vmatrix} - 3\begin{vmatrix} 2 & 5 \\ -2 & 3 \end{vmatrix} + (-7)\begin{vmatrix} 2 & 1 \\ -2 & -3 \end{vmatrix}$$

$$= -2(3+15) - 3(6+10) - 7(-6+2)$$

$$= -2(18) - 3(16) - 7(-4)$$

$$= -36-48+28$$

$$= -56$$

$a = \dfrac{D_a}{D} = \dfrac{-28}{-28} = 1$, $b = \dfrac{D_b}{D} = \dfrac{-28}{-28} = 1$, and

$c = \dfrac{D_c}{D} = \dfrac{-56}{-28} = 2$

The solution is (1, 1, 2).

60.
$$y + 3z = 4$$
$$-x - y + 2z = 0$$
$$x + 2y + z = 1$$

To solve, first calculate D, D_x, D_y, and D_z.

$D = \begin{vmatrix} 0 & 1 & 3 \\ -1 & -1 & 2 \\ 1 & 2 & 1 \end{vmatrix} = 0\begin{vmatrix} -1 & 2 \\ 2 & 1 \end{vmatrix} - (-1)\begin{vmatrix} 1 & 3 \\ 2 & 1 \end{vmatrix} + 1\begin{vmatrix} 1 & 3 \\ -1 & 2 \end{vmatrix}$

$= 0(-1 - 4) + 1(1 - 6) + 1(2 + 3)$

$= 0(-5) + 1(-5) + 1(5)$

$= 0 - 5 + 5$

$= 0$

$D_x = \begin{vmatrix} 4 & 1 & 3 \\ 0 & -1 & 2 \\ 1 & 2 & 1 \end{vmatrix} = 4\begin{vmatrix} -1 & 2 \\ 2 & 1 \end{vmatrix} - 0\begin{vmatrix} 1 & 3 \\ 2 & 1 \end{vmatrix} + 1\begin{vmatrix} 1 & 3 \\ -1 & 2 \end{vmatrix}$

$= 4(-1 - 4) - 0(1 - 6) + 1(2 + 3)$

$= 4(-5) - 0(-5) + 1(5)$

$= -20 + 0 + 5$

$= -15$

Since $D = 0$ and $D_x = -15$, the system is inconsistent and has no solution.

Chapter 9 Practice Test

1. a.

$x + 2y = -6$	$3x + 2y = -12$
$-6 + 2(0) = -6$	$3(-6) + 2(0) = -12$
$-6 = -6$	$-18 = -12$
True	False

Since (–6, 0) does not satisfy both equations, it is not a solution to the system.

b.

$x + 2y = -6$	$3x + 2y = -12$
$-3 + 2\left(-\dfrac{3}{2}\right) = -6$	$3(-3) + 2\left(-\dfrac{3}{2}\right) = -12$
$-6 = -6$	$-12 = -12$
True	True

$\left(-3, -\dfrac{3}{2}\right)$ is a solution to the system.

c.

$x + 2y = -6$	$3x + 2y = -12$
$2 + 2(-4) = -6$	$3(2) + 2(-4) = -12$
$-6 = -6$	$-2 = -12$
True	False

Since (2, –4) does not satisfy both equations, it is not a solution to the system.

2. The system is consistent; it has exactly one solution.

3. The system is inconsistent, it has no solution.

4. The system is dependent; it has an infinite number of solutions.

5. Write both equations in slope-intercept form.

$5x + 2y = 4$	$6x = 3y - 7$
$2y = -5x + 4$	$-3y = -6x - 7$
$y = \dfrac{-5x + 4}{2}$	$y = \dfrac{-6x - 7}{-3}$
$y = -\dfrac{5}{2}x + 2$	$y = 2x + \dfrac{7}{3}$

Since the slope of the first line is $-\dfrac{5}{2}$ and the slope of the second line is 2, the slopes are different so that the lines intersect to produce one solution. This is a consistent system.

6. Write both equations in slope-intercept form.

$5x + 3y = 9$	$2y = -\dfrac{10}{3}x + 6$
$3y = -5x + 9$	
$y = \dfrac{-5x + 9}{3}$	$\dfrac{1}{2}(2y) = \dfrac{1}{2}\left(-\dfrac{10}{3}x + 6\right)$
$y = -\dfrac{5}{3}x + 3$	$y = -\dfrac{5}{3}x + 3$

Since the equations are identical, there are an infinite number of solutions and this is a dependent system.

7. Write both equations in slope-intercept form.

$$5x - 4y = 6 \qquad -10x + 8y = -10$$

$$-4y = -5x + 6 \qquad 8y = 10x - 10$$

$$y = \frac{-5x + 6}{-4} \qquad y = \frac{10x - 10}{8}$$

$$y = \frac{5}{4}x - \frac{3}{2} \qquad y = \frac{5}{4}x - \frac{5}{4}$$

Since the slope of each line is $\frac{5}{4}$, but the y-intercepts are different $\left(b = -\frac{3}{2} \text{ for the first equation, } b = -\frac{5}{4} \text{ for the second equation} \right)$, for the second the two lines are parallel and produce no solution. This is an inconsistent system.

8. Graph the equations $y = 3x - 2$ and $y = -2x + 8$.

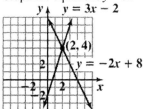

The lines intersect and the point of intersection is (2, 4).

9. Graph the equations $y = -x + 6$ and $y = 2x + 3$.

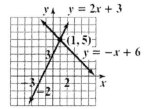

The lines intersect and the point of intersection is (1, 5).

10. $y = 4x - 3$

$y = 5x - 4$

Substitute $4x - 3$ for y in the second equation.

$$4x - 3 = 5x - 4$$

$$-x - 3 = -4$$

$$-x = -1$$

$$x = 1$$

Substitute 1 for x in the first equation.

$$y = 4x - 3$$

$$y = 4(1) - 3 = 4 - 3 = 1$$

The solution is (1, 1).

11. $4a + 7b = 2$

$5a + b = -13$

Solve the second equation for b.

$$5a + b = -13$$

$$b = -5a - 13$$

Substitute $-5a - 13$ for b in the first equation.

$$4a + 7(-5a - 13) = 2$$

$$4a - 35a - 91 = 2$$

$$-31a - 91 = 2$$

$$-31a = 93$$

$$a = -3$$

Substitute -3 for a in the equation $b = -5a - 13$.

$$b = -5(-3) - 13 = 2$$

The solution is (-3, 2).

12.
$$0.3x = 0.2y + 0.4$$

$$-1.2x + 0.8y = -1.6$$

Write the system in standard form.

$$0.3x - 0.2y = 0.4$$

$$-1.2x + 0.8y = -1.6$$

To eliminate x, multiply the first equation by 4 and then add.

$$4[0.3x - 0.2y = 0.4]$$

$$-1.2x + 0.8y = -1.6$$

gives

$$1.2x - 0.8y = 1.6$$

$$\underline{-1.2x + 0.8y = -1.6}$$

Add: $\qquad 0 = 0 \qquad$ True

Since this is a true statement, there are an infinite number of solutions and this is a dependent system.

13. $\dfrac{3}{2}a + b = 6$

$a - \dfrac{5}{2}b = -4$

To clear fractions, multiply both equations by 2.

$$2\left[\frac{3}{2}a + b = 6\right]$$

$$2\left[a - \frac{5}{2}b = -4\right]$$

gives

$$3a + 2b = 12$$

$$2a - 5b = -8$$

Now, to eliminate b, multiply the first equation by 5 and the second equation by 2 and then add.

$5[3a + 2b = 12]$

$2[2a - 5b = -8]$

gives

$$15a + 10b = 60$$
$$\underline{4a - 10b = -16}$$

Add: $19a \quad\quad = 44$

$$a = \frac{44}{19}$$

Substitute $\frac{44}{19}$ for a in the first equation.

$$\frac{3}{2}a + b = 6$$

$$\frac{3}{2}\left(\frac{44}{19}\right) + b = 6$$

$$\frac{66}{19} + b = 6$$

$$b = 6 - \frac{66}{19}$$

$$b = \frac{114}{19} - \frac{66}{19}$$

$$b = \frac{48}{19}$$

The solution is $\left(\frac{44}{19}, \frac{48}{19}\right)$.

14. $\quad x + y + z = 2 \ (1)$

$-2x - y + z = 1 \ (2)$

$x - 2y - z = 1 \ (3)$

To eliminate z between equations (1) and (3) simply add.

$$x + y + z = 2$$
$$\underline{x - 2y - z = 1}$$

Add: $2x - y \quad\quad = 3 \quad (4)$

To eliminate z between equations (2) and (3) simply add.

$$-2x - y + z = 1$$
$$\underline{x - 2y - z = 1}$$

Add: $-x - 3y \quad\quad = 2 \quad (5)$

Equations (4) and (5) are two equations in two unknowns.

$2x - y = 3$

$-x - 3y = 2$

To eliminate x, multiply equation (5) by 2 and then add.

$2x - y = 3$

$2[-x - 3y = 2]$

gives

$$2x - y = 3$$
$$\underline{-2x - 6y = 4}$$

Add: $-7y = 7$

$$y = \frac{7}{-7} = -1$$

Substitute -1 for y in equation (4).

$$2x - y = 3$$
$$2x - (-1) = 3$$
$$2x + 1 = 3$$
$$2x = 2$$
$$x = \frac{2}{2} = 1$$

Finally, substitute 1 for x and -1 for y in equation (1).

$$x + y + z = 2$$
$$1 - 1 + z = 2$$
$$0 + z = 2$$
$$z = 2$$

The solution is $(1, -1, 2)$.

15. $\quad -2x + 3y + 7z = 5$

$3x - 2y + z = -2$

$x - 6y + 9z = -13$

The augmented matrix is

$$\begin{bmatrix} -2 & 3 & 7 & | & 5 \\ 3 & -2 & 1 & | & -2 \\ 1 & -6 & 9 & | & -13 \end{bmatrix}$$

16. $\begin{bmatrix} 6 & -2 & 4 & | & 4 \\ 4 & 3 & 5 & | & 6 \\ 2 & -1 & 4 & | & -3 \end{bmatrix}$

$\begin{bmatrix} 6 & -2 & 4 & | & 4 \\ 0 & 5 & -3 & | & 12 \\ 2 & -1 & 4 & | & -3 \end{bmatrix} -2R_3 + R_2$

17. $2x + 7y = 1$

$3x + 5y = 7$

$\begin{bmatrix} 2 & 7 & | & 1 \\ 3 & 5 & | & 7 \end{bmatrix}$

$\begin{bmatrix} 1 & \frac{7}{2} & | & \frac{1}{2} \\ 3 & 5 & | & 7 \end{bmatrix} \frac{1}{2} R_1$

$\begin{bmatrix} 1 & \frac{7}{2} & | & \frac{1}{2} \\ 0 & -\frac{11}{2} & | & \frac{11}{2} \end{bmatrix} -3R_1 + R_2$

$\begin{bmatrix} 1 & \frac{7}{2} & | & \frac{1}{2} \\ 0 & 1 & | & -1 \end{bmatrix} -\frac{2}{11} R_2$

The system is

$x + \dfrac{7}{2}y = \dfrac{1}{2}$

$y = -1$

Substitute -1 for y in the first equation.

$x + \dfrac{7}{2}(-1) = \dfrac{1}{2}$

$x - \dfrac{7}{2} = \dfrac{1}{2}$

$x = \dfrac{8}{2} = 4$

The solution is $(4, -1)$.

18. $x - 2y + z = 7$

$-2x - y - z = -7$

$4x + 5y - 2z = 3$

$\begin{bmatrix} 1 & -2 & 1 & | & 7 \\ -2 & -1 & -1 & | & -7 \\ 4 & 5 & -2 & | & 3 \end{bmatrix}$

$\begin{bmatrix} 1 & -2 & 1 & | & 7 \\ 0 & -5 & 1 & | & 7 \\ 4 & 5 & -2 & | & 3 \end{bmatrix} 2R_1 + R_2$

$\begin{bmatrix} 1 & -2 & 1 & | & 7 \\ 0 & -5 & 1 & | & 7 \\ 0 & 13 & -6 & | & -25 \end{bmatrix} -4R_1 + R_3$

$\begin{bmatrix} 1 & -2 & 1 & | & 7 \\ 0 & 1 & -\frac{1}{5} & | & -\frac{7}{5} \\ 0 & 13 & -6 & | & -25 \end{bmatrix} -\frac{1}{5}R_2$

$\begin{bmatrix} 1 & -2 & 1 & | & 7 \\ 0 & 1 & -\frac{1}{5} & | & -\frac{7}{5} \\ 0 & 0 & -\frac{17}{5} & | & -\frac{34}{5} \end{bmatrix} -13R_2 + R_3$

$\begin{bmatrix} 1 & -2 & 1 & | & 7 \\ 0 & 1 & -\frac{1}{5} & | & -\frac{7}{5} \\ 0 & 0 & 1 & | & 2 \end{bmatrix} -\frac{5}{17}R_3$

The system is

$x - 2y + z = 7$

$y - \dfrac{1}{5}z = -\dfrac{7}{5}$

$z = 2$

Substitute 2 for z in the second equation.

$y - \dfrac{1}{5}z = -\dfrac{7}{5}$

$y - \dfrac{1}{5}(2) = -\dfrac{7}{5}$

$y - \dfrac{2}{5} = -\dfrac{7}{5}$

$y = -1$

Substitute -1 for y and 2 for z in the first equation.

$x - 2y + z = 7$

$x - 2(-1) + 2 = 7$

$x + 2 + 2 = 7$

$x + 4 = 7$

$x = 3$

The solution is $(3, -1, 2)$.

19. $\begin{vmatrix} 3 & -1 \\ 5 & -2 \end{vmatrix} = (3)(-2) - (5)(-1)$

$= -6 - (-5)$

$= -6 + 5$

$= -1$

20. $\begin{vmatrix} 8 & 2 & -1 \\ 3 & 0 & 5 \\ 6 & -3 & 4 \end{vmatrix} = 8\begin{vmatrix} 0 & 5 \\ -3 & 4 \end{vmatrix} - 3\begin{vmatrix} 2 & -1 \\ -3 & 4 \end{vmatrix} + 6\begin{vmatrix} 2 & -1 \\ 0 & 5 \end{vmatrix}$

$= 8(0 + 15) - 3(8 - 3) + 6(10 - 0)$

$= 8(15) - 3(5) + 6(10)$

$= 120 - 15 + 60$

$= 165$

21. $4x + 3y = -6$

$-2x + 5y = 16$

To solve, first calculate D, D_x, and D_y.

$D = \begin{vmatrix} 4 & 3 \\ -2 & 5 \end{vmatrix} = (4)(5) - (-2)(3) = 20 + 6 = 26$

$D_x = \begin{vmatrix} -6 & 3 \\ 16 & 5 \end{vmatrix} = (-6)(5) - (16)(3) = -30 - 48 = -78$

$D_y = \begin{vmatrix} 4 & -6 \\ -2 & 16 \end{vmatrix} = (4)(16) - (-2)(-6) = 64 - 12 = 52$

$x = \dfrac{D_x}{D} = \dfrac{-78}{26} = -3$ and $y = \dfrac{D_y}{D} = \dfrac{52}{26} = 2$.

The solution is $(-3, 2)$.

22. $2r - 4s + 3t = -1$

$-3r + 5s - 4t = 0$

$-2r + s - 3t = -2$

To solve, first calculate D, D_r, D_s, and D_t.

$D = \begin{vmatrix} 2 & -4 & 3 \\ -3 & 5 & -4 \\ -2 & 1 & -3 \end{vmatrix}$

$= 2\begin{vmatrix} 5 & -4 \\ 1 & -3 \end{vmatrix} - (-4)\begin{vmatrix} -3 & -4 \\ -2 & -3 \end{vmatrix} + 3\begin{vmatrix} -3 & 5 \\ -2 & 1 \end{vmatrix}$

$= 2(-15 + 4) + 4(9 - 8) + 3(-3 + 10)$

$= 2(-11) + 4(1) + 3(7)$

$= -22 + 4 + 21$

$= 3$

$D_r = \begin{vmatrix} -1 & -4 & 3 \\ 0 & 5 & -4 \\ -2 & 1 & -3 \end{vmatrix}$

$= -1\begin{vmatrix} 5 & -4 \\ 1 & -3 \end{vmatrix} - (-4)\begin{vmatrix} 0 & -4 \\ -2 & -3 \end{vmatrix} + 3\begin{vmatrix} 0 & 5 \\ -2 & 1 \end{vmatrix}$

$= -1(-15 + 4) + 4(0 - 8) + 3(0 + 10)$

$= -1(-11) + 4(-8) + 3(10)$

$= 11 - 32 + 30$

$= 9$

$D_s = \begin{vmatrix} 2 & -1 & 3 \\ -3 & 0 & -4 \\ -2 & -2 & -3 \end{vmatrix}$

$= 2\begin{vmatrix} 0 & -4 \\ -2 & -3 \end{vmatrix} - (-1)\begin{vmatrix} -3 & -4 \\ -2 & -3 \end{vmatrix} + 3\begin{vmatrix} -3 & 0 \\ -2 & -2 \end{vmatrix}$

$= 2(0 - 8) + 1(9 - 8) + 3(6 - 0)$

$= 2(-8) + 1(1) + 3(6)$

$= -16 + 1 + 18$

$= 3$

$D_t = \begin{vmatrix} 2 & -4 & -1 \\ -3 & 5 & 0 \\ -2 & 1 & -2 \end{vmatrix}$

$= 2\begin{vmatrix} 5 & 0 \\ 1 & -2 \end{vmatrix} - (-4)\begin{vmatrix} -3 & 0 \\ -2 & -2 \end{vmatrix} + (-1)\begin{vmatrix} -3 & 5 \\ -2 & 1 \end{vmatrix}$

$= 2(-10 - 0) + 4(6 - 0) - 1(-3 + 10)$

$= 2(-10) + 4(6) - 1(7)$

$= -20 + 24 - 7$

$= -3$

$r = \dfrac{D_r}{D} = \dfrac{9}{3} = 3$, $s = \dfrac{D_s}{D} = \dfrac{3}{3} = 1$, $t = \dfrac{D_t}{D} = \dfrac{-3}{3} = -1$

The solution is $(3, 1, -1)$.

23. Let x = the number of pounds of sunflower seeds and y = the number of pounds of bird seed.

$x + y = 20$

$0.49x + 0.89y = 0.73(20)$

Solve the first equation for y.

$x + y = 20$

$y = -x + 20$

Substitute $-x + 20$ for y in the second equation.

$0.49x + 0.89(-x + 20) = 0.73(20)$

$0.49x - 0.89x + 17.8 = 14.6$

$-0.4x = -3.2$

$x = 8$

Substitute 8 for x in the equation for y.

$y = -x + 20$

$y = -(8) + 20$

$y = 12$

Thus, 8 pounds of sunflower seeds should be mixed with 12 pounds of the gourmet bird seed to obtain the desired mixture.

24. Let x = amount of 6% solution

y = amount of 15% solution

$x + y = 10$

$0.06x + 0.15y = 0.09(10)$

The system can be written as

$x + y = 10$

$6x + 15y = 90$

Solve the first equation for y.

$y = 10 - x$

Substitute $10 - x$ for y in the second equation.

$$6x + 15y = 90$$
$$6x + 15(10 - x) = 90$$
$$6x + 150 - 15x = 90$$
$$-9x = -60$$
$$x = \frac{-60}{-9} = \frac{20}{3} = 6\frac{2}{3}$$

Substitute $6\frac{2}{3}$ for x into $y = 10 - x$

$$y = 10 - 6\frac{2}{3}$$

$$y = 3\frac{1}{3}$$

She should mix $6\frac{2}{3}$ liters of 6% solution and $3\frac{1}{3}$ liters of 15% solution.

25. Let x = smallest number
y = remaining number
z = largest number
$$x + y + z = 29$$
$$z = 4x$$
$$y = 2x + 1$$

Substitute $2x + 1$ for y and $4x$ for z in the first equation.
$$x + y + z = 29$$
$$x + 2x + 1 + 4x = 29$$
$$7x = 28$$
$$x = 4$$

Substitute 4 for x in the third equation.
$$y = 2(4) + 1$$
$$y = 9$$

Substitute 4 for x in the second equation.
$$z = 4(4)$$
$$z = 16$$

The three numbers are 4, 9, and 16.

Chapter 9 Cumulative Review Test

1. a. 9 and 1 are natural numbers.

b. $\frac{1}{2}$, −4, 9, 0, −4.63, and 1 are rational numbers.

c. $\frac{1}{2}$, −4, 9, 0, $\sqrt{3}$, −4.63, and 1 are real numbers.

2. $16 \div \left\{ 4\left[3 + \left(\frac{5 + 10}{5} \right)^2 \right] - 32 \right\}$

$= 16 \div \left\{ 4\left[3 + \left(\frac{15}{5} \right)^2 \right] - 32 \right\}$

$= 16 \div \left\{ 4\left[3 + (3)^2 \right] - 32 \right\}$

$= 16 \div \left\{ 4[3 + 9] - 32 \right\}$

$= 16 \div \left\{ 4[12] - 32 \right\}$

$= 16 \div \left\{ 48 - 32 \right\}$

$= 16 \div \left\{ 16 \right\}$

$= 1$

3. $-7(3 - x) = 4(x + 2) - 3x$
$$-21 + 7x = 4x + 8 - 3x$$
$$-21 + 7x = x + 8$$
$$6x = 29$$
$$x = \frac{29}{6}$$

4. $4x - 8y = 16$
Find the x- and y-intercepts:

x-intercept	y-intercept
Let $y = 0$.	Let $x = 0$
$4x - 8(0) = 16$	$4(0) - 8y = 16$
$4x = 16$	$-8y = 16$
$x = 4$	$y = -2$

The intercepts are $(4, 0)$ and $(0, -2)$.

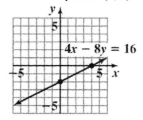

5. $y - y_1 = m(x - x_1)$

$$y - 1 = \frac{2}{5}\left[x - (-3) \right]$$

$$y - 1 = \frac{2}{5}(x + 3)$$

$$y - 1 = \frac{2}{5}x + \frac{6}{5}$$

$$y = \frac{2}{5}x + \frac{6}{5} + 1$$

$$y = \frac{2}{5}x + \frac{11}{5}$$

6. The line passes through the points $(1, 1)$ and $(0, -2)$. The y-intercept is $(0, -2)$, so $b = -2$.

The slope is $m = \dfrac{y_2 - y_1}{x_2 - x_1} = \dfrac{-2 - 1}{0 - 1} = \dfrac{-3}{-1} = 3$.

Thus, the equation of the line is:
$$y = mx + b$$
$$y = 3x - 2$$

7. $\dfrac{4a^3 b^{-5}}{28a^8 b} = \dfrac{a^{3-8} b^{-5-1}}{7} = \dfrac{a^{-5} b^{-6}}{7} = \dfrac{1}{7a^5 b^6}$

8. $3x^3 + 4x^2 + 6x + 8 = x^2(3x + 4) + 2(3x + 4)$
$$= (3x + 4)(x^2 + 2)$$

9. Note that $(-14)(-2) = 28$ and $-14 + (-2) = -16$.
$$x^2 - 16x + 28 = (x - 14)(x - 2)$$

10. $x^2 - 5x = 0$
$$x(x - 5) = 0$$
$$x = 0 \quad \text{or} \quad x - 5 = 0$$
$$x = 5$$
The solutions are 0 and 5.

11. $\dfrac{y - 5}{8} + \dfrac{2y + 7}{8} = \dfrac{y - 5 + 2y + 7}{8} = \dfrac{3y + 2}{8}$

12. $\dfrac{5}{y + 2} = \dfrac{3}{2y - 7}$
$$5(2y - 7) = 3(y + 2)$$
$$10y - 35 = 3y + 6$$
$$7y = 41$$
$$y = \dfrac{41}{7}$$

13. a. It is a function since it passes the vertical line test.

b. It is a function since it passes the vertical line test.

c. It is not a function since it fails the vertical line test.

14. $f(x) = \dfrac{x + 3}{x^2 - 9}$

a. $f(-4) = \dfrac{(-4) + 3}{(-4)^2 - 9} = \dfrac{-1}{16 - 9} = -\dfrac{1}{7}$

b. $f(h) = \dfrac{h + 3}{h^2 - 9}$

c. $f(3) = \dfrac{(3) + 3}{(3)^2 - 9} = \dfrac{6}{9 - 9} = \dfrac{6}{0}$ undefined

15. $3x + y = 6$
$$y = 4x - 1$$
Substitute $4x - 1$ for y in the first equation.
$$3x + y = 6$$
$$3x + 4x - 1 = 6$$
$$7x - 1 = 6$$
$$7x = 7$$
$$x = \dfrac{7}{7} = 1$$
Substitute 1 for x in the second equation.
$$y = 4x - 1$$
$$y = 4(1) - 1$$
$$y = 4 - 1$$
$$y = 3$$
The solution is $(1, 3)$.

16. $2p + 3q = 11$
$$-3p - 5q = -16$$
To eliminate p, multiply the first equation by 3 and the second equation by 2 and then add.
$$3[2p + 3q = 11]$$
$$2[-3p - 5q = -16]$$
gives
$$6p + 9q = 33$$
$$\underline{-6p - 10q = -32}$$
Add: $\qquad -q = 1$
$$q = -1$$
Substitute -1 for q in the first equation.
$$2p + 3(-1) = 11$$
$$2p - 3 = 11$$
$$2p = 14$$
$$p = 7$$
The solution is $(7, -1)$.

17. $x - 2y = 0$ (1)

$2x + z = 7$ (2)

$y - 2z = -5$ (3)

To eliminate z between equations (2) and (3), multiply equation (2) by 2 and then add.

$2[2x + z = 7]$

$y - 2z = -5$

gives

$4x \quad + 2z = 14$

$\underline{\quad\quad y - 2z = -5}$

Add: $4x + y \quad = 9$ (4)

Equations (4) and (1) are two equations in two unknowns:

$x - 2y = 0$

$4x + y = 9$

To eliminate y, multiply equation (4) by 2 and then add.

$x - 2y = 0$

$2[4x + y = 9]$

gives

$x - 2y = 0$

$\underline{8x + 2y = 18}$

Add: $9x \quad = 18$

$x = \dfrac{18}{9} = 2$

Substitute 2 for x in equation (4).

$4x + y = 9$

$4(2) + y = 9$

$8 + y = 9$

$y = 1$

Finally, substitute 2 for x in equation (2).

$2x + z = 7$

$2(2) + z = 7$

$4 + z = 7$

$z = 3$

The solution is $(2, 1, 3)$.

18. Let x be the measure of the smallest angle. Then $9x$ is the measure of the largest angle and $x + 70$ is the measure of the remaining angle. The sum of the measures of the three angles is $180°$.

$x + (x + 70) + 9x = 180$

$11x + 70 = 180$

$11x = 110$

$x = 10$

$x + 70 = 10 + 70 = 80; \quad 9x = 9(10) = 90$

The three angles are $10°$, $80°$, and $90°$.

19. Let t be the time for Judy to catch up to Dawn.

	rate	time	distance
Judy	6	t	$6t$
Mark	4	$t + \dfrac{1}{2}$	$4\left(t + \dfrac{1}{2}\right)$

$6t = 4\left(t + \dfrac{1}{2}\right)$

$6t = 4t + 2$

$2t = 2$

$t = \dfrac{2}{2} = 1$

It takes 1 hour for Judy to catch up to Mark.

20. Let x = the number of \$20 tickets sold and y = the number of \$16 tickets sold.

The system is

$x + y = 1000$

$20x + 16y = 18,400$

Solve the first equation for y.

$x + y = 1000$

$y = -x + 1000$

Substitute $-x + 1000$ for y in the second equation.

$20x + 16(-x + 1,000) = 18,400$

$20x - 16x + 16,000 = 18,400$

$4x + 16,000 = 18,400$

$4x = 2400$

$x = \dfrac{2400}{4} = 600$

Substitute 600 for x in the equation for y.

$y = -x + 1000$

$y = -(600) + 1000$

$y = 400$

Thus, 600 tickets sold at \$20 and 400 tickets sold at \$16 for the concert.

Chapter 10

Exercise Set 10.1

1. It is necessary to change the direction of the inequality symbol when multiplying or dividing by a negative number.

3. a. Use open circles when the endpoints are not included.

   b. Use closed circles when the endpoints are included.

   c. Answers may vary. One possible answer is $x > 4$.

   d. Answers may vary. One possible answer is $x \geq 4$.

5. $a < x < b$ means $a < x$ and $x < b$.

7. a.

-2

   b. $(-2, \infty)$

   c. $\{x \mid x > -2\}$

9. a.

π

   b. $(-\infty, \pi]$

   c. $\{w \mid w \leq \pi\}$

11. a.

$-3 \qquad \dfrac{4}{5}$

   b. $\left(-3, \dfrac{4}{5}\right]$

   c. $\left\{q \,\middle|\, -3 < q \leq \dfrac{4}{5}\right\}$

13. a.

$-7 \qquad -4$

   b. $(-7, -4]$

   c. $\{x \mid -7 < x \leq -4\}$

15. $x - 9 > -6$

$x > 3$

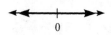
3

17. $3 - x < -4$

$-x < -7$

Reverse the inequality

$\dfrac{-x}{-1} > \dfrac{-7}{-1}$

$x > 7$

7

19. $\qquad 4.7x - 5.48 \geq 11.44$

$4.7x - 5.48 + 5.48 \geq 11.44 + 5.48$

$\qquad\qquad 4.7x \geq 16.92$

$\qquad\qquad \dfrac{4.7x}{4.7} \geq \dfrac{16.92}{4.7}$

$\qquad\qquad\quad x \geq 3.6$

3.6

21. $4(x + 2) \leq 4x + 8$

$4x + 8 \leq 4x + 8$

$\qquad 8 \leq 8$

Since this is a true statement, the solution is the entire real number line.

0

23. $5b - 6 \geq 3(b + 3) + 2b$

$5b - 6 \geq 3b + 9 + 2b$

$5b - 6 \geq 5b + 9$

$\quad -6 \geq 9$

Since this is a false statement, there is no solution.

0

25.
$$2y - 6y + 8 \leq 2(-2y + 9)$$
$$-4y + 8 \leq -4y + 18$$
$$8 \leq 18$$

Since this is a true statement, the solution is all real numbers.

27.
$$4 + \frac{4x}{3} < 6$$
$$\frac{4x}{3} < 2$$
$$3\left(\frac{4x}{3}\right) < 3(2)$$
$$4x < 6$$
$$\frac{4x}{4} < \frac{6}{4}$$
$$x < \frac{3}{2}$$
$$\left(-\infty, \frac{3}{2}\right)$$

29.
$$\frac{v - 5}{3} - v \geq -3(v - 1)$$
$$3\left[\frac{v - 5}{3} - v\right] \geq 3\left[-3(v - 1)\right]$$
$$v - 5 - 3v \geq -9(v - 1)$$
$$-5 - 2v \geq -9v + 9$$
$$7v - 5 \geq 9$$
$$7v \geq 14$$
$$v \geq 2 \quad \Rightarrow \quad [2, \infty)$$

31.
$$\frac{t}{3} - t + 7 \leq -\frac{4t}{3} + 8$$
$$3\left(\frac{t}{3} - t + 7\right) \leq 3\left(-\frac{4t}{3} + 8\right)$$
$$t - 3t + 21 \leq -4t + 24$$
$$-2t + 21 \leq -4t + 24$$
$$2t \leq 3$$
$$\frac{2t}{2} \leq \frac{3}{2}$$
$$t \leq \frac{3}{2} \quad \Rightarrow \quad \left(-\infty, \frac{3}{2}\right]$$

33.
$$-3x + 1 < 3\left[(x + 2) - 2x\right] - 1$$
$$-3x + 1 < 3\left[x + 2 - 2x\right] - 1$$
$$-3x + 1 < 3\left[2 - x\right] - 1$$
$$-3x + 1 < 6 - 3x - 1$$
$$-3x + 1 < 5 - 3x$$
$$1 < 5 \quad \Rightarrow \quad \text{a true statement}$$

The solution set is $(-\infty, \infty)$.

35. $A = \{5, 6, 7\}$; $B = \{6, 7, 8\}$
$$A \cup B = \{5, 6, 7, 8\}$$
$$A \cap B = \{6, 7\}$$

37. $A = \{-2, -4, -5\}$; $B = \{-1, -2, -4, -6\}$
$$A \cup B = \{-1, -2, -4, -5, -6\}$$
$$A \cap B = \{-2, -4\}$$

39. $A = \{\ \}$; $B = \{0, 1, 2, 3\}$
$$A \cup B = \{0, 1, 2, 3\}$$
$$A \cap B = \{\ \}$$

41. $A = \{0, 2, 4, 6, 8\}$; $B = \{1, 3, 5, 7\}$
$$A \cup B = \{0, 1, 2, 3, 4, 5, 6, 7, 8\}$$
$$A \cap B = \{\ \}$$

43. $A = \{0.1, 0.2, 0.3\}$; $B = \{0.2, 0.3, 0.4, 0.5, \ldots\}$
$$A \cup B = \{0.1, 0.2, 0.3, 0.4, 0.5, \ldots\}$$
$$A \cap B = \{0.2, 0.3\}$$

45.
$$-2 \leq t + 3 < 4$$
$$-2 - 3 \leq t + 3 - 3 < 4 - 3$$
$$-5 \leq t < 1$$
$$[-5, 1)$$

47.
$$-15 \leq -3z \leq 12$$

Divide by -3 and reverse inequalities.
$$\frac{-15}{-3} \geq \frac{-3z}{-3} \geq \frac{12}{-3}$$
$$5 \geq z \geq -4$$
$$-4 \leq z \leq 5$$
$$[-4, 5]$$

49. $4 \le 2x - 4 < 7$

$4 + 4 \le 2x - 4 + 4 < 7 + 4$

$8 \le 2x < 11$

$\dfrac{8}{2} \le \dfrac{2x}{2} < \dfrac{11}{2}$

$4 \le x < \dfrac{11}{2}$

$\left[4, \dfrac{11}{2}\right)$

51. $14 \le 2 - 3g < 15$

$14 - 2 \le 2 - 3g - 2 < 15 - 2$

$12 \le -3g < 13$

Divide by -3 and reverse inequalities.

$\dfrac{12}{-3} \ge \dfrac{-3g}{-3} > \dfrac{13}{-3}$

$-4 \ge g > -\dfrac{13}{3}$

$-\dfrac{13}{3} < g \le -4$

$\left(-\dfrac{13}{3}, -4\right]$

53. $5 \le \dfrac{3x + 1}{2} < 11$

$2(5) \le 2\left(\dfrac{3x + 1}{2}\right) < 2(11)$

$10 \le 3x + 1 < 22$

$10 - 1 \le 3x + 1 - 1 < 22 - 1$

$9 \le 3x < 21$

$\dfrac{9}{3} \le \dfrac{3x}{3} < \dfrac{21}{3}$

$3 \le x < 7$

$\{x | 3 \le x < 7\}$

55. $-6 \le -3(2x - 4) < 12$

$-6 \le -6x + 12 < 12$

$-6 - 12 \le -6x + 12 - 12 < 12 - 12$

$-18 \le -6x < 0$

Divide by -6 and reverse inequalities

$\dfrac{-18}{-6} \ge \dfrac{-6x}{-6} > \dfrac{0}{-6}$

$3 \ge x > 0$

$0 < x \le 3$

$\{x | 0 < x \le 3\}$

57. $0 \le \dfrac{3(u - 4)}{7} \le 1$

$7(0) \le 7\left(\dfrac{3(u - 4)}{7}\right) \le 7(1)$

$0 \le 3(u - 4) \le 7$

$0 \le 3u - 12 \le 7$

$0 + 12 \le 3u - 12 + 12 \le 7 + 12$

$12 \le 3u \le 19$

$\dfrac{12}{3} \le \dfrac{3u}{3} \le \dfrac{19}{3}$

$4 \le u \le \dfrac{19}{3}$

$\left\{u \,\middle|\, 4 \le u \le \dfrac{19}{3}\right\}$

59. $c \le 1$

$c > -3$

$c \le 1$ and $c > -3 \Rightarrow -3 < c \le 1$

$\{c | -3 < c \le 1\}$

61. $x < 2$

$x > 4$

$x < 2$ and $x > 4$

There is no overlap so the solution is the empty set, $\varnothing$.

63. $x+1<3$ and $x+1>-4$

$\quad\quad x<2$ and $\quad x>-5$

$x>-5$

$\quad\quad\quad -5$

$x<2$

$\quad\quad\quad\quad 2$

$x<-2$ and $x>-5$ which is $-5<x<2$ or

$\quad -5 \quad\quad 2$

$\{x|-5<x<2\}$

65. $2s+3<7$ or $-3s+4\le-17$

$\quad\quad 2s<4$ or $\quad\quad -3s\le-21$

$\quad \dfrac{2s}{2}<\dfrac{4}{2}$ or $\dfrac{-3s}{-3}\ge\dfrac{-21}{-3}$

$\quad\quad s<2$ $\quad\quad\quad s\ge7$

$s<2$ or $s\ge7$ which is $\left(-\infty,2\right)\cup\left[7,\infty\right)$.

67. $4x+5\ge5$ and $3x-7\le-1$

$\quad\quad 4x\ge0$ and $\quad 3x\le6$

$\quad\quad x\ge0$ and $\quad\quad x\le2$

$x\ge0$

$\quad\quad 0$

$x\le2$

$\quad\quad\quad 2$

$x\ge0$ and $x\le2$ which is $0\le x\le2$

$\quad\quad 0 \quad\quad 2$

In interval notation: $[0, 2]$

69. $4-r<-2$ or $3r-1<-1$

$\quad\quad -r<-6$ $\quad\quad\quad 3r<0$

$\quad\quad r>6$ $\quad\quad\quad\quad r<0$

$r>6$

$\quad\quad\quad 6$

$r<0$

$\quad\quad 0$

$r>6$ or $r<0$

$\quad\quad 0 \quad\quad 6$

In interval notation: $(-\infty,0)\cup(6,\infty)$

71. $2k+5>-1$ and $7-3k\le7$

$\quad\quad 2k>-6$ $\quad\quad\quad -3k\le0$

$\quad \dfrac{2k}{2}>\dfrac{-6}{2}$ $\quad\quad \dfrac{-3k}{-3}\ge\dfrac{0}{-3}$

$\quad\quad k>-3$ $\quad\quad\quad k\ge0$

$k>-3$ and $k\ge0 \Rightarrow k\ge0$

In interval notation: $\left[0,\infty\right)$

73. a. $l+g\le130$

b. $g=2w+2d$

$\quad\quad l+g\le130$

$\quad\quad l+2w+2d\le130$

c. $l=40,\ w=20.5$

$\quad\quad l+2w+2d\le130$

$\quad 40+2(20.5)+2d\le130$

$\quad\quad 40+41+2d\le130$

$\quad\quad\quad 81+2d\le130$

$\quad\quad\quad\quad 2d\le49$

$\quad\quad\quad\quad d\le24.5$

The maximum depth is 24.5 inches.

75. Let x be the maximum number of boxes.

$70x\le800$

$\quad x\le\dfrac{800}{70}$

$\quad x\le11.43$

The maximum number of boxes is 11.

77. Let x = the number of minutes she talks beyond the first 20 minutes.

$0.99+0.07x\le5.00$

$\quad\quad 0.07x\le4.01$

$\quad \dfrac{0.07x}{0.07}\le\dfrac{4.01}{0.07}$

$\quad\quad\quad x\le57$ (to nearest whole number)

She can talk for 57 minutes beyond the first 20 minutes for a total of 77 minutes.

79. To make a profit, the cost must be less than the revenue: cost < revenue.
$$10,025 + 1.09x < 6.42x$$
$$10,025 < 5.33x$$
$$\frac{10,025}{5.33} < x$$
$$1880.86 < x$$
She needs to sell a minimum of 1881 books to make a profit.

81. Let x = the number of additional ounces beyond the first ounce.
$$0.39 + 0.24x \leq 10.00$$
$$0.24x \leq 9.61$$
$$\frac{0.24x}{0.24} \leq \frac{9.61}{0.24}$$
$$x \leq 40 \ \text{(rounded down)}$$
The maximum weight is 41 ounces (the first ounce plus up to 40 additional ounces).

83. Let x be the amount of sales in dollars.
$$300 + 0.10x > 400 + 0.08x$$
$$0.10x > 100 + 0.08x$$
$$0.02x > 100$$
$$x > \frac{100}{0.02}$$
$$x > 5000$$
She will earn more by plan 1 if her weekly sales total more than $5000.

85. Let x be the minimum score for the sixth exam.
$$\frac{66 + 72 + 90 + 49 + 59 + x}{6} \geq 60$$
$$\frac{336 + x}{6} \geq 60$$
$$6\left(\frac{336 + x}{6}\right) \geq 6(60)$$
$$336 + x \geq 360$$
$$x \geq 24$$
She must make a 24 or higher on the sixth exam to pass the course.

87. Let x be the score on the fifth exam.
$$80 \leq \frac{85 + 92 + 72 + 75 + x}{5} < 90$$
$$80 \leq \frac{324 + x}{5} < 90$$
$$5(80) \leq 5\left(\frac{324 + x}{5}\right) < 5(90)$$
$$400 \leq 324 + x < 450$$
$$76 \leq x < 126$$
To receive a final grade of B, Ms. Mahoney must score 76 or higher on the fifth exam. That is, the score must be
$76 \leq x \leq 100$ (maximum grade is 100).

89. a. The taxable income of $78,221 places a married couple filing jointly in the 25% tax bracket. The tax is $8,180.00 plus 25% of the taxable income over $59,400.
The tax is
$$\$8,180.00 + 0.25(\$78,221 - \$59,400)$$
$$\$8,180.00 + 0.25(\$18,821)$$
$$\$8,180.00 + \$4,705.25$$
$$\$12,885.25$$
They will owe $12,885.25 in taxes.

b. The taxable income of $301,233 places a married couple filing jointly in the 33% tax bracket. The tax is $40,915.50 plus 33% of the taxable income over $182,800.
The tax is
$$\$40,915.5 + 0.33(\$301,233 - \$182,800)$$
$$\$40,915.5 + 0.33(\$118,433.00)$$
$$\$40,915.5 + \$39,082.89$$
$$\$79,998.39$$
They will owe $79,998.39 in taxes.

91. a.
$$v(t) \geq 0$$
$$-32t + 96 \geq 0$$
$$-32t \geq -96$$
$$t \leq 3$$
The object is traveling upward on the interval $[0, 3]$.

b. $v(t) \leq 0$

$-32t + 96 \leq 0$

$-32t \leq -96$

$t \geq 3$

The object is traveling downward on the interval $[3, 10]$.

93. a. $v(t) \geq 0$

$-9.8t + 49 \geq 0$

$-9.8t \geq -49$

$t \leq 5$

The object is traveling upward on the interval $[0, 5]$.

b. $v(t) \leq 0$

$-9.8t + 49 \leq 0$

$-9.8t \leq -49$

$t \geq 5$

The object is traveling downward on the interval $[5, 13]$.

95. a. $v(t) \geq 0$

$-32t + 320 \geq 0$

$-32t \geq -320$

$t \leq 10$

The object is traveling upward on the interval $[0, 8]$ (note: we restricted t to the interval $[0, 8]$).

b. From part (a) we saw that the object was moving upward for all values of t on the interval $[0, 8]$. Therefore, there are no values for t in that interval where the object is traveling downward.

97. Let x be the value of the third reading.

$$7.2 < \frac{7.48 + 7.15 + x}{3} < 7.8$$

$$7.2 < \frac{14.63 + x}{3} < 7.8$$

$$3(7.2) < 3\left(\frac{14.63 + x}{3}\right) < 3(7.8)$$

$$21.6 < 14.63 + x < 23.4$$

$$6.97 < x < 8.77$$

Any value between 6.97 and 8.77 would result in a normal pH reading.

99. a. The goal is greater than 6000 for all five months shown and the enlisted number is greater than 4000 for all five months shown except for April. Thus, the goal is greater than 6000 and the number enlisted is greater than 4000 for the months January, February, March, and May.

b. March, April, May

c. April

101. Answers may vary. The three components cover all possible total medical costs.

First piece: 0   if $c \leq \$100$
There is a deductible of $100 so for total costs less than or equal to $100, the patient covers all costs.

Second piece: $0.80(c - 100)$ if $\$100 < c \leq \$2100$

If total costs are more than $100, but no more than $2100, the patient pays the $100 deductible and Blue Cross/Blue Shield pays 80% of the remaining amount.

Third piece: $c - 500$ if $c > \$2100$
The maximum out-of-pocket cost for the patient is $500. For the second piece, the maximum cost to the patient occurs when $c = \$2100$. Since $0.80(2100 - 100) = 500$, this corresponds to the maximum amount a patient will have to pay. Thus, for any total medical cost over $2100, the patient must pay the $500 limit and Blue Cross/Blue Shield will pay the remaining amount.

103. a. From the chart, the 10th percentile is approximately 17.5 pounds and the 90th percentile is approximately 23.5 pounds. Therefore, 80% of the weights for 9 month old boys are in the interval $[17.5, 23.5]$ (in pounds).

b. From the chart, the 10th percentile is approximately 23.5 pounds and the 90th percentile is approximately 31 pounds. Therefore, 80% of the weights for 21 month old boys are in the interval $[23.5, 31]$ (in pounds).

c. From the chart, the 10th percentile is approximately 27.2 pounds and the 90th percentile is approximately 36.5 pounds. Therefore, 80% of the weights for 36 month old boys are in the interval $[27.2, 36.5]$ (in pounds).

105. First find the average of 82, 90, 74, 76, and 68.
$$\frac{82 + 90 + 74 + 76 + 68}{5} = \frac{390}{5} = 78$$

This represents $\frac{2}{3}$ of the final grade.

Let x be the score from the final exam. Since this represents $\frac{1}{3}$ of the final grade, the inequality is

$$80 \le \frac{2}{3}(78) + \frac{1}{3}x < 90$$

$$3(80) < 3\left[\frac{2}{3}(78) + \frac{1}{3}x\right] < 3(90)$$

$$240 \le 2(78) + x < 270$$
$$240 \le 156 + x < 270$$
$$84 \le x < 114$$

Stephen must score at least 84 points on the final exam to have a final grade of B. The range is $84 \le x \le 100$.

107. a. Answers may vary. One possible answer is: Write $x < 2x + 3 < 2x + 5$ as $x < 2x + 3$ and $2x + 3 < 2x + 5$

b. Solve each of the inequalities.
$$x < 2x + 3 \quad \text{and} \quad 2x + 3 < 2x + 5$$
$$-x < 3 \qquad\qquad\qquad 3 < 5$$
$$x > -3 \qquad\qquad \text{All real numbers}$$
The final answer is $x > -3$ or $(-3, \infty)$.

109. a. 4 is a counting number.

b. 0 and 4 are whole numbers.

c. $-3, 4, \dfrac{5}{2}, 0$ and $-\dfrac{29}{80}$ are rational numbers.

d. $-3, 4, \dfrac{5}{2}, \sqrt{7}, 0$ and $-\dfrac{29}{80}$ are real numbers.

110. Associative property of addition.

111. Commutative property of addition

112.
$$R = L + (V - D)r$$
$$R = L + Vr - Dr$$
$$R - L + Dr = Vr$$
$$\frac{R - L + Dr}{r} = V \text{ or } V = \frac{R - L + Dr}{r}$$

113. a. $A \cup B = \{1, 2, 3, 4, 5, 6, 8, 9\}$

b. $A \cap B = \{1, 8\}$

Exercise Set 10.2

1. $|x| = a, a > 0$
Set $x = a$ or $x = -a$.

3. $|x| < a, a > 0$
Write $-a < x < a$.

5. $|x| > a, a > 0$
Write $x < -a$ or $x > a$.

7. The solution to $|x| > 0$ is all real numbers except 0.
The absolute value of any real number, except 0, is greater than 0, i.e., positive.

9. Set $x = y$ or $x = -y$.

11. If $a \neq 0$, and $k > 0$,

    a. $|ax + b| = k$ has 2 solutions.

    b. $|ax + b| < k$ has an infinite number of solutions.

    c. $|ax + b| > k$ has an infinite number of solutions.

13. a. $|x| = 5, \{-5, 5\}$, D

    b. $|x| < 5, \{x | -5 < x < 5\}$, B

    c. $|x| > 5, \{x | x < -5 \text{ or } x > 5\}$, E

    d. $|x| \leq 5, \{x | -5 \leq x \leq 5\}$, C

    e. $|x| \geq 5, \{x | x \leq -5 \text{ or } x \geq 5\}$, A

15. $|a| = 2$
$a = 2$ or $a = -2$
The solution set is $\{-2, 2\}$.

17. $|c| = \dfrac{1}{2}$
$c = \dfrac{1}{2}$ or $c = -\dfrac{1}{2}$
The solution set is $\left\{-\dfrac{1}{2}, \dfrac{1}{2}\right\}$.

19. $|d| = -\dfrac{5}{6}$

There is no solution since the right side is a negative number and the absolute value can never be equal to a negative number. The solution set $\varnothing$.

21. $|x + 5| = 8$
$\begin{aligned} x + 5 &= 8 & x + 5 &= -8 \\ x &= 3 \text{ or } & x &= -13 \end{aligned}$
The solution set is $\{-13, 3\}$.

23. $|4.5q + 31.5| = 0$
$4.5q + 31.5 = 0$
$4.5q = -31.5$
$q = -7$
The solution set is $\{-7\}$.

25. $|5 - 3x| = \dfrac{1}{2}$

$\begin{aligned} 5 - 3x &= \dfrac{1}{2} & \text{or} && 5 - 3x &= -\dfrac{1}{2} \\ -3x &= \dfrac{1}{2} - 5 & && -3x &= -\dfrac{1}{2} - 5 \\ -3x &= -\dfrac{9}{2} & && -3x &= -\dfrac{11}{2} \\ -\dfrac{1}{3}(3x) &= -\dfrac{1}{3}\left(-\dfrac{9}{2}\right) & && -\dfrac{1}{3}(-3x) &= -\dfrac{1}{3}\left(-\dfrac{11}{2}\right) \\ x &= \dfrac{3}{2} & && x &= \dfrac{11}{6} \end{aligned}$

The solution set is $\left\{\dfrac{3}{2}, \dfrac{11}{6}\right\}$.

27. $\left|\dfrac{x - 3}{4}\right| = 5$

$\begin{aligned} \dfrac{x - 3}{4} &= 5 & \text{or} && \dfrac{x - 3}{4} &= -5 \\ 4\left(\dfrac{x - 3}{4}\right) &= 4(5) & && 4\left(\dfrac{x - 3}{4}\right) &= 4(-5) \\ x - 3 &= 20 & && x - 3 &= -20 \\ x &= 23 & && x &= -17 \end{aligned}$

The solution set is $\{-17, 23\}$.

29. $\left|\dfrac{x - 3}{4}\right| + 8 = 8$

$\left|\dfrac{x - 3}{4}\right| = 0$

$\dfrac{x - 3}{4} = 0$

$4\left(\dfrac{x - 3}{4}\right) = 4(0)$

$x - 3 = 0$

$x = 3$

The solution set is $\{3\}$.

31. $|w| < 11$

$-11 < w < 11$

The solution set is $\{w | -11 < w < 11\}$.

33. $|q + 5| \le 8$

$-8 \le q + 5 \le 8$

$-8 - 5 \le q + 5 - 5 \le 8 - 5$

$-13 \le q \le 3$

The solution set is $\left\{ q \middle| -13 \le q \le 3 \right\}$.

35. $|5b - 15| < 10$

$-10 < 5b - 15 < 10$

$-10 + 15 < 5b - 15 + 15 < 10 + 15$

$5 < 5b < 25$

$\dfrac{5}{5} < \dfrac{5b}{5} < \dfrac{25}{5}$

$1 < b < 5$

The solution set is $\left\{ b \middle| 1 < b < 5 \right\}$.

37. $|2x + 3| - 5 \le 10$

$|2x + 3| \le 15$

$-15 \le 2x + 3 \le 15$

$-15 - 3 \le 2x + 3 - 3 \le 15 - 3$

$-18 \le 2x \le 12$

$\dfrac{-18}{2} \le \dfrac{2x}{2} \le \dfrac{12}{2}$

$-9 \le x \le 6$

The solution set is $\{x | -9 \le x \le 6\}$

39. $|3x - 7| + 8 < 14$

$|3x - 7| < 6$

$-6 < 3x - 7 < 6$

$-6 + 7 < 3x - 7 + 7 < 6 + 7$

$1 < 3x < 13$

$\dfrac{1}{3} < \dfrac{3x}{3} < \dfrac{13}{3}$

$\dfrac{1}{3} < x < \dfrac{13}{3}$

The solution set is $\left\{ x \middle| \dfrac{1}{3} < x < \dfrac{13}{3} \right\}$.

41. $|2x - 6| + 5 \le 1$

$|2x - 6| \le -4$

There is no solution since the right side is negative whereas the left side is non-negative; zero or a positive number is never less than a negative number. The solution set is $\varnothing$.

43. $\left| \dfrac{1}{2} j + 4 \right| < 7$

$-7 < \dfrac{1}{2} j + 4 < 7$

$-7 - 4 < \dfrac{1}{2} j + 4 - 4 < 7 - 4$

$-11 < \dfrac{1}{2} j < 3$

$2(-11) < 2\left(\dfrac{1}{2} j \right) < 2(3)$

$-22 < j < 6$

The solution set is $\left\{ j \middle| -22 < j < 6 \right\}$.

45. $\left| \dfrac{x - 3}{2} \right| - 4 \le -2$

$\left| \dfrac{x - 3}{2} \right| \le 2$

$-2 \le \dfrac{x - 3}{2} \le 2$

$2(-2) \le 2\left(\dfrac{x - 3}{2} \right) \le 2(2)$

$-4 \le x - 3 \le 4$

$-4 + 3 \le x - 3 + 3 \le 4 + 3$

$-1 \le x \le 7$

The solution set is $\{x | -1 \le x \le 7\}$.

47. $|y| > 2$

$y < -2$ or $y > 2$

The solution set is $\{y | y < -2 \text{ or } y > 2\}$.

49. $|x + 4| > 5$

$x + 4 < -5$ or $x + 4 > 5$

$x < -9$ $x > 1$

The solution set is $\{x | x < -9 \text{ or } x > 1\}$.

51. $|7-3b|>5$

$7-3b<-5$ or $7-3b>5$

$-3b<-12$ $-3b>-2$

$\dfrac{-3b}{-3}>\dfrac{-12}{-3}$ $\dfrac{-3b}{-3}<\dfrac{-2}{-3}$

$b>4$ $b<\dfrac{2}{3}$

The solution set is $\left\{b\,\middle|\,b<\dfrac{2}{3}\ \text{or}\ b>4\right\}$.

53. $|2h-5|>3$

$2h-5<-3$ or $2h-5>3$

$2h<2$ $2h>8$

$h<\dfrac{2}{2}$ $h>\dfrac{8}{2}$

$h<1$ $h>4$

The solution set is $\left\{h\,\middle|\,h<1\ \text{or}\ h>4\right\}$.

55. $|0.1x-0.4|+0.4>0.6$

$|0.1x-0.4|>0.2$

$0.1x-0.4<-0.2$ or $0.1x-0.4>0.2$

$0.1x<0.2$ $0.1x>0.6$

$x<\dfrac{0.2}{0.1}$ $x>\dfrac{0.6}{0.1}$

$x<2$ $x>6$

The solution set is $\{x\,|\,x<2\ \text{or}\ x>6\}$.

57. $\left|\dfrac{x}{2}+4\right|\ge 5$

$\dfrac{x}{2}+4\le -5$ or $\dfrac{x}{2}+4\ge 5$

$2\left(\dfrac{x}{2}+4\right)\le 2(-5)$ $2\left(\dfrac{x}{2}+4\right)\ge 2(5)$

$x+8\le -10$ $x+8\ge 10$

$x\le -18$ $x\ge 2$

The solution set is $\{x\,|\,x\le -18\ \text{or}\ x\ge 2\}$.

59. $|7w+3|-12\ge -12$

$|7w+3|\ge 0$

Observe that the absolute value of a number is always greater than or equal to 0. Thus, the solution is the set of real numbers, or $\mathbb{R}$.

61. $|4-2x|>0$

$4-2x<0$ or $4-2x>0$

$-2x<-4$ $-2x>-4$

$x>\dfrac{-4}{-2}$ $x<\dfrac{-4}{-2}$

$x>2$ $x<2$

The solution set is $\{x\,|\,x<2\ \text{or}\ x>2\}$.

63. $|3p-5|=|2p+10|$

$3p-5=-(2p+10)$ or $3p-5=2p+10$

$3p-5=-2p-10$ $p-5=10$

$5p-5=-10$ $p=15$

$5p=-5$

$p=-1$

The solution set is $\{-1,\,15\}$.

65. $|6x|=|3x-9|$

$6x=-(3x-9)$ or $6x=3x-9$

$6x=-3x+9$ $3x=-9$

$9x=9$ $x=\dfrac{-9}{3}$

$x=\dfrac{9}{9}$ $x=-3$

$x=1$

The solution set is $\{-3,\,1\}$.

67. $\left|\dfrac{2r}{3}+\dfrac{5}{6}\right|=\left|\dfrac{r}{2}-3\right|$

$\dfrac{2r}{3}+\dfrac{5}{6}=-\left(\dfrac{r}{2}-3\right)$ or $\dfrac{2r}{3}+\dfrac{5}{6}=\dfrac{r}{2}-3$

$\dfrac{2r}{3}+\dfrac{5}{6}=-\dfrac{r}{2}+3$ $6\left(\dfrac{2r}{3}+\dfrac{5}{6}\right)=6\left(\dfrac{r}{2}-3\right)$

$6\left(\dfrac{2r}{3}+\dfrac{5}{6}\right)=6\left(-\dfrac{r}{2}+3\right)$ $4r+5=3r-18$

$4r+5=-3r+18$ $r+5=-18$

$7r+5=18$ $r=-23$

$7r=13$

$r=\dfrac{13}{7}$

The solution set is $\left\{-23,\dfrac{13}{7}\right\}$.

69. $\left|-\dfrac{3}{4}m+8\right| = \left|7-\dfrac{3}{4}m\right|$

$-\dfrac{3}{4}m+8 = -\left(7-\dfrac{3}{4}m\right)$ or $-\dfrac{3}{4}m+8 = 7-\dfrac{3}{4}m$

$-\dfrac{3}{4}m+8 = -7+\dfrac{3}{4}m$ $\qquad$ $-\dfrac{3}{4}m+8 = 7-\dfrac{3}{4}m$

$\qquad -\dfrac{6}{4}m = -15$ $\qquad\qquad\qquad 8 = 7$ False!

$\qquad\qquad m = 10$

The solution set is $\{10\}$.

71. $|h| = 1$

$h = 1$ or $h = -1$

The solution set is $\{-1, 1\}$.

73. $|q+6| > 2$

$q+6 < -2$ or $q+6 > 2$

$\quad q < -8 \qquad\qquad q > -4$

The solution set is $\left\{q \mid q < -8 \text{ or } q > -4\right\}$.

75. $|2w-7| \le 9$

$-9 \le 2w-7 \le 9$

$-9+7 \le 2w-7+7 \le 9+7$

$-2 \le 2w \le 16$

$\dfrac{-2}{2} \le \dfrac{2w}{2} \le \dfrac{16}{2}$

$-1 \le w \le 8$

The solution set is $\left\{w \mid -1 \le w \le 8\right\}$.

77. $|5a-1| = 9$

$5a-1 = -9$ or $5a-1 = 9$

$\quad 5a = -8 \qquad\qquad 5a = 10$

$\quad a = -\dfrac{8}{5} \qquad\qquad a = 2$

The solution set is $\left\{-\dfrac{8}{5}, 2\right\}$.

79. $|5x+2| > 0$

$5+2x < 0$ or $5+2x > 0$

$\quad 2x < -5 \qquad\qquad 2x > -5$

$\quad x < -\dfrac{5}{2} \qquad\qquad x > -\dfrac{5}{2}$

The solution set is $\left\{x \mid x < -\dfrac{5}{2} \text{ or } x > -\dfrac{5}{2}\right\}$.

81. $|4+3x| \le 9$

$-9 \le 4+3x \le 9$

$-13 \le 3x \le 5$

$-\dfrac{13}{3} \le x \le \dfrac{5}{3}$

The solution set is $\left\{x \mid -\dfrac{13}{3} \le x \le \dfrac{5}{3}\right\}$.

83. $|3n+8| - 4 = -10$

$|3n+8| = -6$

Since the right side is negative and the left side is non-negative, there is no solution since the absolute value can never equal a negative number. The solution set is $\varnothing$.

85. $\left|\dfrac{w+4}{3}\right| + 5 < 9$

$\left|\dfrac{w+4}{3}\right| < 4$

$-4 < \dfrac{w+4}{3} < 4$

$3(-4) < 3\left(\dfrac{w+4}{3}\right) < 3(4)$

$-12 < w+4 < 12$

$-16 < w < 8$

The solution set is $\{w \mid -16 < w < 8\}$.

87. $\left|\dfrac{3x-2}{4}\right| - \dfrac{1}{3} \ge -\dfrac{1}{3}$

$\left|\dfrac{3x-2}{4}\right| \ge 0$

Since the absolute value of a number is always greater than or equal to zero, the solution is the set of all real numbers or $\mathbb{R}$.

89. $\left|2x-8\right|=\left|\dfrac{1}{2}x+3\right|$

$$2x-8=-\left(\dfrac{1}{2}x+3\right) \quad \text{or} \quad 2x-8=\dfrac{1}{2}x+3$$

$$2x-8=-\dfrac{1}{2}x-3 \qquad\qquad \dfrac{3}{2}x-8=3$$

$$\dfrac{5}{2}x-8=-3 \qquad\qquad\qquad \dfrac{3}{2}x=11$$

$$\dfrac{5}{2}x=5 \qquad\qquad\qquad \dfrac{2}{3}\left(\dfrac{3}{2}x\right)=\dfrac{2}{3}(11)$$

$$\dfrac{2}{5}\left(\dfrac{5}{2}x\right)=\dfrac{2}{5}(5) \qquad\qquad x=\dfrac{22}{3}$$

$$x=2$$

The solution set is $\left\{2,\dfrac{22}{3}\right\}$.

91. $\left|2-3x\right|=\left|4-\dfrac{5}{3}x\right|$

$$2-3x=-\left(4-\dfrac{5}{3}x\right) \quad \text{or} \quad 2-3x=4-\dfrac{5}{3}x$$

$$2-3x=-4+\dfrac{5}{3}x \qquad\qquad -3x=2-\dfrac{5}{3}x$$

$$-3x=-6+\dfrac{5}{3}x \qquad\qquad -\dfrac{4}{3}x=2$$

$$-\dfrac{14}{3}x=-6 \qquad\qquad -\dfrac{3}{4}\left(-\dfrac{4}{3}x\right)=-\dfrac{3}{2}(2)$$

$$\left(-\dfrac{3}{14}\right)\left(-\dfrac{14}{3}\right)x=\left(-\dfrac{3}{14}\right)(-6) \qquad x=-\dfrac{3}{2}$$

$$x=\dfrac{9}{7}$$

The solution set is $\left\{-\dfrac{3}{2},\dfrac{9}{7}\right\}$.

93. a. $\left|t-0.089\right|\le 0.004$

$$-0.004\le t-0.089\le 0.004$$

$$-0.004+0.089\le t-0.089+0.089\le 0.004+0.089$$

$$0.085\le t\le 0.093$$

The solution is $\left[0.085,0.093\right]$.

b. 0.085 inches

c. 0.093 inches

95. a. $\left|d-160\right|\le 28$

$$-28\le d-160\le 28$$

$$-28+160\le d-160+160\le 28+160$$

$$132\le d\le 188$$

The solution is [132, 188]

b. The submarine can move between 132 feet and 188 feet below sea level, inclusive.

97. $\left\{-5,5\right\}$ is the solution set to the equation $\left|x\right|=5$.

99. $\{x\,|\,x\le -5 \text{ or } x\ge 5\}$ is the solution set of $\left|x\right|\ge 5$.

101. $\left|ax+b\right|\le 0$

$$0\le ax+b\le 0$$

which is the same as

$$ax+b=0$$

$$ax=-b$$

$$x=-\dfrac{b}{a}$$

103. a. Set $ax+b=-c$ or $ax+b=c$ and solve each equation for x.

b.
$$ax+b=-c \qquad \text{or} \quad ax+b=c$$

$$ax=-c-b \qquad\qquad ax=c-b$$

$$x=\dfrac{-c-b}{a} \qquad\qquad x=\dfrac{c-b}{a}$$

The solution is $x=\dfrac{-c-b}{a}$ or $x=\dfrac{c-b}{a}$.

105. a. Write $ax+b<-c$ or $ax+b>c$ and solve each inequality for x.

b.
$$ax+b<-c \qquad \text{or} \quad ax+b>c$$

$$ax<-c-b \qquad\qquad ax>c-b$$

$$x<\dfrac{-c-b}{a} \qquad\qquad x>\dfrac{c-b}{a}$$

The solution is $x<\dfrac{-c-b}{a}$ or $x>\dfrac{c-b}{a}$.

107. $|x-4| = |4-x|$

$x-4 = -(4-x)$ or $x-4 = 4-x$

$x-4 = -4+x$ $2x-4 = 4$

$0 = 0$ $2x = 8$

    True $x = 4$

Since the first statement is always true all real values work. The solution set is $\mathbb{R}$.

109. $|x| = x$

By definition $|x| = \begin{cases} x, & x \geq 0 \\ -x, & x < 0 \end{cases}$

Thus, $|x| = x$ when $x \geq 0$

The solution set is $\{x | x \geq 0\}$.

111. $|x+1| = 2x-1$

$x+1 = -(2x-1)$ or $x+1 = 2x-1$

$x+1 = -2x+1$ $1 = x-1$

$3x+1 = 1$ $2 = x$

$3x = 0$

$x = 0$

Checking both possible solutions, only $x = 2$ checks. The solution set is $\{2\}$.

113. $|x-4| = -(x-4)$

By the definition, $|x-4| = \begin{cases} x-4, & x-4 \geq 0 \\ -(x-4), & x-4 \leq 0 \end{cases}$

$= \begin{cases} x-4, & x \geq 4 \\ -(x-4), & x \leq 4 \end{cases}$

Thus, $|x-4| = -(x-4)$ for $x \leq 4$.

The solution set is $\{x | x \leq 4\}$.

115. $x + |-x| = 8$

For $x \geq 0$, $x + |-x| = 8$

$x + x = 8$

$2x = 8$

$x = 4$

For $x < 0$, $x + |-x| = 8$

$x - x = 8$

$0 = 8$ False

The solution set is $\{4\}$.

117. $x - |x| = 8$

For $x \geq 0$, $x - |x| = 8$

$x - x = 8$

$0 = 8$ False

For $x < 0$, $x - |x| = 8$

$x - (-x) = 8$

$x + x = 8$

$2x = 8$

$x = 4$ Contradicts $x < 0$

There are no values of x, so the solution set is $\varnothing$.

119. $\dfrac{1}{3} + \dfrac{1}{4} \div \dfrac{2}{5}\left(\dfrac{1}{3}\right)^2 = \dfrac{1}{3} + \dfrac{1}{4} \div \dfrac{2}{5} \cdot \dfrac{1}{9}$

$= \dfrac{1}{3} + \dfrac{1}{4} \cdot \dfrac{5}{2} \cdot \dfrac{1}{9}$

$= \dfrac{1}{3} + \dfrac{5}{72}$

$= \dfrac{1}{3} \cdot \dfrac{24}{24} + \dfrac{5}{72}$

$= \dfrac{24}{72} + \dfrac{5}{72}$

$= \dfrac{29}{72}$

120. Substitute 1 for x and 3 for y.

$4(x+3y) - 5xy = 4(1+3\cdot3) - 5(1)(3)$

$= 4(1+9) - 5(1)(3)$

$= 4(10) - 5(1)(3)$

$= 40 - 15$

$= 25$

121. Let x be the time needed to swim across the lake. Then $1.5 - x$ is the time needed to make the return trip.

	Rate	Time	Distance
First Trip	2	x	$2x$
Return Trip	1.6	$1.5 - x$	$1.6(1.5 - x)$

The distances are the same.
$$2x = 1.6(1.5 - x)$$
$$2x = 2.4 - 1.6x$$
$$3.6x = 2.4$$
$$x = \frac{2.4}{3.6} = \frac{2}{3}$$

The total distance across the lake is
$$2x = 2\left(\frac{2}{3}\right) = \frac{4}{3} \text{ or } 1.33 \text{ miles.}$$

122. $3(x - 2) - 4(x - 3) > 2$
$$3x - 6 - 4x + 12 > 2$$
$$-x + 6 > 2$$
$$-x > -4$$
$$\frac{-x}{-1} < \frac{-4}{-1}$$
$$x < 4$$
The solution set is $\{x \mid x < 4\}$

Mid-Chapter Test: 10.1-10.2

1. $x > \dfrac{5}{2}$

2. $-4 \leq w < 5$
Interval notation: $[-4, 5)$

3. $6x - 5 \leq 4x + 61$
$$2x - 5 \leq 61$$
$$2x \leq 66$$
$$x \leq 33$$

4. $5y - 9y + 8 \leq 2(-2y + 1)$
$$-4y + 8 \leq -4y + 2$$
$$8 \leq 2 \text{ false}$$
There is no solution to the inequality.

5. $2.4w - 3.2 < 3.6w - 0.8$
$$-1.2w - 3.2 < -0.8$$
$$-1.2w < 2.4$$
$$w > -2$$

6. $3 - 2x < 6 + 2x + 1$
$$3 - 2x < 2x + 7$$
$$3 < 4x + 7$$
$$-4 < 4x$$
$$-1 < x \text{ or } x > -1$$
Interval notation: $(-1, \infty)$

7. $-8 < p - 3 \leq 4$
$$-5 < p \leq 7$$
Interval notation: $(-5, 7]$

8. $\dfrac{1}{3} \leq 4x + 1 < 5$
$$-\frac{2}{3} \leq 4x < 4$$
$$-\frac{2}{12} \leq x < 1$$
$$-\frac{1}{6} \leq x < 1$$
Interval notation: $\left[-\dfrac{1}{6}, 1\right)$

9. $-3 \leq \dfrac{2x + 1}{5} \leq 1$
$$-15 \leq 2x + 1 \leq 5$$
$$-16 \leq 2x \leq 4$$
$$-8 \leq x \leq 2$$
Solution set: $\{x \mid -8 \leq x \leq 2\}$

10. $m \leq 7$ and $m > -6$

We can rewrite this statement as a compound inequality:

$-6 < m \leq 7$

Solution set: $\{m \mid -6 < m \leq 7\}$

11. $9 - x < 3$ or $7x - 5 \leq -5$

    $-x < -6$ $7x \leq 0$

     $x > 6$ $x \leq 0$

We have $x \leq 0$ or $x > 6$.

Solution set: $\{x \mid x \leq 0 \text{ or } x > 6\}$.

12. $A \cup B = \{7, 8, 9, 10, 12, 15\}$

    $A \cap B = \{8, 10\}$

13. $|t - 4| = 6$

   $t - 4 = 6$ or $t - 4 = -6$

      $t = 10$ $t = -2$

Solution set: $\{-2, 10\}$

14. $\left| \dfrac{w - 5}{3} \right| \leq 7$

   $-7 \leq \dfrac{w - 5}{3} \leq 7$

   $-21 \leq w - 5 \leq 21$

   $-16 \leq w \leq 26$

Solution set: $\{w \mid -16 \leq w \leq 26\}$

15. $|z| \geq 8$

   $z \leq -8$ or $z \geq 8$

Solution set: $\{z \mid z \leq -8 \text{ or } z \geq 8\}$

16. $|5 - 2x| \geq 0$

Since absolute value is always nonnegative, the solution set is all real numbers.

Solution set: $\mathbb{R}$

17. $|4 - 2x| = 5$

   $4 - 2x = -5$ or $4 - 2x = 5$

    $-2x = -9$ $-2x = 1$

      $x = \dfrac{9}{2}$ $x = -\dfrac{1}{2}$

Solution set: $\left\{ -\dfrac{1}{2}, \dfrac{9}{2} \right\}$

18. $|-3m + 4| < 7$

   $-7 < -3m + 4 < 7$

   $-11 < -3m < 3$

   $\dfrac{11}{3} > m > -1$

   $-1 < m < \dfrac{11}{3}$

Solution set: $\left\{ m \,\middle|\, -1 < m < \dfrac{11}{3} \right\}$

19. $|6y + 9| = |4y - 1|$

  $6y + 9 = 4y - 1$ or $6y + 9 = -(4y - 1)$

    $2y = -10$ $6y + 9 = -4y + 1$

     $y = -5$ $10y = -8$

                       $y = -\dfrac{4}{5}$

Solution set: $\left\{ -5, -\dfrac{4}{5} \right\}$

20. $\left| \dfrac{1}{2}z - 3 \right| = \left| \dfrac{3}{2}z + 2 \right|$

  $\dfrac{1}{2}z - 3 = \dfrac{3}{2}z + 2$ or $\dfrac{1}{2}z - 3 = -\left(\dfrac{3}{2}z + 2 \right)$

     $-z = 5$ $\dfrac{1}{2}z - 3 = -\dfrac{3}{2}z - 2$

     $z = -5$ $2z = 1$

                      $z = \dfrac{1}{2}$

Solution set: $\left\{ -5, \dfrac{1}{2} \right\}$

Exercise Set 10.3

1. The inequalities $>$ and $<$ do not include the corresponding equation. The points on the line satisfy only the equation.

3. $(0, 0)$ cannot be used as a test point if the line passes through the origin.

5. Answers will vary.

7. Yes; the point of intersection of the boundary lines is in the solution set if the inequalities are both non-strict.

355

9. $x > 1$

Graph the line $x = 1$ (vertical line) using a dashed line. For the check point, select $(0, 0)$:

$x > 1$

$0 > 1 \leftarrow$ Substitute 0 for x

Since this is a false statement, shade the region which does not contain $(0, 0)$.

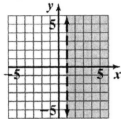

11. $y < -2$

Graph the line $y = -2$ (horizontal line) using a dashed line. For the check point, select $(0, 0)$.

$y < -2$

$0 < -2 \leftarrow$ Substitute 0 for y.

Since this is a false statement, shade the region which does not contain $(0, 0)$.

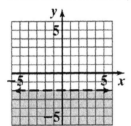

13. $y \geq -\dfrac{1}{2}x$

Graph the line $y = -\dfrac{1}{2}x$ using a solid line. For the check point, select $(0, 2)$.

$y \geq -\dfrac{1}{2}x$

$2 \geq -\dfrac{1}{2}(0) \leftarrow$ Substitute 0 for x and 2 for y

$2 \geq 0$

Since this is a true statement, shade the region which contains the point $(0, 2)$.

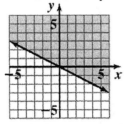

15. $y < 2x + 1$

Graph the line $y = 2x + 1$ using a dashed line. For the check point, select $(0, 0)$.

$y < 2x + 1$

$0 < 2(0) + 1 \leftarrow$ Substitute 0 for x and y

$0 < 1$

Since this is a true statement, shade the region which contains the point $(0, 0)$.

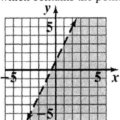

17. $y > 2x - 1$

Graph the line $y = 2x - 1$ using a dashed line. For the check point, select $(0, 0)$.

$y > 2x - 1$

$0 > 2(0) - 1 \leftarrow$ Substitute 0 for x and y

$0 > -1$

Since this is a true statement, shade the region which contains the point $(0, 0)$.

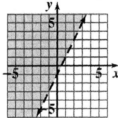

19. $y \geq \frac{1}{2}x - 3$

Graph the line $y = \frac{1}{2}x - 3$ using a solid line. For
the check point, select (0, 0).

$y \geq \frac{1}{2}x - 3$

$0 \geq \frac{1}{2}(0) - 3 \leftarrow$ Substitute 0 for x and y

$0 \geq -3$

Since this is a true statement, shade the region
which contains the point (0, 0).

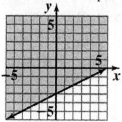

21. $2x + 3y > 6$

Graph the line $2x + 3y = 6$ using a dashed line.
For the check point, select (0, 0).

$2x + 3y > 6$

$2(0) + 3(0) > 6 \leftarrow$ Substitute 0 for x and y

$0 > 6$

Since this is a false statement, shade the region
which does not contain the point (0, 0).

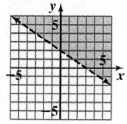

23. $y \leq -3x + 5$

Graph the line $y = -3x + 5$ using a solid line.
For the check point, select (0, 0).

$y \leq -3x + 5$

$0 \leq -3(0) + 5 \leftarrow$ Substitute 0 for x and y

$0 \leq 5$

Since this is a true statement, shade the region
which contains the point (0, 0).

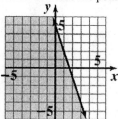

25. $2x + y < 4$

Graph the line $2x + y = 4$ using a dashed line.
For the check point, select (0, 0).

$2x + y < 4$

$2(0) + 0 < 4 \leftarrow$ Substitute 0 for x and y

$0 < 4$

Since this is a true statement, shade the region
which contains the point (0, 0).

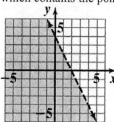

27. $10 \geq 5x - 2y$

Graph the line $10 = 5x - 2y$ using a solid line.

For the check point, select (0, 0).

$10 \geq 5x - 2y$

$10 \geq 5(0) - 2(0) \leftarrow$ Substitute 0 for x and y

$10 \geq 0$

Since this is a true statement, shade the region
which contains the point (0, 0).

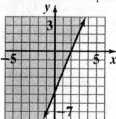

29. $2x - y < 4$

$y \geq -x + 2$

For $2x - y < 4$, graph the line $2x - y = 4$ using a dashed line. For the check point, select $(0, 0)$:

$2x - y < 4$

$2(0) - (0) < 4$

$0 < 4$ True

Since this is a true statement, shade the region which contains the point $(0, 0)$. This is the region "above" the line.

For $y \geq -x + 2$, graph the line

$y = -x + 2$ using a solid line. For the check point, select $(0, 0)$:

$y \geq -x + 2$

$(0) \geq -(0) + 2$

$0 \geq 2$ False

Since this is a false statement, shade the region which does not contain the point $(0, 0)$. This is the region "above" the line. To obtain the final region, take the intersection of the above two regions.

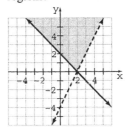

31. $y < 3x - 2$

$y \leq -2x + 3$

For $y < 3x - 2$, graph the line $y = 3x - 2$ using a dashed line. For the check point, select $(0, 0)$:

$y < 3x - 2$

$(0) < 3(0) - 2$

$0 < -2$ False

Since this is a false statement, shade the region which does not contain the point $(0, 0)$. This is the region "below" the line.

For $y \leq -2x + 3$, graph the line $y = -2x + 3$ using a solid line. For the check point, use $(0, 0)$:

$y \leq -2x + 3$

$(0) \leq -2(0) + 3$

$0 \leq 3$ True

Since this is a true statement, shade the region which contains the point $(0, 0)$. This is the region "below" the line. To obtain the final region, take

the intersection of the above two regions.

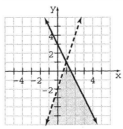

33. $y < x$

$y \geq 3x + 2$

For $y < x$, graph the line $y = x$ using a dashed line. For the check point, select $(1, 0)$:

$y < x$

$0 < 1$ True

Since this is a true statement, shade the region which contains the point $(1, 0)$. This is the region "below" the line.

For $y \geq 3x + 2$, graph the line $y = 3x + 2$ using a solid line. For the check point, select $(0, 0)$:

$y \geq 3x + 2$

$(0) \geq 3(0) + 2$

$0 \geq 2$ False

Since this is a false statement, shade the region which does not contain the point $(0, 0)$. This is the region "above" the line. To obtain the final region, take the intersection of the above two regions.

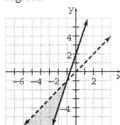

35. $-2x + 3y < -5$

$3x - 8y > 4$

For $-2x + 3y < -5$, graph the line $-2x + 3y = -5$ using a dashed line. For the check point, select $(0, 0)$:

$-2x + 3y < -5$

$-2(0) + 3(0) < -5$

$0 < -5$ False

Since this is a false statement, shade the region which does not contain the point $(0, 0)$. This is the region "below" the line.

For $3x - 8y > 4$, graph the line $3x - 8y = 4$ using

a dashed line. For the check point, select (0, 0):

$3x - 8y > 4$

$3(0) - 8(0) > 4$

$0 > 4$ False

Since this is a false statement, shade the region which does not contain the point (0, 0). This is the region "below" the line. To obtain the final region, take the intersection of the above two regions.

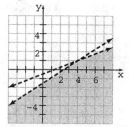

37. $-4x + 5y < 20$

      $x \geq -3$

For $-4x + 5y < 20$, graph the line $-4x + 5y = 20$ using a dashed line. For the check point, select (0, 0):

$-4x + 5y < 20$

$-4(0) + 5(0) < 20$

      $0 < 20$ True

Since this is a true statement, shade the region which contains the point (0, 0). This is the region "below" the line.

For $x \geq -3$, the graph is the line $x = -3$ along with the region to the right of $x = -3$. To obtain the final region, take the intersection of the above two regions.

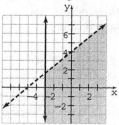

39. $x \leq 4$

      $y \geq -2$

For $x \leq 4$, the graph is the line $x = 4$ along with the region to the left of $x = 4$.

For $y \geq -2$, the graph is the line $y = -2$ along with the region above the line $y = -2$. To obtain the final region, take the intersection of the above two regions.

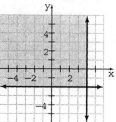

41. $5x + 2y > 10$

      $3x - y > 3$

For $5x + 2y > 10$, graph the line $5x + 2y = 10$ using a dashed line. For the check point, select (0, 0):

$5x + 2y > 10$

$5(0) + 2(0) > 10$

      $0 > 10$ False

Since this is a false statement, shade the region which does not contain the point (0, 0). This is the region "above" the line.

For $3x - y > 3$, graph the line $3x - y = 3$ using a dashed line. For the check point, select (0, 0):

$3x - y > 3$

$3(0) - 0 > 3$

      $0 > 3$ False

Since this is a false statement, shade the region which does not contain the point (0, 0). This is the region "below" the line. To obtain the final region, take the intersection of the above two regions.

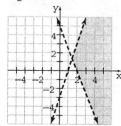

43. $-2x > y + 4$

$$-x < \frac{1}{2}y - 1$$

For $-2x > y + 4$, graph the line $-2x = y + 4$ using a dashed line. For the check point, select $(0, 0)$:

$-2x > y + 4$

$-2(0) > 0 + 4$

$\qquad 0 > 4 \qquad$ False

Since this is a false statement, shade the region which does not contain the point $(0, 0)$. This is the region "below" the line.

For $-x < \frac{1}{2}y - 1$, graph the line $-x = \frac{1}{2}y - 1$ using a dashed line. For the check point, select $(0, 0)$:

$$-x < \frac{1}{2}y - 1$$

$$-0 < \frac{1}{2}(0) - 1$$

$\qquad 0 < -1 \qquad$ False

Since this is a false statement, shade the region which does not contain the point $(0, 0)$. This is the region "above" the line. To obtain the final region take the intersection of the above two regions. Since the regions do not overlap, the final result is the empty set which means there is no solution.

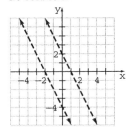

45. $y < 3x - 4$

$6x \geq 2y + 8$

Solve the second inequality for y.

$$6x \geq 2y + 8$$

$6x - 2y > 8$

$\qquad -2y \geq -6x + 8$

$\qquad\quad y \leq 3x - 4$

The second inequality is now identical to the first except that the second inequality includes the line $y = 3x - 4$.

For $y < 3x - 4$, graph the line $y = 3x - 4$ using a dashed line. For the check point, select $(0, 0)$:

$y < 3x - 4$

$0 < 3(0) - 4$

$0 < -4 \qquad$ False

Since this is a false statement, shade the region which does not contain the point $(0, 0)$. This is the region "below" the line.

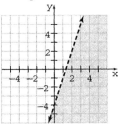

47. $x \geq 0$

$y \geq 0$

$2x + 3y \leq 6$

$4x + y \leq 4$

The first two inequalities indicate that the region must be in the first quadrant. For $2x + 3y \leq 6$, the graph is the line $2x + 3y = 6$ along with the region below this line. For $4x + y \leq 4$, the graph is the line $4x + y = 4$ along with the region below this line.

To obtain the final region, take the intersection of these regions.

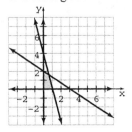

49. $x \geq 0$

$y \geq 0$

$2x + 3y \leq 8$

$4x + 2y \leq 8$

The first two inequalities indicate that the region must be in the first quadrant. For $2x + 3y \leq 8$, the graph is the line $2x + 3y = 8$ along with the region below this line. For $4x + 2y \leq 8$, the graph is the line $4x + 2y = 8$ along with the region below this line. To obtain the final region, take the intersection of these regions.

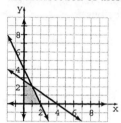

51. $x \geq 0$

$y \geq 0$

$3x + y \leq 9$

$2x + 5y \leq 10$

The first two inequalities indicate that the region must be in the first quadrant. For $3x + y \leq 9$, the graph is the line $3x + y = 9$ along with the region below this line. For $2x + 5y \leq 10$, the graph is the line $2x + 5y = 10$ along with the region below this line. To obtain the final region, take the intersection of these regions.

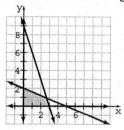

53. $x \geq 0$

$y \geq 0$

$x \leq 4$

$x + y \leq 6$

$x + 2y \leq 8$

The first two inequalities indicate that the region must be in the first quadrant. The third inequality indicates that the region must be on or to the left of the line $x = 4$. For $x + y \leq 6$, the graph is the line $x + y = 6$ along with the region below this

line. For $x + 2y \leq 8$, the graph is the line $x + 2y = 8$ along with the region below the line. To obtain the final region, take the intersection of these regions.

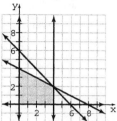

55. $x \geq 0$

$y \geq 0$

$x \leq 15$

$30x + 25y \leq 750$

$10x + 40y \leq 800$

The first two inequalities indicate that the region must be in the first quadrant. The third inequality indicates that the region must be on or to the left of the line $x = 15$. For $30x + 25y \leq 750$ the graph is the line $30x + 25y = 750$ along with the region below this line. For $10x + 40y \leq 800$, the graph is the line $10x + 40y = 800$ along with the region below the line. To obtain the final region, take the intersection of these regions.

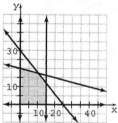

57. $|x| < 2$

For $|x| < 2$, the graph is the region between the dashed lines $x = -2$ and $x = 2$.

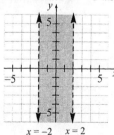

59. $|y-2| \le 4$

$-4 \le y-2 \le 4$

$-2 \le y \le 6$

For $|y-2| \le 4$, the graph is the solid lines $y = -2$ and $y = 6$ along with the region between these lines.

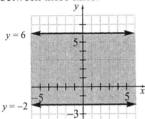

61. $|y| > 2$

$y \le x+3$

For $|y| > 2$, the graph is the region above the dashed line $y = 2$ along with the region below the dashed line $y = -2$. For $y \le x + 3$, the graph is the region below the solid line $y = x + 3$. To obtain the final region, take the intersection of these regions.

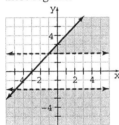

63. $|y| < 4$

$y \ge -2x+2$

For $|y| < 4$, the graph is the region between the dashed lines $y = -4$ and $y = 4$. For $y \ge -2x + 2$, the graph is the region above the solid line $y = -2x + 2$. To obtain the final region, take the intersection of these regions.

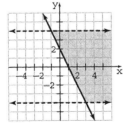

65. $|x+2| < 3$

$|y| > 4$

$|x+2| < 3$ can be written as

$-3 < x+2 < 3$

$-5 < x < 1$

For $|x+2| < 3$, the graph is the region between the dashed lines $x = -5$ and $x = 1$. For $|y| > 4$, the graph is the region above the dashed line $y = 4$ along with the region below the dashed line $y = -4$. To obtain the final region, take the intersection of these regions.

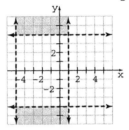

67. $|x-3| \le 4$

$|y+2| \le 1$

$|x-3| \le 4$ can be written as

$-4 \le x-3 \le 4$

$-1 \le x \le 7$

For $|x-3| \le 4$, the graph is the region between the solid lines $x = -1$ and $x = 7$.

$|y+2| \le 1$ can be written as

$-1 \le y+2 \le 1$

$-3 \le y \le -1$

For $|y+2| \le 1$, the graph is the region between the solid lines $y = -3$ and $y = -1$. To obtain the final region, take the intersection of these regions.

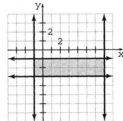

69. a,b.

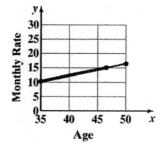

c. The age at which the rate first exceeds $15 per month is 47.

71. a,b.

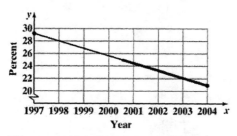

c. The year 2003 is first year in which the percentage of Americans 18 and older who smoke was less than or equal to 23%.

73. a. Region A; we are looking for the region that lies below the graph for the paper returns and above the graph for the electronic returns.

b. Region B; we are looking for the region that lies above the graph for the paper returns and below the graph for the electronic returns.

75. a.

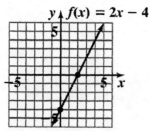

b.

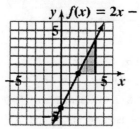

77. Yes; if the boundary lines are parallel, there may be no solution. For example, the system $y < x$ and $y > x + 1$ has no solution.

79. There are no solutions. Opposite sides of the same line are being shaded and only one inequality includes the line.

81. There are an infinite number of solutions. Both inequalities include the line $5x - 2y = 3$.

83. There are an infinite number of solutions. The lines are not parallel or identical.

85. $y \geq x^2$

$y \leq 4$

For $y \geq x^2$, graph the equation $y = x^2$ using a solid line. For the check point, select (0, 1).

$y \geq x^2$

$(1) \geq (0)^2$

$1 \geq 0$ True

Since this is a true statement, shade the region which contains the point (0, 1). This is the region "above" the graph of $y \geq x^2$.

For $y \leq 4$, the graph is the region below the solid line $y = 4$. To obtain the final region, take the intersection of these regions.

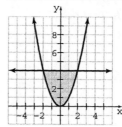

87. $y < |x|$

$y < 4$

For $y < |x|$, graph the equation $y = |x|$ using a dashed line. For the check point, select (0, 3).

$y < |x|$

$3 < |0|$

$3 < 0$ False

Since this is a false statement, shade the region which does not contain the point (0, 3). This is the region below the graph of $y = |x|$.

For $y < 4$, the graph is the region below the dashed line $y = 4$. To obtain the final region, take the intersection of these regions.

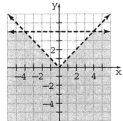

89. $f_1 d_1 + f_2 d_2 = f_3 d_3$

$f_1 d_1 - f_1 d_1 + f_2 d_2 = f_3 d_3 - f_1 d_1$

$f_2 d_2 = f_3 d_3 - f_1 d_1$

$\dfrac{f_2 d_2}{d_2} = \dfrac{f_3 d_3 - f_1 d_1}{d_2}$

$f_2 = \dfrac{f_3 d_3 - f_1 d_1}{d_2}$

90. Domain: {−1, 0, 4, 5}
Range: {−5, −2, 2, 3}

91. Domain: $\mathbb{R}$
Range: $\mathbb{R}$

92. Domain: $\mathbb{R}$
Range: $\left\{ y \mid y \geq -1 \right\}$

Chapter 10 Review Exercises

1. $3z + 9 \leq 15$

$3z \leq 6$

$z \leq 2$

2. $8 - 2w > -4$

$-2w > -12$

$\dfrac{-2w}{-2} < \dfrac{-12}{-2}$

$w < 6$

3. $2x + 1 > 6$

$2x > 5$

$x > \dfrac{5}{2}$

4. $26 \leq 4x + 5$

$21 \leq 4x$

$\dfrac{21}{4} \leq x$

5. $\dfrac{4x + 3}{3} > -5$

$3 \left(\dfrac{4x + 3}{3} \right) > 3(-5)$

$4x + 3 > -15$

$4x > -18$

$x > \dfrac{-18}{4}$

$x > -\dfrac{9}{2}$

6. $2(x - 1) > 3x + 8$

$2x - 2 > 3x + 8$

$2x - 10 > 3x$

$-10 > x$

7. $-4(x - 2) \geq 6x + 8 - 10x$

$-4x + 8 \geq -4x + 8$

$8 \geq 8$ a true statement

The solution is all real numbers.

364

8. $\dfrac{x}{2}+\dfrac{3}{4} > x-\dfrac{x}{2}+1$

$4\left(\dfrac{x}{2}+\dfrac{3}{4}\right) > 4\left(x-\dfrac{x}{2}+1\right)$

$2x+3 > 4x-2x+4$

$2x+3 > 2x+4$

$3 > 4$

This is a contradiction, so the solution is { }.

9. Let x be the maximum number of 40-pound boxes. Since the maximum load is 560 pounds, the total weight of Bob, Kathie, and the boxes must be less than or equal to 560 pounds.

$300+40x \le 560$

$40x \le 260$

$x \le \dfrac{260}{40}$

$x \le 6.5$

The maximum number of boxes that they can carry in the canoe is 6.

10. Let x be the number of additional minutes (beyond 3 minutes) of the phone call.

$4.50+0.95x \le 8.65$

$0.95x \le 4.15$

$x \le \dfrac{4.15}{0.95}$

$x \le 4.4$

The customer can talk for 3 minutes plus an additional 4 minutes for a total of 7 minutes.

11. Let x be the number of weeks (after the first week) needed to lose 27 pounds.

$5+1.5x \ge 27$

$1.5x \ge 22$

$x \ge \dfrac{22}{1.5}$

$x \ge 14\dfrac{2}{3} \approx 14.67$

The number of weeks is about 14.67 plus the initial week for a total of 15.67 weeks.

12. Let x be the grade from the 5th exam. The inequality is

$80 \le \dfrac{94+73+72+80+x}{5} < 90$

$80 \le \dfrac{319+x}{5} < 90$

$5(80) \le 5\left(\dfrac{319+x}{5}\right) < 5(90)$

$400 \le 319+x < 450$

$400-319 \le 319+x < 450-319$

$81 \le x < 131$

(must use 100 here since it is not possible to score 131)

Thus, Patrice needs to score 81 or higher on the 5th exam to receive a B.

$\{x \mid 81 \le x \le 100\}$

13. $A \cup B = \{1,2,3,4,5\}$

$A \cap B = \{2,3,4,5,\}$

14. $A \cup B = \{2,3,4,5,6,7,8,9\}$

$A \cap B = \{\quad\}$

15. $A \cup B = \{1,2,3,4,...\}$

$A \cap B = \{2,4,6,...\}$

16. $A \cup B = \{3,4,5,6,9,10,11,12\}$

$A \cap B = \{9,10\}$

17. $1 < x-4 < 7$

$1+4 < x-4+4 < 7+4$

$5 < x < 11$

$(5, 11)$

18. $8 < p+11 \le 16$

$8-11 < p+11-11 \le 16-11$

$-3 < p \le 5$

$\left(-3, 5\right]$

19. $3 < 2x - 4 < 12$

$3 + 4 < 2x - 4 + 4 < 12 + 4$

$7 < 2x < 16$

$\dfrac{7}{2} < \dfrac{2x}{2} < \dfrac{16}{2}$

$\dfrac{7}{2} < x < 8$

$\left(\dfrac{7}{2}, 8 \right)$

20. $-12 < 6 - 3x < -2$

$-12 - 6 < 6 - 6 - 3x < -2 - 6$

$18 < -3x < -8$

$\dfrac{-18}{-3} > \dfrac{-3x}{-3} > \dfrac{-8}{-3}$

$6 > x > \dfrac{8}{3}$

$\dfrac{8}{3} < x < 6$

$\left(\dfrac{8}{3}, 6 \right)$

21. $-1 < \dfrac{5}{9}x + \dfrac{2}{3} \le \dfrac{11}{9}$

$9(-1) < 9\left(\dfrac{5}{9}x + \dfrac{2}{3} \right) \le 9\left(\dfrac{11}{9} \right)$

$-9 < 5x + 6 \le 11$

$-9 - 6 < 5x + 6 - 6 \le 11 - 6$

$-15 < 5x \le 5$

$\dfrac{-15}{5} < \dfrac{5x}{5} \le \dfrac{5}{5}$

$-3 < x \le 1$

$\left(-3, 1 \right]$

22. $-8 < \dfrac{4 - 2x}{3} < 0$

$3(-8) < 3\left(\dfrac{4 - 2x}{3} \right) < 3(0)$

$-24 < 4 - 2x < 0$

$-24 - 4 < 4 - 4 - 2x < 0 - 4$

$-28 < -2x < -4$

$\dfrac{-28}{-2} > \dfrac{-2x}{-2} > \dfrac{-4}{-2}$

$14 > x > 2$

$2 < x < 14$

$(2, 14)$

23. $h \le 1$ and $7h - 4 > -25$

$h \le 1$ and $7h > -21$

$h \le 1$ and $h > -3$

$h \le 1$

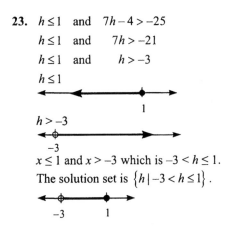

$h > -3$

$x \le 1$ and $x > -3$ which is $-3 < h \le 1$.

The solution set is $\{ h \mid -3 < h \le 1 \}$.

24. $2x - 1 > 5$ or $3x - 2 \le 10$

$2x > 6$ or $3x \le 12$

$x > 3$ or $x \le 4$

$x > 3$

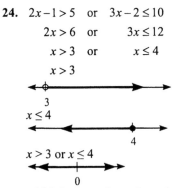

$x \le 4$

$x > 3$ or $x \le 4$

which is the entire real number line or $\mathbb{R}$.

25. $4x - 5 < 11$ and $-3x - 4 \geq 8$

$\quad\quad 4x < 16$ and $\quad -3x \geq 12$

$\quad\quad\;\; x < 4$ and $\quad\quad x \leq -4$

$x \leq 4$

$x \leq -4$

$x \leq -4$ and $x < 4$ which is $x \leq -4$

$\{x \mid x \leq -4\}$

26. $\dfrac{7 - 2g}{3} \leq -5$ or $\dfrac{3 - g}{9} > 1$

$\quad 7 - 2g \leq -15$ or $\quad 3 - g > 9$

$\quad\;\; -2g \leq -22$ or $\quad\; -g > 6$

$\quad\quad\;\; g \geq 11$ or $\quad\quad g < -6$

$g < -6$

$g \geq 11$

$g < -6$ or $g \geq 11$

$\left\{ g \mid g < -6 \;\text{ or }\; g \geq 11 \right\}$

27. $|a| = 2$

$a = 2$ or $a = -2$

The solution set is $\{-2, 2\}$.

28. $|x| < 8$

$-8 < x < 8$

The solution set is $\{x \mid -8 < x < 8\}$.

29. $|x| \geq 9$

$x \leq -9$ or $x \geq 9$

The solution set is $\{x \mid x \leq -9 \text{ or } x \geq 9\}$.

30. $|l + 5| = 13$

$l + 5 = -13$ or $l + 5 = 13$

$\quad\quad l = -18$ $\quad\quad l = 8$

The solution set is $\{-18, 8\}$.

31. $|x - 2| \geq 5$

$x - 2 \leq -5$ or $x - 2 \geq 5$

$\quad\; x \leq -3$ $\quad\quad\quad x \geq 7$

The solution set is $\{x \mid x \leq -3 \text{ or } x \geq 7\}$.

32. $|4 - 2x| = 5$

$4 - 2x = 5$ or $4 - 2x = -5$

$\quad -2x = 1$ $\quad\quad -2x = -9$

$\quad\quad x = \dfrac{1}{-2}$ $\quad\quad\; x = \dfrac{-9}{-2}$

$\quad\quad x = -\dfrac{1}{2}$ $\quad\quad\; x = \dfrac{9}{2}$

The solution set is $\left\{ -\dfrac{1}{2}, \dfrac{9}{2} \right\}$.

33. $|-2q + 9| < 7$

$\quad -7 < -2q + 9 < 7$

$-7 - 9 < -2q + 9 - 9 < 7 - 9$

$\quad\quad -16 < -2q < -2$

$\quad\quad \dfrac{-16}{-2} > \dfrac{-2q}{-2} > \dfrac{-2}{-2}$

$\quad\quad\quad\; 8 > q > 1$

$\quad\quad\quad\; 1 < q < 8$

The solution set is $\left\{ q \mid 1 < q < 8 \right\}$.

34. $\left| \dfrac{2x - 3}{5} \right| = 1$

$\dfrac{2x - 3}{5} = 1$ or $\dfrac{2x - 3}{5} = -1$

$2x - 3 = 5$ $\quad\quad 2x - 3 = -5$

$\quad 2x = 8$ $\quad\quad\quad 2x = -2$

$\quad\;\; x = 4$ $\quad\quad\quad\; x = -1$

The solution set is $\{-1, 4\}$.

35. $\left| \dfrac{x - 4}{3} \right| < 6$

$-6 < \dfrac{x - 4}{3} < 6$

$3(-6) < 3\left(\dfrac{x - 4}{3} \right) < 3(6)$

$-18 < x - 4 < 18$

$-14 < x < 22$

The solution set is $\{x \mid -14 < x < 22\}$.

36. $|4d - 1| = |6d + 9|$

$$4d - 1 = -(6d + 9) \quad \text{or} \quad 4d - 1 = 6d + 9$$
$$4d - 1 = -6d - 9 \qquad\qquad 4d - 10 = 6d$$
$$10d - 1 = -9 \qquad\qquad\quad -10 = 2d$$
$$10d = -8 \qquad\qquad\qquad -5 = d$$
$$d = -\frac{4}{5}$$

The solution set is $\left\{-5, -\dfrac{4}{5}\right\}$.

37. $|2x - 3| + 4 \geq -17$

$$|2x - 3| \geq -21$$

Since the right side is negative and the left side is non-negative, the solution is the entire real number line since the absolute value of a number is always greater than a negative number. The solution set is all real numbers, or $\mathbb{R}$.

38. $|3c + 8| - 6 \leq 1$

$$|3c + 8| \leq 7$$
$$-7 \leq 3c + 8 \leq 7$$
$$-7 - 8 \leq 3c + 8 - 8 \leq 7 - 8$$
$$-15 \leq 3c \leq -1$$
$$\frac{-15}{3} \leq \frac{3c}{3} \leq \frac{-1}{3}$$
$$-5 \leq c \leq -\frac{1}{3}$$
$$\left[-5, -\frac{1}{3}\right]$$

39. $3 < 2x - 5 \leq 11$

$$3 + 5 < 2x - 5 + 5 \leq 11 + 5$$
$$8 < 2x \leq 16$$
$$\frac{8}{2} < \frac{2x}{2} \leq \frac{16}{2}$$
$$4 < x \leq 8$$

The solution is (4, 8].

40. $-6 \leq \dfrac{3 - 2x}{4} < 5$

$$4(-6) \leq 4\left(\frac{3 - 2x}{4}\right) < 4(5)$$
$$-24 \leq 3 - 2x < 20$$
$$-27 \leq -2x < 17$$
$$\frac{-27}{-2} \geq \frac{-2x}{-2} > \frac{17}{-2}$$
$$\frac{27}{2} \geq x > -\frac{17}{2}$$
$$-\frac{17}{2} < x \leq \frac{27}{2}$$

The solution is $\left(-\dfrac{17}{2}, \dfrac{27}{2}\right]$.

41. $2p - 5 < 7 \quad \text{or} \quad 9 - 3p \leq 15$

$$2p < 12 \quad \text{or} \quad -3p \leq 6$$
$$p < 6 \quad \text{or} \quad p \geq -2$$
$$-2 \leq p < 6$$

The solution is $[-2, 6)$.

42. $\quad x - 3 \leq 4 \qquad \text{or} \qquad 2x - 5 > 7$

$$x - 3 + 3 \leq 4 + 3 \qquad 2x - 5 + 5 > 7 + 5$$
$$x \leq 7 \qquad\qquad\qquad 2x > 12$$
$$x > 6$$

The solution is $(-\infty, \infty)$.

43. $-10 < 3(x - 4) \leq 18$

$$-10 < 3x - 12 \leq 18$$
$$-10 + 12 < 3x - 12 + 12 \leq 18 + 12$$
$$2 < 3x \leq 30$$
$$\frac{2}{3} < x \leq 10$$

The solution is $\left(\dfrac{2}{3}, 10\right]$.

44. $y \geq -5$

Graph the line $y = -5$ using a solid line. For the check point, select $(0, 0)$.

$y \geq -5$

$0 \geq -5$ ← Substitute 0 for y

Since this is a true statement, shade the region which contains $(0, 0)$.

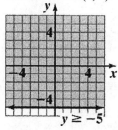

45. $x < 4$

Graph the line $x = 4$ using a dashed line. For the check point, select $(0, 0)$.

$x < 4$

$0 < 4$ ← Substitute 0 for x

Since this is a true statement, shade the region which contains $(0, 0)$.

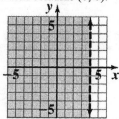

46. $y \leq 4x - 3$

Graph the line $y = 4x - 3$ using a solid line. For the check point, select $(0, 0)$.

$y \leq 4x - 3$

$0 \leq 4(0) - 3$ ← Substitute 0 for x and y

$0 \leq -3$

Since this is a false statement, shade the region which does not contain $(0, 0)$.

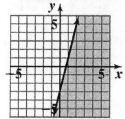

47. $y < \frac{1}{3}x - 2$

Graph the line $y = \frac{1}{3}x - 2$ using a dashed line.

For the check point, select $(0, 0)$.

$y < \frac{1}{3}x - 2$

$0 < \frac{1}{3}(0) - 2$ ← Substitute 0 for x and y

$0 < -2$

Since this is a false statement, shade the region which does not contain $(0, 0)$.

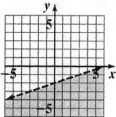

48. $-x + 3y > 6$

$2x - y \leq 2$

For $-x + 3y > 6$, graph the line $-x + 3y = 6$ using a dashed line. For the check point, select $(0, 0)$:

$-x + 3y > 6$

$-0 + 3(0) > 6$

$\quad 0 > 6 \qquad$ False

Since this is a false statement, shade the region which does not contain the point $(0, 0)$.

For $2x - y \leq 2$, graph the line $2x - y = 2$ using a solid line. For the check point, select $(0, 0)$:

$2x - y \leq 2$

$2(0) - 0 \leq 2$

$\quad 0 \leq 2 \qquad$ True

Since this is a true statement, shade the region which contains the point $(0, 0)$.

To obtain the final region, take the intersection of the above two regions.

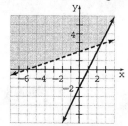

49. $5x - 2y \le 10$

$3x + 2y > 6$

For $5x - 2y \le 10$, graph the line $5x - 2y = 10$ using a solid line.
For the check point, select (0, 0):

$5x - 2y \le 10$

$5(0) - 2(0) \le 10$

$0 \le 10$ True

Since this is a true statement, shade the region which contains the point (0, 0).
For $3x + 2y > 6$, graph the line $3x + 2y = 6$ using a dashed line. For the check point, select (0, 0):

$3x + 2y > 6$

$3(0) + 2(0) > 6$

$0 > 6$ False

Since this is a false statement, shade the region which does not contain the point (0, 0). To obtain the final region, take the intersection of the above two regions.

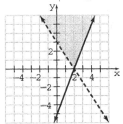

50. $y > 2x + 3$

$y < -x + 4$

For $y > 2x + 3$, graph the line $y = 2x + 3$ using a dashed line. For the check point, select (0, 0):

$y > 2x + 3$

$0 > 2(0) + 3$

$0 > 3$ False

Since this is a false statement, shade the region which does not contain the point (0, 0).
For $y < -x + 4$, graph the line
$y = -x + 4$ using a dashed line. For the check point, select (0, 0):

$y < -x + 4$

$0 < -0 + 4$

$0 < 4$ True

Since this is a true statement, shade the region which contains the point (0, 0). To obtain the final region, take the intersection of the above

two regions.

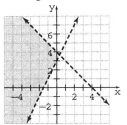

51. $x > -2y + 4$

$y < -\dfrac{1}{2}x - \dfrac{3}{2}$

For $x > -2y + 4$, graph the line $x = -2y + 4$ using a dashed line. For the check point, select (0, 0):

$x > -2y + 4$

$0 > -2(0) + 4$

$0 > 4$ False

Since this is a false statement, shade the region which does not contain the point (0, 0).

For $y < -\dfrac{1}{2}x - \dfrac{3}{2}$, graph the line $y = -\dfrac{1}{2}x - \dfrac{3}{2}$

using a dashed line. For the check point, select (0, 0):

$y < -\dfrac{1}{2}x - \dfrac{3}{2}$

$0 < \dfrac{1}{2}(0) \quad \dfrac{3}{2}$

$0 < -\dfrac{3}{2}$ False

Since this is a false statement, shade the region which does not contain the point (0, 0). To obtain the final region, take the intersection of the above two regions. The regions do not overlap, so there are no solutions.

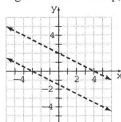

52. $x \geq 0$

$y \geq 0$

$x + y \leq 6$

$4x + y \leq 8$

The first two inequalities indicate that the solution must be in the first quadrant. For $x + y \leq 6$, the graph is the line $x + y = 6$ along with the region below this line. For $4x + y \leq 8$, the graph is the line $4x + y = 8$ along with the region below this line. To obtain the final region, take the intersection of these regions.

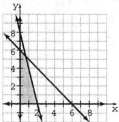

53. $x \geq 0$

$y \geq 0$

$2x + y \leq 6$

$4x + 5y \leq 20$

The first two inequalities indicate that the solution must be in the first quadrant. For $2x + y \leq 6$, graph the line $2x + y = 6$ along with the region below this line. For $4x + 5y \leq 20$, graph the line $4x + 5y = 20$ along with the region below this line. To obtain the final region, take the intersection of these regions.

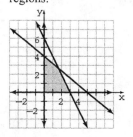

54. $|x| \leq 3$

$|y| > 2$

For $|x| \leq 3$, the graph is the region between the solid lines $x = -3$ and $x = 3$. For $|y| > 2$, the graph is the region above the dashed line $y = 2$ along with the region below the dashed line $y = -2$. To obtain the final region, take the intersection of these regions.

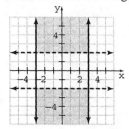

55. $|x| > 4$

$|y - 2| \leq 3$

For $|x| > 4$, the graph is the region to the left of dashed line $x = -4$ along with the region to the right of the dashed line $x = 4$.

$|y - 2| \leq 3$ can be written as

$-3 \leq y - 2 \leq 3$

$-1 \leq y \leq 5$

For $|y - 2| \leq 3$, the graph is the region between the solid lines $y = -1$ and $y = 5$. To obtain the final region, take the intersection of these regions.

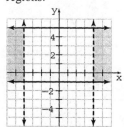

Chapter 10 Practice Test

1. $A \cup B = \{5, 7, 8, 9, 10, 11, 14\}$

$A \cap B = \{8, 10\}$

2. $A \cup B = \{1, 3, 5, 7, \ldots\}$

$A \cap B = \{3, 5, 7, 9, 11\}$

3. $3(2q+4) < 5(q-1)+7$

$\quad 6q+12 < 5q-5+7$

$\quad 6q+12 < 5q+2$

$\quad q+12 < 2$

$\quad\quad q < -10$

$\quad\quad\quad -10$

4. $\dfrac{6-2x}{5} \geq -12$

$\quad 5\left(\dfrac{6-2x}{5}\right) \geq 5(-12)$

$\quad\quad 6-2x \geq -60$

$\quad\quad -2x \geq -66$

$\quad\quad \dfrac{-2x}{-2} \leq \dfrac{-66}{-2}$

$\quad\quad\quad x \leq 33$

$\quad\quad\quad 33$

5. $\quad x-3 \leq 4 \quad$ and $\quad 2x+1 > 10$

$\quad x-3+3 \leq 4+3 \quad\quad 2x+1-1 > 10-1$

$\quad\quad\quad x \leq 7 \quad\quad\quad\quad 2x > 9$

$\quad\quad\quad\quad\quad\quad\quad\quad\quad\quad x > \dfrac{9}{2}$

The solution is $\left(\dfrac{9}{2}, 7\right]$.

6. $\quad 7 \leq \dfrac{2u-5}{3} < 9$

$\quad 3(7) \leq 3\left(\dfrac{2u-5}{3}\right) < 3(9)$

$\quad\quad 21 \leq 2u-5 < 27$

$\quad 21+5 \leq 2u-5+5 < 27+5$

$\quad\quad 26 \leq 2u < 32$

$\quad\quad 13 \leq u < 16$

The solution is $[13, 16)$.

7. $|2b+5| = 9$

$\quad 2b+5 = -9 \quad$ or $\quad 2b+5 = 9$

$\quad 2b = -14 \quad\quad\quad 2b = 4$

$\quad\quad b = -7 \quad\quad\quad\quad b = 2$

The solution set is $\{-7, 2\}$.

8. $|2x-3| = \left|\dfrac{1}{2}x-10\right|$

$\quad 2x-3 = -\left(\dfrac{1}{2}x-10\right) \quad$ or $\quad 2x-3 = \dfrac{1}{2}x-10$

$\quad 2x-3 = -\dfrac{1}{2}x+10 \quad\quad\quad \dfrac{3}{2}x-3 = -10$

$\quad \dfrac{5}{2}x-3 = 10 \quad\quad\quad\quad\quad \dfrac{3}{2}x = -7$

$\quad \dfrac{5}{2}x = 13 \quad\quad\quad\quad \dfrac{2}{3}\left(\dfrac{3}{2}x\right) = \dfrac{2}{3}(-7)$

$\quad \dfrac{2}{5}\left(\dfrac{5}{2}x\right) = \dfrac{2}{5}(13) \quad\quad\quad x = -\dfrac{14}{3}$

$\quad\quad x = \dfrac{26}{5}$

The solution set is $\left\{-\dfrac{14}{3}, \dfrac{26}{5}\right\}$.

9. $|4z+12| = 0$

$\quad 4z+12 = 0$

$\quad\quad 4z = -12$

$\quad\quad\quad z = -3$

The solution set is $\{-3\}$.

10. $|2x-3|+6 > 11$

$\quad\quad |2x-3| > 5$

$\quad 2x-3 < -5 \quad$ or $\quad 2x-3 > 5$

$\quad\quad 2x < -2 \quad\quad\quad\quad 2x > 8$

$\quad\quad\quad x < -1 \quad\quad\quad\quad x > 4$

The solution set is $\{x \mid x < -1 \text{ or } x > 4\}$.

11. $\left|\dfrac{2x-3}{8}\right| \leq \dfrac{1}{4}$

$\quad -\dfrac{1}{4} \leq \dfrac{2x-3}{8} \leq \dfrac{1}{4}$

$\quad 8\left(-\dfrac{1}{4}\right) \leq 8\left(\dfrac{2x-3}{8}\right) \leq 8\left(\dfrac{1}{4}\right)$

$\quad\quad -2 \leq 2x-3 \leq 2$

$\quad\quad\quad 1 \leq 2x \leq 5$

$\quad\quad\quad \dfrac{1}{2} \leq x \leq \dfrac{5}{2}$

The solution set is $\left\{x \mid \dfrac{1}{2} \leq x \leq \dfrac{5}{2}\right\}$.

12. Graph the line $y = 3x - 2$ using a dashed line. For the check point, select $(0, 0)$.

$y < 3x - 2$

$0 < 3 \cdot 0 - 2 \leftarrow$ Substitute 0 for x and y

$0 < -2$

Since the statement is false, shade the region that does not contain $(0, 0)$.

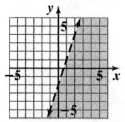

13. $3x + 2y < 9$

$-2x + 5y \leq 10$

For $3x + 2y < 9$, graph the line $3x + 2y = 9$ using a dashed line. For the check point, select $(0, 0)$.

$3x + 2y < 9$

$3(0) + 2(0) < 9$

$\qquad 0 < 9 \qquad$ True

Since this is a true statement, shade the region which contains the point $(0, 0)$. this is the region "below" the line.

For $-2x + 5y \leq 10$, graph the line $-2x + 5y = 10$ using a solid line. For the check point, select $(0, 0)$.

$-2x + 5y \leq 10$

$-2(0) + 5(0) \leq 10$

$\qquad 0 \leq 10 \qquad$ True

Since this is a true statement, shade the region which contains the point $(0, 0)$. this is the region "below" the line. To obtain the final region, take the intersection of the above two regions.

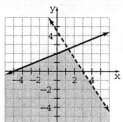

14. $|x| > 3$

$|y| \leq 1$

For $|x| > 3$, the graph is the region to the left of the dashed line $x = -3$ along with the region to the right of the dashed line $x = 3$.

For $|y| \leq 1$, the graph is the region between the solid lines $y = -1$ and $y = 1$. To obtain the final region, take the intersection of these regions.

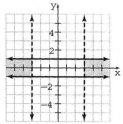

Chapter 10 Cumulative Review Test

1. $-40 - 3(4 - 8)^2$

$-40 - 3(-4)^2$

$-40 - 3(16)$

$-40 - 48$

-88

2. $2(x + 2y) + 4x - 5y$ for $x = 2$ and $y = 3$:

$2(2 + 2(3)) + 4(2) - 5(3)$

$2(2 + 6) + 8 - 15$

$2(8) - 7$

$16 - 7$

9

3. $\dfrac{1}{2}(x + 3) + \dfrac{1}{3}(3x + 6)$

$\dfrac{1}{2}x + \dfrac{3}{2} + x + 2$

$\dfrac{3}{2}x + \dfrac{7}{2}$

4.
$$\frac{a-7}{3} = \frac{a+5}{2} - \frac{7a-1}{6}$$

$$6\left(\frac{a-7}{3}\right) = 6\left(\frac{a+5}{2} - \frac{7a-1}{6}\right)$$

$$2(a-7) = 3(a+5) - (7a-1)$$

$$2a - 14 = 3a + 15 - 7a + 1$$

$$2a - 14 = -4a + 16$$

$$6a = 30$$

$$a = 5$$

5.
$$\frac{40}{5000} = \frac{x}{26,000}$$

$$x = \frac{40(26,000)}{5000} = 208$$

208 pounds of fertilizer are needed.

6. Let x = the amount of assets managed.

$$1000 + 0.01x = 500 + 0.02x$$

$$500 = 0.01x$$

$$\frac{500}{0.01} = x$$

$$50,000 = x$$

The plans will have the same total fee if she manages total assets of \$50,000.

7. Let p – the price of the mixture.

$$\text{Cost}_{\text{mix}} = \text{Cost}_{\text{sunflower}} + \text{Cost}_{\text{corn}}$$

$$3.5p = 1.80(2.5) + 1.40(1)$$

$$3.5p = 4.50 + 1.40$$

$$3.5p = 5.90$$

$$p \approx 1.69$$

The mixture should be sold for \$1.69 per pound.

8.
$$\left(\frac{4a^3b^{-2}}{2a^{-2}b^{-3}}\right)^{-2} = \left(2a^5b^1\right)^{-2} = \frac{1}{\left(2a^5b\right)^2} = \frac{1}{4a^{10}b^2}$$

9.
$$(0.003)(0.00015) = \left(3.0 \times 10^{-3}\right)\left(1.5 \times 10^{-4}\right)$$
$$= (3)(1.5) \times 10^{-3} \cdot 10^{-4}$$
$$= 4.5 \times 10^{-7}$$

10.
$$24p^3q + 16p^2q - 30pq$$
$$= 2pq\left(12p^2 + 8p - 15\right)$$
$$= 2pq(2p+3)(6p-5)$$

11.
$$\frac{2b}{b+1} = 2 - \frac{5}{2b}$$

$$\frac{2b}{b+1} = \frac{4b-5}{2b}$$

$$4b^2 = (b+1)(4b-5)$$

$$4b^2 = 4b^2 - b - 5$$

$$b = -5$$

12.
$$\frac{4}{r+5} + \frac{1}{r+3} = \frac{2}{r^2+8r+15}$$

LCD: $(r+5)(r+3)$ Domain: $x \neq -5, -3$

$$(r+5)(r+3)\left(\frac{4}{r+5} + \frac{1}{r+3}\right) = (r+5)(r+3)\frac{2}{r^2+8r+15}$$

$$4(r+3) + 1(r+5) = 2$$

$$4r + 12 + r + 5 = 2$$

$$5r + 17 = 2$$

$$5r = -15$$

$$r = -3$$

This value makes a denominator equal 0, so it must be excluded. Therefore, there is no solution to the equation.

13. $y = x^2 - 2$

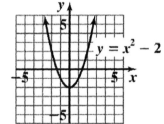

14. a. Function; each input corresponds to exactly one output.

b. Not a function; fails the vertical line test.

c. Function; passes the vertical line test.

15. $3x - 5y = 7$

$-5y = -3x + 7$

$y = \dfrac{3}{5}x - \dfrac{7}{5} \quad \rightarrow \quad \text{slope} = \dfrac{3}{5}$

Since parallel lines have the same slope, the slope of our line must also be $\dfrac{3}{5}$.

Using the slope, $m = \dfrac{3}{5}$, and the y-intercept, $(0, -2)$, the equation of our line is

$y = \dfrac{3}{5}x - 2$.

16. $4x - 3y = 12$

$-3y = -4x + 12$

$y = \dfrac{4}{3}x - 4 \quad \rightarrow \quad \text{slope} = \dfrac{4}{3}$

Since perpendicular lines have opposite-reciprocal slopes, the slope of our line must be $m = -\dfrac{3}{4}$.

$y - y_1 = m(x - x_1)$

$y - (-2) = -\dfrac{3}{4}(x - 3)$

$y + 2 = -\dfrac{3}{4}x + \dfrac{9}{4}$

$y = -\dfrac{3}{4}x + \dfrac{1}{4}$

17. a. $(f \cdot g)(x) = (x^2 - 11x + 30)(x - 5)$

$= x^3 - 5x^2 - 11x^2 + 55x + 30x - 150$

$= x^3 - 16x^2 + 85x - 150$

b. $(f / g)(x) = \dfrac{x^2 - 11x + 30}{x - 5}$

$= \dfrac{(x - 6)(x - 5)}{x - 5}$

$= x - 6$

c. Since $g(x) = 0$ when $x = 5$, this value must be excluded from the domain of (f / g). Therefore, the domain is $\{x \mid x \neq 5\}$.

18. $2a - b - 2c = -1$ (1)

   $a - 2b - c = 1$ (2)

   $a + b + c = 4$ (3)

Multiply equation (2) by -2 and add to equation (1).

$2a - b - 2c = -1$

$\underline{-2a + 4b + 2c = -2}$

       $3b = -3$

        $b = -1$

Substitute -1 for b in equations (2) and (3).

$a - 2(-1) - c = 1 \quad \rightarrow \quad a - c = -1$

$a + (-1) + c = 4 \quad \rightarrow \quad a + c = 5$

Add the resulting equations.

$a - c = -1$

$\underline{a + c = 5}$

  $2a = 4$

   $a = 2$

Substitute -1 for b and 2 for a in equation (3).

$2 + (-1) + c = 4$

       $c = 3$

The solution is $(2, -1, 3)$.

19. $\dfrac{1}{2} < 3x + 4 < 6$

$-\dfrac{7}{2} < 3x < 2$

$-\dfrac{7}{6} < x < \dfrac{2}{3}$

Solution: $\left(-\dfrac{7}{6}, \dfrac{2}{3} \right)$

20. $y < 2x - 3$

Graph the line $y = 2x - 3$ with a dashed line since the inequality is strict. Select a test point, such as $(0, 0)$.

$0 < 2(0) - 3$

$0 < -3$ false

Since this resulted in a false statement, shade the half-plane that does not contain the point $(0, 0)$.

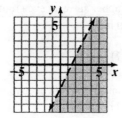

Chapter 11

1. a. Every real number has two square roots; a positive, or principal, square root and a negative square root.

   b. The square roots of 49 are 7 and –7.

   c. When we say square root, we are referring to the principal square root.

   d. $\sqrt{49} = 7$

3. There is no real number which, when squared, results in –81.

5. No. If the number under the radical is negative, the answer is not a real number.

7. a. $\sqrt{(1.3)^2} = \sqrt{1.69} = 1.3$

   b. $\sqrt{(-1.3)^2} = \sqrt{1.69} = 1.3$

9. a. $\sqrt[3]{27} = 3$ since $3^3 = 27$

   b. $-\sqrt[3]{27} = -3$

   c. $\sqrt[3]{-27} = -3$ since $(-3)^3 = -27$

11. $\sqrt{36} = 6$ since $6^2 = 36$

13. $\sqrt[3]{-64} = -4$ since $(-4)^3 = -64$

15. $\sqrt[3]{-125} = -5$ since $(-5)^3 = -125$

17. $\sqrt[5]{-1} = -1$ since $(-1)^5 = -1$

19. $\sqrt[5]{1} = 1$ since $1^5 = 1$

21. $\sqrt[6]{-64}$ is not a real number since an even root of a negative is not a real number.

23. $\sqrt[3]{-343} = -7$ since $(-7)^3 = -343$

25. $\sqrt{-36}$ is not a real number since an even root of a negative is not a real number.

27. $\sqrt{-45.3}$ is not a real number since an even root of a negative is not a real number.

29. $\sqrt{\dfrac{1}{25}} = \dfrac{1}{5}$ since $\left(\dfrac{1}{5}\right)^2 = \dfrac{1}{25}$

31. $\sqrt[3]{\dfrac{1}{8}} = \dfrac{1}{2}$ since $\left(\dfrac{1}{2}\right)^3 = \dfrac{1}{8}$

33. $\sqrt{\dfrac{4}{49}} = \dfrac{2}{7}$ since $\left(\dfrac{2}{7}\right)^2 = \dfrac{4}{49}$

35. $\sqrt[3]{-\dfrac{8}{27}} = -\dfrac{2}{3}$ since $\left(-\dfrac{2}{3}\right)^3 = -\dfrac{8}{27}$

37. $-\sqrt[4]{18.2} \approx -2.07$

39. $\sqrt{7^2} = |7| = 7$

41. $\sqrt{(19)^2} = |19| = 19$

43. $\sqrt{119^2} = |119| = 119$

45. $\sqrt{(235.23)^2} = |235.23| = 235.23$

47. $\sqrt{(0.06)^2} = |0.06| = 0.06$

49. $\sqrt{\left(\dfrac{12}{13}\right)^2} = \left|\dfrac{12}{13}\right| = \dfrac{12}{13}$

51. $\sqrt{(x-4)^2} = |x-4|$

53. $\sqrt{(x-3)^2} = |x-3|$

55. $\sqrt{(3x^2-1)^2} = |3x^2-1|$

57. $\sqrt{(6a^3-5b^4)^2} = |6a^3-5b^4|$

59. $\sqrt{a^{14}} = \sqrt{(a^7)^2} = |a^7|$

61. $\sqrt{z^{32}} = \sqrt{(z^{16})^2} = \left| z^{16} \right| = z^{16}$

63. $\sqrt{a^2 - 8a + 16} = \sqrt{(a-4)^2} = |a-4|$

65. $\sqrt{9a^2 + 12ab + 4b^2} = \sqrt{(3a+2b)^2} = |3a+2b|$

67. $\sqrt{49x^2} = \sqrt{(7x)^2} = 7x$

69. $\sqrt{16c^6} = \sqrt{(4c^3)^2} = 4c^3$

71. $\sqrt{x^2 + 4x + 4} = \sqrt{(x+2)^2} = x+2$

73. $\sqrt{4x^2 + 4xy + y^2} = \sqrt{(2x+y)^2} = 2x+y$

75. $f(x) = \sqrt{5x-6}$

$$f(2) = \sqrt{5 \cdot 2 - 6}$$
$$= \sqrt{10 - 6}$$
$$= \sqrt{4}$$
$$= 2$$

77. $q(x) = \sqrt{76 - 3x}$

$$q(4) = \sqrt{76 - 3 \cdot 4}$$
$$= \sqrt{76 - 12}$$
$$= \sqrt{64}$$
$$= 8$$

79. $t(a) = \sqrt{-15a - 9}$

$$t(-6) = \sqrt{-15(-6) - 9}$$
$$= \sqrt{90 - 9}$$
$$= \sqrt{81}$$
$$= 9$$

81. $g(x) = \sqrt{64 - 8x}$

$$g(-3) = \sqrt{64 - 8(-3)}$$
$$= \sqrt{64 + 24}$$
$$= \sqrt{88}$$
$$\approx 9.381$$

83. $h(x) = \sqrt[3]{9x^2 + 4}$

$$h(4) = \sqrt[3]{9(4)^2 + 4}$$
$$= \sqrt[3]{144 + 4}$$
$$= \sqrt[3]{148}$$
$$\approx 5.290$$

85. $f(x) = \sqrt[3]{-2x^2 + x - 6}$

$$f(-3) = \sqrt[3]{-2(-3)^2 + (-3) - 6}$$
$$= \sqrt[3]{-18 - 3 - 6}$$
$$= \sqrt[3]{-27}$$
$$= -3$$

87. $f(x) = x + \sqrt{x} + 7$

$$f(81) = 81 + \sqrt{81} + 7$$
$$= 81 + 9 + 7$$
$$= 97$$

89. $t(x) = \dfrac{x}{2} + \sqrt{2x - 4}$

$$t(18) = \frac{18}{2} + \sqrt{2(18) - 4}$$
$$= 9 + \sqrt{36 - 4}$$
$$= 9 + 6 - 4$$
$$= 11$$

91. $k(x) = x^2 + \sqrt{\dfrac{x}{2} - 21}$

$$k(8) = (8)^2 + \sqrt{\frac{8}{2} - 21}$$
$$= 64 + \sqrt{4} - 21$$
$$= 64 + 2 - 21$$
$$= 45$$

93. Choose a value for x that will make the expression $2x + 1$ a negative number. For example, select $x = -1$.

$$\sqrt{(2(-1)+1)^2} \ne 2(-1)+1$$
$$\sqrt{(-1)^2} \ne -1$$
$$\sqrt{1} \ne -1$$
$$1 \ne -1$$

Choosing any value for x less than $-\dfrac{1}{2}$ will show this inequality.

95. $\sqrt{(x-1)^2} = x-1$ for all $x \geq 1$. The expression

$\sqrt{(x-1)^2} = x-1$, when $(x-1)$ is equal to zero or a positive number. Therefore, solving for x,
$x-1 \geq 0$
$\quad x \geq 1$.

97. $\sqrt{(2x-6)^2} = 2x-6$ for all $x \geq 3$. The expression

$\sqrt{(2x-6)^2} = 2x-6$, when $(2x-6)$ is positive or equal to 0. Therefore, solving for x,
$2x-6 \geq 0$
$\quad x \geq 3$

99. a. $\sqrt{a^2} = |a|$ for all real values

b. $\sqrt{a^2} = a$ when $a \geq 0$

c. $\sqrt[3]{a^3} = a$ for all real values

101. If n is even, we are finding the even root of a positive number. If n is odd, the expression is a real number even if the radicand is negative.

103. $\dfrac{\sqrt{x+5}}{\sqrt[3]{x+5}}$ The denominator cannot equal zero.

$\sqrt[3]{x+5} \neq 0$
$\quad x \neq -5$
The radicand in the numerator must be greater than or equal to zero.
$x+5 \geq 0$
$\quad x \geq -5$
Therefore the domain is $\{x \mid x > -5\}$.

105. $f(x) = \sqrt{x}$ matches graph d).
The x-intercept is 0 and the domain is $x \geq 0$.

107. $f(x) = \sqrt{x-5}$ matches graph a). The x-intercept is 5 and the domain is $x-5 \geq 0$, or $x \geq 5$.

109. Answers may vary. One answer is
$f(x) = \sqrt{x-8}$

111. $f(x) = -\sqrt{x}$

a. No; since $\sqrt{x} \geq 0$, $-\sqrt{x}$ must be ≤ 0.

b. Yes, if $x = 0$.

c. Yes; since $\sqrt{x} \geq 0$, $-\sqrt{x}$ must be ≤ 0.

113. $V = \sqrt{64.4h}$

a. $V = \sqrt{64.4(20)}$
$\quad = \sqrt{1288}$
$\quad \approx 35.89$
The velocity will be about 35.89 ft/sec.

b. $V = \sqrt{64.4(40)}$
$\quad = \sqrt{2576}$
$\quad \approx 50.75$
The velocity will be about 50.75 ft/sec.

115. $f(x) = \sqrt{x+1}$

x	$f(x)$
-1	0
0	1
3	2
8	3

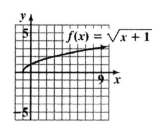

117. $g(x) = \sqrt{x} + 1$

x	$g(x)$
0	1
4	3
9	4

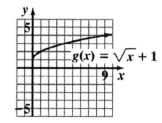

119. $y_1 = \sqrt{x+1}$

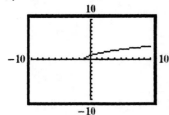

121. $y_1 = \dfrac{\sqrt{x+5}}{\sqrt[3]{x+5}}$

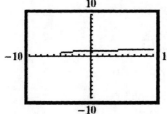

Yes, the domain is $x > -5$.

123. $y = \sqrt[3]{x+4}$

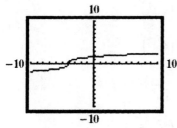

127. $9ax - 3bx + 12ay - 4by$
$= 3x(3a - b) + 4y(3a - b)$
$= (3a - b)(3x + 4y)$

128. $3x^3 - 18x^2 + 24x = 3x(x^2 - 6x + 8)$
$= 3x(x - 4)(x - 2)$

129. $8x^4 + 10x^2 - 3$
$= 8(x^2)^2 + 10(x^2) - 3 \leftarrow$ use y for x^2
$= 8y^2 + 10y - 3$
$= (4y - 1)(2y + 3) \leftarrow$ use x^2 for y
$= (4x^2 - 1)(2x^2 + 3)$
$= ((2x)^2 - 1^2)(2x^2 + 3)$
$= (2x - 1)(2x + 1)(2x^2 + 3)$

130. $x^3 - \dfrac{8}{27}y^3 = x^3 - \left(\dfrac{2}{3}y\right)^3$

$= \left(x - \dfrac{2}{3}y\right)\left(x^2 + \dfrac{2}{3}xy + \left(\dfrac{2}{3}y\right)^2\right)$

$= \left(x - \dfrac{2}{3}y\right)\left(x^2 + \dfrac{2}{3}xy + \dfrac{4}{9}y^2\right)$

Exercise Set 11.2

1. a. $\sqrt[n]{a}$ is a real number when n is even and $a \geq 0$, or n is odd.

b. $\sqrt[n]{a}$ can be expressed with rational exponents as $a^{1/n}$.

3. a. $\sqrt[n]{a^n}$ is always real

b. $\sqrt[n]{a^n} = a$ when $a \geq 0$ and n is even

c. $\sqrt[n]{a^n} = a$ when n is odd

d. $\sqrt[n]{a^n} = |a|$ when n is even and a is any real number

5. a. No, $(xy)^{1/2} = x^{1/2}y^{1/2} \neq xy^{1/2}$

b. No, since $(xy)^{-1/2} = \dfrac{1}{(xy)^{1/2}} = \dfrac{1}{x^{1/2}y^{1/2}}$

but $\dfrac{x^{1/2}}{y^{-1/2}} = x^{1/2}y^{1/2}$

7. $\sqrt{a^3} = a^{3/2}$

9. $\sqrt{9^5} = 9^{5/2}$

11. $\sqrt[3]{z^5} = z^{5/3}$

13. $\sqrt[3]{7^{10}} = 7^{10/3}$

15. $\sqrt[4]{9^7} = 9^{7/4}$

17. $\left(\sqrt[3]{y}\right)^{14} = y^{14/3}$

19. $\sqrt[4]{a^3 b} = (a^3 b)^{1/4}$

21. $\sqrt[4]{x^9 z^5} = (x^9 z^5)^{1/4}$

23. $\sqrt[6]{3a + 8b} = (3a + 8b)^{1/6}$

25. $\sqrt[5]{\dfrac{2x^6}{11y^7}} = \left(\dfrac{2x^6}{11y^7}\right)^{1/5}$

27. $a^{1/2} = \sqrt{a}$

29. $c^{5/2} = \sqrt{c^5}$

31. $18^{5/3} = \sqrt[3]{18^5}$

33. $(24x^3)^{1/2} = \sqrt{24x^3}$

35. $(11b^2 c)^{3/5} = \left(\sqrt[5]{11b^2 c}\right)^3$

37. $(6a + 5b)^{1/5} = \sqrt[5]{6a + 5b}$

39. $(b^3 - d)^{-1/3} = \dfrac{1}{\sqrt[3]{b^3 - d}}$

41. $\sqrt{a^6} = a^{6/2} = a^3$

43. $\sqrt[3]{x^9} = x^{9/3} = x^3$

45. $\sqrt[6]{y^2} = y^{2/6} = y^{1/3} = \sqrt[3]{y}$

47. $\sqrt[6]{y^3} = y^{3/6} = y^{1/2} = \sqrt{y}$

49. $\left(\sqrt{19.3}\right)^2 = (19.3)^{2/2} = (19.3)^1 = 19.3$

51. $\left(\sqrt[3]{xy^2}\right)^{15} = (xy^2)^{15/3}$
$= (xy^2)^5$
$= x^5 y^{10}$

53. $\left(\sqrt[8]{xyz}\right)^4 = (xyz)^{4/8}$
$= (xyz)^{1/2}$
$= \sqrt{xyz}$

55. $\sqrt{\sqrt{x}} = \left(\sqrt{x}\right)^{1/2}$
$= \left(x^{1/2}\right)^{1/2}$
$= x^{1/4}$
$= \sqrt[4]{x}$

57. $\sqrt{\sqrt[4]{y}} = \left(\sqrt[4]{y}\right)^{1/2}$
$= \left(y^{1/4}\right)^{1/2}$
$= y^{1/8}$
$= \sqrt[8]{y}$

59. $\sqrt[3]{\sqrt[3]{x^2 y}} = \left(\sqrt[3]{x^2 y}\right)^{1/3}$
$= (x^2 y)^{1/3 \cdot 1/3}$
$= (x^2 y)^{1/9}$
$= \sqrt[9]{x^2 y}$

61. $\sqrt{\sqrt[5]{a^9}} = \left(\sqrt[5]{a^9}\right)^{1/2}$
$= \left(a^{9/5}\right)^{1/2}$
$= a^{9/10}$
$= \sqrt[10]{a^9}$

63. $25^{1/2} = \sqrt{25} = 5$

65. $64^{1/3} = \sqrt[3]{64} = 4$

67. $64^{2/3} = \left(\sqrt[3]{64}\right)^2 = 4^2 = 16$

69. $(-49)^{1/2} = \sqrt{-49}$ is not a real number.

71. $\left(\dfrac{25}{9}\right)^{1/2} = \sqrt{\dfrac{25}{9}} = \dfrac{5}{3}$

73. $\left(\dfrac{1}{8}\right)^{1/3} = \sqrt[3]{\dfrac{1}{8}} = \dfrac{1}{2}$

75. $-81^{1/2} = -\sqrt{81} = -9$

77. $-64^{1/3} = -\sqrt[3]{64} = -4$

79. $64^{-1/3} = \dfrac{1}{64^{1/3}} = \dfrac{1}{\sqrt[3]{64}} = \dfrac{1}{4}$

81. $16^{-3/2} = \dfrac{1}{16^{3/2}} = \dfrac{1}{\left(\sqrt{16}\right)^3} = \dfrac{1}{4^3} = \dfrac{1}{64}$

83. $\left(\dfrac{64}{27}\right)^{-1/3} = \left(\dfrac{27}{64}\right)^{1/3} = \sqrt[3]{\dfrac{27}{64}} = \dfrac{3}{4}$

85. $(-100)^{3/2} = \left(\sqrt{-100}\right)^3$ is not a real number.

87. $121^{1/2} + 169^{1/2} = \sqrt{121} + \sqrt{169} = 11 + 13 = 24$

89. $343^{-1/3} + 16^{-1/2} = \dfrac{1}{343^{1/3}} + \dfrac{1}{16^{1/2}}$
$$= \dfrac{1}{\sqrt[3]{343}} + \dfrac{1}{\sqrt{16}}$$
$$= \dfrac{1}{7} + \dfrac{1}{4}$$
$$= \dfrac{4}{28} + \dfrac{7}{28}$$
$$= \dfrac{11}{28}$$

91. $x^4 \cdot x^{1/2} = x^{4+1/2} = x^{9/2}$

93. $\dfrac{x^{1/2}}{x^{1/3}} = x^{1/2-1/3} = x^{3/6-2/6} = x^{1/6}$

95. $(x^{1/2})^{-2} = x^{1/2(-2)} = x^{-1} = \dfrac{1}{x}$

97. $(9^{-1/3})^0 = 9^{-1/3(0)} = 9^0 = 1$

99. $\dfrac{5y^{-1/3}}{60y^{-2}} = \dfrac{1}{12}y^{-1/3-(-2)} = \dfrac{1}{12}y^{5/3} = \dfrac{y^{5/3}}{12}$

101. $4x^{5/3} \cdot 3x^{-7/2} = 4 \cdot 3 \cdot x^{5/3} \cdot x^{-7/2} = 12x^{5/3-7/2}$
$$= 12x^{10/6-21/6} = 12x^{-11/6} = \dfrac{12}{x^{11/6}}$$

103. $\left(\dfrac{3}{24x}\right)^{1/3} = \left(\dfrac{1}{8x}\right)^{1/3} = \dfrac{1^{1/3}}{8^{1/3}x^{1/3}} = \dfrac{1}{2x^{1/3}}$

105. $\left(\dfrac{22x^{3/7}}{2x^{1/2}}\right)^2 = (11x^{3/7-1/2})^2 = (11x^{6/14-7/14})^2$
$$= (11x^{-1/14})^2 = (11)^2(x^{-1/14})^2 = 121x^{-1/7}$$
$$= \dfrac{121}{x^{1/7}}$$

107. $\left(\dfrac{a^4}{4a^{-2/5}}\right)^{-3} = \dfrac{a^{-12}}{4^{-3}a^{6/5}}$
$$= 4^3 a^{-12-6/5}$$
$$= 64a^{-66/5}$$
$$= \dfrac{64}{a^{66/5}}$$

109. $\left(\dfrac{x^{3/4}y^{-3}}{x^{1/2}y^2}\right)^4 = \left(x^{3/4-1/2}y^{-3-2}\right)^4$
$$= (x^{1/4}y^{-5})^4$$
$$= (x^{1/4})^4(y^{-5})^4$$
$$= xy^{-20}$$
$$= \dfrac{x}{y^{20}}$$

111. $4z^{-1/2}(2z^4 - z^{1/2}) = 4z^{-1/2} \cdot 2z^4 - 4z^{-1/2}z^{1/2}$
$$= 8z^{-1/2+4} - 4z^{-1/2+1/2}$$
$$= 8z^{7/2} - 4z^0$$
$$= 8z^{7/2} - 4$$

113. $5x^{-1}(x^{-4} + 4x^{-1/2}) = 5x^{-1} \cdot x^{-4} + 5x^{-1} \cdot 4x^{-1/2}$
$$= 5x^{-1-4} + 20x^{-1-1/2}$$
$$= 5x^{-5} + 20x^{-3/2}$$
$$= \dfrac{5}{x^5} + \dfrac{20}{x^{3/2}}$$

115. $-6x^{5/3}(-2x^{1/2} + 3x^{1/3})$
$$= (-6x^{5/3})(-2x^{1/2}) + (-6x^{5/3})(3x^{1/3})$$
$$= 12x^{5/3+1/2} - 18x^{5/3+1/3}$$
$$= 12x^{13/6} - 18x^{6/3}$$
$$= 12x^{13/6} - 18x^2$$

117. $\sqrt{180} \approx 13.42$

119. $\sqrt[5]{402.83} \approx 3.32$

121. $93^{2/3} \approx 20.53$

123. $1000^{-1/2} \approx 0.03$

125. $\sqrt[n]{a^n} = \left(\sqrt[n]{a}\right)^n = a$ when n is odd, or n is even with $a \geq 0$.

127. To show $(a^{1/2} + b^{1/2})^2 \neq a + b$, use $a = 9$ and $b = 16$. Then $(a^{1/2} + b^{1/2})^2$ becomes $(9^{1/2} + 16^{1/2})^2 = (3 + 4)^2 = 7^2 = 49$ whereas $a + b$ becomes $9 + 16 = 25$. Since $49 \neq 25$, then $(a^{1/2} + b^{1/2})^2 \neq a + b$. Answers will vary.

129. To show $(a^{1/3} + b^{1/3})^3 \neq a + b$, use $a = 1$ and $b = 1$. Then $(a^{1/3} + b^{1/3})^3$ becomes $(1^{1/3} + 1^{1/3})^3 = (\sqrt[3]{1} + \sqrt[3]{1})^3 = (1 + 1)^3 = 2^3 = 8$ whereas $a + b$ becomes $1 + 1 = 2$. Since $8 \neq 2$, then $(a^{1/3} + b^{1/3})^3 \neq a + b$. Answers will vary.

131. $x^{3/2} + x^{1/2} = x^{1/2} \cdot x^1 + x^{1/2}$
$= x^{1/2}(x + 1)$

133. $y^{1/3} - y^{7/3} = y^{1/3} - y^{1/3}y^2$
$= y^{1/3}(1 - y^2)$
$= y^{1/3}(1 - y)(1 + y)$

135. $y^{-2/5} + y^{8/5} = y^{-2/5} + y^{-2/5}y^2$
$= y^{-2/5}(1 + y^2)$
$= \dfrac{1 + y^2}{y^{2/5}}$

137. a. $B(t) = 2^{10} \cdot 2^t$
$B(0) = 2^{10} \cdot 2^0$
$= 2^{10} \cdot 1$
$= 2^{10}$
Initially, there are 2^{10} or 1024 bacteria.

b. $B\left(\dfrac{1}{2}\right) = 2^{10} \cdot 2^{1/2}$
$= 2^{10}\sqrt{2}$
≈ 1448.15
After $\dfrac{1}{2}$ hour there are about 1448 bacteria.

139. $A(t) = 2.69t^{3/2}$

a. $t = 200 - 1993 = 7$
$A(7) = 2.69(7)^{3/2}$
≈ 49.82
In 2000, there was about \$49.82 billion in total assets in the U.S. in 401(k) plans.

b. $t = 2009 - 1993 = 16$
$A(16) = 2.69(16)^{3/2}$
$= 172.16$
In 2009, there will be about \$172.16 billion in total assets in the U.S. in 401(k) plans.

141. $(3^{\sqrt{2}})^{\sqrt{2}} = 3^{\sqrt{2} \cdot \sqrt{2}} = 3^2 = 9$

143. $f(x) = (x - 7)^{1/2}(x + 3)^{-1/2}$
$= \dfrac{(x - 7)^{1/2}}{(x + 3)^{1/2}}$
$= \dfrac{\sqrt{x - 7}}{\sqrt{x + 3}}$
The denominator must be greater than zero.
$\sqrt{x + 3} > 0$
$x > -3$
The numerator must be greater than or equal to zero.
$\sqrt{x - 7} \geq 0$
$x \geq 7$
Therefore, the domain is $\{x \mid x \geq 7\}$.

145. a. If n is even: $\sqrt[n]{(x - 6)^{2n}} = \sqrt[n]{\left((x - 6)^2\right)^n}$
$= \left|(x - 6)^2\right|$
$= (x - 6)^2$

b. If n is odd: $\sqrt[n]{(x - 6)^{2n}} = \sqrt[n]{\left((x - 6)^2\right)^n}$
$= (x - 6)^2$

147. Let a be the unknown index in the shaded area.

$$\sqrt[4]{\sqrt[5]{\sqrt[a]{\sqrt[3]{z}}}} = z^{1/120}$$

$$\left(\left(\left(z^{1/3}\right)^{1/a}\right)^{1/5}\right)^{1/4} = z^{1/120}$$

$$z^{1/60a} = z^{1/120}$$

$$\frac{1}{60a} = \frac{1}{120} \leftarrow \text{Equate exponents}$$

$$60a = 120$$

$$a = 2$$

149. $\dfrac{a^{-2} + ab^{-1}}{ab^{-2} - a^{-2}b^{-1}} = \dfrac{\dfrac{1}{a^2} + \dfrac{a}{b}}{\dfrac{a}{b^2} - \dfrac{1}{a^2 b}}$

$$= \frac{a^2 b^2 \left(\dfrac{1}{a^2}\right) + a^2 b^2 \left(\dfrac{a}{b}\right)}{a^2 b^2 \left(\dfrac{a}{b^2}\right) - a^2 b^2 \left(\dfrac{1}{a^2 b}\right)}$$

$$= \frac{b^2 + a^3 b}{a^3 - b}$$

150. $\dfrac{3x-2}{x+4} = \dfrac{2x+1}{3x-2}$

$$(3x-2)(3x-2) = (2x+1)(x+4)$$

$$9x^2 - 12x + 4 = 2x^2 + 9x + 4$$

$$7x^2 - 21x = 0$$

$$7x(x-3) = 0$$

$$7x = 0 \quad \text{or} \quad x - 3 = 0$$

$$x = 0 \qquad \qquad x = 3$$

The solutions are 0 and 3.

151. Let y be the speed of the plane in still air. The table is

	d	r	$t = \dfrac{d}{r}$
With wind	560	$y+25$	$\dfrac{560}{y+25}$
Against wind	500	$y-25$	$\dfrac{500}{y-25}$

Since the time is the same for both parts of the trip. The equation is

$$\frac{560}{y+25} = \frac{500}{y-25}$$

$$560(y-25) = 500(y+25)$$

$$560y - 14,000 = 500y + 12,500$$

$$560y = 500y + 26,500$$

$$60y = 26,500$$

$$y = \frac{26,500}{60}$$

$$\approx 441.67 \text{ mph}$$

The speed of the plane in still air is about 441.67 mph.

152. a. The graph is a relation but not a function because it fails the vertical line test.

b. The graph is a relation but not a function because it fails the vertical line test.

c. The graph is both a relation and a function. It passes the vertical line test.

Exercise Set 11.3

1. a. Square the natural numbers.

b. $1^2 = 1, 2^2 = 4, 3^2 = 9,$
$4^2 = 16, 5^2 = 25, 6^2 = 36$

3. a. Raise natural numbers to the fifth power.

b. $1^5 = 1, 2^5 = 32, 3^5 = 243,$
$4^5 = 1024, 5^5 = 3125$

5. If n is even and a is negative, $\sqrt[n]{a}$ is not a real number and the rule does not apply. Similarly for the case when b is negative.

7. If n is even and a is negative, $\sqrt[n]{a}$ is not a real number and the rule does not apply. Similarly for the case when b is negative.

9. $\sqrt{8} = \sqrt{4 \cdot 2} = \sqrt{4}\sqrt{2} = 2\sqrt{2}$

11. $\sqrt{24} = \sqrt{4 \cdot 6} = 2\sqrt{6}$

13. $\sqrt{32} = \sqrt{16 \cdot 2} = \sqrt{16}\sqrt{2} = 4\sqrt{2}$

15. $\sqrt{50} = \sqrt{25 \cdot 2} = \sqrt{25}\sqrt{2} = 5\sqrt{2}$

17. $\sqrt{75} = \sqrt{25 \cdot 3} = \sqrt{25}\sqrt{3} = 5\sqrt{3}$

19. $\sqrt{40} = \sqrt{4 \cdot 10} = \sqrt{4}\sqrt{10} = 2\sqrt{10}$

21. $\sqrt[3]{16} = \sqrt[3]{8 \cdot 2} = \sqrt[3]{8}\sqrt[3]{2} = 2\sqrt[3]{2}$

23. $\sqrt[3]{54} = \sqrt[3]{27 \cdot 2} = \sqrt[3]{27}\sqrt[3]{2} = 3\sqrt[3]{2}$

25. $\sqrt[3]{32} = \sqrt[3]{8 \cdot 4} = \sqrt[3]{8}\sqrt[3]{4} = 2\sqrt[3]{4}$

27. $\sqrt[3]{40} = \sqrt[3]{8 \cdot 5} = \sqrt[3]{8}\sqrt[3]{5} = 2\sqrt[3]{5}$

29. $\sqrt[4]{48} = \sqrt[4]{16 \cdot 3} = \sqrt[4]{16}\sqrt[4]{3} = 2\sqrt[4]{3}$

31. $-\sqrt[5]{64} = -\sqrt[5]{32 \cdot 2} = -\sqrt[5]{32}\sqrt[5]{2} = -2\sqrt[5]{2}$

33. $\sqrt[3]{b^9} = b^3$

35. $\sqrt[3]{x^6} = x^2$

37. $\sqrt{x^3} = \sqrt{x^2 \cdot x} = \sqrt{x^2}\sqrt{x} = x\sqrt{x}$

39. $\sqrt{a^{11}} = \sqrt{a^{10} \cdot a} = \sqrt{a^{10}}\sqrt{a} = a^5\sqrt{a}$

41. $8\sqrt[3]{z^{32}} = 8\sqrt[3]{z^{30} \cdot z^2} = 8\sqrt[3]{z^{30}}\sqrt[3]{z^2} = 8z^{10}\sqrt[3]{z^2}$

43. $\sqrt[4]{b^{23}} = \sqrt[4]{b^{20} \cdot b^3} = \sqrt[4]{b^{20}}\sqrt[4]{b^3} = b^5\sqrt[4]{b^3}$

45. $\sqrt[6]{x^9} = \sqrt[6]{x^6 \cdot x^3} = \sqrt[6]{x^6}\sqrt[6]{x^3} = x\sqrt[6]{x^3}$ or $x\sqrt{x}$

47. $3\sqrt[5]{y^{23}} = 3\sqrt[5]{y^{20} \cdot y^3} = 3\sqrt[5]{y^{20}}\sqrt[5]{y^3} = 3y^4\sqrt[5]{y^3}$

49. $2\sqrt{50y^9} = 2\sqrt{25 \cdot 2 \cdot y^8 \cdot y}$
$$= 2\sqrt{25y^8 \cdot 2y}$$
$$= 2\sqrt{25y^8}\sqrt{2y}$$
$$= 10y^4\sqrt{2y}$$

51. $\sqrt[3]{x^3y^7} = \sqrt[3]{x^3 \cdot y^6 \cdot y}$
$$= \sqrt[3]{x^3y^6 \cdot y}$$
$$= \sqrt[3]{x^3y^6}\sqrt[3]{y}$$
$$= xy^2\sqrt[3]{y}$$

53. $\sqrt[5]{a^6b^{23}} = \sqrt[5]{a^5b^{20} \cdot ab^3}$
$$= \sqrt[5]{a^5b^{20}} \cdot \sqrt[5]{ab^3}$$
$$= ab^4\sqrt[5]{ab^3}$$

55. $\sqrt{24x^{15}y^{20}z^{27}} = \sqrt{4 \cdot 6 \cdot x^{14} \cdot x \cdot y^{20} \cdot z^{26} \cdot z}$
$$= \sqrt{4x^{14}y^{20}z^{26} \cdot 6xz}$$
$$= \sqrt{4x^{14}y^{20}z^{26}}\sqrt{6xz}$$
$$= 2x^7y^{10}z^{13}\sqrt{6xz}$$

57. $\sqrt[3]{81a^6b^8} = \sqrt[3]{27 \cdot 3 \cdot a^6 \cdot b^6 \cdot b^2}$
$$= \sqrt[3]{27a^6b^6 \cdot 3b^2}$$
$$= \sqrt[3]{27a^6b^6}\sqrt[3]{3b^2}$$
$$= 3a^2b^2\sqrt[3]{3b^2}$$

59. $\sqrt[4]{32x^8y^9z^{19}} = \sqrt[4]{16 \cdot 2 \cdot x^8 \cdot y^8 \cdot y \cdot z^{16} \cdot z^3}$
$$= \sqrt[4]{16x^8y^8z^{16} \cdot 2yz^3}$$
$$= \sqrt[4]{16x^8y^8z^{16}}\sqrt[4]{2yz^3}$$
$$= 2x^2y^2z^4\sqrt[4]{2yz^3}$$

61. $\sqrt[4]{81a^8b^9} = \sqrt[4]{81 \cdot a^8 \cdot b^8 \cdot b}$
$$= \sqrt[4]{81a^8b^8 \cdot b}$$
$$= \sqrt[4]{81a^8b^8}\sqrt[4]{b}$$
$$= 3a^2b^2\sqrt[4]{b}$$

63. $\sqrt[5]{32a^{10}b^{12}} = \sqrt[5]{32 \cdot a^{10} \cdot b^{10} \cdot b^2}$
$$= \sqrt[5]{32a^{10}b^{10} \cdot b^2}$$
$$= \sqrt[5]{32a^{10}b^{10}}\sqrt[5]{b^2}$$
$$= 2a^2b^2\sqrt[5]{b^2}$$

65. $\sqrt{\dfrac{75}{3}} = \sqrt{25} = 5$

67. $\sqrt{\dfrac{81}{100}} = \dfrac{\sqrt{81}}{\sqrt{100}} = \dfrac{9}{10}$

69. $\dfrac{\sqrt{27}}{\sqrt{3}} = \sqrt{\dfrac{27}{3}} = \sqrt{9} = 3$

71. $\dfrac{\sqrt{3}}{\sqrt{48}} = \sqrt{\dfrac{3}{48}} = \sqrt{\dfrac{1}{16}} = \dfrac{1}{4}$

73. $\sqrt[3]{\dfrac{3}{24}} = \sqrt[3]{\dfrac{1}{8}} = \dfrac{\sqrt[3]{1}}{\sqrt[3]{8}} = \dfrac{1}{2}$

75. $\dfrac{\sqrt[3]{3}}{\sqrt[3]{81}} = \sqrt[3]{\dfrac{3}{81}} = \sqrt[3]{\dfrac{1}{27}} = \dfrac{1}{3}$

77. $\sqrt[4]{\dfrac{3}{48}} = \sqrt[4]{\dfrac{1}{16}} = \dfrac{1}{2}$

79. $\sqrt[5]{\dfrac{96}{3}} = \sqrt[5]{32} = 2$

81. $\sqrt{\dfrac{r^4}{4}} = \dfrac{\sqrt{r^4}}{\sqrt{4}} = \dfrac{r^2}{2}$

83. $\sqrt{\dfrac{16x^4}{25y^{10}}} = \dfrac{\sqrt{16x^4}}{\sqrt{25y^{10}}} = \dfrac{4x^2}{5y^5}$

85. $\sqrt[3]{\dfrac{c^6}{64}} = \dfrac{\sqrt[3]{c^6}}{\sqrt[3]{64}} = \dfrac{c^2}{4}$

87. $\sqrt[3]{\dfrac{a^8 b^{12}}{b^{-8}}} = \sqrt[3]{a^8 b^{12+8}}$

$\qquad = \sqrt[3]{a^8 b^{20}}$

$\qquad = \sqrt[3]{a^6 b^{18} \cdot a^2 b^2}$

$\qquad = \sqrt[3]{a^6 b^{18}} \cdot \sqrt[3]{a^2 b^2}$

$\qquad = a^2 b^6 \sqrt[3]{a^2 b^2}$

89. $\dfrac{\sqrt{24}}{\sqrt{3}} = \sqrt{\dfrac{24}{3}} = \sqrt{8} = \sqrt{4 \cdot 2} = 2\sqrt{2}$

91. $\dfrac{\sqrt{27x^6}}{\sqrt{3x^2}} = \sqrt{\dfrac{27x^6}{3x^2}} = \sqrt{9x^4} = 3x^2$

93. $\dfrac{\sqrt{48x^6 y^9}}{\sqrt{6x^2 y^6}} = \sqrt{\dfrac{40x^6 y^9}{5x^2 y^6}}$

$\qquad = \sqrt{8x^4 y^3}$

$\qquad = \sqrt{4x^4 y^2 \cdot 2y}$

$\qquad = 2x^2 y \sqrt{2y}$

95. $\sqrt[3]{\dfrac{5xy}{8x^{13}}} = \sqrt[3]{\dfrac{5y}{8x^{12}}} = \dfrac{\sqrt[3]{5y}}{\sqrt[3]{8x^{12}}} = \dfrac{\sqrt[3]{5y}}{2x^4}$

97. $\sqrt[3]{\dfrac{25x^2 y^9}{5x^8 y^2}} = \sqrt[3]{\dfrac{5y^7}{x^6}}$

$\qquad = \dfrac{\sqrt[3]{5y^7}}{\sqrt[3]{x^6}}$

$\qquad = \dfrac{\sqrt[3]{y^6 \cdot 5y}}{x^2}$

$\qquad = \dfrac{y^2 \sqrt[3]{5y}}{x^2}$

99. $\sqrt[4]{\dfrac{10x^4 y}{81x^{-8}}} = \sqrt[4]{\dfrac{10x^{12} y}{81}}$

$\qquad = \dfrac{\sqrt[4]{10x^{12} y}}{\sqrt[4]{81}}$

$\qquad = \dfrac{\sqrt[4]{x^{12}} \sqrt[4]{10y}}{\sqrt[4]{81}}$

$\qquad = \dfrac{x^3 \sqrt[4]{10y}}{3}$

101. $\sqrt{a \cdot b} = (a \cdot b)^{1/2} = a^{1/2} \cdot b^{1/2} = \sqrt{a} \cdot \sqrt{b}$

103. No, for example $\dfrac{\sqrt{18}}{\sqrt{2}} = \sqrt{\dfrac{18}{2}} = \sqrt{9} = 3$.

105. a. no

$\quad$ **b.** $\dfrac{\sqrt[n]{x}}{\sqrt[n]{x}}$ is equal to 1 when $\sqrt[n]{x}$ is a real number and not equal to 0.

106. $F = \dfrac{9}{5}C + 32$

$\qquad F - 32 = \dfrac{9}{5}C$

$\qquad \dfrac{5}{9}(F - 32) = \dfrac{5}{9}\left(\dfrac{9}{5}C\right)$

$\qquad \dfrac{5}{9}(F - 32) = C$ or $C = \dfrac{5}{9}(F - 32)$

107. $\dfrac{15x^{12} - 5x^9 + 20x^6}{5x^6} = \dfrac{15x^{12}}{5x^6} - \dfrac{5x^9}{5x^6} + \dfrac{20x^6}{5x^6}$

$\qquad = 3x^6 - x^3 + 4$

108. $(x-3)^3 + 8$

$= (x-3)^3 + (2)^3$

$= \left((x-3)+2\right)\left((x-3)^2 - (x-3)(2) + (2)^2\right)$

$= (x-1)\left(x^2 - 6x + 9 - 2x + 6 + 4\right)$

$= (x-1)\left(x^2 - 8x + 19\right)$

109. $(2x-3)(x-2) = 4x - 6$

$2x^2 - 7x + 6 = 4x - 6$

$2x^2 - 11x + 12 = 0$

$(2x-3)(x-4) = 0$

$2x - 3 = 0$ or $x - 4 = 0$

$2x = 3$ $x = 4$

$x = \dfrac{3}{2}$

110. $\left|\dfrac{2x-4}{5}\right| = 12$

$\dfrac{2x-4}{5} = -12$ or $\dfrac{2x-4}{5} = 12$

$2x - 4 = -60$ $2x - 4 = 60$

$2x = -56$ $2x = 64$

$x = -28$ $x = 32$

The solution is $\{-28, 32\}$.

Exercise Set 11.4

1. Like radicals are radicals with the same radicands and index.

3. $\sqrt{3} + 3\sqrt{2} \approx 1.732 + 3(1.414)$

$\approx 1.732 + 4.242$

$\approx 5.974 \text{ or } 5.97$

5. No. To see this, let $a = 16$ and $b = 9$. Then, the left side is $\sqrt{a} + \sqrt{b} = \sqrt{16} + \sqrt{9} = 4 + 3 = 7$ whereas the right side is $\sqrt{a+b} = \sqrt{16+9} = \sqrt{25} = 5$.

7. $\sqrt{3} - \sqrt{3} = 0$

9. $6\sqrt{5} - 2\sqrt{5} = 4\sqrt{5}$

11. $2\sqrt{3} - 2\sqrt{3} - 4\sqrt{3} + 5 = -4\sqrt{3} + 5$

13. $2\sqrt[4]{y} - 9\sqrt[4]{y} = -7\sqrt[4]{y}$

15. $3\sqrt{5} - \sqrt[3]{x} + 6\sqrt{5} + 3\sqrt[3]{x} = 2\sqrt[3]{x} + 9\sqrt{5}$

17. $5\sqrt{x} - 8\sqrt{y} + 3\sqrt{x} + 2\sqrt{y} - \sqrt{x} = 7\sqrt{x} - 6\sqrt{y}$

19. $\sqrt{5} + \sqrt{20} = \sqrt{5} + \sqrt{4} \cdot \sqrt{5}$

$= \sqrt{5} + 2\sqrt{5}$

$= 3\sqrt{5}$

21. $-6\sqrt{75} + 5\sqrt{125} = -6\sqrt{25} \cdot \sqrt{3} + 5\sqrt{25} \cdot \sqrt{5}$

$= -6\left(5\sqrt{3}\right) + 5\left(5\sqrt{5}\right)$

$= -30\sqrt{3} + 25\sqrt{5}$

23. $-4\sqrt{90} + 3\sqrt{40} + 2\sqrt{10}$

$= -4\sqrt{9} \cdot \sqrt{10} + 3\sqrt{4} \cdot \sqrt{10} + 2\sqrt{10}$

$= -4\left(3\sqrt{10}\right) + 3\left(2\sqrt{10}\right) + 2\left(\sqrt{10}\right)$

$= -12\sqrt{10} + 6\sqrt{10} + 2\sqrt{10}$

$= -4\sqrt{10}$

25. $\sqrt{500xy^2} + y\sqrt{320x}$

$= \sqrt{100y^2} \cdot \sqrt{5x} + y\sqrt{64}\sqrt{5x}$

$= 10y\sqrt{5x} + 8y\sqrt{5x}$

$= 18y\sqrt{5x}$

27. $2\sqrt{5x} - 3\sqrt{20x} - 4\sqrt{45x}$

$= 2\sqrt{5x} - 3\sqrt{4} \cdot \sqrt{5x} - 4\sqrt{9} \cdot \sqrt{5x}$

$= 2\sqrt{5x} - 3\left(2\sqrt{5x}\right) - 4\left(3\sqrt{5x}\right)$

$= 2\sqrt{5x} - 6\sqrt{5x} - 12\sqrt{5x}$

$= -16\sqrt{5x}$

29. $3\sqrt{50a^2} - 3\sqrt{72a^2} - 8a\sqrt{18}$

$= 3\sqrt{25a^2} \cdot \sqrt{2} - 3\sqrt{36a^2} \cdot \sqrt{2} - 8a\sqrt{9} \cdot \sqrt{2}$

$= 3\left(5a\sqrt{2}\right) - 3\left(6a\sqrt{2}\right) - 8a\left(3\sqrt{2}\right)$

$= 15a\sqrt{2} - 18a\sqrt{2} - 24a\sqrt{2}$

$= -27a\sqrt{2}$

31. $\sqrt[3]{108} + \sqrt[3]{32} = \sqrt[3]{27} \cdot \sqrt[3]{4} + \sqrt[3]{8} \cdot \sqrt[3]{4}$
$\qquad = 3\sqrt[3]{4} + 2\sqrt[3]{4}$
$\qquad = 5\sqrt[3]{4}$

33. $\sqrt[3]{27} - 5\sqrt[3]{8} = 3 - 5 \cdot 2 = 3 - 10 = -7$

35. $2\sqrt[3]{a^4 b^2} + 4a\sqrt[3]{ab^2} = 2\sqrt[3]{a^3} \cdot \sqrt[3]{ab^2} + 4a\sqrt[3]{ab^2}$
$\qquad = 2a\sqrt[3]{ab^2} + 4a\sqrt[3]{ab^2}$
$\qquad = 6a\sqrt[3]{ab^2}$

37. $\sqrt{4r^7 s^5} + 3r^2\sqrt{r^3 s^5} - 2rs\sqrt{r^5 s^3}$
$= \sqrt{4r^6 s^4} \cdot \sqrt{rs} + 3r^2\sqrt{r^2 s^4} \cdot \sqrt{rs} - 2rs\sqrt{r^4 s^2} \cdot \sqrt{rs}$
$= 2r^3 s^2\sqrt{rs} + 3r^2(rs^2\sqrt{rs}) - 2rs(r^2 s\sqrt{rs})$
$= 2r^3 s^2\sqrt{rs} + 3r^3 s^2\sqrt{rs} - 2r^3 s^2\sqrt{rs}$
$= 3r^3 s^2\sqrt{rs}$

39. $\sqrt[3]{128x^8 y^{10}} - 2x^2 y\sqrt[3]{16x^2 y^7}$
$= \sqrt[3]{64x^8 y^9}\sqrt[3]{2y} - 2x^2 y\sqrt[3]{8x^2 y^6}\sqrt[3]{2y}$
$= 4x^2 y^3\sqrt[3]{2x^2 y} - 2x^2 y\left(2y^2\sqrt[3]{2x^2 y}\right)$
$= 4x^2 y^3\sqrt[3]{2x^2 y} - 4x^2 y^3\sqrt[3]{2x^2 y}$
$= 0$

41. $\sqrt{3}\sqrt{27} = \sqrt{3 \cdot 27} = \sqrt{81} = 9$

43. $\sqrt[3]{4}\sqrt[3]{14} = \sqrt[3]{56} = \sqrt[3]{8 \cdot 7} = 2\sqrt[3]{7}$

45. $\sqrt{9m^3 n^7}\sqrt{3mn^4} = \sqrt{9m^3 n^7 \cdot 3mn^4}$
$\qquad = \sqrt{27m^4 n^{11}}$
$\qquad = \sqrt{9 \cdot 3 \cdot m^4 \cdot n^{10} \cdot n}$
$\qquad = \sqrt{9m^4 n^{10} \cdot 3n}$
$\qquad = \sqrt{9m^4 n^{10}}\sqrt{3n}$
$\qquad = 3m^2 n^5\sqrt{3n}$

47. $\sqrt[3]{9x^7 y^{10}}\sqrt[3]{6x^4 y^3} = \sqrt[3]{9x^7 y^{10} \cdot 6x^4 y^3}$
$\qquad = \sqrt[3]{54x^{11} y^{13}}$
$\qquad = \sqrt[3]{27 \cdot 2 \cdot x^9 \cdot x^2 \cdot y^{12} \cdot y}$
$\qquad = \sqrt[3]{27x^9 y^{12} \cdot 2x^2 y}$
$\qquad = \sqrt[3]{27x^9 y^{12}}\sqrt[3]{2x^2 y}$
$\qquad = 3x^3 y^4\sqrt[3]{2x^2 y}$

49. $\sqrt[5]{x^{24} y^{30} z^9}\sqrt[5]{x^{13} y^8 z^7}$
$= \sqrt[5]{x^{24} y^{30} z^9 \cdot x^{13} y^8 z^7}$
$= \sqrt[5]{x^{37} y^{38} z^{16}}$
$= \sqrt[5]{x^{35} \cdot x^2 \cdot y^{35} \cdot y^3 \cdot z^{15} \cdot z}$
$= \sqrt[5]{x^{35} y^{35} z^{15} \cdot x^2 y^3 z}$
$= \sqrt[5]{x^{35} y^{35} z^{15}}\sqrt[5]{x^2 y^3 z}$
$= x^7 y^7 z^3\sqrt[5]{x^2 y^3 z}$

51. $\left(\sqrt[3]{2x^3 y^4}\right)^2 = \sqrt[3]{(2x^3 y^4)^2}$
$\qquad = \sqrt[3]{4x^6 y^8}$
$\qquad = \sqrt[3]{4 \cdot x^6 \cdot y^6 \cdot y^2}$
$\qquad = \sqrt[3]{x^6 y^6 \cdot 4y^2}$
$\qquad = \sqrt[3]{x^6 y^6}\sqrt[3]{4y^2}$
$\qquad = x^2 y^2\sqrt[3]{4y^2}$

53. $\sqrt{5}\left(\sqrt{5} - \sqrt{3}\right) = \left(\sqrt{5}\right)\left(\sqrt{5}\right) - \left(\sqrt{5}\right)\left(\sqrt{3}\right)$
$\qquad = \sqrt{25} - \sqrt{15}$
$\qquad = 5 - \sqrt{15}$

55. $\sqrt[3]{y}\left(2\sqrt[3]{y} - \sqrt[3]{y^8}\right)$
$= \left(\sqrt[3]{y}\right)\left(2\sqrt[3]{y}\right) - \left(\sqrt[3]{y}\right)\left(\sqrt[3]{y^8}\right)$
$= 2\sqrt[3]{y^2} - \sqrt[3]{y^9}$
$= 2\sqrt[3]{y^2} - y^3$

57. $2\sqrt[3]{x^4y^5}\left(\sqrt[3]{8x^{12}y^4}+\sqrt[3]{16xy^9}\right)$

$\quad =\left(2\sqrt[3]{x^4y^5}\right)\left(\sqrt[3]{8x^{12}y^4}\right)+\left(2\sqrt[3]{x^4y^5}\right)\left(\sqrt[3]{16xy^9}\right)$

$\quad =2\sqrt[3]{8x^{16}y^9}+2\sqrt[3]{16x^5y^{14}}$

$\quad =2\sqrt[3]{8x^{15}y^9}\sqrt[3]{x}+2\sqrt[3]{8x^3y^{12}}\sqrt[3]{2x^2y^2}$

$\quad =2\cdot2x^5y^3\sqrt[3]{x}+2\cdot2xy^4\sqrt[3]{2x^2y^2}$

$\quad =4x^5y^3\sqrt[3]{x}+4xy^4\sqrt[3]{2x^2y^2}$

59. $(8+\sqrt5)(8-\sqrt5)=8^2-(\sqrt5)^2$

$\quad\quad\quad\quad\quad\quad\quad\quad =64-5$

$\quad\quad\quad\quad\quad\quad\quad\quad =59$

61. $\left(\sqrt6+x\right)\left(\sqrt6-x\right)=\left(\sqrt6\right)^2-x^2$

$\quad\quad\quad\quad\quad\quad\quad\quad =6-x^2$

63. $\left(\sqrt7-\sqrt z\right)\left(\sqrt7+\sqrt z\right)=\left(\sqrt7\right)^2-\left(\sqrt z\right)^2$

$\quad\quad\quad\quad\quad\quad\quad\quad\quad =7-z$

65. $\left(\sqrt3+4\right)\left(\sqrt3+5\right)=\sqrt9+5\sqrt3+4\sqrt3+20$

$\quad\quad\quad\quad\quad\quad\quad\quad =3+9\sqrt3+20$

$\quad\quad\quad\quad\quad\quad\quad\quad =23+9\sqrt3$

67. $\left(3-\sqrt2\right)\left(4-\sqrt8\right)=12-3\sqrt8-4\sqrt2+\sqrt{16}$

$\quad\quad\quad\quad\quad\quad\quad\quad =12-3\cdot2\sqrt2-4\sqrt2+4$

$\quad\quad\quad\quad\quad\quad\quad\quad =12-6\sqrt2-4\sqrt2+4$

$\quad\quad\quad\quad\quad\quad\quad\quad =16-10\sqrt2$

69. $\left(4\sqrt3+\sqrt2\right)\left(\sqrt3-\sqrt2\right)$

$\quad =4\sqrt9-4\sqrt6+\sqrt6-\sqrt4$

$\quad =4\cdot3-3\sqrt6-2$

$\quad =12-3\sqrt6-2$

$\quad =10-3\sqrt6$

71. $\left(2\sqrt5-3\right)^2=\left(2\sqrt5-3\right)\left(2\sqrt5-3\right)$

$\quad\quad\quad\quad\quad =4\sqrt{25}-6\sqrt5-6\sqrt5+9$

$\quad\quad\quad\quad\quad =4\cdot5-12\sqrt5+9$

$\quad\quad\quad\quad\quad =20+9-12\sqrt5$

$\quad\quad\quad\quad\quad =29-12\sqrt5$

73. $\left(2\sqrt{3x}-\sqrt y\right)\left(3\sqrt{3x}+\sqrt y\right)$

$\quad =6\left(\sqrt{3x}\right)^2+2\sqrt{3x}\sqrt y-3\sqrt{3x}\sqrt y-\left(\sqrt y\right)^2$

$\quad =6(3x)+2\sqrt{3xy}-3\sqrt{3xy}-y$

$\quad =18x-\sqrt{3xy}-y$

75. $\left(\sqrt[3]4-\sqrt[3]6\right)\left(\sqrt[3]2-\sqrt[3]{36}\right)$

$\quad =\sqrt[3]4\sqrt[3]2-\sqrt[3]4\sqrt[3]{36}-\sqrt[3]6\sqrt[3]2+\sqrt[3]6\sqrt[3]{36}$

$\quad =\sqrt[3]8-\sqrt[3]{144}-\sqrt[3]{12}+\sqrt[3]{216}$

$\quad =2-2\sqrt[3]{18}-\sqrt[3]{12}+6$

$\quad =8-2\sqrt[3]{18}-\sqrt[3]{12}$

77. $(f\cdot g)(x)=f(x)\cdot g(x)$

$\quad\quad\quad\quad\quad =\sqrt{2x}(\sqrt{8x}-\sqrt{32})$

$\quad\quad\quad\quad\quad =\sqrt{2x}\cdot\sqrt{8x}-\sqrt{2x}\cdot\sqrt{32}$

$\quad\quad\quad\quad\quad =\sqrt{16x^2}-\sqrt{64x}$

$\quad\quad\quad\quad\quad =4x-8\sqrt x$

79. $(f\cdot g)(x)=f(x)\cdot g(x)$

$\quad\quad\quad\quad\quad =\sqrt[3]x\left(\sqrt[3]{x^5}+\sqrt[3]{x^4}\right)$

$\quad\quad\quad\quad\quad =\sqrt[3]x\,\sqrt[3]{x^5}+\sqrt[3]x\,\sqrt[3]{x^4}$

$\quad\quad\quad\quad\quad =\sqrt[3]{x^6}+\sqrt[3]{x^5}$

$\quad\quad\quad\quad\quad =\sqrt[3]{x^6}+\sqrt[3]{x^3\cdot x^2}$

$\quad\quad\quad\quad\quad =x^2+x\sqrt[3]{x^2}$

81. $(f \cdot g)(x) = f(x) \cdot g(x)$

$\qquad = \sqrt[4]{3x^2}\left(\sqrt[4]{9x^4} - \sqrt[4]{x^7}\right)$

$\qquad = \sqrt[4]{3x^2}\,\sqrt[4]{9x^4} - \sqrt[4]{3x^2}\,\sqrt[4]{x^7}$

$\qquad = \sqrt[4]{27x^6} - \sqrt[4]{3x^9}$

$\qquad = \sqrt[4]{x^4}\sqrt[4]{27x^2} - \sqrt[4]{x^8}\sqrt[4]{3x}$

$\qquad = x\sqrt[4]{27x^2} - x^2\sqrt[4]{3x}$

83. $\sqrt{24} = \sqrt{4 \cdot 6} = 2\sqrt{6}$

85. $\sqrt{125} - \sqrt{20} = \sqrt{25 \cdot 5} - \sqrt{4 \cdot 5}$

$\qquad = 5\sqrt{5} - 2\sqrt{5}$

$\qquad = 3\sqrt{5}$

87. $\left(3\sqrt{2} - 4\right)\left(\sqrt{2} + 5\right)$

$\qquad = 3\left(\sqrt{2}\right)^2 + 15\sqrt{2} - 4\sqrt{2} - 20$

$\qquad = 6 + 11\sqrt{2} - 20$

$\qquad = -14 + 11\sqrt{2}$

89. $\sqrt{6}\left(5 - \sqrt{2}\right) = \sqrt{6} \cdot 5 - \sqrt{6} \cdot \sqrt{2}$

$\qquad = 5\sqrt{6} - \sqrt{12}$

$\qquad = 5\sqrt{6} - 2\sqrt{3}$

91. $\sqrt{150}\sqrt{3} = \sqrt{450} = \sqrt{225 \cdot 2} = 15\sqrt{2}$

93. $\sqrt[3]{80x^{11}} = \sqrt[3]{8x^9 \cdot 10x^2} = 2x^3\sqrt[3]{10x^2}$

95. $\sqrt[6]{128ab^{17}c^9} = \sqrt[6]{64b^{12}c^6 \cdot 2ab^5c^3}$

$\qquad = 2b^2c\sqrt[6]{2ab^5c^3}$

97. $2b\sqrt[4]{a^4b} + ab\sqrt[4]{16b} = 2b\sqrt[4]{a^4} \cdot \sqrt[4]{b} + ab\sqrt[4]{16} \cdot \sqrt[4]{b}$

$\qquad = 2b\left(a\sqrt[4]{b}\right) + ab\left(2\sqrt[4]{b}\right)$

$\qquad = 2ab\sqrt[4]{b} + 2ab\sqrt[4]{b}$

$\qquad = 4ab\sqrt[4]{b}$

99. $\left(\sqrt[3]{x^2} - \sqrt[3]{y}\right)\left(\sqrt[3]{x} - 2\sqrt[3]{y^2}\right)$

$\qquad = \sqrt[3]{x^2}\sqrt[3]{x} - 2\sqrt[3]{x^2}\sqrt[3]{y^2} - \sqrt[3]{y}\sqrt[3]{x} + 2\sqrt[3]{y}\sqrt[3]{y^2}$

$\qquad = \sqrt[3]{x^3} - 2\sqrt[3]{x^2y^2} - \sqrt[3]{xy} + 2\sqrt[3]{y^3}$

$\qquad = x - 2\sqrt[3]{x^2y^2} - \sqrt[3]{xy} + 2y$

101. $\sqrt[3]{3ab^2}\left(\sqrt[3]{4a^4b^3} - \sqrt[3]{8a^5b^4}\right)$

$\qquad = \left(\sqrt[3]{3ab^2}\right)\left(\sqrt[3]{4a^4b^3}\right) - \left(\sqrt[3]{3ab^2}\right)\left(\sqrt[3]{8a^5b^4}\right)$

$\qquad = \sqrt[3]{12a^5b^5} - \sqrt[3]{24a^6b^6}$

$\qquad = \sqrt[3]{a^3b^3 \cdot 12a^2b^2} - \sqrt[3]{8a^6b^6 \cdot 3}$

$\qquad = ab\sqrt[3]{12a^2b^2} - 2a^2b^2\sqrt[3]{3}$

103. $f(x) = \sqrt{2x - 5}\sqrt{2x - 5} = 2x - 5$

No absolute value needed since $x \geq \dfrac{5}{2}$.

105. $h(r) = \sqrt{4r^2 - 32r + 64}$

$\qquad = \sqrt{4\left(r^2 - 8r + 16\right)}$

$\qquad = \sqrt{4\left(r - 4\right)^2}$

$\qquad = 2\left|r - 4\right|$

107. Perimeter $= \sqrt{45} + \sqrt{45} + \sqrt{80} + \sqrt{80}$

$\qquad = 2\sqrt{45} + 2\sqrt{80}$

$\qquad = 2\sqrt{9}\sqrt{5} + 2\sqrt{16}\sqrt{5}$

$\qquad = 2\left(3\sqrt{5}\right) + 2\left(4\sqrt{5}\right)$

$\qquad = 6\sqrt{5} + 8\sqrt{5}$

$\qquad = 14\sqrt{5}$

Area $= \sqrt{45}\sqrt{80}$

$\qquad = 3\sqrt{5} \cdot 4\sqrt{5}$

$\qquad = 12\left(\sqrt{5}\right)^2$

$\qquad = 12 \cdot 5$

$\qquad = 60$

109. Perimeter $= \sqrt{245} + \sqrt{180} + \sqrt{80}$
$$= \sqrt{49}\sqrt{5} + \sqrt{36}\sqrt{5} + \sqrt{16}\sqrt{5}$$
$$= 7\sqrt{5} + 6\sqrt{5} + 4\sqrt{5}$$
$$= 17\sqrt{5}$$

Area $= \frac{1}{2}\sqrt{245}\sqrt{45}$
$$= \frac{1}{2}\sqrt{49}\sqrt{5}\sqrt{9}\sqrt{5}$$
$$= \frac{1}{2} \cdot 7 \cdot 3\left(\sqrt{5}\right)^2$$
$$= \frac{21}{2} \cdot 5$$
$$= 52.5$$

111. No, for example $-\sqrt{2} + \sqrt{2} = 0$

113. a. $s = \sqrt{30FB}$
$$s = \sqrt{30(0.85)(80)} \approx 45.17$$
The car's speed was about 45.17 mph.

b. $s = \sqrt{30FB}$
$$s = \sqrt{30(0.52)(80)} \approx 35.33$$
The car's speed was about 35.33 mph.

115. $f(t) = 3\sqrt{t} + 19$

a. $f(36) = 3\sqrt{36} + 19$
$$= 3(6) + 19$$
$$= 18 + 19$$
$$= 37$$
The length at 36 months is 37 inches.

b. $f(40) = 3\sqrt{40} + 19$
$$= 3\sqrt{4}\sqrt{10} + 19$$
$$= 3 \cdot 2\sqrt{10} + 19$$
$$= 6\sqrt{10} + 19$$
$$\approx 37.97$$
The length at 40 months is about 37.97 inches.

117. a. $f(x) = \sqrt{x}$
$g(x) = 2$
$(f + g)(x) = f(x) + g(x) = \sqrt{x} + 2$

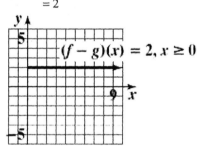

b. It raises the graph 2 units.

119. a. $(f - g)(x) = f(x) - g(x)$
$$= \sqrt{x} - \left(\sqrt{x} - 2\right)$$
$$= \sqrt{x} - \sqrt{x} + 2$$
$$= 2$$

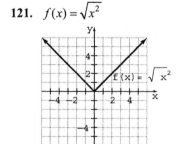

b. $\sqrt{x} \geq 0$, so $x \geq 0$
The domain is $\{x \mid x \geq 0\}$.

121. $f(x) = \sqrt{x^2}$

123. A rational number is a quotient of two integers with a nonzero denominator.

124. A real number is a number that can represented on a real number line.

125. An irrational number is a real number that cannot be expressed as the quotient of two integers.

126. $|a| = \begin{cases} a, & a \geq 0 \\ -a, & a < 0 \end{cases}$

127. $E = \frac{1}{2}mv^2$

$2E = 2\left(\frac{1}{2}mv^2\right)$

$2E = mv^2$

$\frac{2E}{v^2} = m$ or $m = \frac{2E}{v^2}$

128. a. $-4 < 2x - 3 \leq 7$

$-4 + 3 < 2x - 3 + 3 \leq 7 + 3$

$-1 < 2x \leq 10$

$-\frac{1}{2} < x \leq 5$

b. $\left(-\frac{1}{2}, 5\right]$

c. $\left\{x \mid -\frac{1}{2} < x \leq 5\right\}$

Mid-Chapter Test: 11.1 – 11.4

1. $\sqrt{121} = \sqrt{(11)^2} = 11$

2. $\sqrt[3]{-\frac{27}{64}} = \sqrt[3]{-\left(\frac{3}{4}\right)^3} = -\frac{3}{4}$

3. $\sqrt{(-16.3)^2} = |-16.3| = 16.3$

4. $\sqrt{\left(3a^2 - 4b^3\right)^2} = \left|3a^2 - 4b^3\right|$

5. $g(x) = \frac{x}{8} + \sqrt{4x} - 7$

$g(16) = \frac{16}{8} + \sqrt{4(16)} - 7$

$= 2 + \sqrt{64} - 7$

$= 2 + 8 - 7$

$= 3$

6. $\sqrt[5]{7a^4b^3} = \left(7a^4b^3\right)^{1/5}$

7. $-49^{1/2} + 81^{3/4} = -\sqrt{49} + \left(\sqrt[4]{81}\right)^3$

$= -7 + (3)^3$

$= -7 + 27$

$= 20$

8. $\left(\sqrt[4]{a^2b^3c}\right)^{20} = \left[\left(a^2b^3c\right)^{1/4}\right]^{20} = \left(a^2b^3c\right)^{(1/4)\cdot 20}$

$= \left(a^2b^3c\right)^5 = a^{2\cdot 5}b^{3\cdot 5}c^{1\cdot 5}$

$= a^{10}b^{15}c^5$

9. $7x^{-5/2} \cdot 2x^{3/2} = 7 \cdot 2 \cdot x^{-5/2} \cdot x^{3/2}$

$= 14x^{-5/2 + 3/2}$

$= 14x^{-2/2}$

$= 14x^{-1}$

$= \frac{14}{x}$

10. $8x^{-2}\left(x^3 + 2x^{-1/2}\right)$

$= 8x^{-2}\left(x^3\right) + 8x^{-2}\left(2x^{-1/2}\right)$

$= 8x^{-2+3} + 16x^{-2 + (-1/2)}$

$= 8x^1 + 16x^{-5/2}$

$= 8x + \frac{16}{x^{5/2}}$

11. $\sqrt{32x^4y^9} = \sqrt{16x^4y^8} \cdot \sqrt{2y}$

$= 4x^2y^4\sqrt{2y}$

12. $\sqrt[6]{64a^{13}b^{23}c^{15}} = \sqrt[6]{64a^{12}b^{18}c^{12}} \cdot \sqrt[6]{ab^5c^3}$

$= 2a^2b^3c^2\sqrt[6]{ab^5c^3}$

13. $\frac{\sqrt[3]{3}}{\sqrt[3]{81}} = \sqrt[3]{\frac{3}{81}} = \sqrt[3]{\frac{1}{27}} = \frac{1}{3}$

14. $\frac{\sqrt{20x^5y^{12}}}{\sqrt{180x^{15}y^7}} = \sqrt{\frac{20x^5y^{12}}{180x^{15}y^7}} = \sqrt{\frac{y^5}{9x^{10}}}$

$= \frac{\sqrt{y^4} \cdot \sqrt{y}}{\sqrt{9x^{10}}} = \frac{y^2\sqrt{y}}{3x^5}$

15. $2\sqrt{x} - 3\sqrt{y} + 9\sqrt{x} + 15\sqrt{y}$

$= 2\sqrt{x} + 9\sqrt{x} - 3\sqrt{y} + 15\sqrt{y}$

$= 11\sqrt{x} + 12\sqrt{y}$

16. $2\sqrt{90x^2 y} + 3x\sqrt{490 y}$

$= 2\sqrt{9x^2} \cdot \sqrt{10 y} + 3x\sqrt{49} \cdot \sqrt{10 y}$

$= 2(3x)\sqrt{10 y} + 3x(7)\sqrt{10 y}$

$= 6x\sqrt{10 y} + 21x\sqrt{10 y}$

$= 27x\sqrt{10 y}$ (assuming $x \geq 0$)

17. $\left(x + \sqrt{5}\right)\left(2x - 3\sqrt{5}\right)$

$= x(2x) - x\left(3\sqrt{5}\right) + \sqrt{5}(2x) - \left(\sqrt{5}\right)\left(3\sqrt{5}\right)$

$= 2x^2 - 3x\sqrt{5} + 2x\sqrt{5} - 3(5)$

$= 2x^2 - x\sqrt{5} - 15$

18. $2\sqrt{3a}\left(\sqrt{27a^2} - 5\sqrt{4a}\right)$

$= 2\sqrt{3a} \cdot \sqrt{27a^2} - 2\sqrt{3a} \cdot 5\sqrt{4a}$

$= 2\sqrt{81a^3} - 10\sqrt{12a^2}$

$= 2\sqrt{81a^2} \cdot \sqrt{a} - 10\sqrt{4a^2} \cdot \sqrt{3}$

$= 2(9a)\sqrt{a} - 10(2a)\sqrt{3}$

$= 18a\sqrt{a} - 20a\sqrt{3}$

19. $3b\sqrt[4]{a^5 b} + 2ab\sqrt[4]{16ab}$

$= 3b\sqrt[4]{a^4} \cdot \sqrt[4]{ab} + 2ab\sqrt[4]{16} \cdot \sqrt[4]{ab}$

$= 3b(a)\sqrt[4]{ab} + 2ab(2)\sqrt[4]{ab}$

$= 3ab\sqrt[4]{ab} + 4ab\sqrt[4]{ab}$

$= 7ab\sqrt[4]{ab}$ (assuming $a \geq 0$)

20. a. $\sqrt{(x-3)^2} = |x - 3|$

The absolute value is needed here because we do not know the domain for x. Since the expression $x - 3$ is negative for $x < 3$, the absolute value is required.

b. $\sqrt{64x^2} = \sqrt{(8x)^2} = 8x$

Since $x \geq 0$, we know that $8x \geq 0$. Since $8x$ is never negative, the absolute value is not needed.

Exercise Set 11.5

1. a. The conjugate of a binomial is a binomial with the same two terms as the original but the sign of the second term is changed. The conjugate of $a + b$ is $a - b$. Also, the conjugate of $a - b$ is $a + b$.

b. The conjugate of $x - \sqrt{3}$ is $x + \sqrt{3}$.

3. a. Answers will vary. Possible answer: Multiply the numerator and denominator by a quantity that will result in no radicals in the denominator.

b. $\dfrac{4}{\sqrt{3y}} = \dfrac{4}{\sqrt{3y}} \cdot \dfrac{\sqrt{3y}}{\sqrt{3y}} = \dfrac{4\sqrt{3y}}{\sqrt{9y^2}} = \dfrac{4\sqrt{3y}}{3y}$

5. (1) No perfect powers are factors of any radicand.

(2) No radicand contains fractions.

(3) No radicals are in any denominator.

7. $\dfrac{1}{\sqrt{3}} = \dfrac{1}{\sqrt{3}} \cdot \dfrac{\sqrt{3}}{\sqrt{3}} = \dfrac{\sqrt{3}}{3}$

9. $\dfrac{4}{\sqrt{5}} = \dfrac{4}{\sqrt{5}} \cdot \dfrac{\sqrt{5}}{\sqrt{5}} = \dfrac{4\sqrt{5}}{5}$

11. $\dfrac{6}{\sqrt{6}} = \dfrac{6}{\sqrt{6}} \cdot \dfrac{\sqrt{6}}{\sqrt{6}} = \dfrac{6\sqrt{6}}{6} = \sqrt{6}$

13. $\dfrac{1}{\sqrt{z}} = \dfrac{1}{\sqrt{z}} \cdot \dfrac{\sqrt{z}}{\sqrt{z}} = \dfrac{\sqrt{z}}{z}$

15. $\dfrac{p}{\sqrt{2}} = \dfrac{p}{\sqrt{2}} \cdot \dfrac{\sqrt{2}}{\sqrt{2}} = \dfrac{p\sqrt{2}}{2}$

17. $\dfrac{\sqrt{y}}{\sqrt{7}} = \dfrac{\sqrt{y}}{\sqrt{7}} \cdot \dfrac{\sqrt{7}}{\sqrt{7}} = \dfrac{\sqrt{7y}}{7}$

19. $\dfrac{6\sqrt{3}}{\sqrt{6}} = \dfrac{6\sqrt{3}}{\sqrt{6}} \cdot \dfrac{\sqrt{6}}{\sqrt{6}} = \dfrac{6\sqrt{18}}{6} = \dfrac{6 \cdot 3\sqrt{2}}{6} = 3\sqrt{2}$

21. $\dfrac{\sqrt{x}}{\sqrt{y}} = \dfrac{\sqrt{x}}{\sqrt{y}} \cdot \dfrac{\sqrt{y}}{\sqrt{y}} = \dfrac{\sqrt{xy}}{y}$

23. $\sqrt{\dfrac{5m}{8}} = \dfrac{\sqrt{5m}}{\sqrt{8}}$

$\qquad = \dfrac{\sqrt{5m}}{2\sqrt{2}}$

$\qquad = \dfrac{\sqrt{5m}}{2\sqrt{2}} \cdot \dfrac{\sqrt{2}}{\sqrt{2}}$

$\qquad = \dfrac{\sqrt{10m}}{2\cdot 2} = \dfrac{\sqrt{10m}}{4}$

25. $\dfrac{2n}{\sqrt{18n}} = \dfrac{2n}{\sqrt{9}\sqrt{2n}}$

$\qquad = \dfrac{2n}{3\sqrt{2n}}$

$\qquad = \dfrac{2n}{3\sqrt{2n}} \cdot \dfrac{\sqrt{2n}}{\sqrt{2n}}$

$\qquad = \dfrac{2n\sqrt{2n}}{3\cdot 2n} = \dfrac{\sqrt{2n}}{3}$

27. $\sqrt{\dfrac{18x^4 y^3}{2z^3}} = \sqrt{\dfrac{9x^4 y^3}{z^3}} = \dfrac{\sqrt{9x^4 y^3}}{\sqrt{z^2 \cdot z}}$

$\qquad = \dfrac{\sqrt{9x^4 y^2}\sqrt{y}}{\sqrt{z^2}\sqrt{z}} = \dfrac{3x^2 y\sqrt{y}}{z\sqrt{z}}$

$\qquad = \dfrac{3x^2 y\sqrt{y}}{z\sqrt{z}} \cdot \dfrac{\sqrt{z}}{\sqrt{z}}$

$\qquad = \dfrac{3x^2 y\sqrt{yz}}{z^2}$

29. $\sqrt{\dfrac{20y^4 z^3}{3xy^{-4}}} = \sqrt{\dfrac{20y^8 z^3}{3x}} = \dfrac{\sqrt{20y^8 z^3}}{\sqrt{3x}}$

$\qquad = \dfrac{\sqrt{4y^8 z^2}\sqrt{5z}}{\sqrt{3x}} = \dfrac{2y^4 z\sqrt{5z}}{\sqrt{3x}}$

$\qquad = \dfrac{2y^4 z\sqrt{5z}}{\sqrt{3x}} \cdot \dfrac{\sqrt{3x}}{\sqrt{3x}}$

$\qquad = \dfrac{2y^4 z\sqrt{15xz}}{3x}$

31. $\sqrt{\dfrac{48x^6 y^5}{3z^3}} = \sqrt{\dfrac{16x^6 y^5}{z^3}} = \dfrac{\sqrt{16x^6 y^5}}{\sqrt{z^2 \cdot z}}$

$\qquad = \dfrac{\sqrt{16x^6 y^4}\sqrt{y}}{\sqrt{z^2}\sqrt{z}} = \dfrac{4x^3 y^2\sqrt{y}}{z\sqrt{z}}$

$\qquad = \dfrac{4x^3 y^2\sqrt{y}}{z\sqrt{z}} \cdot \dfrac{\sqrt{z}}{\sqrt{z}}$

$\qquad = \dfrac{4x^3 y^2\sqrt{yz}}{z^2}$

33. $\dfrac{1}{\sqrt[3]{2}} = \dfrac{1}{\sqrt[3]{2}} \cdot \dfrac{\sqrt[3]{4}}{\sqrt[3]{4}} = \dfrac{\sqrt[3]{4}}{\sqrt[3]{8}} = \dfrac{\sqrt[3]{4}}{2}$

35. $\dfrac{8}{\sqrt[3]{y}} = \dfrac{8}{\sqrt[3]{y}} \cdot \dfrac{\sqrt[3]{y^2}}{\sqrt[3]{y^2}} = \dfrac{8\sqrt[3]{y^2}}{y}$

37. $\dfrac{1}{\sqrt[4]{3}} = \dfrac{1}{\sqrt[4]{3}} \cdot \dfrac{\sqrt[4]{27}}{\sqrt[4]{27}} = \dfrac{\sqrt[4]{27}}{\sqrt[4]{27}} = \dfrac{\sqrt[4]{27}}{3}$

39. $\dfrac{a}{\sqrt[4]{8}} = \dfrac{a}{\sqrt[4]{8}} \cdot \dfrac{\sqrt[4]{2}}{\sqrt[4]{2}} = \dfrac{a\sqrt[4]{2}}{\sqrt[4]{16}} = \dfrac{a\sqrt[4]{2}}{2}$

41. $\dfrac{5}{\sqrt[4]{z^2}} = \dfrac{5}{\sqrt[4]{z^2}} \cdot \dfrac{\sqrt[4]{z^2}}{\sqrt[4]{z^2}} = \dfrac{5\sqrt[4]{z^2}}{\sqrt[4]{z^4}} = \dfrac{5\sqrt[4]{z^2}}{z}$

43. $\dfrac{10}{\sqrt[5]{y^3}} = \dfrac{10}{\sqrt[5]{y^3}} \cdot \dfrac{\sqrt[5]{y^2}}{\sqrt[5]{y^2}} = \dfrac{10\sqrt[5]{y^2}}{\sqrt[5]{y^5}} = \dfrac{10\sqrt[5]{y^2}}{y}$

45. $\dfrac{2}{\sqrt[7]{a^4}} = \dfrac{2}{\sqrt[7]{a^4}} \cdot \dfrac{\sqrt[7]{a^3}}{\sqrt[7]{a^3}} = \dfrac{2\sqrt[7]{a^3}}{\sqrt[7]{a^7}} = \dfrac{2\sqrt[7]{a^3}}{a}$

47. $\sqrt[3]{\dfrac{1}{2x}} = \dfrac{\sqrt[3]{1}}{\sqrt[3]{2x}} = \dfrac{1}{\sqrt[3]{2x}} \cdot \dfrac{\sqrt[3]{4x^2}}{\sqrt[3]{4x^2}} = \dfrac{\sqrt[3]{4x^2}}{2x}$

49. $\dfrac{5m}{\sqrt[4]{2}} = \dfrac{5m}{\sqrt[4]{2}} \cdot \dfrac{\sqrt[4]{2^3}}{\sqrt[4]{2^3}} = \dfrac{5m\sqrt[4]{8}}{2}$

51. $\sqrt[4]{\dfrac{5}{3x^3}} = \dfrac{\sqrt[4]{5}}{\sqrt[4]{3x^3}} \cdot \dfrac{\sqrt[4]{3^3 x}}{\sqrt[4]{3^3 x}} = \dfrac{\sqrt[4]{135x}}{3x}$

53. $\sqrt[3]{\dfrac{3x^2}{2y^2}} = \dfrac{\sqrt[3]{3x^2}}{\sqrt[3]{2y^2}} \cdot \dfrac{\sqrt[3]{4y}}{\sqrt[3]{4y}} = \dfrac{\sqrt[3]{12x^2 y}}{2y}$

55. $\sqrt[3]{\dfrac{14xy^2}{2z^2}} = \sqrt[3]{\dfrac{7xy^2}{z^2}}$

$= \dfrac{\sqrt[3]{7xy^2}}{\sqrt[3]{z^2}}$

$= \dfrac{\sqrt[3]{7xy^2}}{\sqrt[3]{z^2}} \cdot \dfrac{\sqrt[3]{z}}{\sqrt[3]{z}}$

$= \dfrac{\sqrt[3]{7xy^2 z}}{z}$

57. $\left(5 - \sqrt{6}\right)\left(5 + \sqrt{6}\right) = 5^2 - \left(\sqrt{6}\right)^2$

$= 25 - 6$

$= 19$

59. $\left(8 + \sqrt{2}\right)\left(8 - \sqrt{2}\right) = 8^2 - \left(\sqrt{2}\right)^2 = 64 - 2 = 62$

61. $\left(2 - \sqrt{10}\right)\left(2 + \sqrt{10}\right) = 2^2 - \left(\sqrt{10}\right)^2 = 4 - 10 = -6$

63. $\left(\sqrt{a} - \sqrt{b}\right)\left(\sqrt{a} + \sqrt{b}\right) = \left(\sqrt{a}\right)^2 - \left(\sqrt{b}\right)^2 = a - b$

65. $\left(2\sqrt{x} - 3\sqrt{y}\right)\left(2\sqrt{x} + 3\sqrt{y}\right) = \left(2\sqrt{x}\right)^2 - \left(3\sqrt{y}\right)^2$

$= 4x - 9y$

67. $\dfrac{2}{\sqrt{3} + 1} = \dfrac{2}{\sqrt{3} + 1} \cdot \dfrac{\left(\sqrt{3} - 1\right)}{\left(\sqrt{3} - 1\right)}$

$= \dfrac{2\left(\sqrt{3} - 1\right)}{\left(\sqrt{3}\right)^2 - 1^2}$

$= \dfrac{2\left(\sqrt{3} - 1\right)}{3 - 1}$

$= \dfrac{2\left(\sqrt{3} - 1\right)}{2}$

$= \sqrt{3} - 1$

69. $\dfrac{1}{2 + \sqrt{3}} = \dfrac{1}{2 + \sqrt{3}} \cdot \dfrac{\left(2 - \sqrt{3}\right)}{\left(2 - \sqrt{3}\right)}$

$= \dfrac{2 - \sqrt{3}}{2^2 - \left(\sqrt{3}\right)^2}$

$= \dfrac{2 - \sqrt{3}}{4 - 3}$

$= \dfrac{2 - \sqrt{3}}{1}$

$= 2 - \sqrt{3}$

71. $\dfrac{5}{\sqrt{2} - 7} = \dfrac{5}{\sqrt{2} - 7} \cdot \dfrac{\left(\sqrt{2} + 7\right)}{\left(\sqrt{2} + 7\right)}$

$= \dfrac{5\sqrt{2} + 35}{2 - 49}$

$= \dfrac{5\sqrt{2} + 35}{-47}$

$= \dfrac{-5\sqrt{2} - 35}{47}$

73. $\dfrac{\sqrt{5}}{2\sqrt{5} - \sqrt{6}} = \dfrac{\sqrt{5}}{2\sqrt{5} - \sqrt{6}} \cdot \dfrac{\left(2\sqrt{5} + \sqrt{6}\right)}{\left(2\sqrt{5} + \sqrt{6}\right)}$

$= \dfrac{10 + \sqrt{30}}{20 - 6}$

$= \dfrac{10 + \sqrt{30}}{14}$

75. $\dfrac{3}{6 + \sqrt{x}} = \dfrac{3}{6 + \sqrt{x}} \cdot \dfrac{\left(6 - \sqrt{x}\right)}{\left(6 - \sqrt{x}\right)} = \dfrac{18 - 3\sqrt{x}}{36 - x}$

77. $\dfrac{4\sqrt{x}}{\sqrt{x} - y} = \dfrac{4\sqrt{x}}{\sqrt{x} - y} \cdot \dfrac{\left(\sqrt{x} + y\right)}{\left(\sqrt{x} + y\right)} = \dfrac{4x + 4y\sqrt{x}}{x - y^2}$

79. $\dfrac{\sqrt{2}-2\sqrt{3}}{\sqrt{2}+4\sqrt{3}} = \dfrac{\sqrt{2}-2\sqrt{3}}{\sqrt{2}+4\sqrt{3}} \cdot \dfrac{\left(\sqrt{2}-4\sqrt{3}\right)}{\left(\sqrt{2}-4\sqrt{3}\right)}$

$\qquad = \dfrac{2-4\sqrt{6}-2\sqrt{6}+8\cdot 3}{2-16\cdot 3}$

$\qquad = \dfrac{26-6\sqrt{6}}{-46}$

$\qquad = \dfrac{-13+3\sqrt{6}}{23}$

81. $\dfrac{\sqrt{a^3}+\sqrt{a^7}}{\sqrt{a}} = \dfrac{\sqrt{a^3}+\sqrt{a^7}}{\sqrt{a}} \cdot \dfrac{\left(\sqrt{a}\right)}{\left(\sqrt{a}\right)}$

$\qquad = \dfrac{\sqrt{a^4}+\sqrt{a^8}}{a}$

$\qquad = \dfrac{a^2+a^4}{a}$

$\qquad = a+a^3$

83. $\dfrac{4}{\sqrt{x+2}-3} = \dfrac{4}{\sqrt{x+2}-3} \cdot \dfrac{\left(\sqrt{x+2}+3\right)}{\left(\sqrt{x+2}+3\right)}$

$\qquad = \dfrac{4\sqrt{x+2}+12}{\left(\sqrt{x+2}\right)^2-3^2}$

$\qquad = \dfrac{4\sqrt{x+2}+12}{x+2-9}$

$\qquad = \dfrac{4\sqrt{x+2}+12}{x-7}$

85. $\sqrt{\dfrac{x}{16}} = \dfrac{\sqrt{x}}{\sqrt{16}} = \dfrac{\sqrt{x}}{4}$

87. $\sqrt{\dfrac{2}{9}} = \dfrac{\sqrt{2}}{\sqrt{9}} = \dfrac{\sqrt{2}}{3}$

89. $\left(\sqrt{7}+\sqrt{6}\right)\left(\sqrt{7}-\sqrt{6}\right) = \left(\sqrt{7}\right)^2-\left(\sqrt{6}\right)^2$

$\qquad\qquad\qquad\qquad = 7-6$

$\qquad\qquad\qquad\qquad = 1$

91. $\sqrt{\dfrac{24x^3y^6}{5z}} = \dfrac{\sqrt{24x^3y^6}}{\sqrt{5z}}$

$\qquad = \dfrac{\sqrt{4x^2y^6}\sqrt{6x}}{\sqrt{5z}}$

$\qquad = \dfrac{2xy^3\sqrt{6x}}{\sqrt{5z}} \cdot \dfrac{\sqrt{5z}}{\sqrt{5z}}$

$\qquad = \dfrac{2xy^3\sqrt{30xz}}{5z}$

93. $\sqrt{\dfrac{28xy^4}{2x^3y^4}} = \sqrt{\dfrac{14}{x^2}} = \dfrac{\sqrt{14}}{\sqrt{x^2}} = \dfrac{\sqrt{14}}{x}$

95. $\dfrac{1}{\sqrt{a}+7} = \dfrac{1}{\sqrt{a}+7} \cdot \dfrac{\sqrt{a}-7}{\sqrt{a}-7}$

$\qquad = \dfrac{\sqrt{a}-7}{\left(\sqrt{a}+7\right)\left(\sqrt{a}-7\right)}$

$\qquad = \dfrac{\sqrt{a}-7}{\left(\sqrt{a}\right)^2-\left(7\right)^2}$

$\qquad = \dfrac{\sqrt{a}-7}{a-49}$

97. $-\dfrac{7\sqrt{x}}{\sqrt{98}} = -\dfrac{7\sqrt{x}}{7\sqrt{2}}$

$\qquad = -\dfrac{\sqrt{x}}{\sqrt{2}} \cdot \dfrac{\sqrt{2}}{\sqrt{2}}$

$\qquad = -\dfrac{\sqrt{2x}}{\sqrt{4}}$

$\qquad = -\dfrac{\sqrt{2x}}{2}$

99. $\sqrt[4]{\dfrac{3y^2}{2x}} = \dfrac{\sqrt[4]{3y^2}}{\sqrt[4]{2x}} \cdot \dfrac{\sqrt[4]{8x^3}}{\sqrt[4]{8x^3}} = \dfrac{\sqrt[4]{24x^3y^2}}{\sqrt[4]{16x^4}} = \dfrac{\sqrt[4]{24x^3y^2}}{2x}$

101. $\sqrt[3]{\dfrac{32y^{12}z^{10}}{2x}} = \sqrt[3]{\dfrac{16y^{12}z^{10}}{x}}$

$= \dfrac{\sqrt[3]{16y^{12}z^{10}}}{\sqrt[3]{x}}$

$= \dfrac{\sqrt[3]{8y^{12}z^{9}}\sqrt[3]{2z}}{\sqrt[3]{x}}$

$= \dfrac{2y^{4}z^{3}\sqrt[3]{2z}}{\sqrt[3]{x}} \cdot \dfrac{\sqrt[3]{x^{2}}}{\sqrt[3]{x^{2}}}$

$= \dfrac{2y^{4}z^{3}\sqrt[3]{2x^{2}z}}{\sqrt[3]{x^{3}}}$

$= \dfrac{2y^{4}z^{3}\sqrt[3]{2x^{2}z}}{x}$

103. $\dfrac{\sqrt{ar}}{\sqrt{a}-2\sqrt{r}} \cdot \dfrac{\left(\sqrt{a}+2\sqrt{r}\right)}{\left(\sqrt{a}+2\sqrt{r}\right)} = \dfrac{\sqrt{ar}\left(\sqrt{a}+2\sqrt{r}\right)}{\left(\sqrt{a}\right)^{2}-\left(2\sqrt{r}\right)^{2}}$

$= \dfrac{a\sqrt{r}+2r\sqrt{a}}{a-4r}$

105. $\dfrac{\sqrt[3]{6x}}{\sqrt[3]{5xy}} = \sqrt[3]{\dfrac{6x}{5xy}}$

$= \sqrt[3]{\dfrac{6}{5y}}$

$= \dfrac{\sqrt[3]{6}}{\sqrt[3]{5y}} \cdot \dfrac{\sqrt[3]{25y^{2}}}{\sqrt[3]{25y^{2}}}$

$= \dfrac{\sqrt[3]{150y^{2}}}{5y}$

107. $\sqrt[4]{\dfrac{2x^{7}y^{12}z^{4}}{3x^{9}}} = \sqrt[4]{\dfrac{2y^{12}z^{4}}{3x^{2}}}$

$= \dfrac{\sqrt[4]{2y^{12}z^{4}}}{\sqrt[4]{3x^{2}}}$

$= \dfrac{\sqrt[4]{y^{12}z^{4}}\sqrt[4]{2}}{\sqrt[4]{3x^{2}}}$

$= \dfrac{y^{3}z\sqrt[4]{2}}{\sqrt[4]{3x^{2}}} \cdot \dfrac{\sqrt[4]{27x^{2}}}{\sqrt[4]{27x^{2}}}$

$= \dfrac{y^{3}z\sqrt[4]{54x^{2}}}{\sqrt[4]{81x^{4}}}$

$= \dfrac{y^{3}z\sqrt[4]{54x^{2}}}{3x}$

109. $\dfrac{1}{\sqrt{2}}+\dfrac{\sqrt{2}}{2} = \dfrac{1}{\sqrt{2}} \cdot \dfrac{\sqrt{2}}{\sqrt{2}}+\dfrac{\sqrt{2}}{2}$

$= \dfrac{\sqrt{2}}{\sqrt{4}}+\dfrac{\sqrt{2}}{2}$

$= \dfrac{\sqrt{2}}{2}+\dfrac{\sqrt{2}}{2}$

$= \dfrac{2\sqrt{2}}{2} = \sqrt{2}$

111. $\sqrt{5}-\dfrac{2}{\sqrt{5}} = \sqrt{5}-\dfrac{2}{\sqrt{5}} \cdot \dfrac{\sqrt{5}}{\sqrt{5}}$

$= \sqrt{5}-\dfrac{2\sqrt{5}}{5}$

$= \dfrac{5\sqrt{5}}{5}-\dfrac{2\sqrt{5}}{5}$

$= \dfrac{5\sqrt{5}-2\sqrt{5}}{5}$

$= \dfrac{3\sqrt{5}}{5}$

113. $4\sqrt{\dfrac{1}{6}} + \sqrt{24} = \dfrac{4}{\sqrt{6}} + 2\sqrt{6}$

$$= \dfrac{4}{\sqrt{6}} \cdot \dfrac{\sqrt{6}}{\sqrt{6}} + 2\sqrt{6}$$

$$= \dfrac{4\sqrt{6}}{6} + 2\sqrt{6}$$

$$= \dfrac{2\sqrt{6}}{3} + \dfrac{(2\sqrt{6})3}{3}$$

$$= \dfrac{2\sqrt{6}}{3} + \dfrac{6\sqrt{6}}{3}$$

$$= \dfrac{8\sqrt{6}}{3}$$

115. $5\sqrt{2} - \dfrac{2}{\sqrt{8}} + \sqrt{50} = 5\sqrt{2} - \dfrac{2}{\sqrt{8}} \cdot \dfrac{\sqrt{8}}{\sqrt{8}} + \sqrt{25}\sqrt{2}$

$$= 3\sqrt{2} - \dfrac{2\sqrt{4}\sqrt{2}}{\sqrt{64}} + 5\sqrt{2}$$

$$= \dfrac{5\sqrt{2}}{1} - \dfrac{4\sqrt{2}}{8} + \dfrac{5\sqrt{2}}{1}$$

$$= \dfrac{10\sqrt{2}}{2} - \dfrac{\sqrt{2}}{2} + \dfrac{10\sqrt{2}}{2}$$

$$= \dfrac{(10-1+10)\sqrt{2}}{2}$$

$$= \dfrac{19\sqrt{2}}{2}$$

117. $\sqrt{\dfrac{1}{2}} + 7\sqrt{2} + \sqrt{18} = \dfrac{\sqrt{1}}{\sqrt{2}} \cdot \dfrac{\sqrt{2}}{\sqrt{2}} + 7\sqrt{2} + \sqrt{9} \cdot \sqrt{2}$

$$= \dfrac{\sqrt{2}}{\sqrt{4}} + 7\sqrt{2} + 3\sqrt{2}$$

$$= \dfrac{\sqrt{2}}{2} + 10\sqrt{2}$$

$$= \dfrac{\sqrt{2}}{2} + \dfrac{20\sqrt{2}}{2}$$

$$= \dfrac{21\sqrt{2}}{2}$$

119. $\dfrac{2}{\sqrt{50}} - 3\sqrt{50} - \dfrac{1}{\sqrt{8}}$

$$= \dfrac{2}{5\sqrt{2}} - 3(5\sqrt{2}) - \dfrac{1}{2\sqrt{2}}$$

$$= \dfrac{2}{5\sqrt{2}} \cdot \dfrac{\sqrt{2}}{\sqrt{2}} - 15\sqrt{2} - \dfrac{1}{2\sqrt{2}} \cdot \dfrac{\sqrt{2}}{\sqrt{2}}$$

$$= \dfrac{2\sqrt{2}}{10} - 15\sqrt{2} - \dfrac{\sqrt{2}}{4}$$

$$= \dfrac{\sqrt{2}}{5} - 15\sqrt{2} - \dfrac{\sqrt{2}}{4}$$

$$= \dfrac{4\sqrt{2}}{20} - \dfrac{300\sqrt{2}}{20} - \dfrac{5\sqrt{2}}{20}$$

$$= -\dfrac{301\sqrt{2}}{20}$$

121. $\sqrt{\dfrac{3}{8}} + \sqrt{\dfrac{3}{2}} = \dfrac{\sqrt{3}}{\sqrt{8}} + \dfrac{\sqrt{3}}{\sqrt{2}}$

$$= \dfrac{\sqrt{3}}{2\sqrt{2}} + \dfrac{\sqrt{3}}{\sqrt{2}}$$

$$= \dfrac{\sqrt{3}}{2\sqrt{2}} \cdot \dfrac{\sqrt{2}}{\sqrt{2}} + \dfrac{\sqrt{3}}{\sqrt{2}} \cdot \dfrac{2\sqrt{2}}{2\sqrt{2}}$$

$$= \dfrac{\sqrt{6}}{4} + \dfrac{2\sqrt{6}}{4}$$

$$= \dfrac{3\sqrt{6}}{4}$$

123. $-2\sqrt{\dfrac{x}{y}} + 3\sqrt{\dfrac{y}{x}} = -2\dfrac{\sqrt{x}}{\sqrt{y}} + 3\dfrac{\sqrt{y}}{\sqrt{x}}$

$$= -2\dfrac{\sqrt{x}}{\sqrt{y}} \cdot \dfrac{\sqrt{y}}{\sqrt{y}} + 3\dfrac{\sqrt{y}}{\sqrt{x}} \cdot \dfrac{\sqrt{x}}{\sqrt{x}}$$

$$= -2\dfrac{\sqrt{xy}}{y} + 3\dfrac{\sqrt{xy}}{x}$$

$$= \left(-\dfrac{2}{y} + \dfrac{3}{x}\right)\sqrt{xy}$$

125. $\dfrac{3}{\sqrt{a}} - \sqrt{\dfrac{9}{a}} + 2\sqrt{a} = \dfrac{3}{\sqrt{a}} - \dfrac{\sqrt{9}}{\sqrt{a}} + 2\sqrt{a}$

$$= \dfrac{3}{\sqrt{a}}\left(\dfrac{\sqrt{a}}{\sqrt{a}}\right) - \dfrac{\sqrt{9}}{\sqrt{a}}\left(\dfrac{\sqrt{a}}{\sqrt{a}}\right) + 2\sqrt{a}$$

$$= \dfrac{3\sqrt{a}}{a} - \dfrac{3\sqrt{a}}{a} + 2\sqrt{a}$$

$$= 2\sqrt{a}$$

127. $\dfrac{\sqrt{(a+b)^4}}{\sqrt[3]{a+b}} = \dfrac{(a+b)^{4/2}}{(a+b)^{1/3}}$

$= (a+b)^{6/3-1/3}$

$= (a+b)^{5/3}$

$= \sqrt[3]{(a+b)^5}$

129. $\dfrac{\sqrt[5]{(a+2b)^4}}{\sqrt[3]{(a+2b)^2}} = \dfrac{(a+2b)^{4/5}}{(a+2b)^{2/3}}$

$= (a+2b)^{4/5-2/3}$

$= (a+2b)^{2/15}$

$= \sqrt[15]{(a+2b)^2}$

131. $\dfrac{\sqrt[3]{r^2 s^4}}{\sqrt{rs}} = \dfrac{(r^2 s^4)^{1/3}}{(rs)^{1/2}}$

$= \dfrac{r^{2/3} s^{4/3}}{r^{1/2} s^{1/2}}$

$= r^{2/3-1/2} s^{4/3-1/2}$

$= r^{1/6} s^{5/6}$

$= (rs^5)^{1/6}$

$= \sqrt[6]{rs^5}$

133. $\dfrac{\sqrt[5]{x^4 y^6}}{\sqrt[3]{(xy)^2}} = \dfrac{(x^4 y^6)^{1/5}}{(xy)^{2/3}}$

$= \dfrac{x^{4/5} y^{6/5}}{x^{2/3} y^{2/3}}$

$= x^{4/5-2/3} y^{6/5-2/3}$

$= x^{2/15} y^{8/15}$

$= (x^2 y^8)^{1/15}$

$= \sqrt[15]{x^2 y^8}$

135. $d = \sqrt{\dfrac{72}{I}}$

$d = \sqrt{\dfrac{72}{5.3}} \approx 3.69$

The person is about 3.69 m from the light source.

137. $r = \sqrt[3]{\dfrac{3V}{4\pi}}$

$r = \sqrt[3]{\dfrac{3(7238.23)}{4\pi}} = 12$

The radius of the tank is 12 inches.

139. $N(t) = \dfrac{6.21}{\sqrt[4]{t}}$

a. $t = 1960 - 1959 = 1$

$N(1) = \dfrac{6.21}{\sqrt[4]{1}} = 6.21$

The number of farms in 1960 was 6.21 million.

b. $t = 2008 - 1959 = 49$

$N(49) = \dfrac{6.21}{\sqrt[4]{49}} \approx 2.35$

The number of farms in 2008 will be about 2.35 million.

141. $\dfrac{2}{\sqrt{2}} = \dfrac{2}{\sqrt{2}} \cdot \dfrac{\sqrt{2}}{\sqrt{2}} = \dfrac{2\sqrt{2}}{\sqrt{4}} = \dfrac{2\sqrt{2}}{2} = \sqrt{2}$

$\dfrac{3}{\sqrt{3}} = \dfrac{3}{\sqrt{3}} \cdot \dfrac{\sqrt{3}}{\sqrt{3}} = \dfrac{3\sqrt{3}}{\sqrt{9}} = \dfrac{3\sqrt{3}}{3} = \sqrt{3}$

Since $3 > 2$, then $\sqrt{3} > \sqrt{2}$ and we conclude that $\dfrac{3}{\sqrt{3}} > \dfrac{2}{\sqrt{2}}$.

143. $\dfrac{1}{\sqrt{3}+2} = \dfrac{1}{\sqrt{3}+2} \cdot \dfrac{\sqrt{3}-2}{\sqrt{3}-2}$

$= \dfrac{\sqrt{3}-2}{\left(\sqrt{3}\right)^2 - 2^2}$

$= \dfrac{\sqrt{3}-2}{3-4}$

$= \dfrac{\sqrt{3}-2}{-1}$

$= -\sqrt{3}+2$

$= 2-\sqrt{3}$

$2+\sqrt{3} > 2-\sqrt{3}$

Therefore, $2+\sqrt{3} > \dfrac{1}{\sqrt{3}+2}$.

145. $f(x) = x^{a/2}$, $g(x) = x^{b/3}$

a. $x^{4/2} = x^2$

$x^{12/2} = x^6$

$x^{8/2} = x^4$

Therefore $x^{a/2}$ is a perfect square when $a = 4, 8, 12$.

b. $x^{9/3} = x^3$

$x^{18/3} = x^6$

$x^{27/3} = x^9$

Therefore, $x^{b/3}$ is a perfect cube when $b = 9, 18, 27$.

c. $(f \cdot g)(x) = f(x) \cdot g(x)$

$= x^{a/2} \cdot x^{b/3}$

$= x^{a/2 + b/3}$

$= x^{3a/6 + 2b/6}$

$= x^{(3a+2b)/6}$

d. $\left(\dfrac{f}{g}\right)(x) = \dfrac{f(x)}{g(x)}$

$= \dfrac{x^{a/2}}{x^{b/3}}$

$= x^{a/2 - b/3}$

$= x^{3a/6 - 2b/6}$

$= x^{(3a-2b)/6}$

147. $\dfrac{3}{\sqrt{2a-3b}} = \dfrac{3}{\sqrt{2a-3b}} \cdot \dfrac{\sqrt{2a-3b}}{\sqrt{2a-3b}} = \dfrac{3\sqrt{2a-3b}}{2a-3b}$

149. $\dfrac{5-\sqrt{5}}{6} = \dfrac{5-\sqrt{5}}{6} \cdot \dfrac{5+\sqrt{5}}{5+\sqrt{5}}$

$= \dfrac{25-5}{30+6\sqrt{5}} = \dfrac{20}{2(15+3\sqrt{5})}$

$= \dfrac{10}{15+3\sqrt{5}}$

151. $\dfrac{\sqrt{x+h}-\sqrt{x}}{h} = \dfrac{\sqrt{x+h}-\sqrt{x}}{h} \cdot \dfrac{\sqrt{x+h}+\sqrt{x}}{\sqrt{x+h}+\sqrt{x}}$

$= \dfrac{x+h-x}{h\left(\sqrt{x+h}+\sqrt{x}\right)}$

$= \dfrac{h}{h\left(\sqrt{x+h}+\sqrt{x}\right)}$

$= \dfrac{1}{\sqrt{x+h}+\sqrt{x}}$

154. $A = \dfrac{1}{2}h(b_1 + b_2)$

$2A = h(b_1 + b_2)$

$\dfrac{2A}{h} = b_1 + b_2$

$\dfrac{2A}{h} - b_1 = b_2$

$b_2 = \dfrac{2A}{h} - b_1$

155. Let r be the rate of the slower car and $r + 10$ be the rate of the faster.
Distance the first traveled plus distance the second traveled is 270 miles.

$3r + 3(r+10) = 270$

$3r + 3r + 30 = 270$

$6r + 30 = 270$

$6r = 240$

$r = 40$

The rate of the slower car is 40 mph and the rate of the faster car is $r + 10 = 50$ mph.

156.

$$
\begin{array}{r}
4x^2 + 9x - 2 \\
x - 2 \\
\hline
-8x^2 - 18x + 4 \\
4x^3 + 9x^2 - 2x \\
\hline
4x^3 + x^2 - 20x + 4
\end{array}
$$

157. $\dfrac{x}{2} - \dfrac{4}{x} = -\dfrac{7}{2}$

$2x\left(\dfrac{x}{2} - \dfrac{4}{x}\right) = 2x\left(-\dfrac{7}{2}\right)$

$x^2 - 8 = -7x$

$x^2 + 7x - 8 = 0$

$(x+8)(x-1) = 0$

$x = -8 \text{ or } 1$

Exercise Set 11.6

1. a. Answers will vary.

b.
$$\sqrt{2x+26} - 2 = 4$$
$$\sqrt{2x+26} - 2 + 2 = 4 + 2$$
$$\sqrt{2x+26} = 6$$
$$\left(\sqrt{2x+26}\right)^2 = 6^2$$
$$2x + 26 = 36$$
$$2x = 10$$
$$x = \frac{10}{2}$$
$$x = 5$$

3. 0 is the only solution to the equation. For all other values, the left side of the equation is negative and the right side is positive.

5. Answers will vary. Possible answer:
The equation has no solution since the left side is a positive number whereas the right side is 0. A positive number is never equal to 0.

Also, the equation can be written as $\sqrt{x-3} = -4$ for which the left side is positive and the right side is negative. It is impossible for $\sqrt{x-3}$ to equal a negative number.

7. One solution, $x = 25$. Note that the domain of x is all nonnegative numbers. Therefore we have
$$\left(\sqrt{x}\right)^2 = 5^2$$
$$x = 25$$

9.
$$\sqrt{x} = 4$$
$$\left(\sqrt{x}\right)^2 = 4^2$$
$$x = 16$$

Check: $\sqrt{16} = 4$
$$4 = 4 \text{ True}$$

11. $\sqrt{x} = -9$ No real solution. The principal square root is never negative.

13.
$$\sqrt[3]{x} = -4$$
$$\left(\sqrt[3]{x}\right)^3 = (-4)^3$$
$$x = -64$$

Check: $\sqrt[3]{-64} = -4$
$$\sqrt[3]{(-4)^3} = -4$$
$$-4 = -4 \text{ True}$$

15.
$$\sqrt{2x+3} = 5$$
$$\left(\sqrt{2x+3}\right)^2 = (5)^2$$
$$2x + 3 = 25$$
$$2x = 22$$
$$x = 11$$

Check: $\sqrt{2x+3} = 5$
$$\sqrt{2(11)+3} = 5$$
$$\sqrt{22+3} = 5$$
$$\sqrt{25} = 5$$
$$5 = 5 \text{ True}$$

17.
$$\sqrt[3]{3x} + 4 = 7$$
$$\sqrt[3]{3x} = 3$$
$$\left(\sqrt[3]{3x}\right)^3 = (3)^3$$
$$3x = 27$$
$$x = 9$$

Check: $\sqrt[3]{3(9)} + 4 = 7$
$$\sqrt[3]{27} + 4 = 7$$
$$3 + 4 = 7$$
$$7 = 7 \text{ True}$$

19.
$$\sqrt[3]{2x+29} = 3$$
$$\left(\sqrt[3]{2x+29}\right)^3 = 3^3$$
$$2x + 29 = 27$$
$$2x = -2$$
$$x = -1$$

Check: $\sqrt[3]{2(-1)+29} = 3$
$$\sqrt[3]{-2+29} = 3$$
$$\sqrt[3]{27} = 3$$
$$3 = 3 \text{ True}$$

21. $\sqrt[4]{x} = 3$

$\left(\sqrt[4]{x}\right)^4 = 3^4$

$x = 81$

Check: $\sqrt[4]{81} = 3$

$\sqrt[4]{3^4} = 3$

$3 = 3$ True

23. $\sqrt[4]{x+10} = 3$

$\left(\sqrt[4]{x+10}\right)^4 = 3^4$

$x + 10 = 81$

$x = 71$

Check: $\sqrt[4]{71+10} = 3$

$\sqrt[4]{81} = 3$

$\sqrt[4]{3^4} = 3$

$3 = 3$ True

25. $\sqrt[4]{2x+1} + 6 = 2$

$\sqrt[4]{2x+1} = -4$

No real solution exists because an even root of a real number is never negative.

27. $\sqrt{x+8} = \sqrt{x-8}$

$\left(\sqrt{x+8}\right)^2 = \left(\sqrt{x-8}\right)^2$

$x + 8 = x - 8$

$8 = -8$ False

A contradiction results so the equation has no solution.

29. $2\sqrt[3]{x-1} = \sqrt[3]{x^2 + 2x}$

$\left(2\sqrt[3]{x-1}\right)^3 = \left(\sqrt[3]{x^2+2x}\right)^3$

$8(x-1) = x^2 + 2x$

$8x - 8 = x^2 + 2x$

$x^2 - 6x + 8 = 0$

$(x-4)(x-2) = 0$

$x = 4$ or $x = 2$

A check will show that both 2 and 4 are solutions to the equation.

31. $\sqrt[4]{x+8} = \sqrt[4]{2x}$

$\left(\sqrt[4]{x+8}\right)^4 = \left(\sqrt[4]{2x}\right)^4$

$x + 8 = 2x$

$x = 8$

A check will show that the solution is 8.

33. $\sqrt{5x+1} - 6 = 0$

$\sqrt{5x+1} = 6$

$\left(\sqrt[2]{5x+1}\right)^2 = (6)^2$

$5x + 1 = 36$

$5x = 35$

$x = 7$

A check will show that the solution is 7.

35. $\sqrt{m^2 + 6m - 4} = m$

$\left(\sqrt{m^2+6m-4}\right)^2 = (m)^2$

$m^2 + 6m - 4 = m^2$

$6m - 4 = 0$

$6m = 4$

$m = \dfrac{2}{3}$

Check: $\sqrt{\left(\dfrac{2}{3}\right)^2 + 6\left(\dfrac{2}{3}\right) - 4} = \dfrac{2}{3}$

$\sqrt{\dfrac{4}{9} + 4 - 4} = \dfrac{2}{3}$

$\sqrt{\dfrac{4}{9}} = \dfrac{2}{3}$

$\dfrac{2}{3} = \dfrac{2}{3}$ True

37. $\sqrt{5c+1} - 9 = 0$

$\sqrt{5c+1} = 9$

$\left(\sqrt{5c+1}\right)^2 = 9^2$

$5c + 1 = 81$

$5c = 80$

$c = 16$

A check will show that the solution is 16.

39.
$$\sqrt{z^2 + 5} = z + 1$$
$$\left(\sqrt{z^2 + 5}\right)^2 = (z+1)^2$$
$$z^2 + 5 = z^2 + 2z + 1$$
$$5 = 2z + 1$$
$$4 = 2z$$
$$2 = z$$
A check will show that the solution is 2.

41.
$$\sqrt{2y + 5} + 5 - y = 0$$
$$\sqrt{2y + 5} = y - 5$$
$$\left(\sqrt{2y + 5}\right)^2 = (y-5)^2$$
$$2y + 5 = y^2 - 10y + 25$$
$$0 = y^2 - 12y + 20$$
$$0 = (y - 10)(y - 2)$$
$$y = 10 \text{ or } y = 2$$

Check:
$$\sqrt{2y + 5} + 5 - y = 0$$
$$\sqrt{2(10) + 5} + 5 - (10) = 0$$
$$\sqrt{25} - 5 = 0$$
$$0 = 0 \text{ True}$$

$$\sqrt{2y + 5} + 5 - y = 0$$
$$\sqrt{2(2) + 5} + 5 - (2) = 0$$
$$\sqrt{9} + 3 = 0$$
$$6 = 0 \text{ False}$$

This check shows that 10 is the only solution to this equation.

43.
$$\sqrt{5x + 6} = 2x - 6$$
$$\left(\sqrt{5x + 6}\right)^2 = (2x - 6)^2$$
$$5x + 6 = 4x^2 - 24x + 36$$
$$0 = 4x^2 - 29x + 30$$
$$0 = (4x - 5)(x - 6)$$
$$x = \frac{5}{4} \text{ or } x = 6$$

Check:
$$\sqrt{5x + 6} = 2x - 6 \qquad \sqrt{5x + 6} = 2x - 6$$
$$\sqrt{5\left(\frac{5}{4}\right) + 6} = 2\left(\frac{5}{4}\right) - 6 \qquad \sqrt{5(6) + 6} = 2(6) - 6$$
$$\qquad \qquad \qquad \qquad \sqrt{36} = 12 - 6$$
$$\sqrt{\frac{49}{4}} = -\frac{14}{4} \qquad \qquad 6 = 6 \text{ True}$$
$$\frac{7}{2} = -\frac{7}{2} \text{ False}$$

This check shows that 6 is the only solution to this equation.

45.
$$(2a + 9)^{1/2} - a + 3 = 0$$
$$(2a + 9)^{1/2} = a - 3$$
$$[(2a + 9)^{1/2}]^2 = (a - 3)^2$$
$$2a + 9 = a^2 - 6a + 9$$
$$0 = a^2 - 8a$$
$$0 = a(a - 8)$$
$$a = 0 \text{ or } a = 8$$

Check: $(2a + 9)^{1/2} - a + 3 = 0$
$$(2 \cdot 0 + 9)^{1/2} - 0 + 3 = 0$$
$$3 + 3 = 0$$
$$6 = 0 \text{ False}$$

Check: $(2a + 9)^{1/2} - a + 3 = 0$
$$(2 \cdot 8 + 9)^{1/2} - 8 + 3 = 0$$
$$5 - 8 + 3 = 0$$
$$0 = 0 \text{ True}$$

This check shows that 8 is the only solution to the equation.

47.
$$(2x^2 + 4x + 9)^{1/2} = \sqrt{2x^2 + 9}$$
$$[(2x^2 + 4x + 9)^{1/2}]^2 = \left(\sqrt{2x^2 + 9}\right)^2$$
$$2x^2 + 4x + 9 = 2x^2 + 9$$
$$4x = 0$$
$$x = 0$$

Check: $\left(2(0)^2 + 4(0) + 9\right)^{1/2} = \sqrt{2(0)^2 + 9}$
$$(9)^{1/2} = \sqrt{9}$$
$$3 = 3 \text{ True}$$

49.
$$(r+4)^{1/3} = (3r+10)^{1/3}$$
$$[(r+4)^{1/3}]^3 = [(3r+10)^{1/3}]^3$$
$$r+4 = 3r+10$$
$$4 = 2r+10$$
$$-6 = 2r$$
$$-3 = r$$

Check: $(-3+4)^{1/3} = [3(-3)+10]^{1/3}$
$$(1)^{1/3} = (1)^{1/3}$$
$$1 = 1 \quad \text{True}$$

51.
$$(5x+7)^{1/4} = (9x+1)^{1/4}$$
$$[(5x+7)^{1/4}]^4 = [(9x+1)^{1/4}]^4$$
$$5x+7 = 9x+1$$
$$7 = 4x+1$$
$$6 = 4x$$
$$\frac{3}{2} = x$$

Check: $(5x+7)^{1/4} = (9x+1)^{1/4}$
$$\left(5 \cdot \frac{3}{2}+7\right)^{1/4} = \left(9 \cdot \frac{3}{2}+1\right)^{1/4}$$
$$\left(\frac{15}{2}+7\right)^{1/4} = \left(\frac{27}{2}+1\right)^{1/4}$$
$$\left(\frac{29}{2}\right)^{1/4} = \left(\frac{29}{2}\right)^{1/4} \quad \text{True}$$

53.
$$\sqrt[4]{x+5} = -2$$
$$\left(\sqrt[4]{x+5}\right)^4 = (-2)^4$$
$$x+5 = 16$$
$$x = 11$$

Check: $\sqrt[4]{x+5} = -2$
$$\sqrt[4]{11+5} = -2$$
$$\sqrt[4]{16} = -2$$
$$2 = -2 \quad \text{False}$$

Thus, 11 is not a solution to this equation and we conclude that there is no real solution.
Note: We could have determined that there was no real solution immediately by noting that an even root of a real number is never negative.

55.
$$\sqrt{4x+1} = \sqrt{2x}+1$$
$$\left(\sqrt{4x+1}\right)^2 = \left(\sqrt{2x}+1\right)^2$$
$$4x+1 = \left(\sqrt{2x}\right)^2 + 2 \cdot 1 \cdot \sqrt{2x} + 1^2$$
$$4x+1 = 2x + 2\sqrt{2x} + 1$$
$$2x = 2\sqrt{2x}$$
$$x = \sqrt{2x}$$
$$x^2 = \left(\sqrt{2x}\right)^2$$
$$x^2 = 2x$$
$$x^2 - 2x = 0$$
$$x(x-2) = 0$$
$$x = 0 \quad \text{or} \quad x-2 = 0$$
$$x = 2$$
A check will show that both 0 and 2 are solutions to the equation.

57.
$$\sqrt{3a+1} = \sqrt{a-4}+3$$
$$\left(\sqrt{3a+1}\right)^2 = \left(\sqrt{a-4}+3\right)^2$$
$$3a+1 = \left(\sqrt{a-4}\right)^2 + 2 \cdot 3\sqrt{a-4} + 3^2$$
$$3a+1 = a-4 + 6\sqrt{a-4} + 9$$
$$2a-4 = 6\sqrt{a-4}$$
$$(2a-4)^2 = \left(6\sqrt{a-4}\right)^2$$
$$4a^2 - 16a + 16 = 36(a-4)$$
$$4a^2 - 52a + 160 = 0$$
$$4(a^2 - 13a + 40) = 0$$
$$4(a-8)(a-5) = 0$$
$$a = 8 \quad \text{or} \quad a = 5$$
A check will show that both 5 and 8 are solutions to the equation.

59. $\sqrt{x+3} = \sqrt{x} - 3$

$\left(\sqrt{x+3}\right)^2 = \left(\sqrt{x} - 3\right)^2$

$x + 3 = x - 6\sqrt{x} + 9$

$-6 = -6\sqrt{x}$

$1 = \sqrt{x}$

$(1)^2 = \left(\sqrt{x}\right)^2$

$1 = x$

Check: $\sqrt{x+3} = \sqrt{x} - 3$

$\sqrt{1+3} = \sqrt{1} - 3$

$\sqrt{4} = 1 - 3$

$2 = -2$ False

Thus, 1 is not a solution to this equation and we conclude that there is no real solution.

61. $\sqrt{x+7} = 6 - \sqrt{x-5}$

$\left(\sqrt{x+7}\right)^2 = \left(6 - \sqrt{x-5}\right)^2$

$x + 7 = 36 - 12\sqrt{x-5} + x - 5$

$12\sqrt{x-5} = 24$

$\sqrt{x-5} = 2$

$x - 5 = 4$

$x = 9$

A check will show that the solution is 9.

63. $\sqrt{4x-3} = 2 + \sqrt{2x-5}$

$\left(\sqrt{4x-3}\right)^2 = \left(2 + \sqrt{2x-5}\right)^2$

$4x - 3 = 4 + 4\sqrt{2x-5} + 2x - 5$

$2x - 2 = 4\sqrt{2x-5}$

$x - 1 = 2\sqrt{2x-5}$

$(x-1)^2 = \left(2\sqrt{2x-5}\right)^2$

$x^2 - 2x + 1 = 4(2x-5)$

$x^2 - 2x + 1 = 8x - 20$

$x^2 - 10x + 21 = 0$

$(x-7)(x-3) = 0$

$x - 7 = 0$ or $x - 3 = 0$

$x = 7$ $x = 3$

A check will show that both 3 and 7 are solutions to the equation.

65. $\sqrt{y+1} = \sqrt{y+10} - 3$

$\left(\sqrt{y+1}\right)^2 = \left(\sqrt{y+10} - 3\right)^2$

$y + 1 = \left(\sqrt{y+10}\right)^2 - 6\sqrt{y+10} + 9$

$y + 1 = y + 10 - 6\sqrt{y+10} + 9$

$6\sqrt{y+10} = 18$

$\sqrt{y+10} = 3$

$\left(\sqrt{y+10}\right)^2 = 3^2$

$y + 10 = 9$

$y = -1$

A check will show that the solution is -1.

67. $f(x) = g(x)$

$\sqrt{x+8} = \sqrt{2x+1}$

$\left(\sqrt{x+8}\right)^2 = \left(\sqrt{2x+1}\right)^2$

$x + 8 = 2x + 1$

$7 = x$

A check will show that $x = 7$ is the solution to the equation $f(x) = g(x)$.

69. $f(x) = g(x)$

$\sqrt[3]{5x-19} = \sqrt[3]{6x-23}$

$\left(\sqrt[3]{5x-19}\right)^3 = \left(\sqrt[3]{6x-23}\right)^3$

$5x - 19 = 6x - 23$

$x = 4$

A check will show that $x = 4$ is the solution to the equation $f(x) = g(x)$.

71. $f(x) = g(x)$

$2(8x+24)^{1/3} = 4(2x-2)^{1/3}$

$[2(8x+24)^{1/3}]^3 = [4(2x-2)^{1/3}]^3$

$8(8x+24) = 64(2x-2)$

$64x + 192 = 128x - 128$

$64x = 320$

$x = 5$

A check will show that $x = 5$ is the solution to the equation $f(x) = g(x)$.

73.

$$p = \sqrt{2v}$$

$$p^2 = \left(\sqrt{2v}\right)^2$$

$$p^2 = 2v$$

$$\frac{p^2}{2} = v \text{ or } v = \frac{p^2}{2}$$

75.

$$v = \sqrt{2gh}$$

$$v^2 = \left(\sqrt{2gh}\right)^2$$

$$v^2 = 2gh$$

$$g = \frac{v^2}{2h}$$

77.

$$v = \sqrt{\frac{FR}{M}}$$

$$v^2 = \left(\sqrt{\frac{FR}{M}}\right)^2$$

$$v^2 = \frac{FR}{M}$$

$$Mv^2 = FR$$

$$F = \frac{Mv^2}{R}$$

79.

$$x = \sqrt{\frac{m}{k}}V_0$$

$$x^2 = \left(\sqrt{\frac{m}{k}}V_0\right)^2$$

$$x^2 = \frac{mV_0^2}{k}$$

$$x^2 k = mV_0^2$$

$$m = \frac{x^2 k}{V_0^2}$$

81.

$$r = \sqrt{\frac{A}{\pi}}$$

$$r^2 = \left(\sqrt{\frac{A}{\pi}}\right)^2$$

$$r^2 = \frac{A}{\pi}$$

$$\pi r^2 = A \text{ or } A = \pi r^2$$

83.

$$a^2 + b^2 = c^2$$

$$\left(\sqrt{6}\right)^2 + 9^2 = x^2$$

$$6 + 81 = x^2$$

$$87 = x^2 \Rightarrow x = \sqrt{87}$$

85.

$$a^2 + b^2 = c^2$$

$$x^2 + 5^2 = \left(\sqrt{65}\right)^2$$

$$x^2 + 25 = 65$$

$$x^2 = 40$$

$$x = \sqrt{40} \Rightarrow x = 2\sqrt{10}$$

87.

$$\sqrt{x+5} - \sqrt{x} = \sqrt{x-3}$$

$$\left(\sqrt{x+5} - \sqrt{x}\right)^2 = \left(\sqrt{x-3}\right)^2$$

$$x + 5 - 2\sqrt{x(x+5)} + x = x - 3$$

$$x + 8 = 2\sqrt{x^2 + 5x}$$

$$(x+8)^2 = \left(2\sqrt{x^2 + 5x}\right)^2$$

$$x^2 + 16x + 64 = 4(x^2 + 5x)$$

$$x^2 + 16x + 64 = 4x^2 + 20x$$

$$3x^2 + 4x - 64 = 0$$

$$(3x+16)(x-4) = 0 \Rightarrow x = -\frac{16}{3} \text{ or } x = 4$$

A check will show that $-\dfrac{16}{3}$ is an extraneous solution. The solution to the equation is 4.

89.

$$\sqrt{4y+6} + \sqrt{y+5} = \sqrt{y+1}$$

$$\left(\sqrt{4y+6} + \sqrt{y+5}\right)^2 = \left(\sqrt{y+1}\right)^2$$

$$4y + 6 + 2\sqrt{(4y+6)(y+5)} + y + 5 = y + 1$$

$$2\sqrt{(4y+6)(y+5)} = -4y - 10$$

$$\left(2\sqrt{4y^2 + 26y + 30}\right)^2 = (-4y-10)^2$$

$$4(4y^2 + 26y + 30) = 16y^2 + 80y + 100$$

$$16y^2 + 104y + 120 = 16y^2 + 80y + 100$$

$$24y = -20 \Rightarrow y = -\frac{5}{6}$$

Upon checking, this value does not satisfy the equation. There is no solution.

91. $\sqrt{c+1} + \sqrt{c-2} = \sqrt{3c}$

$\left(\sqrt{c+1} + \sqrt{c-2}\right)^2 = \left(\sqrt{3c}\right)^2$

$c+1 + 2\sqrt{(c+1)(c-2)} + c - 2 = 3c$

$2\sqrt{(c+1)(c-2)} = c + 1$

$\left(2\sqrt{c^2-c-2}\right)^2 = (c+1)^2$

$4(c^2 - c - 2) = c^2 + 2c + 1$

$4c^2 - 4c - 8 = c^2 + 2c + 1$

$3c^2 - 6c - 9 = 0$

$3(c+1)(c-3) = 0 \implies c = -1$ or $c = 3$

Upon checking, only $c = 3$ satisfies the equation.

93. $\sqrt{a+2} - \sqrt{a-3} = \sqrt{a-6}$

$\left(\sqrt{a+2} - \sqrt{a-3}\right)^2 = \left(\sqrt{a-6}\right)^2$

$a+2 - 2\sqrt{(a+2)(a-3)} + a - 3 = a - 6$

$a + 5 = 2\sqrt{(a+2)(a-3)}$

$(a+5)^2 = \left(2\sqrt{a^2-a-6}\right)^2$

$a^2 + 10a + 25 = 4(a^2 - a - 6)$

$a^2 + 10a + 25 = 4a^2 - 4a - 24$

$3a^2 - 14a - 49 = 0$

$(3a+7)(a-7) = 0 \implies a = -\dfrac{7}{3}$ or $a = 7$

Upon checking, only $a = 7$ satisfies the equation.

95. $\sqrt{2 - \sqrt{x}} = \sqrt{x}$

$\left(\sqrt{2-\sqrt{x}}\right)^2 = \left(\sqrt{x}\right)^2$

$2 - \sqrt{x} = x$

$2 - x = \sqrt{x}$

$(2-x)^2 = \left(\sqrt{x}\right)^2$

$4 - 4x + x^2 = x$

$x^2 - 5x + 4 = 0$

$(x-4)(x-1) = 0 \implies x = 4$ or $x = 1$

Upon checking, only $x = 1$ satisfies the equation.

97. $\sqrt{2 + \sqrt{x+1}} = \sqrt{7-x}$

$\left(\sqrt{2+\sqrt{x+1}}\right)^2 = \left(\sqrt{7-x}\right)^2$

$2 + \sqrt{x+1} = 7 - x$

$\sqrt{x+1} = 5 - x$

$\left(\sqrt{x+1}\right)^2 = (5-x)^2$

$x + 1 = 25 - 10x + x^2$

$x^2 - 11x + 24 = 0$

$(x-8)(x-3) = 0 \implies x = 8$ or $x = 3$

Upon checking, only $x = 3$ satisfies the equation.

99. $c = \sqrt{90^2 + 90^2} = \sqrt{2(90)^2} = 90\sqrt{2} \approx 127.28$

Second base is about 127.28 feet from home plate.

101. $s = \sqrt{A}$

$s = \sqrt{169}$

$\quad = 13$ feet

103. $T = 2\pi\sqrt{\dfrac{l}{32}}$

a. Let $l = 8$

$T = 2\pi\sqrt{\dfrac{l}{32}}$

$\quad = 2\pi\sqrt{\dfrac{8}{32}} = 2\pi\sqrt{\dfrac{1}{4}}$

$\quad = 2\pi\dfrac{1}{2} = \pi$

$\quad \approx 3.14$ seconds

b. Replace l with $2l$:

$T_D = 2\pi\sqrt{\dfrac{2l}{32}}$

$\quad = 2\pi\sqrt{2}\sqrt{\dfrac{l}{32}}$

$\quad = \sqrt{2}\left(2\pi\sqrt{\dfrac{l}{32}}\right)$

$\quad = \sqrt{2} \cdot T$

c. This part must be solved in two phases. First, we need to find the length of the pendulum:

$$T = 2\pi\sqrt{\frac{l}{g}}$$

$$2 = 2\pi\sqrt{\frac{l}{32}}$$

$$\frac{1}{\pi} = \sqrt{\frac{l}{32}}$$

$$\left(\frac{1}{\pi}\right)^2 = \frac{l}{32}$$

$$l = \frac{32}{\pi^2}$$

Now, find T using $g = \frac{32}{6}$ and $l = \frac{32}{\pi^2}$

$$T = 2\pi\sqrt{\frac{l}{g}}$$

$$= 2\pi\sqrt{\frac{\frac{32}{\pi^2}}{\frac{32}{6}}} = 2\pi\sqrt{\frac{6}{\pi^2}}$$

$$= 2\pi\frac{\sqrt{6}}{\pi} = 2\sqrt{6}$$

$$\approx 4.90 \text{ seconds}$$

105.
$$r = \sqrt[4]{\frac{8\mu l}{\pi R}}$$

$$r^4 = \left(\sqrt[4]{\frac{8\mu l}{\pi R}}\right)^4$$

$$r^4 = \frac{8\mu l}{\pi R^4}$$

$$\pi R r^4 = 8\mu l$$

$$R = \frac{8\mu l}{\pi r^4}$$

107. $N = 0.2\left(\sqrt{R}\right)^3$

$$N = 0.2\left(\sqrt{149.4}\right)^3$$

$$= 0.2(12.223)^3$$

$$= 0.2(1826.106)$$

$$\approx 365.2 \text{ days}$$

109. $R = \sqrt{F_1^2 + F_2^2}$

$$R = \sqrt{60^2 + 80^2}$$

$$= \sqrt{10,000}$$

$$= 100 \text{ lb}$$

111. $c = \sqrt{gH} = \sqrt{32 \cdot 10} = \sqrt{320} \approx 17.89$ ft/sec

113. The diagonal and the two given sides form a right triangle. Use the Pythagorean formula to solve for the diagonal.

$$a^2 + b^2 = c^2$$

$$25^2 + 32^2 = c^2$$

$$625 + 1024 = c^2$$

$$1649 = c^2 \quad \Rightarrow \quad c = \sqrt{1649} \approx 40.61$$

The diagonal is about 40.61 meters in length.

115. $x = \dfrac{-b \pm \sqrt{b^2 - 4ac}}{2a}$

$$x = \frac{-0 \pm \sqrt{0^2 - 4(1)(-4)}}{2(1)} = \frac{\pm\sqrt{16}}{2} = \frac{\pm 4}{2} = \pm 2$$

Now, $x = 2$ or $x = -2$.

117. $x = \dfrac{-b \pm \sqrt{b^2 - 4ac}}{2a}$

$$x = \frac{-4 \pm \sqrt{4^2 - 4(-1)(5)}}{2(-1)}$$

$$= \frac{-4 \pm \sqrt{36}}{-2}$$

$$= \frac{-4 \pm 6}{-2}$$

Now, $x = \dfrac{-4+6}{-2} = \dfrac{2}{-2} = -1$ or

$$x = \frac{-4-6}{-2} = \frac{-10}{-2} = 5$$

119. $f(x) = \sqrt{x-5}$

$$5 = \sqrt{x-5}$$

$$5^2 = \left(\sqrt{x-5}\right)^2$$

$$25 = x - 5$$

$$30 = x$$

121. $f(x) = \sqrt{3x^2 - 11} + 7$

$15 = \sqrt{3x^2 - 11} + 7$

$8 = \sqrt{3x^2 - 11}$

$8^2 = \left(\sqrt{3x^2 - 11}\right)^2$

$64 = 3x^2 - 11$

$75 = 3x^2$

$25 = x^2$

$\pm 5 = x$

123. a. $y = \sqrt{4x - 12}$, $y = x - 3$

The points of intersection are (3, 0) and (7, 4). The x-values are 3 and 7.

b. $\sqrt{4x - 12} = x - 3$

For $x = 3$:	For $x = 7$:
$\sqrt{4 \cdot 3 - 12} = 3 - 3$	$\sqrt{4x - 12} = x - 3$
$\sqrt{12 - 12} = 0$	$\sqrt{4 \cdot 7 - 12} = 7 - 3$
$\sqrt{0} = 0$	$\sqrt{16} = 4$
$0 = 0$ True	$4 = 4$ True

Both values are solutions to the equation.

c. $\sqrt{4x - 12} = x - 3$

$\left(\sqrt{4x - 12}\right)^2 = (x - 3)^2$

$4x - 12 = x^2 - 6x + 9$

$0 = x^2 - 10x + 21$

$0 = (x - 3)(x - 7)$

$x - 3 = 0$ or $x - 7 = 0$

$x = 3 \qquad x = 7$

The answers agree.

125. At $x = 4$, $g(x) = 0$ or $y = 0$.

Therefore, the graph must have an x-intercept at 4.

127. $L_1 = p - 1.96\sqrt{\dfrac{p(1-p)}{n}}$

$L_1 = 0.60 - 1.96\sqrt{\dfrac{0.60(1 - 0.60)}{36}}$

$\approx 0.60 - 0.16$

≈ 0.44

$L_2 = p + 1.96\sqrt{\dfrac{p(1-p)}{n}}$

$L_2 = 0.60 + 1.96\sqrt{\dfrac{0.60(1 - 0.60)}{36}}$

$= 0.60 + 0.16$

≈ 0.76

129. $\sqrt{x^2 + 49} = (x^2 + 49)^{1/2}$

$\left(\sqrt{x^2 + 49}\right)^2 = [(x^2 + 49)^{1/2}]^2$

$x^2 + 49 = x^2 + 49$

$49 = 49$

All real numbers, x, satisfy this equation.

131. Graph:

$y_1 = \sqrt{x + 8}$

$y_2 = \sqrt{3x + 5}$

Intersection X=1.5   Y=3.082207

$-10, 10, 1, -10, 10, 1$

The graphs of the equations intersect when $x = 1.5$.

133. Graph:

$y = \sqrt[3]{5x^2 - 6} - 4$

The graph of the equation crosses the x-axis at $x \approx -3.7$ and $x \approx 3.7$.

135.
$$\sqrt{\sqrt{x+25}-\sqrt{x}} = 5$$
$$\left(\sqrt{\sqrt{x+25}-\sqrt{x}}\right)^2 = 5^2$$
$$\sqrt{x+25}-\sqrt{x} = 25$$
$$\sqrt{x+25} = 25+\sqrt{x}$$
$$\left(\sqrt{x+25}\right)^2 = \left(25+\sqrt{x}\right)^2$$
$$x+25 = 625+50\sqrt{x}+x$$
$$-600 = 50\sqrt{x}$$
$$-12 = \sqrt{x}$$
$$(-12)^2 = \left(\sqrt{x}\right)^2$$
$$144 = x$$

Check:
$$\sqrt{\sqrt{x+25}-\sqrt{x}} = 5$$
$$\sqrt{\sqrt{144+25}-\sqrt{144}} = 5$$
$$\sqrt{\sqrt{169}-\sqrt{144}} = 5$$
$$\sqrt{13-12} = 5$$
$$\sqrt{1} = 5$$
$$1 = 5 \text{ False}$$

Thus, 144 is not a solution and we conclude that there is no solution.

137.
$$z = \frac{\overline{x}-\mu}{\dfrac{\sigma}{\sqrt{n}}}$$
$$z\left(\frac{\sigma}{\sqrt{n}}\right) = \overline{x}-\mu$$
$$\left(z\frac{\sigma}{\sqrt{n}}\right)^2 = \left(\overline{x}-\mu\right)^2$$
$$\frac{z^2\sigma^2}{n} = \left(\overline{x}-\mu\right)^2$$
$$z^2\sigma^2 = n(\overline{x}-\mu)^2$$
$$\frac{z^2\sigma^2}{(\overline{x}-\mu)^2} = n \text{ or } n = \frac{z^2\sigma^2}{(\overline{x}-\mu)^2}$$

140. $P_1P_2 - P_1P_3 = P_2P_3$ Solve for P_2.
$$P_1P_2 - P_1P_2 - P_1P_3 = P_2P_3 - P_1P_2$$
$$-P_1P_3 = P_2P_3 - P_1P_2$$
$$-P_1P_3 = P_2(P_3 - P_1)$$
$$\frac{-P_1P_3}{P_3 - P_1} = \frac{P_2(P_3 - P_1)}{P_3 - P_1}$$
$$\frac{P_1P_3}{P_1 - P_3} = P_2 \text{ or } P_2 = \frac{P_1P_3}{P_1 - P_3}$$

141.
$$\frac{x(x-5)+x(x-2)}{2x-7} = \frac{x^2-5x+x^2-2x}{2x-7}$$
$$= \frac{2x^2-7x}{2x-7}$$
$$= \frac{x(2x-7)}{2x-7}$$
$$= x$$

142.
$$\frac{4a^2-9b^2}{4a^2+12ab+9b^2} \cdot \frac{6a^2b}{8a^2b^2-12ab^3}$$
$$= \frac{(2a-3b)(2a+3b)}{(2a+3b)(2a+3b)} \cdot \frac{6a^2b}{4ab^2(2a-3b)}$$
$$= \frac{3a}{2b(2a+3b)}$$

143.
$$(t^2-2t-15) \div \frac{t^2-9}{t^2-3t}$$
$$= (t^2-2t-15) \cdot \frac{t^2-3t}{t^2-9}$$
$$= (t-5)(t+3) \cdot \frac{t(t-3)}{(t+3)(t-3)}$$
$$= \frac{t(t-3)(t-5)(t+3)}{(t+3)(t-3)}$$
$$= \frac{t\,\cancel{(t-3)}\,(t-5)\,\cancel{(t+3)}}{\cancel{(t+3)}\,\cancel{(t-3)}}$$
$$= t(t-5)$$

144. $\dfrac{2}{x+3} - \dfrac{1}{x-3} + \dfrac{2x}{x^2-9}$

$= \dfrac{2}{x+3} - \dfrac{1}{x-3} + \dfrac{2x}{(x+3)(x-3)}$

$= \dfrac{2}{x+3} \cdot \dfrac{x-3}{x-3} - \dfrac{1}{x-3} \cdot \dfrac{x+3}{x+3} + \dfrac{2x}{(x+3)(x-3)}$

$= \dfrac{2(x-3)}{(x+3)(x-3)} - \dfrac{x+3}{(x+3)(x-3)} + \dfrac{2x}{(x+3)(x-3)}$

$= \dfrac{2x-6}{(x+3)(x-3)} - \dfrac{x+3}{(x+3)(x-3)} + \dfrac{2x}{(x+3)(x-3)}$

$= \dfrac{2x-6-(x+3)+2x}{(x+3)(x-3)}$

$= \dfrac{2x-6-x-3+2x}{(x+3)(x-3)}$

$= \dfrac{3x-9}{(x+3)(x-3)}$

$= \dfrac{3(x-3)}{(x+3)(x-3)}$

$= \dfrac{3}{x+3}$

145.

$2 + \dfrac{3x}{x-1} = \dfrac{8}{x-1}$

$(x-1)(2) + (x-1)\left(\dfrac{3x}{x-1}\right) = (x-1)\left(\dfrac{8}{x-1}\right)$

$2(x-1) + 3x = 8$

$2x - 2 + 3x = 8$

$5x - 2 = 8$

$5x = 10$

$x = 2$

Exercise Set 11.7

1. a. $i = \sqrt{-1}$

b. $i^2 = -1$

3. Yes, all the numbers listed are complex numbers.

5. Yes

7. The conjugate of $a + bi$ is $a - bi$.

9. Answers will vary. Possible answers:
   a. $\sqrt{2}$ **b.** 1 **c.** $\sqrt{-3}$ or $2i$ **d.** 6
   e. Every number we have studied is a complex number.

11. $7 = 7 + 0i$

13. $\sqrt{25} = 5 = 5 + 0i$

15. $21 - \sqrt{-36} = 21 - \sqrt{36}\sqrt{-1}$
$= 21 - 6i$

17. $\sqrt{-24} = \sqrt{4}\sqrt{-1}\sqrt{6} = 2i\sqrt{6} = 0 + 2i\sqrt{6}$

19. $8 - \sqrt{-12} = 8 - \sqrt{12}\sqrt{-1}$
$= 8 - \sqrt{4}\sqrt{3}\sqrt{-1}$
$= 8 - 2i\sqrt{3}$

21. $3 + \sqrt{-98} = 3 + \sqrt{98}\sqrt{-1}$
$= 3 + \sqrt{49}\sqrt{2}\sqrt{-1}$
$= 3 + 7i\sqrt{2}$

23. $12 - \sqrt{-25} = 12 - \sqrt{25}\sqrt{-1}$
$= 12 - 5i$

25. $7i - \sqrt{-45} = 0 + 7i - \sqrt{9}\sqrt{5}\sqrt{-1}$
$= 0 + 7i - 3i\sqrt{5}$
$= 0 + \left(7 - 3\sqrt{5}\right)i$

27. $(19 - i) + (2 + 9i) = 19 - i + 2 + 9i$
$= 21 + 8i$

29. $\left(8 - 3i\right) + \left(-8 + 3i\right) = 8 - 3i - 8 + 3i$
$= 0$

31. $\left(1 + \sqrt{-1}\right) + \left(-18 - \sqrt{-169}\right)$
$= 1 + \sqrt{-1} - 18 - \sqrt{-169}$
$= 1 + i - 18 - 13i$
$= -17 - 12i$

33. $\left(\sqrt{3} + \sqrt{2}\right) + \left(3\sqrt{2} - \sqrt{-8}\right)$
$= \sqrt{3} + \sqrt{2} + 3\sqrt{2} - \sqrt{-8}$
$= \sqrt{3} + \sqrt{2} + 3\sqrt{2} - \sqrt{-4 \cdot 2}$
$= \sqrt{3} + \sqrt{2} + 3\sqrt{2} - 2i\sqrt{2}$
$= \sqrt{3} + 4\sqrt{2} - 2i\sqrt{2}$
$= \left(\sqrt{3} + 4\sqrt{2}\right) - 2i\sqrt{2}$

35. $\left(5-\sqrt{-72}\right)+\left(6+\sqrt{-8}\right)$

$= 5 - \sqrt{-72} + 6 + \sqrt{-8}$

$= 5 - \sqrt{36}\sqrt{2}\sqrt{-1} + 6 + \sqrt{4}\sqrt{2}\sqrt{-1}$

$= 5 - 6i\sqrt{2} + 6 + 2i\sqrt{2}$

$= 11 - 4i\sqrt{2}$

37. $\left(\sqrt{4}-\sqrt{-45}\right)+\left(-\sqrt{25}+\sqrt{-5}\right)$

$= \sqrt{4} - \sqrt{-45} - \sqrt{25} + \sqrt{-5}$

$= \sqrt{4} - \sqrt{9}\sqrt{5}\sqrt{-1} - \sqrt{25} + \sqrt{-1}\sqrt{5}$

$= 2 - 3i\sqrt{5} - 5 + i\sqrt{5}$

$= -3 - 2i\sqrt{5}$

39. $2(3-i) = 6-2i$

41. $i(4+9i) = 4i + 9i^2 = 4i + 9(-1) = -9 + 4i$

43. $\sqrt{-9}(6+11i) = 3i(6+11i)$

$= 18i + 33i^2$

$= 18i + 33(-1)$

$= -33 + 18i$

45. $\sqrt{-16}\left(\sqrt{3}-7i\right) = 4i\left(\sqrt{3}-7i\right)$

$= 4i\sqrt{3} - 28i^2$

$= 4i\sqrt{3} - 28(-1)$

$= 4i\sqrt{3} + 28$

$= 28 + 4i\sqrt{3}$

47. $\sqrt{-27}\left(\sqrt{3}-\sqrt{-3}\right) = \sqrt{-9}\sqrt{3}\left(\sqrt{3}-\sqrt{-3}\right)$

$= 3i\sqrt{3}\left(\sqrt{3}-i\sqrt{3}\right)$

$= 3i \cdot 3 - 3i^2 \cdot 3$

$= 9i - 3(-1) \cdot 3$

$= 9 + 9i$

49. $(3+2i)(1+i)$

$= 3(1) + 3(i) + 2i(1) + 2i(i)$

$= 3 + 3i + 2i + 2i^2$

$= 3 + 3i + 2i + 2(-1)$

$= 3 + 3i + 2i - 2$

$= 1 + 5i$

51. $(10-3i)(10+3i) = 100 + 30i - 30i - 9i^2$

$= 100 + 30i - 30i - 9(-1)$

$= 100 + 9$

$= 109$

53. $\left(7+\sqrt{-2}\right)\left(5-\sqrt{-8}\right)$

$= \left(7+i\sqrt{2}\right)\left(5-2i\sqrt{2}\right)$

$= 35 - 14i\sqrt{2} + 5i\sqrt{2} - 2i^2 \cdot 2$

$= 35 - 14i\sqrt{2} + 5i\sqrt{2} - 2(-1) \cdot 2$

$= 35 - 14i\sqrt{2} + 5i\sqrt{2} + 4$

$= 39 - 9i\sqrt{2}$

55. $\left(\dfrac{1}{2}-\dfrac{1}{3}i\right)\left(\dfrac{1}{4}+\dfrac{2}{3}i\right)$

$= \dfrac{1}{8} + \dfrac{1}{3}i - \dfrac{1}{12}i - \dfrac{2}{9}i^2$

$= \dfrac{1}{8} + \dfrac{1}{3}i - \dfrac{1}{12}i - \dfrac{2}{9}(-1)$

$= \dfrac{1}{8} + \dfrac{1}{3}i - \dfrac{1}{12}i + \dfrac{2}{9}$

$= \dfrac{1}{8} + \dfrac{2}{9} + \left(\dfrac{1}{3} - \dfrac{1}{12}\right)i$

$= \dfrac{25}{72} + \dfrac{1}{4}i$

57. $\dfrac{8}{3i} = \dfrac{8}{3i} \cdot \dfrac{-i}{-i} = \dfrac{-8i}{-3i^2} = \dfrac{-8i}{-3(-1)} = -\dfrac{8i}{3}$

59. $\dfrac{2+3i}{2i} = \dfrac{2+3i}{2i} \cdot \dfrac{-i}{-i}$

$= \dfrac{(2+3i)(-i)}{-2i^2}$

$= \dfrac{-2i-3i^2}{-2i^2}$

$= \dfrac{-2i-3(-1)}{-2(-1)}$

$= \dfrac{-2i+3}{2}$

$= \dfrac{3-2i}{2}$

61. $\dfrac{6}{2-i} = \dfrac{6}{2-i} \cdot \dfrac{2+i}{2+i}$

$\quad = \dfrac{6(2+i)}{(2-i)(2+i)}$

$\quad = \dfrac{12+6i}{4+2i-2i-i^2}$

$\quad = \dfrac{12+6i}{4+2i-2i-(-1)}$

$\quad = \dfrac{12+6i}{5}$

63. $\dfrac{3}{1-2i} = \dfrac{3}{1-2i} \cdot \dfrac{1+2i}{1+2i}$

$\quad = \dfrac{3(1+2i)}{(1-2i)(1+2i)}$

$\quad = \dfrac{3+6i}{1+2i-2i-4i^2}$

$\quad = \dfrac{3+6i}{1+2i-2i-4(-1)}$

$\quad = \dfrac{3+6i}{5}$

65. $\dfrac{6-3i}{4+2i} = \dfrac{6-3i}{4+2i} \cdot \dfrac{4-2i}{4-2i}$

$\quad = \dfrac{(6-3i)(4-2i)}{16-4i^2}$

$\quad = \dfrac{24-12i-12i+6i^2}{16-4i^2}$

$\quad = \dfrac{24-12i-12i-6}{16+4}$

$\quad = \dfrac{18-24i}{20}$

$\quad = \dfrac{2(9-12i)}{20}$

$\quad = \dfrac{9-12i}{10}$

67. $\dfrac{4}{6-\sqrt{-4}} = \dfrac{4}{6-\sqrt{4}\sqrt{-1}}$

$\quad = \dfrac{4}{6-2i} \cdot \dfrac{6+2i}{6+2i}$

$\quad = \dfrac{4(6+2i)}{36-4i^2}$

$\quad = \dfrac{24+8i}{36-4(-1)}$

$\quad = \dfrac{24+8i}{36+4}$

$\quad = \dfrac{8(3+i)}{40} = \dfrac{3+i}{5}$

69. $\dfrac{\sqrt{2}}{5+\sqrt{-12}} = \dfrac{\sqrt{2}}{5+\sqrt{4}\sqrt{3}\sqrt{-1}}$

$\quad = \dfrac{\sqrt{2}}{5+2i\sqrt{3}} \cdot \dfrac{5-2i\sqrt{3}}{5-2i\sqrt{3}}$

$\quad = \dfrac{\sqrt{2}\left(5-2i\sqrt{3}\right)}{25-4i^2\sqrt{3}^2}$

$\quad = \dfrac{5\sqrt{2}-2i\sqrt{6}}{25-4(-1)(3)}$

$\quad = \dfrac{5\sqrt{2}-2i\sqrt{6}}{25+12}$

$\quad = \dfrac{5\sqrt{2}-2i\sqrt{6}}{37}$

71. $\dfrac{\sqrt{10}+\sqrt{-3}}{5-\sqrt{-20}}$

$\quad = \dfrac{\sqrt{10}+\sqrt{3}\sqrt{-1}}{5-\sqrt{4}\sqrt{5}\sqrt{-1}}$

$\quad = \dfrac{\sqrt{10}+i\sqrt{3}}{5-2i\sqrt{5}} \cdot \dfrac{5+2i\sqrt{5}}{5+2i\sqrt{5}}$

$\quad = \dfrac{(\sqrt{10}+i\sqrt{3})(5+2i\sqrt{5})}{5^2-4i^2\sqrt{5}^2}$

$\quad = \dfrac{5\sqrt{10}+2i\sqrt{50}+5i\sqrt{3}+2i^2\sqrt{15}}{5^2-4(-1)(5)}$

$\quad = \dfrac{5\sqrt{10}+2i\sqrt{25}\sqrt{2}+5i\sqrt{3}+2(-1)\sqrt{15}}{25+20}$

$\quad = \dfrac{(5\sqrt{10}-2\sqrt{15})+(10\sqrt{2}+5\sqrt{3})i}{45}$

73. $\dfrac{\sqrt{-75}}{\sqrt{-3}} = \dfrac{\sqrt{25}\sqrt{3}\sqrt{-1}}{\sqrt{3}\sqrt{-1}} = \dfrac{5i\sqrt{3}}{i\sqrt{3}} = 5$

75. $\dfrac{\sqrt{-32}}{\sqrt{-18}\sqrt{8}} = \dfrac{\sqrt{32}\sqrt{-1}}{\sqrt{144}\sqrt{-1}} = \dfrac{i\sqrt{16}\sqrt{2}}{i\sqrt{144}} = \dfrac{4i\sqrt{2}}{12i} = \dfrac{\sqrt{2}}{3}$

77. $(9-2i)+(3-5i) = 9+3-2i-5i = 12-7i$

79. $\left(\sqrt{50}-\sqrt{2}\right)-\left(\sqrt{-12}-\sqrt{-48}\right)$

$= \left(5\sqrt{2}-\sqrt{2}\right)-\left(2i\sqrt{3}-4i\sqrt{3}\right)$

$= 4\sqrt{2}-\left(-2i\sqrt{3}\right)$

$= 4\sqrt{2}+2i\sqrt{3}$

81. $5.2(4-3.2i) = 5.2(4)-5.2(3.2i) = 20.8-16.64i$

83. $(9+2i)(3-5i)$

$= 27-45i+6i-10i^2$

$= 27-39i-10(-1)$

$= 27+10-39i$

$= 37-39i$

85. $\dfrac{11+4i}{2i} = \dfrac{11+4i}{2i}\cdot\dfrac{-2i}{-2i}$

$= \dfrac{(11+4i)(-2i)}{-4i^2}$

$= \dfrac{-22i-8i^2}{-4i^2}$

$= \dfrac{-22i-8(-1)}{-4(-1)}$

$= \dfrac{8-22i}{4}$

$= \dfrac{4-11i}{2}$

87. $\dfrac{6}{\sqrt{3}-\sqrt{-4}} = \dfrac{6}{\sqrt{3}-2i}$

$= \dfrac{6}{\sqrt{3}-2i}\cdot\dfrac{\sqrt{3}+2i}{\sqrt{3}+2i}$

$= \dfrac{6(\sqrt{3}+2i)}{(\sqrt{3})^2-(2i)^2}$

$= \dfrac{6\sqrt{3}+12i}{3-4i^2}$

$= \dfrac{6\sqrt{3}+12i}{3-4(-1)}$

$= \dfrac{6\sqrt{3}+12i}{7}$

89. $\left(11-\dfrac{5}{9}i\right)-\left(4-\dfrac{3}{5}i\right) = 11-\dfrac{5}{9}i-4+\dfrac{3}{5}i$

$= 7-\dfrac{5}{9}i+\dfrac{3}{5}i$

$= 7-\dfrac{25}{45}i+\dfrac{27}{45}i$

$= 7+\dfrac{2}{45}i$

91. $\left(\dfrac{2}{3}-\dfrac{1}{5}i\right)\left(\dfrac{3}{5}-\dfrac{3}{4}i\right)$

$= \left(\dfrac{2}{3}\right)\left(\dfrac{3}{5}\right)-\dfrac{2}{3}\left(\dfrac{3}{4}i\right)-\left(\dfrac{1}{5}i\right)\left(\dfrac{3}{5}\right)+\left(\dfrac{1}{5}i\right)\left(\dfrac{3}{4}i\right)$

$= \dfrac{2}{5}-\dfrac{1}{2}i-\dfrac{3}{25}i+\dfrac{3}{20}i^2$

$= \dfrac{2}{5}-\dfrac{1}{2}i-\dfrac{3}{25}i+\dfrac{3}{20}(-1)$

$= \left(\dfrac{2}{5}-\dfrac{3}{20}\right)+\left(-\dfrac{1}{2}-\dfrac{3}{25}\right)i$

$= \left(\dfrac{8}{20}-\dfrac{3}{20}\right)+\left(-\dfrac{25}{50}-\dfrac{6}{50}\right)i$

$= \dfrac{5}{20}-\dfrac{31}{50}i$

$= \dfrac{1}{4}-\dfrac{31}{50}i$

93. $\dfrac{\sqrt{-48}}{\sqrt{-12}} = \dfrac{\sqrt{48}\sqrt{-1}}{\sqrt{12}\sqrt{-1}}$

$= \dfrac{i\sqrt{48}}{i\sqrt{12}}$

$= \dfrac{\sqrt{48}}{\sqrt{12}}$

$= \sqrt{\dfrac{48}{12}}$

$= \sqrt{4}$

$= 2$

95. $(5.23 - 6.41i) - (9.56 + 4.5i)$
$= 5.23 - 6.41i - 9.56 - 4.5i$
$= -4.33 - 10.91i$

97. $i^6 = i^4 \cdot i^2 = 1 \cdot (-1) = -1$

99. $i^{160} = (i^4)^{40} = 1^{40} = 1$

101. $i^{93} = i^{92} \cdot i^1 = (i^4)^{23} i = 1^{23} \cdot i = 1(i) = i$

103. $i^{811} = i^{808} \cdot i^3$

$= (i^4)^{202} \cdot i^3$

$= 1^{202} \cdot i^3$

$= 1 \cdot (i^3)$

$= i^3$

$= -i$

105. a. The additive inverse of $2 + 3i$ is its opposite, $-2 - 3i$. Note that
$(2 + 3i) + (-2 - 3i) = 0$.

b. The multiplicative inverse of $2 + 3i$ is its reciprocal, $\dfrac{1}{2 + 3i}$. To simplify this, multiply the numerator and denominator by the conjugate of the denominator.

$\dfrac{1}{2+3i} = \dfrac{1}{2+3i} \cdot \dfrac{2-3i}{2-3i}$

$= \dfrac{2-3i}{(2+3i)(2-3i)}$

$= \dfrac{2-3i}{4-6i+6i-9i^2}$

$= \dfrac{2-3i}{13}$

107. True. The product of two pure imaginary numbers is always a real number. Consider two pure imaginary numbers bi and di where b, d are non-zero real numbers whose product
$(bi)(di) = bdi^2$
$= bd(-1)$
$= -bd$
which is a real number.

109. False. The product of two complex numbers is not always a real number. For example,
$(1+i)(1+i) = 1 + i + i + i^2$
$= 1 + 2i + (-1)$
$= 0 + 2i$
which is not a real number.

111. Even values of n will result in i^n being a real number. $i^2 = -1, i^{2n} = (i^2)^n = (-1)^n$

113. $f(x) = x^2$

$f(2i) = (2i)^2 = 4i^2 = 4(-1) = -4$

115. $f(x) = x^4 - 2x$

$f(2i) = (2i)^4 - 2(2i)$

$= 2^4 i^4 - 4i$

$= 16(1) - 4i$

$= 16 - 4i$

117. $f(x) = x^2 + 2x$

$f(3+i) = (3+i)^2 + 2(3+i)$

$= 9 + 6i + i^2 + 6 + 2i$

$= 9 + 6i - 1 + 6 + 2i$

$= 14 + 8i$

119. $x^2 - 2x + 5$

$= (1+2i)^2 - 2(1+2i) + 5$

$= 1^2 + 2(1)(2i) + (2i)^2 - 2 - 4i + 5$

$= 1 + 4i - 4 - 2 - 4i + 5$

$= 0$

121. $x^2 + 2x + 7$
$= (-1 + i\sqrt{5})^2 + 2(-1 + i\sqrt{5}) + 7$
$= (-1)^2 - 2(1)(i\sqrt{5}) + (i\sqrt{5})^2 - 2 + 2i\sqrt{5} + 7$
$= 1 - 2i\sqrt{5} - 5 - 2 + 2i\sqrt{5} + 7$
$= 1 + 0i$
$= 1$

123.
$$x^2 - 4x + 5 = 0$$
$$(2 - i)^2 - 4(2 - i) + 5 = 0$$
$$2^2 - 2(2)(i) + (i)^2 - 8 + 4i + 5 = 0$$
$$4 - 4i - 1 - 8 + 4i + 5 = 0$$
$$0 + 0i = 0$$
$$0 = 0 \quad \text{True}$$

$2 - i$ is a solution.

125.
$$x^2 - 6x + 11 = 0$$
$$(-3 + i\sqrt{3})^2 - 6(-3 + i\sqrt{3}) + 11 = 0$$
$$(-3)^2 - 2(3)(i\sqrt{3}) + (i\sqrt{3})^2 + 18 - 6i\sqrt{3} + 11 = 0$$
$$9 - 6i\sqrt{3} - 3 + 18 - 6i\sqrt{3} + 11 = 0$$
$$35 - 12i\sqrt{3} = 0$$

False, $-3 + i\sqrt{3}$ is not a solution.

127. $Z = \dfrac{V}{I}$

$Z = \dfrac{1.8 + 0.5i}{0.6i}$

$= \dfrac{1.8 + 0.5i}{0.6i} \cdot \dfrac{-0.6i}{-0.6i}$

$= \dfrac{-1.08i - 0.3i^2}{-0.36i^2}$

$= \dfrac{-1.08i - 0.3(-1)}{-0.36(-1)}$

$= \dfrac{0.3 - 1.08i}{0.36}$

$\approx 0.83 - 3i$

129. $Z_T = \dfrac{Z_1 Z_2}{Z_1 + Z_2}$

$= \dfrac{(2 - i)(4 + i)}{(2 - i) + (4 + i)}$

$= \dfrac{8 + 2i - 4i - i^2}{6}$

$= \dfrac{8 - 2i - (-1)}{6}$

$= \dfrac{9 - 2i}{6}$

$\approx 1.5 - 0.33i$

131. $i^{-1} = \dfrac{1}{i} = \dfrac{1}{i} \cdot \dfrac{i}{i} = \dfrac{i}{i^2} = \dfrac{i}{-1} = -i$

133. $x^2 - 2x + 6 = 0$
$a = 1, b = -2, c = 6$

$x = \dfrac{-(-2) \pm \sqrt{(-2)^2 - 4(1)(6)}}{2(1)}$

$= \dfrac{2 \pm \sqrt{4 - 24}}{2}$

$= \dfrac{2 \pm \sqrt{-20}}{2}$

$= \dfrac{2 \pm 2i\sqrt{5}}{2}$

$= \dfrac{2\left(1 \pm i\sqrt{5}\right)}{2}$

$= 1 \pm i\sqrt{5}$

135. $a + b = 5 + 2i\sqrt{3} + 1 + i\sqrt{3}$
$= 5 + 1 + 2i\sqrt{3} + i\sqrt{3}$
$= 6 + 3i\sqrt{3}$

137. $ab = (5 + 2i\sqrt{3})(1 + i\sqrt{3})$
$= 5(1) + (5)(i\sqrt{3}) + (2i\sqrt{3})(1) + (2i\sqrt{3})(i\sqrt{3})$
$= 5 + 5i\sqrt{3} + 2i\sqrt{3} + 2i^2(\sqrt{3})^2$
$= 5 + 5i\sqrt{3} + 2i\sqrt{3} - 6$
$= -1 + 7i\sqrt{3}$

139.

$$4c+9 \overline{\smash{\big)}\, 8c^2 + 6c - 35} \qquad \overset{2c-3}{}$$

$$\underline{8c^2 + 18c} \qquad \leftarrow 2c(4c+9)$$
$$-12c - 35$$
$$\underline{-12c - 27} \qquad \leftarrow -3(4c+9)$$
$$-8 \qquad \leftarrow \text{Remainder}$$

Thus, $\dfrac{8c^2 + 6c - 35}{4c + 9} = 2c - 3 - \dfrac{8}{4c + 9}$

140.

$$\frac{b}{a-b} + \frac{a+b}{b} = \frac{b}{a-b} \cdot \frac{b}{b} + \frac{a+b}{b} \cdot \frac{a-b}{a-b}$$

$$= \frac{b^2}{b(a-b)} + \frac{(a+b)(a-b)}{b(a-b)}$$

$$= \frac{b^2 + (a+b)(a-b)}{b(a-b)}$$

$$= \frac{b^2 + a^2 - b^2}{b(a-b)}$$

$$= \frac{a^2}{b(a-b)}$$

141.

$$\frac{x}{4} + \frac{1}{2} = \frac{x-1}{2}$$

$$4\left(\frac{x}{4} + \frac{1}{2}\right) = 4\left(\frac{x-1}{2}\right)$$

$$x + 2 = 2(x-1)$$

$$x + 2 = 2x - 2$$

$$x = 4$$

142. This problem can be solved using a single variable. To do this, let x be the amount that is $5.50 per pound. Then $40 - x$ is the amount that is $6.30 per pound and the equation is

$$5.50(x) + 6.30(40 - x) = 6(40)$$

$$5.5x + 252 - 6.3x = 240$$

$$252 - 0.8x = 240$$

$$-0.8x = -12$$

$$x = 15$$

Thus, combine 15 lb of the $5.50 per pound coffee with $40 - 15 = 25$ lb of the $6.30 per pound coffee to obtain 40 lb of $6.00 per pound coffee.

Chapter 11 Review Exercises

1. $\sqrt{100} = \sqrt{10^2} = 10$

2. $\sqrt[3]{-27} = \sqrt[3]{(-3)^3} = -3$

3. $\sqrt[3]{-125} = \sqrt[3]{(-5)^3} = -5$

4. $\sqrt[4]{256} = \sqrt[4]{4^4} = 4$

5. $\sqrt{(-8)^2} = |-8| = 8$

6. $\sqrt{(38.2)^2} = 38.2$

7. $\sqrt{x^2} = |x|$

8. $\sqrt{(x-3)^2} = |x-3|$

9. $\sqrt{(x-y)^2} = |x-y|$

10. $\sqrt{\left(x^2 - 4x + 12\right)^2} = |x^2 - 4x + 12|$

(Note: absolute value would not actually be required because $x^2 - 4x + 12 > 0$ for all x.)

11. $f(x) = \sqrt{10x + 9}$

$$f(4) = \sqrt{10(4) + 9}$$
$$= \sqrt{40 + 9}$$
$$= \sqrt{49}$$
$$= 7$$

12. $k(x) = 2x + \sqrt{\dfrac{x}{3}}$

$$k(27) = 2(27) + \sqrt{\dfrac{27}{3}}$$
$$= 54 + \sqrt{9}$$
$$= 54 + 3$$
$$= 57$$

13. $g(x) = \sqrt[3]{2x+3}$

$g(4) = \sqrt[3]{2(4)+3}$

$ = \sqrt[3]{11}$

$ \approx 2.2$

14. $\text{Area} = (\text{side})^2$

$144 = s^2$

$\pm 12 = s$

Disregard the negative value since lengths must be positive. The length of each side is 12 m.

15. $\sqrt{x^7} = x^{7/2}$

16. $\sqrt[3]{x^5} = x^{5/3}$

17. $\left(\sqrt[4]{y}\right)^{13} = y^{13/4}$

18. $\sqrt[7]{6^{-2}} = 6^{-2/7}$

19. $x^{1/2} = \sqrt{x}$

20. $a^{4/5} = \sqrt[5]{a^4}$

21. $(8m^2n)^{7/4} = \left(\sqrt[4]{8m^2n}\right)^7$

22. $(x+y)^{-5/3} = \dfrac{1}{\left(\sqrt[3]{x+y}\right)^5}$

23. $\sqrt[3]{4^6} = 4^{6/3} = 4^2 = 16$

24. $\sqrt{x^{12}} = x^{12/2} = x^6$

25. $\left(\sqrt[4]{9}\right)^8 = 9^{8/4} = 9^2 = 81$

26. $\sqrt[20]{a^5} = a^{5/20} = a^{1/4} = \sqrt[4]{a}$

27. $-36^{1/2} = -\sqrt{36} = -6$

28. $(-36)^{1/2}$ is not a real number.

29. $\left(\dfrac{64}{27}\right)^{-1/3} = \left(\dfrac{27}{64}\right)^{1/3} = \sqrt[3]{\dfrac{27}{64}} = \dfrac{\sqrt[3]{27}}{\sqrt[3]{64}} = \dfrac{3}{4}$

30. $64^{-1/2} + 8^{-2/3} = \dfrac{1}{64^{1/2}} + \dfrac{1}{8^{2/3}}$

$\phantom{64^{-1/2} + 8^{-2/3}} = \dfrac{1}{\sqrt{64}} + \dfrac{1}{\left(\sqrt[3]{8}\right)^2}$

$\phantom{64^{-1/2} + 8^{-2/3}} = \dfrac{1}{8} + \dfrac{1}{2^2}$

$\phantom{64^{-1/2} + 8^{-2/3}} = \dfrac{1}{8} + \dfrac{1}{4}$

$\phantom{64^{-1/2} + 8^{-2/3}} = \dfrac{1}{8} + \dfrac{2}{8}$

$\phantom{64^{-1/2} + 8^{-2/3}} = \dfrac{3}{8}$

31. $x^{3/5}x^{-1/3} = x^{3/5-1/3} = x^{9/15-5/15} = x^{4/15}$

32. $\left(\dfrac{64}{y^9}\right)^{1/3} = \sqrt[3]{\dfrac{64}{y^9}} = \dfrac{\sqrt[3]{64}}{\sqrt[3]{y^9}} = \dfrac{4}{y^3}$

33. $\left(\dfrac{a^{-6/5}}{a^{2/5}}\right)^{2/3} = \dfrac{a^{-6/5 \cdot 2/3}}{a^{2/5 \cdot 2/3}}$

$\phantom{\left(\dfrac{a^{-6/5}}{a^{2/5}}\right)^{2/3}} = \dfrac{a^{-12/15}}{a^{4/15}}$

$\phantom{\left(\dfrac{a^{-6/5}}{a^{2/5}}\right)^{2/3}} = a^{-12/15 - (4/15)}$

$\phantom{\left(\dfrac{a^{-6/5}}{a^{2/5}}\right)^{2/3}} = a^{-16/15}$

$\phantom{\left(\dfrac{a^{-6/5}}{a^{2/5}}\right)^{2/3}} = \dfrac{1}{a^{16/15}}$

34. $\left(\dfrac{20x^5y^{-3}}{4y^{1/2}}\right)^2 = \left(\dfrac{5x^5}{y^{7/2}}\right)^2$

$\phantom{\left(\dfrac{20x^5y^{-3}}{4y^{1/2}}\right)^2} = \dfrac{5^2 x^{5 \cdot 2}}{y^{(7/2) \cdot 2}}$

$\phantom{\left(\dfrac{20x^5y^{-3}}{4y^{1/2}}\right)^2} = \dfrac{25x^{10}}{y^7}$

35. $a^{1/2}\left(5a^{3/2}-3a^2\right)=a^{1/2}\left(5a^{3/2}\right)-a^{1/2}\left(3a^2\right)$

$\qquad = 5a^{1/2+3/2}-3a^{1/2+2}$

$\qquad = 5a^{4/2}-3a^{1/2+4/2}$

$\qquad = 5a^2-3a^{5/2}$

36. $4x^{-2/3}\left(x^{-1/2}+\dfrac{11}{4}x^{2/3}\right)$

$\qquad = 4x^{-2/3}\left(x^{-1/2}\right)+4x^{-2/3}\left(\dfrac{11}{4}x^{2/3}\right)$

$\qquad = 4x^{-2/3+(-1/2)}+11x^{-2/3+(2/3)}$

$\qquad = 4x^{-4/6+(-3/6)}+11x^{0/3}$

$\qquad = 4x^{-7/6}+11x^0$

$\qquad = \dfrac{4}{x^{7/6}}+11$

37. $x^{2/5}+x^{7/5}=x^{2/5}+x^{2/5}\cdot x^1$

$\qquad = x^{2/5}\left(1+x\right)$

38. $a^{-1/2}+a^{3/2}=a^{-1/2}+a^{-1/2}\cdot a^{4/2}$

$\qquad = a^{-1/2}(1+a^2)$

$\qquad = \dfrac{1+a^2}{a^{1/2}}$

39. $f(x)=\sqrt{6x-11}$

$\quad f(6)=\sqrt{6(6)-11}$

$\qquad = \sqrt{36-11}$

$\qquad = \sqrt{25}$

$\qquad = 5$

40. $g(x)=\sqrt[3]{9x-17}$

$\quad g(4)=\sqrt[3]{9(4)-17}$

$\qquad = \sqrt[3]{36-17}$

$\qquad = \sqrt[3]{19}$

$\qquad \approx 2.668$

41. $f(x)=\sqrt{x}$

x	$f(x)$
0	0
1	1
4	2
9	3

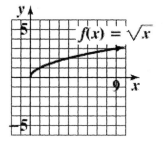

42. $f(x)=\sqrt{x}-4$

x	$f(x)$
0	−4
1	−3
4	−2
9	−1
1	0

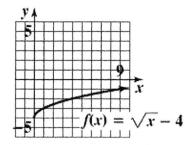

43. $\sqrt{48}=\sqrt{16}\sqrt{3}=4\sqrt{3}$

44. $\sqrt[3]{128}=\sqrt[3]{64}\sqrt[3]{2}=4\sqrt[3]{2}$

45. $\sqrt{\dfrac{49}{9}}=\dfrac{\sqrt{49}}{\sqrt{9}}=\dfrac{7}{3}$

46. $\sqrt[3]{\dfrac{8}{125}}=\dfrac{\sqrt[3]{8}}{\sqrt[3]{125}}=\dfrac{2}{5}$

47. $-\sqrt{\dfrac{81}{49}} = -\dfrac{\sqrt{81}}{\sqrt{49}} = -\dfrac{9}{7}$

48. $\sqrt[3]{-\dfrac{27}{125}} = \dfrac{\sqrt[3]{-27}}{\sqrt[3]{125}} = \dfrac{-3}{5} = -\dfrac{3}{5}$

49. $\sqrt{32}\sqrt{2} = \sqrt{32 \cdot 2} = \sqrt{64} = 8$

50. $\sqrt[3]{32} \cdot \sqrt[3]{2} = \sqrt[3]{32 \cdot 2} = \sqrt[3]{64} = 4$

51. $\sqrt{18x^2 y^3 z^4} = \sqrt{9x^2 y^2 z^4}\sqrt{2y} = 3xyz^2\sqrt{2y}$

52. $\sqrt{75x^3 y^7} = \sqrt{25x^2 y^6}\sqrt{3xy} = 5xy^3\sqrt{3xy}$

53. $\sqrt[3]{54a^7 b^{10}} = \sqrt[3]{27a^6 b^9}\sqrt[3]{2ab} = 3a^2 b^3\sqrt[3]{2ab}$

54. $\sqrt[3]{125x^8 y^9 z^{16}} = \sqrt[3]{125x^6 y^9 z^{15}}\sqrt[3]{x^2 z}$
$$= 5x^2 y^3 z^5\sqrt[3]{x^2 z}$$

55. $\left(\sqrt[6]{x^2 y^3 z^5}\right)^{42} = (x^2 y^3 z^5)^{42/6}$
$$= (x^2 y^3 z^5)^7$$
$$= x^{14} y^{21} z^{35}$$

56. $\left(\sqrt[5]{2ab^4 c^6}\right)^{15} = \left(2ab^4 c^6\right)^{15/5}$
$$= \left(2ab^4 c^6\right)^3$$
$$= 8a^3 b^{12} c^{18}$$

57. $\sqrt{5x}\sqrt{8x^5} = \sqrt{40x^6} = \sqrt{4x^6}\sqrt{10} = 2x^3\sqrt{10}$

58. $\sqrt[3]{2x^2 y}\,\sqrt[3]{4x^9 y^4} = \sqrt[3]{8x^{11} y^5} = 2x^3 y\sqrt[3]{x^2 y^2}$

59. $\sqrt[3]{2x^4 y^5}\,\sqrt[3]{16x^4 y^4} = \sqrt[3]{32x^8 y^9}$
$$= \sqrt[3]{8x^6 y^9}\sqrt[3]{4x^2}$$
$$= 2x^2 y^3\sqrt[3]{4x^2}$$

60. $\sqrt[4]{4x^4 y^7}\,\sqrt[4]{4x^5 y^9} = \sqrt[4]{16x^9 y^{16}}$
$$= \sqrt[4]{16x^8 y^{16}}\sqrt[4]{x}$$
$$= 2x^2 y^4\sqrt[4]{x}$$

61. $\sqrt{3x}(\sqrt{12x} - \sqrt{20}) = \sqrt{36x^2} - \sqrt{60x}$
$$= \sqrt{36x^2} - \sqrt{4}\sqrt{15x}$$
$$= 6x - 2\sqrt{15x}$$

62. $\sqrt[3]{2x^2 y}\left(\sqrt[3]{4x^4 y^7} + \sqrt[3]{9x}\right)$
$$= \sqrt[3]{8x^6 y^8} + \sqrt[3]{18x^3 y}$$
$$= \sqrt[3]{8x^6 y^6}\sqrt[3]{y^2} + \sqrt[3]{x^3}\sqrt[3]{18y}$$
$$= 2x^2 y^2\sqrt[3]{y^2} + x\sqrt[3]{18y}$$

63. $\sqrt{\sqrt{a^3 b^2}} = \left(\sqrt{a^3 b^2}\right)^{1/2}$
$$= \left[\left(a^3 b^2\right)^{1/2}\right]^{1/2}$$
$$= \left(a^3 b^2\right)^{1/4}$$
$$= \sqrt[4]{a^3 b^2}$$

64. $\sqrt{\sqrt[3]{x^5 y^2}} = \left(\sqrt[3]{x^5 y^2}\right)^{1/2}$
$$= \left[\left(x^5 y^2\right)^{1/3}\right]^{1/2}$$
$$= \left(x^5 y^2\right)^{1/6}$$
$$= \sqrt[6]{\left(x^5 y^2\right)}$$

65. $\left(\dfrac{4r^2 p^{1/3}}{r^{1/2} p^{4/3}}\right)^3 = (4r^{2-1/2} p^{1/3-4/3})^3$
$$= (4r^{3/2} p^{-1})^3$$
$$= 4^3 (r^{3/2})^3 (p^{-1})^3$$
$$= 64r^{9/2} p^{-3}$$
$$= \dfrac{64r^{9/2}}{p^3}$$

66. $\left(\dfrac{6y^{2/5}z^{1/3}}{x^{-1}y^{3/5}}\right)^{-1} = \left(\dfrac{6y^{2/5-3/5}z^{1/3}}{x^{-1}}\right)^{-1}$

$\quad = \left(\dfrac{6y^{-1/5}z^{1/3}}{x^{-1}}\right)^{-1}$

$\quad = \left(\dfrac{6xz^{1/3}}{y^{1/5}}\right)^{-1}$

$\quad = \dfrac{y^{1/5}}{6xz^{1/3}}$

67. $\sqrt{\dfrac{3}{5}} = \dfrac{\sqrt{3}}{\sqrt{5}} \cdot \dfrac{\sqrt{5}}{\sqrt{5}} = \dfrac{\sqrt{15}}{5}$

68. $\sqrt[3]{\dfrac{7}{9}} = \dfrac{\sqrt[3]{7}}{\sqrt[3]{9}} = \dfrac{\sqrt[3]{7}}{\sqrt[3]{9}} \cdot \dfrac{\sqrt[3]{3}}{\sqrt[3]{3}} = \dfrac{\sqrt[3]{21}}{\sqrt[3]{27}} = \dfrac{\sqrt[3]{21}}{3}$

69. $\sqrt[4]{\dfrac{5}{4}} = \dfrac{\sqrt[4]{5}}{\sqrt[4]{4}} = \dfrac{\sqrt[4]{5}}{\sqrt[4]{4}} \cdot \dfrac{\sqrt[4]{2^2}}{\sqrt[4]{2^2}} = \dfrac{\sqrt[4]{20}}{\sqrt[4]{2^4}} = \dfrac{\sqrt[4]{20}}{2}$

70. $\dfrac{x}{\sqrt{10}} = \dfrac{x}{\sqrt{10}} \cdot \dfrac{\sqrt{10}}{\sqrt{10}} = \dfrac{x\sqrt{10}}{10}$

71. $\dfrac{8}{\sqrt{x}} = \dfrac{8}{\sqrt{x}} \cdot \dfrac{\sqrt{x}}{\sqrt{x}} = \dfrac{8\sqrt{x}}{x}$

72. $\dfrac{m}{\sqrt[3]{25}} = \dfrac{m}{\sqrt[3]{5^2}} \cdot \dfrac{\sqrt[3]{5}}{\sqrt[3]{5}} = \dfrac{m\sqrt[3]{5}}{\sqrt[3]{5^3}} = \dfrac{m\sqrt[3]{5}}{5}$

73. $\dfrac{10}{\sqrt[3]{y^2}} = \dfrac{10}{\sqrt[3]{y^2}} \cdot \dfrac{\sqrt[3]{y}}{\sqrt[3]{y}} = \dfrac{10\sqrt[3]{y}}{\sqrt[3]{y^3}} = \dfrac{10\sqrt[3]{y}}{y}$

74. $\dfrac{9}{\sqrt[4]{z}} = \dfrac{9}{\sqrt[4]{z}} \cdot \dfrac{\sqrt[4]{z^3}}{\sqrt[4]{z^3}} = \dfrac{9\sqrt[4]{z^3}}{\sqrt[4]{z^4}} = \dfrac{9\sqrt[4]{z^3}}{z}$

75. $\sqrt[3]{\dfrac{x^3}{27}} = \dfrac{\sqrt[3]{x^3}}{\sqrt[3]{27}} = \dfrac{x}{3}$

76. $\dfrac{\sqrt[3]{2x^{10}}}{\sqrt[3]{16x^7}} = \sqrt[3]{\dfrac{2x^{10}}{16x^7}} = \sqrt[3]{\dfrac{x^3}{8}} = \dfrac{\sqrt[3]{x^3}}{\sqrt[3]{8}} = \dfrac{x}{2}$

77. $\sqrt{\dfrac{32x^2y^5}{2x^8y}} = \sqrt{\dfrac{16y^4}{x^6}} = \dfrac{\sqrt{16y^4}}{\sqrt{x^6}} = \dfrac{4y^2}{x^3}$

78. $\sqrt[4]{\dfrac{48x^9y^{15}}{3xy^3}} = \sqrt[4]{16x^8y^{12}} = 2x^2y^3$

79. $\sqrt{\dfrac{6x^4}{y}} = \dfrac{\sqrt{6x^4}}{\sqrt{y}}$

$\quad = \dfrac{\sqrt{x^4}\sqrt{6}}{\sqrt{y}}$

$\quad = \dfrac{x^2\sqrt{6}}{\sqrt{y}} \cdot \dfrac{\sqrt{y}}{\sqrt{y}}$

$\quad = \dfrac{x^2\sqrt{6y}}{y}$

80. $\sqrt{\dfrac{12a}{7b}} = \dfrac{\sqrt{12a}}{\sqrt{7b}}$

$\quad = \dfrac{2\sqrt{3a}}{\sqrt{7b}}$

$\quad = \dfrac{2\sqrt{3a}}{\sqrt{7b}} \cdot \dfrac{\sqrt{7b}}{\sqrt{7b}}$

$\quad = \dfrac{2\sqrt{21ab}}{7b}$

81. $\sqrt{\dfrac{18x^4y^5}{3z}} = \dfrac{\sqrt{18x^4y^5}}{\sqrt{3z}}$

$\quad = \dfrac{\sqrt{9x^4y^4}\sqrt{2y}}{\sqrt{3z}}$

$\quad = \dfrac{3x^2y^2\sqrt{2y}}{\sqrt{3z}}$

$\quad = \dfrac{3x^2y^2\sqrt{2y}}{\sqrt{3z}} \cdot \dfrac{\sqrt{3z}}{\sqrt{3z}}$

$\quad = \dfrac{3x^2y^2\sqrt{6yz}}{3z}$

$\quad = \dfrac{x^2y^2\sqrt{6yz}}{z}$

82. $\sqrt{\dfrac{125x^2y^5}{3z}} = \dfrac{\sqrt{125x^2y^5}}{\sqrt{3z}}$

$= \dfrac{\sqrt{25x^2y^4}\,\sqrt{5y}}{\sqrt{3z}}$

$= \dfrac{5xy^2\sqrt{5y}}{\sqrt{3z}} \cdot \dfrac{\sqrt{3z}}{\sqrt{3z}}$

$= \dfrac{5xy^2\sqrt{15yz}}{3z}$

83. $\sqrt[3]{\dfrac{108x^3y^7}{2y^3}} = \sqrt[3]{54x^3y^4} = 3xy\sqrt[3]{2y}$

84. $\sqrt[3]{\dfrac{3x}{5y}} = \dfrac{\sqrt[3]{3x}}{\sqrt[3]{5y}} \cdot \dfrac{\sqrt[3]{25y^2}}{\sqrt[3]{25y^2}} = \dfrac{\sqrt[3]{75xy^2}}{5y}$

85. $\sqrt[3]{\dfrac{9x^5y^3}{x^6}} = \sqrt[3]{\dfrac{9y^3}{x}}$

$= \dfrac{\sqrt[3]{9y^3}}{\sqrt[3]{x}}$

$= \dfrac{\sqrt[3]{y^3}\,\sqrt[3]{9}}{\sqrt[3]{x}}$

$= \dfrac{y\sqrt[3]{9}}{\sqrt[3]{x}} \cdot \dfrac{\sqrt[3]{x^2}}{\sqrt[3]{x^2}}$

$= \dfrac{y\sqrt[3]{9x^2}}{x}$

86. $\sqrt[3]{\dfrac{y^6}{5x^2}} = \dfrac{\sqrt[3]{y^6}}{\sqrt[3]{5x^2}} = \dfrac{y^2}{\sqrt[3]{5x^2}} \cdot \dfrac{\sqrt[3]{25x}}{\sqrt[3]{25x}} = \dfrac{y^2\sqrt[3]{25x}}{5x}$

87. $\sqrt[4]{\dfrac{2a^2b^{11}}{a^5b}} = \sqrt[4]{\dfrac{2b^{10}}{a^3}}$

$= \dfrac{\sqrt[4]{2b^{10}}}{\sqrt[4]{a^3}} \cdot \dfrac{\sqrt[4]{a}}{\sqrt[4]{a}}$

$= \dfrac{\sqrt[4]{2ab^{10}}}{\sqrt[4]{a^4}}$

$= \dfrac{b^2\sqrt[4]{2ab^2}}{a}$

88. $\sqrt[4]{\dfrac{3x^2y^6}{8x^3}} = \sqrt[4]{\dfrac{3y^6}{8x}}$

$= \dfrac{\sqrt[4]{3y^6}}{\sqrt[4]{8x}}$

$= \dfrac{y\sqrt[4]{3y^2}}{\sqrt[4]{8x}}$

$= \dfrac{y\sqrt[4]{3y^2}}{\sqrt[4]{8x}} \cdot \dfrac{\sqrt[4]{2x^3}}{\sqrt[4]{2x^3}}$

$= \dfrac{y\sqrt[4]{6x^3y^2}}{2x}$

89. $\left(3-\sqrt{2}\right)\left(3+\sqrt{2}\right) = 3^2 - \left(\sqrt{2}\right)^2 = 9-2 = 7$

90. $\left(\sqrt{x}+y\right)\left(\sqrt{x}-y\right) = \left(\sqrt{x}\right)^2 - y^2 = x - y^2$

91. $\left(x-\sqrt{y}\right)\left(x+\sqrt{y}\right) = x^2 - \left(\sqrt{y}\right)^2 = x^2 - y$

92. $\left(\sqrt{3}+2\right)^2 = \left(\sqrt{3}\right)^2 + 2(2)\left(\sqrt{3}\right) + 2^2$

$= 3 + 4\sqrt{3} + 4$

$= 7 + 4\sqrt{3}$

93. $\left(\sqrt{x}-\sqrt{3y}\right)\left(\sqrt{x}+\sqrt{5y}\right)$

$= \left(\sqrt{x}\right)^2 + \sqrt{x}\sqrt{5y} - \sqrt{x}\sqrt{3y} - \sqrt{3y}\sqrt{5y}$

$= x + \sqrt{5xy} - \sqrt{3xy} - \sqrt{15y^2}$

$= x + \sqrt{5xy} - \sqrt{3xy} - y\sqrt{15}$

94. $\left(\sqrt[3]{2x}-\sqrt[3]{3y}\right)\left(\sqrt[3]{3x}-\sqrt[3]{2y}\right)$

$= \sqrt[3]{2x}\left(\sqrt[3]{3x}\right) - \left(\sqrt[3]{2x}\right)\left(\sqrt[3]{2y}\right)$

$\quad - \sqrt[3]{3y}\left(\sqrt[3]{3x}\right) + \sqrt[3]{3y}\sqrt[3]{2y}$

$= \sqrt[3]{6x^2} - \sqrt[3]{4xy} - \sqrt[3]{9xy} + \sqrt[3]{6y^2}$

95. $\dfrac{6}{2+\sqrt{5}} = \dfrac{6}{2+\sqrt{5}} \cdot \dfrac{2-\sqrt{5}}{2-\sqrt{5}}$

$\phantom{\dfrac{6}{2+\sqrt{5}}} = \dfrac{6\left(2-\sqrt{5}\right)}{2^2 - \left(\sqrt{5}\right)^2}$

$\phantom{\dfrac{6}{2+\sqrt{5}}} = \dfrac{12-6\sqrt{5}}{4-5}$

$\phantom{\dfrac{6}{2+\sqrt{5}}} = \dfrac{12-6\sqrt{5}}{-1}$

$\phantom{\dfrac{6}{2+\sqrt{5}}} = -12+6\sqrt{5}$

96. $\dfrac{x}{4+\sqrt{x}} = \dfrac{x}{4+\sqrt{x}} \cdot \dfrac{4-\sqrt{x}}{4-\sqrt{x}}$

$\phantom{\dfrac{x}{4+\sqrt{x}}} = \dfrac{x\left(4-\sqrt{x}\right)}{4^2 - \left(\sqrt{x}\right)^2}$

$\phantom{\dfrac{x}{4+\sqrt{x}}} = \dfrac{4x-x\sqrt{x}}{16-x}$

97. $\dfrac{a}{4-\sqrt{b}} = \dfrac{a}{4-\sqrt{b}} \cdot \dfrac{4+\sqrt{b}}{4+\sqrt{b}}$

$\phantom{\dfrac{a}{4-\sqrt{b}}} = \dfrac{a\left(4+\sqrt{b}\right)}{\left(4-\sqrt{b}\right)\left(4+\sqrt{b}\right)}$

$\phantom{\dfrac{a}{4-\sqrt{b}}} = \dfrac{4a+a\sqrt{b}}{16+4\sqrt{b}-4\sqrt{b}-b}$

$\phantom{\dfrac{a}{4-\sqrt{b}}} = \dfrac{4a+a\sqrt{b}}{16-b}$

98. $\dfrac{x}{\sqrt{y}-7} = \dfrac{x}{\sqrt{y}-7} \cdot \dfrac{\sqrt{y}+7}{\sqrt{y}+7}$

$\phantom{\dfrac{x}{\sqrt{y}-7}} = \dfrac{x\left(\sqrt{y}+7\right)}{\left(\sqrt{y}-7\right)\left(\sqrt{y}+7\right)}$

$\phantom{\dfrac{x}{\sqrt{y}-7}} = \dfrac{x\sqrt{y}+7x}{y+7\sqrt{y}-7\sqrt{y}-49}$

$\phantom{\dfrac{x}{\sqrt{y}-7}} = \dfrac{x\sqrt{y}+7x}{y-49}$

99. $\dfrac{\sqrt{x}}{\sqrt{x}+\sqrt{y}} = \dfrac{\sqrt{x}}{\sqrt{x}+\sqrt{y}} \cdot \dfrac{\sqrt{x}-\sqrt{y}}{\sqrt{x}-\sqrt{y}}$

$\phantom{\dfrac{\sqrt{x}}{\sqrt{x}+\sqrt{y}}} = \dfrac{\sqrt{x}\left(\sqrt{x}-\sqrt{y}\right)}{\left(\sqrt{x}\right)^2 - \left(\sqrt{y}\right)^2}$

$\phantom{\dfrac{\sqrt{x}}{\sqrt{x}+\sqrt{y}}} = \dfrac{\sqrt{x^2}-\sqrt{xy}}{x-y}$

$\phantom{\dfrac{\sqrt{x}}{\sqrt{x}+\sqrt{y}}} = \dfrac{x-\sqrt{xy}}{x-y}$

100. $\dfrac{\sqrt{x}-3\sqrt{y}}{\sqrt{x}-\sqrt{y}} = \dfrac{\sqrt{x}-3\sqrt{y}}{\sqrt{x}-\sqrt{y}} \cdot \dfrac{\sqrt{x}+\sqrt{y}}{\sqrt{x}+\sqrt{y}}$

$\phantom{\dfrac{\sqrt{x}-3\sqrt{y}}{\sqrt{x}-\sqrt{y}}} = \dfrac{x+\sqrt{xy}-3\sqrt{xy}-3y}{\left(\sqrt{x}\right)^2 - \left(\sqrt{y}\right)^2}$

$\phantom{\dfrac{\sqrt{x}-3\sqrt{y}}{\sqrt{x}-\sqrt{y}}} = \dfrac{x-2\sqrt{xy}-3y}{x-y}$

101. $\dfrac{2}{\sqrt{a-1}-2} = \dfrac{2}{\sqrt{a-1}-2} \cdot \dfrac{\sqrt{a-1}+2}{\sqrt{a-1}+2}$

$\phantom{\dfrac{2}{\sqrt{a-1}-2}} = \dfrac{2\left(\sqrt{a-1}+2\right)}{\left(\sqrt{a-1}-2\right)\left(\sqrt{a-1}+2\right)}$

$\phantom{\dfrac{2}{\sqrt{a-1}-2}} = \dfrac{2\sqrt{a-1}+4}{a-1+2\sqrt{a-1}-2\sqrt{a-1}-4}$

$\phantom{\dfrac{2}{\sqrt{a-1}-2}} = \dfrac{2\sqrt{a-1}+4}{a-5}$

102. $\dfrac{5}{\sqrt{y+2}-3} = \dfrac{5}{\sqrt{y+2}-3} \cdot \dfrac{\sqrt{y+2}+3}{\sqrt{y+2}+3}$

$\phantom{\dfrac{5}{\sqrt{y+2}-3}} = \dfrac{5\sqrt{y+2}+15}{\left(\sqrt{y+2}\right)^2 - 3^2}$

$\phantom{\dfrac{5}{\sqrt{y+2}-3}} = \dfrac{5\sqrt{y+2}+15}{y+2-9}$

$\phantom{\dfrac{5}{\sqrt{y+2}-3}} = \dfrac{5\sqrt{y+2}+15}{y-7}$

103. $\sqrt[3]{x}+10\sqrt[3]{x}-2\sqrt[3]{x}=9\sqrt[3]{x}$

104. $\sqrt{3} + \sqrt{27} - \sqrt{192} = \sqrt{3} + 3\sqrt{3} - \sqrt{64}\sqrt{3}$

$$= \sqrt{3} + 3\sqrt{3} - 8\sqrt{3}$$

$$= -4\sqrt{3}$$

105. $\sqrt[3]{16} - 5\sqrt[3]{54} + 3\sqrt[3]{64}$

$$= \sqrt[3]{8}\sqrt[3]{2} - 5\sqrt[3]{27}\sqrt[3]{2} + 3\sqrt[3]{64}$$

$$= 2\sqrt[3]{2} - 5\left(3\sqrt[3]{2}\right) + 3(4)$$

$$= 2\sqrt[3]{2} - 15\sqrt[3]{2} + 12$$

$$= 12 - 13\sqrt[3]{2}$$

106. $\sqrt{2} - \dfrac{3}{\sqrt{32}} + \sqrt{50}$

$$= \sqrt{2} - \frac{3}{4\sqrt{2}} + 5\sqrt{2}$$

$$= \sqrt{2} - \frac{3}{4\sqrt{2}} \cdot \frac{\sqrt{2}}{\sqrt{2}} + 5\sqrt{2}$$

$$= \sqrt{2} - \frac{3\sqrt{2}}{8} + 5\sqrt{2}$$

$$= \frac{8}{8}\left(\sqrt{2}\right) - \frac{3\sqrt{2}}{8} + \left(\frac{8}{8}\right)5\sqrt{2}$$

$$= \frac{8\sqrt{2}}{8} - \frac{3\sqrt{2}}{8} + \frac{40\sqrt{2}}{8}$$

$$= \frac{45\sqrt{2}}{8}$$

107. $9\sqrt{x^5 y^6} - \sqrt{16x^7 y^8}$

$$= 9\sqrt{x^4 y^6}\sqrt{x} - \sqrt{16x^6 y^8}\sqrt{x}$$

$$= 9x^2 y^3 \sqrt{x} - 4x^3 y^4 \sqrt{x}$$

$$= \left(9x^2 y^3 - 4x^3 y^4\right)\sqrt{x}$$

108. $8\sqrt[3]{x^7 y^8} - \sqrt[3]{x^4 y^2} + 3\sqrt[3]{x^{10} y^2}$

$$= 8\sqrt[3]{x^6 y^6}\sqrt[3]{xy^2} - \sqrt[3]{x^3}\sqrt[3]{xy^2} + 3\sqrt[3]{x^9}\sqrt[3]{xy^2}$$

$$= 8x^2 y^2 \sqrt[3]{xy^2} - x\sqrt[3]{xy^2} + 3x^3 \sqrt[3]{xy^2}$$

$$= \left(8x^2 y^2 - x + 3x^3\right)\sqrt[3]{xy^2}$$

109. $(f \cdot g)(x) = f(x) \cdot g(x)$

$$= \sqrt{3x} \cdot \left(\sqrt{6x} - \sqrt{15}\right)$$

$$= \sqrt{3x}\sqrt{6x} - \sqrt{3x}\sqrt{15}$$

$$= \sqrt{18x^2} - \sqrt{45x}$$

$$= \sqrt{9x^2}\sqrt{2} - \sqrt{9}\sqrt{5x}$$

$$= 3x\sqrt{2} - 3\sqrt{5x}$$

110. $(f \cdot g)(x) = f(x) \cdot g(x)$

$$= \sqrt[3]{2x^2}\left(\sqrt[3]{4x^4} + \sqrt[3]{16x^5}\right)$$

$$= \sqrt[3]{2x^2}\sqrt[3]{4x^4} + \sqrt[3]{2x^2}\sqrt[3]{16x^5}$$

$$= \sqrt[3]{8x^6} + \sqrt[3]{32x^7}$$

$$= \sqrt[3]{8x^6} + \sqrt[3]{8x^6}\sqrt[3]{4x}$$

$$= 2x^2 + 2x^2\sqrt[3]{4x}$$

111. $f(x) = \sqrt{2x+7}\sqrt{2x+7}, \quad x \geq -\dfrac{7}{2}$

$$= \sqrt{(2x+7)^2}$$

$$= |2x+7|$$

$$= 2x + 7 \quad \text{since } x \geq -\frac{7}{2}$$

112. $g(a) = \sqrt{20a^2 + 100a + 125}$

$$= \sqrt{5(4a^2 + 20a + 25)}$$

$$= \sqrt{5(2a+5)^2}$$

$$= \sqrt{5}\,|2a+5|$$

113. $\dfrac{\sqrt[3]{(x+5)^5}}{\sqrt{(x+5)^3}} = \dfrac{(x+5)^{5/3}}{(x+5)^{3/2}}$

$$= (x+5)^{5/3 - 3/2}$$

$$= (x+5)^{1/6}$$

$$= \sqrt[6]{x+5}$$

114. $\dfrac{\sqrt[3]{a^3b^2}}{\sqrt[4]{a^4b}} = \dfrac{a\sqrt[3]{b^2}}{a\sqrt[4]{b}}$

$\qquad = \dfrac{\sqrt[3]{b^2}}{\sqrt[4]{b}}$

$\qquad = \dfrac{b^{2/3}}{b^{1/4}}$

$\qquad = b^{2/3-1/4}$

$\qquad = b^{5/12}$

$\qquad = \sqrt[12]{b^5}$

115. a. $P = 2l + 2w$

$\qquad P = 2\sqrt{48} + 2\sqrt{12}$

$\qquad = 2\sqrt{16 \cdot 3} + 2\sqrt{4 \cdot 3}$

$\qquad = 8\sqrt{3} + 4\sqrt{3}$

$\qquad = 12\sqrt{3}$

b. $A = lw$

$\qquad A = \left(\sqrt{48}\right)\left(\sqrt{12}\right)$

$\qquad = \sqrt{576}$

$\qquad = 24$

116. a. $P = s_1 + s_2 + s_3$

$\qquad P = \sqrt{125} + \sqrt{45} + \sqrt{130}$

$\qquad = 5\sqrt{5} + 3\sqrt{5} + \sqrt{130}$

$\qquad = 8\sqrt{5} + \sqrt{130}$

b. $A = \dfrac{1}{2}bh$

$\qquad A = \dfrac{1}{2}\left(\sqrt{130}\right)\left(\sqrt{40}\right)$

$\qquad = \dfrac{1}{2}\sqrt{5200}$

$\qquad = \dfrac{1}{2}\sqrt{400 \cdot 13}$

$\qquad = \dfrac{20}{2}\sqrt{13}$

$\qquad = 10\sqrt{13}$

117. a. $f(x) = \sqrt{x} + 2$

$\qquad g(x) = -3$

$\qquad (f+g)(x) = f(x) + g(x)$

$\qquad\qquad = \sqrt{x} + 2 - 3$

$\qquad\qquad = \sqrt{x} - 1$

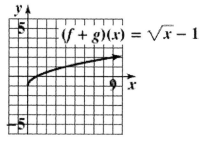

b. $\sqrt{x} \geq 0, x \geq 0$

The domain is $\left\{x \mid x \geq 0\right\}$.

118. a. $f(x) = -\sqrt{x}$

$\qquad g(x) = \sqrt{x} + 2$

$\qquad (f+g)(x) = f(x) + g(x)$

$\qquad\qquad = -\sqrt{x} + \sqrt{x} + 2$

$\qquad\qquad = 2$

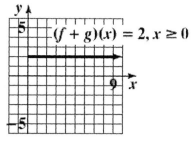

b. $\sqrt{x} \geq 0, x \geq 0$

The domain is $\left\{x \mid x \geq 0\right\}$.

119. $\sqrt{x} = 9$

$\qquad \left(\sqrt{x}\right)^2 = 9^2$

$\qquad x = 81$

Check: $\sqrt{81} = 9$ True

81 is the solution.

120. $\sqrt{x} = -4$

$\left(\sqrt{x}\right)^2 = \left(-4\right)^2$

$x = 16$

Check: $\sqrt{16} = -4$

$\qquad\qquad 4 = -4$ False

no solution

121. $\sqrt[3]{x} = 4$

$\left(\sqrt[3]{x}\right)^3 = 4^3$

$\qquad x = 64$

Check: $\sqrt[3]{x} = 4$

$\qquad\qquad \sqrt[3]{64} = 4$

$\qquad\qquad\qquad 4 = 4$ True

The solution is 64.

122. $\sqrt[3]{x} = -5$

$\left(\sqrt[3]{x}\right)^3 = \left(-5\right)^3$

$\qquad x = -125$

Check: $\sqrt[3]{x} = -5$

$\qquad\qquad \sqrt[3]{-125} = -5$

$\qquad\qquad\qquad -5 = -5$ True

-125 is the solution.

123. $7 + \sqrt{x} = 10$

$\qquad \sqrt{x} = 3$

$\qquad \left(\sqrt{x}\right)^2 = 3^2$

$\qquad\qquad x = 9$

Check: $7 + \sqrt{x} = 10$

$\qquad\qquad 7 + \sqrt{9} = 10$

$\qquad\qquad 7 + 3 = 10$

$\qquad\qquad\qquad 10 = 10$ True

9 is the solution.

124. $7 + \sqrt[3]{x} = 12$

$\qquad \sqrt[3]{x} = 5$

$\qquad \left(\sqrt[3]{x}\right)^3 = 5^3$

$\qquad\qquad x = 125$

Check: $7 + \sqrt[3]{x} = 12$

$\qquad\qquad 7 + \sqrt[3]{125} = 12$

$\qquad\qquad\qquad 7 + 5 = 12$

$\qquad\qquad\qquad\qquad 12 = 12$ True

125 is the solution.

125 $\sqrt{3x+4} = \sqrt{5x+14}$

$\left(\sqrt{3x+4}\right)^2 = \left(\sqrt{5x+14}\right)^2$

$3x + 4 = 5x + 14$

$-10 = 2x$

$-5 = x$

Check: $\sqrt{3x+4}$ becomes

$\sqrt{3(-5)+4} = \sqrt{-15+4} = \sqrt{-11}$

which is not a real number. -5 is not a solution so there is no real solution.

126. $\sqrt{x^2 + 2x - 8} = x$

$\left(\sqrt{x^2 + 2x - 8}\right)^2 = (x)^2$

$x^2 + 2x - 8 = x^2$

$2x - 8 = 0$

$2x = 8$

$x = 4$

Check: $\sqrt{2^2 + 2 \cdot 4 - 8} = 2$

$\sqrt{4 + 8 - 8} = 2$

$\sqrt{4} = 2$

$2 = 2$ True

4 is the solution.

127. $\sqrt[3]{x-9} = \sqrt[3]{5x+3}$

$\left(\sqrt[3]{x-9}\right)^3 = \left(\sqrt[3]{5x+3}\right)^3$

$x - 9 = 5x + 3$

$-4x = 12$

$x = -3$

Check: $\sqrt[3]{-3-9} = \sqrt[3]{5(-3)+3}$

$\sqrt[3]{-12} = \sqrt[3]{-15+3}$

$\sqrt[3]{-12} = \sqrt[3]{-12}$ True

-3 is the solution.

128. $(x^2 + 7)^{1/2} = x + 1$

$\left[(x^2+7)^{1/2}\right]^2 = (x+1)^2$

$x^2 + 7 = x^2 + 2x + 1$

$7 = 2x + 1$

$6 = 2x$

$x = 3$

Check: $(3^2 + 7)^{1/2} = 3 + 1$

$(9 + 7)^{1/2} = 4$

$16^{1/2} = 4$

$4 = 4$ True

3 is the solution.

129. $\sqrt{x} + 3 = \sqrt{3x+9}$

$\left(\sqrt{x}+3\right)^2 = \left(\sqrt{3x+9}\right)^2$

$\left(\sqrt{x}+3\right)\left(\sqrt{x}+3\right) = 3x+9$

$x + 6\sqrt{x} + 9 = 3x + 9$

$6\sqrt{x} = 2x$

$\left(6\sqrt{x}\right)^2 = (2x)^2$

$36x = 4x^2$

$4x^2 - 36x = 0$

$4x(x-9) = 0$

$4x = 0$ or $x - 9 = 0$

$x = 0$ $\qquad$ $x = 9$

A check shows that 0 and 9 are both solutions.

130. $\sqrt{6x-5} - \sqrt{2x+6} - 1 = 0$

$\sqrt{6x-5} = \sqrt{2x+6} + 1$

$\left(\sqrt{6x-5}\right)^2 = \left(\sqrt{2x+6}+1\right)^2$

$6x - 5 = 2x + 6 + 2\sqrt{2x+6} + 1$

$6x - 5 = 2x + 7 + 2\sqrt{2x+6}$

$4x - 12 = 2\sqrt{2x+6}$

$\dfrac{4x}{2} - \dfrac{12}{2} = \dfrac{2}{2}\sqrt{2x+6}$

$2x - 6 = \sqrt{2x+6}$

$(2x-6)^2 = \left(\sqrt{2x+6}\right)^2$

$4x^2 - 24x + 36 = 2x + 6$

$4x^2 - 26x + 30 = 0$

$2x^2 - 13x + 15 = 0$

Express the middle term, $-13x$, as $-10x - 3x$.

$2x^2 - 10x - 3x + 15 = 0$

$2x(x-5) - 3(x-5) = 0$

$(2x-3)(x-5) = 0$

$2x - 3 = 0$ or $x - 5 = 0$

$x = \dfrac{3}{2}$ $\qquad$ $x = 5$

The solution $x = 5$ checks in the original equation but $x = \dfrac{3}{2}$ does not check. Therefore the only solution is $x = 5$.

131. $f(x) = g(x)$

$\sqrt{3x+4} = 2\sqrt{2x-4}$

$\left(\sqrt{3x+4}\right)^2 = \left(2\sqrt{2x-4}\right)^2$

$3x + 4 = 4(2x-4)$

$3x + 4 = 8x - 16$

$20 = 5x$

$4 = x$

132. $f(x) = g(x)$

$(4x+5)^{1/3} = (6x-7)^{1/3}$

$[(4x+5)^{1/3}]^3 = [(6x-7)^{1/3}]^3$

$4x + 5 = 6x - 7$

$12 = 2x$

$6 = x$

133. $V = \sqrt{\dfrac{2L}{w}}$ Solve for L.

$$V^2 = \dfrac{2L}{w}$$

$$V^2 w = 2L$$

$$\dfrac{V^2 w}{2} = L \text{ or } L = \dfrac{V^2 w}{2}$$

134. $r = \sqrt{\dfrac{A}{\pi}}$ Solve for A.

$$r^2 = \left(\sqrt{\dfrac{A}{\pi}}\right)^2$$

$$r^2 = \dfrac{A}{\pi}$$

$$\pi r^2 = A \text{ or } A = \pi r^2$$

135. Pythagorean Theorem: $a^2 + b^2 = c^2$

$$\left(\sqrt{20}\right)^2 + 6^2 = x^2$$

$$20 + 36 = x^2$$

$$56 = x^2$$

$$\sqrt{56} = x \text{ or } x = 2\sqrt{14}$$

136. Pythagorean Theorem: $a^2 + b^2 = c^2$

$$\left(\sqrt{26}\right)^2 + x^2 = \left(\sqrt{101}\right)^2$$

$$26 + x^2 = 101$$

$$x^2 = 75$$

$$x = \sqrt{75} \text{ or } x = 5\sqrt{3}$$

137. $l = \sqrt{a^2 + b^2}$

$$= \sqrt{5^2 + 2^2}$$

$$= \sqrt{29}$$

$$\approx 5.39 \text{ m}$$

138. $v = \sqrt{2gh}$

$$= \sqrt{2(32)(20)}$$

$$= \sqrt{1280} \approx 35.78 \text{ ft/sec}$$

139. $T = 2\pi \sqrt{\dfrac{L}{32}}$

$$= 2\pi \sqrt{\dfrac{64}{32}}$$

$$= 2\pi \sqrt{2}$$

$$\approx 8.89 \text{ sec}$$

140. $V = \sqrt{\dfrac{2K}{m}}$

$$= \sqrt{\dfrac{2(45)}{0.145}}$$

$$\approx 24.91 \text{ meters per second}$$

141. $m = \dfrac{m_0}{\sqrt{1 - \dfrac{v^2}{c^2}}}$

$$= \dfrac{m_0}{\sqrt{1 - \dfrac{(0.98c)^2}{c^2}}}$$

$$= \dfrac{m_0}{\sqrt{1 - \dfrac{0.9604c^2}{c^2}}}$$

$$= \dfrac{m_0}{\sqrt{1 - 0.9604}}$$

$$= \dfrac{m_0}{\sqrt{0.0396}}$$

$$\approx 5m_0$$

It is about 5 times its original mass.

142. $5 = 5 + 0i$

143. $-8 = -8 + 0i$

144. $7 - \sqrt{-256} = 7 - \sqrt{-1}\sqrt{256}$

$$= 7 - 16i$$

145. $9 + \sqrt{-16} = 9 + \sqrt{16}\sqrt{-1}$

$$= 9 + 4i$$

146. $\left(3 + 2i\right) + \left(10 - i\right) = 3 + 2i + 10 - i = 13 + i$

147. $\left(9 - 6i\right) - \left(3 - 4i\right) = 9 - 6i - 3 + 4i = 6 - 2i$

148. $\left(\sqrt{3}+\sqrt{-5}\right)+\left(11\sqrt{3}-\sqrt{-7}\right)$

$=\sqrt{3}+\sqrt{5}\sqrt{-1}+11\sqrt{3}-\sqrt{7}\sqrt{-1}$

$=\sqrt{3}+i\sqrt{5}+11\sqrt{3}-i\sqrt{7}$

$=12\sqrt{3}+\left(\sqrt{5}-\sqrt{7}\right)i$

149. $\sqrt{-6}\left(\sqrt{6}+\sqrt{-6}\right)=\sqrt{6}\sqrt{-1}\left(\sqrt{6}+\sqrt{6}\sqrt{-1}\right)$

$=i\sqrt{6}\left(\sqrt{6}+i\sqrt{6}\right)$

$=i\sqrt{36}+i^2\sqrt{36}$

$=6i+6(-1)=-6+6i$

150. $(4+3i)(2-3i)=8-12i+6i-9i^2$

$=8-6i-9(-1)$

$=8-6i+9$

$=17-6i$

151. $\left(6+\sqrt{-3}\right)\left(4-\sqrt{-15}\right)$

$=\left(6+\sqrt{3}\sqrt{-1}\right)\left(4-\sqrt{15}\sqrt{-1}\right)$

$=\left(6+i\sqrt{3}\right)\left(4-i\sqrt{15}\right)$

$=24-6i\sqrt{15}+4i\sqrt{3}-i^2\sqrt{45}$

$=24-6i\sqrt{15}+4i\sqrt{3}-(-1)\sqrt{9}\sqrt{5}$

$=\left(24+3\sqrt{5}\right)+\left(4\sqrt{3}-6\sqrt{15}\right)i$

152. $\dfrac{8}{3i}=\dfrac{8}{3i}\cdot\dfrac{-3i}{-3i}$

$=\dfrac{-24i}{-9i^2}$

$=\dfrac{-24i}{-9(-1)}$

$=-\dfrac{24i}{9}$

$=-\dfrac{8i}{3}$

153. $\dfrac{2+\sqrt{3}}{2i}=\dfrac{2+\sqrt{3}}{2i}\cdot\dfrac{-2i}{-2i}$

$=\dfrac{-4i-2i\sqrt{3}}{-4i^2}$

$=\dfrac{-4i-2i\sqrt{3}}{-4(-1)}$

$=\dfrac{2\left(-2i-i\sqrt{3}\right)}{4}$

$=\dfrac{-2i-i\sqrt{3}}{2}$

$=\dfrac{\left(-2-\sqrt{3}\right)i}{2}$

154. $\dfrac{4}{3+2i}=\dfrac{4}{3+2i}\cdot\dfrac{3-2i}{3-2i}$

$=\dfrac{4(3-2i)}{9-4i^2}$

$=\dfrac{12-8i}{9-4(-1)}$

$=\dfrac{12-8i}{9+4}$

$=\dfrac{12-8i}{13}$

155. $\dfrac{\sqrt{3}}{5-\sqrt{-6}}=\dfrac{\sqrt{3}}{5-i\sqrt{6}}$

$=\dfrac{\sqrt{3}}{\left(5-i\sqrt{6}\right)}\cdot\dfrac{5+i\sqrt{6}}{5+i\sqrt{6}}$

$=\dfrac{5\sqrt{3}+i\sqrt{18}}{(5)^2-\left(i\sqrt{6}\right)^2}$

$=\dfrac{5\sqrt{3}+3i\sqrt{2}}{(5)^2+\left(\sqrt{6}\right)^2}$

$=\dfrac{5\sqrt{3}+3i\sqrt{2}}{25+6}$

$=\dfrac{5\sqrt{3}+3i\sqrt{2}}{31}$

156. x^2-2x+9

$=\left(1+2i\sqrt{2}\right)^2-2\left(1+2i\sqrt{2}\right)+9$

$=1^2+2(1)\left(2i\sqrt{2}\right)+\left(2i\sqrt{2}\right)^2-2-4i\sqrt{2}+9$

$=1+4i\sqrt{2}-8-2-4i\sqrt{2}+9$

$=0+0i$

$=0$

157. $x^2 - 2x + 12$

$= (1 - 2i)^2 - 2(1 - 2i) + 12$

$= 1^2 - 2(1)(2i) + (2i)^2 - 2 + 4i + 12$

$= 1 - 4i - 4 - 2 + 4i + 12$

$= 7 + 0i$

$= 7$

158. $i^{33} = i^{32}i = (i^4)^8 = 1^8 \cdot i = i$

159. $i^{59} = i^{56}i^3 = (i^4)^{14}i^3 = 1^{14} \cdot i^3 = 1(i^3) = i^3 = -i$

160. $i^{404} = (i^4)^{101} = 1^{101} = 1$

161. $i^{802} = i^{800}i^2$

$= (i^4)^{200}i^2$

$= 1^{200} \cdot i^2$

$= 1(i^2)$

$= i^2$

$= -1$

Chapter 11 Practice Test

1. $\sqrt{(5x-3)^2} = |5x - 3|$

2. $\left(\dfrac{x^{2/5} \cdot x^{-1}}{x^{3/5}}\right)^2 = \left(x^{2/5 - 3/5 - 1}\right)^2$

$= \left(x^{2/5 - 3/5 - 5/5}\right)^2$

$= \left(x^{-6/5}\right)^2$

$= x^{-12/5}$

$= \dfrac{1}{x^{12/5}}$

3. $x^{-2/3} + x^{4/3} = x^{-2/3}(1) + x^{-2/3}\left(x^{6/3}\right)$

$= x^{-2/3}\left(1 + x^{6/3}\right)$

$= x^{-2/3}\left(1 + x^2\right)$

$= \dfrac{1 + x^2}{x^{2/3}}$

4. $g(x) = \sqrt{x} + 1$

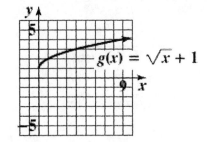

5. $\sqrt{54x^7y^{10}} = \sqrt{9x^6y^{10}}\sqrt{6x}$

$= 3x^3y^5\sqrt{6x}$

6. $\sqrt[3]{25x^5y^2}\sqrt[3]{10x^6y^8} = \sqrt[3]{250x^{11}y^{10}}$

$= \sqrt[3]{125x^9y^9} \cdot \sqrt[3]{2x^2y}$

$= 5x^3y^3\sqrt[3]{2x^2y}$

7. $\sqrt{\dfrac{7x^6y^3}{8z}} = \dfrac{\sqrt{7x^6y^3}}{\sqrt{8z}}$

$= \dfrac{\sqrt{x^6y^2}\sqrt{7y}}{\sqrt{4}\sqrt{2z}}$

$= \dfrac{x^3y\sqrt{7y}}{2\sqrt{2z}} \cdot \dfrac{\sqrt{2z}}{\sqrt{2z}}$

$= \dfrac{x^3y\sqrt{14yz}}{2(2z)}$

$= \dfrac{x^3y\sqrt{14yz}}{4z}$

8. $\dfrac{9}{\sqrt[3]{x}} = \dfrac{9}{\sqrt[3]{x}} \cdot \dfrac{\sqrt[3]{x^2}}{\sqrt[3]{x^2}} = \dfrac{9\sqrt[3]{x^2}}{x}$

9. $\dfrac{\sqrt{3}}{3+\sqrt{27}} = \dfrac{\sqrt{3}}{3+\sqrt{27}} \cdot \dfrac{3-\sqrt{27}}{3-\sqrt{27}}$

$\qquad = \dfrac{\sqrt{3}\left(3-\sqrt{27}\right)}{\left(3+\sqrt{27}\right)\left(3-\sqrt{27}\right)}$

$\qquad = \dfrac{3\sqrt{3}-\sqrt{81}}{9-3\sqrt{27}+3\sqrt{27}-27}$

$\qquad = \dfrac{3\sqrt{3}-9}{-18}$

$\qquad = \dfrac{\sqrt{3}-3}{-6} \ \text{ or } \ \dfrac{3-\sqrt{3}}{6}$

10. $2\sqrt{24}-6\sqrt{6}+3\sqrt{54}$

$\qquad = 2\sqrt{4}\sqrt{6}-6\sqrt{6}+3\sqrt{9}\sqrt{6}$

$\qquad = 4\sqrt{6}-6\sqrt{6}+9\sqrt{6}$

$\qquad = 7\sqrt{6}$

11. $\sqrt[3]{8x^3y^5}+4\sqrt[3]{x^6y^8}$

$\qquad = \sqrt[3]{8x^3y^3}\,\sqrt[3]{y^2}+4\sqrt[3]{x^6y^6}\,\sqrt[3]{y^2}$

$\qquad = 2xy\sqrt[3]{y^2}+4x^2y^2\sqrt[3]{y^2}$

$\qquad = (2xy+4x^2y^2)\sqrt[3]{y^2}$

12. $\left(\sqrt{3}-2\right)\left(6-\sqrt{8}\right) = \sqrt{3}(6)-\sqrt{3}\sqrt{8}-2(6)+2\sqrt{8}$

$\qquad = 6\sqrt{3}-\sqrt{24}-12+2\sqrt{8}$

$\qquad = 6\sqrt{3}-\sqrt{4\cdot 6}-12+2\sqrt{4\cdot 2}$

$\qquad = 6\sqrt{3}-2\sqrt{6}-12+4\sqrt{2}$

13. $\sqrt[4]{\sqrt{x^5y^3}} = \sqrt[4]{\left(x^5y^3\right)^{1/2}}$

$\qquad = \left[\left(x^5y^3\right)^{1/2}\right]^{1/4}$

$\qquad = \left(x^5y^3\right)^{1/8}$

$\qquad = \sqrt[8]{x^5y^3}$

14. $\dfrac{\sqrt[4]{(7x+2)^5}}{\sqrt[3]{(7x+2)^2}} = \dfrac{(7x+2)^{5/4}}{(7x+2)^{2/3}}$

$\qquad = (7x+2)^{5/4-2/3}$

$\qquad = (7x+2)^{7/12}$

$\qquad = \sqrt[12]{(7x+2)^7}$

15. $\qquad \sqrt{2x+19} = 3$

$\qquad \left(\sqrt{2x+19}\right)^2 = 3^2$

$\qquad\qquad 2x+19 = 9$

$\qquad\qquad\quad 2x = -10$

$\qquad\qquad\quad\ x = -5$

Check: $\qquad \sqrt{2x+19} = 3$

$\qquad\qquad \sqrt{2(-5)+19} = 3$

$\qquad\qquad\qquad \sqrt{9} = 3$

$\qquad\qquad\qquad\quad 3 = 3 \ \text{ True}$

This check confirms that -5 is the solution to the equation.

16. $\qquad \sqrt{x^2-x-12} = x+3$

$\qquad \left(\sqrt{x^2-x-12}\right)^2 = (x+3)^2$

$\qquad\qquad x^2-x-12 = x^2+6x+9$

$\qquad\qquad\quad -x-12 = 6x+9$

$\qquad\qquad\qquad -12 = 7x+9$

$\qquad\qquad\qquad -21 = 7x$

$\qquad\qquad\qquad\quad x = -3$

Check:

$\qquad \sqrt{(-3)^2-(-3)-12} = -3+3$

$\qquad\qquad \sqrt{9+3-12} = -3+3$

$\qquad\qquad\qquad \sqrt{0} = 0$

$\qquad\qquad\qquad\quad 0 = 0$

This check confirms that -3 is the solution to the equation.

17. $\sqrt{a-8} = \sqrt{a} - 2$

$\left(\sqrt{a-8}\right)^2 = \left(\sqrt{a}-2\right)^2$

$a-8 = a - 4\sqrt{a} + 4$

$-12 = -4\sqrt{a}$

$\sqrt{a} = 3$

$a = 3^2 = 9$

Check:

$\sqrt{a-8} = \sqrt{a} - 2$

$\sqrt{9-8} = \sqrt{9} - 2$

$\sqrt{1} = 3 - 2$

$1 = 1$ True

This check confirms that 9 is the solution.

18. $f(x) = g(x)$

$(9x+37)^{1/3} = 2(2x+2)^{1/3}$

$[(9x+37)^{1/3}]^3 = [2(2x+2)^{1/3}]^3$

$9x+37 = 8(2x+2)$

$9x+37 = 16x+16$

$21 = 7x$

$3 = x$

19. $w = \dfrac{\sqrt{2gh}}{4}$ Solve for g.

$4w = \sqrt{2gh}$

$\left(4w\right)^2 = 2gh$

$\dfrac{16w^2}{2h} = \dfrac{2gh}{2h}$

$\dfrac{8w^2}{h} = g$

20. $V = \sqrt{64.4h}$

$V = \sqrt{64.4(200)}$

$= \sqrt{12,880}$

≈ 113.49 ft/sec

21. Let x be the length of the ladder.

$x = \sqrt{12^2 + 5^2}$

$= 169$

$= 13$ feet

22. $T = 2\pi\sqrt{\dfrac{m}{k}}$

$T = 2\pi\sqrt{\dfrac{1400}{65,000}}$

≈ 0.92 sec

23. $\left(6-\sqrt{-4}\right)\left(2+\sqrt{-16}\right) = (6-2i)(2+4i)$

$= 12 + 24i - 4i - 8i^2$

$= 12 + 20i - 8(-1)$

$= 12 + 20i + 8$

$= 20 + 20i$

24. $\dfrac{5-i}{7+2i} = \dfrac{5-i}{7+2i} \cdot \dfrac{7-2i}{7-2i}$

$= \dfrac{(5-i)(7-2i)}{(7+2i)(7-2i)}$

$= \dfrac{35-10i-7i+2i^2}{49-14i+14i-4i^2}$

$= \dfrac{35-17i+2(-1)}{49-4(-1)}$

$= \dfrac{35-17i-2}{49+4}$

$= \dfrac{33-17i}{53}$

25. $x^2 + 6x + 12$

$= (-3+i)^2 + 6(-3+i) + 12$

$= (-3)^2 - 2(3)(i) + (i)^2 - 18 + 6i + 12$

$= 9 - 6i - 1 - 18 + 6i + 12$

$= 2 + 0i$

$= 2$

Chapter 11 Cumulative Review Test

1.
$$\frac{1}{5}(x-3) = \frac{3}{4}(x+3) - x$$
$$20\left[\frac{1}{5}(x-3)\right] = 20\left[\frac{3}{4}(x+3)\right] - 20x$$
$$4(x-3) = 5(3)(x+3) - 20x$$
$$4x - 12 = 15x + 45 - 20x$$
$$4x - 12 = -5x + 45$$
$$9x - 12 = 45$$
$$9x = 57$$
$$x = \frac{57}{9}$$

2. $3(x-4) = 6x - (4-5x)$
$$3x - 12 = 6x - 4 + 5x$$
$$3x - 12 = 11x - 4$$
$$-8x = 8$$
$$x = -1$$

3. Let x be the original price of the sweater.
$$x - 60\%x = 16$$
$$x - 0.60x = 16$$
$$0.40x = 16$$
$$x = \frac{16}{0.40}$$
$$x = 40$$
The original price of the sweater is $40.

4. $y = \frac{3}{2}x - 3$

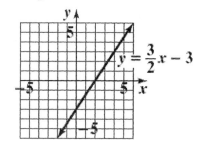

5. $(5xy - 3)(5xy + 3)$
$$= 25x^2 y^2 + 15xy - 15xy - 9$$
$$= 25x^2 y^2 - 9$$

6. $V = lwh$
$$6r^3 + 5r^2 + r = (3r+1)(w)(r)$$
$$\frac{6r^3 + 5r^2 + r}{(3r+1)(r)} = \frac{(3r+1)(w)(r)}{(3r+1)(r)}$$
$$\frac{r(6r^2 + 5r + 1)}{r(3r+1)} = w$$
$$\frac{r(3r+1)(2r+1)}{r(3r+1)} = w$$
$$w = 2r + 1$$

7. $4x^3 - 9x^2 + 5x = x(4x^2 - 9x + 5)$
$$= x(4x - 5)(x - 1)$$

8. $(x+1)^3 - 27 = (x+1)^3 - 3^3$
$$= (x+1-3)\left((x+1)^2 + 3(x+1) + 3^2\right)$$
$$= (x-2)(x^2 + 2x + 1 + 3x + 3 + 9)$$
$$= (x-2)(x^2 + 5x + 13)$$

9. $8x^2 - 3 = -10x$
$$8x^2 + 10x - 3 = 0$$
$$(4x-1)(2x+3) = 0$$
$$4x - 1 = 0 \quad \text{or} \quad 2x + 3 = 0$$
$$4x = 1 \qquad\qquad 2x = -3$$
$$x = \frac{1}{4} \qquad\qquad x = -\frac{3}{2}$$

10. $\dfrac{4x+4y}{x^2 y} \cdot \dfrac{y^3}{12x} = \dfrac{4(x+y)}{x^2 y} \cdot \dfrac{y^3}{12x}$
$$= \frac{x+y}{x^2} \cdot \frac{y^2}{3x}$$
$$= \frac{(x+y)y^2}{3x^3}$$

11. $\dfrac{x-4}{x-5} - \dfrac{3}{x+5} - \dfrac{10}{x^2-25}$

$= \dfrac{x-4}{x-5} - \dfrac{3}{x+5} - \dfrac{10}{(x+5)(x-5)}$

$= \dfrac{x-4}{x-5} \cdot \dfrac{x+5}{x+5} - \dfrac{3}{x+5} \cdot \dfrac{x-5}{x-5} - \dfrac{10}{(x+5)(x-5)}$

$= \dfrac{(x-4)(x+5)}{(x+5)(x-5)} - \dfrac{3(x-5)}{(x+5)(x-5)} - \dfrac{10}{(x+5)(x-5)}$

$= \dfrac{(x-4)(x+5)-3(x-5)-10}{(x+5)(x-5)}$

$= \dfrac{x^2+x-20-3x+15-10}{(x+5)(x-5)}$

$= \dfrac{x^2-2x-15}{(x+5)(x-5)}$

$= \dfrac{(x-5)(x+3)}{(x+5)(x-5)}$

$= \dfrac{x+3}{x+5}$

12. $\dfrac{4}{x} - \dfrac{1}{6} = \dfrac{1}{x}$

$6x\left(\dfrac{4}{x} - \dfrac{1}{6} = \dfrac{1}{x}\right)$

$24 - x = 6$

$-x = -18$

$x = 18$

Upon checking, this value satisfies the equation. The solution is 18.

13. $y = 3x-8 \;\Rightarrow\; m_1 = 3$

$6y = 18x+12 \;\Rightarrow\; y = 3x+2 \;\Rightarrow\; m_2 = 3$

The slope of both lines is 3. Since the slopes of the two lines are the same, and the y-intercepts are different, the lines are parallel.

14. First find the slope of the given line.

$3x - 2y = 6$

$-2y = -3x + 6$

$y = \dfrac{3}{2}x - 3 \;\Rightarrow\;$ The slope is $\dfrac{3}{2}$.

The slope of any line perpendicular to this line must have an opposite reciprocal slope. Therefore, the slope of the line perpendicular to the given line is $-\dfrac{2}{3}$. Finally, use the point-slope formula to find the equation of the line.

$y - y_1 = m(x - x_1)$ with $m = -\dfrac{2}{3}$ and $(1, -4)$

$y - (-4) = -\dfrac{2}{3}(x - 1)$

$y + 4 = -\dfrac{2}{3}x + \dfrac{2}{3}$

$y = -\dfrac{2}{3}x - \dfrac{10}{3}$

15. $f(x) = x^2 - 3x + 4$ and $g(x) = 2x - 9$

$(g - f)(x) = (2x - 9) - (x^2 - 3x + 4)$

$\qquad = 2x - 9 - x^2 + 3x - 4$

$\qquad = -x^2 + 5x - 13$

16.
$$x + 2y = 12 \quad (1)$$
$$4x = 8 \quad (2)$$
$$3x - 4y + 5z = 20 \quad (3)$$

Using equation (2), solve for x.

$4x = 8 \;\Rightarrow\; x = \dfrac{8}{4} \;\Rightarrow\; x = 2$

Substitute 2 for x in equation (1) in order to solve for y.

$2 + 2y = 12 \;\Rightarrow\; 2y = 10 \;\Rightarrow\; y = 5$

Substitute 2 for x and 5 for y in equation (3) in order to solve for z.

$3(2) - 4(5) + 5z = 20$

$6 - 20 + 5z = 20$

$-14 + 5z = 20$

$5z = 34$

$z = \dfrac{34}{5}$

The solution is $\left(2, 5, \dfrac{34}{5}\right)$.

17. $\begin{vmatrix} 3 & -6 & -1 \\ 2 & 1 & -2 \\ 1 & 3 & 1 \end{vmatrix}$

$= 3\begin{vmatrix} 1 & -2 \\ 3 & 1 \end{vmatrix} - (-6)\begin{vmatrix} 2 & -2 \\ 1 & 1 \end{vmatrix} + (-1)\begin{vmatrix} 2 & 1 \\ 1 & 3 \end{vmatrix}$

$= 3\big(1-(-6)\big) + 6\big(2-(-2)\big) - (6-1)$

$= 3(1+6) + 6(2+2) - (6-1)$

$= 3(7) + 6(4) - (5)$

$= 21 + 24 - 5$

$= 40$

18. $\qquad |3-2x| < 5$

$\qquad -5 < 3-2x < 5$

$\quad -5-3 < 3-2x-3 < 5-3$

$\qquad -8 < -2x < 2$

$\qquad \dfrac{-8}{-2} > \dfrac{-2x}{-2} > \dfrac{2}{-2}$

$\qquad 4 > x > -1 \ \text{ or } \ -1 < x < 4$

The solution set is $\{x | -1 < x < 4\}$

19. $\sqrt{2x^2+7} + 3 = 8$

$\sqrt{2x^2+7} = 5$

$\left(\sqrt{2x^2+7}\right)^2 = 5^2$

$2x^2 + 7 = 25$

$2x^2 = 18$

$x^2 = 9$

$x = 3 \ \text{ or } \ -3$

Check:

$\sqrt{2(3)^2+7} + 3 = 8 \ \text{ or } \ \sqrt{2(-3)^2+7} + 3 = 8$

$\qquad \sqrt{25} = 5 \qquad\qquad\qquad \sqrt{25} = 5$

$\qquad\quad 5 = 5 \qquad\qquad\qquad\quad 5 = 5$

Both values check. The solutions are 3 and –3.

20. $d = kt^2$

$16 = k(1)^2 \ \Rightarrow \ k = 16$

$d = 16(5)^2$

$\quad = 16 \cdot 25$

$\quad = 400$

The object will fall 400 feet in 5 seconds.

Chapter 12

Exercise Set 12.1

1. The two square roots of 36 are $\pm\sqrt{36} = \pm 6$.

3. The square root property is: If $x^2 = a$, where a is a real number, then $x = \pm\sqrt{a}$.

5. A trinomial, $x^2 + bx + c$, is a perfect square trinomial if $\left(\dfrac{b}{2}\right)^2 = c$.

7. **a.** Yes, $x = 4$ is the solution to the equation. It is the only real number that satisfies the equation.

   b. No, $x = 2$ is not the solution. Both -2 and 2 satisfy the equation.

9. Multiply the equation by $\dfrac{1}{2}$ to obtain a leading coefficient of 1.

11. You should add the square of half the coefficient of the first degree term: $\left(\dfrac{-6}{2}\right)^2 = (-3)^2 = 9$.

13. $x^2 - 25 = 0$
    $$x^2 = 25$$
    $$x = \pm\sqrt{25} = \pm 5$$

15. $x^2 + 49 = 0$
    $$x^2 = -49$$
    $$x = \pm\sqrt{-49} = \pm 7i$$

17. $y^2 + 24 = 0$
    $$y^2 = -24$$
    $$y = \pm\sqrt{-24} = \pm 2i\sqrt{6}$$

19. $y^2 + 10 = -51$
    $$y^2 = -61$$
    $$y = \pm\sqrt{-61} = \pm i\sqrt{61}$$

21. $(p-4)^2 = 16$
    $$p - 4 = \pm\sqrt{16}$$
    $$p - 4 = \pm 4$$
    $$p = 4 \pm 4$$
    $$p = 4 + 4 \quad \text{or} \quad p = 4 - 4$$
    $$p = 8 \qquad\qquad p = 0$$

23. $(x+3)^2 + 25 = 0$
    $$(x+3)^2 = -25$$
    $$x + 3 = \pm\sqrt{-25}$$
    $$x + 3 = \pm\sqrt{-25}$$
    $$x = -3 \pm 5i$$

25. $(a-2)^2 + 45 = 0$
    $$(a-2)^2 = -45$$
    $$a - 2 = \pm\sqrt{-45}$$
    $$a - 2 = \pm 3i\sqrt{5}$$
    $$a = 2 \pm 3i\sqrt{5}$$

27. $\left(b+\dfrac{1}{3}\right)^2 = \dfrac{4}{9}$
    $$b + \frac{1}{3} = \pm\sqrt{\frac{4}{9}}$$
    $$b + \frac{1}{3} = \pm\frac{2}{3}$$
    $$b = -\frac{1}{3} \pm \frac{2}{3}$$
    $$b = -\frac{1}{3} + \frac{2}{3} \quad \text{or} \quad b = -\frac{1}{3} - \frac{2}{3}$$
    $$b = \frac{1}{3} \qquad\qquad b = -\frac{3}{3}$$
    $$\qquad\qquad\qquad\qquad b = -1$$

29. $\left(b - \dfrac{2}{3}\right)^2 + \dfrac{4}{9} = 0$

$\left(b - \dfrac{2}{3}\right)^2 = -\dfrac{4}{9}$

$b - \dfrac{2}{3} = \pm\sqrt{-\dfrac{4}{9}}$

$b = \dfrac{2}{3} \pm \dfrac{2}{3}i \quad \text{or} \quad b = \dfrac{2 \pm 2i}{3}$

31. $\left(x + 0.8\right)^2 = 0.81$

$x + 0.8 = \pm\sqrt{0.81}$

$x + 0.8 = \pm 0.9$

$x = -0.8 \pm 0.9$

$x = -0.8 + 0.9 \quad \text{or} \quad x = -0.8 - 0.9$

$x = 0.1 \qquad\qquad x = -1.7$

33. $\left(2a - 5\right)^2 = 18$

$2a - 5 = \pm\sqrt{18}$

$2a - 5 = \pm 3\sqrt{2}$

$2a = 5 \pm 3\sqrt{2}$

$a = \dfrac{5 \pm 3\sqrt{3}}{2}$

35. $\left(2y + \dfrac{1}{2}\right)^2 = \dfrac{4}{25}$

$2y + \dfrac{1}{2} = \pm\sqrt{\dfrac{4}{25}}$

$2y + \dfrac{1}{2} = \pm\dfrac{2}{5}$

$2y + \dfrac{1}{2} = \dfrac{2}{5} \quad \text{or} \quad 2y + \dfrac{1}{2} = -\dfrac{2}{5}$

$2y = -\dfrac{1}{2} + \dfrac{2}{5} \qquad 2y = -\dfrac{1}{2} - \dfrac{2}{5}$

$2y = -\dfrac{1}{10} \qquad\qquad 2y = -\dfrac{9}{10}$

$y = -\dfrac{1}{20} \qquad\qquad y = -\dfrac{9}{20}$

37. $x^2 + 3x - 4 = 0$

$x^2 + 3x = 4$

$x^2 + 3x + \dfrac{9}{4} = 4 + \dfrac{9}{4}$

$x^2 + 3x + \dfrac{9}{4} = \dfrac{16}{4} + \dfrac{9}{4}$

$\left(x + \dfrac{3}{2}\right)^2 = \dfrac{25}{4}$

$x + \dfrac{3}{2} = \pm\sqrt{\dfrac{25}{4}}$

$x + \dfrac{3}{2} = \pm\dfrac{5}{2}$

$x = -\dfrac{3}{2} \pm \dfrac{5}{2}$

$x = -\dfrac{3}{2} + \dfrac{5}{2} \quad \text{or} \quad x = -\dfrac{3}{2} - \dfrac{5}{2}$

$x = \dfrac{2}{2} \qquad\qquad x = \dfrac{-8}{2}$

$x = 1 \qquad\qquad x = -4$

39. $x^2 + 2x - 15 = 0$

$x^2 + 2x = 15$

$x^2 + 2x + 1 = 15 + 1$

$\left(x + 1\right)^2 = 16$

$x + 1 = \pm\sqrt{16}$

$x + 1 = \pm 4$

$x = -1 \pm 4$

$x = -1 + 4 \quad \text{or} \quad x = -1 - 4$

$x = 3 \qquad\qquad x = -5$

41. $x^2 + 6x + 8 = 0$

$x^2 + 6x = -8$

$x^2 + 6x + 9 = -8 + 9$

$x^2 + 6x + 9 = 1$

$\left(x + 3\right)^2 = 1$

$x + 3 = \pm\sqrt{1}$

$x + 3 = \pm 1$

$x = -3 \pm 1$

$x = -3 + 1 \quad \text{or} \quad x = -3 - 1$

$x = -2 \qquad\qquad x = -4$

43.
$$x^2 - 7x + 6 = 0$$
$$x^2 - 7x = -6$$
$$x^2 - 7x + \frac{49}{4} = -6 + \frac{49}{4}$$
$$x^2 - 7x + \frac{49}{4} = -\frac{24}{4} + \frac{49}{4}$$
$$x^2 - 7x + \frac{49}{4} = \frac{25}{4}$$
$$\left(x - \frac{7}{2}\right)^2 = \frac{25}{4}$$
$$x - \frac{7}{2} = \pm\sqrt{\frac{25}{4}}$$
$$x - \frac{7}{2} = \pm\frac{5}{2}$$
$$x = \frac{7}{2} \pm \frac{5}{2}$$
$$x = \frac{7}{2} + \frac{5}{2} \quad \text{or} \quad x = \frac{7}{2} - \frac{5}{2}$$
$$x = \frac{12}{2} \qquad\qquad x = \frac{2}{2}$$
$$x = 6 \qquad\qquad x = 1$$

45.
$$2x^2 + x - 1 = 0$$
$$\frac{1}{2}\left(2x^2 + x - 1\right) = \frac{1}{2}(0)$$
$$x^2 + \frac{1}{2}x - \frac{1}{2} = 0$$
$$x^2 + \frac{1}{2}x = \frac{1}{2}$$
$$x^2 + \frac{1}{2}x + \frac{1}{16} = \frac{1}{2} + \frac{1}{16}$$
$$\left(x + \frac{1}{4}\right)^2 = \frac{9}{16}$$
$$x + \frac{1}{4} = \pm\sqrt{\frac{9}{16}}$$
$$x + \frac{1}{4} = \pm\frac{3}{4}$$
$$x = -\frac{1}{4} \pm \frac{3}{4}$$
$$x = -\frac{1}{4} + \frac{3}{4} \quad \text{or} \quad x = -\frac{1}{4} - \frac{3}{4}$$
$$x = \frac{2}{4} \qquad\qquad x = -\frac{4}{4}$$
$$x = \frac{1}{2} \qquad\qquad x = -1$$

47.
$$2z^2 - 7z - 4 = 0$$
$$\frac{1}{2}\left(2z^2 - 7z - 4\right) = \frac{1}{2}(0)$$
$$z^2 - \frac{7}{2}z - 2 = 0$$
$$z^2 - \frac{7}{2}z = 2$$
$$z^2 - \frac{7}{2}z + \frac{49}{16} = 2 + \frac{49}{16}$$
$$\left(z - \frac{7}{4}\right)^2 = \frac{81}{16}$$
$$z - \frac{7}{4} = \pm\sqrt{\frac{81}{16}}$$
$$z - \frac{7}{4} = \pm\frac{9}{4}$$
$$z = \frac{7}{4} \pm \frac{9}{4}$$
$$z = \frac{7}{4} + \frac{9}{4} \quad \text{or} \quad z = \frac{7}{4} - \frac{9}{4}$$
$$z = \frac{16}{4} \qquad\qquad z = -\frac{2}{4}$$
$$z = 4 \qquad\qquad z = -\frac{1}{2}$$

49.
$$x^2 - 13x + 40 = 0$$
$$x^2 - 13x = -40$$
$$x^2 - 13x + \frac{169}{4} = -40 + \frac{169}{4}$$
$$\left(x - \frac{13}{2}\right)^2 = \frac{9}{4}$$
$$x - \frac{13}{2} = \pm\sqrt{\frac{9}{4}}$$
$$x - \frac{13}{2} = \pm\frac{3}{2}$$
$$x = \frac{13}{2} \pm \frac{3}{2}$$
$$x = \frac{13}{2} + \frac{3}{2} \quad \text{or} \quad x = \frac{13}{2} - \frac{3}{2}$$
$$x = \frac{16}{2} \qquad\qquad x = \frac{10}{2}$$
$$x = 8 \qquad\qquad x = 5$$

51.
$$-x^2 + 6x + 7 = 0$$
$$-1(-x^2 + 6x + 7) = -1(0)$$
$$x^2 - 6x - 7 = 0$$
$$x^2 - 6x = 7$$
$$x^2 - 6x + 9 = 7 + 9$$
$$(x-3)^2 = 16$$
$$x - 3 = \pm\sqrt{16}$$
$$x - 3 = \pm 4$$
$$x = 3 \pm 4$$
$$x = 3 + 4 \quad \text{or} \quad x = 3 - 4$$
$$x = 7 \qquad\qquad x = -1$$

53.
$$-z^2 + 9z - 20 = 0$$
$$-1(-z^2 + 9z - 20) = -1(0)$$
$$z^2 - 9z + 20 = 0$$
$$z^2 - 9z = -20$$
$$z^2 - 9z + \frac{81}{4} = -20 + \frac{81}{4}$$
$$\left(z - \frac{9}{2}\right)^2 = \frac{1}{4}$$
$$z - \frac{9}{2} = \pm\sqrt{\frac{1}{4}}$$
$$z - \frac{9}{2} = \pm\frac{1}{2}$$
$$z = \frac{9}{2} \pm \frac{1}{2}$$
$$z = \frac{9}{2} + \frac{1}{2} \quad \text{or} \quad z = \frac{9}{2} - \frac{1}{2}$$
$$z = \frac{10}{2} \qquad\qquad z = \frac{8}{2}$$
$$z = 5 \qquad\qquad z = 4$$

55.
$$b^2 = 3b + 28$$
$$b^2 - 3b = 28$$
$$b^2 - 3b + \frac{9}{4} = \frac{112}{4} + \frac{9}{4}$$
$$\left(b - \frac{3}{2}\right)^2 = \frac{121}{4}$$
$$b - \frac{3}{2} = \pm\sqrt{\frac{121}{4}}$$
$$b - \frac{3}{2} = \pm\frac{11}{2}$$
$$b = \frac{3}{2} \pm \frac{11}{2}$$

$$b = \frac{3}{2} + \frac{11}{2} \quad \text{or} \quad b = \frac{3}{2} - \frac{11}{2}$$
$$b = \frac{14}{2} \qquad\qquad b = -\frac{8}{2}$$
$$b = 7 \qquad\qquad b = -4$$

57.
$$x^2 + 10x = 11$$
$$x^2 + 10x + 25 = 11 + 25$$
$$(x + 5)^2 = 36$$
$$x + 5 = \pm\sqrt{36}$$
$$x + 5 = \pm 6$$
$$x = -5 \pm 6$$
$$x = \frac{11}{2} - \frac{9}{2} \quad \text{or} \quad x = -\frac{11}{2} - \frac{9}{2}$$
$$x = \frac{2}{2} \qquad\qquad x = -\frac{20}{2}$$
$$x = 1 \qquad\qquad x = -10$$

59.
$$x^2 - 4x - 10 = 0$$
$$x^2 - 4x = 10$$
$$x^2 - 4x + 4 = 10 + 4$$
$$(x - 2)^2 = 14$$
$$x - 2 = \pm\sqrt{14}$$
$$x = 2 \pm \sqrt{14}$$

61.
$$r^2 + 8r + 5 = 0$$
$$r^2 + 8r = -5$$
$$r^2 + 8r + 16 = -5 + 16$$
$$(r + 4)^2 = 11$$
$$r + 4 = \pm\sqrt{11}$$
$$r = -4 \pm \sqrt{11}$$

63.
$$c^2 - c - 3 = 0$$
$$c^2 - c = 3$$
$$c^2 - c + \frac{1}{4} = 3 + \frac{1}{4}$$
$$\left(c - \frac{1}{2}\right)^2 = \frac{13}{4}$$
$$c - \frac{1}{2} = \pm\sqrt{\frac{13}{4}}$$
$$c = \frac{1}{2} \pm \frac{\sqrt{13}}{2} = \frac{1 \pm \sqrt{13}}{2}$$

65. $x^2 + 3x + 6 = 0$

$x^2 + 3x = -6$

$x^2 + 3x + \dfrac{9}{4} = -6 + \dfrac{9}{4}$

$\left(x + \dfrac{3}{2}\right)^2 = \dfrac{-15}{4}$

$x + \dfrac{3}{2} = \pm\sqrt{\dfrac{-15}{4}}$

$x + \dfrac{3}{2} = \pm\dfrac{i\sqrt{15}}{2}$

$x = -\dfrac{3}{2} \pm \dfrac{i\sqrt{15}}{2} = \dfrac{-3 \pm i\sqrt{15}}{2}$

67. $9x^2 - 9x = 0$

$\dfrac{1}{9}(9x^2 - 9x) = \dfrac{1}{9}(0)$

$x^2 - x = 0$

$x^2 - x + \dfrac{1}{4} = \dfrac{1}{4}$

$\left(x - \dfrac{1}{2}\right)^2 = \dfrac{1}{4}$

$x - \dfrac{1}{2} = \pm\dfrac{1}{2}$

$x = \dfrac{1}{2} \pm \dfrac{1}{2}$

$x = \dfrac{1}{2} + \dfrac{1}{2}$ or $x = \dfrac{1}{2} - \dfrac{1}{2}$

$x = 1$ $\qquad$ $x = 0$

69. $-\dfrac{3}{4}b^2 - \dfrac{1}{2}b = 0$

$-\dfrac{4}{3}\left(-\dfrac{3}{4}b^2 - \dfrac{1}{2}b\right) = -\dfrac{4}{3}(0)$

$b^2 + \dfrac{2}{3}b = 0$

$b^2 + \dfrac{2}{3}b + \dfrac{1}{9} = 0 + \dfrac{1}{9}$

$\left(b + \dfrac{1}{3}\right)^2 = \dfrac{1}{9}$

$b + \dfrac{1}{3} = \pm\sqrt{\dfrac{1}{9}}$

$b + \dfrac{1}{3} = \pm\dfrac{1}{3}$

$b = -\dfrac{1}{3} \pm \dfrac{1}{3}$

$b = -\dfrac{1}{3} + \dfrac{1}{3}$ or $b = -\dfrac{1}{3} - \dfrac{1}{3}$

$b = 0$ $\qquad$ $b = -\dfrac{2}{3}$

71. $36z^2 - 6z = 0$

$\dfrac{1}{36}\left(36z^2 - 6z\right) = \dfrac{1}{36}(0)$

$z^2 - \dfrac{1}{6}z = 0$

$z^2 - \dfrac{1}{6}z + \dfrac{1}{144} = 0 + \dfrac{1}{144}$

$\left(z - \dfrac{1}{12}\right)^2 = \dfrac{1}{144}$

$z - \dfrac{1}{12} = \pm\sqrt{\dfrac{1}{144}}$

$z - \dfrac{1}{12} = \pm\dfrac{1}{12}$

$z = \dfrac{1}{12} \pm \dfrac{1}{12}$

$z = \dfrac{1}{12} + \dfrac{1}{12}$ or $z = \dfrac{1}{12} - \dfrac{1}{12}$

$z = \dfrac{2}{12}$ $\qquad$ $z = 0$

$z = \dfrac{1}{6}$

73. $-\dfrac{1}{2}p^2 - p + \dfrac{3}{2} = 0$

$-2\left(-\dfrac{1}{2}p^2 - p + \dfrac{3}{2}\right) = -2(0)$

$p^2 + 2p - 3 = 0$

$p^2 + 2p = 3$

$p^2 + 2p + 1 = 3 + 1$

$(p + 1)^2 = 4$

$p + 1 = \pm\sqrt{4}$

$p + 1 = \pm 2$

$p = -1 \pm 2$

$p = -1 + 2$ or $p = -1 - 2$

$p = 1$ $\qquad$ $p = -3$

75.

$$2x^2 = 8x + 64$$

$$\frac{1}{2}(2x^2) = \frac{1}{2}(8x + 64)$$

$$x^2 = 4x + 32$$

$$x^2 - 4x = 32$$

$$x^2 - 4x + 4 = 32 + 4$$

$$(x - 2)^2 = 36$$

$$x - 2 = \pm\sqrt{36}$$

$$x - 2 = \pm 6$$

$$x = 2 \pm 6$$

$$x = 2 + 6 \quad \text{or} \quad x = 2 - 6$$

$$x = 8 \qquad \text{or} \quad x = -4$$

77.

$$2x^2 + 18x + 4 = 0$$

$$\frac{1}{2}(2x^2 + 18x + 4) = \frac{1}{2}(0)$$

$$x^2 + 9x + 2 = 0$$

$$x^2 + 9x = -2$$

$$x^2 + 9x + \frac{81}{4} = -2 + \frac{81}{4}$$

$$\left(x + \frac{9}{2}\right)^2 = -\frac{8}{4} + \frac{81}{4}$$

$$\left(x + \frac{9}{2}\right)^2 = \frac{73}{4}$$

$$x + \frac{9}{2} = \pm\frac{\sqrt{73}}{2}$$

$$x = -\frac{9}{2} \pm \frac{\sqrt{73}}{2}$$

$$x = \frac{-9 \pm \sqrt{73}}{2}$$

79.

$$\frac{3}{4}w^2 + \frac{1}{2}w - \frac{1}{4} = 0$$

$$\frac{4}{3}\left(\frac{3}{4}w^2 + \frac{1}{2}w - \frac{1}{4}\right) = \frac{4}{3}(0)$$

$$w^2 + \frac{2}{3}w - \frac{1}{3} = 0$$

$$w^2 + \frac{2}{3}w = \frac{1}{3}$$

$$w^2 + \frac{2}{3}w + \frac{1}{9} = \frac{1}{3} + \frac{1}{9}$$

$$\left(w + \frac{1}{3}\right)^2 = \frac{4}{9}$$

$$w + \frac{1}{3} = \pm\sqrt{\frac{4}{9}}$$

$$w + \frac{1}{3} = \pm\frac{2}{3}$$

$$w = -\frac{1}{3} \pm \frac{2}{3}$$

$$w = -\frac{1}{3} + \frac{2}{3} \quad \text{or} \quad w = -\frac{1}{3} - \frac{2}{3}$$

$$w = \frac{1}{3} \qquad\qquad w = -1$$

81.

$$2x^2 - x = -5$$

$$\frac{1}{2}(2x^2 - x) = \frac{1}{2}(-5)$$

$$x^2 - \frac{1}{2}x = -\frac{5}{2}$$

$$x^2 - \frac{1}{2}x + \frac{1}{16} = -\frac{40}{16} + \frac{1}{16}$$

$$\left(x - \frac{1}{4}\right)^2 = -\frac{39}{16}$$

$$x - \frac{1}{4} = \pm\frac{i\sqrt{39}}{4}$$

$$x = \frac{1}{4} \pm \frac{i\sqrt{39}}{4}$$

$$x = \frac{1 \pm i\sqrt{39}}{4}$$

83.
$$-3x^2 + 6x = 6$$
$$-\frac{1}{3}\left(-3x^2 + 6x\right) = -\frac{1}{3}(6)$$
$$x^2 - 2x = -2$$
$$x^2 - 2x + 1 = -2 + 1$$
$$(x-1)^2 = -1$$
$$x - 1 = \pm\sqrt{-1}$$
$$x - 1 = \pm i$$
$$x = 1 \pm i$$

85. a. $21 = (x+2)(x-2)$

b.
$$21 = (x+2)(x-2)$$
$$21 = x^2 - 2x + 2x - 4$$
$$0 = x^2 - 25$$
$$0 = (x+5)(x-5)$$
$$x + 5 = 0 \quad \text{or} \quad x - 5 = 0$$
$$x = -5 \qquad\qquad x = 5$$
Disregard the negative answer since x represents a distance. $x = 5$.

87. a. $18 = (x+4)(x+2)$

b.
$$18 = (x+4)(x+2)$$
$$18 = x^2 + 2x + 4x + 8$$
$$0 = x^2 + 6x - 10$$
$$x^2 + 6x = 10$$
$$x^2 + 6x + 9 = 10 + 9$$
$$(x+3)^2 = 19$$
$$x + 3 = \pm\sqrt{19}$$
$$x = -3 \pm \sqrt{19}$$
Disregard the negative answer since x represents a distance. $x = -3 + \sqrt{19}$.

89.
$$d = \frac{1}{6}x^2$$
$$150 = \frac{1}{6}x^2$$
$$6 \cdot 150 = x^2$$
$$900 = x^2$$
$$x = \pm\sqrt{900} = \pm 30$$
Disregard the negative answer since speed must be positive. The car's speed was about 30 mph.

91. Let x be the first integer. Then $x + 2$ is the next consecutive odd integer.
$$x(x+2) = 35$$
$$x^2 + 2x = 35$$
$$x^2 + 2x + 1 = 35 + 1$$
$$(x+1)^2 = 36$$
$$x + 1 = \pm\sqrt{36}$$
$$x + 1 = \pm 6$$
$$x = -1 \pm 6$$
$$x = -1 + 6 \quad \text{or} \quad x = -1 - 6$$
$$x = 5 \qquad\qquad x = -7$$
Since it was given that the integers are positive, one integer is 5 and the other is $5 + 2 = 7$.

93. Let x be the width of the rectangle. Then $2x + 2$ is the length. Use length $\cdot$ width = area.
$$(2x+2)x = 60$$
$$2x^2 + 2x = 60$$
$$x^2 + x = 30$$
$$x^2 + x + \frac{1}{4} = 30 + \frac{1}{4}$$
$$\left(x + \frac{1}{2}\right)^2 = \frac{120}{4} + \frac{1}{4}$$
$$\left(x + \frac{1}{2}\right)^2 = \frac{121}{4}$$
$$x + \frac{1}{2} = \pm\sqrt{\frac{121}{4}}$$
$$x + \frac{1}{2} = \pm\frac{11}{2}$$
$$x = -\frac{1}{2} \pm \frac{11}{2}$$
$$x = -\frac{1}{2} + \frac{11}{2} \quad \text{or} \quad x = -\frac{1}{2} - \frac{11}{2}$$
$$x = \frac{10}{2} \qquad\qquad x = -\frac{12}{2}$$
$$x = 5 \qquad\qquad x = -6$$
Since the width cannot be negative, the width is 5 ft.
Length $= 2(5) + 2 = 10 + 2 = 12$ ft.
The rectangle is 5 ft by 12 ft.

95. Let s be the length of the side. Then $s + 6$ is the length of the diagonal (d). Use $s^2 + s^2 = d^2$.

$$2s^2 = (s+6)^2$$
$$2s^2 = s^2 + 12s + 36$$
$$s^2 = 12s + 36$$
$$s^2 - 12s = 36$$
$$s^2 - 12s + 36 = 36 + 36$$
$$(s-6)^2 = 72$$
$$s - 6 = \pm\sqrt{72}$$
$$s - 6 = \pm 6\sqrt{2}$$
$$s = 6 \pm 6\sqrt{2}$$

Length is never negative. Thus,
$s = 6 + 6\sqrt{2} \approx 14.49$.
The patio is about 14.49 ft by 14.49 ft.

97. Since the radius is 10 inches, the diameter (d) is 20 inches. Use the formula $s^2 + s^2 = d^2$ to find the length (s) of the other two sides.

$$s^2 + s^2 = d^2$$
$$2s^2 = 20^2$$
$$2s^2 = 400$$
$$s^2 = 200$$
$$s = \pm\sqrt{200} = \pm 10\sqrt{2}$$

Length is never negative.
Thus, $s = 10\sqrt{2} \approx 14.14$ inches.

99. $A = \pi r^2$
$$24\pi = \pi r^2$$
$$24 = r^2$$
$$r = \pm\sqrt{24} = \pm 2\sqrt{6}$$
Length is never negative.
Thus, $r = 2\sqrt{6} \approx 4.90$ feet.

101. $A = P\left(1 + \dfrac{r}{n}\right)^{nt}$

$$540.80 = 500\left(1 + \frac{r}{1}\right)^{1(2)}$$
$$540.80 = 500(1+r)^2$$
$$1.0816 = (1+r)^2$$
$$1 + r = \pm\sqrt{1.0816}$$
$$1 + r = \pm 1.04$$
$$r = -1 \pm 1.04$$
$$r = -1 + 1.04 \quad\text{or}\quad r = -1 - 1.04$$
$$r = 0.04 \qquad\text{or}\quad r = -2.04$$

Interest rates are never negative, so $r = 0.04 = 4\%$.

103. $A = P\left(1 + \dfrac{r}{n}\right)^{nt}$

$$1432.86 = 1200\left(1 + \frac{r}{2}\right)^{2(3)}$$
$$1432.86 = 1200\left(1 + \frac{r}{2}\right)^{6}$$
$$1.19405 = \left(1 + \frac{r}{2}\right)^{6}$$
$$1 + \frac{r}{2} \approx \pm 1.03$$
$$\frac{r}{2} \approx -1 \pm 1.03$$
$$r \approx -2 \pm 2.06$$
$$r \approx -2 + 2.06 \quad\text{or}\quad r \approx -2 - 2.06$$
$$r \approx 0.06 \qquad\text{or}\quad r \approx -4.06$$

An interest rate is never negative. Thus, Steve Rodi's annual interest rate is about 6%.

105. a. To find the surface area, we must first determine the radius. Use $V = \pi r^2 h$ with $V = 160$ and $h = 10$ to get

$$160 = \pi r^2 (10)$$
$$16 = \pi r^2$$
$$\frac{16}{\pi} = r^2$$
$$r = \pm\frac{4}{\sqrt{\pi}}$$

The length cannot be negative, so $r = \dfrac{4}{\sqrt{\pi}}$.

Use the formula $S = 2\pi r^2 + 2\pi r h$ to calculate the surface area.

$$S = 2\pi\left(\frac{4}{\sqrt{\pi}}\right)^2 + 2\pi\left(\frac{4}{\sqrt{\pi}}\right)(10)$$
$$= 2\pi\left(\frac{16}{\pi}\right) + \frac{80\pi}{\sqrt{\pi}}$$
$$= 32 + 80\sqrt{\pi} \approx 173.80$$

The surface area is about 173.80 square inches.

b. Use $V = \pi r^2 h$ with $V = 160$ and $h = 10$ to obtain $160 = \pi r^2 (10)$. In part (a) this was solved for r to get

$$r = \frac{4}{\sqrt{\pi}} = \frac{4}{\sqrt{\pi}} \cdot \frac{\sqrt{\pi}}{\sqrt{\pi}} = \frac{4\sqrt{\pi}}{\pi} \approx 2.26$$

The radius is about 2.26 inches.

c. Use $S = 2\pi r^2 + 2\pi rh$ with $S = 160$ and $h = 10$.

$$160 = 2\pi r^2 + 2\pi r(10)$$

$$160 = 2\pi r^2 + 20\pi r$$

$$\frac{160}{2\pi} = \frac{2\pi r^2}{2\pi} + \frac{20\pi r}{2\pi}$$

$$\frac{80}{\pi} = r^2 + 10r$$

$$\frac{80}{\pi} + 25 = r^2 + 10r + 25$$

$$\frac{80 + 25\pi}{\pi} = (r+5)^2$$

$$r + 5 = \pm\sqrt{\frac{80 + 25\pi}{\pi}}$$

$$r = -5 \pm \sqrt{\frac{80 + 25\pi}{\pi}}$$

The radius is never negative so $r \approx 2.1$ in.

107. $-4(2z - 6) = -3(z - 4) + z$

$$-8z + 24 = -3z + 12 + z$$

$$-8z + 24 = -2z + 12$$

$$-6z = -12$$

$$z = 2$$

108. Let $x =$ the amount invested at 7%. Then the amount invested at $6\frac{1}{4}\%$ will be $10,000 - x$.

The interest earned at 7% will be $0.07x$. and the amount of interest earned at 6.25% will be $.0625(10,000 - x)$. The total interest earned is \$656.50.

$$0.07x + 0.0625(10,000 - x) = 656.50$$

$$0.07x + 625 - 0.0625x = 656.50$$

$$0.0075x = 31.5$$

$$x = 4200$$

Thus, \$4200 was invested at 7% and

$\$10,000 - \$4200 = \$5800$ was invested at $6\frac{1}{4}\%$

109. $m = \dfrac{y_2 - y_1}{x_2 - x_1} = \dfrac{5 - 5}{0 - (-2)} = \dfrac{0}{0 + 2} = \dfrac{0}{2} = 0$

110.

$$\begin{array}{r} 4x^2 + 9x - 3 \\ \underline{x - 2} \\ -8x^2 - 18x + 6 \\ \underline{4x^3 + 9x^2 - 3x} \\ 4x^3 + x^2 - 21x + 6 \end{array}$$

111. $|x + 3| = |2x - 7|$

$$x + 3 = 2x - 7 \quad \text{or} \quad x + 3 = -(2x - 7)$$

$$\begin{array}{ll} -x = -10 & x + 3 = -2x + 7 \\ x = 10 & 3x = 4 \\ & x = \dfrac{4}{3} \end{array}$$

Exercise Set 12.2

1. For a quadratic equation in standard form, $ax^2 + bx + c = 0$, the quadratic formula is

$$x = \frac{-b \pm \sqrt{b^2 - 4ac}}{2a}.$$

3. $\quad 6x - 3x^2 + 8 = 0$

$$-3x^2 + 6x + 8 = 0$$

$$a = -3, \ b = 6, \text{ and } c = 8$$

5. Yes, multiply both sides of the equation

$$-6x^2 + \frac{1}{2}x - 5 = 0 \ \text{ by } -1 \text{ to obtain}$$

$6x^2 - \dfrac{1}{2}x + 5 = 0$. The equations are equivalent

so they will have the same solutions.

7. a. For a quadratic equation in standard form, $ax^2 + bx + c = 0$, the discriminant is the expression under the square root symbol in the quadratic formula, $b^2 - 4ac$.

b. $3x^2 - 6x + 10 = 0$, $a = 3$, $b = -6$, and $c = 10$.
$b^2 - 4ac = (-6)^2 - 4(3)(10) = 36 - 120 = -84$

c. If $b^2 - 4ac > 0$, then the quadratic equation will have two distinct real solutions. Since there is a positive number under the radical sign in the quadratic formula, the value of the radical will be real and there will be two real solutions. If $b^2 - 4ac = 0$, then the equation has the single real solution $\dfrac{-b}{2a}$. If $b^2 - 4ac < 0$, then the expression under the radical sign in the quadratic formula is negative. Thus, the equation will have no real solution.

9. $x^2 + 3x + 1 = 0$

   $b^2 - 4ac = (3)^2 - 4(1)(1) = 9 - 4 = 5$

   Since $5 > 0$, there are two real solutions.

11. $4z^2 + 6z + 5 = 0$

    $b^2 - 4ac = 6^2 - 4(4)(5) = 36 - 80 = -44$

    Since $-44 < 0$, there is no real solution.

13. $5p^2 + 3p - 7 = 0$

    $b^2 - 4ac = 3^2 - 4(5)(-7) = 9 + 140 = 149$

    Since $149 > 0$, there are two real solutions.

15. $-5x^2 + 5x - 8 = 0$

    $b^2 - 4ac = 5^2 - 4(-5)(-8) = 25 - 160 = -135$

    Since $-135 < 0$, there is no real solution.

17. $x^2 + 10.2x + 26.01 = 0$

    $$b^2 - 4ac = (10.2)^2 - 4(1)(26.01)$$
    $$= 104.04 - 104.04$$
    $$= 0$$

    Since the discriminant is 0, there is one real solution.

19. $b^2 = -3b - \dfrac{9}{4}$

    $b^2 + 3b + \dfrac{9}{4} = 0$

    $b^2 - 4ac = 3^2 - 4(1)\left(\dfrac{9}{4}\right) = 9 - 9 = 0$

    Since the discriminant is 0, there is one real solution.

21. $x^2 - 9x + 18 = 0$

    $$x = \frac{-(-9) \pm \sqrt{(-9)^2 - 4(1)(18)}}{2(1)}$$
    $$= \frac{9 \pm \sqrt{81 - 72}}{2}$$
    $$= \frac{9 \pm \sqrt{9}}{2}$$
    $$= \frac{9 \pm 3}{2}$$

    $x = \dfrac{9+3}{2}$ or $x = \dfrac{9-3}{2}$

    $= \dfrac{12}{2}$ $= \dfrac{6}{2}$

    $= 6$ $= 3$

    The solutions are 6 and 3.

23. $a^2 - 6a + 8 = 0$

    $$a = \frac{6 \pm \sqrt{(-6)^2 - 4(1)(8)}}{2(1)}$$
    $$= \frac{6 \pm \sqrt{36 - 32}}{2}$$
    $$= \frac{6 \pm \sqrt{4}}{2}$$
    $$= \frac{6 \pm 2}{2}$$

    $a = \dfrac{6-2}{2}$ or $a = \dfrac{6+2}{2}$

    $= \dfrac{4}{2}$ $= \dfrac{8}{2}$

    $= 2$ $= 4$

    The solutions are 2 and 4.

25. $\qquad x^2 = -6x + 7$

    $x^2 + 6x - 7 = 0$

    $$x = \frac{-6 \pm \sqrt{6^2 - 4(1)(-7)}}{2(1)}$$
    $$= \frac{-6 \pm \sqrt{36 + 28}}{2}$$
    $$= \frac{-6 \pm \sqrt{64}}{2}$$
    $$= \frac{-6 \pm 8}{2}$$

    $x = \dfrac{-6+8}{2}$ or $x = \dfrac{-6-8}{2}$

    $= \dfrac{2}{2}$ $= \dfrac{-14}{2}$

    $= 1$ $= -7$

    The solutions are 1 and –7.

27. $\qquad -b^2 = 4b - 20$

    $b^2 + 4b - 20 = 0$

    $$b = \frac{-4 \pm \sqrt{(4)^2 - 4(1)(-20)}}{2(1)}$$
    $$= \frac{-4 \pm \sqrt{16 + 80}}{2}$$
    $$= \frac{-4 \pm \sqrt{96}}{2}$$
    $$= \frac{-4 \pm 4\sqrt{6}}{2}$$
    $$= -2 \pm \sqrt{6}$$

    The solutions are $-2 + 2\sqrt{6}$ and $-2 - 2\sqrt{6}$.

29. $b^2 - 64 = 0$

$$b = \frac{0 \pm \sqrt{0^2 - 4(1)(-64)}}{2(1)}$$

$$= \frac{\pm\sqrt{256}}{2}$$

$$= \frac{\pm 16}{2}$$

$$= \pm 8$$

The solutions are 8 and –8.

31. $3w^2 - 4w + 5 = 0$

$$w = \frac{-(-4) \pm \sqrt{(-4)^2 - 4(3)(5)}}{2(3)}$$

$$= \frac{4 \pm \sqrt{16 - 60}}{6}$$

$$= \frac{4 \pm \sqrt{-44}}{6}$$

$$= \frac{4 \pm 2i\sqrt{11}}{6}$$

$$= \frac{2(2 \pm i\sqrt{11})}{6}$$

$$= \frac{2 \pm i\sqrt{11}}{3}$$

The solutions are $\dfrac{2 - i\sqrt{11}}{3}$ and $\dfrac{2 + i\sqrt{11}}{3}$.

33. $c^2 - 5c = 0$

$$c = \frac{-(-5) \pm \sqrt{(-5)^2 - 4(1)(0)}}{2(1)}$$

$$= \frac{5 \pm \sqrt{25}}{2}$$

$$= \frac{5 \pm 5}{2}$$

$$c = \frac{5 + 5}{2} \quad \text{or} \quad c = \frac{5 - 5}{2}$$

$$= \frac{10}{2} \qquad\qquad = \frac{0}{2}$$

$$= 5 \qquad\qquad\quad = 0$$

The solutions are 5 and 0.

35. $4s^2 - 8s + 6 = 0$

$$\frac{1}{2}(4s^2 - 8s + 6) = 0$$

$$2s^2 - 4s + 3 = 0$$

$$s = \frac{-(-4) \pm \sqrt{(-4)^2 - 4(2)(3)}}{2(2)}$$

$$= \frac{4 \pm \sqrt{16 - 24}}{4}$$

$$= \frac{4 \pm \sqrt{-8}}{4}$$

$$= \frac{4 \pm 2i\sqrt{2}}{4}$$

$$= \frac{2 \pm i\sqrt{2}}{2}$$

The solutions are $\dfrac{2 - i\sqrt{2}}{2}$ and $\dfrac{2 + i\sqrt{2}}{2}$.

37. $a^2 + 2a + 1 = 0$

$$a = \frac{-2 \pm \sqrt{2^2 - 4(1)(1)}}{2(1)}$$

$$= \frac{-2 \pm \sqrt{4 - 4}}{2}$$

$$= \frac{-2 \pm \sqrt{0}}{2}$$

$$= \frac{-2}{2}$$

$$= -1$$

The solution is –1.

39. $16x^2 - 8x + 1 = 0$

$$x = \frac{-(-8) \pm \sqrt{(-8)^2 - 4(16)(1)}}{2(16)}$$

$$= \frac{8 \pm \sqrt{64 - 64}}{32}$$

$$= \frac{8 \pm \sqrt{0}}{32}$$

$$= \frac{8}{32}$$

$$= \frac{1}{4}$$

The solution is $\dfrac{1}{4}$.

41. $x^2 - 2x - 1 = 0$

$$x = \frac{-(-2) \pm \sqrt{(-2)^2 - 4(1)(-1)}}{2(1)}$$

$$= \frac{2 \pm \sqrt{4 + 4}}{2}$$

$$= \frac{2 \pm \sqrt{8}}{2}$$

$$= \frac{2 \pm 2\sqrt{2}}{2}$$

$$= 1 \pm \sqrt{2}$$

The solutions are $1 - \sqrt{2}$ and $1 + \sqrt{2}$.

43. $-n^2 = 3n + 6$

$$0 = n^2 + 3n + 6$$

$$n = \frac{-3 \pm \sqrt{3^2 - 4(1)(6)}}{2(1)}$$

$$= \frac{-3 \pm \sqrt{9 - 24}}{2}$$

$$= \frac{-3 \pm \sqrt{-15}}{2}$$

$$= \frac{-3 \pm i\sqrt{15}}{2}$$

The solutions are $\dfrac{-3 + i\sqrt{15}}{2}$ and $\dfrac{-3 - i\sqrt{15}}{2}$.

45. $2x^2 + 5x - 3 = 0$

$$x = \frac{-5 \pm \sqrt{5^2 - 4(2)(-3)}}{2(2)}$$

$$= \frac{-5 \pm \sqrt{25 + 24}}{4}$$

$$= \frac{-5 \pm \sqrt{49}}{4}$$

$$= \frac{-5 \pm 7}{4}$$

$$x = \frac{-5 + 7}{4} \quad \text{or} \quad x = \frac{-5 - 7}{4}$$

$$= \frac{2}{4} \qquad\qquad = \frac{-12}{4}$$

$$= \frac{1}{2} \qquad\qquad = -3$$

The solutions are $\dfrac{1}{2}$ and -3.

47. $(2a + 3)(3a - 1) = 2$

$$6a^2 + 7a - 3 = 2$$

$$6a^2 + 7a - 5 = 0$$

$$a = \frac{-(7) \pm \sqrt{(7)^2 - 4(6)(-5)}}{2(6)}$$

$$= \frac{-7 \pm \sqrt{49 + 120}}{12}$$

$$= \frac{-7 \pm \sqrt{169}}{12}$$

$$= \frac{-7 \pm 13}{12}$$

$$a = \frac{-7 + 13}{12} \quad \text{or} \quad a = \frac{-7 - 13}{12}$$

$$= \frac{6}{12} \qquad\qquad = \frac{-20}{12}$$

$$= \frac{1}{2} \qquad\qquad = -\frac{5}{3}$$

The solutions are $\dfrac{1}{2}$ and $-\dfrac{5}{3}$.

49. $\dfrac{1}{2}t^2 + t - 12 = 0$

$$2\left(\frac{1}{2}t^2 + t - 12\right) = 2(0)$$

$$t^2 + 2t - 24 = 0$$

$$t = \frac{-(2) \pm \sqrt{(2)^2 - 4(1)(-24)}}{2(1)}$$

$$= \frac{-2 \pm \sqrt{4 + 96}}{2}$$

$$= \frac{-2 \pm \sqrt{100}}{2}$$

$$= \frac{-2 \pm 10}{2}$$

$$t = \frac{-2 + 10}{2} \quad \text{or} \quad t = \frac{-2 - 10}{2}$$

$$= \frac{8}{2} \qquad\qquad = \frac{-12}{2}$$

$$= 4 \qquad\qquad = -6$$

The solutions are 4 and –6.

51. $9r^2 + 3r - 2 = 0$

$$r = \frac{-3 \pm \sqrt{3^2 - 4(9)(-2)}}{2(9)}$$

$$= \frac{-3 \pm \sqrt{9 + 72}}{18}$$

$$= \frac{-3 \pm \sqrt{81}}{18}$$

$$= \frac{-3 \pm 9}{18}$$

$$r = \frac{-3 + 9}{18} \quad \text{or} \quad r = \frac{-3 - 9}{18}$$

$$= \frac{6}{18} \qquad\qquad = \frac{-12}{18}$$

$$= \frac{1}{3} \qquad\qquad = -\frac{2}{3}$$

The solutions are $\dfrac{1}{3}$ and $-\dfrac{2}{3}$.

53. $\dfrac{1}{2}x^2 + 2x + \dfrac{2}{3} = 0$

$$6\left(\frac{1}{2}x^2 + 2x + \frac{2}{3}\right) = 6(0)$$

$$3x^2 + 12x + 4 = 0$$

$$x = \frac{-12 \pm \sqrt{(12)^2 - 4(3)(4)}}{2(3)}$$

$$= \frac{-12 \pm \sqrt{144 - 48}}{6}$$

$$= \frac{-12 \pm \sqrt{96}}{6}$$

$$= \frac{-12 \pm 4\sqrt{6}}{6}$$

$$= \frac{2(-6 \pm 2\sqrt{6})}{2(3)}$$

$$= \frac{-6 \pm 2\sqrt{6}}{3}$$

The solutions are $\dfrac{-6 + 2\sqrt{6}}{3}$ and $\dfrac{-6 - 2\sqrt{6}}{3}$.

55. $a^2 - \dfrac{a}{5} - \dfrac{1}{3} = 0$

$$15\left(a^2 - \frac{a}{5} - \frac{1}{3}\right) = 15(0)$$

$$15a^2 - 3a - 5 = 0$$

$$a = \frac{-(-3) \pm \sqrt{(-3)^2 - 4(15)(-5)}}{2(15)}$$

$$= \frac{3 \pm \sqrt{9 + 300}}{30}$$

$$= \frac{3 \pm \sqrt{309}}{30}$$

The solutions are $\dfrac{3 - \sqrt{309}}{30}$ and $\dfrac{3 + \sqrt{309}}{30}$.

57. $c = \dfrac{c - 6}{4 - c}$

$$c(4 - c) = c - 6$$

$$4c - c^2 = c - 6$$

$$0 = c^2 - 3c - 6$$

$$c = \frac{-(-3) \pm \sqrt{(-3)^2 - 4(1)(-6)}}{2(1)}$$

$$= \frac{3 \pm \sqrt{9 + 24}}{2}$$

$$= \frac{3 \pm \sqrt{33}}{2}$$

The solutions are $\dfrac{3 + \sqrt{33}}{2}$ and $\dfrac{3 - \sqrt{33}}{2}$.

59. $2x^2 - 4x + 5 = 0$

$$x = \frac{-(-4) \pm \sqrt{(-4)^2 - 4(2)(5)}}{2(2)}$$

$$= \frac{4 \pm \sqrt{16 - 40}}{4}$$

$$= \frac{4 \pm \sqrt{-24}}{4}$$

$$= \frac{4 \pm 2i\sqrt{6}}{4}$$

$$= \frac{2 \pm i\sqrt{6}}{2}$$

The solutions are $\dfrac{2 + i\sqrt{6}}{2}$ and $\dfrac{2 - i\sqrt{6}}{2}$.

61.
$$y^2 + \frac{y}{2} = -\frac{3}{2}$$
$$2\left(y^2 + \frac{y}{2}\right) = 2\left(-\frac{3}{2}\right)$$
$$2y^2 + y = -3$$
$$2y^2 + y + 3 = 0$$
$$y = \frac{-1 \pm \sqrt{(1)^2 - 4(2)(3)}}{2(2)}$$
$$= \frac{-1 \pm \sqrt{1 - 24}}{4}$$
$$= \frac{-1 \pm \sqrt{-23}}{4}$$
$$= \frac{-1 \pm i\sqrt{23}}{4}$$

The solutions are $\dfrac{-1 + i\sqrt{23}}{4}$ and $\dfrac{-1 - i\sqrt{23}}{4}$.

63.
$$0.1x^2 + 0.6x - 1.2 = 0$$
$$10(0.1x^2 + 0.6x - 1.2) = 10(0)$$
$$x^2 + 6x - 12 = 0$$
$$x = \frac{-6 \pm \sqrt{6^2 - 4(1)(-12)}}{2(1)}$$
$$= \frac{-6 \pm \sqrt{36 + 48}}{2}$$
$$= \frac{-6 \pm \sqrt{84}}{2}$$
$$= \frac{-6 \pm 2\sqrt{21}}{2}$$
$$= -3 \pm \sqrt{21}$$

The solutions are $-3 + \sqrt{21}$ or $-3 - \sqrt{21}$.

65. $f(x) = x^2 - 2x + 5$, $f(x) = 5$
$$x^2 - 2x + 5 = 5$$
$$x^2 - 2x = 0$$
$$x = \frac{2 \pm \sqrt{(-2)^2 - 4(1)(0)}}{2(1)} = \frac{2 \pm \sqrt{4}}{2} = \frac{2 \pm 2}{2}$$
$$x = \frac{2+2}{2} \quad \text{or} \quad x = \frac{2-2}{2}$$
$$= \frac{4}{2} \qquad\qquad = \frac{0}{2}$$
$$= 2 \qquad\qquad = 0$$

The values of x are 2 and 0.

67. $k(x) = x^2 - x - 15$, $k(x) = 15$
$$x^2 - x - 15 = 15$$
$$x^2 - x - 30 = 0$$
$$x = \frac{1 \pm \sqrt{(-1)^2 - 4(1)(-30)}}{2(1)}$$
$$= \frac{1 \pm \sqrt{1 + 120}}{2}$$
$$= \frac{1 \pm \sqrt{121}}{2}$$
$$= \frac{1 \pm 11}{2}$$
$$x = \frac{1+11}{2} \quad \text{or} \quad x = \frac{1-11}{2}$$
$$= \frac{12}{2} \qquad\qquad = \frac{-10}{2}$$
$$= 6 \qquad\qquad = -5$$

The values of x are 6 and –5.

69. $h(t) = 2t^2 - 7t + 6$, $h(t) = 2$
$$2t^2 - 7t + 6 = 2$$
$$2t^2 - 7t + 4 = 0$$
$$t = \frac{7 \pm \sqrt{(-7)^2 - 4(2)(4)}}{2(2)}$$
$$= \frac{7 \pm \sqrt{49 - 32}}{4}$$
$$= \frac{7 \pm \sqrt{17}}{4}$$

The values of t are $\dfrac{7 + \sqrt{17}}{4}$ and $\dfrac{7 - \sqrt{17}}{4}$.

71. $g(a) = 2a^2 - 3a + 16$, $g(a) = 14$
$$2a^2 - 3a + 16 = 14$$
$$2a^2 - 3a + 2 = 0$$
$$a = \frac{3 \pm \sqrt{(-3)^2 - 4(2)(2)}}{2(2)}$$
$$= \frac{3 \pm \sqrt{9 - 16}}{4}$$
$$= \frac{3 \pm \sqrt{-7}}{4}$$

There are no real values of a for which $g(a) = 14$.

73. $x = 2$ and $x = 5$
$x - 2 = 0$ and $x - 5 = 0$
$(x - 2)(x - 5) = 0$
$x^2 - 5x - 2x + 10 = 0$
$x^2 - 7x + 10 = 0$

75. $x = 1$ and $x = -9$
$x - 1 = 0$ and $x + 9 = 0$
$(x - 1)(x + 9) = 0$
$x^2 + 9x - x - 9 = 0$
$x^2 + 8x - 9 = 0$

77. $x = -\dfrac{3}{5}$ and $x = \dfrac{2}{3}$
$5x = -3$ and $3x = 2$
$5x + 3 = 0$ and $3x - 2 = 0$
$(5x + 3)(3x - 2) = 0$
$15x^2 - 10x + 9x - 6 = 0$
$15x^2 - x - 6 = 0$

79. $x = \sqrt{2}$ and $x = -\sqrt{2}$
$x - \sqrt{2} = 0$ and $x + \sqrt{2} = 0$.
$\left(x - \sqrt{2}\right)\left(x + \sqrt{2}\right) = 0$
$x^2 + x\sqrt{2} - x\sqrt{2} - 2 = 0$
$x^2 - 2 = 0$

81. $x = 3i$ and $x = -3i$
$x - 3i = 0$ and $x + 3i = 0$.
$(x - 3i)(x + 3i) = 0$
$x^2 + 3ix - 3ix - 9i^2 = 0$
$x^2 - 9i^2 = 0$
$x^2 - 9(-1) = 0$
$x^2 + 9 = 0$

83. $x = 3 + \sqrt{2}$ and $x = 3 - \sqrt{2}$
$x - 3 - \sqrt{2} = 0$ and $x - 3 + \sqrt{2} = 0$
$(x - 3 - \sqrt{2})(x - 3 + \sqrt{2}) = 0$
$\left[(x - 3) - \sqrt{2}\right]\left[(x - 3) + \sqrt{2}\right] = 0$
$(x - 3)^2 - \left(\sqrt{2}\right)^2 = 0$
$x^2 - 6x + 9 - 2 = 0$
$x^2 - 6x + 7 = 0$

85. $x = 2 + 3i$ and $x = 2 - 3i$
$x - 2 - 3i = 0$ and $x - 2 + 3i = 0$
$(x - 2 - 3i)(x - 2 + 3i) = 0$
$\left[(x - 2) - 3i\right]\left[(x - 2) + 3i\right] = 0$
$(x - 2)^2 - (3i)^2 = 0$
$x^2 - 4x + 4 - 9i^2 = 0$
$x^2 - 4x + 4 + 9 = 0$
$x^2 - 4x + 13 = 0$

87. a. $n(10 - 0.02n) = 450$

 b. $\qquad n(10 - 0.02n) = 450$
$\qquad 10n - 0.02n^2 = 450$
$0.02n^2 - 10n + 450 = 0$
$n = \dfrac{-(-10) \pm \sqrt{(-10)^2 - 4(0.02)(450)}}{2(0.02)}$
$n = \dfrac{10 \pm \sqrt{100 - 36}}{0.04}$
$n = \dfrac{10 \pm \sqrt{64}}{0.04}$
$n = \dfrac{10 \pm 8}{0.04}$
$n = 450$ or $n = 50$
Since $n \le 65$, the number of lamps that must be sold is 50.

89. a. $n(50 - 0.2n) = 1680$

 b. $\qquad n(50 - 0.4n) = 660$
$\qquad 50n - 0.4n^2 = 660$
$0.4n^2 - 50n + 660 = 0$
$n = \dfrac{-(-50) \pm \sqrt{(-50)^2 - 4(0.4)(660)}}{2(0.4)}$
$n = \dfrac{50 \pm \sqrt{2500 - 1056}}{0.8}$
$n = \dfrac{50 \pm \sqrt{1444}}{0.8}$
$n = \dfrac{50 \pm 38}{0.8}$
$n = 110$ or $n = 15$
Since $n \le 50$, the number of chairs that must be sold is 15.

91. Any quadratic equation for which the discriminant is a non-negative perfect square can be solved by factoring. Any quadratic equation for which the discriminant is a positive number but not a perfect square can be solved by the quadratic formula but not by factoring over the set of integers.

93. Yes. If the discriminant is a perfect square, the simplified expression will not contain a radical and the quadratic equation can be solved by factoring.

95. Let x be the number.

$$2x^2 + 3x = 27$$

$$2x^2 + 3x - 27 = 0$$

$$x = \frac{-3 \pm \sqrt{3^2 - 4(2)(-27)}}{2(2)}$$

$$x = \frac{-3 \pm \sqrt{9 + 112}}{4}$$

$$x = \frac{-3 \pm \sqrt{121}}{4}$$

$$x = \frac{-3 \pm 11}{4}$$

$$x = \frac{8}{4} \text{ or } x = \frac{-14}{4}$$

$$x = 2 \text{ or } x = -\frac{7}{2}$$

Since the number must be positive, it is 3.

97. Let x be the width. Then $3x - 1$ is the length.
Use $A = (\text{length})(\text{width})$.

$$24 = (3x - 1)(x)$$

$$24 = 3x^2 - x$$

$$0 = 3x^2 - x - 24$$

$$x = \frac{-(-1) \pm \sqrt{(-1)^2 - 4(3)(-24)}}{2(3)}$$

$$= \frac{1 \pm \sqrt{1 + 288}}{6}$$

$$= \frac{1 \pm \sqrt{289}}{6}$$

$$= \frac{1 \pm 17}{6}$$

Since width is positive, use

$$x = \frac{1 + 17}{6} = \frac{18}{6} = 3$$

$$3x - 1 = 3(3) - 1 = 9 - 1 = 8$$

The width is 3 feet and the length is 8 feet.

99. Let x be the amount by which each side is to be reduced.
Then $6 - x$ is the new width
and $8 - x$ is the new length

$$\text{new area} = \frac{1}{2}(\text{old area}) = \frac{1}{2}(6 \cdot 8) = 24$$

new area = (new width)(new length)

$$24 = (6 - x)(8 - x)$$

$$0 = 48 - 14x + x^2 - 24$$

$$0 = x^2 - 14x + 24$$

$$x = \frac{-(-14) \pm \sqrt{(-14)^2 - 4(1)(24)}}{2(1)}$$

$$= \frac{14 \pm \sqrt{196 - 96}}{2}$$

$$= \frac{14 \pm \sqrt{100}}{2}$$

$$= \frac{14 \pm 10}{2}$$

$$x = \frac{14 + 10}{2} \text{ or } x = \frac{14 - 10}{2}$$

$$= \frac{24}{2} \qquad\qquad = \frac{4}{2}$$

$$= 12 \qquad\qquad = 2$$

We reject $x = 12$, since this would give negative values for width and length.
The only meaningful value is $x = 2$ inches since this gives positive values for the new width and length.

101. Substitute $h = 0$ in the formula $h = -16t^2 + 308$.

$$0 = -16t^2 + 308$$

$$16t^2 = 308$$

$$t^2 = \frac{308}{16}$$

$$t^2 = \frac{77}{4}$$

$$t = \pm\sqrt{\frac{77}{7}} = \pm\frac{\sqrt{77}}{2}$$

Time must be positive, so $t = \frac{\sqrt{77}}{2} \approx 4.39$. It take approximately 4.39 seconds for the water to reach the bottom of the falls.

103. a. $h = \frac{1}{2}at^2 + v_0t + h_0$

$$20 = \frac{1}{2}(-32)t^2 + 60t + 80$$

$$20 = -16t^2 + 60t + 80$$

$$16t^2 - 60t - 60 = 0$$

$$t = \frac{-(-60) \pm \sqrt{(-60)^2 - 4(16)(-60)}}{2(16)}$$

$$= \frac{60 \pm \sqrt{7440}}{32}$$

Since time must be positive, use

$$t = \frac{60 + \sqrt{7440}}{32} \approx 4.57$$

The horseshoe is 20 feet from the ground after about 4.57 seconds.

b. $0 = \frac{1}{2}(-32)t^2 + 60t + 80$

$$0 = -16t^2 + 60t + 80$$

$$16t^2 - 60t - 80 = 0$$

$$t = \frac{-(-60) \pm \sqrt{(-60)^2 - 4(16)(-80)}}{2(16)}$$

$$= \frac{60 \pm \sqrt{8720}}{32}$$

Since time must be positive, use

$$t = \frac{60 + \sqrt{8720}}{32} \approx 4.79$$

The horseshoes strike the ground after about 4.79 seconds.

105. $x^2 - \sqrt{5}x - 10 = 0, \ a = 1, b = -\sqrt{5}, c = -10$

$$x = \frac{-(-\sqrt{5}) \pm \sqrt{(-\sqrt{5})^2 - 4(1)(-10)}}{2(1)}$$

$$= \frac{\sqrt{5} \pm \sqrt{5 + 40}}{2}$$

$$= \frac{\sqrt{5} \pm \sqrt{45}}{2}$$

$$= \frac{\sqrt{5} \pm 3\sqrt{5}}{2}$$

$$x = \frac{\sqrt{5} + 3\sqrt{5}}{2} \quad \text{or} \quad x = \frac{\sqrt{5} - 3\sqrt{5}}{2}$$

$$= \frac{4\sqrt{5}}{2} \qquad\qquad = \frac{-2\sqrt{5}}{2}$$

$$= 2\sqrt{5} \qquad\qquad = -\sqrt{5}$$

The solutions are $2\sqrt{5}$ and $-\sqrt{5}$.

107. Let s be the length of the side of the original cube. Then $s + 0.2$ is the length of the side of the expanded cube.

$$(s + 0.2)^3 = s^3 + 6$$

$$s^3 + 0.6s^2 + 0.12s + 0.008 = s^3 + 6$$

$$0.6s^2 + 0.12s + 0.008 = 6$$

$$0.6s^2 + 0.12s - 5.992 = 0$$

$$s = \frac{-0.12 \pm \sqrt{(0.12)^2 - 4(0.6)(-5.992)}}{2(0.6)}$$

$$= \frac{-0.12 \pm \sqrt{0.0144 + 14.3803}}{1.2}$$

$$= \frac{-0.12 \pm \sqrt{14.3952}}{1.2}$$

Use the positive value since a length cannot be negative.

$$s = \frac{-0.12 + \sqrt{14.3952}}{1.2}$$

$$\approx \frac{-0.12 + 3.7941}{1.2}$$

$$\approx 3.0618$$

The original side was about 3.0618 mm long.

109. a. $h = \frac{1}{2}at^2 + v_0t + h_0$

$$0 = \frac{1}{2}(-32)t^2 + 0t + 60$$

$$0 = -16t^2 + 0t + 60$$

$$t = \frac{-(0) \pm \sqrt{0^2 - 4(-16)(60)}}{2(-16)}$$

$$t = \frac{0 \pm \sqrt{3840}}{-32}$$

$t \approx -1.94 \ \text{or} \ t \approx 1.94$

Since time must be positive, use 1.94 sec.

b. $h = \frac{1}{2}at^2 + v_0t + h_0$

$$0 = \frac{1}{2}(-32)t^2 + 0t + 120$$

$$0 = -16t^2 + 0t + 120$$

$$t = \frac{-(0) \pm \sqrt{0^2 - 4(-16)(120)}}{2(-16)}$$

$$t = \frac{0 \pm \sqrt{7680}}{-32}$$

$t \approx -2.74 \ \text{or} \ t \approx 2.74$

Since time must be positive, use 2.74 sec.

c. The height of Travis' rock is given by $h(t) = -16t^2 + 100t + 60$. Set the height equal to 0 to determine the how long it will take Travis' rock to strike the ground.

$$-16t^2 + 100t + 60 = 0$$
$$4t^2 - 25t - 15 = 0$$
$$t = \frac{-(-25) \pm \sqrt{(-25)^2 - 4(4)(-15)}}{2(4)}$$
$$= \frac{25 \pm \sqrt{625 + 240}}{8}$$
$$= \frac{25 \pm \sqrt{865}}{8}$$
$$t \approx 6.80 \text{ or } t \approx -0.55$$

Time must be positive, so it will take about 6.80 seconds for Travis' rock to strike the ground.

The height of Courtney's rock is given by $h(t) = -16t^2 + 60t + 120$. Set the height equal to 0 to determine the how long it will take Courtney's rock to strike the ground.

$$-16t^2 + 60t + 120 = 0$$
$$4t^2 - 15t - 30 = 0$$
$$t = \frac{-(-15) \pm \sqrt{(-15)^2 - 4(4)(-30)}}{2(4)}$$
$$= \frac{15 \pm \sqrt{225 + 480}}{8}$$
$$= \frac{15 \pm \sqrt{705}}{8}$$
$$t \approx 5.19 \text{ or } t \approx -1.44$$

Time must be positive, so it will take about 5.19 seconds for Travis' rock to strike the ground.

Thus, Courtney's rock will strike the ground sooner than Travis'.

d. The height of Travis' rock is given by $h(t) = -16t^2 + 100t + 60$. The height of Courtney's rock is given by $h(t) = -16t^2 + 60t + 120$. We want to know when these will be equal.

$$-16t^2 + 100t + 60 = -16t^2 + 60t + 120$$
$$100t + 60 = 60t + 120$$
$$40t = 60$$
$$t = \frac{60}{40} = 1.5$$

The rocks will be the same distance above the ground after 1.5 seconds.

110. $\dfrac{5.55 \times 10^3}{1.11 \times 10^1} = \dfrac{5.55}{1.11} \times \dfrac{10^3}{10^1} = 5 \times 10^2 \text{ or } 500$

111. $f(x) = x^2 + 2x - 8$
$$f(3) = (3)^2 + 2(3) - 8$$
$$= 9 + 6 - 8$$
$$= 7$$

112. $3x + 4y = 2$
$2x = -5y - 1$
Rewrite the system in standard form.
$3x + 4y = 2$
$2x + 5y = -1$
To eliminate the x variable, multiply the first equation by 2 and the second equation by -3, and then add.
$$6x + 8y = 4$$
$$\underline{-6x - 15y = 3}$$
$$-7y = 7$$
$$y = -1$$
Substitute -1 for y in the first equation to find x.
$$3x + 4(-1) = 2$$
$$3x - 4 = 2$$
$$3x = 6$$
$$x = 2$$
The solution is $(2, -1)$.

113. $\dfrac{x + \sqrt{y}}{x - \sqrt{y}} \cdot \dfrac{x + \sqrt{y}}{x + \sqrt{y}} = \dfrac{x^2 + x\sqrt{y} + x\sqrt{y} + (\sqrt{y})^2}{x^2 + x\sqrt{y} - x\sqrt{y} - (\sqrt{y})^2}$
$$= \dfrac{x^2 + 2x\sqrt{y} + y}{x^2 - y}$$

114. $\sqrt{x^2 - 6x - 4} = x$
$$x^2 - 6x - 4 = x^2$$
$$-6x - 4 = 0$$
$$-6x = 4$$
$$x = \frac{4}{-6} = -\frac{2}{3}$$
Upon checking, this value does not satisfy the equation. There is no real solution.

Exercise Set 12.3

1. Answers will vary.

3. $A = s^2$, for s.

$\sqrt{A} = s$

5. $d = 4.9t^2$, for t

$\dfrac{d}{4.9} = t^2$

$\sqrt{\dfrac{d}{4.9}} = t$

7. $E = i^2 r$, for i

$\dfrac{E}{r} = i^2$

$\sqrt{\dfrac{E}{r}} = i$

9. $d = 16t^2$, for t

$\dfrac{d}{16} = t^2$

$\sqrt{\dfrac{d}{16}} = t \;\Rightarrow\; t = \dfrac{\sqrt{d}}{4}$

11. $E = mc^2$, for c

$\dfrac{E}{m} = c^2$

$\sqrt{\dfrac{E}{m}} = c$

13. $V = \dfrac{1}{3}\pi r^2 h$, for r

$3V = \pi r^2 h$

$\dfrac{3V}{\pi h} = r^2$

$\sqrt{\dfrac{3V}{\pi h}} = r$

15. $d = \sqrt{L^2 + W^2}$, for W

$d^2 = L^2 + W^2$

$d^2 - L^2 = W^2$

$\sqrt{d^2 - L^2} = W$

17. $a^2 + b^2 = c^2$, for b

$b^2 = c^2 - a^2$

$b = \sqrt{c^2 - a^2}$

19. $d = \sqrt{L^2 + W^2 + H^2}$, for H

$d^2 = L^2 + W^2 + H^2$

$d^2 - L^2 - W^2 = H^2$

$\sqrt{d^2 - L^2 - W^2} = H$

21. $h = -16t^2 + s_0$, for t

$16t^2 = s_0 - h$

$t^2 = \dfrac{s_0 - h}{16}$

$t = \sqrt{\dfrac{s_0 - h}{16}}$

$t = \dfrac{\sqrt{s_0 - h}}{4}$

23. $E = \dfrac{1}{2}mv^2$, for v

$2E = mv^2$

$\dfrac{2E}{m} = v^2$

$\sqrt{\dfrac{2E}{m}} = v$

25. $a = \dfrac{v_2^2 - v_1^2}{2d}$, for v_1

$2ad = v_2^2 - v_1^2$

$2ad + v_1^2 = v_2^2$

$v_1^2 = v_2^2 - 2ad$

$v_1 = \sqrt{v_2^2 - 2ad}$

27. $v' = \sqrt{c^2 - v^2}$, for c

$\left(v'\right)^2 = c^2 - v^2$

$\left(v'\right)^2 + v^2 = c^2$

$\sqrt{\left(v'\right)^2 + v^2} = c$

29. a. $P(n) = 2.7n^2 + 9n - 3$

$P(5) = 2.7(5)^2 + 9(5) - 3 = 109.5$

The profit would be \$10,950.

b. $P(n) = 2.7n^2 + 9n - 3$

$200 = 2.7n^2 + 9n - 3$

$0 = 2.7n^2 + 9n - 203$

$x = \dfrac{-9 \pm \sqrt{9^2 - 4(2.7)(-203)}}{2(2.7)}$

$x = \dfrac{-9 \pm \sqrt{2273.4}}{5.4}$

$x \approx 7 \text{ or } x \approx -10$

We disregard the negative answer. Thus 7 tractors must be sold.

31. $T = 6.2t^2 + 12t + 32$

a. When the car is turned on, $t = 0$.

$T = 6.2(0)^2 + 12(0) + 32 = 32$

The temperature is $32°F$.

b. $T = 6.2(2)^2 + 12(2) + 32 = 80.8$

The temperature after 2 minutes is $80.8°F$.

c. $120 = 6.2t^2 + 12t + 32$

$0 = 6.2t^2 + 12t - 88$

$t = \dfrac{-12 \pm \sqrt{12^2 - 4(6.2)(-88)}}{2(6.2)}$

$t \approx \dfrac{-12 \pm 48.23}{12.4}$

$t \approx 2.92 \text{ or } t \approx -4.86$

The radiator temperature will reach $120°F$ about 2.92 min. after the engine is started.

33. a. In 2006, $t = 4$.

$D(4) = 0.04(4)^2 - 0.03(4) + 0.01 = 0.53$

The number of downloads in 2006 was about 0.53 billion.

b. $D(t) = 0.04t^2 - 0.03t + 0.01,\ D(t) = 1$

$1 = 0.04t^2 - 0.03t + 0.01$

$0 = 0.04t^2 - 0.03t - 0.99$

$t = \dfrac{-(-0.03) \pm \sqrt{(-0.03)^2 - 4(0.04)(-0.99)}}{2(0.04)}$

$= \dfrac{0.03 \pm \sqrt{0.1593}}{0.08}$

$\approx \dfrac{0.03 \pm 0.3991}{0.08}$

$t \approx \dfrac{0.03 + 0.3991}{0.08} \quad \text{or} \quad t \approx \dfrac{0.03 - 0.3991}{0.08}$

$\approx \dfrac{0.4291}{0.08} \qquad\qquad \approx \dfrac{-0.3691}{0.08}$

$\approx 5.36 \qquad\qquad\qquad \approx -4.61$

We disregard the negative answer. The number of downloads will be 1 billion 5.36 years after 2002, which is the year 2007.

35. a. In 2003, $t = 3$.

$Y(3) = 0.66(3)^2 - 2.49(3) + 12.93 = 11.4$

In 2003, the yield was about 11.4 tons per acre.

b. $0.66t^2 - 2.49t + 12.93 = 13$

$0.66t^2 - 2.49t - 0.07 = 0$

$t = \dfrac{-(-2.49) \pm \sqrt{(-2.49)^2 - 4(0.66)(-0.07)}}{2(0.66)}$

$= \dfrac{2.49 \pm \sqrt{6.3849}}{2(0.66)}$

$= \dfrac{2.49 \pm 2.5268}{1.32}$

$t = \dfrac{2.49 + 2.5268}{1.32} \quad \text{or} \quad t = \dfrac{2.49 - 2.5268}{1.32}$

$= \dfrac{5.0168}{1.32} \qquad\qquad = \dfrac{-0.0368}{1.32}$

$\approx 3.80 \qquad\qquad\qquad \approx -0.03$

We disregard the negative answer. Thus the yield will be 13 tons per acre about 3.8 years after 2000, which would be in the year 2003.

37. a. $M = -0.00434t^2 + 0.142t + 0.315$

In 2007, $t = 10$.

$M(1) = -0.00434(10)^2 + 0.142(10) + 0.315$

$= 1.301$

In 2007, about 1.301 million motorcycles was sold in the United States.

b. $M = -0.00434t^2 + 0.142t + 0.315,\ M = 1.4$

$1.4 = -0.00434t^2 + 0.142t + 0.315$

$0.00434t^2 - 0.142t + 1.085 = 0$

$t = \dfrac{-(-0.142) \pm \sqrt{(-0.142)^2 - 4(0.00434)(1.085)}}{2(0.00434)}$

$= \dfrac{0.142 \pm \sqrt{0.0013284}}{0.00868}$

$\approx \dfrac{0.142 \pm 0.036447}{0.00868}$

$$t \approx \frac{0.142 + 0.036447}{0.00868} = \frac{0.178447}{0.00868} \approx 20.56$$

or

$$t \approx \frac{0.142 - 0.036447}{0.00868} = \frac{0.105553}{0.00868} \approx 12.16$$

The number of motorcycles sold in the U.S. will be 1.4 million 12.16 years after 1997, which is the year 2009, and 20.56 years after 1997, which is the year 2017.

39. Let x be the width of the playground. Then the length is given by $x + 10$.

Area = length × width

$$600 = (x+10)x$$
$$0 = x^2 + 10x - 600$$
$$0 = (x-20)(x+30)$$
$$x - 20 = 0 \quad \text{or} \quad x + 30 = 0$$
$$x = 20 \quad \text{or} \qquad x = -30$$

Disregard the negative value. The width of the playground is 20 meters and the length is $20 + 10 = 30$ meters.

41. Let r be the rate at which the present equipment drills.

	d	r	$t = \dfrac{d}{r}$
Present equipment	64	r	$\dfrac{64}{r}$
New equipment	64	$r+1$	$\dfrac{64}{r+1}$

They would have hit water in 3.2 hours less time with the new equipment.

$$\frac{64}{r+1} = \frac{64}{r} - 3.2$$
$$r(r+1)\left(\frac{64}{r+1}\right) = r(r+1)\left(\frac{64}{r}\right) - r(r+1)(3.2)$$
$$64r = 64(r+1) - 3.2r(r+1)$$
$$64r = 64r + 64 - 3.2r^2 - 3.2r$$
$$3.2r^2 + 3.2r - 64 = 0$$
$$r^2 + r - 20 = 0$$
$$(r+5)(r-4) = 0$$
$$r + 5 = 0 \quad \text{or} \quad r - 4 = 0$$
$$r = -5 \qquad\qquad r = 4$$

Disregard the negative answer. Thus the present equipment drills at a rate of 4 ft/hr.

43. Let x be Latoya's rate going uphill so $x + 2$ is her rate going downhill. Using $\dfrac{d}{r} = t$ gives

$$t_{\text{uphill}} + t_{\text{downhill}} = 1.75$$
$$\frac{6}{x} + \frac{6}{x+2} = 1.75$$
$$x(x+2)\left(\frac{6}{x}\right) + x(x+2)\left(\frac{6}{x+2}\right) = x(x+2)(1.75)$$
$$6(x+2) + 6x = 1.75x(x+2)$$
$$6x + 12 + 6x = 1.75x^2 + 3.5x$$
$$0 = 1.75x^2 - 8.5x - 12$$
$$x = \frac{-(-8.5) \pm \sqrt{(-8.5)^2 - 4(1.75)(-12)}}{2(1.75)}$$
$$= \frac{8.5 \pm \sqrt{156.25}}{3.5} = \frac{8.5 \pm 12.5}{3.5}$$
$$x = 6 \quad \text{or} \quad x \approx -1.14$$

Since the time must be positive, Latoya's uphill rate is 6 mph and her downhill rate is $x + 2 = 8$ mph.

45. Let x be the Bonita's time, then $x + 1$ is the Pamela's time.

$$\frac{6}{x} + \frac{6}{x+1} = 1$$
$$x(x+1)\left(\frac{6}{x}\right) + x(x+1)\left(\frac{6}{x+1}\right) = x(x+1)(1)$$
$$6(x+1) + 6x = x^2 + x$$
$$6x + 6 + 6x = x^2 + x$$
$$0 = x^2 - 11x - 6$$
$$x = \frac{-(-11) \pm \sqrt{(-11)^2 - 4(1)(-6)}}{2(1)}$$
$$= \frac{11 \pm \sqrt{121 + 24}}{2}$$
$$= \frac{11 \pm \sqrt{145}}{2}$$
$$x = \frac{11 + \sqrt{145}}{2} \quad \text{or} \quad x = \frac{11 - \sqrt{145}}{2}$$
$$\approx 11.52 \qquad\qquad \approx -0.52$$

Since the time must be positive, it takes Bonita about 11.52 hours and Pamela about 12.52 hours to rebuild the engine.

47. Let r be the speed of the plane in still air.

	d	r	$t = \dfrac{d}{r}$
With wind	80	$r + 30$	$\dfrac{80}{r+30}$
Against wind	80	$r - 30$	$\dfrac{80}{r-30}$

The total time is 1.3 hours

$$\frac{80}{r+30} + \frac{80}{r-30} = 1.3$$

$$(r+30)(r-30)\left(\frac{80}{r+30} + \frac{80}{r-30} = 1.3\right)$$

$$80(r-3) + 80(r+30) = 1.3(r^2 - 900)$$

$$80r - 240 + 80r + 240 = 1.3r^2 - 1170$$

$$160r = 1.3r^2 - 1170$$

$$0 = 1.3r^2 - 160r - 1170$$

$$r = \frac{-(-160) \pm \sqrt{(-160)^2 - 4(1.3)(-1170)}}{2(1.3)}$$

$$= \frac{160 \pm \sqrt{25,600 + 6084}}{2.6}$$

$$= \frac{160 \pm \sqrt{31,684}}{2.6}$$

$$= \frac{160 \pm 178}{2.6}$$

$$r = \frac{160 + 178}{2.6} \quad \text{or} \quad r = \frac{160 - 178}{2.6}$$

$$= \frac{338}{2.6} \qquad\qquad = \frac{-18}{2.6}$$

$$= 130 \qquad\qquad \approx -6.92$$

Since speed must be positive, the speed of the plane in still air is 130 mph.

49. Let t be the number of hours for Chris to clean alone. Then $t + 0.5$ is the number of hours for John to clean alone.

	Rate of work	Time worked	Part of Task completed
Chris	$\dfrac{1}{t}$	6	$\dfrac{6}{t}$
John	$\dfrac{1}{t+0.5}$	6	$\dfrac{6}{t+0.5}$

$$1 = \frac{6}{t} + \frac{6}{t+0.5}$$

$$t(t+0.5)(1) = t(t+0.5)\left(\frac{6}{t}\right) + t(t+0.5)\left(\frac{6}{t+0.5}\right)$$

$$t(t+0.5) = 6(t+0.5) + 6t$$

$$t^2 + 0.5t = 6t + 3 + 6t$$

$$t^2 + 0.5t = 12t + 3$$

$$t^2 - 11.5t - 3 = 0$$

$$t = \frac{11.5 \pm \sqrt{(-11.5)^2 - 4(1)(-3)}}{2(1)}$$

$$= \frac{11.5 \pm \sqrt{132.25 + 12}}{2}$$

$$= \frac{11.5 \pm \sqrt{144.25}}{2}$$

$$t = \frac{11.5 + \sqrt{144.25}}{2} \quad \text{or} \quad t = \frac{11.5 - \sqrt{144.25}}{2}$$

$$\approx 11.76 \qquad\qquad \approx -0.26$$

Since the time must be positive, it takes Chris about 11.76 hours and John about $11.76 + 0.5 = 12.26$ hours to clean alone

51. Let x be the speed of the trip from Lubbock to Plainview. Then $x - 10$ is the speed from Plainview to Amarillo.

	d	r	t
First part	75	x	$\dfrac{75}{x}$
Second part	195	$x - 10$	$\dfrac{195}{x-10}$

Including the 2 hours she spent in Plainview, the entire trip took Lisa 6 hours.

$$\frac{75}{x} + 2 + \frac{195}{x-10} = 6$$

$$\frac{75}{x} + \frac{195}{x-10} = 4$$

$$x(x-10)\left(\frac{75}{x} + \frac{195}{x-10}\right) = x(x-10)4$$

$$75(x-10) + 195x = 4x^2 - 40x$$

$$75x - 750 + 195x = 4x^2 - 40x$$

$$270x - 750 = 4x^2 - 40x$$

$$0 = 4x^2 - 310x + 750$$

$$0 = 2(2x^2 - 155x + 375)$$

$$0 = 2(2x - 5)(x - 75)$$

$$2x - 5 = 0 \quad \text{or} \quad x - 75 = 0$$

$$x = \frac{5}{2} \qquad\qquad x = 75$$

We disregard $x = \dfrac{5}{2}$ since it would result in a negative speed for the second part of the trip. Shywanda's average speed from San Antonio to Austin was 75 mph.

53. From the figure, $16x =$ the length and $9x =$ the height.

$$(16x)^2 + (9x)^2 = 40^2$$
$$256x^2 + 81x^2 = 1600$$
$$337x^2 = 1600$$
$$x^2 = \frac{1600}{337}$$
$$x = \pm\sqrt{\frac{1600}{337}} \approx \pm 2.179$$

Disregard the negative answer.
The length is $16(2.179) \approx 34.86$ inches and the height is $9(2.179) \approx 19.61$ inches.

55. Answers will vary.

57. Let $l =$ original length and $w =$ original width. A system of equations that describes this situation is
$$l \cdot w = 18$$
$$(l+2)(w+3) = 48$$

Solve the first equation for l to obtain $l = \dfrac{18}{w}$.

Substitute $\dfrac{18}{w}$ for l in the second equation. The result is an equation in only one variable which can be solved.

$$(l+2)(w+3) = 48$$
$$\left(\frac{18}{w}+2\right)(w+3) = 48$$
$$18 + \frac{54}{w} + 2w + 6 = 48$$
$$2w - 24 + \frac{54}{w} = 0$$
$$w\left(2w - 24 + \frac{54}{w}\right) = w(0)$$
$$2w^2 - 24w + 54 = 0$$
$$2(w-3)(w-9) = 0$$
$$w = 3 \text{ or } w = 9$$

If $w = 3$, then $l = \dfrac{18}{3} = 6$. One possible set of

dimensions for the original rectangle is 6 meters by 3 meters. If $w = 9$, then $l = \dfrac{18}{9} = 2$. Another possible set of dimensions for the original rectangle is 2 meters by 9 meters.

59. $-\left[4(5-3)^3\right] + 2^4 = -\left[4(2)^3\right] + 2^4$
$$= -[4(8)] + 16$$
$$= -32 + 16$$
$$= -16$$

60. $IR + Ir = E, \text{ for } R$
$$IR = E - Ir$$
$$R = \frac{E - Ir}{I}$$

61. $\dfrac{r}{r-4} - \dfrac{r}{r+4} + \dfrac{32}{r^2-16}$

$$= \frac{r}{r-4} \cdot \frac{r+4}{r+4} - \frac{r}{r+4} \cdot \frac{r-4}{r-4} + \frac{32}{(r+4)(r-4)}$$
$$= \frac{r(r+4) - r(r-4) + 32}{(r+4)(r-4)}$$
$$= \frac{r^2 + 4r - r^2 + 4r + 32}{(r+4)(r-4)}$$
$$= \frac{8r + 32}{(r+4)(r-4)}$$
$$= \frac{8(r+4)}{(r+4)(r-4)}$$
$$= \frac{8}{r-4}$$

62. $\left(\dfrac{x^{3/4}y^{-2}}{x^{1/2}y^2}\right)^8 = \left(x^{(3/4)-(1/2)}y^{-2-2}\right)^8$

$$= \left(x^{1/4}y^{-4}\right)^8$$
$$= x^2 y^{-32}$$
$$= \frac{x^2}{y^{32}}$$

63. $\sqrt{x^2 + 3x + 12} = x$
$$x^2 + 3x + 12 = x^2$$
$$3x + 12 = 0$$
$$3x = -12$$
$$x = -4$$

Upon checking, this value does not satisfy the equation. There is no real solution.

Mid-Chapter Test: 12.1-12.3

1. $x^2 - 12 = 86$

   $x^2 = 98$

   $x = \pm\sqrt{98} = \pm 7\sqrt{2}$

2. $(a-3)^2 + 20 = 0$

   $(a-3)^2 = -20$

   $a - 3 = \pm\sqrt{-20}$

   $a - 3 = \pm 2i\sqrt{5}$

   $a = 3 \pm 2i\sqrt{5}$

3. $(2m+7)^2 = 36$

   $2m + 7 = \pm\sqrt{36}$

   $2m + 7 = \pm 6$

   $2m = -7 \pm 6$

   $m = \dfrac{-7 \pm 6}{2}$

   $m = \dfrac{-7+6}{2}$ or $m = \dfrac{-7-6}{2}$

   $m = -\dfrac{1}{2}$ $\qquad$ $m = -\dfrac{13}{2}$

4. $y^2 + 4y - 12 = 0$

   $y^2 + 4y = 12$

   $y^2 + 4y + 4 = 12 + 4$

   $(y+2)^2 = 16$

   $y + 2 = \pm\sqrt{16}$

   $y + 2 = \pm 4$

   $y = -2 \pm 4$

   $y = -2 + 4$ or $y = -2 - 4$

   $y = 2$ $\qquad$ $y = -6$

5. $3a^2 - 12a - 30 = 0$

   $\dfrac{1}{3}(3a^2 - 12a - 30) = \dfrac{1}{3}(0)$

   $a^2 - 4a - 10 = 0$

   $a^2 - 4a = 10$

   $a^2 - 4a + 4 = 10 + 4$

   $(a-2)^2 = 14$

   $a - 2 = \pm\sqrt{14}$

   $y = 2 \pm \sqrt{14}$

6. $\quad 4c^2 + c = -9$

   $\dfrac{1}{4}(4c^2 + c) = \dfrac{1}{4}(-9)$

   $c^2 + \dfrac{1}{4}c = -\dfrac{9}{4}$

   $c^2 + \dfrac{1}{4}c + \dfrac{1}{64} = -\dfrac{9}{4} + \dfrac{1}{64}$

   $\left(c + \dfrac{1}{8}\right)^2 = -\dfrac{143}{64}$

   $\left(c + \dfrac{1}{8}\right)^2 = -\dfrac{143}{64}$

   $c + \dfrac{1}{8} = \sqrt{-\dfrac{143}{64}}$

   $c + \dfrac{1}{8} = \pm\dfrac{i\sqrt{143}}{8}$

   $c = -\dfrac{1}{8} \pm \dfrac{i\sqrt{143}}{8}$

   $c = \dfrac{-1 \pm i\sqrt{143}}{8}$

7. Let $x =$ the length of one side of the patio. Then $x + 6$ is the length of the diagonal.

   $x^2 + x^2 = (x+6)^2$

   $2x^2 = x^2 + 12x + 36$

   $x^2 - 12x = 36$

   $x^2 - 12x + 36 = 36 + 36$

   $(x-6)^2 = 72$

   $x - 6 = \pm\sqrt{72}$

   $x - 6 = \pm 6\sqrt{2}$

   $x = 6 \pm 6\sqrt{2}$

   The length of one side is $\left(6 \pm 6\sqrt{2}\right)$ meters.

8. **a.** $b^2 - 4ac$

   b. If $b^2 - 4ac > 0$, then the quadratic equation will have two distinct real solutions. If $b^2 - 4ac = 0$, then the equation has the single real solution $\dfrac{-b}{2a}$. If $b^2 - 4ac < 0$, then the expression under the radical sign in the quadratic formula is negative. Thus, the equation will have no real solution.

9. $2b^2 - 6b - 11 = 0$

   $b^2 - 4ac = (-6)^2 - 4(2)(-11) = 36 + 88 = 124$

   Since the discriminant is positive, the equation will have two distinct real solutions.

10.
$$6n^2 + n = 15$$
$$6n^2 + n - 15 = 0$$
$$n = \frac{-b \pm \sqrt{b^2 - 4ac}}{2a}$$
$$= \frac{-1 \pm \sqrt{1^2 - 4(6)(-15)}}{2(6)}$$
$$= \frac{-1 \pm \sqrt{1 + 360}}{12}$$
$$= \frac{-1 \pm \sqrt{361}}{12}$$
$$= \frac{-1 \pm 19}{12}$$
$$n = \frac{-1 + 19}{12} \quad \text{or} \quad n = \frac{-1 - 19}{12}$$
$$= \frac{18}{12} \qquad\qquad = \frac{-20}{12}$$
$$= \frac{3}{2} \qquad\qquad = -\frac{5}{3}$$

11.
$$p^2 = -4p + 8$$
$$p^2 + 4p - 8 = 0$$
$$p = \frac{-b \pm \sqrt{b^2 - 4ac}}{2a}$$
$$= \frac{-4 \pm \sqrt{4^2 - 4(1)(-8)}}{2(1)}$$
$$= \frac{-4 \pm \sqrt{16 + 32}}{2}$$
$$= \frac{-4 \pm \sqrt{48}}{2}$$
$$= \frac{-4 \pm 4\sqrt{3}}{2} = -2 \pm 2\sqrt{3}$$

12.
$$3d^2 - 2d + 5 = 0$$
$$d = \frac{-b \pm \sqrt{b^2 - 4ac}}{2a}$$
$$= \frac{-(-2) \pm \sqrt{(-2)^2 - 4(3)(5)}}{2(3)}$$
$$= \frac{2 \pm \sqrt{4 - 60}}{6}$$
$$= \frac{2 \pm \sqrt{-56}}{6}$$
$$= \frac{2 \pm 2i\sqrt{14}}{6} = \frac{1 \pm i\sqrt{14}}{3}$$

13.
$$x = 7 \text{ and } x = -2$$
$$x - 7 = 0 \text{ and } x + 2 = 0$$
$$(x - 7)(x + 2) = 0$$
$$x^2 + 2x - 7x - 14 = 0$$
$$x^2 - 5x - 14 = 0$$

14.
$$x = 2 + \sqrt{5} \text{ and } x = 2 - \sqrt{5}$$
$$x - 2 - \sqrt{5} = 0 \text{ and } x - 2 + \sqrt{5} = 0$$
$$\left(x - 2 - \sqrt{5}\right)\left(x - 2 + \sqrt{5}\right) = 0$$
$$\left[(x - 2) - \sqrt{5}\right]\left[(x - 2) + \sqrt{5}\right] = 0$$
$$(x - 2)^2 - \left(\sqrt{5}\right)^2 = 0$$
$$x^2 - 4x + 4 - 5 = 0$$
$$x^2 - 4x - 1 = 0$$

15. The revenue function is
$$R(n) = n(60 - 0.5n) = 60n - 0.5n^2$$
Let $R(n) = 550$.
$$550 = 60n - 0.5n^2$$
$$0.5n^2 - 60n + 550 = 0$$
$$2(0.5n^2 - 60n + 550) = 2(0)$$
$$n^2 - 120n + 1100 = 0$$
$$n = \frac{-b \pm \sqrt{b^2 - 4ac}}{2a}$$
$$= \frac{-(-120) \pm \sqrt{(-120)^2 - 4(1)(1100)}}{2(1)}$$
$$= \frac{120 \pm \sqrt{14,400 - 4400}}{2}$$
$$= \frac{120 \pm \sqrt{10,000}}{2}$$
$$= \frac{120 \pm 100}{2}$$
$$n = \frac{120 + 100}{2} \quad \text{or} \quad n = \frac{120 - 100}{2}$$
$$= \frac{220}{2} \qquad\qquad = \frac{20}{2}$$
$$= 110 \qquad\qquad = 10$$

We have a restriction of $n \le 20$, so disregard 110. Thus 10 lamps must be sold.

16. $y = x^2 - r^2$, for r

$$r^2 = x^2 - y$$

$$r = \sqrt{x^2 - y}$$

17. $A = \dfrac{1}{3}kx^2$, for r

$$3(A) = 3\left(\dfrac{1}{3}kx^2\right)$$

$$3A = kx^2$$

$$\dfrac{3A}{k} = x^2$$

$$\sqrt{\dfrac{3A}{k}} = x$$

18. $D = \sqrt{x^2 + y^2}$, for y

$$D^2 = x^2 + y^2$$

$$D^2 - x^2 = y^2$$

$$\sqrt{D^2 - x^2} = y$$

19. Let w = the width of the rectangle. Then $2w + 2$ = the length.

$$w(2w + 2) = 60$$

$$2w^2 + 2w - 60 = 0$$

$$\dfrac{1}{2}(2w^2 + 2w - 60) = \dfrac{1}{2}(0)$$

$$w^2 + w - 30 = 0$$

$$w = \dfrac{-1 \pm \sqrt{1^2 - 4(1)(-30)}}{2(1)}$$

$$= \dfrac{-1 \pm \sqrt{1 + 120}}{2}$$

$$= \dfrac{-1 \pm \sqrt{121}}{2}$$

$$= \dfrac{-1 \pm 11}{2}$$

$$w = \dfrac{-1 + 11}{2} \quad \text{or} \quad w = \dfrac{-1 - 11}{2}$$

$$= \dfrac{10}{2} \qquad\qquad = \dfrac{-12}{2}$$

$$= 5 \qquad\qquad\quad = -6$$

The width must be positive, so disregard the negative answer. The width is 5 feet and the length is $2(5) + 2 = 12$ feet.

20. Note: $2000 = 20$ hundreds

$$p(n) = 2n^2 + n - 35; \quad p(n) = 20$$

$$20 = 2n^2 + n - 35$$

$$0 = 2n^2 + n - 55$$

$$0 = 2n^2 + n - 55$$

$$n = \dfrac{-1 \pm \sqrt{1^2 - 4(2)(-55)}}{2(2)}$$

$$= \dfrac{-1 \pm \sqrt{1 + 440}}{4}$$

$$= \dfrac{-1 \pm \sqrt{441}}{4}$$

$$= \dfrac{-1 \pm 21}{4}$$

$$n = \dfrac{-1 + 21}{4} \quad \text{or} \quad n = \dfrac{-1 - 21}{4}$$

$$= \dfrac{20}{4} \qquad\qquad = \dfrac{-22}{4}$$

$$= 5 \qquad\qquad\quad = -\dfrac{11}{2}$$

Disregard the negative answer. Thus, 5 clocks must be sold.

Exercise Set 12.4

1. A given equation can be expressed as an equation in quadratic form if the equation can be written in the form $au^2 + bu + c = 0$.

3. Let $u = x^2$. Then

$$3x^4 - 5x^2 + 1 = 0$$

$$3(x^2)^2 - 5x^2 + 1 = 0$$

$$3u^2 - 5u + 1 = 0$$

5. Let $u = z^{-1}$. Then

$$z^{-2} - z^{-1} = 56$$

$$(z^{-1})^2 - z^{-1} = 56$$

$$u^2 - u = 56$$

7. Let $u = x^2$.

$$x^4 + x^2 - 6 = (x^2)^2 + x^2 - 6$$

$$= u^2 + u - 6$$

$$= (u + 3)(u - 2)$$

Back substitute x^2 for u.

$$= (x^2 + 3)(x^2 - 2)$$

9. Let $u = x^2$.
$$x^4 + 5x^2 + 6 = \left(x^2\right)^2 + 5x^2 + 6$$
$$= u^2 + 5u + 6$$
$$= (u+2)(u+3)$$
Back substitute x^2 for u.
$$= (x^2 + 2)(x^2 + 3)$$

11. Let $u = a^2$.
$$6a^4 + 5a^2 - 25 = 6\left(a^2\right)^2 + 5a^2 - 25$$
$$= 6u^2 + 5u - 25$$
$$= (2u+5)(3u-5)$$
Back substitute a^2 for u.
$$= \left(2a^2 + 5\right)\left(3a^2 - 5\right)$$

13. Let $u = x+1$.
$$4(x+1)^2 + 8(x+1) + 3$$
$$= 4u^2 + 8u + 3$$
$$= (2u+3)(2u+1)$$
Back substitute $x+1$ for u.
$$= [2(x+1)+3][2(x+1)+1]$$
$$= (2x+2+3)(2x+2+1)$$
$$= (2x+5)(2x+3)$$

15. Let $u = a+2$.
$$6(a+2)^2 - 7(a+2) - 5$$
$$= 6u^2 - 7u - 5$$
$$= (2u+1)(3u-5)$$
Back substitute $a+2$ for u.
$$= [2(a+2)+1][3(a+2)-5]$$
$$= (2a+4+1)(3a+6-5)$$
$$= (2a+5)(3a+1)$$

17. Let $u = ab$.
$$a^2b^2 + 8ab + 15 = (ab)^2 + 8ab + 15$$
$$= u^2 + 8u + 15$$
$$= (u+3)(u+5)$$
Back substitute ab for u.
$$= (ab+3)(ab+5)$$

19. Let $u = xy$.
$$3x^2y^2 - 2xy - 5 = 3(xy)^2 - 2xy - 5$$
$$= 3u^2 - 2u - 5$$
$$= (3u-5)(u+1)$$
Back substitute xy for u.
$$= (3xy-5)(xy+1)$$

21. Factor out $(5-a)$.
$$2a^2(5-a) - 7a(5-a) + 5(5-a)$$
$$= (5-a)(2a^2 - 7a + 5)$$
$$= (5-a)(2a-5)(a-1)$$

23. Factor out $(x-3)$.
$$2x^2(x-3) + 7x(x-3) + 6(x-3)$$
$$= (x-3)(2x^2 + 7x + 6)$$
$$= (x-3)(2x+3)(x+2)$$

25. Let $u = y^2$.
$$y^4 + 13y^2 + 30 = \left(y^2\right)^2 + 13y^2 + 30$$
$$= u^2 + 13u + 30$$
$$= (u+10)(u+3)$$
Back substitute y^2 for u.
$$= \left(y^2 + 10\right)\left(y^2 + 3\right)$$

27. Factor out $(x+3)$.
$$x^2(x+3) + 3x(x+3) + 2(x+3)$$
$$= (x+3)(x^2 + 3x + 2)$$
$$= (x+3)(x+1)(x+2)$$

29. Factor out a^3b^2.
$$5a^5b^2 - 8a^4b^3 + 3a^3b^4$$
$$= a^3b^2\left(5a^2 - 8ab + 3b^2\right)$$
$$= a^3b^2(5a-3b)(a-b)$$

31.
$$x^4 - 10x^2 + 9 = 0$$
$$\left(x^2\right)^2 - 10x^2 + 9 = 0$$
Let $u = x^2$.
$$u^2 - 10u + 9 = 0$$
$$(u-9)(u-1) = 0$$
$$u - 9 = 0 \quad \text{or} \quad u - 1 = 0$$
$$u = 9 \qquad\qquad u = 1$$
Substitute x^2 for u.
$$x^2 = 9 \qquad \text{or} \quad x^2 = 1$$
$$x = \pm\sqrt{9} = \pm 3 \qquad x = \pm\sqrt{1} = \pm 1$$
The solutions are 3, –3, 1, and –1.

33. $x^4 + 17x^2 + 16 = 0$

$(x^2)^2 + 17x^2 + 16 = 0$

Let $u = x^2$.

$u^2 + 17u + 16 = 0$

$(u + 16)(u + 1) = 0$

$u + 16 = 0$ or $u + 1 = 0$

     $u = -16$ $u = -1$

Substitute x^2 for u.

$x^2 = -16$ or $x^2 = -1$

  $x = \pm\sqrt{-16} = \pm 4i$ $x = \pm\sqrt{-1} = \pm i$

The solutions are $4i$, $-4i$, i, and $-i$.

35. $x^4 - 13x^2 + 36 = 0$

$(x^2)^2 - 13x^2 + 36 = 0$

Let $u = x^2$.

$u^2 - 13u + 36 = 0$

$(u - 9)(u - 4) = 0$

$u - 9 = 0$ or $u - 4 = 0$

   $u = 9$ $u = 4$

Substitute x^2 for u.

$x^2 = 9$ or $x^2 = 4$

  $x = \pm\sqrt{9} = \pm 3$ $x = \pm\sqrt{4} = \pm 2$

The solutions are 3, -3, 2, and -2.

37. $a^4 - 7a^2 + 12 = 0$

$(a^2)^2 - 7a^2 + 12 = 0$

Let $u = a^2$.

$u^2 - 7u + 12 = 0$

$(u - 4)(u - 3) = 0$

$u - 4 = 0$ or $u - 3 = 0$

   $u = 4$ $u = 3$

Substitute a^2 for u.

$a^2 = 4$ or $a^2 = 3$

  $a = \pm\sqrt{4} = \pm 2$ $a = \pm\sqrt{3}$

The solutions are 2, -2, $\sqrt{3}$, and $-\sqrt{3}$.

39. $4x^4 - 17x^2 + 4 = 0$

$4(x^2)^2 - 17x^2 + 4 = 0$

Let $u = x^2$.

$4u^2 - 17u + 4 = 0$

$(4u - 1)(u - 4) = 0$

$4u - 1 = 0$ or $u - 4 = 0$

    $u = \dfrac{1}{4}$ $u = 4$

Substitute x^2 for u.

$x^2 = \dfrac{1}{4}$ or $x^2 = 4$

$x = \pm\sqrt{\dfrac{1}{4}} = \pm\dfrac{1}{2}$ $x = \pm\sqrt{4} = \pm 4$

The solutions are $\dfrac{1}{2}$, $-\dfrac{1}{2}$, 2, and -2.

41. $r^4 - 8r^2 = -15$

$r^4 - 8r^2 + 15 = 0$

$(r^2)^2 - 8r^2 + 15 = 0$

Let $u = r^2$.

$u^2 - 8u + 15 = 0$

$(u - 3)(u - 5) = 0$

$u - 3 = 0$ or $u - 5 = 0$

   $u = 3$ $u = 5$

Substitute r^2 for u.

$r^2 = 3$ or $r^2 = 5$

  $r = \pm\sqrt{3}$ $r = \pm\sqrt{5}$

The solutions are $\sqrt{3}, -\sqrt{3}, \sqrt{5}$, and $-\sqrt{5}$.

43. $z^4 - 7z^2 = 18$

$z^4 - 7z^2 - 18 = 0$

$(z^2)^2 - 7z^2 - 18 = 0$

Let $u = z$.

$u^2 - 7u - 18 = 0$

$(u - 9)(u + 2) = 0$

$u - 9 = 0$ or $u + 2 = 0$

   $u = 9$ $u = -2$

Substitute z^2 for u.

$z^2 = 9$ or $z^2 = -2$

  $z = \pm 3$ $z = \pm\sqrt{-2} = \pm i\sqrt{2}$

The solutions are $3, -3, i\sqrt{2}$, and $-i\sqrt{2}$.

45. $-c^4 = 4c^2 - 5$

$0 = c^4 + 4c^2 - 5$

$0 = \left(c^2\right)^2 + 4c^2 - 5$

Let $u = c$.

$u^2 + 4u - 5 = 0$

$(u-1)(u+5) = 0$

$u - 1 = 0$ or $u + 5 = 0$

$u = 1$ $u = -5$

Substitute c^2 for u.

$c^2 = 1$ or $c^2 = -5$

$c = \pm 1$ $c = \pm\sqrt{-5} = \pm i\sqrt{5}$

The solutions are $1, -1, i\sqrt{5}$, and $-i\sqrt{5}$.

47. $\sqrt{x} = 2x - 6$

$0 = 2x - \sqrt{x} - 6$

$0 = 2\left(x^{1/2}\right)^2 - x^{1/2} - 6$

Let $u = x^{1/2}$.

$2u^2 - u - 6 = 0$

$(2u+3)(u-2) = 0$

$2u + 3 = 0$ or $u - 2 = 0$

$u = -\dfrac{3}{2}$ $u = 2$

Substitute $x^{1/2}$ for u.

$x^{1/2} = -\dfrac{3}{2}$ or $x^{1/2} = 2$

Not real $x = 2^2 = 4$

The solution is 4.

49. $x - \sqrt{x} = 6$

$x - \sqrt{x} - 6 = 0$

$\left(x^{1/2}\right)^2 - x^{1/2} - 6 = 0$

Let $u = x^{1/2}$.

$u^2 - u - 6 = 0$

$(u+2)(u-3) = 0$

$u + 2 = 0$ or $u - 3 = 0$

$u = -2$ or $u = 3$

Substitute $x^{1/2}$ for u.

$x^{1/2} = -2$ or $x^{1/2} = 3$

Not real $x = 3^2 = 9$

The solution is 9.

51. $9x + 3\sqrt{x} = 2$

$9x + 3\sqrt{x} - 2 = 0$

$9\left(x^{1/2}\right)^2 + 3x^{1/2} - 2 = 0$

Let $u = x^{1/2}$.

$9u^2 + 3u - 2 = 0$

$(3u+2)(3u-1) = 0$

$3u + 2 = 0$ or $3u - 1 = 0$

$u = -\dfrac{2}{3}$ $u = \dfrac{1}{3}$

Substitute $x^{1/2}$ for u.

$x^{1/2} = -\dfrac{2}{3}$ or $x^{1/2} = \dfrac{1}{3}$

Not real $x = \left(\dfrac{1}{3}\right)^2 = \dfrac{1}{9}$

The solution is $\dfrac{1}{9}$.

53. $(x+3)^2 + 2(x+3) = 24$

$(x+3)^2 + 2(x+3) - 24 = 0$

Let $u = x + 3$.

$u^2 + 2u - 24 = 0$

$(u-4)(u+6) = 0$

$u - 4 = 0$ or $u + 6 = 0$

$u = 4$ $u = -6$

Substitute $x + 3$ for u.

$x + 3 = 4$ or $x + 3 = -6$

$x = 1$ $x = -9$

The solutions are 1 and -9.

55. $6(a-2)^2 = -19(a-2) - 10$

$6(a-2)^2 + 19(a-2) + 10 = 0$

Let $u = a - 2$.

$6u^2 + 19u + 10 = 0$

$(3u+2)(2u+5) = 0$

$3u + 2 = 0$ or $2u + 5 = 0$

$u = -\dfrac{2}{3}$ $u = -\dfrac{5}{2}$

Substitute $a - 2$ for u.

$a - 2 = -\dfrac{2}{3}$ or $a - 2 = -\dfrac{5}{2}$

$a = \dfrac{4}{3}$ $a = -\dfrac{1}{2}$

The solutions are $\dfrac{4}{3}$ and $-\dfrac{1}{2}$.

57. $(x^2-3)^2-(x^2-3)-6=0$

Let $u=x^2-3$.

$u^2-u-6=0$

$(u+2)(u-3)=0$

$u+2=0$ or $u-3=0$

$u=-2$ $u=3$

Substitute x^2-1 for u.

$x^2-3=-2$ or $x^2-3=3$

$x^2=1$ $x^2=6$

$x=\pm\sqrt{1}=\pm1$ $x=\pm\sqrt{6}$

The solutions are $1,-1,\sqrt{6}$, and $-\sqrt{6}$.

59. $2(b+3)^2+5(b+3)-3=0$

Let $u=b+3$.

$2u^2+5u-3=0$

$(u+3)(2u-1)=0$

$u+3=0$ or $2u-1=0$

$u=-3$ $u=\dfrac{1}{2}$

Substitute $b+3$ for u.

$b+3=-3$ or $b+3=\dfrac{1}{2}$

$b=-6$ $b=-\dfrac{5}{2}$

The solutions are -6 and $-\dfrac{5}{2}$.

61. $18(x^2-5)^2+27(x^2-5)+10=0$

Let $u=x^2-5$.

$18u^2+27u+10=0$

$(3u+2)(6u+5)=0$

$3u+2=0$ or $6u+5=0$

$u=-\dfrac{2}{3}$ $u=-\dfrac{5}{6}$

Substitute x^2-5 for u.

or $x^2-5=-\dfrac{5}{6}$

$x^2-5=-\dfrac{2}{3}$ or $x^2-5=-\dfrac{5}{6}$

$x^2=\dfrac{13}{3}$ $x^2=\dfrac{25}{6}$

$x=\pm\sqrt{\dfrac{13}{3}}$ $x=\pm\sqrt{\dfrac{25}{6}}$

$=\pm\dfrac{\sqrt{13}}{\sqrt{3}}\cdot\dfrac{\sqrt{3}}{\sqrt{3}}$ $=\pm\dfrac{5}{\sqrt{6}}\cdot\dfrac{\sqrt{6}}{\sqrt{6}}$

$=\pm\dfrac{\sqrt{39}}{3}$ $=\pm\dfrac{5\sqrt{6}}{6}$

The solutions are $\dfrac{\sqrt{39}}{3},-\dfrac{\sqrt{39}}{3},\dfrac{5\sqrt{6}}{6}$, and $-\dfrac{5\sqrt{6}}{6}$.

63. $a^{-2}+4a^{-1}+4=0$

$\left(a^{-1}\right)^2+4\left(a^{-1}\right)+4=0$

Let $u=a^{-1}$.

$u^2+4u+4=0$

$(u+2)^2=0$

$u+2=0$

$u=-2$

Substitute a^{-1} for u.

$a^{-1}=-2$

$a=-\dfrac{1}{2}$

The solution is $-\dfrac{1}{2}$.

65. $12b^{-2}-7b^{-1}+1=0$

$12\left(b^{-1}\right)^2-7\left(b^{-1}\right)+1=0$

Let $u=b^{-1}$.

$12u^2-7u+1=0$

$(4u-1)(3u-1)=0$

$4u-1=0$ or $3u-1=0$

$u=\dfrac{1}{4}$ $u=\dfrac{1}{3}$

Substitute b^{-1} for u.

$b^{-1}=\dfrac{1}{4}$ or $b^{-1}=\dfrac{1}{3}$

$b=4$ $b=3$

The solutions are 4 and 3.

67.
$$2b^{-2} = 7b^{-1} - 3$$
$$2b^{-2} - 7b^{-1} + 3 = 0$$
$$2(b^{-1})^2 - 7(b^{-1}) + 3 = 0$$
Let $u = b^{-1}$.
$$2u^2 - 7u + 3 = 0$$
$$(2u - 1)(u - 3) = 0$$
$$2u - 1 = 0 \quad \text{or} \quad u - 3 = 0$$
$$u = \frac{1}{2} \qquad u = 3$$
Substitute b^{-1} for u.
$$b^{-1} = \frac{1}{2} \quad \text{or} \quad b^{-1} = 3$$
$$b = 2 \qquad b = \frac{1}{3}$$

The solutions are 2 and $\frac{1}{3}$.

69.
$$x^{-2} + 9x^{-1} = 10$$
$$x^{-2} + 9x^{-1} - 10 = 0$$
$$(x^{-1})^2 + 9(x^{-1}) - 10 = 0$$
Let $u = x^{-1}$.
$$u^2 + 9u - 10 = 0$$
$$(u + 10)(u - 1) = 0$$
$$u + 10 = 0 \quad \text{or} \quad u - 1 = 0$$
$$u = -10 \qquad u = 1$$
Substitute x^{-1} for u.
$$x^{-1} = -10 \quad \text{or} \quad x^{-1} = 1$$
$$x = -\frac{1}{10} \qquad x = 1$$

The solutions are $-\frac{1}{10}$ and 1.

71.
$$x^{-2} = 4x^{-1} + 12$$
$$x^{-2} - 4x^{-1} - 12 = 0$$
$$(x^{-1})^2 - 4(x^{-1}) - 12 = 0$$
Let $u = x^{-1}$.
$$u^2 - 4u - 12 = 0$$
$$(u + 2)(u - 6) = 0$$
$$u + 2 = 0 \quad \text{or} \quad u - 6 = 0$$
$$u = -2 \qquad u = 6$$
Substitute x^{-1} for u.
$$x^{-1} = -2 \quad \text{or} \quad x^{-1} = 6$$
$$x = -\frac{1}{2} \qquad x = \frac{1}{6}$$

The solutions are $-\frac{1}{2}$ and $\frac{1}{6}$.

73.
$$x^{2/3} - 4x^{1/3} = -3$$
$$(x^{1/3})^2 - 4x^{1/3} + 3 = 0$$
Let $u = x^{1/3}$.
$$u^2 - 4u + 3 = 0$$
$$(u - 1)(u - 3) = 0$$
$$u - 1 = 0 \quad \text{or} \quad u - 3 = 0$$
$$u = 1 \qquad u = 3$$
Substitute $x^{1/3}$ for u.
$$x^{1/3} = 1 \qquad \text{or} \quad x^{1/3} = 3$$
$$x = 1^3 = 1 \qquad x = 3^3 = 27$$
The solutions are 1 and 27.

75.
$$b^{2/3} - 9b^{1/3} + 18 = 0$$
$$(b^{1/3})^2 - 9b^{1/3} + 18 = 0$$
Let $u = b^{1/3}$.
$$u^2 - 9u + 18 = 0$$
$$(u - 6)(u - 3) = 0$$
$$u - 6 = 0 \quad \text{or} \quad u - 3 = 0$$
$$u = 6 \qquad u = 3$$
Substitute $b^{1/3}$ for u.
$$b^{1/3} = 6 \qquad \text{or} \quad b^{1/3} = 3$$
$$b = 6^3 = 216 \qquad b = 3^3 = 27$$
The solutions are 216 and 27.

77.
$$-2a - 5a^{1/2} + 3 = 0$$
$$-2(a^{1/2})^2 - 5a^{1/2} + 3 = 0$$
Let $u = a^{1/2}$.
$$-2u^2 - 5u + 3 = 0$$
$$2u^2 + 5u - 3 = 0$$
$$(2u - 1)(u + 3) = 0$$
$$2u - 1 = 0 \quad \text{or} \quad u + 3 = 0$$
$$u = \frac{1}{2} \qquad u = -3$$
Substitute $a^{1/2}$ for u.
$$a^{1/2} = 2 \qquad \text{or} \quad a^{1/2} = -3$$
$$a = \left(\frac{1}{2}\right)^2 = \frac{1}{4} \qquad \text{not real}$$

The solution is $\frac{1}{4}$.

79. $c^{2/5} + 3c^{1/5} + 2 = 0$

$(c^{1/5})^2 + 3c^{1/5} + 2 = 0$

Let $u = c^{1/5}$.

$u^2 + 3u + 2 = 0$

$(u+2)(u+1) = 0$

$u+2 = 0$ or $u+1 = 0$

$u = -2$ $u = -1$

Substitute $c^{1/5}$ for u.

$c^{1/5} = -2$ or $c^{1/5} = -1$

$c = (-2)^5 = -32$ $c = (-1)^5 = -1$

The solutions are -32 and -1.

81. $f(x) = x - 5\sqrt{x} + 4$, $f(x) = 0$

$0 = (x^{1/2})^2 - 5x^{1/2} + 4$

Let $u = x^{1/2}$.

$0 = u^2 - 5u + 4$

$0 = (u-1)(u-4)$

$u-1 = 0$ or $u-4 = 0$

$u = 1$ $u = 4$

Substitute $x^{1/2}$ for u.

$x^{1/2} = 1$ or $x^{1/2} = 4$

$x = 1^2 = 1$ $x = 4^2 = 16$

The x-intercepts are $(1, 0)$ and $(16, 0)$.

83. $h(x) = x + 14\sqrt{x} + 45$, $h(x) = 0$

$0 = (x^{1/2})^2 + 14x^{1/2} + 45$

Let $u = x^{1/2}$.

$0 = u^2 + 14u + 45$

$0 = (u+9)(u+5)$

$u+9 = 0$ or $u+5 = 0$

$u = -9$ $u = -5$

Substitute $x^{1/2}$ for u.

$x^{1/2} = -9$ or $x^{1/2} = -5$

There are no values of x for which $x^{1/2} = -9$ or $x^{1/2} = -5$. There are no x-intercepts.

85. $p(x) = 4x^{-2} - 19x^{-1} - 5$, $p(x) = 0$

$0 = 4(x^{-1})^2 - 19x^{-1} - 5$

Let $u = x^{-1}$.

$0 = 4u^2 - 19u - 5$

$0 = (4u+1)(u-5)$

$4u+1 = 0$ or $u-5 = 0$

$u = -\dfrac{1}{4}$ $u = 5$

Substitute x^{-1} for u.

$x^{-1} = -\dfrac{1}{4}$ or $x^{-1} = 5$

$x = -4$ $x = \dfrac{1}{5}$

The x-intercepts are $(-4, 0)$ and $\left(\dfrac{1}{5}, 0\right)$.

87. $f(x) = x^{2/3} - x^{1/3} - 6$, $f(x) = 0$

$0 = (x^{1/3})^2 - x^{1/3} - 6$

Let $u = x^{1/3}$.

$0 = u^2 - u - 6$

$0 = (u-3)(u+2)$

$u-3 = 0$ or $u+2 = 0$

$u = 3$ $u = -2$

Substitute $x^{1/3}$ for u.

$x^{1/3} = 3^3 = 27$ or $x^{1/3} = (-2)^3 = -8$

The x-intercepts are $(27, 0)$ and $(-8, 0)$.

89. $g(x) = (x^2 - 3x)^2 + 2(x^2 - 3x) - 24$, $g(x) = 0$

Let $u = x^2 - 3x$.

$0 = u^2 + 2u - 24$

$0 = (u+6)(u-4)$

$u+6 = 0$ or $u-4 = 0$

$u = -6$ $u = 4$

Substitute $(x^2 - 3x)$ for u.

$x^2 - 3x = -6$ or $x^2 - 3x = 4$

$x^2 - 3x + 6 = 0$ $x^2 - 3x - 4 = 0$

$(x-4)(x+1) = 0$

$x = 4$ or $x = -1$

There are no x-intercepts for $x^2 - 3x + 6 = 0$ since $b^2 - 4ac = (-3)^2 - 4(1)(6) = -15$ (that is, since the discriminant is negative). The x-intercepts are $(4, 0)$ or $(-1, 0)$.

91. $f(x) = x^4 - 29x + 100,\ f(x) = 0$

$0 = (x^2)^2 - 29x^2 + 100$

Let $u = x^2$.

$0 = u^2 - 29u + 100$

$0 = (u - 25)(u - 4)$

$u - 25 = 0 \quad$ or $\quad u - 4 = 0$

$\qquad u = 25 \qquad\qquad u = 4$

Substitute x^2 for u.

$x^2 = 25 \qquad$ or $\quad x^2 = 4$

$x = \pm\sqrt{25} = \pm 5 \qquad x = \pm\sqrt{4} = \pm 2$

The x-intercepts are $(5, 0)$, $(-5, 0)$, $(2, 0)$, and $(-2, 0)$.

93. When solving an equation of the form $ax^4 + bx^2 + c = 0$, let $u = x^2$.

95. When solving an equation of the form $ax^{-2} + bx^{-1} + c = 0$, let $u = x^{-1}$.

97. If the solutions are ± 2 and ± 1, the factors must be $(x - 2), (x + 2), (x - 1),$ and $(x + 1)$.

$(x - 2)(x + 2)(x - 1)(x + 1) = 0$

$\quad (x^2 - 4)(x^2 - 1) = 0$

$\qquad x^4 - 5x^2 + 4 = 0$

99. If the solutions are $\pm\sqrt{2}$ and $\pm\sqrt{5}$, the factors must be $(x + \sqrt{2}), (x - \sqrt{2}), (x + \sqrt{5}), (x - \sqrt{5})$.

$(x + \sqrt{2})(x - \sqrt{2})(x + \sqrt{5})(x - \sqrt{5}) = 0$

$\qquad (x^2 - 2)(x^2 - 5) = 0$

$\qquad\qquad x^4 - 7x^2 + 10 = 0$

101. No. An equation of the form $ax^4 + bx^2 + c = 0$ can have no imaginary solutions, two imaginary solutions, or four imaginary solutions.

103. a. $\qquad \dfrac{3}{x^2} - \dfrac{3}{x} = 60 \quad$ The LCD is x^2

$x^2\left(\dfrac{3}{x^2}\right) - x^2\left(\dfrac{3}{x}\right) = x^2(60)$

$\qquad\qquad 3 - 3x = 60x^2$

$\qquad\qquad 0 = 60x^2 + 3x - 3$

$\qquad\qquad 0 = 3(20x^2 + x - 1)$

$\qquad\qquad 0 = 3(5x - 1)(4x + 1)$

$5x - 1 = 0 \quad$ or $\quad 4x + 1 = 0$

$\quad x = \dfrac{1}{5} \qquad\qquad x = -\dfrac{1}{4}$

The solutions are $\dfrac{1}{5}$ and $-\dfrac{1}{4}$.

b. $\qquad\qquad \dfrac{3}{x^2} - \dfrac{3}{x} = 60$

$\qquad\qquad 3x^{-2} - 3x^{-1} = 60$

$\qquad 3(x^{-1})^2 - 3x^{-1} - 60 = 0$

Let $u = x^{-1}$.

$\qquad 3u^2 - 3u - 60 = 0$

$\qquad 3(u^2 - u - 20) = 0$

$\qquad 3(u - 5)(u + 4) = 0$

$u - 5 = 0 \quad$ or $\quad u + 4 = 0$

$\qquad u = 5 \qquad\qquad u = -4$

Substitute x^{-1} for u.

$x^{-1} = 5 \quad$ or $\quad x^{-1} = -4$

$\quad x = \dfrac{1}{5} \qquad\qquad x = -\dfrac{1}{4}$

The solutions are $\dfrac{1}{5}$ and $-\dfrac{1}{4}$.

105. $\qquad 15(r + 2) + 22 = -\dfrac{8}{r + 2}$

$15(r + 2)(r + 2) + 22(r + 2) = -\dfrac{8}{r + 2}(r + 2)$

$\qquad 15(r + 2)^2 + 22(r + 2) = -8$

$\qquad 15(r + 2)^2 + 22(r + 2) + 8 = 0$

Let $u = r + 2$.

$\qquad 15u^2 + 22u + 8 = 0$

$\qquad (5u + 4)(3u + 2) = 0$

$5u + 4 = 0 \quad$ or $\quad 3u + 2 = 0$

$\quad u = -\dfrac{4}{5} \qquad\qquad u = -\dfrac{2}{3}$

Substitute $r + 2$ for u.

$r + 2 = -\dfrac{4}{5} \quad$ or $\quad r + 2 = -\dfrac{2}{3}$

$\quad r = -\dfrac{14}{5} \qquad\qquad r = -\dfrac{8}{3}$

The solutions are $-\dfrac{14}{5}$ and $-\dfrac{8}{3}$.

107.
$$4-(x-1)^{-1}=3(x-1)^{-2}$$
$$3(x-1)^{-2}+(x-1)^{-1}-4=0$$
$$3\left[(x-1)^{-1}\right]^{2}+(x-1)^{-1}-4=0$$
Let $u=(x-1)^{-1}$
$$3u^{2}+u-4=0$$
$$(3u+4)(u-1)=0$$
$$3u+4=0 \quad \text{or} \quad u-1=0$$
$$u=-\frac{4}{3} \qquad u=1$$
Substitute $(x-1)^{-1}$ for u.
$$(x-1)^{-1}=-\frac{4}{3} \quad \text{or} \quad (x-1)^{-1}=1$$
$$x-1=-\frac{3}{4} \qquad x-1=1$$
$$x=\frac{1}{4} \qquad x=2$$
The solutions are $\frac{1}{4}$ and 2.

109.
$$x^{6}-9x^{3}+8=0$$
$$\left(x^{3}\right)^{2}-9x^{3}+8=0$$
Let $u=x^{3}$.
$$u^{2}-9u+8=0$$
$$(u-8)(u-1)=0$$
$$u-8=0 \quad \text{or} \quad u-1=0$$
$$u=8 \qquad u=1$$
Substitute x^{3} for u.
$$x^{3}=8 \qquad \text{or} \quad x^{3}=1$$
$$x=\sqrt[3]{8}=2 \qquad x=\sqrt[3]{1}=1$$
The solutions are 2 and 1.

111. $(x^{2}+2x-2)^{2}-7(x^{2}+2x-2)+6=0$
Let $u=x^{2}+2x-2$.
$$u^{2}-7u+6=0$$
$$(u-6)(u-1)=0$$
$$u=6 \quad \text{or} \quad u=1$$
Substitute $x^{2}+2x-2$ for u.
$$x^{2}+2x-2=6 \quad \text{or} \quad x^{2}+2x-2=1$$
$$x^{2}+2x-8=0 \qquad x^{2}+2x-3=0$$
$$(x+4)(x-2)=0 \qquad (x+3)(x-1)=0$$
$$x=-4 \text{ or } x=2 \qquad x=-3 \text{ or } x=1$$
The solutions are –4, 2, –3, and 1.

113.
$$2n^{4}-6n^{2}-3=0$$
$$2\left(n^{2}\right)^{2}-6n^{2}-3=0$$
Let $u=n^{2}$.
$$2u^{2}-6u-3=0$$
$$u=\frac{6\pm\sqrt{(-6)^{2}-4(2)(-3)}}{2(2)}$$
$$=\frac{6\pm\sqrt{60}}{4}=\frac{6\pm2\sqrt{15}}{4}=\frac{3\pm\sqrt{15}}{2}$$
Substitute n^{2} for u.
$$n^{2}=\frac{3\pm\sqrt{15}}{2}$$
$$n=\pm\sqrt{\frac{3\pm\sqrt{15}}{2}}$$

115.
$$\frac{4}{5}-\left(\frac{3}{4}-\frac{2}{3}\right)=\frac{4}{5}-\left(\frac{9}{12}-\frac{8}{12}\right)$$
$$=\frac{4}{5}-\frac{1}{12}$$
$$=\frac{48}{60}-\frac{5}{60}$$
$$=\frac{43}{60}$$

116. $3(x+2)-2(3x+3)=-3$
$$3x+6-6x-6=-3$$
$$-3x=-3$$
$$x=1$$

117. $y=(x-3)^{2}$
Domain : $\mathbb{R}$
Range: $\left\{y\,|\,y\geq0\right\}$

118. $\sqrt[3]{16x^{3}y^{6}}=\sqrt[3]{8\cdot2\cdot x^{3}\left(y^{2}\right)^{3}}=2xy^{2}\sqrt[3]{2}$

119. $\sqrt{75}+\sqrt{48}=\sqrt{25\cdot3}+\sqrt{16\cdot3}$
$$=5\sqrt{3}+4\sqrt{3}$$
$$=9\sqrt{3}$$

Exercise Set 12.5

1. The graph of a quadratic equation is called a parabola.

3. The axis of symmetry of a parabola is the line where, if the graph is folded, the two sides overlap.

5. For $f(x) = ax^2 + bx + c$, the vertex of the graph is $\left(-\dfrac{b}{2a}, \dfrac{4ac - b^2}{4}\right)$.

7. **a.** For $f(x) = ax^2 + bx + c$, $f(x)$ will have a minimum if $a > 0$ since the graph opens upward.

   b. For $f(x) = ax^2 + bx + c$, $f(x)$ will have a maximum if $a < 0$ since the graph opens downward.

9. To find the y-intercepts of the graph of a quadratic function, set $x = 0$ and solve for y.

11. **a.** For $f(x) = ax^2$, the general shape of $f(x)$ if $a > 0$ is

   b. For $f(x) = ax^2$, the general shape of $f(x)$ if $a < 0$ is

13. Since $a = 3$ is greater than 0, the graph opens upward, and therefore has a minimum value.

15. $P(x) = x^2 - 6x + 4$

   $P(2) = (2)^2 - 6(2) + 4 = 4 - 12 + 4 = -8 + 4 = -4$

17. $P(x) = 2x^2 - 3x - 6$

   $P\left(\dfrac{1}{2}\right) = 2\left(\dfrac{1}{2}\right)^2 - 3\left(\dfrac{1}{2}\right) - 6$

   $= 2\left(\dfrac{1}{4}\right) - 3\left(\dfrac{1}{2}\right) - 6$

   $= \dfrac{1}{2} - \dfrac{3}{2} - \dfrac{12}{2}$

   $= -\dfrac{14}{2}$

   $= -7$

19. $P(x) = 0.2x^3 + 1.6x^2 - 2.3$

   $P(0.4) = 0.2(0.4)^3 + 1.6(0.4)^2 - 2.3$

   $= 0.2(0.064) + 1.6(0.16) - 2.3$

   $= 0.0128 + 0.256 - 2.3$

   $= 0.2688 - 2.3$

   $= -2.0312$

21. $f(x) = x^2 + 8x + 15$

   a. Since $a = 1$, the parabola opens upward.

   b. $y = (0)^2 + 8(0) + 15 = 15$
   The y-intercept is $(0, 15)$.

   c. $x = -\dfrac{b}{2a} = -\dfrac{8}{2(1)} = -\dfrac{8}{2} = -4$

   $y = \dfrac{4ac - b^2}{4a}$

   $= \dfrac{4(1)(15) - 8^2}{4(1)}$

   $= \dfrac{60 - 64}{4}$

   $= \dfrac{-4}{4}$

   $= -1$
   The vertex is $(-4, -1)$.

   d. $x^2 + 8x + 15 = 0$
   $(x + 5)(x + 3) = 0$
   $x + 5 = 0 \quad$ or $\quad x + 3 = 0$
   $\quad x = -5 \qquad\qquad x = -3$
   The x-intercepts are $(-5, 0)$ and $(-3, 0)$.

   e.

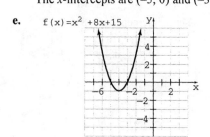

23. $f(x) = x^2 - 4x + 3$

    a. Since $a = 1$, the parabola opens upward.

    b. $y = 0^2 - 4(0) + 3 = 3$
        The y-intercept is $(0, 3)$.

    c. $x = -\dfrac{b}{2a} = -\dfrac{-4}{2(1)} = \dfrac{4}{2} = 2$

        $y = \dfrac{4ac - b^2}{4a}$

        $= \dfrac{4(1)(3) - (-4)^2}{4(1)} = \dfrac{12 - 16}{4} = \dfrac{-4}{4} = -1$

        The vertex is $(2, -1)$.

    d. $x^2 - 4x + 3 = 0$
        $(x - 3)(x - 1) = 0$
        $x - 3 = 0 \quad$ or $\quad x - 1 = 0$
        $x = 3 \qquad\qquad x = 1$
        The x-intercepts are $(3, 0)$ and $(1, 0)$.

    e.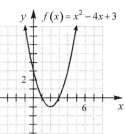

25. $f(x) = -x^2 - 2x + 8$

    a. Since $a = -1$, the parabola opens downward.

    b. $y = -(0)^2 - 2(0) + 8 = 8$
        The y-intercept is $(0, 8)$.

    c. $x = -\dfrac{b}{2a} = -\dfrac{-2}{2(-1)} = -\dfrac{2}{2} = -1$

        $y = \dfrac{4ac - b^2}{4a}$

        $= \dfrac{4(-1)(8) - (-2)^2}{4(-1)} = \dfrac{-32 - 4}{-4} = \dfrac{-36}{-4} = 9$

        The vertex is $(-1, 9)$.

    d. $-x^2 - 2x + 8 = 0$
        $x^2 + 2x - 8 = 0$
        $(x + 4)(x - 2) = 0$
        $x + 4 = 0 \quad$ or $\quad x - 2 = 0$
        $x = -4 \qquad\qquad x = 2$
        The x-intercepts are $(-4, 0)$ and $(2, 0)$.

    e.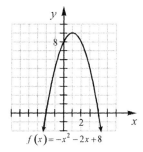

        $f(x) = -x^2 - 2x + 8$

27. $g(x) = -x^2 + 4x + 5$

    a. Since $a = -1$, the parabola opens downward.

    b. $y = -(0)^2 + 4(0) + 5 = 8$
        The y-intercept is $(0, 5)$.

    c. $x = -\dfrac{b}{2a} = -\dfrac{4}{2(-1)} = \dfrac{4}{2} = 2$

        $y = \dfrac{4ac - b^2}{4a}$

        $= \dfrac{4(-1)(5) - (4)^2}{4(-1)}$

        $= \dfrac{-20 - 16}{-4} = \dfrac{-36}{-4} = 9$

        The vertex is $(2, 9)$.

    d. $-x^2 + 4x + 5 = 0$
        $x^2 - 4x - 5 = 0$
        $(x - 5)(x + 1) = 0$
        $x - 5 = 0 \quad$ or $\quad x + 1 = 0$
        $x = 5 \qquad\qquad x = -1$
        The x-intercepts are $(5, 0)$ and $(-1, 0)$.

    e.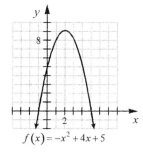

        $f(x) = -x^2 + 4x + 5$

29. $t(x) = -x^2 + 4x - 5$

a. Since $a = -1$, the parabola opens downward.

b. $y = -0^2 + 4(0) - 5 = -5$; The y-intercept is $(0, -5)$.

c. $x = -\dfrac{b}{2a} = -\dfrac{4}{2(-1)} = -\dfrac{4}{-2} = 2$

$y = \dfrac{4ac - b^2}{4a} = \dfrac{4(-1)(-5) - (4)^2}{4(-1)}$

$= \dfrac{20 - 16}{-4} = \dfrac{4}{-4} = -1$

The vertex is $(2, -1)$.

d. $0 = -x^2 + 4x - 5$

Since this is not factorable, check the discriminant.

$b^2 - 4ac = 4^2 - 4(-1)(-5) = 16 - 20 = -4$

Since $-4 < 0$ there are no real roots. Thus, there are no x-intercepts.

e.

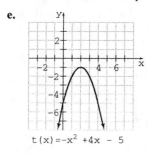

$t(x) = -x^2 + 4x - 5$

31. $f(x) = x^2 - 4x + 4$

a. Since $a = 1$, the parabola opens upward.

b. $y = 0^2 - 4(0) + 4 = 4$

The y-intercept is $(0, 4)$.

c. $x = -\dfrac{b}{2a} = -\dfrac{-4}{2(1)} = \dfrac{4}{2} = 2$

$y = \dfrac{4ac - b^2}{4a} = \dfrac{4(1)(4) - (-4)^2}{4(1)}$

$= \dfrac{16 - 16}{4} = \dfrac{0}{4} = 0$

The vertex is $(2, 0)$.

d. $x^2 - 4x + 4 = 0$

$(x - 2)^2 = 0$

$x - 2 = 0$

$x = 2$

The x-intercept $(2, 0)$.

e.

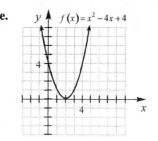

$f(x) = x^2 - 4x + 4$

33. $r(x) = x^2 + 2$

a. Since $a = 1$ the parabola opens upward.

b. $y = 0^2 + 2 = 2$

The y-intercept is $(0, 2)$.

c. $x = -\dfrac{b}{2a} = -\dfrac{0}{2(1)} = \dfrac{0}{2} = 0$

$y = \dfrac{4ac - b^2}{4a} = \dfrac{4(1)(2) - 0^2}{4(1)} = \dfrac{8}{4} = 2$

The vertex is $(0, 2)$.

d. $0 = x^2 + 2$

Since this is not factorable, check the discriminant.

$b^2 - 4ac = 0 - 4(1)2 = -8$

There are no real roots. Thus, there are no x-intercepts.

e.

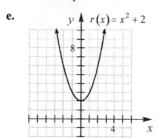

$r(x) = x^2 + 2$

35. $l(x) = -x^2 + 5$

a. Since $a = -1$, the parabola opens downward.

b. $y = -(0)^2 + 5 = 5$

The y-intercept is $(0, 5)$.

c. $x = -\dfrac{b}{2a} = -\dfrac{0}{2(-1)} = \dfrac{0}{2} = 0$

$y = \dfrac{4ac - b^2}{4a} = \dfrac{4(-1)(5) - (0)^2}{4(-1)}$

$= \dfrac{-20 - 0}{-4} = \dfrac{-20}{-4} = 5$

The vertex is $(0, 5)$.

d. $-x^2 + 5 = 0$

$-x^2 = -5$

$x^2 = 5$

$x = \pm\sqrt{5}$

The x-intercepts are $\left(\sqrt{5}, 0\right)$ and $\left(-\sqrt{5}, 0\right)$.

e.

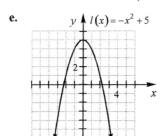

37. $y = -2x^2 + 4x - 8$

a. Since $a = -2$ the parabola opens downward.

b. $y = -2(0)^2 + 4(0) - 8 = -8$

The y-intercept is $(0, -8)$.

c. $x = -\dfrac{b}{2a} = -\dfrac{4}{2(-2)} = -\dfrac{4}{-4} = 1$

$y = \dfrac{4ac - b^2}{4a} = \dfrac{4(-2)(-8) - (4)^2}{4(-2)}$

$= \dfrac{64 - 16}{-8} = \dfrac{48}{-8} = -6$

The vertex is $(1, -6)$.

d. $-2x^2 + 4x - 8 = 0$

$-2(x^2 - 2x + 4) = 0$

Since this is not factorable, check the discriminant.

$b^2 - 4ac = 4^2 - 4(-2)(-8) = 16 - 64 = -48$

Since $-48 < 0$, there are no real roots. Thus, there are no x-intercepts.

e.

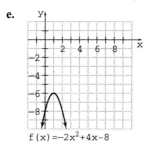

$f(x) = -2x^2 + 4x - 8$

39. $m(x) = 3x^2 + 4x + 3$

a. Since $a = 3$ the parabola opens upward.

b. $y = 3(0) + 4(0) + 3 = 3$

The y-intercept is $(0, 3)$.

c. $x = -\dfrac{b}{2a} = -\dfrac{4}{2(3)} = -\dfrac{4}{6} = -\dfrac{2}{3}$

$y = \dfrac{4ac - b^2}{4a} = \dfrac{4(3)(3) - 4^2}{4(3)}$

$= \dfrac{36 - 16}{12} = \dfrac{20}{12} = \dfrac{5}{3}$

The vertex is $\left(-\dfrac{2}{3}, \dfrac{5}{3}\right)$.

d. $0 = 3x^2 + 4x + 3$

Since this is not factorable, check the discriminant.

$b^2 - 4ac = 4^2 - 4(3)(3) = 16 - 36 = -20$

Since $-20 < 0$ there are no real roots. Thus, there are no x-intercepts.

e.

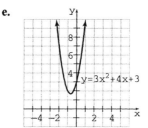

$y = 3x^2 + 4x + 3$

41. $y = 3x^2 + 4x - 6$

a. Since $a = 3$ the parabola opens upward.

b. $y = 3(0)^2 + 4(0) - 6 = -6$

The y-intercept is $(0, -6)$.

c. $x = -\dfrac{b}{2a} = -\dfrac{4}{2(3)} = -\dfrac{4}{6} = -\dfrac{2}{3}$

$y = \dfrac{4ac - b^2}{4a} = \dfrac{4(3)(-6) - 4^2}{4(3)}$

$= \dfrac{-72 - 16}{12} = -\dfrac{22}{3}$

The vertex is $\left(-\dfrac{2}{3}, -\dfrac{22}{3}\right)$.

d. $0 = 3x^2 + 4x - 6$

Since this is not factorable, check the discriminant.

$b^2 - 4ac = 4^2 - 4(3)(-6) = 88$

Since $88 > 0$ there are two real roots.

$x = \dfrac{-b \pm \sqrt{b^2 - 4ac}}{2a} = \dfrac{-4 \pm \sqrt{88}}{2(3)}$

$= \dfrac{-4 \pm 2\sqrt{22}}{6} = \dfrac{-2 \pm \sqrt{22}}{3}$

The x-intercepts are

$\left(\dfrac{-2 + \sqrt{22}}{3}, 0\right)$ and $\left(\dfrac{-2 - \sqrt{22}}{3}, 0\right)$.

e.

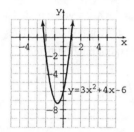

d. $0 = x^2 - 6x + 4$

Since this is not factorable check the discriminant.

$b^2 - 4ac = (-6)^2 - 4(1)(4) = 36 - 16 = 20$

Since $20 > 0$ there are two real roots.

$x = \dfrac{-b \pm \sqrt{b^2 - 4ac}}{2a} = \dfrac{-(-6) \pm \sqrt{20}}{2(1)}$

$= \dfrac{6 \pm 2\sqrt{5}}{2} = 3 \pm \sqrt{5}$

The x-intercepts are

$\left(3 + \sqrt{5}, 0\right)$ and $\left(3 - \sqrt{5}, 0\right)$.

e.

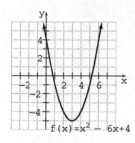

43. $y = 2x^2 - x - 6$

a. Since $a = 2$ the parabola opens upward.

b. $y = 2(0)^2 - 0 - 6 = -6$

The y-intercept is $(0, -6)$.

c. $x = -\dfrac{b}{2a} = \dfrac{-1}{2(2)} = \dfrac{1}{4}$

$y = \dfrac{4ac - b^2}{4a} = \dfrac{4(2)(-6) - (-1)^2}{4(2)}$

$= \dfrac{-48 - 1}{8} = -\dfrac{49}{8}$

The vertex is $\left(\dfrac{1}{4}, -\dfrac{49}{8}\right)$.

d. $2x^2 - x - 6 = 0$

$(2x + 3)(x - 2) = 0$

$2x + 3 = 0 \quad$ or $\quad x - 2 = 0$

$x = -\dfrac{3}{2} \qquad\qquad x = 2$

The x-intercepts are $\left(-\dfrac{3}{2}, 0\right)$ and $(2, 0)$.

e.

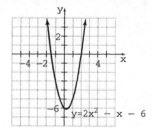

45. $f(x) = -x^2 + 3x - 5$

a. Since $a = -1$ the parabola opens downward.

b. $y = -0^2 + 3(0) - 5 = -5$

The y-intercept is $(0, -5)$.

c. $x = -\dfrac{b}{2a} = -\dfrac{3}{2(-1)} = -\dfrac{3}{-2} = \dfrac{3}{2}$

$y = \dfrac{4ac - b^2}{4a} = \dfrac{4(-1)(-5) - 3^2}{4(-1)} = -\dfrac{11}{4}$

The vertex is $\left(\dfrac{3}{2}, -\dfrac{11}{4}\right)$.

d. $0 = -x^2 + 3x - 5$

Since this is not factorable, check the discriminant.

$b^2 - 4ac = 3^2 - 4(-1)(-5) = 9 - 20 = -11$

Since $-11 < 0$ there are no real roots. Thus, there are no x-intercepts.

e.

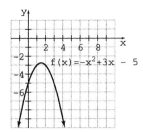

47. In the function $f(x) = (x-3)^2$, h has a value of 3. The graph of $f(x)$ is the graph of $g(x) = x^2$ shifted 3 units to the right.

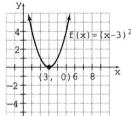

49. In the function $f(x) = (x+1)^2$, h has a value of -1. The graph of $f(x)$ is the graph of $g(x) = x^2$ shifted 1 unit to the left.

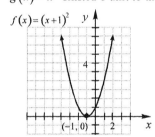

51. In the function $f(x) = x^2 + 3$, k has a value of 3. The graph $f(x)$ will be the graph of $g(x) = x^2$ shifted 3 units up.

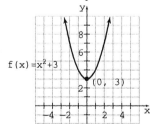

53. In the function $f(x) = x^2 - 1$, k has a value of -1. The graph $f(x)$ will be the graph of $g(x) = x^2$ shifted 1 units down.

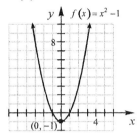

55. In the function $f(x) = (x-2)^2 + 3$, h has a value of 2 and k has a value of 3. The graph of $f(x)$ will be the graph of $g(x) = x^2$ shifted 2 units to the right and 3 units up.

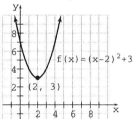

57. In the function $f(x) = (x+4)^2 + 4$, h has a value of -4 and k has a value of 4. The graph of $f(x)$ will be the graph of $g(x) = x^2$ shifted 4 units to the left and 4 units up.

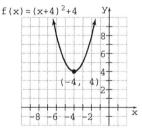

59. In the function $g(x) = -(x+3)^2 - 2$, a has the value -1, h has the value -3, and k has the value -2. Since $a < 0$, the parabola opens downward. The graph of $g(x)$ will be the graph of $f(x) = -x^2$ shifted 3 units to the left and 2 units down.

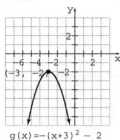

$g(x) = -(x+3)^2 - 2$

61. In the function $y = -2(x-2)^2 + 2$, a has a value of -2, h has a value of 2, and k has a value of 2. The graph of y will be the graph of $g(x) = -2x^2$ shifted 2 units to the right and 2 units up.

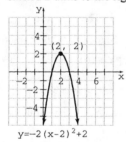

$y = -2(x-2)^2 + 2$

63. In the function $h(x) = -2(x+1)^2 - 3$, a has a value of -2, h has a value of -1, and k has a value of -3. Since $a < 0$, the parabola opens downward. The graph of $h(x)$ will be the graph of $f(x) = -2x^2$ shifted 1 unit left and 3 units down.

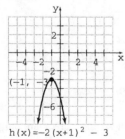

$h(x) = -2(x+1)^2 - 3$

65. a. $f(x) = x^2 - 6x + 8$
$f(x) = (x^2 - 6x + 9) + 8 - 9$
$f(x) = (x-3)^2 - 1$

b. Since $h = 3$ and $k = -1$, the vertex is $(3, -1)$.

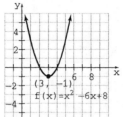

67. a. $g(x) = x^2 - x - 3$

$g(x) = \left(x^2 - x + \dfrac{1}{4}\right) - 3 - \dfrac{1}{4}$

$g(x) = \left(x - \dfrac{1}{2}\right)^2 - \dfrac{13}{4}$

b. Since $h = \dfrac{1}{2}$ and $k = -\dfrac{13}{4}$, the vertex is

$\left(\dfrac{1}{2}, -\dfrac{13}{4}\right)$.

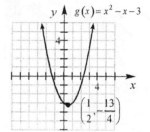

69. a. $f(x) = -x^2 - 4x - 6$
$f(x) = -(x^2 + 4x) - 6$
$f(x) = -(x^2 + 4x + 4) - 6 - (-4)$
$f(x) = -(x+2)^2 - 2$

b. Since $h = -2$ and $k = -2$, the vertex is $(-2, -2)$. Since $a < 0$, the parabola opens downward.

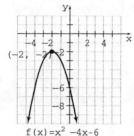

$f(x) = x^2 - 4x - 6$

71. a. $g(x) = x^2 - 4x - 1$

$g(x) = (x^2 - 4x + 4) - 1 - 4$

$g(x) = (x - 2)^2 - 5$

b. Since $h = 2$ and $k = -5$, the vertex is $(2, -5)$.

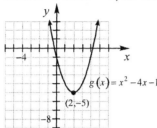

73. a. $f(x) = 2x^2 + 5x - 3$

$f(x) = 2\left(x^2 + \dfrac{5}{2}x\right) - 3$

$f(x) = 2\left(x^2 + \dfrac{5}{2}x + \dfrac{25}{16}\right) - 3 - 2\left(\dfrac{25}{16}\right)$

$f(x) = 2\left(x^2 + \dfrac{5}{2}x + \dfrac{25}{16}\right) - 3 - \dfrac{25}{8}$

$f(x) = 2\left(x + \dfrac{5}{4}\right)^2 - \dfrac{49}{8}$

b. Since $h = -\dfrac{5}{4}$ and $k = -\dfrac{49}{8}$, the vertex is $\left(-\dfrac{5}{4}, -\dfrac{49}{8}\right)$. The graph of $f(x)$ will be the graph of $g(x) = 2x^2$ shifted $\dfrac{5}{4}$ units left and $\dfrac{49}{8}$ units down.

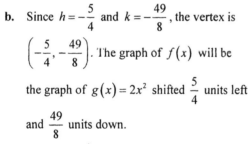

75. $y = x^2 + 3x - 4$

$x = -\dfrac{b}{2a} = -\dfrac{3}{2}$

The parabola opens up and the x-coordinate of the vertex is $-\dfrac{3}{2}$. The graph is (c).

77. $y = -x^3 + 2x^2 - 6$

The leading coefficient is negative. The function will decrease as x increases. The y-intercept is $(0, -6)$. The graph is (c).

79. a. $y_1 = x^3$

$y_2 = x^3 - 3x^2 - 3$

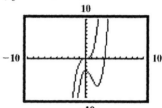

b. As x increases the functions increase.

c. Answers will vary.

d. As x decreases, the functions decrease.

e. Answers will vary.

81. $f(x) = 2(x + 3)^2 - 1$. The vertex is $(h, k) = (-3, -1)$. Since $a > 0$, the parabola opens up. The graph is (d).

83. $f(x) = 2(x - 1)^2 + 3$ The vertex is $(h, k) = (1, 3)$. Since $a > 0$, the parabola opens up. The graph is (b).

85. a. $f(x) = (x + 4)(18 - x)$

$= 18x - x^2 + 72 - 4x$

$= -x^2 + 14x + 72$

Since $a = -1$, the graph of this function is a parabola that opens downward and thus has a maximum value at its vertex.

$x = -\dfrac{b}{2a} = -\dfrac{14}{2(-1)} = 7$

b. $y = \dfrac{4ac - b^2}{4a} = \dfrac{4(-1)(72) - (14)^2}{4(-1)}$

$= \dfrac{-288 - 196}{-4} = \dfrac{-484}{-4} = 121$

The maximum area is 121 square units.

87. a. $f(x) = (x+5)(26-x)$

$$= 26x - x^2 + 130 - 5x$$

$$= -x^2 + 21x + 130$$

Since $a = -1$, the graph of this function is a parabola that opens downward and thus has a maximum value at its vertex.

$$x = -\frac{b}{2a} = -\frac{21}{2(-1)} = 10.5$$

b. $y = \frac{4ac - b^2}{4a} = \frac{4(-1)(130) - (21)^2}{4(-1)}$

$$= \frac{-520 - 441}{-4} = 240.25$$

The maximum area is 240.25 square units.

89. a. $R(n) = -0.02n^2 + 8n$

Since $a = -0.02$, the graph of this function is a parabola that opens downward and thus has a maximum value at its vertex.

$$x = -\frac{b}{2a} = -\frac{8}{2(-0.02)} = 200$$

The maximum revenue will be achieved when 200 batteries are sold.

b. $y = \frac{4ac - b^2}{4a} = \frac{4(-0.02)(0) - (8)^2}{4(-0.02)}$

$$= \frac{0 - 64}{-0.08} = \frac{-64}{-0.08} = 800$$

The maximum revenue is $800.

91. $N(t) = -0.043t^2 + 1.82t + 46.0$

Since $a = -0.043$, the graph of this function is a parabola that opens downward and thus has a maximum value at its vertex.

$$x = -\frac{b}{2a} = -\frac{1.82}{2(-0.043)} \approx 21.2$$

The maximum enrollment will be obtained about 21 years after 1989 which is the year 2010.

93. For $f(x) = (x-2)^2 + \frac{5}{2}$, the vertex is $\left(2, \frac{5}{2}\right)$.

For $g(x) = (x-2)^2 - \frac{3}{2}$, the vertex is $\left(2, -\frac{3}{2}\right)$.

These points are on the vertical line $x = 2$. The distance between the two points is

$$\frac{5}{2} - \left(-\frac{3}{2}\right) = \frac{5}{2} + \frac{3}{2} = \frac{8}{2} = 4 \text{ units.}$$

95. For $f(x) = 2(x+4)^2 - 3$, the vertex is $(-4, -3)$.

For $g(x) = -(x+1)^2 - 3$, the vertex is $(-1, -3)$.

These points are on the horizontal line $y = -3$. The distance between the two points is $-1 - (-4) = -1 + 4 = 3$ units.

97. A function that has the shape of $f(x) = 2x^2$ will have the form $f(x) = 2(x-h)^2 + k$. If $(h, k) = (3, -2)$, the function is

$$f(x) = 2(x-3)^2 - 2.$$

99. A function that has the shape of $f(x) = -4x^2$ will have the form $f(x) = -4(x-h)^2 + k$. If $(h, k) = \left(-\frac{3}{5}, -\sqrt{2}\right)$, the function is

$$f(x) = -4\left(x + \frac{3}{5}\right)^2 - \sqrt{2}.$$

101. a. The graphs will have the same x-intercepts but $f(x) = x^2 - 8x + 12$ will open up and $g(x) = -x^2 + 8x - 12$ will open down.

b. Yes, because the x-intercepts are located by setting $x^2 - 8x + 12$ and $-x^2 + 8x - 12$ equal to zero. They have the same solution set, therefore the x-intercepts are equal. The x-intercepts for both are $(6, 0)$, and $(2, 0)$.

c. No. The vertex for $f(x) = x^2 - 8x + 12$ is $(4, -4)$ and the vertex for $g(x) = -x^2 + 8x - 12$ is $(4, 4)$.

d.

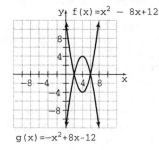

$$477$$

103. a. The vertex $x = -\dfrac{b}{2a} = -\dfrac{24}{2(-1)} = 12$

$I = -(12)^2 + 24(12) - 44$

$= -144 + 288 - 44$

$= 100$

The vertex is at (12, 100). To find the roots set $I = 0$.

$0 = -x^2 + 24x - 44$

$= -1(x^2 - 24x + 44)$

$= -1(x - 2)(x - 22)$

$x - 2 = 0$ or $x - 22 = 0$

$x = 2 \qquad\qquad x = 22$

The roots are 2 and 22.

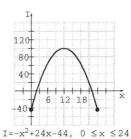

$I = -x^2 + 24x - 44, \ 0 \le x \le 24$

b. The minimum cost will be $2 since the smaller root is 2.

c. The maximum cost is $22 since the larger root is 22.

d. The maximum value will occur at the vertex of the parabola, (12, 100). Therefore, they should charge $12.

e. The maximum value will occur at the vertex of the parabola, (12, 100). Since I is in hundreds of dollars the maximum income is 100($100) = $10,000.

105. a. The number of bird feeders sold for the maximum profit will be the x-coordinate of the vertex.

$f(x) = -0.4x^2 + 80x - 200$

$x = -\dfrac{b}{2a} = -\dfrac{80}{2(-0.4)} = -\dfrac{80}{-0.8} = 100$

The company must sell 100 bird feeders for maximum profit.

b. The maximum profit will be the y-coordinate of the vertex, $y = f(100)$.

$f(100) = -0.4(100)^2 + 80(100) - 200$

$= -0.4(10,000) + 8000 - 200$

$= -4000 + 8000 - 200$

$= 3800$

The maximum profit will be $3800.

107. a. The maximum height is h, the y-coordinate of the vertex.

$h(t) = -4.9t^2 + 24.5t + 9.8$

$h = \dfrac{4ac - b^2}{4a} = \dfrac{4(-4.9)(9.8) - (24.5)^2}{4(-4.9)}$

$= \dfrac{-192.08 - 600.25}{-19.6} = \dfrac{-792.33}{-19.6} = 40.425$

The maximum height obtained by the cannonball is 40.425 meters.

b. The time the cannonball reaches the maximum height is t, the x-coordinate of the vertex,. $t = -\dfrac{b}{2a} = -\dfrac{24.5}{2(-4.9)} = 2.5$

The cannonball will reach the maximum height after 2.5 seconds.

c. $f(x) = 0$ when the cannonball hits the ground.

$h(t) = -4.9t^2 + 24.5t + 9.8$

$0 = -4.9t^2 + 24.5t + 9.8$

$t = \dfrac{-24.5 \pm \sqrt{(24.5)^2 - 4(-4.9)(9.8)}}{2(-4.9)}$

$= \dfrac{-24.5 \pm \sqrt{792.33}}{-9.8}$

$\approx \dfrac{-24.5 \pm 28.14836}{-9.8}$

$t \approx -0.37$ or $t \approx 5.37$

Disregard the negative value. The cannonball will hit the ground after about 5.37 seconds.

109. a. The year 2007 is represented by $t = 13$.

$r(t) = -2.723t^2 + 35.273t + 579$

$r(13) = -2.723(13)^2 + 35.273(13) + 579 \approx 577$

The average monthly rent should be about $577.

b. The function will reach a maximum at it's vertex: $t = -\dfrac{b}{2a} = -\dfrac{35.273}{2(-2.723)} \approx 6.5$

The monthly rent for an apartment reached a maximum about 6.5 years after 1994, which was during the year 2000.

111. If the perimeter of the room is 8 ft., then $80 = 2l + 2w$, where l is the length and w is the width. Then $80 = 2(l + w)$ and $40 = l + w$. Therefore, $l = 40 - w$. The area of the room is $A = lw = (40 - w)w = 40w - w^2 = -w^2 + 40w$. The maximum area is the y-coordinate of the vertex.

$$A = \frac{4ac - b^2}{4a} = \frac{4(-1)(0) - 40^2}{4(-1)}$$

$$= \frac{0 - 1600}{-4} = \frac{-1600}{-4} = 400$$

The maximum area is 400 square feet.

113. If two numbers differ by 8 and x is one of the numbers, then $x + 8$ is the other number. The product is $f(x) = x(x + 8) = x^2 + 8x$. The maximum product is the y-coordinate of the vertex.

$$x = -\frac{b}{2a} = -\frac{8}{2(1)} = -4$$

$$y = f(-4) = (-4)^2 + 8(-4) = -16$$

The maximum product is –16. The numbers are –4 and –4 + 8 = 4.

115. If two numbers add to 60 and x is one of the numbers, then $60 - x$ is the other number. The product is

$$f(x) = x(60 - x) = 60x - x^2 = -x^2 + 60x.$$

The maximum product is the y-coordinate of the vertex.

$$x = -\frac{b}{2a} = -\frac{60}{2(-1)} = -\frac{60}{-2} = 30$$

$$y = f(30) = -30^2 + 60(30) = -900 + 1800 = 900$$

The maximum product is 900. The numbers are 30 and 60 – 30 = 30.

117. $C(x) = 2000 + 40x$, $R(x) = 800x - x^2$

$$P(x) = R(x) - C(x)$$

$$P(x) = (800x - x^2) - (2000 + 40x)$$

$$= 800x - x^2 - 2000 - 40x$$

$$= -x^2 + 760x - 2000$$

The maximum profit is $P(x)$, the y-coordinate of the vertex. The number of items that must be produced and sold to obtain maximum profit is the x coordinate of the vertex.

$$x = -\frac{b}{2a} = -\frac{760}{2(-1)} = 380$$

a. $P(380) = -380^2 + 760(380) - 2000$

$$= -144,400 + 288800 - 2000$$

$$= 142,400$$

The maximum profit is \$142,400.

b. The number of items that must be produced and sold to obtain maximum profit is 380.

119. The y-intercept is $(0, -5)$. As x increases, the function decreases because of the leading coefficient. The graph is (b).

121. a. $f(t) = -16t^2 + 52t + 3$

$$= -16(t^2 - 3.25t) + 3$$

$$= -16\left[t^2 - 3.25t + \left(\frac{3.25}{2}\right)^2\right] + 3 + 16\left(\frac{3.25}{2}\right)^2$$

$$= -16(t - 1.625)^2 + 3 + 42.25$$

$$= -16(t - 1.625)^2 + 45.25$$

b. The maximum height was 45.25 feet obtained at 1.625 seconds.

c. It is the same answer.

125. The radius of the outer circle is $r = 15$ ft. The area $A = \pi r^2 = \pi(15^2) = 225\pi$ ft^2. The radius of the inner circle is $r = 5$ ft. The area is $A = \pi r^2 = \pi(5^2) = 25\pi$ ft^2. The blue shaded area is 225π ft$^2 - 25\pi$ ft$^2 = 200\pi$ ft^2.

126.

$$(x - 3) \div \frac{x^2 + 3x - 18}{x} = \frac{x - 3}{1} \cdot \frac{x}{x^2 + 3x - 18}$$

$$= \frac{x - 3}{1} \cdot \frac{x}{(x + 6)(x - 3)}$$

$$= \frac{x}{x + 6}$$

127.

$$\begin{aligned} x - y &= -5 \quad (1) \\ 2x + 2y - z &= 0 \quad (2) \\ x + y + z &= 3 \quad (3) \end{aligned}$$

First eliminate the variable z from equations (2) and (3) by adding these equations.

$$\begin{aligned} 2x + 2y - z &= 0 \\ \underline{x + y + z} &= 3 \\ 3x + 3y &= 3 \quad (4) \end{aligned}$$

Equations (1) and (4) form a system of equations in two variables. Multiply equation (1) by 3 and add the result to equation (4) to eliminate the variable y.

$3(x-y) = 3(-5) \implies 3x - 3y = -15$

$3x + 3y = 3 \implies \underline{3x + 3y = 3}$

$6x = -12$

$x = -2$

Substitute -2 for x in equation (1) to find y.

$(-2) - y = -5$

$-y = -3$

$y = 3$

Substitute -2 for x and 3 for y in equation (3) to find z.

$(-2) + (3) + z = 3$

$1 + z = 3$

$z = 2$

The solution is $(-2, 3, 2)$.

128. $\begin{vmatrix} 1 & 3 \\ 2 & \\ 2 & -4 \end{vmatrix} = \left(\frac{1}{2}\right)(-4) - (2)(3) = -2 - 6 = -8$

129. $y \le \frac{2}{3}x + 3$

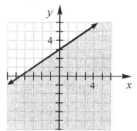

Exercise Set 12.6

1. a. For $f(x) = x^2 - 7x + 10$, $f(x) > 0$ when the graph is above the x-axis. The solution is $x < 2$ or $x > 5$.

b. For $f(x) = x^2 - 7x + 10$ $f(x) < 0$ when the graph is below the x-axis. The solution is $2 < x < 5$.

3. Yes. The boundary values 5 and -3 are included in the solution set since this is a greater than *or equal to* inequality. These values make the expression equal to zero.

5. The boundary values -2 and 1 are included in the solution set since this is a less than *or equal to* inequality. These values make the expression equal to zero. However, the boundary value -1 is not included in the solution set since it would result in a zero in the denominator, which is undefined.

7. $x^2 - 2x - 8 \ge 0$

$(x+2)(x-4) \ge 0$

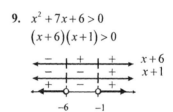

9. $x^2 + 7x + 6 > 0$

$(x+6)(x+1) > 0$

11. $n^2 - 6n + 9 \ge 0$

$(n-3)^2 \ge 0$

13. $x^2 - 16 < 0$

$(x+4)(x-4) < 0$

15. $2x^2 + 5x - 3 \ge 0$

$(2x-1)(x+3) \ge 0$

17. $5x^2 + 6x \le 8$

$5x^2 + 6x - 8 \le 0$

$(x+2)(5x-4) \le 0$

480

19. $2x^2 - 12x + 9 \leq 0$

$2x^2 - 12x + 9 = 0$

$x = \dfrac{-(-12) \pm \sqrt{(-12)^2 - 4(2)(9)}}{2(2)}$

$= \dfrac{12 \pm \sqrt{144 - 72}}{4}$

$= \dfrac{12 \pm \sqrt{72}}{4} = \dfrac{12 \pm 6\sqrt{2}}{4} = \dfrac{6 \pm 3\sqrt{2}}{2}$

$\dfrac{6 - 3\sqrt{2}}{2} \qquad \dfrac{6 + 3\sqrt{2}}{2}$

21. $(x-2)(x+1)(x+5) \geq 0$

$x - 2 = 0 \quad x + 1 = 0 \quad x + 5 = 0$

$x = 2 \qquad x = -1 \qquad x = -5$

$[-5, -1] \cup [2, \infty)$

23. $(a-3)(a+2)(a+4) < 0$

$a - 3 = 0 \quad a + 2 = 0 \quad a + 4 = 0$

$a = 3 \qquad a = -2 \qquad a = -4$

$(-\infty, -4) \cup (-2, 3)$

25. $(2c+5)(3c-6)(c+6) > 0$

$2c + 5 = 0 \qquad 3c - 6 = 0 \quad c + 6 = 0$

$2c = -5 \qquad 3c = 6 \qquad c = -6$

$c = -\dfrac{5}{2} \qquad c = 2$

$\left(-6, -\dfrac{5}{2}\right) \cup (2, \infty)$

27. $(3x+5)(x-3)(x+1) > 0$

$3x + 5 = 0 \qquad x - 3 = 0 \quad x + 1 = 0$

$3x = -5 \qquad x = 3 \qquad x = -1$

$x = -\dfrac{5}{3}$

$\left(-\dfrac{5}{3}, -1\right) \cup (3, \infty)$

29. $(x+2)(x+2)(3x-8) \geq 0$

$x + 2 = 0 \qquad 3x - 8 = 0$

$x = -2 \qquad x = \dfrac{8}{3}$

$\{-2\} \cup \left[\dfrac{8}{3}, \infty\right)$

31. $x^3 - 6x^2 + 9x < 0$

$x(x^2 - 6x + 9) < 0$

$x(x-3)^2 < 0$

$x = 0 \quad x - 3 = 0$

$x = 3$

$(-\infty, 0)$

33. $f(x) = x^2 - 6x, \; f(x) \geq 0$

$x^2 - 6x \geq 0$

$x(x-6) \geq 0$

$x = 0 \quad x - 6 = 0$

$x = 6$

35. $f(x) = x^2 + 4x, \ f(x) > 0$

$x^2 + 4x > 0$

$x(x+4) > 0$

$x = 0 \quad x+4 = 0$

$\qquad\qquad x = -4$

$$
\begin{array}{ccc}
- & - & + & x \\
- & + & + & x+4 \\
+ & - & + & \\
\hline
& -4 \quad\quad 0 &
\end{array}
$$

37. $f(x) = x^2 - 14x + 48, \ f(x) < 0$

$x^2 - 14x + 48 < 0$

$(x-6)(x-8) < 0$

$x - 6 = 0 \quad x - 8 = 0$

$x = 6 \qquad x = 8$

$$
\begin{array}{ccc}
- & + & + & x-6 \\
- & - & + & x-8 \\
+ & - & + & \\
\hline
& 6 \quad\quad 8 &
\end{array}
$$

39. $f(x) = 2x^2 + 9x - 1, \ f(x) \le 5$

$2x^2 + 9x - 1 \le 5$

$2x^2 + 9x - 6 \le 0$

$x = \dfrac{-9 \pm \sqrt{9^2 - 4(2)(-6)}}{2(2)}$

$= \dfrac{-9 \pm \sqrt{129}}{4}$

$x = \dfrac{-9 - \sqrt{129}}{4} \quad x = \dfrac{-98 + \sqrt{129}}{4}$

$$
\begin{array}{ccc}
- & + & + & x - \left(\frac{-9-\sqrt{129}}{4}\right) \\
- & - & + & x - \left(\frac{-9+\sqrt{129}}{4}\right) \\
+ & - & + & \\
\hline
& \frac{-9-\sqrt{129}}{4} \quad \frac{-9+\sqrt{129}}{4} &
\end{array}
$$

41. $f(x) = 2x^3 + 9x^2 - 35x, \ f(x) \ge 0$

$2x^3 + 9x^2 - 35x \ge 0$

$x(2x^2 + 9x - 35) \ge 0$

$x(2x-5)(x+7) \ge 0$

$x = 0 \quad 2x - 5 = 0 \quad x + 7 = 0$

$\qquad\qquad x = \dfrac{5}{2} \qquad x = -7$

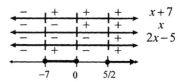

43. $\dfrac{x+2}{x-4} > 0$

$x \ne 4$

$$
\begin{array}{ccc}
- & + & + & x+2 \\
- & - & + & x-4 \\
+ & - & + & \\
\hline
& -2 \quad\quad 4 &
\end{array}
$$

$\{x \,|\, x < -2 \text{ or } x > 4\}$

45. $\dfrac{x-1}{x+5} < 0$

$x \ne -5$

$$
\begin{array}{ccc}
- & + & + & x+5 \\
- & - & + & x-1 \\
+ & - & + & \\
\hline
& -5 \quad\quad 1 &
\end{array}
$$

$\{x \,|\, -5 < x < 1\}$

47. $\dfrac{x+3}{x-2} \ge 0$

$x \ne 2$

$$
\begin{array}{ccc}
- & + & + & x+3 \\
- & - & + & x-2 \\
+ & - & + & \\
\hline
& -3 \quad\quad 2 &
\end{array}
$$

$\{x \,|\, x \le -3 \text{ or } x > 2\}$

49. $\dfrac{a-9}{a+5} < 0$

$a \ne -5$

$$
\begin{array}{ccc}
- & + & + & a+5 \\
- & - & + & a-9 \\
+ & - & + & \\
\hline
& -5 \quad\quad 9 &
\end{array}
$$

$\{a \,|\, -5 < a < 9\}$

51. $\dfrac{c-10}{c-4} > 0$

$c \neq 4$

$\{c \mid c < 4 \text{ or } c > 10\}$

53. $\dfrac{3y+6}{y+4} \leq 0$

$y \neq -4,$
$3y+6 = 0 \implies y = -2$

$\{y \mid -4 < y \leq -2\}$

55. $\dfrac{5a+10}{3a-1} \geq 0$

$a \neq \dfrac{1}{3},$

$5a+10 = 0 \implies a = -2$

$\left\{a \mid a \leq -2 \text{ or } a > \dfrac{1}{3}\right\}$

57. $\dfrac{3x+4}{2x-1} < 0$

$x \neq \dfrac{1}{2}$

$\left\{x \mid -\dfrac{4}{3} < x < \dfrac{1}{2}\right\}$

59. $\dfrac{3x+8}{x-2} \leq 0$

$x \neq 2$

$\left\{x \mid -\dfrac{8}{3} \leq x < 2\right\}$

61. $\dfrac{(x+1)(x-6)}{x+3} < 0$

$x \neq -3$

$(-\infty, -3) \cup (-1, 6)$

63. $\dfrac{(x-2)(x+3)}{x-5} > 0$

$x \neq 5$

$(-3, 2) \cup (5 \; \infty)$

65. $\dfrac{(a-1)(a-7)}{a+2} \geq 0$

$a \neq -2$

$(-2, 1] \cup [7, \infty)$

67. $\dfrac{c}{(c-3)(c+8)} \leq 0$

$c \neq 3, \; c \neq -8$

$(-\infty, -8) \cup [0, 3)$

69. $\dfrac{x-6}{(x+4)(x-1)} \le 0$

$x \ne -4, \ x \ne 1$

$(-\infty, -4) \cup (1, 6]$

71. $\dfrac{(x-3)(2x+5)}{x-4} \ge 0$

$x \ne 4$

$\left[-\dfrac{5}{2}, 3\right] \cup (4, \infty)$

73. $\dfrac{2}{x-4} \ge 1$

$\dfrac{2}{x-4} - 1 \ge 0$

$\dfrac{2}{x-4} + \dfrac{-1(x-4)}{x-4} \ge 0$

$\dfrac{2-x+4}{x-4} \ge 0$

$\dfrac{6-x}{x-4} \ge 0$

$x \ne 4$

75. $\dfrac{3}{x-1} > -1$

$\dfrac{3}{x-1} + 1 > 0$

$\dfrac{3}{x-1} + \dfrac{1(x-1)}{x-1} > 0$

$\dfrac{3+x-1}{x-1} > 0$

$\dfrac{x+2}{x-1} > 0$

$x \ne 1$

77. $\dfrac{5}{x+2} \le 1$

$\dfrac{5}{x+2} - 1 \le 0$

$\dfrac{5}{x+2} + \dfrac{-1(x+2)}{x+2} \le 0$

$\dfrac{5-x-2}{x+2} \le 0$

$\dfrac{3-x}{x+2} \le 0$

$x \ne -2$

79. $\dfrac{2p-5}{p-4} \le 1$

$\dfrac{2p-5}{p-4} - 1 \le 0$

$\dfrac{2p-5}{p-4} - \dfrac{1(p-4)}{p-4} \le 0$

$\dfrac{2p-5-p+4}{p-4} \le 0$

$\dfrac{p-1}{p-4} \le 0, \quad p \ne 4$

81.
$$\frac{4}{x+2} \ge 2$$

$$\frac{4}{x+2} - 2 \ge 0$$

$$\frac{4}{x+2} - \frac{2(x+2)}{x+2} \ge 0$$

$$\frac{4-2x-4}{x+2} \ge 0$$

$$\frac{-2x}{x+2} \ge 0, \quad x \ne -2$$

```
      −   |  +  |  +    x+2
    +   |  +  |  −      −2x
    −   |  +  |  −
  ←———○———●———→
     −2      0
```

83.
$$\frac{w}{3w-2} > -2$$

$$\frac{w}{3w-2} + 2 > 0$$

$$\frac{w}{3w-2} + \frac{2(3w-2)}{3w-2} > 0$$

$$\frac{w+6w-4}{3w-2} > 0$$

$$\frac{7w-4}{3w-2} > 0$$

$$w \ne \frac{2}{3}$$

```
     −   |  +  |  +    7w−4
    −   |  −  |  +      3w−2
    +   |  −  |  +
  ←———○———○———→
     4/7    2/3
```

85. a. $y = \dfrac{x^2 - 4x + 4}{x-4} > 0$ where the graph of y is above the x-axis, on the interval $(4, \infty)$.

b. $y = \dfrac{x^2 - 4x + 4}{x-4} < 0$ where the graph of y is below the x-axis, on the interval $(-\infty, 2) \cup (2, 4)$.

87. A quadratic inequality with the union of the two outer regions of the number line as its solution, not including the boundary values, will be of the form $ax^2 + bx + c > 0$ with $a > 0$. Since the boundary values are $x = -4$ and $x = 2$, the factors are $x + 4$ and $x - 2$. Therefore one quadratic inequality is $(x+4)(x-2) > 0$ or

$$x^2 + 2x - 8 > 0.$$

89. Since the solution set is $x \le -3$ and $x > 4$, the factors are $x + 3$ and $x - 4$. Because -3 is included in the solution set, $x + 3$ is the numerator. Since 4 is not included in the solution set, $x - 4$ is the denominator. The inequality symbol will be $\ge$ because the union of the outer regions of number line is the solution set.

Therefore, the rational inequality is $\dfrac{x+3}{x-4} \ge 0$.

91. $(x+3)^2(x-1)^2 \ge 0$

The solution is all real numbers since any nonzero number squared is positive and zero squared is zero.

93. $\dfrac{x^2}{(x+2)^2} \ge 0$

This solution is all real numbers, except -2, since any nonzero number squared is negative. It is undefined when $x = -2$. Therefore, -2 is not a solution.

95. If $f(x) = ax^2 + bx + c$ and $a > 0$, the graph of $f(x)$ opens upward. If the discriminant is negative, the graph of $f(x)$ has no x-intercepts. Therefore, the graph lies above the x-axis and $f(x) < 0$ has no solution.

97. $(x+1)(x-3)(x+5)(x+8) \ge 0$

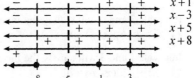

99. One possible answer is: Use a parabola that opens upward and has x-intercepts of $(0, 0)$ and $(3, 0)$. The x-values for which the parabola lies above the x-axis are $(-\infty, 0) \cup (3, \infty)$.

$$x^2 - 3x > 0$$

101. One possible answer is: Use a parabola that opens upward and has its vertex on or above the x-axis. Then there are no x-values for which the parabola lies below the x-axis.

$$x^2 < 0$$

103. $x^4 - 10x^2 + 9 > 0$

$\left(x^2\right)^2 - 10x^2 + 9 > 0$

Let $u = x^2$.

$u^2 - 10u + 9 > 0$

$(u-9)(u-1) > 0$

Substitute x^2 for u.

$\left(x^2 - 9\right)\left(x^2 - 1\right) > 0$

$(x+3)(x-3)(x+1)(x-1) > 0$

$(-\infty, -3) \cup (-1, 1) \cup (3, \infty)$

105. $x^3 + x^2 - 4x - 4 \geq 0$

$x^2(x+1) - 4(x+1) \geq 0$

$(x^2 - 4)(x+1) \geq 0$

$(x+2)(x-2)(x+1) \geq 0$

$[-2, -1] \cup [2, \infty)$

a. Intervals 2 and 4 will satisfy the inequality $(x-a)(x-b)(x-c) > 0$.

b. Intervals 1 and 3 will satisfy the inequality $(x-a)(x-b)(x-c) < 0$.

109. Let x = the number of quarts of 100% antifreeze added. The equation below describes the situation.

$100\%(x) + 20\%(10) = 50\%(x+10)$

$x + 0.2(10) = 0.5(x+10)$

$x + 2 = 0.5x + 5$

$0.5x = 3$

$x = 6$

Paul should add 6 quarts of 100% antifreeze.

110. $\left(6r + 5s - t\right) + \left(-3r - 2s - 8t\right)$

$= 6r + 5s - t - 3r - 2s - 8t$

$= 3r + 3s - 9t$

111. $\dfrac{1 + \dfrac{x}{x+1}}{\dfrac{2x+1}{x-3}} = \dfrac{\dfrac{1}{1} \cdot \dfrac{x+1}{x+1} + \dfrac{x}{x+1}}{\dfrac{2x+1}{x-3}}$

$= \dfrac{\dfrac{2x+1}{x+1}}{\dfrac{2x+1}{x-3}}$

$= \dfrac{2x+1}{x+1} \cdot \dfrac{x-3}{2x+1}$

$= \dfrac{x-3}{x+1}$

112. $h(x) = \dfrac{x^2 + 4x}{x+9}$

$h(-3) = \dfrac{(-3)^2 + 4(-3)}{(-3)+9} = \dfrac{9-12}{6} = \dfrac{-3}{6} = -\dfrac{1}{2}$

113. $(3-4i)(6+5i) = 18 + 15i - 24i - 20i^2$

$= 18 - 9i - 20(-1)$

$= 18 - 9i + 20$

$= 38 - 9i$

Chapter 12 Review Exercises

1. $(x-5)^2 = 24$

$x - 5 = \pm\sqrt{24}$

$x - 5 = \pm 2\sqrt{6}$

$x = 5 \pm 2\sqrt{6}$

$x = 5 + 2\sqrt{6}$ or $x = 5 - 2\sqrt{6}$

2. $(2x+1)^2 = 60$

$2x + 1 = \pm\sqrt{60}$

$2x + 1 = \pm 2\sqrt{15}$

$2x = -1 \pm 2\sqrt{15}$

$x = \dfrac{-1 \pm 2\sqrt{15}}{2}$

$x = \dfrac{-1 + 2\sqrt{15}}{2}$ or $x = \dfrac{-1 - 2\sqrt{15}}{2}$

3. $\left(x-\dfrac{1}{3}\right)^2=\dfrac{4}{9}$

$$x-\dfrac{1}{3}=\pm\sqrt{\dfrac{4}{9}}$$

$$x-\dfrac{1}{3}=\pm\dfrac{2}{3}$$

$$x=\dfrac{1}{3}\pm\dfrac{2}{3}$$

$x=\dfrac{1}{3}+\dfrac{2}{3}$ or $x=\dfrac{1}{3}-\dfrac{2}{3}$

$\quad=\dfrac{3}{3}\qquad\qquad=-\dfrac{1}{3}$

$\quad=1$

$x=1$ or $x=-\dfrac{1}{3}$

4. $\left(2x-\dfrac{1}{2}\right)^2=4$

$$2x-\dfrac{1}{2}=\pm\sqrt{4}$$

$$2x-\dfrac{1}{2}=\pm2$$

$$2x=\dfrac{1}{2}\pm2$$

$$2x=\dfrac{1\pm4}{2}$$

$$x=\dfrac{1\pm4}{4}$$

$x=\dfrac{1+4}{4}$ or $x=\dfrac{1-4}{4}$

$\quad=\dfrac{5}{4}\qquad\qquad=-\dfrac{3}{4}$

5. $x^2-7x+12=0$

$$x^2-7x=-12$$

$$x^2-7x+\dfrac{49}{4}=-\dfrac{48}{4}+\dfrac{49}{4}$$

$$x^2-7x+\dfrac{49}{4}=\dfrac{1}{4}$$

$$\left(x-\dfrac{7}{2}\right)^2=\dfrac{1}{4}$$

$$x-\dfrac{7}{2}=\pm\dfrac{1}{2}$$

$$x=\dfrac{7}{2}\pm\dfrac{1}{2}$$

$x=\dfrac{7}{2}+\dfrac{1}{2}$ or $x=\dfrac{7}{2}-\dfrac{1}{2}$

$\quad=4\qquad\qquad=3$

6. $x^2+4x-32=0$

$$x^2+4x=32$$

$$x^2+4x+4=32+4$$

$$(x+2)^2=36$$

$$x+2=\pm\sqrt{36}$$

$$x+2=\pm6$$

$$x=-2\pm6$$

$$x=4 \text{ or } x=-8$$

7. $a^2+2a-9=0$

$$a^2+2a=9$$

$$a^2+2a+1=9+1$$

$$(a+1)^2=10$$

$$a+1=\pm\sqrt{10}$$

$$a=-1\pm\sqrt{10}$$

$$a=-1+\sqrt{10} \text{ or } a=-1-\sqrt{10}$$

8. $z^2+6z=12$

$$z^2+6z+9=12+9$$

$$(z+3)^2=21$$

$$z+3=\pm\sqrt{21}$$

$$z=-3\pm\sqrt{21}$$

$$z=-3+\sqrt{21} \text{ or } z=-3-\sqrt{21}$$

9. $x^2-2x+10=0$

$$x^2-2x=-10$$

$$x^2-2x+1=-10+1$$

$$(x-1)^2=-9$$

$$(x-1)^2=\sqrt{-9}$$

$$x-1=\pm3i$$

$$x=1\pm3i$$

$$x=1+3i \text{ or } x=1-3i$$

10. $2r^2-8r=-64$

$$r^2-4r=-32$$

$$r^2-4r+4=-32+4$$

$$(r-2)^2=-28$$

$$r-2=\pm\sqrt{-28}$$

$$r=2\pm\sqrt{4}\sqrt{7}\sqrt{-1}$$

$$r=2\pm2i\sqrt{7}$$

$$r=2+2i\sqrt{7} \text{ or } r=2-2i\sqrt{7}$$

11. a. Area = length × width
$$32 = (x+5)(x+1)$$

b. $32 = (x+5)(x+1)$
$$32 = x^2 + x + 5x + 5$$
$$0 = x^2 + 6x - 27$$
$$0 = (x-3)(x+9)$$
$$x = 3 \quad \text{or} \quad x = -9$$
Disregard the negative value. $x = 3$.

12. a. Area = length × width
$$63 = (x+2)(x+4)$$

b. $63 = (x+2)(x+4)$
$$63 = x^2 + 4x + 2x + 8$$
$$0 = x^2 + 6x - 55$$
$$0 = (x-5)(x+11)$$
$$x = 5 \quad \text{or} \quad x = -11$$
Disregard the negative value. $x = 5$.

13. Let x = the smaller integer. The larger will then be $x + 1$. Their product is 42.
$$x(x+1) = 42$$
$$x^2 + x = 42$$
$$x^2 + x - 42 = 0$$
$$(x+7)(x-6) = 0$$
$$x + 7 = 0 \quad \text{or} \quad x - 6 = 0$$
$$x = -7 \qquad\qquad x = 6$$
Since the integers must be positive, disregard the negative value. The smaller integer is 6 and the larger is 7.

14. Let x = the length of side of the square room. The diagonal can then be described by $x + 7$. Two of the adjacent sides and the diagonal make up a right triangle. Use the Pythagorean theorem to solve the problem.
$$a^2 + b^2 = c^2$$
$$x^2 + x^2 = (x+7)^2$$
$$2x^2 = x^2 + 14x + 49$$
$$x^2 - 14x - 49 = 0$$
$$x = \frac{-(-14) \pm \sqrt{(-14)^2 - 4(1)(-49)}}{2(1)} = \frac{14 \pm \sqrt{392}}{2}$$
$$x \approx 16.90 \quad \text{or} \quad x \approx -2.90$$
Disregard the negative value. The room is about 16.90 feet by 16.90 feet.

15. $2x^2 - 5x - 1 = 0$
$a = 2,\ b = -5,\ c = -1$
$b^2 - 4ac = (-5)^2 - 4(2)(-1) = 25 + 8 = 33$
Since the discriminant is positive, this equation has two distinct real solutions.

16. $3x^2 + 2x = -6$
$3x^2 + 2x + 6 = 0$
$a = 3,\ b = 2,\ c = 6$
$b^2 - 4ac = (2)^2 - 4(3)(6) = 4 - 72 = -68$
Since the discriminant is negative, this equation has no real solutions.

17. $r^2 + 16r = -64$
$r^2 + 16r + 64 = 0$
$a = 1,\ b = 16,\ c = 64$
$b^2 - 4ac = (16)^2 - 4(1)(64) = 256 - 256 = 0$
Since the discriminant is 0, the equation has one real solution.

18. $5x^2 - x + 2 = 0$
$a = 5,\ b = -1,\ c = 2$
$b^2 - 4ac = (-1)^2 - 4(5)(2) = 1 - 40 = -39$
Since the discriminant is negative, this equation has no real solutions.

19. $a^2 - 14n = -49$
$a^2 - 14n + 49 = 0$
$a = 1,\ b = -14,\ c = 49$
$b^2 - 4ac = (-14)^2 - 4(1)(49) = 196 - 196 = 0$
Since the discriminant is 0, the equation has one real solution.

20. $\dfrac{1}{2}x^2 - 3x = 8$
$\dfrac{1}{2}x^2 - 3x - 8 = 0$
$a = \dfrac{1}{2},\ b = -3,\ c = -8$
$b^2 - 4ac = (-3)^2 - 4\left(\dfrac{1}{2}\right)(-8) = 9 + 16 = 25$
Since the discriminant is positive, this equation has two real solutions.

21. $3x^2 + 4x = 0$

$a = 3, b = 4, c = 0$

$x = \dfrac{-b \pm \sqrt{b^2 - 4ac}}{2a}$

$x = \dfrac{-4 \pm \sqrt{4^2 - 4(3)(0)}}{2(3)}$

$= \dfrac{-4 \pm \sqrt{16}}{6}$

$= \dfrac{-4 \pm 4}{6}$

$x = \dfrac{-4 + 4}{6}$ or $x = \dfrac{-4 - 4}{6}$

$x = \dfrac{0}{6}$ $\quad x = \dfrac{-8}{6}$

$x = 0$ $\quad\quad x = -\dfrac{4}{3}$

22. $x^2 - 11x = -18$

$x^2 - 11x + 18 = 0$

$a = 1, b = -11, c = 18$

$x = \dfrac{-b \pm \sqrt{b^2 - 4ac}}{2a}$

$x = \dfrac{-(-11) \pm \sqrt{(-11)^2 - 4(1)(18)}}{2(1)}$

$= \dfrac{11 \pm \sqrt{121 - 72}}{2}$

$= \dfrac{11 \pm \sqrt{49}}{2}$

$= \dfrac{11 \pm 7}{2}$

$x = \dfrac{11 + 7}{2}$ or $x = \dfrac{11 - 7}{2}$

$= \dfrac{18}{2}$ $\quad\quad = \dfrac{4}{2}$

$= 9$ $\quad\quad\quad = 2$

23. $r^2 = 3r + 40$

$r^2 - 3r - 40 = 0$

$a = 1, b = -3, c = -40$

$r = \dfrac{-b \pm \sqrt{b^2 - 4ac}}{2a}$

$r = \dfrac{-(-3) \pm \sqrt{(-3)^2 - 4(1)(-40)}}{2(1)}$

$= \dfrac{3 \pm \sqrt{9 + 160}}{2}$

$= \dfrac{3 \pm \sqrt{169}}{2}$

$= \dfrac{3 \pm 13}{2}$

$r = \dfrac{3 + 13}{2}$ or $r = \dfrac{3 - 13}{2}$

$= \dfrac{16}{2}$ $\quad\quad = \dfrac{-10}{2}$

$= 8$ $\quad\quad\quad = -5$

24. $7x^2 = 9x$

$7x^2 - 9x = 0$

$a = 7, b = -9, c = 0$

$x = \dfrac{-b \pm \sqrt{b^2 - 4ac}}{2a}$

$x = \dfrac{-(-9) \pm \sqrt{(-9)^2 - 4(7)(0)}}{2(7)}$

$= \dfrac{9 \pm \sqrt{81 - 0}}{14}$

$= \dfrac{9 \pm \sqrt{81}}{14}$

$= \dfrac{9 \pm 9}{14}$

$x = \dfrac{9 + 9}{14}$ or $x = \dfrac{9 - 9}{14}$

$= \dfrac{18}{14}$ $\quad\quad = \dfrac{0}{14}$

$= \dfrac{9}{7}$ $\quad\quad\quad = 0$

25. $6a^2 + a - 15 = 0$

$a = 6, b = 1, c = -15$

$$a = \frac{-b \pm \sqrt{b^2 - 4ac}}{2a}$$

$$a = \frac{-1 \pm \sqrt{1^2 - 4(6)(-15)}}{2(6)}$$

$$= \frac{-1 \pm \sqrt{1 + 360}}{12}$$

$$= \frac{-1 \pm \sqrt{361}}{12}$$

$$= \frac{-1 \pm 19}{12}$$

$$a = \frac{-1 + 19}{12} \quad \text{or} \quad a = \frac{-1 - 19}{12}$$

$$= \frac{18}{12} \qquad\qquad = \frac{-20}{12}$$

$$= \frac{3}{2} \qquad\qquad = -\frac{5}{3}$$

26. $4x^2 + 11x = 3$

$4x^2 + 11x - 3 = 0$

$a = 4, b = 11, c = -3$

$$x = \frac{-b \pm \sqrt{b^2 - 4ac}}{2a}$$

$$x = \frac{-(11) \pm \sqrt{(11)^2 - 4(4)(-3)}}{2(4)}$$

$$= \frac{-11 \pm \sqrt{121 + 48}}{8}$$

$$= \frac{-11 \pm \sqrt{169}}{8}$$

$$= \frac{-11 \pm 13}{8}$$

$$x = \frac{-11 + 13}{8} \quad \text{or} \quad x = \frac{-11 - 13}{8}$$

$$= \frac{2}{8} \qquad\qquad = \frac{-24}{8}$$

$$= \frac{1}{4} \qquad\qquad = -3$$

27. $x^2 + 8x + 5 = 0$

$a = 1, b = 8, c = 5$

$$x = \frac{-b \pm \sqrt{b^2 - 4ac}}{2a}$$

$$x = \frac{-8 \pm \sqrt{8^2 - 4(1)(5)}}{2(1)}$$

$$= \frac{-8 \pm \sqrt{64 - 20}}{2}$$

$$= \frac{-8 \pm \sqrt{44}}{2}$$

$$= \frac{-8 \pm 2\sqrt{11}}{2} = -4 \pm \sqrt{11}$$

$$x = -4 + \sqrt{11} \quad \text{or} \quad x = -4 - \sqrt{11}$$

28. $b^2 + 4b = 8$

$b^2 + 4b - 8 = 0$

$a = 1, b = 4, c = -8$

$$b = \frac{-b \pm \sqrt{b^2 - 4ac}}{2a}$$

$$b = \frac{-(4) \pm \sqrt{(4)^2 - 4(1)(-8)}}{2(1)}$$

$$= \frac{-4 \pm \sqrt{16 + 32}}{2}$$

$$= \frac{-4 \pm \sqrt{48}}{2}$$

$$= \frac{-4 \pm 4\sqrt{3}}{2} = -2 \pm 2\sqrt{3}$$

$$x = -2 + 2\sqrt{3} \quad \text{or} \quad x = -2 - 2\sqrt{3}$$

29. $2x^2 + 4x - 3 = 0$

$a = 2, b = 4, c = -3$

$$x = \frac{-b \pm \sqrt{b^2 - 4ac}}{2a}$$

$$x = \frac{-4 \pm \sqrt{4^2 - 4(2)(-3)}}{2(2)}$$

$$= \frac{-4 \pm \sqrt{16 + 24}}{4}$$

$$= \frac{-4 \pm \sqrt{40}}{4}$$

$$= \frac{-4 \pm 2\sqrt{10}}{4} = \frac{-2 \pm \sqrt{10}}{2}$$

$$x = \frac{-2 + \sqrt{10}}{2} \quad \text{or} \quad x = \frac{-2 - \sqrt{10}}{2}$$

30. $3y^2 - 6y = 8$

$3y^2 - 6y - 8 = 0$

$a = 3, b = -6, c = -8$

$y = \dfrac{-b \pm \sqrt{b^2 - 4ac}}{2a}$

$y = \dfrac{-(-6) \pm \sqrt{(-6)^2 - 4(3)(-8)}}{2(3)}$

$= \dfrac{6 \pm \sqrt{36 + 96}}{6}$

$= \dfrac{6 \pm \sqrt{132}}{6}$

$= \dfrac{6 \pm 2\sqrt{33}}{6}$

$= \dfrac{2\left(3 \pm \sqrt{33}\right)}{6}$

$= \dfrac{3 \pm \sqrt{33}}{3}$

$y = \dfrac{3 + \sqrt{33}}{3}$ or $y = \dfrac{3 - \sqrt{33}}{3}$

31. $x^2 - x + 13 = 0$

$a = 1, b = -1, c = 13$

$x = \dfrac{-b \pm \sqrt{b^2 - 4ac}}{2a}$

$x = \dfrac{-(-1) \pm \sqrt{(-1)^2 - 4(1)(13)}}{2(1)}$

$= \dfrac{1 \pm \sqrt{1 - 52}}{2}$

$= \dfrac{1 \pm \sqrt{-51}}{2}$

$= \dfrac{1 \pm i\sqrt{51}}{2}$

$x = \dfrac{1 + i\sqrt{51}}{2}$ or $x = \dfrac{1 - i\sqrt{51}}{2}$

32. $x^2 - 2x + 11 = 0$

$a = 1, b = -2, c = 11$

$x = \dfrac{-b \pm \sqrt{b^2 - 4ac}}{2a}$

$x = \dfrac{-(-2) \pm \sqrt{(-2)^2 - 4(1)(11)}}{2(1)}$

$= \dfrac{2 \pm \sqrt{4 - 44}}{2}$

$= \dfrac{2 \pm \sqrt{-40}}{2}$

$= \dfrac{2 \pm 2i\sqrt{10}}{2}$

$= 1 \pm i\sqrt{10}$

$x = 1 + i\sqrt{10}$ or $x = 1 - i\sqrt{10}$

33. $2x^2 - \dfrac{5}{3}x = \dfrac{25}{3}$

$3\left(2x^2 - \dfrac{5}{3}x\right) = 3\left(\dfrac{25}{3}\right)$

$6x^2 - 5x = 25$

$6x^2 - 5x - 25 = 0$

$a = 6, b = -5, c = -25$

$x = \dfrac{-b \pm \sqrt{b^2 - 4ac}}{2a}$

$x = \dfrac{-(-5) \pm \sqrt{(-5)^2 - 4(6)(-25)}}{2(6)}$

$x = \dfrac{5 \pm \sqrt{25 + 600}}{12}$

$= \dfrac{5 \pm \sqrt{625}}{12}$

$= \dfrac{5 \pm 25}{12}$

$x = \dfrac{5 + 25}{12}$ or $x = \dfrac{5 - 25}{12}$

$= \dfrac{30}{12}$ $= \dfrac{-20}{12}$

$= \dfrac{5}{2}$ $= -\dfrac{5}{3}$

34. $4x^2 + 5x - \dfrac{3}{2} = 0$

$2\left(4x^2 + 5x - \dfrac{3}{2}\right) = 2(0)$

$8x^2 + 10x - 3 = 0$

$a = 8,\ b = 10,\ c = -3$

$x = \dfrac{-b \pm \sqrt{b^2 - 4ac}}{2a}$

$= \dfrac{-10 \pm \sqrt{10^2 - 4(8)(-3)}}{2(8)}$

$= \dfrac{-10 \pm \sqrt{100 + 96}}{16}$

$= \dfrac{-10 \pm \sqrt{196}}{16}$

$= \dfrac{-10 \pm 14}{16}$

$x = \dfrac{-10 + 14}{16}$ or $x = \dfrac{-10 - 14}{16}$

$\quad = \dfrac{1}{4} \qquad\qquad\quad = -\dfrac{3}{2}$

35. $f(x) = x^2 - 4x - 35,\ f(x) = 25$

$x^2 - 4x - 35 = 25$

$x^2 - 4x - 60 = 0$

$(x - 10)(x + 6) = 0$

$x - 10 = 0$ or $x + 6 = 0$

$\quad x = 10 \qquad\qquad x = -6$

The solutions are 10 and -6.

36. $g(x) = 6x^2 + 5x,\ g(x) = 6$

$6x^2 + 5x = 6$

$6x^2 + 5x - 6 = 0$

$(2x + 3)(3x - 2) = 0$

$2x + 3 = 0$ or $3x - 2 = 0$

$\quad x = -\dfrac{3}{2} \qquad\qquad x = \dfrac{2}{3}$

The solutions are $-\dfrac{3}{2}$ and $\dfrac{2}{3}$.

37. $h(r) = 5r^2 - 7r - 10,\ h(r) = -8$

$5r^2 - 7r - 10 = -8$

$5r^2 - 7r - 2 = 0$

$r = \dfrac{7 \pm \sqrt{(-7)^2 - 4(5)(-2)}}{2(5)}$

$= \dfrac{7 \pm \sqrt{49 + 40}}{10}$

$= \dfrac{7 \pm \sqrt{89}}{10}$

The solutions are $\dfrac{7 + \sqrt{89}}{10}$ and $\dfrac{7 - \sqrt{89}}{10}$.

38. $f(x) = -2x^2 + 6x + 7,\ f(x) = -2$

$-2x^2 + 6x + 7 = -2$

$-2x^2 + 6x + 9 = 0$

$x = \dfrac{-6 \pm \sqrt{6^2 - 4(-2)(9)}}{2(-2)}$

$= \dfrac{-6 \pm \sqrt{36 + 72}}{-4}$

$= \dfrac{-6 \pm \sqrt{108}}{-4}$

$= \dfrac{-6 \pm 6\sqrt{3}}{-4}$

$= \dfrac{3 \pm 3\sqrt{3}}{2}$

The solutions are $\dfrac{3 + 3\sqrt{3}}{2}$ and $\dfrac{3 - 3\sqrt{3}}{2}$.

39. Solutions are $x = 3$ and $x = -1$.

$x - 3 = 0$ and $x + 1 = 0$.

$(x - 3)(x + 1) = 0$

$x^2 - 2x - 3 = 0$

40. Solutions are $x = \dfrac{2}{3}$ and $x = -2$.

$3x = 2$ and $x + 2 = 0$

$3x - 2 = 0$

$(3x - 2)(x + 2) = 0$

$3x^2 + 4x - 4 = 0$

41. Solutions are $x = -\sqrt{11}$ and $x = \sqrt{11}$.
$$x + \sqrt{11} = 0 \text{ and } x - \sqrt{11} = 0.$$
$$\left(x + \sqrt{11}\right)\left(x - \sqrt{11}\right) = 0$$
$$x^2 - 11 = 0$$

42. Solutions are $x = 3 - 2i$ and $x = 3 + 2i$.
$$x - (3 - 2i) = 0 \text{ and } x - (3 + 2i) = 0$$
$$x - 3 + 2i = 0 \qquad x - 3 - 2i = 0$$
$$(x - 3 + 2i)(x - 3 - 2i) = 0$$
$$[(x - 3) + 2i][(x - 3) - 2i] = 0$$
$$(x - 3)^2 - (2i)^2 = 0$$
$$x^2 - 6x + 9 - 4i^2 = 0$$
$$x^2 - 6x + 9 + 4 = 0$$
$$x^2 - 6x + 13 = 0$$

43. Let x = the width of the garden. Then the length is $x + 4$.
Area = length × width
$$96 = (x + 4)x$$
$$96 = x^2 + 4x$$
$$0 = x^2 + 4x - 96$$
$$0 = (x + 12)(x - 8)$$
$$x + 12 = 0 \quad \text{or} \quad x - 8 = 0$$
$$x = -12 \qquad x = 8$$
Disregard the negative value. The width is 8 feet and the length is $8 + 4 = 12$ feet.

44. Using the Pythagorean Theorem, $a^2 + b^2 = c^2$
$$8^2 + 8^2 = x^2$$
$$64 + 64 = x^2$$
$$128 = x^2$$
$$\sqrt{128} = x$$
$$x = 8\sqrt{2} \approx 11.31$$

45. $A = P\left(1 + \dfrac{r}{n}\right)^{nt}$
$$1081.60 = 1000\left(1 + \frac{r}{1}\right)^{1(2)}$$
$$1081.60 = 1000(1 + r)^2$$
$$1.08160 = (1 + r)^2$$
$$\pm 1.04 = 1 + r$$
$$r = -1 \pm 1.04$$
The rate must be positive, so
$r = -1 + 1.04 = 0.04$.
The annual interest is 4%.

46. Let x be the smaller positive number. Then $x + 4$ is the larger positive number.
$$x(x + 4) = 77$$
$$x^2 + 4x - 77 = 0$$
$$(x + 11)(x - 7) = 0$$
$$x = -11 \quad \text{or} \quad x = 7$$
Since x must be positive, $x = 7$ and $7 + 2 = 9$.

47. Let x be the width. Then $2x - 4$ is the length and the equation is $A = lw$.
$$96 = (2x - 4)(x)$$
$$96 = 2x^2 - 4x$$
$$0 = 2x^2 - 4x - 96$$
$$0 = (2x + 12)(x - 8)$$
$$0 = 2x + 12 \quad \text{or} \quad 0 = x - 8$$
$$-12 = 2x \qquad\qquad 8 = x$$
$$-6 = x$$
Since the width must be positive, $x = 8$.
The width is 8 inches and the length $2x - 4$ is $2(8) - 4 = 16 - 4 = 12$ inches.

48. $V = 12d - 0.05d^2$, $d = 60$
$$V = 12(60) - 0.05(60)^2 = 720 - 180 = 540$$
The value is $540.

49. a. Note that 2004 is represented by $t = 3$.
$$E(t) = 7t^2 - 7.8t + 82.2$$
$$E(3) = 7(3)^2 - 7.8(3) + 82.2 = 121.8$$
The expenditure of oil companies in 2004 was about $121.8 billion.

b. $E(t) = 579$
$$579 = 7t^2 - 7.8t + 82.2$$
$$0 = 7t^2 - 7.8t - 496.8$$
$$0 = 7t^2 - 7.8t - 496.8$$
$$t = \frac{-(-7.8) \pm \sqrt{(-7.8)^2 - 4(7)(-496.8)}}{2(7)}$$
$$= \frac{7.8 \pm \sqrt{13,971.24}}{14} = \frac{7.8 \pm 118.2}{14}$$
$$t = \frac{7.8 + 118.2}{14} \quad \text{or} \quad t = \frac{7.8 - 118.2}{14}$$
$$= \frac{126}{14} = 9 \qquad\qquad = -\frac{110.4}{14}$$
Since the time must be positive, $t = 9$, which represents the year 2010. If the trend continues, the expenditure by oil companies will be $579 billion in the year 2010.

50. $d = -16t^2 + 784$

  a. $d = -16(2)^2 + 784 = -64 + 784 = 720$

  The object is 720 feet from the ground 2 seconds after being dropped.

  b. $0 = -16t^2 + 784$

  $16t^2 = 784$

  $t^2 = 49$

  $t = \pm\sqrt{49} = \pm 7$

  Since the time must be positive, $t = 7$ seconds.

51. a. $L(t) = 0.0004t^2 + 0.16t + 20$

  $L(100) = 0.0004(100)^2 + 0.16(100) + 20 = 40$

  40 milliliters will leak out at 100°C.

  b. $53 = 0.0004t^2 + 0.16t + 20$

  $0 = 0.0004t^2 + 0.16t - 33$

  $t = \dfrac{-0.16 \pm \sqrt{(0.16)^2 - 4(0.0004)(-33)}}{2(0.0004)}$

  $= \dfrac{-0.16 \pm \sqrt{0.0784}}{0.0008}$

  $= \dfrac{-0.16 \pm 0.28}{0.0008}$

  $t = \dfrac{-0.16 + 0.28}{0.0008}$ or $t = \dfrac{-0.16 - 0.28}{0.0008}$

  $= 150$ $= -550$

  Since the temperature must be positive, $t = 150$. The operating temperature is 150°C.

52. Let x be the time in which the smaller machine can do the job then $x - 1$ is the time for the larger machine.

$$\frac{12}{x} + \frac{12}{x-1} = 1$$

$$x(x-1)\left(\frac{12}{x}\right) + x(x-1)\left(\frac{12}{x-1}\right) = x(x-1)$$

$$12(x-1) + 12x = x^2 - x$$

$$12x - 12 + 12x = x^2 - x$$

$$-12 + 24x = x^2 - x$$

$$0 = x^2 - 25x + 12$$

$$x = \frac{-(-25) \pm \sqrt{(-25)^2 - 4(1)(12)}}{2(1)} = \frac{25 \pm \sqrt{577}}{2}$$

$$x = \frac{25 + \sqrt{577}}{2} \approx 24.51 \text{ or } x = \frac{25 - \sqrt{577}}{2} \approx 0.49$$

x cannot equal 0.49 since this would mean the smaller machine could do the work in 0.49 hours and the larger can do the work in $x - 1$ or -0.51 hours. Therefore the smaller machine does the work in 24.51 hours and the larger machine does the work in 23.51 hours.

53. Let x be the speed (in miles per hour) for the first 25 miles. Then, the speed for the next 65 miles is $x + 15$. The time for the first 20 miles is $\dfrac{d}{r} = \dfrac{25}{x}$ and the time for the next 30 miles is $\dfrac{d}{r} = \dfrac{65}{x+15}$. The total time is 1.5 hours.

$$\frac{25}{x} + \frac{65}{x+15} = 1.5$$

$$x(x+15)\left(\frac{25}{x} + \frac{65}{x+15}\right) = x(x+15)(1.5)$$

$$25(x+15) + x(65) = (x^2 + 15x)(1.5)$$

$$90x + 375 = 1.5x^2 + 22.5x$$

$$0 = 1.5x^2 - 67.5x - 375$$

$$0 = 3x^2 - 135x - 750$$

$$0 = 3(x - 50)(x + 5)$$

$$x - 50 = 0 \quad \text{or} \quad x + 5 = 0$$

$$x = 50 \quad \text{or} \quad x = -5$$

Since the speed must be positive, $x = 50$. Thus, the speed was 50 mph.

54. Let r be the speed (in miles per hour) of the canoe in still water. For the trip downstream, the rate is $r + 0.4$ and the distance 3 miles so that the time is $t = \dfrac{3}{r + 0.4}$. For the trip upstream the rate is $r - 0.4$ and the distance is 3 miles so that the time is $t = \dfrac{3}{r - 0.4}$. The total time is 4 hours.

$$\frac{3}{r+0.4} + \frac{3}{r-0.4} = 4$$

$$(r+0.4)(r-0.4)\left[\frac{3}{r+0.4} + \frac{3}{r-0.4} = 4\right]$$

$$3(r-0.4) + 3(r+0.4) = 4(r+0.4)(r-0.4)$$

$$3r - 1.2 + 3r + 1.2 = 4(r^2 - 0.16)$$

$$6r = 4r^2 - 0.64$$

$$0 = 4r^2 - 6r - 0.64$$

$$0 = 2r^2 - 3r - 0.32$$

$$r = \frac{-(-3) \pm \sqrt{(-3)^2 - 4(2)(-0.32)}}{2(2)}$$

$$= \frac{3 \pm \sqrt{9 + 2.56}}{4}$$

$$= \frac{3 \pm \sqrt{11.56}}{4}$$

$$= \frac{3 \pm 3.4}{4}$$

$$r = \frac{3 + 3.4}{4} \quad \text{or} \quad r = \frac{3 - 3.4}{4}$$

$$= \frac{6.4}{4} \qquad\qquad = \frac{-0.4}{4}$$

$$= 1.6 \qquad\qquad = -0.1$$

Since the rate must be positive, $r = 1.6$. Rachel canoes 1.6 miles per hour in still water.

55. Let x be the length. The width is $x - 2$.
Area = length × width

$$80 = x(x - 2)$$

$$80 = x^2 - 2x$$

$$0 = x^2 - 2x - 80$$

$$x = \frac{-(-2) \pm \sqrt{(-2)^2 - 4(1)(-80)}}{2(1)}$$

$$= \frac{2 \pm \sqrt{324}}{2} = \frac{2 \pm 18}{2}$$

$$x = \frac{20}{2} = 10 \quad \text{or} \quad x = \frac{-16}{2} = -8$$

Since the width must be positive, $x = 10$. The length is 10 units and the width is $10 - 2 = 8$ units.

56. If the business sells n tables at a price of $(60 - 0.3n)$ dollars per table, then the revenue is given by $R(n) = n(60 - 0.3n)$ with $n \le 40$. Set this equal to 1080 and solve for n.

$$R(n) = n(60 - 0.3n)$$

$$1080 = n(60 - 0.3n)$$

$$1080 = 60n - 0.3n^2$$

$$0.3n^2 - 60n + 1080 = 0$$

$$n^2 - 200n + 3600 = 0$$

$$(n - 20)(n - 180) = 0$$

$n = 20$ or $n = 180$
Disregard $n = 180$ since $n \le 40$. The business must sell 20 tables.

57. $a^2 + b^2 = c^2$

$$a^2 = c^2 - b^2$$

$$\sqrt{a^2} = \pm\sqrt{c^2 - b^2}$$

$$a = \pm\sqrt{c^2 - b^2}$$

Since a refers to a length, it cannot be negative. Therefore disregard the negative sign.

$$a = \sqrt{c^2 - b^2}$$

58.
$$h = -4.9t^2 + c$$

$$h - c = -4.9t^2$$

$$\frac{h - c}{-4.9} = \frac{-4.9t^2}{-4.9}$$

$$\frac{h - c}{-4.9} = t^2$$

$$\pm\sqrt{\frac{h - c}{-4.9}} = t \quad \text{or} \quad t = \pm\sqrt{\frac{c - h}{4.9}}$$

Disregard the negative value for time:

$$t = \sqrt{\frac{c - h}{4.9}}$$

59. $v_x^2 + v_y^2 = v^2$ for v_y

$$v_y^2 = v^2 - v_x^2$$

$$v_y = \sqrt{v^2 - v_x^2}$$

60. $a = \dfrac{v_2^2 - v_1^2}{2d}$ for v_2

$$2ad = v_2^2 - v_1^2$$

$$2ad + v_1^2 = v_2^2$$

$$v_2 = \sqrt{v_1^2 + 2ad}$$

61. Let $u = x^2$.

$$x^4 - x^2 - 20 = (x^2)^2 - x^2 - 20$$

$$= u^2 - u - 20$$

$$= (u + 4)(u - 5)$$

Back substitute x^2 for u.

$$= (x^2 + 4)(x^2 - 5)$$

62. Let $u = x^2$.

$$4x^4 + 4x^2 - 3 = 4(x^2)^2 + 4x^2 - 3$$

$$= 4u^2 + 4u - 3$$

$$= (2u - 1)(2u + 3)$$

Back substitute x^2 for u.

$$= (2x^2 - 1)(2x^2 + 3)$$

63. Let $u = x+5$.

$(x+5)^2 + 10(x+5) + 24$

$= u^2 + 10u + 24$

$= (u+4)(u+6)$

Back substitute $x+5$ for u.

$= [(x+5)+4][(x+5)+6]$

$= (x+9)(x+11)$

64. Let $u = 2x+3$.

$4(2x+3)^2 - 12(2x+3) + 5$

$= 4u^2 - 12u + 5$

$= (2u-5)(2u-1)$

Back substitute $2x+3$ for u.

$= [2(2x+3)-5][2(2x+3)-1]$

$= (4x+6-5)(4x+6-1)$

$= (4x+1)(4x+5)$

65. $x^4 - 13x^2 + 36 = 0$

$(x^2)^2 - 13x^2 + 36 = 0$

Let $u = x^2$

$u^2 - 13u + 36 = 0$

$(u-9)(u-4) = 0$

$u-9 = 0$ or $u-4 = 0$

$u = 9 \qquad u = 4$

Substitute x^2 for u.

$x^2 = 9$ or $x^2 = 4$

$x = \pm 3 \qquad x = \pm 2$

The solutions are ± 3 and ± 2.

66. $x^4 - 21x^2 + 80 = 0$

$(x^2)^2 - 21x^2 + 80 = 0$

Let $u = x^2$

$u^2 - 21u + 80 = 0$

$(u-16)(u-5) = 0$

$u-16 = 0$ or $u-5 = 0$

$u = 16 \qquad u = 5$

Substitute x^2 for u.

$x^2 = 16$ or $x^2 = 5$

$x = \pm 4 \qquad x = \pm\sqrt{5}$

The solutions are ± 4 and $\pm\sqrt{5}$.

67. $a^4 = 5a^2 + 24$

$a^4 - 5a^2 - 24 = 0$

$(a^2)^2 - 5a^2 - 24 = 0$

Let $u = a^2$

$u^2 - 5u - 24 = 0$

$(u-8)(u+3) = 0$

$u-8 = 0$ or $u+3 = 0$

$u = 8 \qquad u = -3$

Substitute a^2 for u.

$a^2 = 8$ or $a^2 = -3$

$a = \pm\sqrt{8} \qquad a = \pm\sqrt{-3}$

$= \pm 2\sqrt{2} \qquad = \pm i\sqrt{3}$

The solutions are $\pm 2\sqrt{2}$ and $\pm i\sqrt{3}$.

68. $3y^{-2} + 16y^{-1} = 12$

$3(y^{-1})^2 + 16(y^{-1}) - 12 = 0$

Let $u = y^{-1}$.

$3u^2 + 16u - 12 = 0$

$(3u-2)(u+6) = 0$

$3u-2 = 0$ or $u+6 = 0$

$u = \dfrac{2}{3} \qquad u = -6$

Substitute y^{-1} for u.

$y^{-1} = \dfrac{2}{3}$ or $y^{-1} = -6$

$y = \dfrac{3}{2} \qquad y = -\dfrac{1}{6}$

The solutions are $\dfrac{3}{2}$ and $-\dfrac{1}{6}$.

69.

$$3r + 11\sqrt{r} - 4 = 0$$

$$3\left(r^{1/2}\right)^2 + 11r^{1/2} - 4 = 0$$

Let $u = r^{1/2}$

$$3u^2 + 11u - 4 = 0$$

$$(3u - 1)(u + 4) = 0$$

$$3u - 1 = 0 \quad \text{or} \quad u + 4 = 0$$

$$u = \frac{1}{3} \qquad u = -4$$

Substitute $r^{1/2}$ for u.

$$r^{1/2} = \frac{1}{3} \qquad \text{or} \qquad r^{1/2} = -6$$

$$\text{not real}$$

$$r = \left(\frac{1}{3}\right)^2 = \frac{1}{9}$$

There are no solutions for $r^{1/2} = -6$ since there is no real number x for which $r^{1/2} = -6$.

The solution is $\dfrac{1}{9}$.

70.

$$2p^{2/3} - 7p^{1/3} + 6 = 0$$

$$2\left(p^{1/3}\right)^2 - 7p^{1/3} + 6 = 0$$

Let $u = p^{1/3}$

$$2u^2 - 7u + 6 = 0$$

$$(2u - 3)(u - 2) = 0$$

$$2u - 3 = 0 \quad \text{or} \quad u - 2 = 0$$

$$u = \frac{3}{2} \qquad u = 2$$

Substitute $p^{1/3}$ for u.

$$p^{1/3} = \frac{3}{2} \qquad p^{1/3} = 2$$

$$p = \left(\frac{3}{2}\right)^3 \qquad p = 2^3$$

$$= \frac{27}{8} \qquad = 8$$

The solutions are $\dfrac{27}{8}$ and 8.

71.

$$6(x - 2)^{-2} = -13(x - 2)^{-1} + 8$$

$$6\left[(x - 2)^{-1}\right]^2 + 13(x - 2)^{-1} - 8 = 0$$

Let $u = (x - 2)^{-1}$

$$6u^2 + 13u - 8 = 0$$

$$(2u - 1)(3u + 8) = 0$$

$$2u - 1 = 0 \quad \text{or} \quad 3u + 8 = 0$$

$$u = \frac{1}{2} \qquad u = -\frac{8}{3}$$

Substitute $(x - 2)^{-1}$ for u.

$$(x - 2)^{-1} = \frac{1}{2} \quad \text{or} \quad (x - 2)^{-1} = -\frac{8}{3}$$

$$x - 2 = 2 \qquad x - 2 = -\frac{3}{8}$$

$$x = 4 \qquad x = \frac{13}{8}$$

The solutions are 4 and $\dfrac{13}{8}$.

72.

$$10(r + 1) = \frac{12}{r + 1} - 7$$

$$(r + 1)\left[10(r + 1)\right] = (r + 1)\left[\frac{12}{r + 1} - 7\right]$$

$$10(r + 1)^2 + 7(r + 1) = 12$$

$$10(r + 1)^2 + 7(r + 1) - 12 = 0$$

Let $u = r + 1$.

$$10u^2 + 7u - 12 = 0$$

$$(5u - 4)(2u + 3) = 0$$

$$5u - 4 = 0 \quad \text{or} \quad 2u + 3 = 0$$

$$u = \frac{4}{5} \qquad u = -\frac{3}{2}$$

Substitute $(r + 1)$ for u.

$$r + 1 = \frac{4}{5} \quad \text{or} \quad r + 1 = -\frac{3}{2}$$

$$r = -\frac{1}{5} \qquad r = -\frac{5}{2}$$

The solutions are $-\dfrac{1}{5}$ and $-\dfrac{5}{2}$.

73. $f(x) = x^4 - 82x^2 + 81$

To find the *x*-intercepts, set $f(x) = 0$.

$0 = x^4 - 82x^2 + 81$

$0 = \left(x^2\right)^2 - 82x^2 + 81$

Let $u = x^2$.

$0 = u^2 - 82u + 81$

$0 = (u - 81)(u - 1)$

$u - 81 = 0 \quad$ or $\quad u - 1 = 0$

$\qquad u = 81 \qquad\qquad u = 1$

Substitute x^2 for *u*.

$x^2 = 81 \quad$ or $\quad x^2 = 1$

$\quad x = \pm 9 \qquad\qquad x = \pm 1$

The *x*-intercepts are (9, 0), (–9, 0), (1, 0), and (–1, 0).

74. $f(x) = 30x + 13\sqrt{x} - 10$

To find the *x*-intercepts, set $f(x) = 0$.

$0 = 30x + 13\sqrt{x} - 10$

$0 = 30\left(\sqrt{x}\right)^2 + 13\sqrt{x} - 10$

Let $u = \sqrt{x}$.

$0 = 30u^2 + 13u - 10$

$0 = (6u + 5)(5u - 2)$

$6u + 5 = 0 \quad$ or $\quad 5u - 2 = 0$

$\qquad u = -\dfrac{5}{6} \qquad\qquad u = \dfrac{2}{5}$

Substitute $\sqrt{x}$ for *u*.

$\sqrt{x} = -\dfrac{5}{6} \quad$ or $\quad \sqrt{x} = \dfrac{2}{5}$

$\quad$ not real $\qquad\qquad x = \dfrac{4}{25}$

Since $\sqrt{x}$ cannot be negative, the solution is $\dfrac{4}{25}$. The only *x*-intercept is $\left(\dfrac{4}{25}, 0\right)$.

75. $f(x) = x - 6\sqrt{x} + 12$

To find the *x*-intercepts, set $f(x) = 0$.

$0 = x - 6\sqrt{x} + 12$

$0 = \left(\sqrt{x}\right)^2 - 6\sqrt{x} + 12$

Let $u = \sqrt{x}$.

$0 = u^2 - 6u + 12$

$u = \dfrac{-(-6) \pm \sqrt{(-6)^2 - 4(1)(10)}}{2(1)}$

$\quad = \dfrac{6 \pm \sqrt{-4}}{2} = \dfrac{6 \pm 2i}{2} = 3 \pm i$

Substitute $\sqrt{x}$ for *u*.

$\sqrt{x} = 3 \pm i$

Since *x*-intercepts must be real numbers, this function has no *x*-intercepts.

76. $g(x) = \left(x^2 - 6x\right)^2 - 5\left(x^2 - 6x\right) - 24$

To find the *x*-intercepts, set $g(x) = 0$.

$0 = \left(x^2 - 6x\right)^2 - 5\left(x^2 - 6x\right) - 24$

Let $u = x^2 - 6x$.

$0 = u^2 - 5u - 24$

$0 = (u + 3)(u - 8)$

$u + 3 = 0 \quad$ or $\quad u - 8 = 0$

$\qquad u = -3 \qquad\qquad u = 8$

Substitute $\left(x^2 - 6x\right)$ for *u*.

$\quad x^2 - 6x = -3 \quad$ or $\quad x^2 - 6x = 8$

$x^2 - 6x + 3 = 0 \qquad x^2 - 6x - 8 = 0$

$\dfrac{6 \pm \sqrt{(-6)^2 - 4(1)(3)}}{2(1)} \quad \dfrac{6 \pm \sqrt{(-6)^2 - 4(1)(-8)}}{2(1)}$

$\qquad \dfrac{6 \pm \sqrt{24}}{2} \qquad\qquad \dfrac{6 \pm \sqrt{68}}{2}$

$\qquad \dfrac{6 \pm 2\sqrt{6}}{2} \qquad\qquad \dfrac{6 \pm 2\sqrt{17}}{2}$

$\qquad 3 \pm \sqrt{6} \qquad\qquad\quad 3 \pm \sqrt{17}$

The *x*-intercepts are $\left(3 + \sqrt{6}, 0\right), \left(3 - \sqrt{6}, 0\right)$, $\left(3 + \sqrt{17}, 0\right)$ and $\left(3 - \sqrt{17}, 0\right)$.

77. $f(x) = x^2 + 5x$

   a. Since $a = 1$ the parabola opens upward.

   b. $y = f(0) = 0^2 + 5(0) = 0$

      The y-intercept is $(0, 0)$.

   c. $x = -\dfrac{b}{2a} = -\dfrac{5}{2(1)} = -\dfrac{5}{2}$

      $y = \dfrac{4ac - b^2}{4a} = \dfrac{4(1)(0) - 5^2}{4(1)} = -\dfrac{25}{4}$

      The vertex is $\left(-\dfrac{5}{2}, -\dfrac{25}{4}\right)$.

   d. $0 = x^2 + 5x$

      $0 = x(x + 5)$

      $0 = x$ or $0 = x + 5$

      $x = 0$ $x = -5$

      The x-intercepts are $(0, 0)$ and $(-5, 0)$.

   e.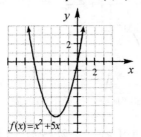

78. $f(x) = x^2 - 2x - 8$

   a. Since $a = 1$ the parabola opens upward.

   b. $y = f(0) = (0)^2 - 2(0) - 8 = -8$

      The y-intercept is $(0, -8)$.

   c. $x = -\dfrac{b}{2a} = -\dfrac{-2}{2(1)} = \dfrac{2}{2} = 1$

      $y = \dfrac{4ac - b^2}{4a} = \dfrac{4(1)(-8) - (-2)^2}{4(1)} = \dfrac{-36}{4} = -9$

      The vertex is $(1, -9)$.

   d. $0 = x^2 - 2x - 8$

      $0 = (x - 4)(x + 2)$

      $0 = x - 4$ or $0 = x + 2$

      $4 = x$ or $-2 = x$

      The x-intercepts are $(4, 0)$ and $(-2, 0)$.

   e.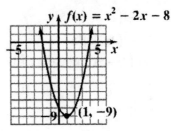

79. $g(x) = -x^2 - 2$

   a. Since $a = -1$ the parabola opens downward.

   b. $y = g(0) = -(0)^2 - 2 = -2$

      The y-intercept is $(0, -2)$.

   c. $x = -\dfrac{b}{2a} = -\dfrac{0}{2(-1)} = -\dfrac{0}{-2} = 0$

      $y = \dfrac{4ac - b^2}{4a} = \dfrac{4(-1)(-2) - 0^2}{4(-1)} = \dfrac{8}{-4} = -2$

      The vertex is $(0, -2)$.

   d. $0 = -x^2 - 2$

      $x^2 = -2$

      $x = \pm\sqrt{-2} = \pm i\sqrt{2}$

      There are no real roots. Thus, there are no x-intercepts.

   e.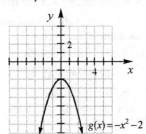

80. $g(x) = -2x^2 - x + 15$

   a. Since $a = -2$ the parabola opens downward.

   b. $y = -2(0)^2 - 0 + 15 = 15$

      The y-intercept is $(0, 15)$.

   c. $x = -\dfrac{b}{2a} = -\dfrac{-1}{2(-2)} = \dfrac{1}{-4} = -\dfrac{1}{4}$

      $y = \dfrac{4ac - b^2}{4a} = \dfrac{4(-2)(15) - (-1)^2}{4(-2)} = \dfrac{121}{8}$

      The vertex is $\left(-\dfrac{1}{4}, \dfrac{121}{8}\right)$.

d. $0 = -1\left(2x^2 + x - 15\right)$

$0 = -1\left(2x - 5\right)\left(x + 3\right)$

$0 = \left(2x - 5\right)$ or $0 = x + 3$

$5 = 2x$

$\dfrac{5}{2} = x$ $\qquad\qquad -3 = x$

The *x*-intercepts are $(-3, 0)$ and $\left(\dfrac{5}{2}, 0\right)$.

e.

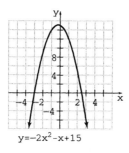

$y = -2x^2 - x + 15$

81. a. $I = -x^2 + 22x - 45$, $2 \le x \le 20$
The *x*-coordinate of the vertex will be the cost per ticket to maximize profit.

$x = -\dfrac{b}{2a} = -\dfrac{22}{2(-1)} = 11$

They should charge $11 per ticket.

b. The maximum profit in hundreds is the *y*-coordinate of the vertex.

$I(11) = -11^2 + 22(11) - 45$

$= -121 + 242 - 45$

$= 76$

The maximum profit is $76 hundred or $7600.

82. a. $s(t) = -16t^2 + 80t + 75$
The ball will attain maximum height at the *x*-coordinate of the vertex.

$t = -\dfrac{b}{2a} = -\dfrac{80}{2(-16)} = -\dfrac{80}{-32} = 2.5$

The ball will attain maximum height 2.5 seconds after being thrown.

b. The maximum height is the *y*-coordinate of the vertex.

$s(2.5) = -16(2.5)^2 + 80(2.5) + 74$

$= -100 + 200 + 75$

$= 175$

The maximum height is 175 feet.

83. The graph of $f(x) = (x - 3)^2$ has vertex $(3, 0)$. The graph will be $g(x) = x^2$ shifted right 3 units.

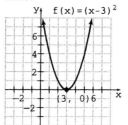

84. The graph of $f(x) = -(x + 2)^2 - 3$ has vertex $(-2, -3)$. Since $a < 0$, the parabola opens downward. The graph will be $g(x) = -x^2$ shifted left 2 units and down 3 units.

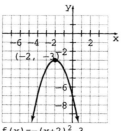

$f(x) = -(x + 2)^2 - 3$

85. The graph of $g(x) = -2(x + 4)^2 - 1$ has vertex $(-4, -1)$. Since $a < 0$, the parabola opens downward. The graph will be $f(x) = -2x^2$ shifted left 4 units and down 1 unit.

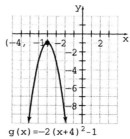

$g(x) = -2(x + 4)^2 - 1$

86. The graph of $h(x) = \dfrac{1}{2}(x-1)^2 + 3$ has vertex $(1, 3)$. The graph will be $f(x) = \dfrac{1}{2}x^2$ shifted right 1 unit and up 3 units.

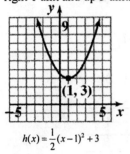

$$h(x) = \frac{1}{2}(x-1)^2 + 3$$

87. $x^2 + 4x + 3 \geq 0$
$(x+1)(x+3) \geq 0$

88. $x^2 + 3x - 10 \leq 0$
$(x+5)(x-2) \leq 0$

89. $x^2 \leq 11x - 20$
$x^2 - 11x + 20 \leq 0$
$x^2 - 11x + 20 = 0$

$$x = \frac{-(-11) \pm \sqrt{(-11)^2 - 4(1)(20)}}{2(1)}$$

$$= \frac{11 \pm \sqrt{121 - 80}}{2}$$

$$= \frac{11 \pm \sqrt{41}}{2}$$

$$\frac{11 - \sqrt{41}}{2} \qquad \frac{11 + \sqrt{41}}{2}$$

90. $3x^2 + 8x > 16$
$3x^2 + 8x - 16 > 0$
$(3x - 4)(x + 4) > 0$
$3x - 4 = 0 \quad$ or $\quad x + 4 = 0$

$$x = \frac{4}{3} \qquad\qquad x = -4$$

91. $4x^2 - 9 \leq 0$
$(2x - 3)(2x + 3) \leq 0$
$2x - 3 = 0 \quad$ or $\quad 2x + 3 = 0$

$$x = \frac{3}{2} \qquad\qquad x = -\frac{3}{2}$$

92. $6x^2 - 30 > 0$
$6(x^2 - 5) > 0$
$6(x + \sqrt{5})(x - \sqrt{5}) > 0$
$x + \sqrt{5} = 0 \qquad$ or $\quad x - \sqrt{5} = 0$
$x = -\sqrt{5} \qquad\qquad x = \sqrt{5}$

93. $\dfrac{x+1}{x-5} > 0$
$x \neq 5$

$$\{x \mid x < -1 \text{ or } x > 5\}$$

501

94. $\dfrac{x-3}{x+2} \le 0$

$x = -2$

$$\begin{array}{cc} - \;\; + \;\; + & x+2 \\ - \;\; - \;\; + & x-3 \\ + \;\; - \;\; + & \end{array}$$

$\circ\!-\!-\!\bullet$ $-2 \quad 3$

$\{x \mid -2 < x \le 3\}$

95. $\dfrac{2x-4}{x+3} \ge 0$

$\dfrac{2(x-2)}{x+3} \ge 0$

$x \ne -3$

$$\begin{array}{cc} - \;\; - \;\; + & 2x-4 \\ - \;\; + \;\; + & x+3 \\ + \;\; - \;\; + & \end{array}$$

$-3 \quad 2$

$\{x \mid x < -3 \text{ or } x \ge 2\}$

96. $\dfrac{3x+5}{x-6} < 0$

$x \ne 6$

$$\begin{array}{cc} - \;\; + \;\; + & 3x+5 \\ - \;\; - \;\; + & x-6 \\ + \;\; - \;\; + & \end{array}$$

$-5/3 \quad 6$

$\left\{x \mid -\dfrac{5}{3} < x < 6\right\}$

97. $(x+4)(x+1)(x-2) > 0$

$$\begin{array}{cc} - \;\; + \;\; + \;\; + & x+4 \\ - \;\; - \;\; + \;\; + & x+1 \\ - \;\; - \;\; - \;\; + & x-2 \\ - \;\; + \;\; - \;\; + & \end{array}$$

$-4 \quad -1 \quad 2$

$\{x \mid -4 < x < -1 \text{ or } x > 2\}$

98. $x(x-3)(x-6) \le 0$

$$\begin{array}{cc} - \;\; + \;\; + \;\; + & x \\ - \;\; - \;\; + \;\; + & x-3 \\ - \;\; - \;\; - \;\; + & x-6 \\ - \;\; + \;\; - \;\; + & \end{array}$$

$0 \quad 3 \quad 6$

$\{x \mid x \le 0 \text{ or } 3 \le x \le 6\}$

99. $(3x+4)(x-1)(x-3) \ge 0$

$$\begin{array}{cc} - \;\; + \;\; + \;\; + & 3x+4 \\ - \;\; - \;\; + \;\; + & x-1 \\ - \;\; - \;\; - \;\; + & x-3 \\ - \;\; + \;\; - \;\; + & \end{array}$$

$-4/3 \quad 1 \quad 3$

$\left[-\dfrac{4}{3}, 1\right] \cup [3, \infty)$

100. $2x(x+2)(x+4) < 0$

$$\begin{array}{cc} - \;\; - \;\; - \;\; + & 2x \\ - \;\; - \;\; + \;\; + & x+2 \\ - \;\; + \;\; + \;\; + & x+4 \\ - \;\; + \;\; - \;\; + & \end{array}$$

$-4 \quad -2 \quad 0$

$(-\infty, -4) \cup (-2, 0)$

101. $\dfrac{x(x-4)}{x+2} > 0$

$x \ne -2$

$$\begin{array}{cc} - \;\; - \;\; + \;\; + & x \\ - \;\; - \;\; - \;\; + & x-4 \\ - \;\; + \;\; + \;\; + & x+2 \\ - \;\; + \;\; - \;\; + & \end{array}$$

$-2 \quad 0 \quad 4$

$(-2, 0) \cup (4, \infty)$

102. $\dfrac{(x-2)(x-8)}{x+3} < 0$

$x \ne -3$

$$\begin{array}{cc} - \;\; + \;\; + \;\; + & x+3 \\ - \;\; - \;\; - \;\; + & x-8 \\ - \;\; - \;\; + \;\; + & x-2 \\ - \;\; + \;\; - \;\; + & \end{array}$$

$-3 \quad 2 \quad 8$

$(-\infty, -3) \cup (2, 8)$

103. $\dfrac{x-3}{(x+2)(x-7)} \ge 0$

$x \ne -2, x \ne 7$

$$\begin{array}{cc} - \;\; + \;\; + \;\; + & x+2 \\ - \;\; - \;\; + \;\; + & x-3 \\ - \;\; - \;\; - \;\; + & x-7 \\ - \;\; + \;\; - \;\; + & \end{array}$$

$-2 \quad 3 \quad 7$

$(-2, 3] \cup (7, \infty)$

104. $\dfrac{x(x-6)}{x+3} \le 0$

$x \ne -3$

$(-\infty, -3) \cup [0, 6]$

105. $\dfrac{5}{x+4} \ge -1$

$\dfrac{5}{x+4} + 1 \ge 0$

$\dfrac{5 + 1(x+4)}{x+4} \ge 0$

$\dfrac{5 + x + 4}{x+4} \ge 0$

$\dfrac{x+9}{x+4} \ge 0$

$x \ne -4$

106. $\dfrac{2x}{x-2} \le 1$

$\dfrac{2x}{x-2} - 1 \le 0$

$\dfrac{2x}{x-2} - \dfrac{1(x-2)}{x-2} \le 0$

$\dfrac{2x - x + 2}{x-2} \le 0$

$\dfrac{x+2}{x-2} \le 0$

$x \ne 2$

107. $\dfrac{2x+3}{3x-5} < 4$

$\dfrac{2x+3}{3x-5} - 4 < 0$

$\dfrac{2x+3 - 4(3x-5)}{3x-5} < 0$

$\dfrac{2x+3 - 12x + 20}{3x-5} < 0$

$\dfrac{-10x + 23}{3x-5} < 0$

Chapter 12 Practice Test

1. $x^2 + 2x - 15 = 0$

$x^2 + 2x = 15$

$x^2 + 2x + 1 = 15 + 1$

$(x+1)^2 = 16$

$x + 1 = \pm 4$

$x = -1 \pm 4$

$x = 3 \ \text{ or } \ x = -5$

2. $a^2 + 7 = 6a$

$a^2 - 6a = -7$

$a^2 - 6a + 9 = -7 + 9$

$(a-3)^2 = 2$

$a - 3 = \pm\sqrt{2}$

$a = 3 \pm \sqrt{2}$

$a = 3 + \sqrt{2} \ \text{ or } \ a = 3 - \sqrt{2}$

3. $x^2 - 6x - 16 = 0$, $a = 1$, $b = -6$, $c = -16$

$$x = \frac{-b \pm \sqrt{b^2 - 4ac}}{2a}$$

$$x = \frac{-(-6) \pm \sqrt{(-6)^2 - 4(1)(-16)}}{2(1)}$$

$$= \frac{6 \pm \sqrt{36 + 64}}{2}$$

$$= \frac{6 \pm \sqrt{100}}{2}$$

$$= \frac{6 \pm 10}{2}$$

$$x = \frac{6 + 10}{2} \quad \text{or} \quad x = \frac{6 - 10}{2}$$

$$= \frac{16}{2} \qquad\qquad = \frac{-4}{2}$$

$$= 8 \qquad\qquad = -2$$

4. $x^2 - 4x = -11$

$x^2 - 4x + 11 = 0$

$a = 1$, $b = -4$, $c = 11$

$$x = \frac{-b \pm \sqrt{b^2 - 4ac}}{2a}$$

$$x = \frac{-(-4) \pm \sqrt{(-4)^2 - 4(1)(11)}}{2(1)}$$

$$= \frac{4 \pm \sqrt{16 - 44}}{2}$$

$$= \frac{4 \pm \sqrt{-28}}{2}$$

$$= \frac{4 \pm 2i\sqrt{7}}{2} = 2 \pm i\sqrt{7}$$

$$x = 2 + i\sqrt{7} \quad \text{or} \quad x = 2 - i\sqrt{7}$$

5. $3r^2 + r = 2$

$3r^2 + r - 2 = 0$

$(3r - 2)(r + 1) = 0$

$3r - 2 = 0 \quad \text{or} \quad r + 1 = 0$

$\quad 3r = 2 \qquad\qquad r = -1$

$$r = \frac{2}{3}$$

6. $\quad p^2 + 4 = -7p$

$p^2 + 7p + 4 = 0$

$$p = \frac{-7 \pm \sqrt{(7)^2 - 4(1)(4)}}{2(1)}$$

$$= \frac{-7 \pm \sqrt{49 - 16}}{2}$$

$$= \frac{-7 \pm \sqrt{33}}{2}$$

$$p = \frac{-7 + \sqrt{33}}{2} \quad \text{or} \quad p = \frac{-7 - \sqrt{33}}{2}$$

7. $x = 4$ and $x = -\dfrac{2}{5}$

$x - 4 = 0$ and $5x + 2 = 0$

$(x - 4)(5x + 2) = 0$

$\quad 5x^2 - 18x - 8 = 0$

8. $K = \dfrac{1}{2}mv^2$ for v

$2K = mv^2$

$$\frac{2K}{m} = v^2$$

$$v = \sqrt{\frac{2K}{m}}$$

9. a. $c(s) = -0.01s^2 + 78s + 22{,}000$

$c(1600) = -0.01(1600)^2 + 78(1600) + 22{,}000$

$\qquad\qquad = -25{,}600 + 124{,}800 + 22{,}000$

$\qquad\qquad = 121{,}200$

The cost is about $121,200.

b. $160{,}000 = -0.01s^2 + 78s + 22{,}000$

$0 = -0.01s^2 + 78s - 138{,}000$

$$s = \frac{-78 \pm \sqrt{78^2 - 4(-0.01)(-138{,}000)}}{2(-0.01)}$$

$$= \frac{-78 \pm \sqrt{564}}{-0.02}$$

$$s = \frac{-78 + \sqrt{564}}{-0.02} \approx 2712.57$$

$$s = \frac{-78 - \sqrt{564}}{-0.02} \approx 5087.43$$

Since $1300 \le s \le 3900$, the house should have about 2712.57 square feet.

10. The formula $d = rt$ can be written $t = \dfrac{d}{r}$.

Let r = his actual rate.

	distance	rate	time $= \dfrac{d}{r}$
Actual trip	520	r	$\dfrac{520}{r}$
Faster trip	520	$r + 15$	$\dfrac{520}{r+15}$

The faster trip would have taken 2.4 hours less time than the actual trip.

$$\frac{520}{r+15} = \frac{520}{r} - 2.4$$

$$r(r+15)\left(\frac{520}{r+15}\right) = r(r+15)\left(\frac{520}{r} - 2.4\right)$$

$$520r = 520(r+15) - 2.4r(r+15)$$

$$520r = 520r + 7800 - 2.4r^2 - 36r$$

$$0 = -2.4r^2 - 36r + 7800$$

$$0 = r^2 + 15r - 3250$$

$$0 = (r+65)(r-50)$$

$$r + 65 = 0 \qquad r - 50 = 0$$

$$r = -65 \qquad r = 50$$

Since speed is never negative, Tom drove an average speed of 50 mph.

11.
$$2x^4 + 15x^2 - 50 = 0$$

$$2\left(x^2\right)^2 + 15x^2 - 50 = 0$$

Let $u = x^2$.

$$2u^2 + 15u - 50 = 0$$

$$(u+10)(2u-5) = 0$$

$$u + 10 = 0 \quad \text{or} \quad 2u - 5 = 0$$

$$u = -10 \qquad u = \frac{5}{2}$$

Substitute x^2 for u.

$$x^2 = -10 \quad \text{or} \quad x^2 = \frac{5}{2}$$

$$x = \pm\sqrt{-10} \qquad x = \pm\sqrt{\frac{5}{2}}$$

$$= \pm i\sqrt{10} \qquad = \pm\frac{\sqrt{5}}{\sqrt{2}} \cdot \frac{\sqrt{2}}{\sqrt{2}}$$

$$= \pm\frac{\sqrt{10}}{2}$$

12.
$$3r^{2/3} + 11r^{1/3} - 42 = 0$$

$$3\left(r^{1/3}\right)^2 + 11r^{1/3} - 42 = 0$$

Let $u = r^{1/3}$.

$$3u^2 + 11u - 42 = 0$$

$$(3u - 7)(u + 6) = 0$$

$$3u - 7 = 0 \quad \text{or} \quad u + 6 = 0$$

$$u = \frac{7}{3} \qquad u = -6$$

$$r^{1/3} = \frac{7}{3} \quad \text{or} \quad r^{1/3} = -6$$

$$r = \left(\frac{7}{3}\right)^3 \qquad r = (-6)^3$$

$$= \frac{343}{27} \qquad r = -216$$

13. $f(x) = 16x - 24\sqrt{x} + 9$

$$0 = 16\left(\sqrt{x}\right)^2 - 24\sqrt{x} + 9$$

Let $u = \sqrt{x}$.

$$0 = 16u^2 - 24u + 9$$

$$0 = (4u - 3)^2$$

$$4u - 3 = 0$$

$$u = \frac{3}{4}$$

Substitute $\sqrt{x}$ for u.

$$\sqrt{x} = \frac{3}{4}$$

$$x = \left(\frac{3}{4}\right)^2 = \frac{9}{16}$$

The x-intercept is $\left(\dfrac{9}{16}, 0\right)$.

14. $f(x) = (x-3)^2 + 2$

The vertex is (3, 2). The graph will be the graph of $g(x) = x^2$ shifted right 3 units and up 2 units.

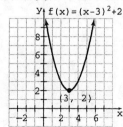

15. $h(x) = -\dfrac{1}{2}(x-2)^2 - 2$

The vertex is $(2, -2)$. The graph will be the graph of $g(x) = -\dfrac{1}{2}x^2$ shifted right 2 units and down 2 units.

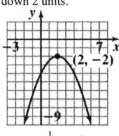

$h(x) = -\dfrac{1}{2}(x-2)^2 - 2$

16. $6x^2 = 2x + 3$

$6x^2 - 2x - 3 = 0$

$a = 6,\ b = -2,\ c = -3$

The discriminant is

$b^2 - 4ac = (-2)^2 - 4(6)(-3) = 4 + 72 = 76$

Since the discriminant is greater than 0, the quadratic equation has two distinct real solutions.

17. $y = x^2 + 2x - 8$

a. Since $a = 1$ the parabola opens upward.

b. Let $x = 0$: $y = 0^2 + 2(0) - 8 = -8$

The y-intercept is $(0, -8)$.

c. $x = -\dfrac{b}{2a} = -\dfrac{2}{2(1)} = -1$

$y = (-1)^2 + 2(-1) - 8 = -9$

The vertex is $(-1, -9)$.

d. The x-intercepts occur when $y = 0$.

$0 = x^2 + 2x - 8$

$0 = (x+4)(x-2)$

$x + 4 = 0$ or $x - 2 = 0$

$x = -4$ $\qquad$ $x = 2$

The x-intercepts are $(2, 0)$ and $(-4, 0)$.

e.

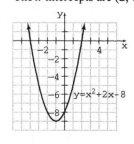

$y = x^2 + 2x - 8$

18. $x = -7$ and $x = \dfrac{1}{2}$

$x + 7 = 0$ and $2x - 1 = 0$

$(x+7)(2x-1) = 0$

$2x^2 + 13x - 7 = 0$

19. $x^2 - x \geq 42$

$x^2 - x - 42 \geq 0$

$(x-7)(x+6) \geq 0$

$x - 7 = 0$ or $x + 6 = 0$

$x = 7$ $\qquad$ $x = -6$

$$\begin{array}{ccccc}
- & | & + & | & + \\
\hline
\end{array} \quad x+6$$

$$\begin{array}{ccccc}
- & | & - & | & + \\
\hline
\end{array} \quad x-7$$

$$\begin{array}{ccccc}
+ & | & - & | & + \\
\hline
\end{array}$$

$\qquad -6 \qquad\quad 7$

20. $\dfrac{(x+5)(x-4)}{x+1} \geq 0$

$x \neq -1$

$x + 5 = 0$ or $x - 4 = 0$ or $x + 1 = 0$

$x = -5$ $\qquad$ $x = 4$ $\qquad$ $x = -1$

$$\begin{array}{ccccccc}
- & | & + & | & + & | & + \\
\hline
\end{array} \quad x+5$$

$$\begin{array}{ccccccc}
- & | & - & | & - & | & + \\
\hline
\end{array} \quad x-4$$

$$\begin{array}{ccccccc}
- & | & - & | & + & | & + \\
\hline
\end{array} \quad x+1$$

$$\begin{array}{ccccccc}
- & | & + & | & - & | & + \\
\hline
\end{array}$$

$\qquad -5 \qquad -1 \qquad 4$

21. $\dfrac{x+3}{x+2} \leq -1$

$\dfrac{x+3}{x+2} + 1 \leq 0$

$\dfrac{x+3}{x+2} + \dfrac{x+2}{x+2} \leq 0$

$\dfrac{2x+5}{x+2} \leq 0$

$x \neq -2$

$$\begin{array}{ccccc}
- & | & + & | & + \\
\hline
\end{array} \quad 2x+5$$

$$\begin{array}{ccccc}
- & | & - & | & + \\
\hline
\end{array} \quad x+2$$

$$\begin{array}{ccccc}
+ & | & - & | & + \\
\hline
\end{array}$$

$\qquad -5/2 \qquad -2$

a. $\left[-\dfrac{5}{2}, -2\right)$

b. $\left\{ x \,\middle|\, -\dfrac{5}{2} \leq x < -2 \right\}$

22. Let x be the width of the carpet. Then $2x + 3$ is the length $A = lw$.

$$65 = x(2x+3)$$
$$65 = 2x^2 + 3x$$
$$0 = 2x^2 + 3x - 65$$
$$0 = (2x+13)(x-5)$$
$$2x+13 = 0 \quad \text{or} \quad x-5 = 0$$
$$x = -\frac{13}{2} \qquad x = 5$$

Disregard the negative answer. The width is 5 feet and the length is $2 \cdot 5 + 3 = 10 + 3 = 13$ feet.

23. $d = -16t^2 + 80t + 96$

$d = 0$ when the ball strikes the ground.

$$0 = -16t^2 + 80t + 96$$
$$0 = -16(t^2 - 5t - 6)$$
$$0 = -16(t-6)(t+1)$$
$$t-6 = 0 \quad \text{or} \quad t+1 = 0$$
$$t = 6 \qquad t = -1$$

The time must be positive, so $t = 6$.
Thus, the ball strikes the ground in 6 seconds.

24. a. $f(x) = -1.4x^2 + 56x - 70$

$$x = -\frac{b}{2a} = -\frac{56}{2(-1.4)} = -\frac{56}{-2.8} = 20$$

The company must sell 20 carvings.

b. $f(20) = -1.4(20)^2 + 56(20) - 70$
$$= -560 + 1120 - 70$$
$$= 490$$

The maximum weekly profit is $490.

25. If the business sells n brooms at a price of $(10 - 0.1n)$ dollars per broom, then the revenue is given by $R(n) = n(10 - 0.1n)$ with $n \le 32$. Set this equal to 160 and solve for n.

$$R(n) = n(10 - 0.1n)$$
$$210 = n(10 - 0.1n)$$
$$210 = 10n - 0.1n^2$$
$$0.1n^2 - 10n + 210 = 0$$
$$n^2 - 100n + 2100 = 0$$
$$(n-30)(n-70) = 0$$
$$n = 30 \quad \text{or} \quad n = 70$$

Disregard $n = 70$ since $n \le 32$. The business must sell 30 brooms.

Chapter 12 Cumulative Review Test

1. $-4 \div (-2) + 18 - \sqrt{49} = -4 \div (-2) + 18 - 7$
$$= 2 + 18 - 7$$
$$= 20 - 7$$
$$= 13$$

2. Evaluate $2x^2 + 3x + 4$ when $x = 2$.
$$2(2)^2 + 3(2) + 4 = 8 + 6 + 4 = 18$$

3. $6x - \{3 - [2(x-2) - 5x]\}$
$$= 6x - \{3 - [2x - 4 - 5x]\}$$
$$= 6x - \{3 - [-4 - 3x]\}$$
$$= 6x - \{3 + 4 + 3x\}$$
$$= 6x - \{7 + 3x\}$$
$$= 6x - 7 - 3x$$
$$= 3x - 7$$

4. $-\frac{1}{2}(4x - 6) = \frac{1}{3}(3 - 6x) + 2$
$$-2x + 3 = 1 - 2x + 2$$
$$-2x + 3 = -2x + 3$$

This is an identity. The solution is all real numbers.

5. a. $x = -4$ is a vertical line.

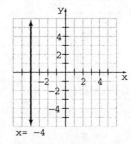

b. $y = 2$ is a horizontal line.

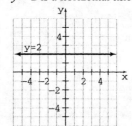

6. $m = \dfrac{y_2 - y_1}{x_2 - x_1} = \dfrac{3-5}{4-6} = \dfrac{-2}{-2} = 1$

   $y - y_1 = m(x - x_1)$
   $y - 3 = 1(x - 4)$
   $y - 3 = x - 4$
   $y = x - 1$

7. $2,540,000 = 2.54 \times 10^6$

8. $\dfrac{x+2}{x^2 - x - 6} + \dfrac{x-3}{x^2 - 8x + 15}$

   $= \dfrac{x+2}{(x-3)(x+2)} + \dfrac{x-3}{(x-5)(x-3)}$

   $= \dfrac{1}{(x-3)} + \dfrac{1}{(x-5)}$

   $= \dfrac{1}{x-3} \cdot \dfrac{x-5}{x-5} + \dfrac{1}{x-5} \cdot \dfrac{x-3}{x-3}$

   $= \dfrac{x-5+x-3}{(x-3)(x-5)}$

   $= \dfrac{2x-8}{(x-3)(x-5)}$ or $\dfrac{2(x-4)}{(x-3)(x-5)}$

9. $\dfrac{1}{a-2} = \dfrac{4a-1}{a^2 + 5a - 14} + \dfrac{2}{a+7}$

   $\dfrac{1}{a-2} = \dfrac{4a-1}{(a+7)(a-2)} + \dfrac{2}{a+7}$

   $(a+7)(a-2)\left[\dfrac{1}{a-2} = \dfrac{4a-1}{(a+7)(a-2)} + \dfrac{2}{a+7}\right]$

   $a+7 = 4a-1+2(a-2)$
   $a+7 = 4a-1+2a-4$
   $a+7 = 6a-5$
   $-5a+7 = -5$
   $-5a = -12$
   $a = \dfrac{12}{5}$

10. $w = kI^2 R$, $w = 12$, $I = 2$, $R = 100$
    $12 = k(2)^2(100)$
    $12 = 400k$
    $k = \dfrac{12}{400} = \dfrac{3}{100}$
    $w = \dfrac{3}{100}I^2 R$, $I = 0.8$, $R = 600$
    $w = \dfrac{3}{100}(0.8)^2(600) = 11.52$
    The wattage is 11.52 watts.

11. $N(x) = -0.2x^2 + 40x$
    $N(50) = -0.2(50)^2 + 40(50)$
    $= -500 + 2000$
    $= 1500$
    50 trees would produce about 1500 baskets of apples.

12. **a.** No, the graph is not a function since each x-value does not have a unique y-value.

    b. The domain is the set of x-values,
    Domain : $\{x \mid x \geq -2\}$.

    The range is the set of y-values,
    Range: $\mathbb{R}$.

13. $4x - 3y = 10$ (1)
    $2x + y = 5$ (2)
    To eliminate the y variable, multiply equation (2) by 3 and add the result to equation (1).
    $4x - 3y = 10$ $\qquad$ $4x - 3y = 10$
    $3(2x + y = 5)$ $\Rightarrow$ $\dfrac{6x + 3y = 15}{10x = 25}$

    $\qquad\qquad\qquad\qquad x = \dfrac{5}{2}$

    Substitute $\dfrac{5}{2}$ for x in equation (2) and then solve for y.
    $2\left(\dfrac{5}{2}\right) + y = 5$
    $5 + y = 5$
    $y = 0$
    The solution is $\left(\dfrac{5}{2}, 0\right)$.

14. $\begin{vmatrix} 4 & 0 & -2 \\ 3 & 5 & 1 \\ 1 & -1 & 7 \end{vmatrix} = 4\begin{vmatrix} 5 & 1 \\ -1 & 7 \end{vmatrix} - 0\begin{vmatrix} 3 & 1 \\ 1 & 7 \end{vmatrix} + (-2)\begin{vmatrix} 3 & 5 \\ 1 & -1 \end{vmatrix}$

    $= 4(35 + 1) - 0(21 - 1) - 2(-3 - 5)$
    $= 4(36) - 0(20) - 2(-8)$
    $= 144 - 0 + 16$
    $= 160$

15. $-4 < \dfrac{x+4}{2} < 6$
    $-8 < x + 4 < 12$
    $-12 < x < 8$
    In interval notation the solution is $(-12, 8)$.

16. $|4-2x| = 5$

$4-2x = 5$ or $4-2x = -5$

$-2x = 1$ $\qquad -2x = -9$

$x = -\dfrac{1}{2} \qquad x = \dfrac{9}{2}$

The solution set is $\left\{-\dfrac{1}{2}, \dfrac{9}{2}\right\}$

17. $\dfrac{3-4i}{2+5i} = \left(\dfrac{3-4i}{2+5i}\right)\left(\dfrac{2-5i}{2-5i}\right)$

$\qquad = \dfrac{6-23i+20i^2}{4-25i^2}$

$\qquad = \dfrac{6-23i-20}{4+25}$

$\qquad = \dfrac{-14-23i}{29}$

18. $4x^2 = -3x - 12$

$4x^2 + 3x + 12 = 0$

Using the quadratic formula, $a = 4$, $b = 3$, $c = 12$.

$x = \dfrac{-b \pm \sqrt{b^2 - 4ac}}{2a}$

$\quad = \dfrac{-(3) \pm \sqrt{(3)^2 - 4(4)(12)}}{2(4)}$

$\quad = \dfrac{-3 \pm \sqrt{9 - 192}}{8}$

$\quad = \dfrac{-3 \pm \sqrt{-183}}{8}$

$\quad = \dfrac{-3 \pm i\sqrt{183}}{8}$

19. $V = \dfrac{1}{3}\pi r^2 h$

$3(V) = 3\left(\dfrac{1}{3}\pi r^2 h\right)$

$3V = \pi r^2 h$

$\dfrac{3V}{\pi h} = \dfrac{\pi r^2 h}{\pi h}$

$\dfrac{3V}{\pi h} = r^2$

$\sqrt{\dfrac{3V}{\pi h}} = \sqrt{r^2}$

$\sqrt{\dfrac{3V}{\pi h}} = r$

20. $(x+3)^2 + 10(x+3) + 24$

$= [(x+3)+4][(x+3)+6]$

$= (x+7)(x+9)$

Chapter 13

Exercise Set 13.1

1. To find $(f \circ g)(x)$, substitute $g(x)$ for x in $f(x)$.

3. **a.** Each y has a unique x in a one-to-one function.

   b. Use the horizontal line test to determine whether a graph is one-to-one.

5. **a.** Yes; each first coordinate is paired with only one second coordinate.

   b. Yes; each second coordinate is paired with only one first coordinate.

   c. $\{(5, 3), (2, 4), (3, -1), (-2, 0)\}$
   Reverse each ordered pair.

7. The domain of f is the range of f^{-1} and the range of f is the domain of f^{-1}.

9. $f(x) = x^2 + 1$, $g(x) = x + 2$

   a. $(f \circ g)(x) = (x+2)^2 + 1$
   $\qquad = x^2 + 4x + 4 + 1$
   $\qquad = x^2 + 4x + 5$

   b. $(f \circ g)(4) = 4^2 + 4(4) + 5 = 37$

   c. $(g \circ f)(x) = (x^2 + 1) + 2 = x^2 + 3$

   d. $(g \circ f)(4) = 4^2 + 3 = 19$

11. $f(x) = x + 3$, $g(x) = x^2 + x - 4$

   a. $(f \circ g)(x) = (x^2 + x - 4) + 3 = x^2 + x - 1$

   b. $(f \circ g)(4) = 4^2 + 4 - 1 = 19$

   c. $(g \circ f)(x) = (x+3)^2 + (x+3) - 4$
   $\qquad = x^2 + 6x + 9 + x + 3 - 4$
   $\qquad = x^2 + 7x + 8$

   d. $(g \circ f)(4) = 4^2 + 7(4) + 8 = 52$

13. $f(x) = \dfrac{1}{x}$, $g(x) = 2x + 3$

   a. $(f \circ g)(x) = \dfrac{1}{2x+3}$

   b. $(f \circ g)(4) = \dfrac{1}{2(4)+3} = \dfrac{1}{11}$

   c. $(g \circ f)(x) = 2\left(\dfrac{1}{x}\right) + 3 = \dfrac{2}{x} + 3$

   d. $(g \circ f)(4) = \dfrac{2}{4} + 3 = 3\dfrac{1}{2}$

15. $f(x) = 3x + 1$, $g(x) = \dfrac{3}{x}$

   a. $(f \circ g)(x) = 3\left(\dfrac{3}{x}\right) + 1 = \dfrac{9}{x} + 1$

   b. $(f \circ g)(4) = \dfrac{9}{4} + 1 = 3\dfrac{1}{4}$

   c. $(g \circ f)(x) = \dfrac{3}{3x+1}$

   d. $(g \circ f)(4) = \dfrac{3}{3(4)+1} = \dfrac{3}{13}$

17. $f(x) = x^2 + 1$, $g(x) = x^2 + 5$

   a. $(f \circ g)(x) = (x^2 + 5)^2 + 1$
   $\qquad = x^4 + 10x^2 + 25 + 1$
   $\qquad = x^4 + 10x^2 + 26$

   b. $(f \circ g)(4) = 4^4 + 10(4)^2 + 26 = 442$

   c. $(g \circ f)(x) = (x^2 + 1)^2 + 5$
   $\qquad = x^4 + 2x^2 + 1 + 5$
   $\qquad = x^4 + 2x^2 + 6$

   d. $(g \circ f)(4) = 4^4 + 2(4)^2 + 6 = 294$

19. $f(x) = x - 4$, $g(x) = \sqrt{x+5}$, $x \geq -5$

   a. $(f \circ g)(x) = \sqrt{x+5} - 4$

   b. $(f \circ g)(4) = \sqrt{4+5} - 4$
$$= \sqrt{9} - 4$$
$$= 3 - 4$$
$$= -1$$

   c. $(g \circ f)(x) = \sqrt{(x-4)+5} = \sqrt{x+1}$

   d. $(g \circ f)(4) = \sqrt{4+1} = \sqrt{5}$

21. This function is not a one-to-one function since it does not pass the horizontal line test.

23. This function is a one-to-one function since it passes the horizontal line test.

25. Yes, the ordered pairs represent a one-to-one function. For each value of x there is a unique value for y and each y-value has a unique x-value.

27. No, the ordered pairs do not represent a one-to-one function. For each value of x there is a unique y, but for each y-value there is not a unique x since $(-4, 2)$ and $(0, 2)$ are ordered pairs in the given set.

29. $y = 2x + 5$ is a line with a slope of 2 and having a y-intercept of 5. It is a one-to-one function since it passes both the vertical line test and the horizontal line test.

31. $y = x^2 - 1$ is a parabola with vertex at $(0, -1)$. It is not a one-to-one function since it does not pass the horizontal line test. Horizontal lines above $y = -1$ intersect the graph at 2 different points.

33. $y = x^2 - 2x + 5$ is a parabola with vertex at $(1, 4)$. It is not a one-to-one function since it does not pass the horizontal line test. Horizontal lines above $y = 4$ intersect the graph at two different points.

35. $y = x^2 - 9$, $x \geq 0$ is the right side of a parabola. It is a one-to-one function since it passes both the vertical line test and the horizontal line test.

37. $y = \sqrt{x}$ is a one-to-one function since it passes both the vertical line test and the horizontal line test.

39. $y = |x|$ is not a one-to-one function since it does not pass the vertical line test and the horizontal line test. Horizontal lines above $y = 0$ intersect the graph at two different points.

41. $y = \sqrt[3]{x}$ is a one-to-one function since it passes both the vertical line test and the horizontal line test.

43. For $f(x)$: Domain: $\{-2, -1, 2, 4, 8\}$
Range: $\{0, 4, 6, 7, 9\}$
For $f^{-1}(x)$: Domain: $\{0, 4, 6, 7, 9\}$
Range: $\{-2, -1, 2, 4, 8\}$

45. For $f(x)$: Domain: $\{-1, 1, 2, 4\}$
Range: $\{-3, -1, 0, 2\}$
For $f^{-1}(x)$: Domain: $\{-3, -1, 0, 2\}$
Range: $\{-1, 1, 2, 4\}$

47. For $f(x)$: Domain: $\{x \mid x \geq 2\}$
Range: $\{y \mid y \geq 0\}$
For $f^{-1}(x)$: Domain: $\{x \mid x \geq 0\}$
Range: $\{y \mid y \geq 2\}$

49. a. Yes, $f(x) = x - 2$ is a one-to-one function.

   b.
$$y = x - 2$$
$$x = y - 2$$
$$x + 2 = y$$
$$y = x + 2$$
$$f^{-1}(x) = x + 2$$

51. a. Yes, $h(x) = 4x$ is a one-to-one function.

   b.
$$y = 4x$$
$$x = 4y$$
$$y = \frac{x}{4}$$
$$h^{-1}(x) = \frac{x}{4}$$

53. a. No, $p(x) = 3x^2$ is not a one-to-one function.

  b. Does not exist

55. a. No; $t(x) = x^2 + 3$ is not a one-to-one function.

  b. Does not exist

57. a. Yes; $g(x) = \dfrac{1}{x}$ is a one-to-one function.

  b.
$$y = \frac{1}{x}$$
$$x = \frac{1}{y}$$
$$y = \frac{1}{x}$$
$$g^{-1}(x) = \frac{1}{x}$$

59. a. No; $f(x) = x^2 + 10$ is not a one-to-one function.

  b. Does not exist

61. a. Yes, $g(x) = x^3 - 6$ is a one-to-one function.

  b.
$$y = x^3 - 6$$
$$x = y^3 - 6$$
$$x + 6 = y^3$$
$$\sqrt[3]{x+6} = y$$
$$g^{-1}(x) = \sqrt[3]{x+6}$$

63. a. Yes, $g(x) = \sqrt{x+2}$, $x \geq -2$ is a one-to-one function.

  b.
$$y = \sqrt{x+2}$$
$$x = \sqrt{y+2}$$
$$x^2 = y + 2$$
$$x^2 - 2 = y$$
$$g^{-1}(x) = x^2 - 2, x \geq 0$$

65. a. Yes, $h(x) = x^2 - 4$, $x \geq 0$ is one-to-one.

  b.
$$y = x^2 - 4$$
$$x = y^2 - 4$$
$$x + 4 = y^2$$
$$y = \sqrt{x+4}$$
$$h^{-1}(x) = \sqrt{x+4}, x \geq -4$$

67. $f(x) = 2x + 8$

  a.
$$y = 2x + 8$$
$$x = 2y + 8$$
$$x - 8 = 2y$$
$$\frac{x-8}{2} = y$$
$$f^{-1}(x) = \frac{x-8}{2}$$

  b.

x	$f(x)$
0	8
–4	0

x	$f^{-1}(x)$
0	–4
8	0

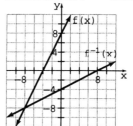

69. $f(x) = \sqrt{x}, x \geq 0$

    a. $y = \sqrt{x}$

             $x = \sqrt{y}$

             $x^2 = \left(\sqrt{y}\right)^2$

             $x^2 = y$

             $f^{-1}(x) = x^2$ for $x \geq 0$

    b.

x	$f(x)$
0	0
1	1
4	2

x	$f^{-1}(x)$
0	0
1	1
2	4

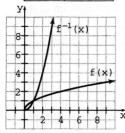

71. $f(x) = \sqrt{x-1}, x \geq 1$

    a. $y = \sqrt{x-1}$

             $x = \sqrt{y-1}$

             $x^2 = \left(\sqrt{y-1}\right)^2$

             $x^2 = y-1$

             $x^2 + 1 = y$

             $f^{-1}(x) = x^2 + 1$ for $x \geq 0$

    b.

x	$f(x)$
1	0
2	1
5	2

x	$f^{-1}(x)$
0	1
1	2
2	5

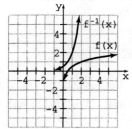

73. $f(x) = \sqrt[3]{x}$

    a. $y = \sqrt[3]{x}$

             $x = \sqrt[3]{y}$

             $x^3 = \left(\sqrt[3]{y}\right)^3$

             $x^3 = y$

             $f^{-1}(x) = x^3$

    b.

x	$f(x)$
−8	−2
−1	−1
0	0
1	1
8	2

x	$f^{-1}(x)$
−2	−8
−1	−1
0	0
1	1
2	8

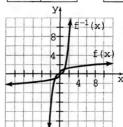

75. $f(x) = \dfrac{1}{x}, x > 0$

    a. $y = \dfrac{1}{x}$

             $x = \dfrac{1}{y}$

             $xy = 1$

             $y = \dfrac{1}{x}$

             $f^{-1}(x) = \dfrac{1}{x}, x > 0$

    b.

x	$f(x)$
$\dfrac{1}{2}$	2
1	1
3	$\dfrac{1}{3}$

x	$f^{-1}(x)$
2	$\dfrac{1}{2}$
1	1
$\dfrac{1}{3}$	3

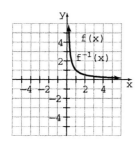

77. $(f \circ f^{-1})(x) = (x+8)-8 = x$

$(f^{-1} \circ f)(x) = (x-8)+8 = x$

79. $(f \circ f^{-1})(x) = \dfrac{1}{2}(2x-6)+3$

$\qquad = x-3+3$

$\qquad = x$

$(f^{-1} \circ f)(x) = 2\left(\dfrac{1}{2}x+3\right)-6$

$\qquad = x+6-6$

$\qquad = x$

81. $(f \circ f^{-1})(x) = \sqrt[3]{(x^3+2)-2}$

$\qquad = \sqrt[3]{x^3}$

$\qquad = x$

$(f^{-1} \circ f)(x) = \left(\sqrt[3]{x-2}\right)^3+2$

$\qquad = x-2+2$

$\qquad = x$

83. $(f \circ f^{-1})(x) = \dfrac{3}{\frac{3}{x}} = 3 \cdot \dfrac{x}{3} = x$

$(f^{-1} \circ f)(x) = \dfrac{3}{\frac{3}{x}} = 3 \cdot \dfrac{x}{3} = x$

85. No, composition of functions is not commutative.
Let $f(x) = x^2$ and $g(x) = x+1$.
Then $(f \circ g)(x) = (x+1)^2 = x^2+2x+1$ while
$(g \circ f)(x) = x^2+1$.

87. a. $(f \circ g)(x) = f[g(x)]$

$\qquad = \left(\sqrt[3]{x-2}\right)^3+2$

$\qquad = x-2+2$

$\qquad = x$

$(g \circ f)(x) = g[f(x)]$

$\qquad = \sqrt[3]{\left(x^3+2\right)-2}$

$\qquad = \sqrt[3]{x^3}$

$\qquad = x$

b. The domain of f is all real numbers and the domain of g is all real numbers. The domains of $(f \circ g)(x)$ and $(g \circ f)(x)$ are also all real numbers.

89. The range of $f^{-1}(x)$ is the domain of $f(x)$.

91. $f(x) = 3x$ converts yards, x, into feet, y.
$y = 3x$

$x = 3y$

$\dfrac{x}{3} = y$

$f^{-1}(x) = \dfrac{x}{3}$

Here, x is feet and $f^{-1}(x)$ is yards. The inverse function converts feet to yards.

93. $f(x) = \dfrac{5}{9}(x-32)$ where x is degrees Fahrenheit

and $f(x)$ is degrees Celsius.

$y = \dfrac{5}{9}(x-32)$

$x = \dfrac{5}{9}(y-32)$

$\dfrac{9}{5}x = \dfrac{9}{5}\left[\dfrac{5}{9}(y-32)\right]$

$\dfrac{9}{5}x = y-32$

$\dfrac{9}{5}x+32 = y$

$f^{-1}(x) = \dfrac{9}{5}x+32$

Here, x is degrees Celsius and $f^{-1}(x)$ is degrees Fahrenheit. The inverse function converts Celsius to Fahrenheit.

95. $f(x) = 16x$; $g(x) = 28.35x$

$(f \circ g)(x) = 16(28.35x) = 453.6x$

In this composition, x represents pounds and $(f \circ g)(x)$ represents grams. The composition converts pounds to grams.

97. $f(x) = 3x$; $g(x) = 0.305x$

$(f \circ g)(x) = 3(0.305x) = 0.915x$

In this composition, x represents yards and $(f \circ g)(x)$ represents meters. The composition converts yards to meters.

99.

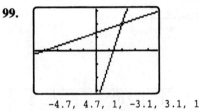

$-4.7, \ 4.7, \ 1, \ -3.1, \ 3.1, \ 1$

Yes, the functions are inverses.

101.

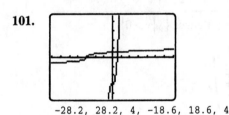

$-28.2, \ 28.2, \ 4, \ -18.6, \ 18.6, \ 4$

Yes, the functions are inverses.

103. a. $r(3) = 2(3) = 6$

The radius is 6 feet.

b. $A = \pi r^2$

$A = \pi(6)^2$

$A = 36\pi \approx 113.10$

The surface area is $36\pi \approx 113.10$ square feet.

c. $(A \circ r)(t) = \pi(2t)^2 = \pi(4t^2) = 4\pi t^2$

d. $4\pi(3)^2 = 4\pi(9) = 36\pi$

e. The answers should agree.

106. First find the slope of the given line.

$2x + 3y - 9 = 0$

$3y = -2x + 9$

$y = -\dfrac{2}{3}x + 3 \ \Rightarrow \ m = -\dfrac{2}{3}$

Now use this slope together with the given point

$\left(\dfrac{1}{2}, 3\right)$ to find the equation.

point-slope form:

$y - y_1 = m(x - x_1)$

$y - 3 = -\dfrac{2}{3}\left(x - \dfrac{1}{2}\right)$

$y - 3 = -\dfrac{2}{3}x + \dfrac{1}{3}$

$3\left(y - 3 = -\dfrac{2}{3}x + \dfrac{1}{3}\right)$

$3y - 9 = -2x + 1$

$2x + 3y = 10$

107. $x + 2 = x - (-2) \rightarrow c = -2$

$$
\begin{array}{r|rrrr}
-2 & 1 & 6 & 6 & -8 \\
   &   & -2 & -8 & 4 \\
\hline
   & 1 & 4 & -2 & -4
\end{array}
$$

$\dfrac{x^3 + 6x^2 + 6x - 8}{x + 2} = x^2 + 4x - 2 - \dfrac{4}{x + 2}$

108. $\dfrac{\dfrac{3}{x^2} - \dfrac{2}{x}}{\dfrac{x}{6}} = \dfrac{6x^2\left(\dfrac{3}{x^2} - \dfrac{2}{x}\right)}{6x^2\left(\dfrac{x}{6}\right)} = \dfrac{18 - 12x}{x^3}$

109. $\sqrt{\dfrac{24x^3y^2}{3xy^3}} = \sqrt{\dfrac{8x^2y}{y^2}} = \dfrac{\sqrt{8x^2y}}{\sqrt{y^2}}$

$= \dfrac{\sqrt{4x^2 \cdot 2y}}{\sqrt{y^2}} = \dfrac{2x\sqrt{2y}}{y}$

110. $x^2 + 2x - 6 = 0$

$x^2 + 2x = 6$

$x^2 + 2x + 1 = 6 + 1$

$(x + 1)^2 = 7$

$x + 1 = \pm\sqrt{7}$

$x = -1 \pm \sqrt{7}$

Exercise Set 13.2

1. Exponential functions are functions of the form $f(x) = a^x, a > 0, a \neq 1$.

3. **a.** $y = \left(\dfrac{1}{2}\right)^x$; as x increases, y decreases.

   b. No, y can never be zero because $\left(\dfrac{1}{2}\right)^x$ can never be 0.

   c. No, y can never be negative because $\left(\dfrac{1}{2}\right)^x$ is never negative.

5. $y = 2^x$ and $y = 3^x$

   a. Let $x = 0$
   $y = 2^0 \quad y = 3^0$
   $y = 1 \quad y = 1$
   They have the same y-intercept at $(0, 1)$.

   b. $y = 3^x$ will be steeper than $y = 2^x$ for $x > 0$ because it has a larger base.

7. $y = 2^x$

x	-2	-1	0	1	2
y	$\frac{1}{4}$	$\frac{1}{2}$	1	2	4

Domain: $\mathbb{R}$; Range: $\{y \mid y > 0\}$

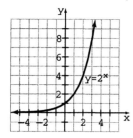

9. $y = \left(\dfrac{1}{2}\right)^x$

x	-2	-1	0	1	2
y	4	2	1	$\frac{1}{2}$	$\frac{1}{4}$

Domain: $\mathbb{R}$; Range: $\{y \mid y > 0\}$

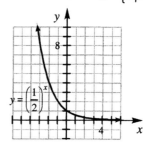

11. $y = 4^x$

x	-2	-1	0	1	2
y	$\frac{1}{16}$	$\frac{1}{4}$	1	4	16

Domain: $\mathbb{R}$; Range: $\{y \mid y > 0\}$

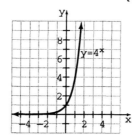

13. $y = \left(\dfrac{1}{4}\right)^x$

x	-2	-1	0	1	2
y	16	4	1	$\frac{1}{4}$	$\frac{1}{16}$

Domain: $\mathbb{R}$; Range: $\{y \mid y > 0\}$

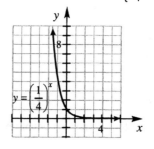

15. $y = 3^{-x} = \dfrac{1}{3^x} = \left(\dfrac{1}{3}\right)^x$

x	-2	-1	0	1	2
y	9	3	1	$\frac{1}{3}$	$\frac{1}{9}$

Domain: $\mathbb{R}$; Range: $\left\{y \,\middle|\, y > 0\right\}$

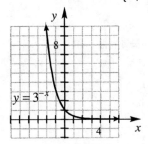

17. $y = \left(\dfrac{1}{3}\right)^{-x} = 3^x$

x	-2	-1	0	1	2
y	$\frac{1}{9}$	$\frac{1}{3}$	1	3	9

Domain: $\mathbb{R}$; Range: $\left\{y \,\middle|\, y > 0\right\}$

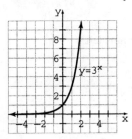

19. $y = 2^{x-1}$

x	-2	0	2	4	6
y	$\frac{1}{8}$	$\frac{1}{2}$	2	8	32

Domain: $\mathbb{R}$; Range: $\left\{y \,\middle|\, y > 0\right\}$

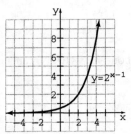

21. $y = \left(\dfrac{1}{3}\right)^{x+1}$

x	-3	-2	-1	0	1
y	9	3	1	$\frac{1}{3}$	$\frac{1}{9}$

Domain: $\mathbb{R}$; Range: $\left\{y \,\middle|\, y > 0\right\}$

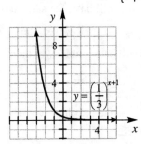

23. $y = 2^x + 1$

x	-2	-1	0	1	2	3
y	$\frac{5}{4}$	$\frac{3}{2}$	2	3	5	9

Domain: $\mathbb{R}$; Range: $\left\{y \,\middle|\, y > 1\right\}$

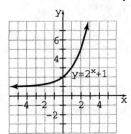

25. $y = 3^x - 1$

x	-2	-1	0	1	2
y	$-\frac{8}{9}$	$-\frac{2}{3}$	0	2	8

Domain: $\mathbb{R}$; Range: $\left\{y \,\middle|\, y > -1\right\}$

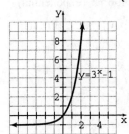

27. **a.** The graph is the horizontal line through $y = 1$.

    b. Yes. A horizontal line will pass the vertical line test.

    c. No. $f(x)$ is not one-to-one and therefore does not have an inverse function.

29. The graph of $y = a^x - k$ will have the same basic shape as the graph of $y = a^x$. However, the graph of $y = a^x - k$ will be k units lower than that of $y = a^x$.

31. The graph of $y = a^{x+2}$ is the graph of $y = a^x$ shifted 2 units to the left.

33. **a.** For 2060 we have $t = 2060 - 1960 = 100$.
    $$f(100) = 0.592(1.042)^{100} \approx 36.232$$
    The function estimates that in 2060 there will be 36.232 million people in the U.S. aged 85 or older.

    b. For 2100 we have $t = 2100 - 1960 = 140$.
    $$f(140) = 0.592(1.042)^{140} \approx 187.846$$
    The function estimates that in 2100 there will be 187.846 million people in the U.S. aged 85 or older.

35. The amount each day is given by the function $A(d) = 2^d$ where d is the number of days.
    $$A(9) = 2^9 = 512$$
    After 9 days, the amount would be $512.

37. **a.** About 14 years

    b. About 10 years

    c. From the graph, the difference is about $25. Using the formulas given, we get:
    $$A = 100(1.07)^{10} = 196.72 \quad \text{(exponential)}$$
    $$A = 100 + 100(.07)(10) = 170 \quad \text{(linear)}$$
    The difference is $26.72.

    d. For daily compounding we would get
    $$A = 100\left(1 + \frac{0.07}{365}\right)^{(365 \cdot 10)} = 201.36$$
    This is about a $5 increase over compounding annually. Daily compounding increases the amount.

39. $N(t) = 5(3)^t, t = 2$
    $$N(2) = 5(3)^2 = 5 \cdot 9 = 45$$
    There will be 45 bacteria in the petri dish after two days.

41. $A = p\left(1 + \dfrac{r}{n}\right)^{nt}$.

    Use $p = 5000$,
    $r = 6\% = 0.06$ and $n = 4$ and $t = 4$.
    $$A = 5000\left(1 + \frac{0.06}{4}\right)^{4 \cdot 4}$$
    $$A = 5000(1 + 0.015)^{16}$$
    $$A = 5000(1.015)^{16}$$
    $$A \approx 5000(1.2689855)$$
    $$A \approx 6344.93$$
    He has $6344.93 after 4 years.

43. $A = A_0 2^{-t/5600}$
    Use $A_0 = 12$ and $t = 1000$.
    $$A = 12(2^{-1000/5600})$$
    $$A \approx 12(2^{-0.18})$$
    $$A \approx 12(0.88)$$
    $$A \approx 10.6 \text{ grams}$$
    There are about 10.6 grams left.

45. $y = 80(2)^{-0.4t}$

    a. $t = 10$
    $$y = 80(2)^{-0.4(10)}$$
    $$y = 80(2)^{-4} = 80\left(\frac{1}{16}\right) = 5$$
    After 10 years, 5 grams remain.

    b. $t = 100$
    $$y = 80(2)^{-0.4(100)}$$
    $$y = 80(2)^{-40}$$
    $$y \approx 80(9.094947 \times 10^{-13})$$
    $$y \approx 7.28 \times 10^{-11}$$
    After 100 years, about 7.28×10^{-11} grams are left.

47. $y = 2000(1.2)^{0.1t}$

    a. $t = 10$

        $y = 2000(1.2)^{0.1(10)}$

        $y = 2000(1.2)^{1}$

        $y = 2400$

        In 10 years, the population is expected to be 2400.

    b. $t = 50$

        $y = 2000(1.2)^{0.1(50)}$

        $y = 2000(1.2)^{5}$

        $y = 2000(2.48832)$

        $y \approx 4977$

        In 50 years, the population is expected to be about 4,977.

49. $V(t) = 24,000(0.82)^{t}$, $t = 4$

    $V(t) = 24,000(0.82)^{4} \approx 10,850.92$

    The SUV will be worth about \$10,850.92 in 4 years.

51. a. Answers will vary. One possible answer is: Since the amount is reduced by 5%, the consumption is 95% of the previous year, or 0.95. Thus, $A(t) = 580,000(0.95)^{t}$.

    b. $t = 2009 - 2005 = 4$

        $A(4) = 580,000(0.95)^{4}$

            $\approx 472,414$

        The expected average use in 2009 is about 472,414 gallons.

53. $A = 41.97(0.996)^{x}$

    $A(389) = 41.97(0.996)^{389}$

    $A \approx 8.83$

    The altitude at the top of Mt. Everest is about 8.83 kilometers.

55. a. $A = p\left(1 + \dfrac{r}{n}\right)^{nt}$

        $A = 100\left(1 + \dfrac{0.07}{365}\right)^{365 \cdot 10}$

        $A \approx 100(1.0001918)^{3650}$

        $A = 201.36$

        The amount is \$201.36.

    b. For simple interest,

        $A = 100 + 100(0.07)t$

        $A = 100 + 100(0.07)(10)$

        $A = 100 + 70$

        $A = 170$

        \$201.36 - \$170 = \$31.36

57. a. $y_1 = 3^{x-5}$

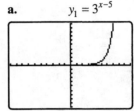

        -10, 10, 1, -10, 10, 1

    b. $4 = 3^{x-5}$ when $x \approx 6.26$.

59. a. Day 15: $2^{15-1} = 2^{14} = \$16,384$

    b. Day 20: $2^{20-1} = 2^{19} = \$524,288$

    c. nth Day: 2^{n-1}

    d. Day 30: $2^{30-1} = \$2^{29} = \$536,870,912$

    e. $2^{0} + 2^{1} + 2^{2} + \cdots + 2^{29}$

61. a. $2.3x^{4}y - 6.2x^{6}y^{2} + 9.2x^{5}y^{2}$

        $= -6.2x^{6}y^{2} + 9.2x^{5}y^{2} + 2.3x^{4}y$

    b. $-6.2x^{6}y^{2}$ is the leading term.

        $6 + 2 = 8$ is the degree of the polynomial.

    c. $-6.2x^{6}y^{2}$ is the leading term, so -6.2 is the leading coefficient.

62. $(f \cdot g)(x) = f(x) \cdot g(x)$

              $= (x + 5)(x^{2} - 2x + 4)$

              $= x^{3} - 2x^{2} + 4x + 5x^{2} - 10x + 20$

              $= x^{3} + 3x^{2} - 6x + 20$

63. $\sqrt{a^{2} - 8a + 16} = \sqrt{(a-4)^{2}} = |a - 4|$

64. $\sqrt[4]{\dfrac{32x^5y^9}{2y^3z}} = \sqrt[4]{\dfrac{16x^5y^6}{z}}$

$$= \dfrac{\sqrt[4]{16x^5y^6}}{\sqrt[4]{z}}$$

$$= \dfrac{\sqrt[4]{16x^4y^4 \cdot xy^2}}{\sqrt[4]{z}}$$

$$= \dfrac{2xy\sqrt[4]{xy^2}}{\sqrt[4]{z}} \cdot \dfrac{\sqrt[4]{z^3}}{\sqrt[4]{z^3}}$$

$$= \dfrac{2xy\sqrt[4]{xy^2z^3}}{\sqrt[4]{z^4}}$$

$$= \dfrac{2xy\sqrt[4]{xy^2z^3}}{z}$$

Exercise Set 13.3

1. $y = \log_a x$

   a. The base a must be positive and must not be equal to one.

   b. The argument x represents a number that is greater than 0. Thus, the domain is $\{x \mid x > 0\}$.

   c. $\mathbb{R}$

3. The functions $f(x) = a^x$ and $g(x) = \log_a x$ are inverse functions. Therefore, some of the points on the function are $g(x) = \log_a x$ are

$$\left(\dfrac{1}{27}, -3\right)\left(\dfrac{1}{9}, -2\right), \left(\dfrac{1}{3}, -1\right)(1, 0), \ (3, 1), \ (9, 2),$$

and $(27, 3)$. These points were obtained by switching the coordinates of the given points for the graph of $f(x) = a^x$.

5. The functions $y = a^x$ and $y = \log_a x$ for $a \neq 1$ are inverses of each other, thus the graphs are symmetric with respect to the line $y = x$. For each ordered pair (x, y) on the graph of $y = a^x$, the ordered pair (y, x) is on the graph of $y = \log_a x$.

7. $y = \log_2 x$

Convert to exponential form.

$2^y = x$

x	$\frac{1}{4}$	$\frac{1}{2}$	1	2	4
y	-2	-1	0	1	2

Domain: $\{x \mid x > 0\}$

Range: $\mathbb{R}$

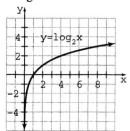

9. $y = \log_{1/2} x$

Convert to exponential form.

$$x = \left(\dfrac{1}{2}\right)^y$$

x	4	2	1	$\frac{1}{2}$	$\frac{1}{4}$
y	-2	-1	0	1	2

Domain: $\{x \mid x > 0\}$
Range: $\mathbb{R}$

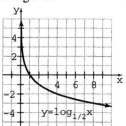

11. $y = \log_5 x$

Convert to the exponential form.

$x = 5^y$

x	$\frac{1}{25}$	$\frac{1}{5}$	1	5	25
y	-2	-1	0	1	2

Domain: $\{x \mid x > 0\}$

Range: $\mathbb{R}$

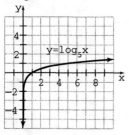

13. $y = \log_{1/5} x$ Convert to exponential form.

$x = \left(\frac{1}{5}\right)^y$

x	25	5	1	$\frac{1}{5}$	$\frac{1}{25}$
y	-2	-1	0	1	2

Domain: $\{x \mid x > 0\}$

Range: $\mathbb{R}$

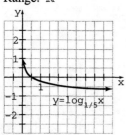

15. $y = 2^x$

x	-2	-1	0	1	2
y	$\frac{1}{4}$	$\frac{1}{2}$	1	2	4

$y = \log_{1/2} x$

Convert to exponential form.

$x = \left(\frac{1}{2}\right)^y$

x	4	2	1	$\frac{1}{2}$	$\frac{1}{4}$
y	-2	-1	0	1	2

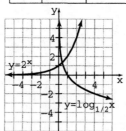

17. $y = 2^x$

x	-2	-1	0	1	2
y	$\frac{1}{4}$	$\frac{1}{2}$	1	2	4

$y = \log_2 x$

Convert to exponential form.

$x = 2^y$

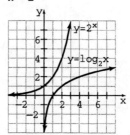

19. $2^3 = 8$

$\log_2 8 = 3$

21. $3^2 = 9$

$\log_3 9 = 2$

23. $16^{1/2} = 4$

$\log_{16} 4 = \dfrac{1}{2}$

25. $8^{1/3} = 2$

$\log_8 2 = \dfrac{1}{3}$

27. $\left(\dfrac{1}{2}\right)^5 = \dfrac{1}{32}$

$\log_{1/2}\left(\dfrac{1}{32}\right) = 5$

29. $2^{-3} = \dfrac{1}{8}$

$\log_2 \dfrac{1}{8} = -3$

31. $4^{-3} = \dfrac{1}{64}$

$\log_4 \dfrac{1}{64} = -3$

33. $64^{1/3} = 4$

$\log_{64} 4 = \dfrac{1}{3}$

35. $8^{-1/3} = \dfrac{1}{2}$

$\log_8 \dfrac{1}{2} = -\dfrac{1}{3}$

37. $81^{-1/4} = \dfrac{1}{3}$

$\log_{81} \dfrac{1}{3} = -\dfrac{1}{4}$

39. $10^{0.8451} = 7$

$\log_{10} 7 = 0.8451$

41. $e^2 = 7.3891$

$\log_e 7.3891 = 2$

43. $a^n = b$

$\log_a b = n$

45. $\log_2 8 = 3$

$2^3 = 8$

47. $\log_{1/3} \dfrac{1}{27} = 3$

$\left(\dfrac{1}{3}\right)^3 = \dfrac{1}{27}$

49. $\log_5 \dfrac{1}{25} = -2$

$5^{-2} = \dfrac{1}{25}$

51. $\log_{49} 7 = \dfrac{1}{2}$

$49^{1/2} = 7$

53. $\log_9 \dfrac{1}{81} = -2$

$9^{-2} = \dfrac{1}{81}$

55. $\log_{10} \dfrac{1}{1000} = -3$

$10^{-3} = \dfrac{1}{1000}$

57. $\log_6 216 = 3$

$6^3 = 216$

59. $\log_{10} 0.62 = -0.2076$

$10^{-0.2076} = 0.62$

61. $\log_e 6.52 = 1.8749$

$e^{1.8749} = 6.52$

63. $\log_w s = -p$

$w^{-p} = s$

65. $\log_4 64 = y$

$\quad 4^y = 64$

$\quad 4^y = 4^3$

$\quad y = 3$

67. $\log_a 125 = 3$

$\quad a^3 = 125$

$\quad a^3 = 5^3$

$\quad a = 5$

69. $\log_3 x = 3$

$\quad 3^3 = x$

$\quad 27 = x$

71. $\log_2 \dfrac{1}{16} = y$

$\quad 2^y = \dfrac{1}{16}$

$\quad 2^y = 2^{-4}$

$\quad y = -4$

73. $\log_{1/2} x = 6$

$\quad \left(\dfrac{1}{2}\right)^6 = x$

$\quad \dfrac{1}{64} = x$

75. $\log_a \dfrac{1}{27} = -3$

$\quad a^{-3} = \dfrac{1}{27}$

$\quad a^{-3} = 3^{-3}$

$\quad a = 3$

77. $\log_{10} 1 = 0$ because $10^0 = 1$

79. $\log_{10} 100 = 2$ because $10^2 = 100$

81. $\log_{10} \dfrac{1}{100} = -2$ because $10^{-2} = \dfrac{1}{10^2} = \dfrac{1}{100}$

83. $\log_{10} 10,000 = 4$ because $10^4 = 10,000$

85. $\log_4 256 = 4$ because $4^4 = 256$

87. $\log_3 \dfrac{1}{81} = -4$ because $3^{-4} = \dfrac{1}{3^4} = \dfrac{1}{81}$

89. $\log_8 \dfrac{1}{64} = -2$ because $8^{-2} = \dfrac{1}{8^2} = \dfrac{1}{64}$

91. $\log_9 1 = 0$ because $9^0 = 1$

93. $\log_9 9 = 1$ because $9^1 = 9$

95. $\log_4 1024 = 5$ because $4^5 = 1024$

97. If $f(x) = 5^x$, then $f^{-1}(x) = \log_5 x$.

99. $\log_3 62$ lies between 3 and 4 since 62 lies between $3^3 = 27$ and $3^4 = 81$.

101. $\log_{10} 425$ lies between 2 and 3 since 425 lies between $10^2 = 100$ and $10^3 = 1000$.

103. For $x > 1$, 2^x will grow faster than $\log_{10} x$. Note that when $x = 10$, $2^x = 1024$ while $\log_{10} x = 1$.

105. $\quad x = \log_{10} 10^6$

$\quad 10^x = 10^6$

$\quad x = 6$

107. $\quad x = \log_b b^8$

$\quad b^x = b^8$

$\quad x = 8$

109. $\quad x = 10^{\log_{10} 3}$

$\quad \log_{10} x = \log_{10} 3$

$\quad x = 3$

111. $\quad x = b^{\log_b 9}$

$\quad \log_b x = \log_b 9$

$\quad x = 9$

113. $\quad R = \log_{10} I$

$\quad 7 = \log_{10} I$

$\quad 10^7 = I$

$\quad I = 10,000,000$

The earthquake is 10,000,000 times more intense than the smallest measurable activity.

115.
$$R = \log_{10} I \qquad R = \log_{10} I$$
$$6 = \log_{10} I \qquad 2 = \log_{10} I$$
$$10^6 = I \qquad 10^2 = I$$
$$1,000,000 = I \qquad 100 = I$$

$$\frac{1,000,000}{100} = 10,000$$

An earthquake that measures 6 is 10,000 times more intense than one that measures 2.

117. $y = \log_2(x-1)$ or $2^y = x-1$

x	$1\frac{1}{4}$	$1\frac{1}{2}$	2	3	5
y	-2	-1	0	1	2

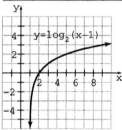

119. $2x^3 - 6x^2 - 36x = 2x\left(x^2 - 3x - 18\right)$
$$= 2x(x+3)(x-6)$$

120. $x^4 - 16 = \left(x^2 - 4\right)\left(x^2 + 4\right)$
$$= (x-2)(x+2)\left(x^2 + 4\right)$$

121. $40x^2 + 52x - 12 = 4\left(10x^2 + 13x - 3\right)$
$$= 4(2x+3)(5x-1)$$

122. $6r^2s^2 + rs - 1 = (3rs - 1)(2rs + 1)$

Exercise Set 13.4

1. Answers will vary.

3. Answers will vary.

5. Yes. This is true because of the product rule for logarithms.

7. $\log_4(3 \cdot 10) = \log_4 3 + \log_4 10$

9. $\log_8 7(x+3) = \log_8 7 + \log_8(x+3)$

11. $\log_2 \dfrac{27}{11} = \log_2 27 - \log_2 11$

13. $\log_{10} \dfrac{\sqrt{x}}{x-9} = \log_{10} \dfrac{x^{1/2}}{x-9}$
$$= \log_{10} x^{1/2} - \log_{10}(x-9)$$
$$= \frac{1}{2} \log_{10} x - \log_{10}(x-9)$$

15. $\log_6 x^7 = 7\log_6 x$

17. $\log_4(r+7)^5 = 5\log_4(r+7)$

19. $\log_4 \sqrt{\dfrac{a^3}{a+2}} = \log_4 \left(\dfrac{a^3}{a+2}\right)^{1/2}$
$$= \frac{1}{2} \log_4 \frac{a^3}{a+2}$$
$$= \frac{1}{2}[\log_4 a^3 - \log_4(a+2)]$$
$$= \frac{1}{2}[3\log_4 a - \log_4(a+2)]$$
$$= \frac{3}{2} \log_4 a - \frac{1}{2}\log_4(a+2)$$

21. $\log_3 \dfrac{d^6}{(a-8)^4} = \log_3 d^6 - \log_3(a-8)^4$
$$= 6\log_3 d - 4\log_3(a-8)$$

23. $\log_8 \dfrac{y(y+4)}{y^3} = \log_8 y + \log_8(y+4) - \log_8 y^3$
$$= \log_8 y + \log_8(y+4) - 3\log_8 y$$
$$= \log_8(y+4) - 2\log_8 y$$

25. $\log_{10} \dfrac{9m}{8n} = \log_{10} 9m - \log_{10} 8n$
$$= \log_{10} 9 + \log_{10} m - (\log_{10} 8 + \log_{10} n)$$
$$= \log_{10} 9 + \log_{10} m - \log_{10} 8 - \log_{10} n$$

27. $\log_5 2 + \log_5 8 = \log_5(2 \cdot 8) = \log_5 16$

29. $\log_2 9 - \log_2 5 = \log_2 \dfrac{9}{5}$

31. $6\log_4 2 = \log_4 2^6 = \log_4 64$

33. $\log_{10} x + \log_{10}(x+3) = \log_{10} x(x+3)$

35. $2\log_9 z - \log_9(z-2) = \log_9 z^2 - \log_9(z-2)$
$$= \log_9 \frac{z^2}{z-2}$$

37. $4(\log_5 p - \log_5 3) = 4\log_5 \dfrac{p}{3}$
$$= \log_5 \left(\frac{p}{3}\right)^4$$

39. $\log_2 n + \log_2(n+4) - \log_2(n-3)$
$\log_2 n(n+4) - \log_2(n-3)$
$\log_2 \dfrac{n(n+4)}{n-3}$

41. $\dfrac{1}{2}\Big[\log_5(x-8) - \log_5 x\Big] = \dfrac{1}{2}\log_5 \dfrac{x-8}{x}$
$$= \log_5 \left[\frac{x-8}{x}\right]^{1/2}$$
$$= \log_5 \sqrt{\frac{x-8}{x}}$$

43. $2\log_9 4 + \dfrac{1}{3}\log_9(r-6) - \dfrac{1}{2}\log_9 r$
$= \log_9 4^2 + \log_9(r-6)^{1/3} - \log_9 r^{1/2}$
$= \log_9 16 + \log_9 \sqrt[3]{r-6} - \log_9 \sqrt{r}$
$= \log_9 16\sqrt[3]{r-6} - \log_9 \sqrt{r}$
$= \log_9 \dfrac{16\sqrt[3]{r-6}}{\sqrt{r}}$

45. $4\log_6 3 - [2\log_6(x+3) + 4\log_6 x]$
$= \log_6 3^4 - [\log_6(x+3)^2 + \log_6 x^4]$
$= \log_6 81 - \log_6(x+3)^2 x^4$
$= \log_6 \dfrac{81}{(x+3)^2 x^4}$

47. $\log_a 10 = \log_a (2)(5)$
$\qquad = \log_a 2 + \log_a (5)$
$\qquad = 0.3010 + 0.6990$
$\qquad = 1$

49. $\log_a 0.4 = \log_a \dfrac{2}{5}$
$\qquad = \log_a 2 - \log_a 5$
$\qquad = 0.3010 - 0.6990$
$\qquad = -0.3980$

51. $\log_a 25 = \log_a 5^2$
$\qquad = 2(\log_a 5)$
$\qquad = 2(0.6990)$
$\qquad = 1.3980$

53. $5^{\log_5 10} = 10$

55. $(2^3)^{\log_8 7} = 8^{\log_8 7} = 7$

57. $\log_3 27 = \log_3 3^3 = 3$

59. $5\left(\sqrt[3]{27}\right)^{\log_3 5} = 5(3)^{\log_3 5}$
$\qquad = 5(5)$
$\qquad = 25$

61. Yes

63. $\log_a \dfrac{x}{y} = \log_a xy^{-1}$
$\qquad = \log_a x + \log_a y^{-1}$
$\qquad = \log_a x + \log_a \dfrac{1}{y}$

65. $\log_a(x^2 - 4) - \log_a(x+2) = \log_a \dfrac{x^2-4}{x+2}$
$\qquad = \log_a \dfrac{(x+2)(x-2)}{x+2}$
$\qquad = \log_a(x-2)$

67. Yes, assuming $x+4 > 0$.
$\log_a(x^2 + 8x + 16) = \log_a(x+4)^2$
$\qquad = 2\log_a(x+4)$

69. $\log_{10} x^2 = 2\log_{10} x = 2(0.4320) = 0.8640$

71. $\log_{10} \sqrt[4]{x} = \log_{10} x^{1/4}$

$$= \frac{1}{4}\log_{10} x = \frac{1}{4}(0.4320) = 0.1080$$

73. $\log_{10} xy = \log_{10} x + \log_{10} y$

$$= 0.5000 + 0.2000 = 0.7000$$

75. No; answers will vary. There is no simplification rule for the log of a sum.

77.
$$\log_2 \frac{\sqrt[4]{xy}\,\sqrt[3]{a}}{\sqrt[5]{a-b}} = \log_2 \sqrt[4]{xy}\,\sqrt[3]{a} - \log_2 \sqrt[5]{a-b}$$

$$= \log_2 (xy)^{1/4} + \log_2 a^{1/3} - \log_2 (a-b)^{1/5}$$

$$= \frac{1}{4}\log_2 xy + \frac{1}{3}\log_2 a - \frac{1}{5}\log_2 (a-b)$$

$$= \frac{1}{4}\log_2 x + \frac{1}{4}\log_2 y + \frac{1}{3}\log_2 a - \frac{1}{5}\log_2 (a-b)$$

79. Let $\log_a x = m$ and $\log_a y = n$. Then

$$a^m = x \text{ and } a^n = y, \text{ so } \frac{x}{y} = \frac{a^m}{a^n} = a^{m-n}.$$

Thus, $\log_a \dfrac{x}{y} = m - n = \log_a x - \log_a y$.

82.
$$\frac{2x+5}{x^2 - 7x + 12} \div \frac{x-4}{2x^2 - x - 15}$$

$$= \frac{2x+5}{(x-4)(x-3)} \div \frac{x-4}{(2x+5)(x-3)}$$

$$= \frac{(2x+5)}{(x-4)(x-3)} \cdot \frac{(2x+5)(x-3)}{(x-4)}$$

$$= \frac{(2x+5)^2}{(x-4)^2}$$

83.
$$\frac{2x+5}{x^2 - 7x + 12} - \frac{x-4}{2x^2 - x - 15}$$

$$= \frac{2x+5}{(x-4)(x-3)} - \frac{x-4}{(2x+5)(x-3)}$$

$$= \frac{(2x+5)(2x+5)}{(x-4)(x-3)(2x+5)} - \frac{(x-4)(x-4)}{(x-4)(x-3)(2x+5)}$$

$$= \frac{(4x^2 + 20x + 25) - (x^2 - 8x + 16)}{(x-4)(x-3)(2x+5)}$$

$$= \frac{4x^2 + 20x + 25 - x^2 + 8x - 16}{(x-4)(x-3)(2x+5)}$$

$$= \frac{3x^2 + 28x + 9}{(x-4)(x-3)(2x+5)}$$

$$= \frac{(3x+1)(x+9)}{(x-4)(x-3)(2x+5)}$$

84.
$$\frac{x}{4} + \frac{x}{5} = 1$$

$$5x + 4x = 20$$

$$9x = 20$$

$$x = \frac{20}{9} = 2\frac{2}{9}$$

It will take them $2\frac{2}{9}$ days to paint the house together.

85.
$$\sqrt[3]{4x^4 y^7} \cdot \sqrt[3]{12x^7 y^{10}} = \sqrt[3]{4x^4 y^7 \cdot 12x^7 y^{10}}$$

$$= \sqrt[3]{48x^{11} y^{17}}$$

$$= \sqrt[3]{8x^9 y^{15} \cdot 6x^2 y^2}$$

$$= 2x^3 y^5 \sqrt[3]{6x^2 y^2}$$

86. $2a - 7\sqrt{a} = 30$

$$2a - 7\sqrt{a} - 30 = 0 \quad \rightarrow \quad \text{let } u = \sqrt{a}$$

$$2u^2 - 7u - 30 = 0$$

$$(2u+5)(u-6) = 0$$

$$2u + 5 = 0 \qquad \text{or} \qquad u - 6 = 0$$

$$2u = -5 \qquad\qquad\qquad u = 6$$

$$2\sqrt{a} = -5 \qquad\qquad\qquad \sqrt{a} = 6$$

$$\text{not possible} \qquad\qquad a = 36$$

The solution is $a = 36$.

Mid-Chapter Test: 13.1 – 13.4

1. **a.** In $f(x)$, replace x by $g(x)$.

   b. $f(x) = 3x + 3$; $g(x) = 2x + 5$
   $$(f \circ g)(x) = 3(2x + 5) + 3$$
   $$= 6x + 15 + 3$$
   $$= 6x + 18$$

2. **a.** $(f \circ g)(x) = \left(\dfrac{6}{x}\right)^2 + 5$
   $$= \dfrac{36}{x^2} + 5$$

   b. $(f \circ g)(3) = \dfrac{36}{3^2} + 5 = \dfrac{36}{9} + 5 = 4 + 5 = 9$

   c. $(g \circ f)(x) = \dfrac{6}{x^2 + 5}$

   d. $(g \circ f)(3) = \dfrac{6}{3^2 + 5} = \dfrac{6}{9 + 5} = \dfrac{6}{14} = \dfrac{3}{7}$

3. **a.** Answers will vary. A function is one-to-one if each input corresponds to exactly one output and each output corresponds to exactly one input.

   b. No, the function is not one-to-one because it fails the horizontal line test.

4. **a.** The function is one-to-one. Each value in the range corresponds to exactly one value in the domain.

   b. $\{(2, -3), (3, 2), (1, 5), (8, 6)\}$

5. **a.** The function is one-to-one. Each value in the range corresponds to exactly one value in the domain.

b.
$$y = \dfrac{1}{3}x - 5$$
$$x = \dfrac{1}{3}y - 5$$
$$x + 5 = \dfrac{1}{3}y$$
$$3(x + 5) = y$$
$$3x + 15 = y \quad \text{or} \quad p^{-1}(x) = 3x + 15$$

6. **a.** The function is one-to-one. Each value in the range corresponds to exactly one value in the domain.

   b.
   $$y = \sqrt{x - 4}$$
   $$x = \sqrt{y - 4}$$
   $$x^2 = y - 4$$
   $$x^2 + 4 = y \quad \text{or} \quad k^{-1}(x) = x^2 + 4, \quad x \geq 0$$

7. $m(x) = -2x + 4$
   $$y = -2x + 4$$
   $$x = -2y + 4$$
   $$x - 4 = -2y$$
   $$\dfrac{x - 4}{-2} = y$$
   $$2 - \dfrac{1}{2}x = y \quad \text{or} \quad m^{-1}(x) = -\dfrac{1}{2}x + 2$$

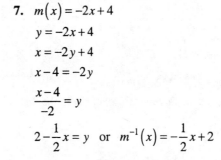

8. $y = 2^x$

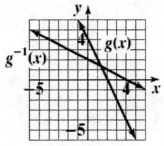

9. $y = 3^{-x}$

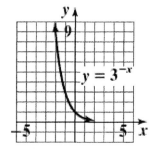

10. $y = \log_2 x$

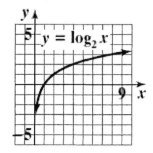

11. a. $N(t) = 5(2)^t$

$N(1) = 5(2)^1 = 5(2) = 10$

After 1 hour there are 10 bacteria in the dish.

b. $N(t) = 5(2)^t$

$N(6) = 5(2)^6 = 5(64) = 320$

After 6 hours there are 320 bacteria in the dish.

12. $27^{2/3} = 9 \quad \Leftrightarrow \quad \log_{27} 9 = \dfrac{2}{3}$

13. $\log_2 \dfrac{1}{64} = -6 \quad \Leftrightarrow \quad 2^{-6} = \dfrac{1}{64}$

14. $\log_5 125 = \log_5 5^3 = 3\log_5 5 = 3 \cdot 1 = 3$

15. $\log_{1/4} \dfrac{1}{16} = x$

$\log_{1/4} \left(\dfrac{1}{4}\right)^2 = x$

$2\log_{1/4} \dfrac{1}{4} = x$

$2 \cdot 1 = x$

$2 = x$

16. $\log_x 64 = 3$

Write the equivalent exponential equation and solve for x.

$x^3 = 64$

$x^3 = 4^3 \quad \Rightarrow \quad x = 4$

17. $\log_9 x^2 (x-5) = \log_9 x^2 + \log_9 (x-5)$

$= 2\log_9 x + \log_9 (x-5)$

18. $\log_5 \dfrac{7m}{\sqrt{n}} = \log_5 (7m) - \log_5 \sqrt{n}$

$= \log_5 (7m) - \log_5 n^{1/2}$

$= \log_5 7 + \log_5 m - \dfrac{1}{2}\log_5 n$

19. $3\log_2 x + \log_2 (x+7) - 4\log_2 (x+1)$

$= \log_2 x^3 + \log_2 (x+7) - \log_2 (x+1)^4$

$= \log_2 x^3 (x+7) - \log_2 (x+1)^4$

$= \log_2 \dfrac{x^3 (x+7)}{(x+1)^4}$

20. $\dfrac{1}{2}\left[\log_7 (x+2) - \log_7 x\right]$

$= \dfrac{1}{2}\left[\log_7 \dfrac{(x+2)}{x}\right]$

$= \log_7 \left[\dfrac{(x+2)}{x}\right]^{1/2}$

$= \log_7 \sqrt{\dfrac{x+2}{x}}$

Exercise Set 13.5

1. Common logarithms are logarithms with base 10.

3. Antilogarithms are numbers obtained by taking the base of the logarithm and raising it to the power that the logarithm is equal to. They are the numbers *inside* the logarithm.

5. $\log 86 = 1.9345$

7. $\log 19{,}200 = 4.2833$

9. $\log 0.0613 = -1.2125$

11. $\log 100 = 2.0000$

13. $\log 3.75 = 0.5740$

15. $\log 0.0173 = -1.7620$

17. antilog $0.2137 = 1.64$

19. antilog $4.6283 = 42,500$

21. antilog $(-1.7086) = 0.0196$

23. antilog $0.0000 = 1.00$

25. antilog $2.7625 = 579$

27. antilog $(-4.1390) = 0.0000726$

29. $\log N = 2.0000$
  $N = $ antilog 2.000
  $N = 100$

31. $\log N = 3.3817$
  $N = $ antilog 3.3817
  $N = 2410$

33. $\log N = 4.1409$
  $N = $ antilog 4.1409
  $N = 13,800$

35. $\log N = -1.06$
  $N = $ antilog (-1.06)
  $N = 0.0871$

37. $\log N = -0.6218$
  $N = $ antilog (-0.6218)
  $N = 0.239$

39. $\log N = -0.1256$
  $N = $ antilog (-0.1256)
  $N = 0.749$

41. $\log 3560 = 3.5514$
  Therefore, $10^{3.5514} \approx 3560$.

43. $\log 0.0727 = -1.1385$
  Therefore, $10^{-1.1385} \approx 0.0727$.

45. $\log 243 = 2.3856$
  Therefore, $10^{2.3856} \approx 243$.

47. $\log 0.00592 = -2.2277$
  Therefore, $10^{-2.2277} \approx 0.00592$.

49. $10^{2.8316} = 679$

51. $10^{-0.5186} = 0.303$

53. $10^{-1.4802} = 0.0331$

55. $10^{1.3503} = 22.4$

57. $\log 1 = x$
  $10^x = 1$
  $10^x = 10^0$
  $x = 0$
  Therefore, $\log 1 = 0$.

59. $\log 0.1 = x$
  $10^x = 0.1$
  $10^x = \dfrac{1}{10}$
  $10^x = 10^{-1}$
  $x = -1$
  Therefore, $\log 0.1 = -1$.

61. $\log 0.01 = x$
  $10^x = 0.01$
  $10^x = \dfrac{1}{100}$
  $10^x = 10^{-2}$
  $x = -2$
  Therefore, $\log 0.01 = -2$.

63. $\log 0.001 = x$
  $10^x = 0.001$
  $10^x = \dfrac{1}{1000}$
  $10^x = 10^{-3}$
  $x = -3$
  Therefore, $\log 0.001 = -3$.

65. $\log 10^7 = 7$

67. $10^{\log 7} = 7$

69. $4 \log 10^{5.2} = 4(5.2) = 20.8$

71. $5(10^{\log 8.3}) = 5(8.3) = 41.5$

73. No; $10^2 = 100$ and since $462 > 100$, $\log 462$ must be greater than 2.

75. No; $10^0 = 1$ and $10^{-1} = 0.1$ and since $0.1 < 0.163 < 1$, $\log 0.163$ must be between 0 and -1.

77. No;

$$\log \frac{y}{4x} = \log y - \log 4x$$
$$= \log y - (\log 4 + \log x)$$
$$= \log y - \log 4 - \log x$$

79. $\log 125 = \log(25 \cdot 5)$
$$= \log 25 + \log 5$$
$$= 1.3979 + 0.6990$$
$$= 2.0969$$

81. $\log \dfrac{1}{5} = \log 5^{-1}$
$$= -\log 5$$
$$= -1(0.6990)$$
$$= -0.6990$$

83. $\log 625 = \log 25^2$
$$= 2 \log 25$$
$$= 2(1.3979)$$
$$= 2.7958$$

85. $R = \log I, R = 3.4$

$3.4 = \log I$

$I = \text{antilog}(3.4)$

$I \approx 2510$

This earthquake is about 2,510 times more intense than the smallest measurable activity.

87. $R = \log I, R = 5.7$

$5.7 = \log I$

$I = \text{antilog}(5.7)$

$I \approx 501,000$

This earthquake is about 501,000 times more intense than the smallest measurable activity.

89. $\log d = 3.7 - 0.2g$

a. $g = 11$
$$\log d = 3.7 - 0.2(11)$$
$$= 3.7 - 2.2$$
$$= 1.5$$
$d = \text{antilog } 1.5 = 31.62$
A planet with absolute magnitude of 11 has a diameter of 31.62 kilometers.

b. $g = 20$
$$\log d = 3.7 - 0.2(20)$$
$$= 3.7 - 4$$
$$= -0.3$$
$d = \text{antilog } (-0.3) = 0.50$
A planet with absolute magnitude of 20 has a diameter of 0.50 kilometers.

c. $d = 5.8$
$$\log 5.8 = 3.7 - 0.2g$$
$$\log 5.8 - 3.7 = -0.2g$$
$$0.76343 - 3.7 = -0.2g$$
$$-2.93657 = -0.2g$$
$$\frac{-2.93657}{-0.2} = g$$
$$14.68 = g$$
A planet with diameter 5.8 kilometers has an absolute magnitude of 14.68.

91. $R(t) = 94 - 46.8 \log(t + 1)$

a. $R(2) = 94 - 46.8 \log(2 + 1)$
$$= 94 - 46.8 \log(3)$$
$$\approx 72$$
After two months, Sammy will remember about 72% of the course material.

b. $R(48) = 94 - 46.8 \log(2 + 48)$
$$= 94 - 46.8 \log(50)$$
$$\approx 15$$
After forty-eight months, Sammy will remember about 15% of the course material.

93. $R = \log I$, $R = 3.8$

$3.8 = \log I$

$I = \text{anti} \log(3.8)$

$I \approx 6310$

This earthquake is about 6310 times more intense than the smallest measurable activity.

95. $\log E = 11.8 + 1.5 m_s$

a. $\log E = 11.8 + 1.5(6)$

$\log E = 20.8$

$10^{20.8} = E$

$E = 6.31 \times 10^{20}$

The energy released is 6.31×10^{20}.

b. $\log(1.2 \times 10^{15}) = 11.8 + 1.5 m_s$

$15.07918125 = 11.8 + 1.5 m_s$

$3.27918125 = 1.5 m_s$

$m_s \approx 2.19$

The surface wave has magnitude 2.19.

97. $M = \dfrac{\log E - 11.8}{1.5}$

$M = \dfrac{\log(1.259 \times 10^{21}) - 11.8}{1.5}$

$= \dfrac{\log 1.259 + \log 10^{21} - 11.8}{1.5}$

$= \dfrac{\log 1.259 + 21 - 11.8}{1.5}$

$= \dfrac{\log 1.259 + 9.2}{1.5}$

$\approx \dfrac{0.1000 + 9.2}{1.5}$

≈ 6.2

The magnitude is about 6.2.

99. $R = \log I$

$\text{antilog}(R) = \text{antilog}(\log I)$

$\text{antilog}(R) = I$

101. $R = 26 - 41.9 \log(t + 1)$

$R - 26 = -41.9 \log(t + 1)$

$\dfrac{R - 26}{-41.9} = \dfrac{-41.9 \log(t + 1)}{-41.9}$

$\dfrac{26 - R}{41.9} = \log(t + 1)$

$\text{antilog}\left(\dfrac{26 - R}{41.9}\right) = \text{antilog}\left(\log(t + 1)\right)$

$\text{antilog}\left(\dfrac{26 - R}{41.9}\right) = t + 1$

$\text{antilog}\left(\dfrac{26 - R}{41.9}\right) - 1 = t$

104. $3r = -4s - 6 \implies 3r + 4s = -6 \quad (1)$

$3s = -5r + 1 \implies 5r + 3s = 1 \quad (2)$

To eliminate the variable r, multiply equation (1) by -5 and equation (2) by 3 then add.

$-5(3r + 4s = -6) \implies -15r - 20s = 30$

$3(5r + 3s = 1) \implies \underline{15r + 9s = 3}$

$-11s = 33$

$s = -3$

Substitute -3 for s in equation (1) and solve for r.

$3r + 4(-3) = -6$

$3r - 12 = -6$

$3r = 6$

$r = 2$

The solution is $(2, -3)$.

105. Let r equal the rate of car 2.

	d	r	t
Car 1	$4(r + 5)$	$r + 5$	4
Car 2	$4r$	r	4

The total distance was 420 miles.

$4(r + 5) + 4r = 420$

$4r + 20 + 4r = 420$

$8r + 20 = 420$

$8r = 400$

$r = 50$

The rate of car 2 is 50 mph and the rate of car 1 is $50 + 5 = 55$ mph.

106. $-3x^2 - 4x - 8 = 0$

$a = -3$, $b = -4$, $c = -8$

$$x = \frac{-(-4) \pm \sqrt{(-4)^2 - 4(-3)(-8)}}{2(-3)}$$

$$= \frac{4 \pm \sqrt{16 - 96}}{-6} = \frac{-4 \pm \sqrt{-80}}{6}$$

$$= \frac{-4 \pm 4i\sqrt{5}}{6} = \frac{-2 \pm 2i\sqrt{5}}{3}$$

107. Let x = the speed of the boat in still water.

Direction	Distance	Rate	Time
downriver	15	$x+5$	$\dfrac{15}{x+5}$
upriver	15	$x-5$	$\dfrac{15}{x-5}$

$$\frac{15}{x+5} + \frac{15}{x-5} = 4$$

$$15(x-5) + 15(x+5) = 4(x^2 - 25)$$

$$15x - 75 + 15x + 75 = 4x^2 - 100$$

$$4x^2 - 30x - 100 = 0$$

$$2x^2 - 15x - 50 = 0$$

$$(2x+5)(x-10) = 0$$

$2x + 5 = 0 \quad$ or $\quad x - 10 = 0$

$2x = -5 \qquad\qquad x = 10$

$x = -\dfrac{5}{2}$

Speed is nonnegative so we discard the negative solution. The speed of the boat in still water is 10 miles per hour.

108. $y = (x-2)^2 + 1$

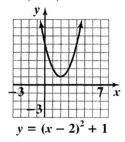

$y = (x - 2)^2 + 1$

109. $\dfrac{2x-3}{5x+10} < 0$

The numerator is 0 when $x = \dfrac{3}{2}$ and the denominator is 0 when $x = -2$. Therefore, our intervals are: $(-\infty, -2)$, $\left(-2, \dfrac{3}{2}\right)$, and $\left(\dfrac{3}{2}, \infty\right)$.

Interval A	Interval B	Interval C
$(-\infty, -2)$	$\left(-2, \dfrac{3}{2}\right)$	$\left(\dfrac{3}{2}, \infty\right)$
test value, -3	test value, 0	test value, 2

$$\frac{2x-3}{5x+10} < 0$$

$$\frac{2(-3)-3}{5(-3)+10} < 0$$

$$\frac{9}{5} < 0 \text{ false}$$

$$\frac{2x-3}{5x+10} < 0$$

$$\frac{2(0)-3}{5(0)+10} < 0$$

$$-\frac{3}{10} < 0 \text{ true}$$

$$\frac{2x-3}{5x+10} < 0$$

$$\frac{2(2)-3}{5(2)+10} < 0$$

$$\frac{1}{20} < 0 \text{ false}$$

Only interval B satisfies the inequality. The inequality is strict so we will not include any boundaries. The solution set is $\left\{x \mid -2 < x < \dfrac{3}{2}\right\}$.

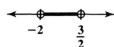

Exercise Set 13.6

1. $c = d$

3. Check for extraneous solutions.

5. log (–2) is not a real number

7. $5^x = 125$

$5^x = 5^3$

$x = 3$

9. $3^x = 81$

$3^x = 3^4$

$x = 4$

11. $64^x = 8$

$(8^2)^x = 8^1$

$8^{2x} = 8^1$

$2x = 1$

$x = \dfrac{1}{2}$

13. $7^{-x} = \dfrac{1}{49}$

$7^{-x} = 7^{-2}$

$-x = -2$

$x = 2$

15. $27^x = \dfrac{1}{3}$

$\left(3^3\right)^x = 3^{-1}$

$3^{3x} = 3^{-1}$

$3x = -1$

$x = -\dfrac{1}{3}$

17. $2^{x+2} = 64$

$2^{x+2} = 2^6$

$x + 2 = 6$

$x = 4$

19. $2^{3x-2} = 128$

$2^{3x-2} = 2^7$

$3x - 2 = 7$

$3x = 9$

$x = 3$

21. $27^x = 3^{2x+3}$

$3^{3x} = 3^{2x+3}$

$3x = 2x + 3$

$x = 3$

23. $7^x = 50$

$\log 7^x = \log 50$

$x \log 7 = \log 50$

$x = \dfrac{\log 50}{\log 7}$

$x \approx 2.01$

25. $4^{x-1} = 35$

$\log 4^{x-1} = \log 35$

$(x-1) \log 4 = \log 35$

$x - 1 = \dfrac{\log 35}{\log 4}$

$x = \dfrac{\log 35}{\log 4} + 1$

$x \approx 3.56$

27. $1.63^{x+1} = 25$

$\log 1.63^{x+1} = \log 25$

$(x+1) \log 1.63 = \log 25$

$x + 1 = \dfrac{\log 25}{\log 1.63}$

$x + 1 \approx 6.59$

$x \approx 5.59$

29. $3^{x+4} = 6^x$

$\log 3^{x+4} = \log 6^x$

$(x+4) \log 3 = x \log 6$

$x \log 3 + 4 \log 3 = x \log 6$

$4 \log 3 = x \log 6 - x \log 3$

$4 \log 3 = x(\log 6 - \log 3)$

$\dfrac{4 \log 3}{\log 6 - \log 3} = x$

$6.34 \approx x$

31. $\log_{36} x = \dfrac{1}{2}$

$36^{1/2} = x$

$\sqrt{36} = x$

$6 = x$

33. $\log_{125} x = \dfrac{1}{3}$

$125^{1/3} = x$

$\sqrt[3]{125} = x$

$5 = x$

35. $\log_2 x = -4$

$2^{-4} = x$

$\dfrac{1}{2^4} = x$

$\dfrac{1}{16} = x$

37. $\log x = 2$

$\log_{10} x = 2$

$10^2 = x$

$100 = x$

39. $\log_2(5-3x) = 3$

$2^3 = 5-3x$

$8 = 5-3x$

$3x = -3$

$x = -1$

41. $\log_5(x+1)^2 = 2$

$(x+1)^2 = 5^2$

$(x+1)^2 = 25$

$x+1 = \pm\sqrt{25}$

$x+1 = \pm 5$

$x+1 = 5 \quad\text{or}\quad x+1 = -5$

$x = 4 \qquad\qquad x = -6$

Both values check. The solutions are 4 and –6.

43. $\log_2(r+4)^2 = 4$

$(r+4)^2 = 2^4$

$r^2 + 8r + 16 = 16$

$r^2 + 8r = 0$

$r(r+8) = 0$

$r = 0 \quad\text{or}\quad r+8 = 0$

$r = -8$

Both values check. The solutions are 0 and –8.

45. $\log(x+8) = 2$

$\log_{10}(x+8) = 2$

$10^2 = x+8$

$100 = x+8$

$x = 92$

47. $\log_2 x + \log_2 5 = 2$

$\log_2 5x = 2$

$5x = 2^2$

$x = \dfrac{4}{5}$

49. $\log(r+2) = \log(3r-1)$

$r+2 = 3r-1$

$3 = 2r$

$\dfrac{3}{2} = r$

51. $\log(2x+1) + \log 4 = \log(7x+8)$

$\log(8x+4) = \log(7x+8)$

$8x+4 = 7x+8$

$x = 4$

53. $\log n + \log(3n-5) = \log 2$

$\log(3n^2 - 5n) = \log 2$

$3n^2 - 5n = 2$

$3n^2 - 5n - 2 = 0$

$(3n+1)(n-2) = 0$

$3n+1 = 0 \quad\text{or}\quad n-2 = 0$

$3n = -1 \qquad\qquad n = 2$

$n = -\tfrac{1}{3}$

Check: $n = -\dfrac{1}{3}$

$\log n + \log(3n-5) = \log 2$

$\log\left(-\dfrac{1}{3}\right) + \log\left[3\left(\dfrac{-1}{3}\right)-5\right] = \log 2$

Logarithms of negative numbers are not real numbers so $-\dfrac{1}{3}$ is an extraneous solution.

Check: $n = 2$

$\log n + \log(3n-5) = \log 2$

$\log 2 + \log[3(2)-5] = \log 2$

$\log 2 + \log 1 = \log 2$

$\log(2 \cdot 1) = \log 2$

$\log 2 = \log 2$

2 is the only solution.

55. $\log 6 + \log y = 0.72$

$\quad\quad \log 6y = 0.72$

$\quad\quad 6y = \text{antilog } 0.72$

$\quad\quad y = \dfrac{\text{antilog } 0.72}{6}$

$\quad\quad y \approx 0.87$

57. $2\log x - \log 9 = 2$

$\quad\quad \log x^2 - \log 9 = 2$

$\quad\quad \log \dfrac{x^2}{9} = 2$

$\quad\quad \dfrac{x^2}{9} = \text{antilog } 2$

$\quad\quad \dfrac{x^2}{9} = 100$

$\quad\quad x^2 = 900$

$\quad\quad x^2 - 900 = 0$

$\quad\quad (x+30)(x-30) = 0$

$\quad\quad x+30 = 0 \quad \text{or} \quad x-30 = 0$

$\quad\quad\quad x = -30 \quad\quad\quad\quad x = 30$

Check: $x = -30$

$\quad\quad 2\log x - \log 9 = 2$

$2\log(-30) - \log 9 = 2$

Logarithms of negative numbers are not real numbers so -30 is an extraneous solution.

Check: $x = 30$

$\quad\quad 2\log x - \log 9 = 2$

$\quad\quad 2\log 30 - \log 9 = 2$

$\quad\quad\quad \log \dfrac{900}{9} = 2$

$\quad\quad\quad \log 100 = 2$

$\quad\quad\quad 100 = \text{antilog } 2$

$\quad\quad\quad 100 = 100$

Thus, 30 is the only solution.

59. $\log x + \log(x-3) = 1$

$\quad\quad \log(x^2 - 3x) = 1$

$\quad\quad x^2 - 3x = \text{antilog } 1$

$\quad\quad x^2 - 3x = 10$

$\quad\quad x^2 - 3x - 10 = 0$

$\quad\quad (x-5)(x+2) = 0$

$\quad x - 5 = 0 \quad \text{or} \quad x+2 = 0$

$\quad\quad x = 5 \quad\quad\quad\quad x = -2$

A check shows that -2 is an extraneous solution so 5 is the only solution.

61. $\log x = \dfrac{1}{3}\log 64$

$\quad\quad \log x = \log 64^{1/3}$

$\quad\quad \log x = \log 4$

$\quad\quad\quad x = 4$

63. $\log_8 x = 4\log_8 2 - \log_8 8$

$\quad\quad \log_8 x = \log_8 2^4 - \log_8 8$

$\quad\quad \log_8 x = \log_8 \dfrac{16}{8}$

$\quad\quad \log_8 x = \log_8 2$

$\quad\quad\quad x = 2$

65. $\log_5(x+3) + \log_5(x-2) = \log_5 6$

$\quad\quad \log_5(x+3)(x-2) = \log_5 6$

$\quad\quad \log_5(x^2 + x - 6) = \log_5 6$

$\quad\quad\quad x^2 + x - 6 = 6$

$\quad\quad\quad x^2 + x - 12 = 0$

$\quad\quad\quad (x+4)(x-3) = 0$

$\quad x+4 = 0 \quad \text{or} \quad x-3 = 0$

$\quad\quad x = -4 \quad\quad\quad\quad x = 3$

Check $x = -4$:

$\log_5\left(-4+3\right) + \log_5\left(-4-2\right) = \log_5 6$

$\log_5\left(-1\right) + \log_5\left(-6\right) = \log_5 6$

Since we cannot take the logarithm of a negative number, -4 is an extraneous solution.

Check $x = 3$:

$\log_5\left(3+3\right) + \log_5\left(3-2\right) = \log_5 6$

$\quad\quad \log_5\left(6\right) + \log_5\left(1\right) = \log_5 6$

$\quad\quad\quad\quad \log_5 6 = \log_5 6$

Therefore, $x = 3$ is the only solution.

67. $\log_2(x+3) - \log_2(x-6) = \log_2 4$

$$\log_2 \frac{x+3}{x-6} = \log_2 4$$

$$\frac{x+3}{x-6} = 4$$

$$x+3 = 4x - 24$$

$$27 = 3x$$

$$9 = x$$

69.
$$50,000 = 4500(2^t)$$

$$\frac{50,000}{4500} = 2^t$$

$$\log \frac{50,000}{4500} = \log 2^t$$

$$\log \frac{50,000}{4500} = t \log 2$$

$$\log 50,000 - \log 4500 = t \log 2$$

$$\frac{\log 50,000 - \log 4500}{\log 2} = t$$

$$3.47 \approx t$$

There are 50,000 bacteria after about 3.47 hours.

71.
$$80 = 200(0.75)^t$$

$$0.4 = (0.75)^t$$

$$\log 0.4 = \log(0.75)^t$$

$$\log 0.4 = t \log 0.75$$

$$\frac{\log 0.4}{\log 0.75} = t$$

$$3.19 \approx t$$

80 grams remain after about 3.19 years.

73.
$$A = P\left(1 + \frac{r}{n}\right)^{nt}$$

$$4600 = 2000\left(1 + \frac{0.05}{1}\right)^{1 \cdot t}$$

$$4600 = 2000(1.05)^t$$

$$\frac{4600}{2000} = 1.05^t$$

$$2.3 = 1.05^t$$

$$\log 2.3 = \log 1.05^t$$

$$\log 2.3 = t \log 1.05$$

$$\frac{\log 2.3}{\log 1.05} = t \quad \Rightarrow \quad t \approx 17.07 \text{ years}$$

The $2000 will grow to $4600 in about 17.07 years.

75. $f(t) = 26 - 12.1 \cdot \log(t+1)$

   a. $x = 1990 - 1960 = 30$
$f(30) = 26 - 12.1 \log (30+1) = 7.95$
In 1990, the rate was about 7.95 deaths per 1000 live births.

   b. $x = 2005 - 1960 = 45$
$f(45) = 26 - 12.1 \log (45+1) \approx 5.88$
In 2005, the rate was about 5.88 deaths per 1000 live births.

77. $c = 50,000, n = 12, r = 0.15.$

$$S = c(1-r)^n$$

$$S = 50,000(1 - 0.15)^{12}$$

$$S = 50,000(0.85)^{12}$$

$$S \approx 7112.09$$

The scrap value is about $7112.09.

79. $P_{out} = 12.6$ and $P_{in} = 0.146$

$$P = 10 \log\left(\frac{12.6}{0.146}\right)$$

$$P \approx 10 \log 86.30137$$

$$P \approx 10(1.936)$$

$$P \approx 19.36$$

The power gain is about 19.36 watts.

81. a. $d = 120$

$$d = 10 \log I$$

$$120 = 10 \log I$$

$$12 = \log I$$

$$I = \text{antilog } 12$$

$$I = 10^{12}$$

$$I = 1,000,000,000,000$$

The intensity is 1,000,000,000,000 times the minimum intensity of audible sound.

   b. $d = 50$

$$d = 10 \log I$$

$$50 = 10 \log I$$

$$5 = \log I$$

$$I = \text{antilog } 5$$

$$I = 10^5$$

$$I = 100,000$$

$$\frac{1,000,000,000,000}{100,000} = 10,000,000$$

The sound of an airplane engine is 10,000,000 times more intense than the noise in a busy city street.

83. $8^x = 16^{x-2}$

$2^{3x} = 2^{4(x-2)}$

$3x = 4(x-2)$

$3x = 4x - 8$

$8 = x$

85. $2^{2x} - 6(2^x) + 8 = 0$

$(2^x)^2 - 6(2^x) + 8 = 0$

$y^2 - 6y + 8 = 0 \leftarrow$ Replace 2^x with y

$(y-4)(y-2) = 0$

$y - 4 = 0$ or $y - 2 = 0$

$\quad y = 4 \qquad\qquad y = 2$

$2^x = 4 \qquad 2^x = 2 \leftarrow$ Replace y with 2^x

$2^x = 2^2 \qquad 2^x = 2^1$

$x = 2 \qquad\quad x = 1$

The solutions are $x = 2$ and $x = 1$.

87. $2^x = 8^y$

$x + y = 4$

The first equation simplifies to

$2^x = (2^3)^y$

$2^x = 2^{3y}$

$x = 3y$

The system becomes

$x = 3y$

$x + y = 4$

Substitute $3y$ for x in the second equation.

$x + y = 4$

$3y + y = 4$

$\quad 4y = 4$

$\quad\quad y = 1$

Now, substitute 1 for y in the first equation.

$x = 3y$

$x = 3(1) = 3$

The solution is $(3, 1)$.

89. $\log(x + y) = 2$

$\quad x - y = 8$

The first equation can be written as

$x + y = 10^2$

$x + y = 100$

The system becomes

$x + y = 100$

$x - y = 8$

Add: $\overline{\;2x\;\; = 108\;}$

$\quad x \;\;\; = 54$

Substitute 54 for x in the first equation.

$54 + y = 100$

$\quad\quad y = 46$

The solution is $(54, 46)$.

91.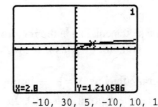

−10, 30, 5, −10, 10, 1

The solution is $x \approx 2.8$.

93.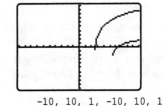

−10, 10, 1, −10, 10, 1

There is no real-number solution.

95. Volume of cylinder:

$$V_1 = \pi r^2 h = \pi\left(\frac{3}{2}\right)^2 \cdot 4 \approx 28.2743 \text{ cubic feet}$$

Volume of box:

$$V_2 = l \cdot w \cdot h = (3)(3)(4) = 36 \text{ cubic feet}$$

The box has the greater volume.

Difference in volumes:

$$V_2 - V_1 \approx 7.73 \text{ cubic feet}$$

96. $f(x) = x^2 - x, \; g(x) = x - 1$

$(g - f)(x) = (x - 1) - (x^2 - x)$

$\qquad = x - 1 - x^2 + x$

$\qquad = -x^2 + 2x - 1$

$(g - f)(3) = -(3)^2 + 2(3) - 1$

$\qquad = -9 + 6 - 1$

$\qquad = -4$

97. $3x - 4y \le 6$

$y > -x + 4$

For $3x - 4y \le 6$, graph the line $3x - 4y = 6$ using a solid line. For the check point, select (0, 0):

$3x - 4y \le 6$

$3(0) - 4(0) \le 6$

$\qquad 0 \le 6 \quad$ True

Since this is a true statement, shade the region which contains the point (0, 0). This is the region "above" the line.

For $y > -x + 4$, graph the line $y = -x + 4$ using a dashed line. For the check point, select (0, 0):

$y > -x + 4$

$(0) > -(0) + 4$

$\qquad 0 > 4 \qquad$ False

Since this is a false statement, shade the region which does not contain the point (0, 0). This is the region "above" the line. To obtain the final region, take the intersection of the above two regions.

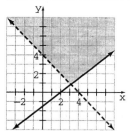

98. $\dfrac{2\sqrt{xy} - \sqrt{xy}}{\sqrt{x} + \sqrt{y}} = \dfrac{2\sqrt{xy} - \sqrt{xy}}{\sqrt{x} + \sqrt{y}} \cdot \dfrac{\sqrt{x} - \sqrt{y}}{\sqrt{x} - \sqrt{y}}$

$\qquad = \dfrac{\left(2\sqrt{xy} - \sqrt{xy}\right) \cdot \left(\sqrt{x} - \sqrt{y}\right)}{\left(\sqrt{x} + \sqrt{y}\right) \cdot \left(\sqrt{x} - \sqrt{y}\right)}$

$\qquad = \dfrac{2\sqrt{x^2 y} - 2\sqrt{xy^2} - \sqrt{x^2 y} + \sqrt{xy^2}}{\sqrt{x^2} - \sqrt{xy} + \sqrt{xy} - \sqrt{y^2}}$

$\qquad = \dfrac{\sqrt{x^2 y} - \sqrt{xy^2}}{\sqrt{x^2} - \sqrt{y^2}}$

$\qquad = \dfrac{x\sqrt{y} - y\sqrt{x}}{x - y}$

99. $E = mc^2$, for c

$\dfrac{E}{m} = \dfrac{mc^2}{m}$

$\dfrac{E}{m} = c^2$

$\sqrt{\dfrac{E}{m}} = c$

(Recall that when using the square root property to solve for a variable in a formula we only use the principal square root because we are generally solving for a variable that cannot be negative.)

100. Use $f(x) = a(x - h)^2 + k$, where $a = 2$ and $(h, k) = (3, -5)$.

$f(x) = 2(x - 3)^2 - 5$

Exercise Set 13.7

1. a. The base in the natural exponential function is e.

    b. The approximate value of e is 2.7183.

3. The domain of $\ln x$ is $\{x \mid x > 0\}$.

5. $\log_a x = \dfrac{\log_b x}{\log_b a}$

7. $\ln e^x = x$

9. The inverse of $\ln x$ is e^x.

11. P decreases when t increases for $k < 0$.

13. $\ln 62 \approx 4.1271$

15. $\ln 0.813 \approx -0.2070$

17. $\ln N = 1.6$

$e^{\ln N} = e^{1.6}$

$N = e^{1.6} \approx 4.95$

19. $\ln N = -2.85$

$e^{\ln N} = e^{-2.85}$

$N = e^{-2.85} \approx 0.0578$

21. $\ln N = -0.0287$

$e^{\ln N} = e^{-0.0287}$

$N = e^{-0.0287} \approx 0.972$

23. $\log_3 56 = \dfrac{\log 56}{\log 3} \approx 3.6640$

25. $\log_2 21 = \dfrac{\log 21}{\log 2} \approx 4.3923$

27. $\log_4 11 = \dfrac{\log 11}{\log 4} \approx 1.7297$

29. $\log_5 82 = \dfrac{\log 82}{\log 5} \approx 2.7380$

31. $\log_6 185 = \dfrac{\log 185}{\log 6} \approx 2.9135$

33. $\ln 51 = \dfrac{\log 51}{\log e} \approx 3.9318$

35. $\log_5 0.463 = \dfrac{\log 0.463}{\log 5} \approx -0.4784$

37. $\ln x + \ln(x-1) = \ln 12$

$\ln x(x-1) = \ln 12$

$e^{\ln[x(x-1)]} = e^{\ln 12}$

$x(x-1) = 12$

$x^2 - x - 12 = 0$

$(x-4)(x+3) = 0$

$x - 4 = 0 \qquad x + 3 = 0$

$x = 4 \qquad x = -3$

Only $x = 4$ checks since $x = -3$ is an extraneous solution. Note that when $x = -3$ we would get $\ln(-3)$ and $\ln(-4)$ which are not real numbers.

39. $\ln x + \ln(x+4) = \ln 5$

$\ln x(x+4) = \ln 5$

$e^{\ln(x^2+4x)} = e^{\ln 5}$

$x^2 + 4x = 5$

$x^2 + 4x - 5 = 0$

$(x+5)(x-1) = 0$

$x + 5 = 0 \quad$ or $\quad x - 1 = 0$

$x = -5 \qquad x = 1$

Only $x = 1$ checks since $x = -5$ is an extraneous solution. Note that when $x = -5$ we would get $\ln(-5)$ and $\ln(-1)$ which are not real numbers.

41. $\ln x = 5 \ln 2 - \ln 8$

$\ln x = \ln 2^5 - \ln 8$

$\ln x = \ln \dfrac{32}{8}$

$\ln x = \ln 4$

$e^{\ln x} = e^{\ln 4}$

$x = 4$

$x = 4$ checks.

43. $\ln(x^2 - 4) - \ln(x+2) = \ln 4$

$\ln(x^2 - 4) - \ln(x+2) = \ln 4$

$\ln\left(\dfrac{x^2 - 4}{x + 2}\right) = \ln 4$

$\ln(x-2) = \ln 4$

$e^{\ln(x-2)} = e^{\ln(4)}$

$x - 2 = 4$

$x = 6$

$x = 6$ checks.

45. $P = 120 e^{(2.3)(1.6)}$

$P = 120 e^{3.68}$

$P \approx 4757.5673$

47. $50 = P_0 e^{-0.5(3)}$

$50 = P_0 e^{-1.5}$

$\dfrac{50}{e^{-1.5}} = P_0$

$P_0 \approx 224.0845$

49. $60 = 20e^{1.4t}$

$3 = e^{1.4t}$

$\ln 3 = \ln e^{1.4t}$

$\ln 3 = 1.4t$

$t = \dfrac{\ln 3}{1.4} \approx 0.7847$

51. $86 = 43e^{k(3)}$

$2 = e^{3k}$

$\ln 2 = \ln e^{3k}$

$\ln 2 = 3k$

$k = \dfrac{\ln 2}{3} \approx 0.2310$

53. $20 = 40e^{k(2.4)}$

$0.5 = e^{2.4k}$

$\ln 0.5 = \ln e^{2.4k}$

$\ln 0.5 = 2.4k$

$k = \dfrac{\ln 0.5}{2.4} \approx -0.2888$

55. $A = 6000e^{-0.08(3)}$

$A = 6000e^{-0.24}$

$A \approx 4719.7672$

57. $V = V_0 e^{kt}$

$\dfrac{V}{e^{kt}} = V_0 \text{ or } V_0 = \dfrac{V}{e^{kt}}$

59. $P = 150e^{7t}$

$\dfrac{P}{150} = e^{7t}$

$\ln \dfrac{P}{150} = \ln e^{7t}$

$\ln \dfrac{P}{150} = 7t$

$\dfrac{\ln P - \ln 150}{7} = t \text{ or } t = \dfrac{\ln P - \ln 150}{7}$

61. $A = A_0 e^{kt}$

$\dfrac{A}{A_0} = e^{kt}$

$\ln \dfrac{A}{A_0} = \ln e^{kt}$

$\ln A - \ln A_0 = kt$

$\dfrac{\ln A - \ln A_0}{t} = k \text{ or } k = \dfrac{\ln A - \ln A_0}{t}$

63. $\ln y - \ln x = 2.3$

$\ln \dfrac{y}{x} = 2.3$

$e^{\ln(y/x)} = e^{2.3}$

$\dfrac{y}{x} = e^{2.3}$

$y = xe^{2.3}$

65. $\ln y - \ln(x + 6) = 5$

$\ln \dfrac{y}{x+6} = 5$

$e^{\ln \frac{y}{x+6}} = e^5$

$\dfrac{y}{x+6} = e^5$

$y = (x+6)e^5$

67. $e^x = 12.183$

Take the natural logarithm of both sides of the equation.

$\ln e^x = \ln 12.183$

$x = \ln 12.183 \approx 2.5000$

69. $P = P_0 e^{kt}$

a. $P = 5000e^{0.06(2)}$

$= 5000e^{0.12} \approx 5637.48$

The amount will be $5637.48.

b. If the amount in the account is to double, then $P = 2(5000) = 10{,}000$.

$10{,}000 = 5000e^{0.06t}$

$2 = e^{0.06t}$

$\ln 2 = \ln e^{0.06t}$

$\ln 2 = 0.06t$

$t = \dfrac{\ln 2}{0.06} \approx 11.55$

It would take about 11.55 years for the value to double.

71. $P = P_0 e^{-0.028t}$

$P = 70e^{-0.028(20)}$

$P = 70e^{-0.56}$

$P \approx 39.98$

After 20 years, about 39.98 grams remain.

73. $f(t) = 1 - e^{-0.04t}$

a. $f(t) = 1 - e^{-0.04(50)} = 1 - e^{-2} \approx 0.8647$

About 86.47% of the target market buys the drink after 50 days of advertising.

b. $0.75 = 1 - e^{-0.04t}$

$-0.25 = -e^{-0.04t}$

$0.25 = e^{-0.04t}$

$\ln 0.25 = \ln e^{-0.04t}$

$\ln 0.25 = -0.04t$

$t = \dfrac{\ln 0.25}{-0.04}$

$t \approx 34.66$

About 34.66 days of advertising are needed if 75% of the target market is to buy the soft drink.

75. $f(P) = 0.37 \ln P + 0.05$

a. $f(972,000) = 0.37 \ln(972,000) + 0.05$

$\approx 5.1012311 + 0.05$

≈ 5.15

The average walking speed in Nashville, Tennessee is 5.15 feet per second.

b. $f(8,567,000) = 0.37 \ln(8,567,000) + 0.05$

$\approx -5.906 + 0.05$

≈ 5.96

The average walking speed in New York City is 5.96 feet per second.

c. $5 = 0.37 \ln P + 0.05$

$4.95 = 0.37 \ln P$

$13.378378 = \ln P$

$e^{13.378378} = e^{\ln P}$

$P = e^{13.378378}$

$P \approx 646,000$

The population is about 646,000.

77. $V(t) = 24e^{0.08t}, \quad t = 2008 - 1626 = 382$

$V(377) = 24e^{0.08(382)}$

$\approx 449,004,412,200,000$

The value of Manhattan in 2003 is about $449,004,412,200,000.

79. $P(t) = 6.30e^{0.013t}$

a. $t = 2010 - 2003 = 7$

$P(7) = 6.30e^{0.013(7)}$

$= 6.30e^{0.091}$

≈ 6.9

The world's population in 2010 is expected to be about 6.9 billion.

b. $2(6.30 \text{ billion}) \implies 12.60 \text{ billion}$

$12.60 = 6.30e^{0.013t}$

$\dfrac{12.60}{6.30} = \dfrac{6.30e^{0.013t}}{6.30}$

$2 = e^{0.013t}$

$\ln 2 = \ln e^{0.013t}$

$\ln 2 = 0.013t$

$t = \dfrac{\ln 2}{0.013} \approx 53$

The world's population will double in about 53 years.

81. $d(t) = 2.19e^{0.0164t}$

a. $t = 2025 - 2005 = 20$

$d(20) = 2.19e^{0.0164(20)} = 2.19e^{0.328} \approx 3.04$

According to the trend, the demand for nurses in 2025 will be approximately 3.04 million.

b. $t = 2040 - 2005 = 35$

$d(35) = 2.19e^{0.0164(35)} = 2.19e^{0.574} \approx 3.89$

According to the trend, the demand for nurses in 2040 will be approximately 3.89 million.

83. $r(t) = 1182.3e^{0.0715t}$

    a. $t = 2006 - 1994 = 12$

$$r(12) = 1182.3e^{0.0715(12)}$$

$$= 1182.3e^{0.858} \approx 2788.38$$

According to the trend, the average annual tax refund was about \$2788.38 in 2006.

    b. $t = 2010 - 1994 = 16$

$$r(16) = 1182.3e^{0.0715(16)}$$

$$= 1182.3e^{1.144} \approx 3711.59$$

According to the trend, the average annual tax refund will be about \$3711.59 in 2010.

85. $y = 15.29 + 5.93 \ln x$

    a. $y(18) = 15.29 + 5.93 \ln(18) \approx 32.43$ in.

    b. $y(30) = 15.29 + 5.93 \ln(30) \approx 35.46$ in.

87. $f(t) = v_0 e^{-0.0001205t}$

    a. Use $f(t) = 9$ and $v_0 = 20$.

$$9 = 20e^{-0.0001205t}$$

$$0.45 = e^{-0.0001205t}$$

$$\ln 0.45 = -0.0001205t$$

$$\frac{\ln 0.45}{-0.0001205} = t$$

$$t \approx 6626.62$$

The bone is about 6626.62 years old.

    b. Let x equal the original amount of carbon 14 then $0.5x$ equals the remaining amount.

$$0.5x = xe^{-0.0001205t}$$

$$\frac{0.5x}{x} = \frac{xe^{-0.0001205t}}{x}$$

$$0.5 = e^{-0.0001205t}$$

$$\ln 0.5 = \ln e^{-0.0001205t}$$

$$\ln 0.5 = -0.0001205t$$

$$\frac{\ln 0.5}{-0.0001205} = t$$

$$t \approx 5752.26$$

If 50% of the carbon 14 remains, the item is about 5752.26 years old.

89. Let P_0 be the initial investment, then $P = 20{,}000$, $k = 0.06$, and $t = 18$.

$$P = P_0 e^{kt}$$

$$20{,}000 = P_0 e^{0.06(18)}$$

$$20{,}000 = P_0 e^{1.08}$$

$$\frac{20{,}000}{e^{1.08}} = P_0$$

$$6791.91 \approx P_0$$

The initial investment should be \$6791.91.

91. a. Strontium 90 has a higher decay rate so it will decompose more quickly.

    b. $P = P_0 e^{-kt}$

$$P = P_0 e^{-0.023(50)}$$

$$= P_0 e^{-1.15}$$

$$\approx P_0 (0.3166)$$

About 31.66% of the original amount will remain.

93. Answers will vary.

95. $e^{x-4} = 12 \ln(x+2)$

$$y_1 = e^{x-4}$$

$$y_2 = 12 \ln(x+2) \cdot$$

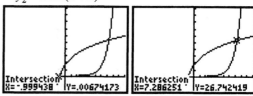

The intersections are approximately $(-0.999, 0.007)$ and $(7.286, 26.742)$. Therefore, $x \approx -0.999$ or $x \approx 7.286$.

97. $3x - 6 = 2e^{0.2x} - 12$

$$y_1 = 3x - 6$$

$$y_2 = 2e^{0.2x} - 12$$

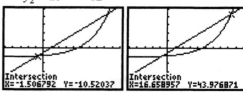

The intersections are approximately $(-1.507, -10.520)$ and $(16.659, 43.977)$. Therefore, $x \approx -1.507$ and 16.659.

99. $x = \dfrac{1}{k}\ln(kv_0t+1)$

$xk = \ln(kv_0t+1)$

$e^{xk} = e^{\ln(kv_0t+1)}$

$e^{xk} = kv_0t+1$

$e^{xk}-1 = kv_0t$

$\dfrac{e^{xk}-1}{kt} = v_0$ or $v_0 = \dfrac{e^{xk}-1}{kt}$

101. $\ln i - \ln I = \dfrac{-t}{RC}$

$\ln\dfrac{i}{I} = \dfrac{-t}{RC}$

$e^{\ln(i/I)} = e^{-t/RC}$

$\dfrac{i}{I} = e^{-t/RC}$

$i = Ie^{-t/RC}$

102. $\left(3xy^2+y\right)\left(4x-3xy\right)$

$= 12x^2y^2 - 9x^2y^3 + 4xy - 3xy^2$

$= -9x^2y^3 + 12x^2y^2 - 3xy^2 + 4xy$

103. $4x^2+bx+25 = \left(2x\right)^2+bx+\left(5\right)^2$ will be

a perfect square trinomial if

$bx = \pm 2\left(2x\right)\left(5\right) \Rightarrow b = \pm 20$

104. $h\left(x\right) = \dfrac{x^2+4x}{x+6}$

a. $h\left(-4\right) = \dfrac{\left(-4\right)^2+4\left(-4\right)}{\left(-4\right)+6} = \dfrac{16-16}{2} = \dfrac{0}{2} = 0$

b. $h\left(\dfrac{2}{5}\right) = \dfrac{\left(\frac{2}{5}\right)^2+4\left(\frac{2}{5}\right)}{\left(\frac{2}{5}\right)+6}$

$= \dfrac{\frac{4}{25}+\frac{40}{25}}{\frac{2}{5}+\frac{30}{5}} = \dfrac{\frac{44}{25}}{\frac{32}{5}}$

$= \dfrac{44}{25}\cdot\dfrac{5}{32} = \dfrac{11}{40}$ or 0.275

105. Let x be the number of adult tickets sold and y be the number of children's tickets sold. The following system describes the situation.

$x + y = 550$

$15x + 11y = 7290$

Solve by elimination.

$-11(x+y=550) \Rightarrow -11x-11y = -6050$

$15x+11y = 7290 \Rightarrow \underline{15x+11y = 7290}$

$4x = 1240$

$x = 310$

Substitute $x = 310$ into the first equation to find y.

$310+y = 550 \Rightarrow y = 240$

Thus, 310 adult tickets and 240 children's tickets must be sold.

106. $\sqrt[3]{x}\left(\sqrt[3]{x^2}+\sqrt[3]{x^5}\right) = \sqrt[3]{x}\cdot\sqrt[3]{x^2}+\sqrt[3]{x}\cdot\sqrt[3]{x^5}$

$= \sqrt[3]{x^3}+\sqrt[3]{x^6}$

$= x+x^2$

Chapter 13 Review Exercises

1. $(f\circ g)(x) = (2x-5)^2 - 3(2x-5)+4$

$= 4x^2 - 20x+25 - 6x+15+4$

$= 4x^2 - 26x+44$

2. $(f\circ g)(x) = 4x^2 - 26x+44$

$(f\circ g)(3) = 4(3)^2 - 26(3)+44$

$= 36-78+44$

$= 2$

3. $(g\circ f)(x) = 2(x^2-3x+4)-5$

$= 2x^2 - 6x+8-5$

$= 2x^2 - 6x+3$

4. $(g\circ f)(x) = 2x^2 - 6x+3$

$(g\circ f)(-3) = 2(-3)^2 - 6(-3)+3$

$= 18+18+3$

$= 39$

5. $(f\circ g)(x) = 6\sqrt{x-3}+7, \; x\geq 3$

6. $(g\circ f)(x) = \sqrt{(6x+7)-3}$

$= \sqrt{6x+4}, \; x\geq -\dfrac{2}{3}$

7. This function is one-to-one since it passes the horizontal line test.

8. The function is not one-to-one since the graph does not pass the horizontal line test.

9. Yes, the ordered pairs represent a one-to-one function. For each value of x, there is a unique value for y and each y-value has a unique x-value.

10. No, the ordered pairs do not represent a one-to-one function since the pairs $(0, -2)$ and $(3, -2)$ have different x-values but the same y-value.

11. Yes, the function $y = \sqrt{x+8}, x \geq -8$, is a one-to-one function since each output, y, corresponds to exactly one intput, x. The graph passes the horizontal line test.

12. No, $y = x^2 - 9$ is a parabola with vertex at $(0, -9)$. It is not a one-to-one function since it does not pass the horizontal line test. Horizontal lines above $y = -9$ intersect the graph at two points.

13. $f(x)$: Domain: $\{-4, -1, 5, 6\}$
Range: $\{-3, 2, 3, 8\}$
$f^{-1}(x)$: Domain: $\{-3, 2, 3, 8\}$
Range: $\{-4, -1, 5, 6\}$

14. $f(x)$: Domain: $\{x \mid x \geq 0\}$
Range: $\{y \mid y \geq 4\}$
$f^{-1}(x)$: Domain: $\{x \mid x \geq 4\}$
Range: $\{y \mid y \geq 0\}$

15. $y = f(x) = 4x - 2$
$x = 4y - 2$
$x + 2 = 4y$
$\dfrac{x+2}{4} = y$ or $y = f^{-1}(x) = \dfrac{x+2}{4}$

x	$f(x)$
0	-2
$\frac{1}{2}$	0

x	$f^{-1}(x)$
0	$\frac{1}{2}$
-2	0

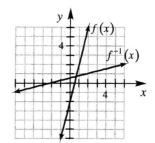

16. $y = f(x) = \sqrt[3]{x-1} = (x-1)^{1/3}$
$x = (y-1)^{1/3}$
$x^3 = [(y-1)^{1/3}]^3$
$x^3 = y - 1$
$x^3 + 1 = y$
$f^{-1}(x) = x^3 + 1$

x	$f(x)$
-7	-2
0	-1
1	0
2	1
9	2

x	$f^{-1}(x)$
-2	-7
-1	0
0	1
1	2
2	9

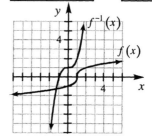

17. $f(x) = 36x \Rightarrow y = 36x$
$x = 36y \Rightarrow y = \dfrac{x}{36} \Rightarrow f^{-1}(x) = \dfrac{x}{36}$
$f^{-1}(x)$ represents yards and x represents inches.

18. $f(x) = 4x \Rightarrow y = 4x$
$x = 4y \Rightarrow y = \dfrac{x}{4} \Rightarrow f^{-1}(x) = \dfrac{x}{4}$
$f^{-1}(x)$ represents gallons and x represents quarts.

19. $y = 2^x$

x	-2	-1	0	1	2	3
y	$\frac{1}{4}$	$\frac{1}{2}$	1	2	4	8

Domain: $\mathbb{R}$

Range: $\left\{ y \mid y > 0 \right\}$

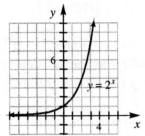

20. $y = \left(\dfrac{1}{2} \right)^x$

x	-2	-1	0	1	2
y	4	2	1	$\frac{1}{2}$	$\frac{1}{4}$

Domain: $\mathbb{R}$

Range: $\{ y \mid y > 0 \}$

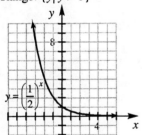

21. $f(t) = 7.02e^{0.365t}$

a. $t = 2003 - 1999 = 4$

$f(4) = 7.02e^{0.365(4)}$

≈ 30.23

The worldwide shipment in 2003 is about 30.23 million.

b. $t = 2005 - 1999 = 6$

$f(6) = 7.02e^{0.365(6)}$

≈ 62.73

The worldwide shipment in 2005 will be about 62.73 million.

c. $t = 2008 - 1999 = 9$

$f(9) = 7.02e^{0.365(9)}$

≈ 187.50

The worldwide shipment in 2008 will be about 187.50 million.

22. $8^2 = 64$

$\log_8 64 = 2$

23. $81^{1/4} = 3$

$\log_{81} 3 = \dfrac{1}{4}$

24. $5^{-3} = \dfrac{1}{125}$

$\log_5 \dfrac{1}{125} = -3$

25. $\log_2 32 = 5$

$2^5 = 32$

26. $\log_{1/4} \dfrac{1}{16} = 2$

$\left(\dfrac{1}{4} \right)^2 = \dfrac{1}{16}$

27. $\log_6 \dfrac{1}{36} = -2$

$6^{-2} = \dfrac{1}{36}$

28. $3 = \log_4 x$

$x = 4^3$

$x = 64$

29. $4 = \log_a 81$

$a^4 = 81$

$a^4 = 3^4$

$a = 3$

30. $-3 = \log_{1/5} x$

$$x = \left(\frac{1}{5}\right)^{-3}$$

$$x = \frac{1}{\left(\frac{1}{5}\right)^3}$$

$$x = \frac{1}{\frac{1}{125}}$$

$$x = 125$$

31. $y = \log_3 x$

$$x = 3^y$$

x	$\frac{1}{9}$	$\frac{1}{3}$	1	3	9	27
y	-2	-1	0	1	2	3

Domain: $\{x \mid x > 0\}$

Range: $\mathbb{R}$

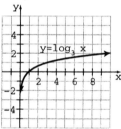

32. $y = \log_{1/2} x$

$$x = \left(\frac{1}{2}\right)^y$$

x	4	2	1	$\frac{1}{2}$	$\frac{1}{4}$
y	-2	-1	0	1	2

Domain: $\{x \mid x > 0\}$

Range: $\mathbb{R}$

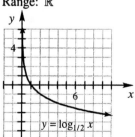

33. $\log_5 17^8 = 8 \log_5 17$

34. $\log_3 \sqrt{x-9} = \log_3 (x-9)^{1/2} = \frac{1}{2} \log_3 (x-9)$

35. $\log \dfrac{6(a+1)}{19} = \log\left[6(a+1)\right] - \log 19$

$$= \log 6 + \log(a+1) - \log 19$$

36. $\log \dfrac{x^4}{7(2x+3)^5} = \log x^4 - \log\left(7(2x+3)^5\right)$

$$= 4\log x - [\log 7 + \log(2x+3)^5]$$
$$= 4\log x - [\log 7 + 5\log(2x+3)]$$
$$= 4\log x - \log 7 - 5\log(2x+3)$$

37. $5\log x - 3\log(x+1) = \log x^5 - \log(x+1)^3$

$$= \log \frac{x^5}{(x+1)^3}$$

38. $4\left(\log 2 + \log x\right) - \log y = 4\left(\log(2x)\right) - \log y$

$$= \log(2x)^4 - \log y$$

$$= \log \frac{(2x)^4}{y} \quad \text{or} \quad \log \frac{16x^4}{y}$$

39. $\dfrac{1}{3}[\ln x - \ln(x+2)] - \ln 2$

$$= \frac{1}{3}\left(\ln \frac{x}{x+2}\right) - \ln 2$$

$$= \ln\left(\frac{x}{x+2}\right)^{1/3} - \ln 2$$

$$= \ln \frac{\sqrt[3]{\frac{x}{x+2}}}{2}$$

40. $3\ln x + \dfrac{1}{2}\ln(x+1) - 6\ln(x+4)$

$$= \ln x^3 + \ln(x+1)^{1/2} - \ln(x+4)^6$$

$$= \ln \frac{x^3 \sqrt{x+1}}{(x+4)^6}$$

41. $8^{\log_8 10} = 10$

42. $\log_4 4^5 = 5$

43. $11^{\log_9 81} = 11^{\log_9 9^2}$

$\quad\quad = 11^2$

$\quad\quad = 121$

44. $9^{\log_8 \sqrt{8}} = 9^{\log_8 8^{1/2}}$

$\quad\quad = 9^{1/2}$

$\quad\quad = \sqrt{9}$

$\quad\quad = 3$

45. $\log 819 = 2.9133$

46. $\ln 0.0281 = -3.5720$

47. antilog $3.159 = 1440$

48. antilog$(-3.157) = 0.000697$

49. $\log N = 4.063$

$\quad\quad N = \text{antilog } 4.063$

$\quad\quad N = 11,600$

50. $\log N = -1.2262$

$\quad\quad N = \text{antilog } (-1.2262)$

$\quad\quad N = 0.0594$

51. $\log 10^5 = 5$

52. $10^{\log 9} = 9$

53. $7 \log 10^{3.2} = 7(3.2) = 22.4$

54. $2\left(10^{\log 4.7}\right) = 2(4.7) = 9.4$

55. $625 = 5^x$

$\quad\quad 5^4 = 5^x$

$\quad\quad 4 = x$

56. $49^x = \dfrac{1}{7}$

$\quad\quad (7^2)^x = 7^{-1}$

$\quad\quad 7^{2x} = 7^{-1}$

$\quad\quad 2x = -1$

$\quad\quad x = -\dfrac{1}{2}$

57. $2^{3x-1} = 32$

$\quad\quad 2^{3x-1} = 2^5$

$\quad\quad 3x - 1 = 5$

$\quad\quad 3x = 6$

$\quad\quad x = 2$

58. $27^x = 3^{2x+5}$

$\quad\quad (3^3)^x = 3^{2x+5}$

$\quad\quad 3^{3x} = 3^{2x+5}$

$\quad\quad 3x = 2x + 5$

$\quad\quad x = 5$

59. $7^x = 152$

$\quad\quad \log 7^x = \log 152$

$\quad\quad x \log 7 = \log 152$

$\quad\quad x = \dfrac{\log 152}{\log 7}$

$\quad\quad x \approx 2.582$

60. $3.1^x = 856$

$\quad\quad \log 3.1^x = \log 856$

$\quad\quad x \log 3.1 = \log 856$

$\quad\quad x = \dfrac{\log 856}{\log 3.1}$

$\quad\quad x \approx 5.968$

61. $12.5^{x+1} = 381$

$\quad\quad \log 12.5^{x+1} = \log 381$

$\quad\quad (x+1) \log 12.5 = \log 381$

$\quad\quad x + 1 = \dfrac{\log 381}{\log 12.5}$

$\quad\quad x = \dfrac{\log 381}{\log 12.5} - 1$

$\quad\quad x \approx 1.353$

62. $3^{x+2} = 8^x$

$\quad\quad \log 3^{x+2} = \log 8^x$

$\quad\quad (x+2) \log 3 = x \log 8$

$\quad\quad x \log 3 + 2 \log 3 = x \log 8$

$\quad\quad 2 \log 3 = x \log 8 - x \log 3$

$\quad\quad 2 \log 3 = x(\log 8 - \log 3)$

$\quad\quad \dfrac{2 \log 3}{\log 8 - \log 3} = x$

$\quad\quad 2.240 \approx x$

63. $\log_7(2x-3) = 2$

$$7^2 = 2x - 3$$
$$49 = 2x - 3$$
$$52 = 2x$$
$$26 = x$$

64. $\log x + \log(4x - 19) = \log 5$

$$\log\big(x(4x-19)\big) = \log 5$$
$$x(4x-19) = 5$$
$$4x^2 - 19x - 5 = 0$$
$$(4x+1)(x-5) = 0$$
$$4x + 1 = 0 \quad \text{or} \quad x - 5 = 0$$
$$x = -\frac{1}{4} \qquad x = 5$$

Only $x = 5$ checks. $x = -\dfrac{1}{4}$ is an extraneous solution since $\log(4x - 19)$ becomes $\log(-20)$ which is not a real number.

65. $\log_3 x + \log_3(2x+1) = 1$

$$\log_3 x(2x+1) = 1$$
$$3^1 = x(2x+1)$$
$$0 = 2x^2 + x - 3$$
$$0 = (2x+3)(x-1)$$
$$2x + 3 = 0 \quad \text{or} \quad x - 1 = 0$$
$$x = -\tfrac{3}{2} \qquad x = 1$$

Only $x = 1$ checks. $x = -\dfrac{3}{2}$ is an extraneous solution since $\log_3 x$ becomes $\log_3\left(-\dfrac{3}{2}\right)$ which is not a real number.

66. $\ln(x+1) - \ln(x-2) = \ln 4$

$$\ln\frac{x+1}{x-2} = \ln 4$$
$$\frac{x+1}{x-2} = 4$$
$$x + 1 = 4(x-2)$$
$$x + 1 = 4x - 8$$
$$1 = 3x - 8$$
$$9 = 3x$$
$$x = 3$$

67. $50 = 25e^{0.6t}$

$$2 = e^{0.6t}$$
$$\ln 2 = \ln e^{0.6t}$$
$$\ln 2 = 0.6t$$
$$\frac{\ln 2}{0.6} = t$$
$$1.155 \approx t$$

68. $100 = A_0 e^{-0.42(3)}$

$$100 = A_0 e^{-1.26}$$
$$\frac{100}{e^{-1.26}} = A_0$$
$$352.542 \approx A_0$$

69. $A = A_0 e^{kt}$

$$\frac{A}{A_0} = e^{kt}$$
$$\ln\frac{A}{A_0} = \ln e^{kt}$$
$$\ln\frac{A}{A_0} = kt$$
$$\frac{\ln\frac{A}{A_0}}{k} = t$$
$$\frac{\ln A - \ln A_0}{k} = t \ \text{ or } \ t = \frac{\ln A - \ln A_0}{k}$$

70. $200 = 800e^{kt}$

$$\frac{200}{800} = e^{kt}$$
$$0.25 = e^{kt}$$
$$\ln 0.25 = \ln e^{kt}$$
$$\ln 0.25 = kt$$
$$\frac{\ln 0.25}{t} = k \ \text{ or } \ k = \frac{\ln 0.25}{t}$$

71. $\ln y - \ln x = 6$

$$\ln\frac{y}{x} = 6$$
$$e^{\ln\frac{y}{x}} = e^6$$
$$\frac{y}{x} = e^6$$
$$y = xe^6$$

72. $\ln(y+1) - \ln(x+8) = \ln 3$

$$\ln \frac{y+1}{x+8} = \ln 3$$

$$\frac{y+1}{x+8} = 3$$

$$y+1 = 3(x+8)$$

$$y = 3(x+8) - 1$$

$$y = 3x + 24 - 1$$

$$y = 3x + 23$$

73. $\log_2 196 = \dfrac{\log 196}{\log 2} \approx 7.6147$

74. $\log_3 47 = \dfrac{\log 47}{\log 3} \approx 3.5046$

75. $A = P(1+r)^n$

$$= 12,000(1+0.06)^8$$

$$= 12,000(1.06)^8$$

$$= 19,126.18$$

The amount is \$19,126.18.

76. $P = P_0 e^{kt}$

$P_0 = 6,000, k = 0.04,$ and $P = 12,000$

$$12,000 = 12,000 e^{(0.04)t}$$

$$2 = e^{0.04t}$$

$$\ln 2 = 0.04t$$

$$t = \frac{\ln 2}{0.04}$$

$$t \approx 17.3$$

It will take about 17.3 years for the \$6,000 to double.

77. $N(t) = 2000(2)^{0.05t}$

a. Let $N(t) = 50,000$.

$$50,000 = 2000(2)^{0.05t}$$

$$\frac{50,000}{2000} = 2^{0.05t}$$

$$25 = 2^{0.05t}$$

$$\log 25 = \log 2^{0.05t}$$

$$\log 25 = 0.05t \log 2$$

$$\frac{\log 25}{0.05 \log 2} = t$$

$$92.88 \approx t$$

The time is 92.88 minutes.

b. Let $N(t) = 120,000$.

$$120,000 = 2000(2)^{0.05t}$$

$$\frac{120,000}{2000} = 2^{0.05t}$$

$$60 = 2^{0.05t}$$

$$\log 60 = \log 2^{0.05t}$$

$$\log 60 = 0.05t \log 2$$

$$\frac{\log 60}{0.05 \log 2} = t$$

$$118.14 \approx t$$

The time is 118.14 minutes.

78. $P = 14.7 e^{-0.00004x}$

$$P = 14.7 e^{-0.00004(8842)}$$

$$P = 14.7 e^{-0.35368}$$

$$P \approx 14.7(0.7021)$$

$$P \approx 10.32$$

The atmospheric pressure is 10.32 pounds per square inch at 8,842 feet above sea level.

79. $A(n) = 72 - 18\log(n+1)$

a. $A(0) = 72 - 18\log(0+1)$

$$= 72 - 18\log(1)$$

$$= 72 - 18(0)$$

$$= 72$$

The original class average was 72.

b. $A(3) = 72 - 18\log(3+1)$

$$= 72 - 18\log(4)$$

$$\approx 72 - 10.8$$

$$= 61.2$$

After 3 months, the class average was 61.2.

c. Let $A(n) = 58.0$.

$$58.0 = 72 - 18\log(n+1)$$

$$-14 = -18\log(n+1)$$

$$\frac{7}{9} = \log(n+1)$$

$$10^{7/9} = 10^{\log(n+1)}$$

$$10^{7/9} = n+1$$

$$10^{7/9} - 1 = n$$

$$5.0 \approx n$$

It takes about 5 months.

Chapter 13 Practice Test

1. a. Yes, $\{(4, 2), (-3, 8), (-1, 3), (6, -7)\}$ is one-to-one.

b. $\{(2, 4), (8, -3), (3, -1), (-7, 6)\}$ is the inverse function.

2. a.
$$(f \circ g)(x) = f[g(x)]$$
$$= f(x+2)$$
$$= (x+2)^2 - 3$$
$$= x^2 + 4x + 4 - 3$$
$$= x^2 + 4x + 1$$

b. $(f \circ g)(6) = 6^2 + 4(6) + 1 = 61$

3. a.
$$(g \circ f)(x) = g[f(x)]$$
$$= g(x^2 + 8)$$
$$= \sqrt{x^2 + 8 - 5}$$
$$= \sqrt{x^2 + 3}$$

b.
$$(g \circ f)(7) = \sqrt{7^2 + 3}$$
$$= \sqrt{49 + 3}$$
$$= \sqrt{52} = 2\sqrt{13}$$

4. a.
$$y = f(x) = -3x - 5$$
$$x = -3y - 5$$
$$x + 5 = -3y$$
$$\frac{x+5}{-3} = y$$
$$-\frac{1}{3}(x+5) = y$$
$$f^{-1}(x) = -\frac{1}{3}(x+5)$$

b.

x	f(x)
0	–5
$-\frac{5}{3}$	0

x	$f^{-1}(x)$
0	$-\frac{5}{3}$
–5	0

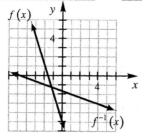

5. a.
$$y = f(x) = \sqrt{x-1}, x \geq 1$$
$$x = (y-1)^{1/2}$$
$$x^2 = [(y-1)^{1/2}]^2$$
$$x^2 = y - 1$$
$$x^2 + 1 = y$$
$$f^{-1}(x) = x^2 + 1, x \geq 0$$

b.

x	f(x)
1	0
2	1
5	2

x	$f^{-1}(x)$
0	1
1	2
2	5

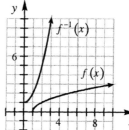

6. The domain of $y = \log_5(x)$ is $\{x \mid x > 0\}$.

7. $\log_4 \dfrac{1}{256} = \log_4 4^{-4} = -4$

8. $y = 3^x$

x	–2	–1	0	2	3
y	$\frac{1}{9}$	$\frac{1}{3}$	1	9	27

Domain: $\mathbb{R}$

Range: $\{y \mid y > 0\}$

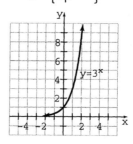

9. $y = \log_2 x$

$x = 2^y$

x	$\frac{1}{4}$	$\frac{1}{2}$	1	2	4
y	-2	-1	0	1	2

Domain: $\{x \mid x > 0\}$

Range: $\mathbb{R}$

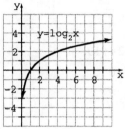

10. $2^{-5} = \dfrac{1}{32}$

$\log_2 \dfrac{1}{32} = -5$

11. $\log_5 125 = 3$

$\qquad 5^3 = 125$

12. $4 = \log_2 (x+3)$

$2^4 = x + 3$

$16 = x + 3$

$13 = x$

13. $y = \log_{64} 16$

$64^y = 16$

$\left(4^3\right)^y = 4^2$

$4^{3y} = 4^2$

$3y = 2$

$y = \dfrac{2}{3}$

14. $\log_2 \dfrac{x^3(x-4)}{x+2}$

$= \log_2 x^3(x-4) - \log_2 (x+2)$

$= \log_2 x^3 + \log_2 (x-4) - \log_2 (x+2)$

$= 3\log_2 x + \log_2 (x-4) - \log_2 (x+2)$

15. $7\log_6 (x-4) + 2\log_6 (x+3) - \dfrac{1}{2}\log_6 x$

$= \log_6 (x-4)^7 + \log_6 (x+3)^2 - \log_6 x^{1/2}$

$= \log_6 \dfrac{(x-4)^7(x+3)^2}{\sqrt{x}}$

16. $10\log_9 \sqrt{9} = 10\log_9 9^{1/2}$

$= 10 \cdot \dfrac{1}{2}$

$= 5$

17. a. $\log 4620 \approx 3.6646$

b. $\ln 0.0692 \approx -2.6708$

18. $3^x = 19$

$\log 3^x = \log 19$

$x\log 3 = \log 19$

$x = \dfrac{\log 19}{\log 3}$

$x \approx 2.68$

19. $\log 4x = \log(x+3) + \log 2$

$\log 4x = \log 2(x+3)$

$4x = 2(x+3)$

$4x = 2x + 6$

$2x = 6$

$x = 3$

20. $\log(x+5) - \log(x-2) = \log 6$

$\log \dfrac{x+5}{x-2} = \log 6$

$\dfrac{x+5}{x-2} = 6$

$x + 5 = 6x - 12$

$17 = 5x$

$\dfrac{17}{5} = x$

21. $\ln N = 2.79$

$e^{2.79} = N$

$16.2810 \approx N$

22. $\log_6 40 = \dfrac{\log 40}{\log 6} \approx 2.0588$

23. $100 = 250e^{-0.03t}$

$\dfrac{100}{250} = e^{-0.03t}$

$\ln \dfrac{100}{250} = -0.03t$

$\dfrac{\ln \frac{100}{250}}{-0.03} = t$

$t \approx 30.5430$

24. $A = p\left(1 + \dfrac{r}{n}\right)^{nt}$

Use $p = 3500$, $r = 0.04$ and $n = 4$

$t = 10$

$A = 3500\left(1 + \dfrac{0.04}{4}\right)^{4 \cdot 10}$

$= 3500(1.01)^{40}$

$= 5211.02$

The amount in the account is $5211.02.

25. $v = v_0 e^{-0.0001205t}$

Use $v = 40$, and $v_0 = 60$.

$40 = 60e^{-0.0001205t}$

$\dfrac{40}{60} = e^{-0.0001205t}$

$\dfrac{2}{3} = e^{-0.0001205t}$

$\ln \dfrac{2}{3} = \ln e^{-0.0001205t}$

$\ln \dfrac{2}{3} = -0.0001205t$

$\dfrac{\ln \frac{2}{3}}{-0.0001205} = t$

$3364.86 \approx t$

The fossil is approximately 3364.86 years old.

Chapter 13 Cumulative Review Test

1. $\dfrac{6 - |-18| \div 3^2 - 6}{4 - |-8| \div 2^2} = \dfrac{6 - 18 \div 3^2 - 6}{4 - 8 \div 2^2}$

$= \dfrac{6 - 18 \div 9 - 6}{4 - 8 \div 4}$

$= \dfrac{6 - 2 - 6}{4 - 2}$

$= \dfrac{-2}{2} = -1$

2. $4 - (6x + 6) = -(-2x + 10)$

$4 - 6x - 6 = 2x - 10$

$-6x - 2 = 2x - 10$

$-8x = -8$

$x = 1$

3. $2x - 3y = 5$

$-3y = 5 - 2x$

$y = \dfrac{5 - 2x}{-3}$ or $y = \dfrac{2x - 5}{3}$

4. Let r be the tax rate.

$92 + 92r = 98.90$

$92r = 6.90$

$r = \dfrac{6.90}{92}$

≈ 0.075

The tax rate is 7.5%.

5. Let $t =$ the time it takes Kendra to catch up to Jason.

Person	Rate	Time	Distance
Jason	4	$t + 0.5$	$4(t + 0.5)$
Kendra	5	t	$5t$

a. $4(t + 0.5) = 5t$

$4t + 2 = 5t$

$2 = t$

Kendra will meet Jason after she has jogged for 2 hours.

b. $5(2) = 10$ miles

6. Two points on the line are $(0, 3)$ and $(-2, -1)$.

$$m = \frac{y_2 - y_1}{x_2 - x_1} = \frac{-1 - 3}{-2 - 0} = \frac{-4}{-2} = 2$$

Use point-slope form with $(0, 3)$ and $m = 2$.

$$y - 3 = 2(x - 0)$$
$$y - 3 = 2x$$
$$y = 2x + 3$$

7. $\left(\dfrac{3x^4 y^{-3}}{6xy^4 z^2}\right)^{-3} = \left(\dfrac{x^3}{2y^7 z^2}\right)^{-3} = \left(\dfrac{2y^7 z^2}{x^3}\right)^3$

$$= \frac{8y^{21} z^6}{x^9}$$

8. $\dfrac{x^3 + 3x^2 + 5x + 4}{x + 1}$

Using synthetic division with $c = -1$:

$$
\begin{array}{r|rrrr}
-1 & 1 & 3 & 5 & 4 \\
   &   & -1 & -2 & -3 \\
\hline
   & 1 & 2 & 3 & 1
\end{array}
$$

$$\frac{x^3 + 3x^2 + 5x + 4}{x + 1} = x^2 + 2x + 3 + \frac{1}{x + 1}$$

9. $12x^2 - 5xy - 3y^2 = (4x - 3y)(3x + y)$

10. $x^2 - 2xy + y^2 - 25 = (x^2 - 2xy + y^2) - 25$

$$= (x - y)^2 - 5^2$$
$$= (x - y - 5)(x - y + 5)$$

11. $\dfrac{x + 1}{x + 2} + \dfrac{x - 2}{x - 3} = \dfrac{x^2 - 4}{x^2 - x - 6}$

LCD: $(x + 2)(x - 3)$

$$(x + 1)(x - 3) + (x - 2)(x + 2) = x^2 - 4$$
$$x^2 - 2x - 3 + x^2 - 4 = x^2 - 4$$
$$x^2 - 2x - 3 = 0$$
$$(x - 3)(x + 1) = 0$$
$$x - 3 = 0 \quad \text{or} \quad x + 1 = 0$$
$$x = 3 \qquad\qquad x = -1$$

We exclude $x = 3$ since it is not in the domain of the variable. The solution is $x = -1$.

12. $L = \dfrac{K}{P^2} = \dfrac{100}{4^2} = \dfrac{100}{16} = 6.25$

13. $h(x) = \dfrac{x^2 + 4x}{x + 6}$

$$h(-3) = \frac{(-3)^2 + 4(-3)}{(-3) + 6}$$

$$= \frac{9 - 12}{3} = \frac{-3}{3} = -1$$

14. $0.4x + 0.6y = 3.2$
$1.4x - 0.3y = 1.6$

Multiply the second equation by 2 and add the equations.

$$0.4x + 0.6y = 3.2$$
$$\underline{2.8x - 0.6y = 3.2}$$
$$3.2x = 6.4$$
$$x = 2$$

Substitute 2 for x in the first equation and solve for y.

$$0.4(2) + 0.6y = 3.2$$
$$0.8 + 0.6y = 3.2$$
$$0.6y = 2.4$$
$$y = 4$$

The solution is $(2, 4)$.

15. $x + y = 6$
$-2x + y = 3$

Write the augmented matrix.

$$\begin{bmatrix} 1 & 1 & | & 6 \\ -2 & 1 & | & 3 \end{bmatrix}$$

Use row transformations to write the matrix in triangular form.

$$\begin{bmatrix} 1 & 1 & | & 6 \\ 0 & 3 & | & 15 \end{bmatrix} 2R_1 + R_2$$

$$\begin{bmatrix} 1 & 1 & | & 6 \\ 0 & 1 & | & 5 \end{bmatrix} \tfrac{1}{3}R_2$$

Write the equivalent system of equations.

$$x + y = 6$$
$$y = 5$$

Substitute 5 for y in the first equation to find x.

$$x + (5) = 6$$
$$x = 1$$

The solution is $(1, 5)$.

16. $\begin{vmatrix} 3 & 0 & -1 \\ 2 & 5 & 3 \\ -1 & 4 & 6 \end{vmatrix}$ Expand across row 1

$= 3\begin{vmatrix} 5 & 3 \\ 4 & 6 \end{vmatrix} - 0\begin{vmatrix} 2 & 3 \\ -1 & 6 \end{vmatrix} + (-1)\begin{vmatrix} 2 & 5 \\ -1 & 4 \end{vmatrix}$

$= 3(5 \cdot 6 - 4 \cdot 3) - 0 - (2 \cdot 4 - (-1) \cdot 5)$

$= 3(30 - 12) - (8 + 5)$

$= 3(18) - 13$

$= 54 - 13$

$= 41$

17. $y \le \dfrac{1}{3}x + 6$

Plot a solid line at $y = \dfrac{1}{3}x + 6$.

Use the point $(0, 0)$ as the check point.

$y \le \dfrac{1}{3}x + 6$

$0 \le \dfrac{1}{3}(0) + 6$

$0 \le 6$ True

Therefore, shade the half-plane containing $(0,0)$.

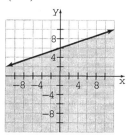

18. $g(x) = x^2 - 4x - 5$

a. $g(x) = x^2 - 4x - 5$

$g(x) = (x^2 - 4x + 4) - 5 - 4$

$g(x) = (x - 2)^2 - 9$

b.

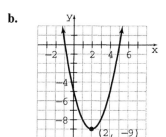

$g(x) = x^2 - 4x - 5$

19. $f(x) = x^3 - 6x^2 + 5x \ge 0$

Factor the left side and set equal to 0.

$x(x^2 - 6x + 5) = 0$

$x(x - 5)(x - 1) = 0$

The left side is 0 when $x = 0$, 5, or 1. Therefore, our intervals are $(-\infty, 0)$, $(0,1)$, $(1,5)$, and $(5,\infty)$.

Interval A   Interval B

test value, -1 test value, 0.5

$f(-1) = -12$ $f(0.5) = \dfrac{9}{8}$

Interval C   Interval D

test value, 2   test value, 6

$f(2) = -6$ $f(6) = 30$

The inequality is satisfied on intervals *B* and *D*. The inequality is non-strict, so we also include the endpoints. The solution set, in interval notation, is $[0,1] \cup [5,\infty)$.

20. a. $A = 600e^{-0.028t}$

When $t = 60$, we get

$A = 600e^{-0.028(60)} \approx 111.82$

After 60 years there will be about 111.82 grams remaining.

b. To find the half-life, we are looking for the time required to have half of the initial amount remaining.

$300 = 600e^{-0.028t}$

$\dfrac{1}{2} = e^{-0.028t}$

$\ln\left(\dfrac{1}{2}\right) = -0.028t$

$\dfrac{\ln\left(\dfrac{1}{2}\right)}{-0.028} = t$ or $t \approx 24.8$

The half-life is about 24.8 years.

Chapter 14

Exercise Set 14.1

1.

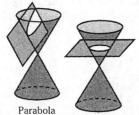  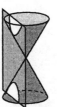

Parabola   Circle   Ellipse   Hyperbola

3. Yes, any parabola in the form $y = a(x-h)^2 + k$ is a function because each value of x corresponds to only one value of y. The domain is $\mathbb{R}$, the set of all real numbers. Since the vertex is at (h, k) and $a > 0$, the range is $\{y \mid y \geq k\}$.

5. The graphs have the same vertex, (3, 4). The first graph opens upward, and the second one opens downward.

7. The distance is always a positive number because both distances are squared and we use the principal square root.

9. A circle is the set of all points in a plane that are the same distance from a fixed point.

11. No, the coefficients of the y^2-term and the x^2-term must both be the same.

13. No, the coefficients of the y^2-term and the x^2-term must both be the same.

15. No, equations of parabolas do not include both x^2- and y^2-terms.

17. $y = (x-2)^2 + 3$

This is a parabola in the form $y = a(x-h)^2 + k$ with $a = 1$, $h = 2$ and $k = 3$. Since $a > 0$, the parabola opens upward. The vertex is (2, 3). The y-intercept is (0, 7). There are no x-intercepts.

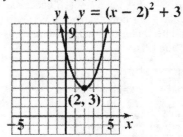

19. $y = (x+3)^2 + 2$

This is a parabola in the form $y = a(x-h)^2 + k$ with $a = 1$, $h = -3$ and $k = 2$. Since $a > 0$, the parabola opens upward. The vertex is (−3, 2). The y-intercept is (0, 11). There are no x-intercepts.

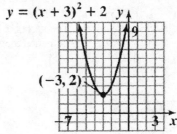

21. $y = (x-2)^2 - 1$

This is a parabola in the form $y = a(x-h)^2 + k$ with $a = 1$, $h = 2$ and $k = -1$. Since $a > 0$, the parabola opens upward. The vertex is $(2, -1)$. The y-intercept is $(0, 3)$. The x-intercepts are $(1, 0)$ and $(3, 0)$.

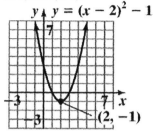

23. $y = -(x-1)^2 + 1$

This is a parabola in the form $y = a(x-h)^2 + k$ with $a = -1$, $h = 1$ and $k = 1$. Since $a < 0$, the parabola opens downward. The vertex is $(1, 1)$. The y-intercept is $(0, 0)$. The x-intercepts are $(0, 0)$ and $(2, 0)$.

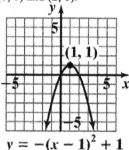

25. $y = -(x+3)^2 + 4$

This is a parabola in the form $y = a(x-h)^2 + k$ with $a = -1$, $h = -3$ and $k = 4$. Since $a < 0$, the parabola opens downward. The vertex is $(-3, 4)$. The y-intercept is $(0, -5)$. The x-intercepts are $(-5, 0)$ and $(-1, 0)$.

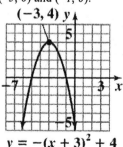

27. $y = -3(x-5)^2 + 3$

This is a parabola in the form $y = a(x-h)^2 + k$ with $a = -3$, $h = 5$ and $k = 3$. Since $a < 0$, the parabola opens downward. The vertex is $(5, 3)$. The y-intercept is $(0, -72)$. The x-intercepts are $(4, 0)$ and $(6, 0)$.

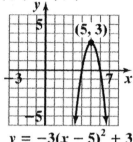

29. $x = (y-4)^2 - 3$

This is a parabola in the form $x = a(y-k)^2 + h$ with $a = 1$, $h = -3$ and $k = 4$. Since $a > 0$, the parabola opens to the right. The vertex is $(-3, 4)$. The y-intercepts are about $(0, 2.27)$ and $(0, 5.73)$. The x-intercept is $(13, 0)$.

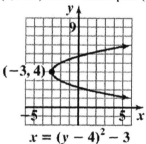

31. $x = -(y-5)^2 + 4$

This is a parabola in the form $x = a(y-k)^2 + h$ with $a = -1$, $h = 4$ and $k = 5$. Since $a < 0$, the parabola opens to the left. The vertex is $(4, 5)$. The y-intercepts are $(0, 3)$ and $(0, 7)$. The x-intercept is $(-21, 0)$.

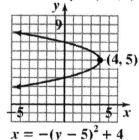

33. $x = -5(y+3)^2 - 6$

This is a parabola in the form $x = a(y-k)^2 + h$ with $a = -5$, $h = -6$ and $k = -3$. Since $a < 0$, the parabola opens to the left. The vertex is $(-6, -3)$. There are no y-intercepts. The x-intercept is $(-51, 0)$

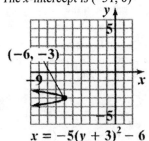

$$x = -5(y+3)^2 - 6$$

35. $y = -2\left(x + \dfrac{1}{2}\right)^2 + 6$

This is a parabola in the form $y = a(x-h)^2 + k$ with $a = -2$, $h = -\dfrac{1}{2}$ and $k = 6$. Since $a < 0$, the parabola opens downward. The vertex is $\left(-\dfrac{1}{2}, 6\right)$. The y-intercept is $\left(0, \dfrac{11}{2}\right)$. The x-intercepts are about $(-2.23, 0)$ and $(1.23, 0)$.

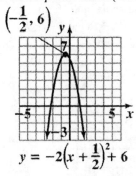

$$y = -2\left(x + \dfrac{1}{2}\right)^2 + 6$$

37. a. $y = x^2 + 2x$

$y = (x^2 + 2x + 1) - 1$

$y = (x+1)^2 - 1$

b. This is a parabola in the form $y = a(x-h)^2 + k$ with $a = 1$, $h = -1$ and $k = -1$. Since $a > 0$, the parabola opens upward. The vertex is $(-1, -1)$. The y-intercept is $(0, 0)$. The x-intercepts are $(-2, 0)$ and $(0, 0)$.

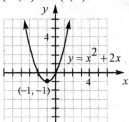

39. a. $y = x^2 + 6x$

$y = (x^2 + 6x + 9) - 9$

$y = (x+3)^2 - 9$

b. This is a parabola in the form $y = a(x-h)^2 + k$ with $a = 1$, $h = -3$ and $k = -9$. Since $a > 0$, the parabola opens upward. The vertex is $(-3, -9)$. The y-intercept is $(0, 0)$. The x-intercepts are $(-6, 0)$ and $(0, 0)$.

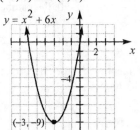

41. a. $x = y^2 + 4y$

$x = (y^2 + 4y + 4) - 4$

$x = (y + 2)^2 - 4$

b. This is a parabola in the form
$x = a(y - k)^2 + h$ with $a = 1$, $h = -4$ and
$k = -2$. Since $a > 0$, the parabola opens to
the right. The vertex is $(-4, -2)$. The
y-intercepts are $(0, -4)$ and $(0, 0)$. The
x-intercept is $(0, 0)$.

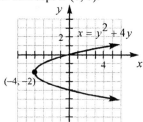

43. a. $y = x^2 + 7x + 10$

$y = \left(x^2 + 7x + \dfrac{49}{4} \right) - \dfrac{49}{4} + 10$

$y = \left(x + \dfrac{7}{2} \right)^2 - \dfrac{9}{4}$

b. This is a parabola in the form
$y = a(x - h)^2 + k$ with $a = 1$, $h = -\dfrac{7}{2}$ and

$k = -\dfrac{9}{4}$. Since $a > 0$, the parabola opens

upward. The vertex is $\left(-\dfrac{7}{2}, -\dfrac{9}{4} \right)$. The

y-intercept is $(0, 10)$. The x-intercepts are
$(-5, 0)$ and $(-2, 0)$.

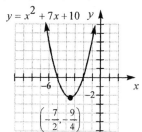

45. a. $x = -y^2 + 6y - 9$

$x = -(y^2 - 6y) - 9$

$x = -(y^2 - 6y + 9) + 9 - 9$

$x = -(y - 3)^2$

b. This is a parabola in the form
$x = a(y - k)^2 + h$ with $a = -1$, $h = 0$ and
$k = 3$. Since $a < 0$, the parabola opens to the
left. The vertex is $(0, 3)$. The y-intercept is
$(0, 3)$. The x-intercept is
$(-9, 0)$.

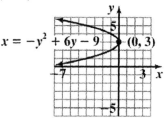

47. a. $y = -x^2 + 4x - 4$

$y = -(x^2 - 4x) - 4$

$y = -(x^2 - 4x + 4) + 4 - 4$

$y = -(x - 2)^2$

b. This is a parabola in the form
$y = a(x - h)^2 + k$ with $a = -1$, $h = 2$ and
$k = 0$. Since $a < 0$, the parabola opens
downward. The vertex is $(2, 0)$. The
y-intercept is $(0, -4)$. The x-intercept is
$(2, 0)$.

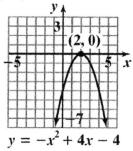

49. a.

$$x = -y^2 + 3y - 4$$

$$x = -\left(y^2 - 3y\right) - 4$$

$$x = -\left(y^2 - 3y + \frac{9}{4}\right) + \frac{9}{4} - 4$$

$$x = -\left(y - \frac{3}{2}\right)^2 - \frac{7}{4}$$

b. This is a parabola in the form

$x = a(y - k)^2 + h$ with $a = -1$, $h = -\dfrac{7}{4}$ and

$k = \dfrac{3}{2}$. Since $a < 0$, the parabola opens to

the left. The vertex is $\left(-\dfrac{7}{4}, \dfrac{3}{2}\right)$. There are

no y-intercepts.
The x-intercept is $(-4, 0)$.

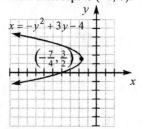

51.

$$d = \sqrt{(x_2 - x_1)^2 + (y_2 - y_1)^2}$$

$$= \sqrt{(5 - 5)^2 + \left[-6 - (-1)\right]^2}$$

$$= \sqrt{0^2 + (-5)^2}$$

$$= \sqrt{0 + 25}$$

$$= \sqrt{25}$$

$$= 5$$

53.

$$d = \sqrt{(x_2 - x_1)^2 + (y_2 - y_1)^2}$$

$$= \sqrt{\left[8 - (-1)\right]^2 + (6 - 6)^2}$$

$$= \sqrt{9^2 + 0^2}$$

$$= \sqrt{81 + 0}$$

$$= \sqrt{81}$$

$$= 9$$

55.

$$d = \sqrt{(x_2 - x_1)^2 + (y_2 - y_1)^2}$$

$$= \sqrt{\left[4 - (-1)\right]^2 + \left[9 - (-3)\right]^2}$$

$$= \sqrt{5^2 + 12^2}$$

$$= \sqrt{25 + 144}$$

$$= \sqrt{169}$$

$$= 13$$

57.

$$d = \sqrt{(x_2 - x_1)^2 + (y_2 - y_1)^2}$$

$$= \sqrt{\left[5 - (-4)\right]^2 + \left[-2 - (-5)\right]^2}$$

$$= \sqrt{9^2 + 3^2}$$

$$= \sqrt{81 + 9}$$

$$= \sqrt{90} \approx 9.49$$

59.

$$d = \sqrt{(x_2 - x_1)^2 + (y_2 - y_1)^2}$$

$$= \sqrt{\left(\frac{1}{2} - 3\right)^2 + \left[4 - (-1)\right]^2}$$

$$= \sqrt{\left(-\frac{5}{2}\right)^2 + 5^2}$$

$$= \sqrt{\frac{25}{4} + 25}$$

$$= \sqrt{\frac{125}{4}}$$

$$\approx 5.59$$

61.

$$d = \sqrt{(x_2 - x_1)^2 + (y_2 - y_1)^2}$$

$$= \sqrt{\left[-4.3 - (-1.6)\right]^2 + (-1.7 - 3.5)^2}$$

$$= \sqrt{(-2.7)^2 + (-5.2)^2}$$

$$= \sqrt{7.29 + 27.04}$$

$$= \sqrt{34.33}$$

$$\approx 5.86$$

63. $d = \sqrt{(x_2 - x_1)^2 + (y_2 - y_1)^2}$

$= \sqrt{\left(0 - \sqrt{7}\right)^2 + \left[0 - \sqrt{3}\right]^2}$

$= \sqrt{\left(-\sqrt{7}\right)^2 + \left(\sqrt{3}\right)^2}$

$= \sqrt{7 + 3}$

$= \sqrt{10}$

≈ 3.16

65. Midpoint $= \left(\dfrac{x_1 + x_2}{2}, \dfrac{y_1 + y_2}{2}\right)$

$= \left(\dfrac{1 + 5}{2}, \dfrac{9 + 3}{2}\right)$

$= (3, 6)$

67. Midpoint $= \left(\dfrac{x_1 + x_2}{2}, \dfrac{y_1 + y_2}{2}\right)$

$= \left(\dfrac{-7 + 7}{2}, \dfrac{2 + (-2)}{2}\right)$

$= (0, 0)$

69. Midpoint $= \left(\dfrac{x_1 + x_2}{2}, \dfrac{y_1 + y_2}{2}\right)$

$= \left(\dfrac{-1 + 4}{2}, \dfrac{4 + 6}{2}\right)$

$= \left(\dfrac{3}{2}, 5\right)$

71. Midpoint $= \left(\dfrac{x_1 + x_2}{2}, \dfrac{y_1 + y_2}{2}\right)$

$= \left(\dfrac{3 + 2}{2}, \dfrac{\frac{1}{2} + (-4)}{2}\right)$

$= \left(\dfrac{5}{2}, -\dfrac{7}{4}\right)$

73. Midpoint $= \left(\dfrac{x_1 + x_2}{2}, \dfrac{y_1 + y_2}{2}\right)$

$= \left(\dfrac{\sqrt{3} + \sqrt{2}}{2}, \dfrac{2 + 7}{2}\right)$

$= \left(\dfrac{\sqrt{3} + \sqrt{2}}{2}, \dfrac{9}{2}\right)$

75. $(x - h)^2 + (y - k)^2 = r^2$

$(x - 0)^2 + (y - 0)^2 = 4^2$

$x^2 + y^2 = 16$

77. $(x - h)^2 + (y - k)^2 = r^2$

$(x - 2)^2 + (y - 0)^2 = 5^2$

$(x - 2)^2 + y^2 = 25$

79. $(x - h)^2 + (y - k)^2 = r^2$

$(x - 0)^2 + (y - 5)^2 = 1^2$

$x^2 + (y - 5)^2 = 1$

81. $(x - h)^2 + (y - k)^2 = r^2$

$(x - 3)^2 + (y - 4)^2 = (8)^2$

$(x - 3)^2 + (y - 4)^2 = 64$

83. $(x - h)^2 + (y - k)^2 = r^2$

$(x - 7)^2 + \left[y - (-6)\right]^2 = 10^2$

$(x - 7)^2 + (y + 6)^2 = 100$

85. $(x - h)^2 + (y - k)^2 = r^2$

$(x - 1)^2 + (y - 2)^2 = \left(\sqrt{5}\right)^2$

$(x - 1)^2 + (y - 2)^2 = 5$

87. The center is (0, 0) and the radius is 4.

$(x - h)^2 + (y - k)^2 = r^2$

$(x - 0)^2 + (y - 0)^2 = 4^2$

$x^2 + y^2 = 16$

89. The center is (3, –2) and the radius is 3.

$(x - h)^2 + (y - k)^2 = r^2$

$(x - 3)^2 + \left[y - (-2)\right]^2 = 3^2$

$(x - 3)^2 + (y + 2)^2 = 9$

91. $x^2 + y^2 = 16$

$x^2 + y^2 = 4^2$

The graph is a circle with its center at the origin and radius 4.

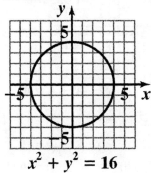

93. $x^2 + y^2 = 10$

$x^2 + y^2 = \left(\sqrt{10}\right)^2$

The graph is a circle with its center at the origin and radius $\sqrt{10}$.

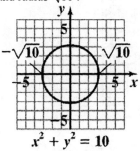

95. $(x+4)^2 + y^2 = 25$

$(x+4)^2 + (y-0)^2 = 5^2$

The graph is a circle with its center at (–4, 0) and radius 5.

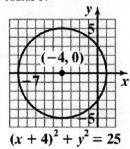

97. $x^2 + (y-3)^2 = 4$

$(x-0)^2 + (y-3)^2 = (2)^2$

The graph is a circle with its center at (0, 3) and radius 2.

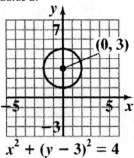

99. $(x+8)^2 + (y+2)^2 = 9$

$(x+8)^2 + (y+2)^2 = 3^2$

The graph is a circle with its center at (–8, –2) and radius 3.

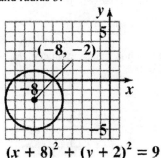

101. $y = \sqrt{25 - x^2}$

If we solve $x^2 + y^2 = 25$ for y, we obtain

$y = \pm\sqrt{25 - x^2}$. Therefore, the graph of

$y = \sqrt{25 - x^2}$ is the upper half ($y \geq 0$) of a

circle with its center at the origin and radius 5.

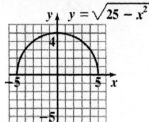

103. $y = -\sqrt{4 - x^2}$

If we solve $x^2 + y^2 = 4$ for y, we obtain

$y = \pm\sqrt{4 - x^2}$. Therefore, the graph of

$y = -\sqrt{4 - x^2}$ is the lower half $(y \le 0)$ of a

circle with its center at the origin and radius 2.

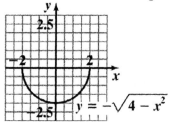

105. a.
$$x^2 + y^2 + 8x + 15 = 0$$
$$x^2 + 8x + y^2 = -15$$
$$(x^2 + 8x + 16) + y^2 = -15 + 16$$
$$(x + 4)^2 + y^2 = 1$$
$$(x + 4)^2 + y^2 = 1^2$$

b. The graph is a circle with center $(-4, 0)$ and radius 1.

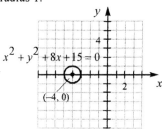

107. a.
$$x^2 + y^2 + 6x - 4y + 4 = 0$$
$$x^2 + 6x + y^2 - 4y = -4$$
$$(x^2 + 6x + 9) + (y^2 - 4y + 4) = -4 + 9 + 4$$
$$(x + 3)^2 + (y - 2)^2 = 9$$
$$(x + 3)^2 + (y - 2)^2 = 3^2$$

b. The graph is a circle with center $(-3, 2)$ and radius 3.

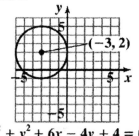

$$x^2 + y^2 + 6x - 4y + 4 = 0$$

109. a.
$$x^2 + y^2 + 6x - 2y + 6 = 0$$
$$x^2 + 6x + y^2 - 2y = -6$$
$$(x^2 + 6x + 9) + (y^2 - 2y + 1) = -6 + 9 + 1$$
$$(x + 3)^2 + (y - 1)^2 = 4$$
$$(x + 3)^2 + (y - 1)^2 = 2^2$$

b. The graph is a circle with center $(-3, 1)$ and radius 2.

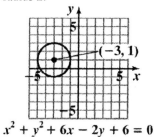

$$x^2 + y^2 + 6x - 2y + 6 = 0$$

111. a.
$$x^2 + y^2 - 8x + 2y + 13 = 0$$
$$x^2 - 8x + y^2 + 2y = -13$$
$$(x^2 - 8x + 16) + (y^2 + 2y + 1) = -13 + 16 + 1$$
$$(x - 4)^2 + (y + 1)^2 = 4$$
$$(x - 4)^2 + (y + 1)^2 = 2^2$$

b. The graph is a circle with center $(4, -1)$ and radius 2.

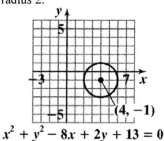

$$x^2 + y^2 - 8x + 2y + 13 = 0$$

113. $(x+4)^2 + y^2 = 25$

$(x+4)^2 + (y-0)^2 = 5^2$

The graph is a circle with its center at (–4, 0) and radius 5.

Area $= \pi r^2 = \pi\left(5\right)^2 = 25\pi \approx 78.5$ sq. units

115. *x*-intercept:

$x = 0^2 - 6(0) - 7$

$x = -7$

The *x*-intercept is (–7, 0)

y-intercepts:

$0 = y^2 - 6y - 7$

$0 = (y+1)(y-7)$

$y = -1$ or $y = 7$

The *y*-intercepts are (0, –1) and (0, 7).

117. *x*-intercept:

$x = 2(0-3)^2 + 6$

$x = 24$

The *x*-intercept is (24, 0).

y-intercepts:

$0 = 2(y-3)^2 + 6$

Since $2(y-3)^2 + 6 \geq 6$ for all real values of *y*, this equation has no real solutions.
There are no *y*-intercepts.

119. No. For example, the origin is the midpoint of both the segment from (1, 1) to (–1, –1) and the segment from (2, 2) to (–2, –2), but these segments have different lengths.

121. The distance from the midpoint (4, –6) to the endpoint (7, –2) is half the length of the line segment.

$\dfrac{d}{2} = \sqrt{(7-4)^2 + \left[-2-(-6)\right]^2}$

$= \sqrt{3^2 + 4^2}$

$= \sqrt{25}$

$= 5$

Since $\dfrac{d}{2} = 5$, $d = 10$. The length is 10 units.

123. Since (–6, 2) is 2 units above the *x*-axis, the radius is 2.

$(x-h)^2 + (y-k)^2 = r^2$

$(x+6)^2 + (y-2)^2 = 2^2$

$(x+6)^2 + (y-2)^2 = 4$

125. a. Diameter $= \sqrt{(x_2 - x_1)^2 + (y_2 - y_1)^2}$

$= \sqrt{(9-5)^2 + (8-4)^2}$

$= \sqrt{4^2 + 4^2}$

$= \sqrt{16+16}$

$= \sqrt{32}$

$= 4\sqrt{2}$

Since the diameter is $4\sqrt{2}$ units, the radius is $2\sqrt{2}$ units.

b. Midpoint $= \left(\dfrac{x_1 + x_2}{2}, \dfrac{y_1 + y_2}{2}\right)$

$= \left(\dfrac{5+9}{2}, \dfrac{4+8}{2}\right)$

$= (7, 6)$

The center is (7, 6).

c. $(x-h)^2 + (y-k)^2 = r^2$

$(x-7)^2 + (y-6)^2 = \left(2\sqrt{2}\right)^2$

$(x-7)^2 + (y-6)^2 = 8$

127. The minimum number is 0 and the maximum number is 4 as shown in the diagrams.

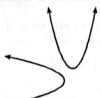

No points of intersection

Four points of intersection

129. a. Since $150 - 2(68.2) = 13.6$, the clearance is 13.6 feet.

b. Since $150 - 68.2 = 81.8$, the center of the wheel is 81.8 feet above the ground.

c. $(x-h)^2+(y-k)^2=r^2$

$(x-0)^2+(y-81.8)^2=68.2^2$

$x^2+(y-81.8)^2=68.2^2$

$x^2+(y-81.8)^2=4651.24$

131. a.   The center of the blue circle is the origin, and the radius is 4.

$x^2+y^2=r^2$

$x^2+y^2=4^2$

$x^2+y^2=16$

b.   The center of the red circle is (2, 0), and the radius is 2.

$(x-h)^2+(y-k)^2=r^2$

$(x-2)^2+(y-0)^2=2^2$

$(x-2)^2+y^2=4$

c.   The center of the green circle is (–2, 0), and the radius is 2.

$(x-h)^2+(y-k)^2=r^2$

$[x-(-2)]^2+(y-0)^2=2^2$

$(x+2)^2+y^2=4$

d.   Shaded area = (blue circle area) – (red circle area) – (green circle area)

$=\pi(4^2)-\pi(2^2)-\pi(2^2)$

$=16\pi-4\pi-4\pi$

$=8\pi$

133.   The radii are 4 and 8, respectively. So, the area between the circles is

$\pi(8)^2-\pi(4)^2=64\pi-16\pi=48\pi$ square units.

136. $\dfrac{6x^{-3}y^4}{18x^{-2}y^3}=\dfrac{6}{18}x^{-3-(-2)}y^{4-3}$

$=\dfrac{1}{3}x^{-3+2}y^1$

$=\dfrac{1}{3}x^{-1}y$ or $\dfrac{y}{3x}$

137. a.   area 1: a^2 area 2: ab

area 3: ab area 4: b^2

b. $a^2+ab+ab+b^2=a^2+2ab+b^2$

$=(a+b)^2$

138. $\begin{vmatrix}4&0&3\\5&2&-1\\3&6&4\end{vmatrix}=4\begin{vmatrix}2&-1\\6&4\end{vmatrix}-5\begin{vmatrix}0&3\\6&4\end{vmatrix}+3\begin{vmatrix}0&3\\2&-1\end{vmatrix}$

$=4(8+6)-5(0-18)+3(0-6)$

$=4(14)-5(-18)+3(-6)$

$=56+90-18$

$=128$

139. $-4<3x-4<17$

$-4+4<3x-4+4<17+4$

$0<3x<21$

$\dfrac{0}{3}<\dfrac{3x}{3}<\dfrac{21}{3}$

$0<x<7$

In interval notation: $(0,7)$

140.   This is a parabola in the form $y=a(x-h)^2+k$ with $a=1$, $h=4$ and $k=1$. Since $a>0$, the parabola opens upward. The vertex is (4, 1). The y-intercept is (0, 17). There are no x-intercepts.

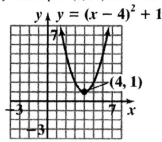

Exercise Set 14.2

1. An ellipse is a set of points in a plane, the sum of whose distances from two fixed points is constant.

3. $\dfrac{(x-h)^2}{a^2}+\dfrac{(y-k)^2}{b^2}=1$

5. If $a = b$, the formula for a circle is obtained.

7. First divide both sides by 180.

9. No, the sign in front of the y^2 is negative.

11. $\dfrac{x^2}{4}+\dfrac{y^2}{1}=1$

Since $a^2 = 4, a = 2.$

Since $b^2 = 1, b = 1.$

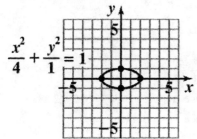

13. $\dfrac{x^2}{4}+\dfrac{y^2}{9}=1$

Since $a^2 = 4, a = 2.$

Since $b^2 = 9, b = 3.$

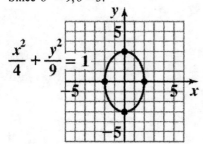

15. $\dfrac{x^2}{25}+\dfrac{y^2}{9}=1$

Since $a^2 = 25, a = 5.$

Since $b^2 = 9, b = 3.$

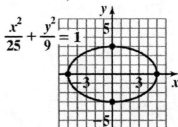

17. $\dfrac{x^2}{16}+\dfrac{y^2}{25}=1$

Since $a^2 = 16, a = 4.$

Since $b^2 = 25, b = 5.$

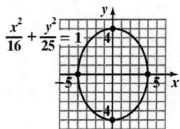

19. $x^2 + 16y^2 = 16$

$\dfrac{x^2}{16}+\dfrac{16y^2}{16}=1$

$\dfrac{x^2}{16}+\dfrac{y^2}{1}=1$

Since $a^2 = 16, a = 4w.$

Since $b^2 = 1, b = 1.$

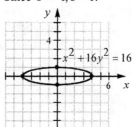

21. $49x^2 + y^2 = 49$

$$\frac{49x^2}{49} + \frac{y^2}{49} = 1$$

$$\frac{x^2}{1} + \frac{y^2}{49} = 1$$

Since $a^2 = 1, a = 1.$

Since $b^2 = 49, b = 7.$

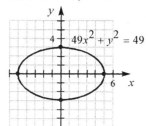

23. $9x^2 + 16y^2 = 144$

$$\frac{9x^2}{144} + \frac{16y^2}{144} = 1$$

$$\frac{x^2}{16} + \frac{y^2}{9} = 1$$

Since $a^2 = 16, a = 4.$

Since $b^2 = 9, b = 3.$

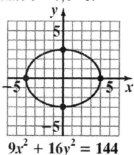

25. $25x^2 + 100y^2 = 400$

$$x^2 + 4y^2 = 16$$

$$\frac{x^2}{16} + \frac{4y^2}{16} = 1$$

$$\frac{x^2}{16} + \frac{y^2}{4} = 1$$

Since $a^2 = 16, a = 4.$

Since $b^2 = 4, b = 2.$

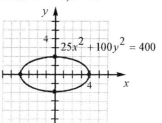

27. $x^2 + 2y^2 = 8$

$$\frac{x^2}{8} + \frac{2y^2}{8} = 1$$

$$\frac{x^2}{8} + \frac{y^2}{4} = 1$$

Since $a^2 = 8, a = \sqrt{8} = 2\sqrt{2} \approx 2.83.$

Since $b^2 = 4, b = 2.$

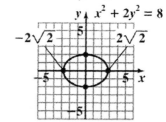

29. $\dfrac{x^2}{16} + \dfrac{(y-2)^2}{9} = 1$

The center is (0, 2).

Since $a^2 = 16, a = 4$.

Since $b^2 = 9, b = 3$.

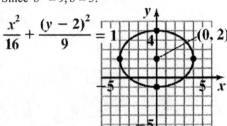

31. $\dfrac{(x-4)^2}{9} + \dfrac{(y+3)^2}{25} = 1$

The center is (4, −3).

Since $a^2 = 9, a = 3$.

Since $b^2 = 25, b = 5$.

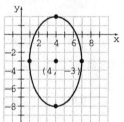

33. $\dfrac{(x+1)^2}{9} + \dfrac{(y-2)^2}{4} = 1$

The center is (−1, 2).

Since $a^2 = 9, a = 3$.

Since $b^2 = 4, b = 2$.

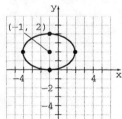

35. $(x+3)^2 + 9(y+1)^2 = 81$

$$\dfrac{(x+3)^2}{81} + \dfrac{9(y+1)^2}{81} = 1$$

$$\dfrac{(x+3)^2}{81} + \dfrac{(y+1)^2}{9} = 1$$

The center is (−3, −1).

Since $a^2 = 81, a = 9$.

Since $b^2 = 9, b = 3$.

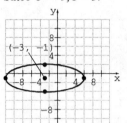

37. $(x-5)^2 + 4(y+4)^2 = 4$

$$\dfrac{(x-5)^2}{4} + \dfrac{4(y+4)^2}{4} = 1$$

$$\dfrac{(x-5)^2}{4} + \dfrac{(y+4)^2}{1} = 1$$

The center is (5, −4).

Since $a^2 = 4, a = 2$.

Since $b^2 = 1, b = 1$.

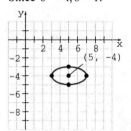

39. $12(x+4)^2 + 3(y-1)^2 = 48$

$$\frac{12(x+4)^2}{48} + \frac{3(y-1)^2}{48} = 1$$

$$\frac{(x+4)^2}{4} + \frac{(y-1)^2}{16} = 1$$

The center is $(-4, 1)$.

Since $a^2 = 4$, $a = 2$.

Since $b^2 = 16$, $b = 4$.

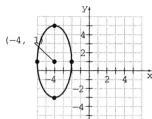

41. $\dfrac{x^2}{4} + \dfrac{y^2}{1} = 1$

Since $a^2 = 4$, $a = 2$; Since $b^2 = 1$, $b = 1$.

Area $= \pi ab$

$ = \pi(2)(1)$

$ = 2\pi \approx 6.3$ square units

43. There is only one point, at $(0, 0)$. Two non-negative numbers can only sum to 0 if they are both 0.

45. The center is the origin, $a = 3$, and $b = 4$.

$$\frac{x^2}{a^2} + \frac{y^2}{b^2} = 1$$

$$\frac{x^2}{3^2} + \frac{y^2}{4^2} = 1$$

$$\frac{x^2}{9} + \frac{y^2}{16} = 1$$

47. The center is the origin, $a = 2$, and $b = 3$.

$$\frac{x^2}{a^2} + \frac{y^2}{b^2} = 1$$

$$\frac{x^2}{2^2} + \frac{y^2}{3^2} = 1$$

$$\frac{x^2}{4} + \frac{y^2}{9} = 1$$

49. There are no points of intersection, because the ellipse with $a = 4$ and $b = 5$ is completely inside the circle of radius 7.

51. $x^2 + 4y^2 + 6x + 16y - 11 = 0$

$$x^2 + 6x + 4y^2 + 16y = 11$$

$$(x^2 + 6x + 9) + 4(y^2 + 4y + 4) = 11 + 9 + 16$$

$$(x+3)^2 + 4(y+2)^2 = 36$$

$$\frac{(x+3)^2}{36} + \frac{(y+2)^2}{9} = 1$$

The center is $(-3, -2)$.

53. Since $90.2 - 20.7 = 69.5$, the distance between the foci is 69.5 feet.

55. a. Consider the ellipse to be centered at the origin, $(0, 0)$. Here, $a = 10$ and $b = 24$. The equation is

$$\frac{x^2}{10^2} + \frac{y^2}{24^2} = 1 \;\Rightarrow\; \frac{x^2}{100} + \frac{y^2}{576} = 1.$$

b. Area $= \pi ab$

$ = \pi(10)(24)$

$ = 240\pi \approx 753.98$ square feet

c. Area of opening is half the area of ellipse

$$= \frac{\pi ab}{2}$$

$$= \frac{\pi(10)(24)}{2}$$

$$= 120\pi \approx 376.99 \text{ square feet}$$

57. Using $a = 3$ and $b = 2$, we may assume that the ellipse has the equation $\dfrac{x^2}{9} + \dfrac{y^2}{4} = 1$ and that the foci are located at $(\pm c, 0)$. Apply the definition of an ellipse using the points $(3, 0)$ and $(0, 2)$. That is, the distance from $(3, 0)$ to $(-c, 0)$ plus the distance from $(3, 0)$ to $(c, 0)$ is the same as the sum of the distance from $(0, 2)$ to $(-c, 0)$ and the distance from $(0, 2)$ to $(c, 0)$.

$$\sqrt{\left[3-(-c)\right]^2 + (0-0)^2} + \sqrt{(3-c)^2 + (0-0)^2}$$
$$= \sqrt{\left[0-(-c)\right]^2 + (2-0)^2} + \sqrt{(0-c)^2 + (2-0)^2}$$
$$\left|3+c\right| + \left|3-c\right| = \sqrt{c^2+4} + \sqrt{(-c)^2+4}$$

Note that the foci are inside the ellipse, so $3 + c > 0$ and $3 - c > 0$. So, $\left|3+c\right| = 3+c$ and $\left|3-c\right| = 3-c$.

$$(3+c) + (3-c) = 2\sqrt{c^2+4}$$
$$6 = 2\sqrt{c^2+4}$$
$$3 = \sqrt{c^2+4}$$
$$9 = c^2 + 4$$
$$5 = c^2$$
$$c = \pm\sqrt{5}$$

The foci are located at $\left(\pm\sqrt{5}, 0\right)$. That is, the foci are $\sqrt{5} \approx 2.24$ feet, in both directions, from the center of the ellipse, along the major axis.

59. Answers will vary.

61. Answers will vary.

63. $h = \dfrac{-3+11}{2} = \dfrac{8}{2} = 4$; $k = \dfrac{5+(-1)}{2} = \dfrac{4}{2} = 2$

Thus, the center is $(4, 2)$.

$$a = \dfrac{11-(-3)}{2} = \dfrac{14}{2} = 7 \; ; \; b = \dfrac{5-(-1)}{2} = \dfrac{6}{2} = 3$$
$$\dfrac{(x-h)^2}{a^2} + \dfrac{(y-k)^2}{b^2} = 1$$
$$\dfrac{(x-4)^2}{7^2} + \dfrac{(y-2)^2}{3^2} = 1$$
$$\dfrac{(x-4)^2}{49} + \dfrac{(y-2)^2}{9} = 1$$

65.
$$S = \dfrac{n}{2}(f+l), \text{ for } l$$
$$S = \dfrac{nf}{2} + \dfrac{nl}{2}$$
$$S - \dfrac{nf}{2} = \dfrac{nl}{2}$$
$$2\left(S - \dfrac{nf}{2}\right) = 2\left(\dfrac{nl}{2}\right)$$
$$2S - nf = nl$$
$$\dfrac{2S - nf}{n} = \dfrac{nl}{n} \;\Rightarrow\; l = \dfrac{2S - nf}{n}$$

66.
$$\begin{array}{r} x + \frac{5}{2} \\ 2x-3 \overline{\smash{\big)}\, 2x^2 + 2x - 7} \\ \underline{2x^2 - 3x} \\ 5x - 7 \\ \underline{5x - \frac{15}{2}} \\ \frac{1}{2} \end{array}$$
$$\dfrac{2x^2 + 2x - 7}{2x-3} = x + \dfrac{5}{2} + \dfrac{1}{2(2x-3)}$$

67.
$$\sqrt{3b-2} = 10 - b$$
$$3b - 2 = (10-b)^2$$
$$3b - 2 = 100 - 20b + b^2$$
$$b^2 - 23b + 102 = 0$$
$$(b-6)(b-17) = 0$$
$$b - 6 = 0 \quad \text{or} \quad b - 17 = 0$$
$$b = 6 \qquad\qquad b = 17$$

Upon checking $b = 17$ is extraneous. The solution is $b = 6$.

68.
$$\dfrac{3x+5}{x-4} \le 0$$
$$3x + 5 = 0 \;\Rightarrow\; x = -\dfrac{5}{3}$$
$$x - 4 = 0 \;\Rightarrow\; x = 4$$
$$\dfrac{3x+5}{x-4} \le 0 \;\Rightarrow\; -\dfrac{5}{3} \le x < 4 \text{ or } \left[-\dfrac{5}{3}, 4\right)$$

69. $\log_8 321 = \dfrac{\ln 321}{\ln 8} \approx 2.7755$

Mid-Chapter Test: 14.1-14.2

1. This is a parabola in the form $y = a(x-h)^2 + k$ with $a = 1$, $h = 2$, and $k = -1$. Since $a > 0$, the parabola opens upward. The vertex is $(2,8)$. The y-intercept is $(0,3)$. The x-intercepts are $(1,0)$ and $(3,0)$.

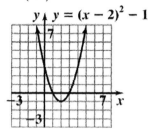

2. This is a parabola in the form $y = a(x-h)^2 + k$ with $a = -1$, $h = -1$, and $k = 3$. Since $a < 0$, the parabola opens downward. The vertex is $(-1,3)$. The y-intercept is $(0,2)$. The x-intercepts are about $(-2.732,0)$ and $(0.732,0)$.

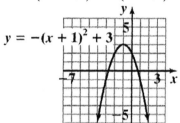

3. This is a parabola in the form $x = a(y-k)^2 + h$ with $a = -1$, $h = 1$, and $k = 4$. Since $a < 0$, the parabola opens to the left. The vertex is $(1,4)$. The x-intercept is $(-15,0)$. The y-intercepts are $(0,5)$ and $(0,3)$.

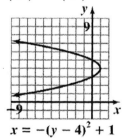

4. This is a parabola in the form $x = a(y-k)^2 + h$ with $a = 2$, $h = -2$, and $k = -3$. Since $a > 0$, the parabola opens to the right. The vertex is $(-2,-3)$. The x-intercept is $(16,0)$. The y-intercepts are $(0,-4)$ and $(0,-2)$.

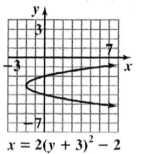

5. $y = x^2 + 6x + 10$
$$= (x^2 + 6x) + 10$$
$$= (x^2 + 6x + 9) + 10 - 9$$
$$= (x+3)^2 + 1$$

This is a parabola in the form $y = a(x-h)^2 + k$ with $a = 1$, $h = -3$, and $k = 1$. Since $a > 0$, the parabola opens upward. The vertex is $(-3,1)$. The y-intercept is $(0,10)$. There are no x-intercepts.

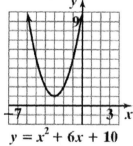

6. $d = \sqrt{(x_2 - x_1)^2 + (y_2 - y_1)^2}$
$$= \sqrt{(-2-(-7))^2 + (-8-4)^2}$$
$$= \sqrt{(5)^2 + (-12)^2}$$
$$= \sqrt{25 + 144}$$
$$= \sqrt{169}$$
$$= 13$$

7. $d = \sqrt{(x_2 - x_1)^2 + (y_2 - y_1)^2}$

$= \sqrt{(2-5)^2 + (9-(-3))^2}$

$= \sqrt{(-3)^2 + (12)^2}$

$= \sqrt{9+144}$

$= \sqrt{153}$

≈ 12.37

8. Midpoint $= \left(\dfrac{x_1 + x_2}{2}, \dfrac{y_1 + y_2}{2} \right)$

$= \left(\dfrac{9 + (-11)}{2}, \dfrac{-1+6}{2} \right)$

$= \left(\dfrac{-2}{2}, \dfrac{5}{2} \right)$

$= \left(-1, \dfrac{5}{2} \right)$

9. Midpoint $= \left(\dfrac{x_1 + x_2}{2}, \dfrac{y_1 + y_2}{2} \right)$

$= \left(\dfrac{-\frac{5}{2} + 8}{2}, \dfrac{7 + \frac{1}{2}}{2} \right)$

$= \left(\dfrac{11}{4}, \dfrac{15}{4} \right)$

10. $(h,k) = (-3,2)$; $r = 5$

$(x-h)^2 + (y-k)^2 = r^2$

$(x-(-3))^2 + (y-2)^2 = 5^2$

$(x+3)^2 + (y-2)^2 = 25$

11. $x^2 + (y-1)^2 = 16$

$x^2 + (y-1)^2 = 4^2$

The graph is a circle with its center at $(0,1)$ and radius 4.

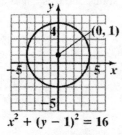

$x^2 + (y-1)^2 = 16$

12. $y = \sqrt{36 - x^2}$

If we solve $x^2 + y^2 = 36$ for y, we obtain

$y = \pm\sqrt{36 - x^2}$. Therefore, the graph of

$y = \sqrt{36 - x^2}$ is the upper half $(y \geq 0)$ of a

circle with its center at the origin and radius 6.

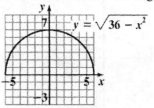

13. $x^2 + y^2 - 2x + 4y - 4 = 0$

Complete the square in both x and y.

$(x^2 - 2x) + (y^2 + 4y) = 4$

$(x^2 - 2x + 1) + (y^2 + 4y + 4) = 4 + 1 + 4$

$(x-1)^2 + (y+2)^2 = 9$

The graph of $x^2 + y^2 - 2x + 4y - 4 = 0$ is a circle

with its center at $(1,-2)$ and radius 3.

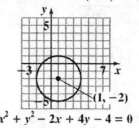

$x^2 + y^2 - 2x + 4y - 4 = 0$

14. A circle is defined to be the set of points that are equidistant from a fixed point. This distance is called the radius and the fixed point is called the center.

15. $\dfrac{x^2}{4} + \dfrac{y^2}{9} = 1$

$a^2 = 4 \to a = 2$; $b^2 = 9 \to b = 3$

The graph of this equation is an ellipse with center at the origin and major axis along the y-axis.

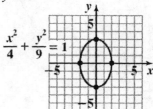

16. $\dfrac{x^2}{81}+\dfrac{y^2}{25}=1$

$a^2=81 \to a=9$; $b^2=25 \to b=5$

The graph of this equation is an ellipse with center at the origin and major axis along the *x*-axis.

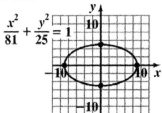

17. $\dfrac{(x-1)^2}{49}+\dfrac{(y+2)^2}{4}=1$

$a^2=49 \to a=7$; $b^2=4 \to b=2$

The graph of this equation is an ellipse with center at $(1,-2)$ and major axis parallel to the *x*-axis.

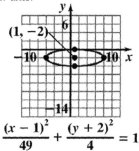

18. $36(x+3)^2+(y-4)^2=36$

$\dfrac{36(x+3)^2}{36}+\dfrac{(y-4)^2}{36}=\dfrac{36}{36}$

$\dfrac{(x+3)^2}{1}+\dfrac{(y-4)^2}{36}=1$

$a^2=1 \to a=1$; $b^2=36 \to b=6$

The graph of this equation is an ellipse with center at $(-3,4)$ and major axis parallel to the *y*-axis.

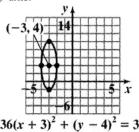

19. $\dfrac{x^2}{4}+\dfrac{y^2}{9}=1$

Since $a^2=4, a=2$; Since $b^2=9, b=3$.

Area $= \pi ab = \pi(2)(3)$
$= 6\pi \approx 18.85$ square units

20. The center of the ellipse is the origin, $(0,0)$. We have $a=8$ and $b=5$ so $a^2=64$ and $b^2=25$. Thus, the equation of the ellipse is

$$\dfrac{(x-0)^2}{8^2}+\dfrac{(y-0)^2}{5^2}=1$$

$$\dfrac{x^2}{64}+\dfrac{y^2}{25}=1$$

Exercise Set 14.3

1. A hyperbola is the set of points in a plane, the difference of whose distances from two fixed points (called foci) is a constant.

3. The graph of $\dfrac{x^2}{a^2}-\dfrac{y^2}{b^2}=1$ is a hyperbola with vertices at $(a,0)$ and $(-a,0)$. Its transverse axis lies along the *x*-axis. The asymptotes are $y=\pm\dfrac{b}{a}x$.

5. No, equations of hyperbolas have one positive square term and one negative square term. This equation has two positive square terms.

7. Yes, equations of hyperbolas have one positive square term and one negative square term. This equation satisfies that condition.

9. The first step is to divide both sides by 81 in order to make the right side equal to 1.

11. a. $\dfrac{x^2}{9}-\dfrac{y^2}{4}=1$

Since $a^2=9$ and $b^2=4$, $a=3$ and $b=2$. The equations of the asymptotes are

$y=\pm\dfrac{b}{a}x$, or $y=\pm\dfrac{2}{3}x$.

b. To graph the asymptotes, plot the points (3, 2), (–3, 2), (3, –2), and (–3, –2). The graph intersects the *x*-axis at (–3, 0) and (3, 0).

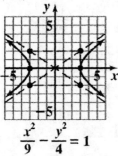

$$\frac{x^2}{9} - \frac{y^2}{4} = 1$$

13. a. $\dfrac{x^2}{4} - \dfrac{y^2}{1} = 1$

Since $a^2 = 4$ and $b^2 = 1$, $a = 2$ and $b = 1$. The equations of the asymptotes are

$$y = \pm\frac{b}{a}x, \text{ or } y = \pm\frac{1}{2}x.$$

b. To graph the asymptotes, plot the points (2, 1), (–2, 1), (2, –1), and (–2, –1). The graph intersects the *x*-axis at (–2, 0) and (2, 0).

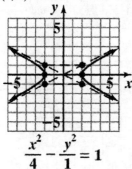

$$\frac{x^2}{4} - \frac{y^2}{1} = 1$$

15. a. $\dfrac{x^2}{9} - \dfrac{y^2}{25} = 1$

Since $a^2 = 9$ and $b^2 = 25$, $a = 3$ and $b = 5$. The equations of the asymptotes are

$$y = \pm\frac{b}{a}x, \text{ or } y = \pm\frac{5}{3}x.$$

b. To graph the asymptotes, plot the points (3, 5), (–3, 5), (3, –5), and (–3, –5). The graph intersects the *x*-axis at (–3, 0) and (3, 0).

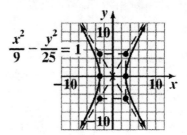

$$\frac{x^2}{9} - \frac{y^2}{25} = 1$$

17. a. $\dfrac{x^2}{25} - \dfrac{y^2}{16} = 1$

Since $a^2 = 25$ and $b^2 = 16$, $a = 5$ and $b = 4$. The equations of the asymptotes are

$$y = \pm\frac{b}{a}x, \text{ or } y = \pm\frac{4}{5}x.$$

b. To graph the asymptotes, plot the points (5, 4), (–5, 4), (5, –4), and (–5, –4). The graph intersects the *x*-axis at (–5, 0) and (5, 0).

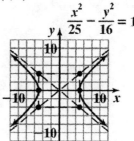

$$\frac{x^2}{25} - \frac{y^2}{16} = 1$$

19. a. $\dfrac{y^2}{25} - \dfrac{x^2}{36} = 1$

Since $a^2 = 36$ and $b^2 = 25$, $a = 6$ and $b = 5$. The equations of the asymptotes are

$$y = \pm\frac{b}{a}x, \text{ or } y = \pm\frac{5}{6}x.$$

b. To graph the asymptotes, plot the points (6, 5), (–6, 5), (6, –5), and (–6, –5). The graph intersects the *y*-axis at (0, –5) and (0, 5).

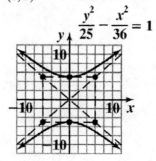

$$\frac{y^2}{25} - \frac{x^2}{36} = 1$$

21. a. $\dfrac{y^2}{9} - \dfrac{x^2}{16} = 1$

Since $a^2 = 16$ and $b^2 = 9$, $a = 4$ and $b = 3$. The equations of the asymptotes are $y = \pm\dfrac{b}{a}x$, or $= y \pm \dfrac{3}{4}x$.

b. To graph the asymptotes, plot the points $(4, 3)$, $(-4, 3)$, $(4, -3)$ and $(-4, -3)$. The graph intersects the y-axis at $(0, -3)$ and $(0, 3)$.

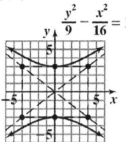

23. a. $\dfrac{y^2}{25} - \dfrac{x^2}{4} = 1$

Since $a^2 = 4$ and $b^2 = 25$, $a = 2$ and $b = 5$. The equations of the asymptotes are $y = \pm\dfrac{b}{a}x$ or $y = \pm\dfrac{5}{2}x$.

b. To graph the asymptotes, plot the points $(2, 5)$, $(-2, 5)$, $(2, -5)$ and $(-2, -5)$. The graph intersects the y-axis at $(0, -5)$ and $(0, 5)$.

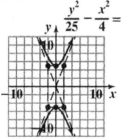

25. a. $\dfrac{x^2}{81} - \dfrac{y^2}{16} = 1$

Since $a^2 = 81$ and $b^2 = 16$, $a = 9$ and $b = 4$. The equations of the asymptotes are $y = \pm\dfrac{b}{a}x$, or $y = \pm\dfrac{4}{9}x$.

b. To graph the asymptotes, plot the points $(9, 4)$, $(-9, 4)$, $(9, -4)$, and $(-9, -4)$. The graph intersects the x-axis at $(-9, 0)$ and $(9, 0)$.

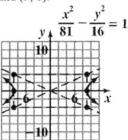

27. a. $x^2 - 25y^2 = 25$

$\dfrac{x^2}{25} - \dfrac{25y^2}{25} = 1$

$\dfrac{x^2}{25} - \dfrac{y^2}{1} = 1$

Since $a^2 = 25$ and $b^2 = 1$, $a = 5$ and $b = 1$. The equations of the asymptotes are $y = \pm\dfrac{b}{a}x$, or $y = \pm\dfrac{1}{5}x$.

b. To graph the asymptotes, plot the points $(5, 1)$, $(-5, 1)$, $(5, -1)$, and $(-5, -1)$. The graph intersects the x-axis at $(-5, 0)$, and $(5, 0)$.

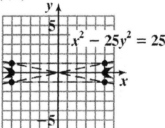

29. a. $4y^2 - 16x^2 = 64$

$\dfrac{4y^2}{64} - \dfrac{16x^2}{64} = 1$

$\dfrac{y^2}{16} - \dfrac{x^2}{4} = 1$

Since $a^2 = 4$ and $b^2 = 16$, $a = 2$ and $b = 4$. The equations of the asymptotes are $y = \pm\dfrac{b}{a}x$, or $y = \pm 2x$.

b. To graph the asymptotes, plot (2, 4), (−2, 4), (2 −4), and (−2, −4). The graph intersects the y-axis at (0, −4) and (0, 4).

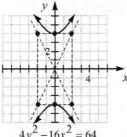

$4y^2 - 16x^2 = 64$

31. a. $9y^2 - x^2 = 9$

$$\frac{9y^2}{9} - \frac{x^2}{9} = 1$$

$$\frac{y^2}{1} - \frac{x^2}{9} = 1$$

Since $a^2 = 9$ and $b^2 = 1$, $a = 3$ and $b = 1$. The equations of the asymptotes are

$$y = \pm\frac{b}{a}x, \text{ or } \pm\frac{1}{3}x.$$

b. To graph the asymptotes, plot the points (3, 1), (−3, 1), (3, −1), and (−3, −1). The graph intersects the y-axis at (0, −1) and (0, 1).

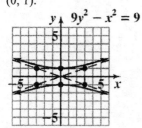

$9y^2 - x^2 = 9$

33. a. $25x^2 - 9y^2 = 225$

$$\frac{25x^2}{225} - \frac{9y^2}{225} = 1$$

$$\frac{x^2}{9} - \frac{y^2}{25} = 1$$

Since $a^2 = 9$ and $b^2 = 25$, $a = 3$ and $b = 5$. The equations of the asymptotes are

$$y = \pm\frac{b}{a}x, \text{ or } y = \pm\frac{5}{3}x.$$

b. To graph the asymptotes, plot the points (3, 5), (−3, 5), (3, −5), and (−3, −5). The graph intersects the x-axis at (−3, 0) and (3, 0).

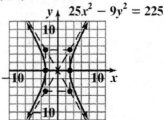

$25x^2 - 9y^2 = 225$

35. a. $4y^2 - 36x^2 = 144$

$$\frac{4y^2}{144} - \frac{36x^2}{144} = \frac{144}{144}$$

$$\frac{y^2}{36} - \frac{x^2}{4} = 1$$

Since $a^2 = 4$ and $b^2 = 36$, $a = 2$ and $b = 6$. The equations of the asymptotes are

$$y = \pm\frac{b}{a}x, \text{ or } y = \pm3x.$$

b. To graph the asymptotes, plot the points (2, 6), (−2, 6), (2, −6), and (−2, −6). The graph intersects the y-axis at (0, −6) and (0, 6).

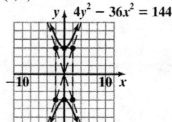

$4y^2 - 36x^2 = 144$

37. $10x^2 + 10y^2 = 40$

$$\frac{10x^2}{10} + \frac{10y^2}{10} = \frac{40}{10}$$

$$x^2 + y^2 = 4$$

The graph is a circle.

39. $x^2 + 16y^2 = 64$

$$\frac{x^2}{64} + \frac{16y^2}{64} = \frac{64}{64}$$

$$\frac{x^2}{64} + \frac{y^2}{4} = 1$$

The graph is an ellipse.

41. $4x^2 - 4y^2 = 29$

$$\frac{4x^2}{29} - \frac{4y^2}{29} = \frac{29}{29}$$

$$\frac{x^2}{\frac{29}{4}} - \frac{y^2}{\frac{29}{4}} = 1$$

The graph is a hyperbola.

43. $2y = 12x^2 - 8x + 16$

$$y = 6x^2 - 4x + 8$$

The graph is a parabola.

45. $6x^2 + 9y^2 = 54$

$$\frac{6x^2}{54} + \frac{9y^2}{54} = \frac{54}{54}$$

$$\frac{x^2}{9} + \frac{y^2}{6} = 1$$

The graph is an ellipse.

47. $3x = -2y^2 + 9y - 15$

$$x = -\frac{2}{3}y^2 + 3y - 5$$

The graph is a parabola.

49. $6x^2 + 6y^2 = 36$

$$\frac{6x^2}{6} + \frac{6y^2}{6} = \frac{36}{6}$$

$$x^2 + y^2 = 6$$

The graph is a circle.

51. $14y^2 = 7x^2 + 35$

$$14y^2 - 7x^2 = 35$$

$$\frac{14y^2}{35} - \frac{7x^2}{35} = \frac{35}{35}$$

$$\frac{y^2}{\frac{5}{2}} - \frac{x^2}{5} = 1$$

The graph is a hyperbola.

53. $x + y = 2y^2 + 6$

$$x = 2y^2 - y + 6$$

The graph is a parabola.

55. $12x^2 = 4y^2 + 48$

$$12x^2 - 4y^2 = 48$$

$$\frac{12x^2}{48} - \frac{4y^2}{48} = \frac{48}{48}$$

$$\frac{x^2}{4} - \frac{y^2}{12} = 1$$

The graph is a hyperbola.

57. $y - x + 4 = x^2$

$$y = x^2 + x - 4$$

The graph is a parabola.

59. $-3x^2 - 3y^2 = -27$

$$\frac{-3x^2}{-3} + \frac{-3y^2}{-3} = \frac{-27}{-3}$$

$$x^2 + y^2 = 9$$

The graph is a circle.

61. Since the vertices are (0, ±2), the hyperbola is of the form $\dfrac{y^2}{b^2} - \dfrac{x^2}{a^2} = 1$ with $b = 2$. Since the asymptotes are $y = \pm\dfrac{1}{2}x$, we have $\dfrac{b}{a} = \dfrac{1}{2}$. Therefore, $\dfrac{2}{a} = \dfrac{1}{2}$, so $a = 4$. The equation of the hyperbola is $\dfrac{y^2}{2^2} - \dfrac{x^2}{4^2} = 1$, or $\dfrac{y^2}{4} - \dfrac{x^2}{16} = 1$.

63. Since the vertices are $\left(0, \pm 3\right)$, the hyperbola is of the form $\dfrac{x^2}{a^2} - \dfrac{y^2}{b^2} = 1$ with $a = 3$. Since the asymptotes are $y = \pm 2x$, we have $\dfrac{b}{a} = 2$. Therefore, $\dfrac{b}{3} = 2$, so $b = 6$. The equation of the hyperbola is $\dfrac{x^2}{3^2} - \dfrac{y^2}{6^2} = 1$, or $\dfrac{x^2}{9} - \dfrac{y^2}{36} = 1$.

65. Since the transverse axis is along the *x*-axis, the equation is of the form $\dfrac{x^2}{a^2} - \dfrac{y^2}{b^2} = 1$ Since the asymptotes are $y = \pm \dfrac{5}{3}x$, we require $\dfrac{b}{a} = \dfrac{5}{3}$.

Using $a = 3$ and $b = 5$, the equation of the hyperbola is $\dfrac{x^2}{3^2} - \dfrac{y^2}{5^2} = 1$, or $\dfrac{x^2}{9} - \dfrac{y^2}{25} = 1$.

No, this is not the only possible answer, because *a* and *b* are not uniquely determined.

$\dfrac{x^2}{18} - \dfrac{y^2}{50} = 1$ and others will also work.

67. No, for each value of *x* with $|x| > a$, there are 2 possible values of *y*.

69. $\dfrac{x^2}{25} - \dfrac{y^2}{4} = 1$. This hyperbola has its transverse axis along the *x*-axis with vertices at $(\pm 5, 0)$.

Domain: $(-\infty, -5] \cup [5, \infty)$

Range: $\mathbb{R}$

71. The equation is changed from $\dfrac{x^2}{a^2} - \dfrac{y^2}{b^2} = 1$ to

$\dfrac{x^2}{b^2} - \dfrac{y^2}{a^2} = 1$. Both graphs have a transverse axis along the *x*-axis. The vertices of the second graph will be closer to the origin, at $(\pm b, 0)$ instead of $(\pm a, 0)$. The second graph will open wider.

73. Answers will vary.

75. The points are $(-6, 4)$ and $(-2, 2)$.

$m = \dfrac{y_2 - y_1}{x_2 - x_1} = \dfrac{2 - 4}{-2 - (-6)} = \dfrac{-2}{4} = -\dfrac{1}{2}$

Use $y - y_1 = m(x - x_1)$, with $m = -\dfrac{1}{2}, (-2, 2)$

$y - 2 = -\dfrac{1}{2}\left(x - (-2)\right)$

$y - 2 = -\dfrac{1}{2}x - 1$

$y = -\dfrac{1}{2}x + 1$

76. $\dfrac{3x}{2x - 3} + \dfrac{2x + 4}{2x^2 + x - 6}$

$= \dfrac{3x}{2x - 3} + \dfrac{2x + 4}{(2x - 3)(x + 2)}$

$= \dfrac{3x}{2x - 3} \cdot \dfrac{x + 2}{x + 2} + \dfrac{2x + 4}{(2x - 3)(x + 2)}$

$= \dfrac{3x^2 + 6x + 2x + 4}{(2x - 3)(x + 2)}$

$= \dfrac{3x^2 + 8x + 4}{(2x - 3)(x + 2)}$

$= \dfrac{(3x + 2)(x + 2)}{(2x - 3)(x + 2)}$

$= \dfrac{3x + 2}{2x - 3}$

77. $f(x) = 3x^2 - x + 5,\ g(x) = 6 - 4x^2$

$(f + g)(x) = \left(3x^2 - x + 5\right) + \left(6 - 4x^2\right)$

$= -x^2 - x + 11$

78. $\begin{array}{rcl} 5(-4x + 9y = 7) & \Rightarrow & -20x + 45y = 35 \\ 4(5x + 6y = -3) & \Rightarrow & \underline{20x + 24y = -12} \\ & & 69y = 23 \\ & & y = \dfrac{1}{3} \end{array}$

$5x + 6\left(\dfrac{1}{3}\right) = -3 \Rightarrow 5x + 2 = -3 \Rightarrow x = -1$

The solution is $\left(-1, \dfrac{1}{3}\right)$.

79. $E = \dfrac{1}{2}mv^2$, for *v*

$2E = mv^2$

$\dfrac{2E}{m} = v^2$

$\sqrt{\dfrac{2E}{m}} = v$ or $v = \sqrt{\dfrac{2E}{m}}$

80. $\log(x+4) = \log 5 - \log x$

$\log(x+4) = \log \dfrac{5}{x}$

$x+4 = \dfrac{5}{x}$

$x(x+4) = 5$

$x^2 + 4x - 5 = 0$

$(x+5)(x-1) = 0$

$x+5 = 0 \quad \text{or} \quad x-1 = 0$

$x = -5 \qquad\qquad x = 1$

Upon checking, $x = -5$ does not satisfy the equation. The solution is $x = 1$.

Exercise Set 14.4

1. A nonlinear system of equations is a system in which at least one equation is nonlinear.

3. Yes

5. Yes

7. $x^2 + y^2 = 18$

$x + y = 0$

Solve $x + y = 0$ for x: $x = -y$.

Substitute $x = -y$ for x in $x^2 + y^2 = 18$.

$x^2 + y^2 = 18$

$(-y)^2 + y^2 = 18$

$y^2 + y^2 = 18$

$2y^2 = 18$

$y^2 = 9$

$y = \pm 3$

$y = 3 \quad \text{or} \quad y = -3$

$x = -3 \qquad\quad x = 3$

The solutions are $(3, -3)$ and $(-3, 3)$.

9. $x^2 + y^2 = 9$

$x + 2y = 3$

Solve $x + 2y = 3$ for x: $x = 3 - 2y$.

Substitute $3 - 2y$ for x in $x^2 + y^2 = 9$.

$x^2 + y^2 = 9$

$(3-2y)^2 + y^2 = 9$

$9 - 12y + 4y^2 + y^2 = 9$

$5y^2 - 12y = 0$

$y(5y - 12) = 0$

$y = 0 \qquad \text{or} \quad y = \dfrac{12}{5}$

$x = 3 - 2y$ $x = 3 - 2y$

$x = 3 - 2(0)$ $x = 3 - 2\left(\dfrac{12}{5}\right)$

$x = 3$ $x = -\dfrac{9}{5}$

The solutions are $(3, 0)$ and $\left(-\dfrac{9}{5}, \dfrac{12}{5}\right)$.

11. $y = x^2 - 5$

$3x + 2y = 10$

Substitute $x^2 - 5$ for y in $3x + 2y = 10$.

$3x + 2y = 10$

$3x + 2(x^2 - 5) = 10$

$3x + 2x^2 - 10 = 10$

$2x^2 + 3x - 20 = 0$

$(x+4)(2x-5) = 0$

$x = -4 \qquad \text{or} \qquad x = \dfrac{5}{2}$

$y = x^2 - 5$ $y = x^2 - 5$

$y = (-4)^2 - 5$ $y = \left(\dfrac{5}{2}\right)^2 - 5$

$y = 11$ $y = \dfrac{5}{4}$

The solutions are $(-4, 11)$ and $\left(\dfrac{5}{2}, \dfrac{5}{4}\right)$.

13. $x^2 + y = 6$

$y = x^2 + 4$

Substitute $x^2 + 4$ for y in $x^2 + y = 6$

$x^2 + x^2 + 4 = 6$

$2x^2 = 2$

$x^2 = 1$

$x = \pm 1$

$y = (1)^2 + 4$ or $y = (-1)^2 + 4$

$y = 1 + 4$ $\qquad$ $y = 1 + 4$

$y = 5$ $\qquad\qquad$ $y = 5$

The solutions are (1, 5) and (–1, 5).

15. $2x^2 + y^2 = 16$

$x^2 - y^2 = -4$

Solve $x^2 - y^2 = -4$ for y^2: $y^2 = x^2 + 4$

Substitute $x^2 + 4$ for y^2 in $2x^2 + y^2 = 16$.

$2x^2 + y^2 = 16$

$2x^2 + (x^2 + 4) = 16$

$3x^2 + 4 = 16$

$3x^2 = 12$

$x^2 = 4$

$x = 2$ $\qquad$ or $\qquad$ $x = -2$

$y^2 = x^2 + 4$	$y^2 = x^2 + 4$
$y^2 = 2^2 + 4$	$y = (-2)^2 + 4$
$y^2 = 8$	$y = 8$
$y = \pm\sqrt{8} = \pm 2\sqrt{2}$	$y = \pm\sqrt{8} = \pm 2\sqrt{2}$

The solutions are $\left(2, 2\sqrt{2}\right), \left(2, -2\sqrt{2}\right),$

$\left(-2, 2\sqrt{2}\right),$ and $\left(-2, -2\sqrt{2}\right).$

17. $x^2 + y^2 = 4$

$y = x^2 - 6 \implies x^2 = y + 6$

Substitute $y + 6$ for x^2 in $x^2 + y^2 = 4$.

$x^2 + y^2 = 4$

$(y + 6) + y^2 = 4$

$y^2 + y + 2 = 0$

$$y = \frac{-1 \pm \sqrt{1^2 - 4(1)(2)}}{2(1)}$$

$$= \frac{-1 \pm \sqrt{-7}}{2}$$

$$= \frac{-1 \pm i\sqrt{31}}{2}$$

There is no real solution.

19. $x^2 + y^2 = 9$

$y = x^2 - 3$

Solve $y = x^2 - 3$ for x^2: $x^2 = y + 3$.

Substitute $y + 3$ for x^2 in $x^2 + y^2 = 9$.

$x^2 + y^2 = 9$

$\left(y + 3\right) + y^2 = 9$

$y^2 + y - 6 = 0$

$\left(y - 2\right)\left(y + 3\right) = 0$

$y = 2$ $\qquad$ or $\qquad$ $y = -3$

$x^2 = y + 3$	$x^2 = y + 3$
$x^2 = 2 + 3$	$x^2 = -3 + 3$
$x^2 = 5$	$x^2 = 0$
$x = \pm\sqrt{5}$	$x = 0$

The solutions are $\left(0, -3\right), \left(\sqrt{5}, 2\right),$ and

$\left(-\sqrt{5}, 2\right).$

21. $2x^2 - y^2 = -8$

$x - y = 6$

Solve the second equation for y: $y = x - 6$.

Substitute $x - 6$ for y in $2x^2 - y^2 = -8$.

$$2x^2 - y^2 = -8$$
$$2x^2 - (x-6)^2 = -8$$
$$2x^2 - (x^2 - 12x + 36) = -8$$
$$2x^2 - x^2 + 12x - 36 = -8$$
$$x^2 + 12x - 28 = 0$$
$$(x-2)(x+14) = 0$$

or

$x = 2$ $\qquad\qquad$ $x = -14$

$y = x - 6$ $\qquad\quad$ $y = x - 6$

$y = 2 - 6$ $\qquad\quad$ $y = -14 - 6$

$y = -4$ $\qquad\qquad$ $y = -20$

The solutions are (2, −4) and (−14, −20).

23. $x^2 - y^2 = 4$

$\underline{2x^2 + y^2 = 8}$

$3x^2 \qquad\ = 12$

$x^2 = 4$

$x = 2$ $\qquad$ or $\qquad$ $x = -2$

$x^2 - y^2 = 4$ $\qquad\qquad$ $x^2 - y^2 = 4$

$2^2 - y^2 = 4$ $\qquad\qquad$ $(-2)^2 - y^2 = 4$

$y^2 = 0$ $\qquad\qquad\qquad$ $y^2 = 0$

$y = 0$ $\qquad\qquad\qquad$ $y = 0$

The solutions are (2, 0) and (−2, 0).

25. $x^2 + y^2 = 16$ (1)

$2x^2 - 5y^2 = 25$ (2)

$-2x^2 - 2y^2 = -32$ (1) multiplied by −2

$\underline{2x^2 - 5y^2 = 25}$ (2)

$-7y^2 = -7$

$y^2 = 1$

$y = 1$ $\qquad$ or $\qquad$ $y = -1$

$x^2 + y^2 = 16$ $\qquad\qquad$ $x^2 + y^2 = 16$

$x^2 + 1^2 = 16$ $\qquad\qquad$ $x^2 + (-1)^2 = 16$

$x^2 = 15$ $\qquad\qquad\qquad$ $x^2 = 15$

$x = \pm\sqrt{15}$ $\qquad\qquad$ $x = \pm\sqrt{15}$

The solutions are $\left(-\sqrt{15}, 1\right)$, $\left(-\sqrt{15}, -1\right)$,

$\left(\sqrt{15}, -1\right)$, and $\left(\sqrt{15}, 1\right)$.

27. $3x^2 - y^2 = 4$ (1)

$x^2 + 4y^2 = 10$ (2)

$12x^2 - 4y^2 = 16$ (1) multiplied by 4

$\underline{x^2 + 4y^2 = 10}$ (2)

$13x^2 = 26$

$x^2 = 2$

$x = \sqrt{2}$ $\qquad$ or $\qquad$ $x = -\sqrt{2}$

$3x^2 - y^2 = 4$ $\qquad\qquad$ $3x^2 - y^2 = 4$

$3(\sqrt{2})^2 - y^2 = 4$ $\qquad$ $3(-\sqrt{2})^2 - y^2 = 4$

$6 - y^2 = 4$ $\qquad\qquad$ $6 - y^2 = 4$

$y^2 = 2$ $\qquad\qquad\qquad$ $y^2 = 2$

$y = \pm\sqrt{2}$ $\qquad\qquad$ $y = \pm\sqrt{2}$

The solutions are $\left(\sqrt{2}, \sqrt{2}\right)$, $\left(\sqrt{2}, -\sqrt{2}\right)$,

$\left(-\sqrt{2}, \sqrt{2}\right)$, and $\left(-\sqrt{2}, -\sqrt{2}\right)$.

29. $4x^2 + 9y^2 = 36$

$\underline{2x^2 - 9y^2 = 18}$

$\qquad 6x^2 = 54$

$\qquad x^2 = 9$

$\qquad x = 3 \quad$ or $\qquad\qquad x = -3$

$4x^2 + 9y^2 = 36 \qquad\qquad 4x^2 + 9y^2 = 36$

$4(3)^2 + 9y^2 = 36 \qquad\quad 4(-3)^2 + 9y^2 = 36$

$\qquad 9y^2 = 0 \qquad\qquad\qquad\quad 9y^2 = 0$

$\qquad y^2 = 0 \qquad\qquad\qquad\quad\; y^2 = 0$

$\qquad y = 0 \qquad\qquad\qquad\quad\;\; y = 0$

The solutions are (3, 0) and (–3, 0).

31. $2x^2 - y^2 = 7 \quad$ (1)

$\quad x^2 + 2y^2 = 6 \quad$ (2)

$\quad 4x^2 - 2y^2 = 14 \quad$ (1) multiplied by 2

$\quad \underline{x^2 + 2y^2 = 6} \quad$ (2)

$\qquad 5x^2 = 20$

$\qquad x^2 = 4$

$\qquad\qquad\qquad$ or

$\qquad x = 2 \qquad\qquad\qquad x = -2$

$2(2)^2 - y^2 = 7 \qquad\quad 2(-2)^2 - y^2 = 7$

$\quad 8 - y^2 = 7 \qquad\qquad\quad 8 - y^2 = 7$

$\qquad y^2 = 1 \qquad\qquad\qquad\quad y^2 = 1$

$\qquad x = \pm 1 \qquad\qquad\qquad\; x = \pm 1$

The solutions are (2, 1), (2, –1), (–2, 1), and (–2, –1).

33. $\quad x^2 + y^2 = 25 \quad$ (1)

$\quad 2x^2 - 3y^2 = -30 \;$ (2)

$\quad 3x^2 + 3y^2 = 75 \quad$ (1) multiplied by 3

$\quad \underline{2x^2 - 3y^2 = -30} \;$ (2)

$\qquad 5x^2 = 45$

$\qquad x^2 = 9$

$\qquad x = 3 \qquad$ or $\qquad\qquad x = -3$

$x^2 + y^2 = 25 \qquad\qquad x^2 + y^2 = 25$

$(3)^2 + y^2 = 25 \qquad\quad (-3)^2 + y^2 = 25$

$\qquad y^2 = 16 \qquad\qquad\qquad y^2 = 16$

$\qquad y = \pm 4 \qquad\qquad\qquad y = \pm 4$

The solutions are (3, 4), (3, –4), (–3, 4), and (–3, –4).

35. $\quad x^2 + y^2 = 9 \quad$ (1)

$\quad 16x^2 - 4y^2 = 64 \;$ (2)

$\quad 4x^2 + 4y^2 = 36 \quad$ (1) multiplied by 4

$\quad \underline{16x^2 - 4y^2 = 64} \;$ (2)

$\qquad 20x^2 = 100$

$\qquad x^2 = 5$

$\qquad x = \sqrt{5} \qquad$ or $\qquad\qquad x = -\sqrt{5}$

$x^2 + y^2 = 9 \qquad\qquad\quad x^2 + y^2 = 9$

$\left(\sqrt{5}\right)^2 + y^2 = 9 \qquad\quad \left(-\sqrt{5}\right)^2 + y^2 = 9$

$\qquad y^2 = 4 \qquad\qquad\qquad\quad y^2 = 4$

$\qquad y = \pm 2 \qquad\qquad\qquad\quad y = \pm 2$

The solutions are $\left(\sqrt{5}, 2\right)$, $\left(\sqrt{5}, -2\right)$, $\left(-\sqrt{5}, 2\right)$,

and $\left(-\sqrt{5}, -2\right)$.

37. $x^2 + y^2 = 4$ (1)

$16x^2 + 9y^2 = 144$ (2)

Multiply the first equation by -16 and add.

$-16x^2 - 16y^2 = -164$

$\underline{16x^2 + 9y^2 = 144}$

$7y^2 = -20$

$y^2 = -\dfrac{20}{7}$

This equation has no real solutions so the system has no real solutions.

39. $x^2 + 4y^2 = 4$

$10y^2 - 9x^2 = 90$

Write the equations in standard form.

$x^2 + 4y^2 = 4$

$-9x^2 + 10y^2 = 90$

Multiply the first equation by -2 and the second equation by 2, then add.

$-5x^2 - 20y^2 = -20$

$\underline{-18x^2 + 20y^2 = 180}$

$-23x^2 = 160$

$x^2 = -\dfrac{160}{23}$

This equation has no real solutions so the system has no real solutions.

41. Answers will vary.

43. Let x = length; y = width

$xy = 440$

$2x + 2y = 84$

Solve $2x + 2y = 84$ for y: $y = 42 - x$.

Substitute $42 - x$ for y in $xy = 440$.

$xy = 440$

$x(42 - x) = 440$

$42x - x^2 = 440$

$x^2 - 42x + 440 = 0$

$(x - 20)(x - 22) = 0$

$x - 20 = 0$ or $x - 22 = 0$

$x = 20$ $\qquad$ $x = 22$

$y = 42 - x$ or $y = 42 - 22$

$y = 42 - 20$ $\qquad$ $y = 42 - 22$

$y = 22$ $\qquad$ $y = 20$

The solutions are (20, 22) and (22, 20).

The dimensions of the floor are 20 m by 22 m.

45. Let x = length

$\quad\;\; y$ = width

$\quad\;\; xy = 270$

$2x + 2y = 78$

Solve $2x + 2y = 78$ for y: $y = 39 - x$.

Substitute $39 - x$ for y in $xy = 270$.

$xy = 270$

$x(39 - x) = 270$

$39x - x^2 = 270$

$x^2 - 39x + 270 = 0$

$(x - 9)(x - 30) = 0$

$x - 9 = 0$ $\qquad$ or $\qquad$ $x - 30 = 0$

$x = 9$ $\qquad\qquad\qquad$ $x = 30$

$y = 39 - x$ $\qquad\qquad$ $y = 39 - x$

$y = 39 - 9$ $\qquad\qquad$ $y = 39 - 30$

$y = 30$ $\qquad\qquad\qquad$ $y = 9$

The solutions are (9, 30) and (30, 9). The dimensions of the garden are 9 ft by 30 ft.

47. Let x = length; $\quad y$ = width

$\qquad xy = 112$

$x^2 + y^2 = \left(\sqrt{260}\right)^2$

Solve $xy = 112$ for y: $y = \dfrac{112}{x}$.

$x^2 + y^2 = 260$

$x^2 + \left(\dfrac{112}{x}\right)^2 = 260$

$x^2 + \dfrac{12,544}{x^2} = 260$

$x^4 + 12,544 = 260x^2$

$x^4 - 260x^2 + 12,544 = 0$

$(x^2 - 64)(x^2 - 196) = 0$

$x^2 - 64 = 0$ or $x^2 - 196 = 0$

$x^2 = 64$ $\qquad$ $x^2 = 196$

$x = \pm 8$ $\qquad$ $x = \pm 14$

Since x must be positive, $x = 8$ or $x = 14$.

If $x = 8$, then $y = \dfrac{112}{8} = 14$.

If $x = 14$, then $y = \dfrac{112}{14} = 8$.

The dimensions of the new bill are 8 cm by 14 cm.

49. Let $x = $ length

$y = $ width

$x^2 + y^2 = 34^2$

$x + y + 34 = 80$

Solve $x + y + 34 = 80$ for y: $y = 46 - x$.

Substitute $46 - x$ for y in $x^2 + y^2 = 34^2$.

$$x^2 + y^2 = 34^2$$

$$x^2 + (46 - x)^2 = 34^2$$

$$x^2 + (2116 - 92x + x^2) = 1156$$

$$2x^2 - 92 + 2116 = 1156$$

$$2x^2 - 92x + 960 = 0$$

$$x^2 - 46x + 480 = 0$$

$$(x - 16)(x - 30) = 0$$

$$x - 16 = 0 \qquad \text{or} \qquad x - 30 = 0$$

$$x = 16 \qquad\qquad\qquad x = 30$$

$$y = 46 - x \qquad\qquad y = 46 - x$$

$$y = 46 - 16 \qquad\qquad y = 46 - 30$$

$$y = 30 \qquad\qquad\qquad y = 16$$

The solutions are (16, 30) and (30, 16).
The dimensions of the piece of wood are 16 in. by 30 in.

51. $d = -16t^2 + 64t$

$d = -16t^2 + 16t + 80$

Substitute $-16t^2 + 64t$ for d in

$d = -16t^2 + 16t + 80$.

$$d = -16t^2 + 16t + 80$$

$$-16t^2 + 64t = -16t^2 + 16t + 80$$

$$64t = 16t + 80$$

$$48t = 80$$

$$t = \frac{80}{48} = \frac{5}{3} \approx 1.67$$

The balls are the same height above the ground at $t \approx 1.67$ sec.

53. Since $t = 1$ year, we may write the formula as
$i = pr$.

$7.50 = pr$

$7.50 = (p + 25)(r - 0.01)$

Rewrite the second equation by multiplying the binomials. Then substitute 7.50 for pr and solve for r.

$$7.50 = (p + 25)(r - 0.01)$$

$$7.50 = pr - 0.01p + 25r - 0.25$$

$$7.50 = 7.50 - 0.01p + 25r - 0.25$$

$$0 = -0.01p + 25r - 0.25$$

$$0.01p + 0.25 = 25r$$

$$\frac{0.01p}{25} + \frac{0.25}{25} = \frac{25r}{25}$$

$$r = 0.0004p + 0.01$$

Substitute $0.0004p + 0.01$ for r in $7.50 = pr$.

$$7.50 = pr$$

$$7.50 = p(0.0004p + 0.01)$$

$$7.50 = 0.0004p^2 + 0.01p$$

$$0 = 0.0004p^2 + 0.01p - 7.50$$

$$0 = p^2 + 25p - 18,750$$

$$0 = (p - 125)(p + 150)$$

$$p - 125 = 0 \qquad \text{or} \qquad p + 150 = 0$$

$$p = 125 \qquad\qquad p = -150$$

Since the principal must be positive, use $p = 125$.

$r = 0.0004p + 0.01$

$r = 0.0004(125) + 0.01$

$r = 0.06$

The principal is \$125 and the interest rate is 6%.

55. $C = 10x + 300$

$R = 30x - 0.1x^2$

$$C = R$$

$$10x + 300 = 30x - 0.1x^2$$

$$0.1x^2 - 20x + 300 = 0$$

$$x = \frac{-b \pm \sqrt{b^2 - 4ac}}{2a}$$

$$= \frac{-(-20) \pm \sqrt{(-20)^2 - 4(0.1)(300)}}{2(0.1)}$$

$$= \frac{20 \pm \sqrt{280}}{0.2}$$

$$x = \frac{20 + \sqrt{280}}{0.2} \approx 183.7$$

or

$$x = \frac{20 - \sqrt{280}}{0.2} \approx 16.3$$

The break-even points are ≈ 16 and ≈ 184.

57. $C = 12.6x + 150$

$R = 42.8x - 0.3x^2$

$$C = R$$

$$12.6x + 150 = 42.8x - 0.3x^2$$

$$0.3x^2 - 30.2x + 150 = 0$$

$$x = \frac{-b \pm \sqrt{b^2 - 4ac}}{2a}$$

$$= \frac{-(-30.2) \pm \sqrt{(-30.2)^2 - 4(0.3)(150)}}{2(0.3)}$$

$$= \frac{30.2 \pm \sqrt{732.04}}{0.6}$$

$$x = \frac{30.2 + \sqrt{732.04}}{0.6} \approx 95.4$$

or

$$x = \frac{30.2 - \sqrt{732.04}}{0.6} \approx 5.2$$

The break-even points are ≈ 5 and ≈ 95.

59. Solve each equation for y.

$3x - 5y = 12 \qquad\qquad x^2 + y^2 = 10$

$-5y = -3x + 12 \qquad\quad y^2 = 10 - x^2$

$y = \frac{3}{5}x - \frac{12}{5} \qquad\qquad y = \pm\sqrt{10 - x^2}$

Use $y_1 = \frac{3}{5}x - \frac{12}{5}$, $y_2 = \sqrt{10 - x^2}$, and

$y_2 = -\sqrt{10 - x^2}$.

$-9.4,\ 9.4,\ 1,\ -6.2,\ 6.2,\ 10$

Approximate solutions: $(-1, -3)$, $(3.12, -0.53)$

61. Let x = length of one leg

$\qquad\quad y$ = length of other leg

$$x^2 + y^2 = 26^2$$

$$\frac{1}{2}xy = 120$$

Solve $\frac{1}{2}xy = 120$ for y: $y = \frac{240}{x}$.

Substitute $\frac{240}{x}$ for y in $x^2 + y^2 = 26^2$

$$x^2 + y^2 = 26^2$$

$$x^2 + \left(\frac{240}{x}\right)^2 = 676$$

$$x^2 + \frac{57,600}{x^2} = 676$$

$$x^4 + 57,600 = 676x^2$$

$$x^4 - 676x^2 + 57,600 = 0$$

$$\left(x^2 - 100\right)\left(x^2 - 576\right) = 0$$

$x^2 - 100 = 0 \qquad$ or $\quad x^2 - 576 = 0$

$\qquad x^2 = 100 \qquad\qquad\quad x^2 = 576$

$\qquad x = \pm 10 \qquad\qquad\quad x = \pm 24$

Since x is a length, x must be positive.

If $x = 10$, then $y = \frac{240}{10} = 24$. If $x = 24$, then

$y = \frac{240}{24} = 10$. The legs have lengths 10 yards

and 24 yards.

63. The operations are evaluated in the following order: parentheses, exponents, multiplication or division, addition or subtraction.

64. $(x+1)^3 + 1$

$$= (x+1)^3 + (1)^3$$

$$= (x+1+1)\left((x+1)^2 - (x+1)(1) + (1)^2\right)$$

$$= (x+2)\left(x^2 + 2x + 1 - x - 1 + 1\right)$$

$$= (x+2)\left(x^2 + x + 1\right)$$

65. $x = \dfrac{k}{P^2}$

$10 = \dfrac{k}{6^2} \;\Rightarrow\; k = 360 \;\Rightarrow\; x = \dfrac{360}{P^2}$

$x = \dfrac{360}{20^2} = \dfrac{360}{400} = \dfrac{9}{10} \;$ or $\; 0.9$

66. $\dfrac{5}{\sqrt{x+2}-3} = \dfrac{5}{\sqrt{x+2}-3} \cdot \dfrac{\sqrt{x+2}+3}{\sqrt{x+2}+3}$

$\quad = \dfrac{5\sqrt{x+2}+15}{x+2+3\sqrt{x+2}-3\sqrt{x+2}-9}$

$\quad = \dfrac{5\sqrt{x+2}+15}{x-7}$

67. $A = A_0 e^{kt}$, for k

$\dfrac{A}{A_0} = e^{kt}$

$\ln \dfrac{A}{A_0} = \ln e^{kt}$

$\ln A - \ln A_0 = kt$

$\dfrac{\ln A - \ln A_0}{t} = k$

Chapter 14 Review Exercises

1. $d = \sqrt{\left(x_2 - x_1\right)^2 + \left(y_2 - y_1\right)^2}$

$\quad = \sqrt{\left(5-0\right)^2 + \left(-12-0\right)^2}$

$\quad = \sqrt{5^2 + \left(-12\right)^2}$

$\quad = \sqrt{25+144}$

$\quad = \sqrt{169}$

$\quad = 13$

Midpoint $= \left(\dfrac{x_1 + x_2}{2}, \dfrac{y_1 + y_2}{2} \right)$

$\quad = \left(\dfrac{0+5}{2}, \dfrac{0+\left(-12\right)}{2} \right)$

$\quad = \left(\dfrac{5}{2}, -6 \right)$

2. $d = \sqrt{\left(x_2 - x_1\right)^2 + \left(y_2 - y_1\right)^2}$

$\quad = \sqrt{\left(-1-\left(-4\right)\right)^2 + \left(5-1\right)^2}$

$\quad = \sqrt{3^2 + 4^2}$

$\quad = \sqrt{9+16}$

$\quad = \sqrt{25}$

$\quad = 5$

Midpoint $= \left(\dfrac{x_1 + x_2}{2}, \dfrac{y_1 + y_2}{2} \right)$

$\quad = \left(\dfrac{-4+\left(-1\right)}{2}, \dfrac{1+5}{2} \right)$

$\quad = \left(-\dfrac{5}{2}, 3 \right)$

3. $d = \sqrt{\left(x_2 - x_1\right)^2 + \left(y_2 - y_1\right)^2}$

$\quad = \sqrt{\left[-1-\left(-9\right)\right]^2 + \left[10-\left(-5\right)\right]^2}$

$\quad = \sqrt{\left(8\right)^2 + 15^2}$

$\quad = \sqrt{64+225}$

$\quad = \sqrt{289}$

$\quad = 17$

Midpoint $= \left(\dfrac{x_1 + x_2}{2}, \dfrac{y_1 + y_2}{2} \right)$

$\quad = \left(\dfrac{-9+\left(-1\right)}{2}, \dfrac{-5+10}{2} \right)$

$\quad = \left(-5, \dfrac{5}{2} \right)$

4. $d = \sqrt{(x_2 - x_1)^2 + (y_2 - y_1)^2}$

$= \sqrt{\left[-2 - (-4)\right]^2 + (5 - 3)^2}$

$= \sqrt{2^2 + 2^2}$

$= \sqrt{4 + 4}$

$= \sqrt{8}$

≈ 2.83

$\text{Midpoint} = \left(\dfrac{x_1 + x_2}{2}, \dfrac{y_1 + y_2}{2} \right)$

$= \left(\dfrac{-4 + (-2)}{2}, \dfrac{3 + 5}{2} \right)$

$= (-3, 4)$

5. $y = (x - 2)^2 + 1$

This is a parabola in the form $y = a(x - h)^2 + k$ with $a = 1$, $h = 2$, and $k = 1$. Since $a > 0$, the parabola opens upward. The vertex is (2, 1). The y-intercept is (0, 5).
There are no x-intercepts.

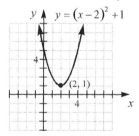

6. $y = (x + 3)^2 - 4$

This is a parabola in the form $y = a(x - h)^2 + k$ with $a = 1$, $h = -3$, and $k = -4$. Since $a > 0$, the parabola opens upward. The vertex is (-3, -4). The y-intercept is (0, 5).
The x-intercepts are about (-5, 0) and (-1, 0).

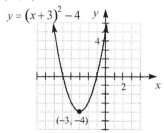

7. $x = (y - 1)^2 + 4$

This is a parabola in the form $x = a(y - k)^2 + h$ with $a = 1$, $h = 4$, and $k = 1$. Since $a > 0$, the parabola opens to the right. The vertex is (4, 1). There are no y-intercepts.
The x-intercept is (5, 0).

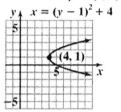

8. $x = -2(y + 4)^2 - 3$

This is a parabola in the form $x = a(y - k)^2 + h$ with $a = -2$, $h = -3$, and $k = -4$. Since $a < 0$, the parabola opens to the left. The vertex is (-3, -4). There are no y-intercepts.
The x-intercept is (-35, 0).

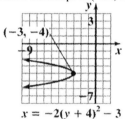

9. a. $y = x^2 - 8x + 22$

$y = \left(x^2 - 8x + 16\right) + 22 - 16$

$y = \left(x - 4\right)^2 + 6$

b. This is a parabola in the form $y = a(x - h)^2 + k$ with $a = 1$, $h = 4$, and $k = 6$. Since $a > 0$, the parabola opens upward. The vertex is (4, 6). The y-intercept is (0, 22). There are no y-intercepts.

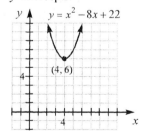

10. a.
$$x = -y^2 - 2y + 5$$
$$x = -(y^2 + 2y) + 5$$
$$x = -(y^2 + 2y + 1) + 1 + 5$$
$$x = -(y+1)^2 + 6$$

b. This is a parabola in the form $x = a(y-k)^2 + h$ with $a = -1$, $h = 6$, and $k = -1$. Since $a < 0$, the parabola opens to the left. The vertex is $(6, -1)$.
The y-intercepts are about $(0, -3.45)$ and $(0, 1.45)$. The x-intercept is $(5, 0)$.

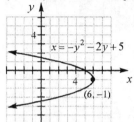

11. a.
$$x = y^2 + 5y + 4$$
$$x = \left(y^2 + 5y + \frac{25}{4}\right) - \frac{25}{4} + 4$$
$$x = \left(y + \frac{5}{2}\right)^2 - \frac{9}{4}$$

b. This is a parabola in the form $x = a(y-k)^2 + h$ with $a = 1$, $h = -\frac{9}{4}$, and $k = -\frac{5}{2}$. Since $a > 0$, the parabola opens to the right. The vertex is $\left(-\frac{9}{4}, -\frac{5}{2}\right)$.
The y-intercepts are $(0, -4)$ and $(0, -1)$. The x-intercept is $(4, 0)$.

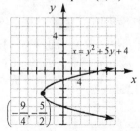

12. a.
$$y = 2x^2 - 8x - 24$$
$$y = 2(x^2 - 4x) - 24$$
$$y = 2(x^2 - 4x + 4) - 8 - 24$$
$$y = 2(x-2)^2 - 32$$

b. This is a parabola in the form $y = a(x-h)^2 + k$ with $a = 2$, $h = 2$, and $k = -32$. Since $a > 0$, the parabola opens upward. The vertex is $(2, -32)$. The y-intercept is $(0, -24)$. The x-intercepts are $(-2, 0)$ and $(6, 0)$.

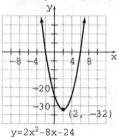

13. a.
$$(x-h)^2 + (y-k)^2 = r^2$$
$$(x-0)^2 + (y-0)^2 = 4^2$$
$$x^2 + y^2 = 4^2$$

b. The graph is a circle with center $(0, 0)$ and radius 4.

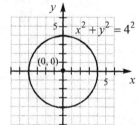

14. a.
$$(x-h)^2 + (y-k)^2 = r^2$$
$$\left[x-(-3)\right]^2 + (y-4)^2 = 1^2$$
$$(x+3)^2 + (y-4)^2 = 1^2$$

b. The graph is a circle with center $(-3,4)$ and radius 1.

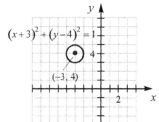

15. a.
$$x^2 + y^2 - 4y = 0$$
$$x^2 + (y^2 - 4y + 4) = 4$$
$$x^2 + (y-2)^2 = 2^2$$

b. The graph is a circle with center $(0, 2)$ and radius 2.

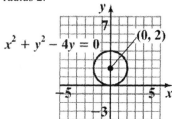

16. a.
$$x^2 + y^2 - 2x + 6y + 1 = 0$$
$$x^2 - 2x + y^2 + 6y = -1$$
$$(x^2 - 2x + 1) + (y^2 + 6y + 9) = -1 + 1 + 9$$
$$(x-1)^2 + (y+3)^2 = 9$$
$$(x-1)^2 + (y+3)^2 = 3^2$$

b. The graph is a circle with center $(1, -3)$ and radius 3.

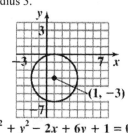

17. a.
$$x^2 - 8x + y^2 - 10y + 40 = 0$$
$$(x^2 - 8x + 16) + (y^2 - 10y + 25) = -40 + 16 + 25$$
$$(x-4)^2 + (y-5)^2 = 1$$
$$(x-4)^2 + (y-5)^2 = 1^2$$

b. The graph is a circle with center $(4, 5)$ and radius 1.

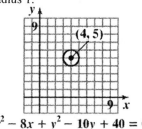

18. a.
$$x^2 + y^2 - 4x + 10y + 17 = 0$$
$$x^2 - 4x + y^2 + 10y = -17$$
$$(x^2 - 4x + 4) + (y^2 + 10y + 25) = -17 + 4 + 25$$
$$(x-2)^2 + (y+5)^2 = 12$$
$$(x-2)^2 + (y+5)^2 = \left(\sqrt{12}\right)^2$$

b. The graph is a circle with center $(2, -5)$ and radius $\sqrt{12} \approx 3.46$.

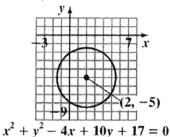

19. $y = \sqrt{9 - x^2}$

If we solve $x^2 + y^2 = 9$ for y, we obtain $y = \pm\sqrt{9 - x^2}$. Therefore, the graph of $y = \sqrt{9 - x^2}$ is the upper half $(y \geq 0)$ of a circle with its center at the origin and radius 4.

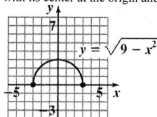

20. $y = -\sqrt{36 - x^2}$

If we solve $x^2 + y^2 = 36$ for y, we obtain

$y = \pm\sqrt{36 - x^2}$. Therefore, the graph of

$y = -\sqrt{36 - x^2}$ is the lower half $(y \le 0)$ of a

circle with its center at the origin and radius 6.

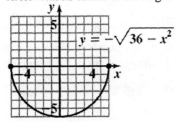

21. The center is $(-1, 1)$ and the radius is 2.

$$(x - h)^2 + (y - k)^2 = r^2$$
$$\left[x - (-1)\right]^2 + (y - 1)^2 = 2^2$$
$$(x + 1)^2 + (y - 1)^2 = 4$$

22. The center is $(5, -3)$ and the radius is 3.

$$(x - h)^2 + (y - k)^2 = r^2$$
$$(x - 5)^2 + \left[y - (-3)\right]^2 = 3^2$$
$$(x - 5)^2 + (y + 3)^2 = 9$$

23. $\dfrac{x^2}{4} + \dfrac{y^2}{9} = 1$

Since $a^2 = 4$, $a = 2$.

Since $b^2 = 9$, $b = 3$.

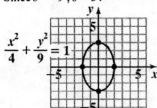

24. $\dfrac{x^2}{36} + \dfrac{y^2}{64} = 1$

Since $a^2 = 36$, $a = 6$.

Since $b^2 = 64$, $b = 8$.

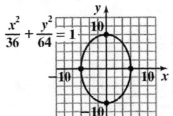

25. $4x^2 + 9y^2 = 36$

$$\frac{4x^2}{36} + \frac{9y^2}{36} = 1$$
$$\frac{x^2}{9} + \frac{y^2}{4} = 1$$

Since $a^2 = 9$, $a = 3$.

Since $b^2 = 4$, $b = 2$.

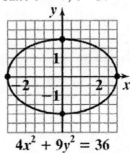

$$4x^2 + 9y^2 = 36$$

26. $9x^2 + 16y^2 = 144$

$$\frac{9x^2}{144} + \frac{16y^2}{144} = 1$$
$$\frac{x^2}{16} + \frac{y^2}{9} = 1$$

Since $a^2 = 16$, $a = 4$.

Since $b^2 = 9$, $b = 3$.

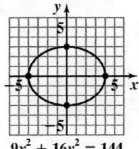

$$9x^2 + 16y^2 = 144$$

27. $\dfrac{(x-3)^2}{16}+\dfrac{(y+2)^2}{4}=1$

The center is $(3, -2)$.

Since $a^2 = 16$, $a = 4$.

Since $b^2 = 4$, $b = 2$.

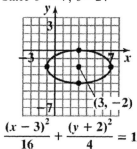

$$\dfrac{(x-3)^2}{16}+\dfrac{(y+2)^2}{4}=1$$

28. $\dfrac{(x+3)^2}{9}+\dfrac{y^2}{25}=1$

The center is $(-3, 0)$.

Since $a^2 = 9$, $a = 3$.

Since $b^2 = 25$, $b = 5$.

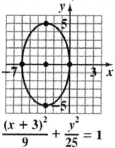

$$\dfrac{(x+3)^2}{9}+\dfrac{y^2}{25}=1$$

29. $25(x-2)^2+9(y-1)^2=225$

$$\dfrac{25(x-2)^2}{225}+\dfrac{9(y-1)^2}{225}=1$$

$$\dfrac{(x-2)^2}{9}+\dfrac{(y-1)^2}{25}=1$$

The center is $(2, 1)$.

Since $a^2 = 9$, $a = 3$.

Since $b^2 = 25$, $b = 5$.

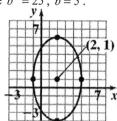

$$25(x-2)^2+9(y-1)^2=225$$

30. $\dfrac{x^2}{4}+\dfrac{y^2}{9}=1$

Since $a^2 = 4$, $a = 2$.

Since $b^2 = 9$, $b = 3$.

Area $= \pi ab = \pi(2)(3) = 6\pi \approx 18.85$ sq. units

31. a. $\dfrac{x^2}{4}-\dfrac{y^2}{16}=1$

Since $a^2 = 4$ and $b^2 = 16$, $a = 2$ and $b = 4$. The equations of the asymptotes are $y = \pm\dfrac{b}{a}x$, or $y = \pm 2x$.

b. To graph the asymptotes, plot the points $(2, 4)$, $(-2, 4)$, $(2, -4)$, and $(-2, -4)$. The graph intersects the x-axis at $(-2, 0)$ and $(2, 0)$.

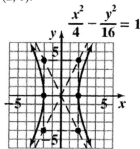

32. a. $\dfrac{x^2}{4}-\dfrac{y^2}{4}=1$

Since $a^2 = 4$ and $b^2 = 4$, $a = 2$ and $b = 2$. The equations of the asymptotes are $y = \pm\dfrac{b}{a}x$, or $y = \pm\dfrac{2}{2}x = \pm x$.

b. To graph the asymptotes, plot the points $(2, 2)$, $(-2, 2)$, $(2, -2)$, and $(-2, -2)$. The graph intersects the x-axis at $(-2, 0)$ and $(2, 0)$.

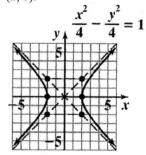

33. a. $\dfrac{y^2}{4} - \dfrac{x^2}{36} = 1$

Since $a^2 = 36$ and $b^2 = 4$, $a = 6$ and $b = 2$. The equations of the asymptotes are $y = \pm\dfrac{b}{a}x$, or $y = \pm\dfrac{1}{3}$.

b. To graph the asymptotes, plot the points $(6, 2)$, $(6, -2)$, $(-6, 2)$, and $(-6, -2)$. The graph intersects the y-axis at $(0, -2)$ and $(0, 2)$.

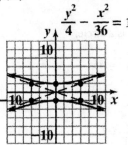

34. a. $\dfrac{y^2}{25} - \dfrac{x^2}{16} = 1$

Since $a^2 = 16$ and $b^2 = 25$, $a = 4$ and $b = 5$. The equations of the asymptotes are $y = \pm\dfrac{b}{a}x$, or $y = \pm\dfrac{5}{4}x$.

b. To graph the asymptotes, plot the points $(4, 5)$, $(4, -5)$, $(-4, 5)$, and $(-4, -5)$. The graph intersects the y-axis at $(0, -5)$ and $(0, 5)$.

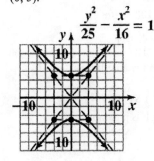

35. a. $x^2 - 9y^2 = 9$

$\dfrac{x^2}{9} - \dfrac{9y^2}{9} = 1$

$\dfrac{x^2}{9} - \dfrac{y^2}{1} = 1$

b. Since $a^2 = 9$ and $b^2 = 1$, $a = 3$ and $b = 1$. The equations of the asymptotes are $y = \pm\dfrac{b}{a}x$, or $y = \pm\dfrac{1}{3}x$.

c. To graph the asymptotes, plot the points $(3, 1)$, $(-3, 1)$, $(3, -1)$, and $(-3, -1)$. The graph intersects the x-axis at $(-3, 0)$ and $(3, 0)$.

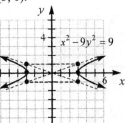

36. a. $25x^2 - 16y^2 = 400$

$\dfrac{25x^2}{400} - \dfrac{16y^2}{400} = 1$

$\dfrac{x^2}{16} - \dfrac{y^2}{25} = 1$

b. Since $a^2 = 16$ and $b^2 = 25$, $a = 4$ and $b = 5$. The equations of the asymptotes are $y = \pm\dfrac{b}{a}x$, or $y = \pm\dfrac{5}{4}x$.

c. To graph the asymptotes, plot the points $(4, 5)$, $(-4, 5)$, $(4, -5)$, and $(-4, -5)$. The graph intersects the x-axis at $(-4, 0)$ and $(4, 0)$.

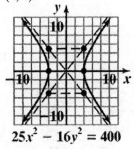

37. a. $4y^2 - 25x^2 = 100$

$\dfrac{4y^2}{100} - \dfrac{25x^2}{100} = 1$

$\dfrac{y^2}{25} - \dfrac{x^2}{4} = 1$

b. Since $a^2 = 4$ and $b^2 = 25$, $a = 2$ and $b = 5$. The equations of the asymptotes are

$$y = \pm \frac{b}{a}x, \text{ or } y = \pm \frac{5}{2}x.$$

c. To graph the asymptotes, plot the points $(2, 5)$, $(2, -5)$, $(-2, 5)$, and $(-2, -5)$. The graph intersects the y-axis at $(0, -5)$ and $(0, 5)$.

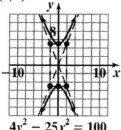

$$4y^2 - 25x^2 = 100$$

38. a. $49y^2 - 9x^2 = 441$

$$\frac{49y^2}{441} - \frac{9x^2}{441} = 1$$

$$\frac{y^2}{9} - \frac{x^2}{49} = 1$$

b. Since $a^2 = 49$ and $b^2 = 9$, $a = 7$ and $b = 3$. The equations of the asymptotes are

$$y = \pm \frac{b}{a}x, \text{ or } y = \pm \frac{3}{7}x.$$

c. To graph the asymptotes, plot the points $(7, 3)$, $(-7, 3)$, $(7, -3)$, and $(-7, -3)$. The graph intersects the y-axis at $(0, -3)$ and $(0, 3)$.

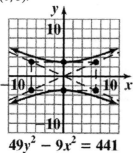

$$49y^2 - 9x^2 = 441$$

39. $\dfrac{x^2}{49} - \dfrac{y^2}{16} = 1$

The graph is a hyperbola.

40. $4x^2 + 8y^2 = 32$

$$\frac{4x^2}{32} + \frac{8y^2}{32} = \frac{32}{32}$$

$$\frac{x^2}{8} + \frac{y^2}{4} = 1$$

The graph is an ellipse.

41. $5x^2 + 5y^2 = 125$

$$\frac{5x^2}{5} + \frac{5y^2}{5} = \frac{125}{5}$$

$$x^2 + y^2 = 25$$

The graph is a circle.

42. $4x^2 - 25y^2 = 25$

$$\frac{4x^2}{25} - \frac{25y^2}{25} = \frac{25}{25}$$

$$\frac{x^2}{6.25} - \frac{y^2}{1} = 1$$

The graph is a hyperbola.

43. $\dfrac{x^2}{18} + \dfrac{y^2}{9} = 1$

The graph is an ellipse.

44. $y = (x - 2)^2 + 1$

The graph is a parabola.

45. $12x^2 + 9y^2 = 108$

$$\frac{12x^2}{108} + \frac{9y^2}{108} = \frac{108}{108}$$

$$\frac{x^2}{9} + \frac{y^2}{12} = 1$$

The graph is an ellipse.

46. $x = -y^2 + 8y - 9$

The graph is a parabola.

47. $x^2 + 2y^2 = 25$

$x^2 - 3y^2 = 25 \implies x^2 = 3y^2 + 25$

Substitute $3y^2 + 25$ for x^2 in $x^2 + 2y^2 = 25$.

$x^2 + 2y^2 = 25$

$3y^2 + 25 + 2y^2 = 25$

$5y^2 = 0$

$y^2 = 0$

Substitute 0 for y^2 in $x^2 - 3y^2 = 25$

$x^2 - 0 = 25 \implies x = \pm 5$

The solutions are (5, 0) and (–5, 0).

48. $x^2 = y^2 + 4$

$x + y = 4$

Solve $x + y = 4$ for y: $y = 4 - x$.

Substitute $4 - x$ for y in $x^2 = y^2 + 4$.

$x^2 = y^2 + 4$

$x^2 = (4 - x)^2 + 4$

$x^2 = (16 - 8x + x^2) + 4$

$8x - 16 = 4$

$8x = 20$

$x = \dfrac{5}{2}$

$y = 4 - x$

$y = 4 - \dfrac{5}{2}$

$y = \dfrac{3}{2}$

The solution is $\left(\dfrac{5}{2}, \dfrac{3}{2} \right)$.

49. $x^2 + y^2 = 9$

$y = 3x + 9$

Substitute $3x + 9$ for y in $x^2 + y^2 = 9$.

$x^2 + y^2 = 9$

$x^2 + (3x + 9)^2 = 9$

$x^2 + 9x^2 + 54x + 81 = 9$

$10x^2 + 54x + 72 = 0$

$5x^2 + 27x + 36 = 0$

$(x + 3)(5x + 12) = 0$

$x + 3 = 0 \qquad$ or $\quad 5x + 12 = 0$

$x = -3 \qquad\qquad\qquad x = -\dfrac{12}{5}$

$y = 3x + 9$

$y = 3(-3) + 9 \qquad\qquad y = 3x + 9$

$y = 0 \qquad\qquad\qquad y = 3\left(-\dfrac{12}{5} \right) + 9$

$\qquad\qquad\qquad\qquad\qquad y = \dfrac{9}{5}$

The solutions are (–3, 0) and $\left(-\dfrac{12}{5}, \dfrac{9}{5} \right)$.

50. $x^2 + 2y^2 = 9$

$x^2 - 6y^2 = 36$

Solve $x^2 + 2y^2 = 9$ for x^2: $x^2 = 9 - 2y^2$.

Substitute $9 - 2y^2$ for x^2 in $x^2 - 6y^2 = 36$.

$x^2 - 6y^2 = 36$

$(9 - 2y^2) - 6y^2 = 36$

$9 - 8y^2 = 36$

$-8y^2 = 27$

$y^2 = -\dfrac{27}{8}$

There is no real solution to this equation so the system has no real solution.

51. $x^2 + y^2 = 36$

$\underline{x^2 - y^2 = 36}$

$2x^2 = 72$

$x^2 = 36$

$x = 6 \qquad$ or $\qquad\qquad x = -6$

$x^2 + y^2 = 36 \qquad\qquad x^2 + y^2 = 36$

$6^2 + y^2 = 36 \qquad\qquad (-6)^2 + y^2 = 36$

$y^2 = 0 \qquad\qquad\qquad y^2 = 0$

$y = 0 \qquad\qquad\qquad y = 0$

The solutions are (6, 0) and (–6, 0).

52.
$$x^2 + y^2 = 25 \ (1)$$
$$x^2 - 2y^2 = -2 \ (2)$$
$$2x^2 + 2y^2 = 50 \ \text{(1) multiplied by 2}$$
$$\underline{x^2 - 2y^2 = -2 \ (2)}$$
$$3x^2 = 48$$
$$x^2 = 16$$
$$x = 4 \quad \text{or} \quad x = -4$$

$$x^2 + y^2 = 25 \qquad x^2 + y^2 = 25$$
$$4^2 + y^2 = 25 \qquad (-4)^2 + y^2 = 25$$
$$y^2 = 9 \qquad\qquad y^2 = 9$$
$$y = \pm 3 \qquad\qquad y = \pm 3$$

The solutions are (4, 3), (4, –3), (–4, 3) and (–4, –3).

53.
$$-4x^2 + y^2 = -15 \ \ (1)$$
$$8x^2 + 3y^2 = -5 \ \ (2)$$
$$-8x^2 + 2y^2 = -30 \ \ \text{(1) multiplied by 2}$$
$$\underline{8x^2 + 3y^2 = -5 \ \ (2)}$$
$$5y^2 = -35$$
$$y^2 = -7$$

This equation has no real solution so there is no real solution to the system.

54.
$$3x^2 + 2y^2 = 6 \quad (1)$$
$$4x^2 + 5y^2 = 15 \quad (2)$$
$$-12x^2 - 8y^2 = -24 \quad \text{(1) multiplied by } -4$$
$$\underline{12x^2 + 15y^2 = 45 \quad \text{(2) multiplied by 3}}$$
$$7y^2 = 21$$
$$y^2 = 3$$
$$y = \pm\sqrt{3}$$
$$3x^2 + 2y^2 = 6$$
$$3x^2 + 2(3) = 6$$
$$3x^2 = 0$$
$$x^2 = 0$$

The solutions are $\left(0, \sqrt{3}\right)$ and $\left(0, -\sqrt{3}\right)$.

55. Let $x =$ length
$$y = \text{width}$$
$$xy = 45$$
$$2x + 2y = 28$$
Solve $2x + 2y = 28$ for y: $y = 14 - x$.
Substitute $14 - x$ for y in $xy = 45$.
$$xy = 45$$
$$x(14 - x) = 45$$
$$14x - x^2 = 45$$
$$x^2 - 14x + 45 = 0$$
$$(x - 5)(x - 9) = 0$$
$$x - 5 = 0 \quad \text{or} \quad x - 9 = 0$$
$$x = 5 \qquad\qquad x = 9$$

$$y = 14 - x \qquad y = 14 - x$$
$$y = 14 - 5 \qquad y = 14 - 9$$
$$y = 9 \qquad\qquad y = 5$$

The solutions are (5, 9) and (9, 5).
The dimensions of the pool table are 5 feet by 9 feet.

56.
$$C = 20.3x + 120$$
$$R = 50.2x - 0.2x^2$$
$$C = R$$
$$20.3x + 120 = 50.2x - 0.2x^2$$
$$0.2x^2 - 29.9 + 120 = 0$$
$$x = \frac{29.9 \pm \sqrt{(-29.9)^2 - 4(0.2)(120)}}{2(0.2)}$$
$$x = \frac{29.9 \pm \sqrt{798.01}}{0.4}$$
$$x \approx 145 \ \text{or} \ 4$$

The company must sell either 4 bottles or 145 bottles to break even.

57. Since $t = 1$ year, we may rewrite the formula $i = prt$ as $i = pr$.

$120 = pr$

$120 = (p+2000)(r-0.01)$

Rewrite the second equation by multiplying the binomials. Then substitute 120 for pr and solve for r.

$$120 = pr - 0.01p + 2000r - 20$$
$$120 = 120 - 0.01p + 2000r - 20$$
$$0 = -0.01p + 2000r - 20$$
$$0.01p + 20 = 2000r$$
$$\frac{0.01p}{2000} + \frac{20}{2000} = r$$
$$r = 0.000005p + 0.01$$

Substitute $0.000005p + 0.01$ for r in $120 = pr$.

$$120 = pr$$
$$120 = p(0.000005p + 0.01)$$
$$120 = 0.000005p^2 + 0.01p$$
$$0 = 0.000005p^2 + 0.01p - 120$$
$$0 = p^2 + 2000p - 24,000,000$$
$$0 = (p-4000)(p+6000)$$
$$p - 4000 = 0 \quad \text{or} \quad p + 6000 = 0$$
$$p = 4000 \qquad\qquad p = -6000$$

The principal must be positive, so use $p = 4000$.

$$r = 0.000005p + 0.01$$
$$r = 0.000005(4000) + 0.01$$
$$r = 0.03$$

The principal is \$4000 and the rate is 3%.

Chapter 14 Practice Test

1. They are formed by cutting a cone or pair of cones.

2. $d = \sqrt{(x_2 - x_1)^2 + (y_2 - y_1)^2}$

$ = \sqrt{[6-(-1)]^2 + (7-8)^2}$

$ = \sqrt{7^2 + (-1)^2}$

$ = \sqrt{49 + 1}$

$ = \sqrt{50}$

The length is $\sqrt{50} \approx 7.07$ units.

3. Midpoint $= \left(\dfrac{x_1 + x_2}{2}, \dfrac{y_1 + y_2}{2} \right)$

$\phantom{\text{Midpoint }} = \left(\dfrac{-9+7}{2}, \dfrac{4+(-1)}{2} \right)$

$\phantom{\text{Midpoint }} = \left(-1, \dfrac{3}{2} \right)$

4. $y = -2(x+3)^2 + 1$

This is a parabola in the form $y = a(x-h)^2 + k$ with $a = -2$, $h = -3$, and $k = 1$. Since $a < 0$, the parabola opens downward. The vertex is $(-3, 1)$. The y-intercept is $(0, -17)$. The x-intercepts are about $(-3.71, 0)$ and $(-2.29, 0)$.

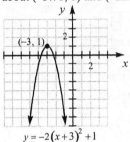

$y = -2(x+3)^2 + 1$

5. $x = y^2 - 2y + 4$

$x = (y^2 - 2y + 1) - 1 + 4$

$x = (y-1)^2 + 3$

This is a parabola in the form $x = a(y-k)^2 + h$ with $a = 1$, $h = 3$ and $k = 1$. Since $a > 0$, the parabola opens to the right. The vertex is $(3, 1)$. There is no y-intercept. The x-intercept is $(4, 0)$.

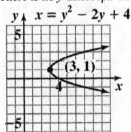

$x = y^2 - 2y + 4$

6. $x = -y^2 - 4y - 5$

$x = -(y^2 + 4y) - 5$

$x = -(y^2 + 4y + 4) + 4 - 5$

$x = -(y + 2)^2 - 1$

This is a parabola in the form $x = a(y - k)^2 + h$ with $a = -1$, $h = -1$, and $k = -2$. Since $a < 0$, the parabola opens to the left. The vertex is $(-1, -2)$. There are no y-intercepts. The x-intercept is $(-5, 0)$.

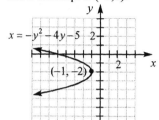

7. $(x - h)^2 + (y - k)^2 = r^2$

$[x - 2]^2 + [y - 4]^2 = 3^2$

$(x - 2)^2 + (y - 4)^2 = 9$

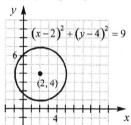

8. $(x + 2)^2 + (y - 8)^2 = 9$. The graph of this equation is a circle with center $(-2, 8)$ and radius 3.

Area $= \pi r^2$

$= \pi 3^2 = 9\pi \approx 28.27$ sq. units

9. The center is $(3, -1)$ and the radius is 4.

$(x - h)^2 + (y - k)^2 = r^2$

$(x - 3)^2 + [y - (-1)]^2 = 4^2$

$(x - 3)^2 + (y + 1)^2 = 16$

10. $y = -\sqrt{16 - x^2}$

If we solve $x^2 + y^2 = 16$ for y, we obtain $y = \pm\sqrt{16 - x^2}$. Therefore, the graph of $y = -\sqrt{16 - x^2}$ is the lower half $(y \le 0)$ of a circle with its center at the origin and radius 4.

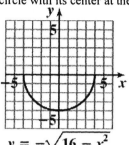

11. $\qquad x^2 + y^2 + 2x - 6y + 1 = 0$

$x^2 + 2x + y^2 - 6y = -1$

$(x^2 + 2x + 1) + (y^2 - 6y + 9) = -1 + 1 + 9$

$(x + 1)^2 + (y - 3)^2 = 9$

The graph is a circle with center $(1, 3)$ and radius 3.

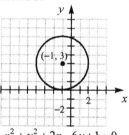

12. $4x^2 + 25y^2 = 100$

$\dfrac{4x^2}{100} + \dfrac{25y^2}{100} = 1$

$\dfrac{x^2}{25} + \dfrac{y^2}{4} = 1$

Since $a^2 = 25$, $a = 5$

Since $b^2 = 4$, $b = 2$.

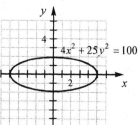

13. The center is $(-2, -1)$, $a = 4$, and $b = 2$.

$$\frac{(x-h)^2}{a^2} + \frac{(y-k)^2}{b^2} = 1$$

$$\frac{[x-(-2)]^2}{4^2} + \frac{[y-(-1)]^2}{2^2} = 1$$

$$\frac{(x+2)^2}{16} + \frac{(y+1)^2}{4} = 1$$

The values of a^2 and b^2 are switched, so this is not the graph of the given equation. The major axis should be along the y-axis.

14. $4(x-4)^2 + 36(y+2)^2 = 36$

$$\frac{4(x-4)^2}{36} + \frac{36(y+2)^2}{36} = 1$$

$$\frac{(x-4)^2}{9} + \frac{(y+2)^2}{1} = 1$$

The center is $(4, -2)$. Since $a^2 = 9$, $a = 3$

Since $b^2 = 1$, $b = 1$

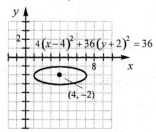

15. $3(x-8)^2 + 6(y+7)^2 = 18$

$$\frac{3(x-8)^2}{18} + \frac{6(y+7)^2}{18} = \frac{18}{18}$$

$$\frac{(x-8)^2}{6} + \frac{(y+7)^2}{3} = 1$$

The center is $(8, -7)$.

16. The transverse axis lies along the axis corresponding to the positive term of the equation in standard form.

17. $\dfrac{x^2}{16} - \dfrac{y^2}{49} = 1$

Since $a^2 = 16$ and $b^2 = 49$, $a = 4$ and $b = 7$. The equations of the asymptotes are

$$y = \pm\frac{b}{a}x, \text{ or } y = \pm\frac{7}{4}x.$$

18. $\dfrac{y^2}{25} - \dfrac{x^2}{1} = 1$

Since $a^2 = 1$ and $b^2 = 25$, $a = 1$ and $b = 5$. The equations of the asymptotes are

$$y = \pm\frac{b}{a}x, \text{ or } y = \pm 5x.$$

To graph the asymptotes, plot the points $(1, 5)$, $(-1, 5)$, $(1, -5)$, and $(-1, -5)$.
The graph intersects the y-axis at $(0, -5)$ and $(0, 5)$.

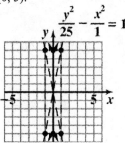

19. $\dfrac{x^2}{4} - \dfrac{y^2}{9} = 1$

Since $a^2 = 4$ and $b^2 = 9$, $a = 2$ and $b = 3$. The equations of the asymptotes are

$$y = \pm\frac{b}{a}x, \text{ or } y = \pm\frac{3}{2}x.$$

To graph the asymptotes, plot the points $(2, 3)$, $(-2, 3)$, $(2, -3)$, and $(-2, -3)$. The graph intersects the x-axis at $(-2, 0)$ and $(2, 0)$.

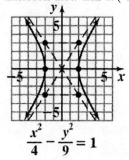

20. $4x^2 - 15y^2 = 30$

$$\frac{4x^2}{30} - \frac{15y^2}{30} = \frac{30}{30}$$

$$\frac{x^2}{\frac{15}{2}} - \frac{y^2}{2} = 1$$

Since the equation is of the form

$$\frac{x^2}{a^2} - \frac{y^2}{b^2} = 1, \text{ the graph is a hyperbola.}$$

21. $25x^2 + 4y^2 = 100$

$$\frac{25x^2}{100} + \frac{4y^2}{100} = \frac{100}{100}$$

$$\frac{x^2}{4} + \frac{y^2}{25} = 1$$

Since the equation is of the form

$\frac{x^2}{a^2} + \frac{y^2}{b^2} = 1$, the graph is an ellipse.

22. $x^2 + y^2 = 7 \quad \overset{\times 3}{\Rightarrow} \quad 3x^2 + 3y^2 = 21$

$\quad 2x^2 - 3y^2 = -1 \qquad \underline{2x^2 - 3y^2 = -1}$

$\qquad\qquad\qquad\qquad\qquad 5x^2 \quad\;\; = 20$

$\qquad\qquad\qquad\qquad\qquad\;\; x^2 = \;\; 4$

$\qquad x = 2 \qquad$ or $\qquad\qquad x = -2$

$\quad x^2 + y^2 = 7 \qquad\qquad\quad x^2 + y^2 = 7$

$\quad (2)^2 + y^2 = 7 \qquad\qquad (-2)^2 + y^2 = 7$

$\quad 4 + y^2 = 7 \qquad\qquad\quad 4 + y^2 = 7$

$\qquad\quad y^2 = 3 \qquad\qquad\qquad\quad y^2 = 3$

$\qquad\quad y = \pm\sqrt{3} \qquad\qquad\qquad y = \pm\sqrt{3}$

The solutions are $\left(2, \sqrt{3}\right), \left(2, -\sqrt{3}\right),$

$\left(-2, \sqrt{3}\right),$ and $\left(-2, -\sqrt{3}\right).$

23. $\quad x + y = 8$

$\quad x^2 + y^2 = 4$

Solve $x + y = 8$ for y: $y = 8 - x$.

Substitute $8 - x$ for y in $x^2 + y^2 = 4$.

$$x^2 + y^2 = 4$$

$$x^2 + (8 - x)^2 = 4$$

$$x^2 + 64 - 16x + x^2 = 4$$

$$2x^2 - 16x + 60 = 0$$

$$x^2 - 8x + 30 = 0$$

$$x = \frac{-(-8) \pm \sqrt{(-8)^2 - 4(1)(30)}}{2(1)}$$

$$= \frac{8 \pm \sqrt{-56}}{2} = 4 \pm i\sqrt{14}$$

There is no real solution.

24. Let x = length, y = width.

$\qquad xy = 1500$

$\quad 2x + 2y = 160$

Solve $2x + 2y = 160$ for y: $y = 80 - x$.

Substitute $80 - x$ for y in $xy = 1500$.

$$xy = 1500$$

$$x(80 - x) = 1500$$

$$80x - x^2 = 1500$$

$x^2 - 80x + 1500 = 0$

$(x - 30)(x - 50) = 0$

$x - 30 = 0 \qquad$ or $\quad x - 50 = 0$

$\qquad x = 30 \qquad\qquad\qquad x = 50$

$\qquad y = 80 - 30 \qquad\qquad y = 80 - 50$

$\qquad y = 50 \qquad\qquad\qquad y = 30$

The solutions are (30, 50) and (50, 30).

The dimensions are 30 m by 50 m.

25. Let x = length, $\quad y$ = width

$\qquad xy = 60$

$\quad x^2 + y^2 = 13^2$

Solve $xy = 60$ for y: $y = \dfrac{60}{x}$.

Substitute $\dfrac{60}{x}$ for y in $x^2 + y^2 = 13^2$.

$$x^2 + y^2 = 13^2$$

$$x^2 + \left(\frac{60}{x}\right)^2 = 169$$

$$x^2 + \frac{3600}{x^2} = 169$$

$$x^4 + 3600 = 169x^2$$

$$x^4 - 169x^2 + 3600 = 0$$

$$\left(x^2 - 25\right)\left(x^2 - 144\right) = 0$$

$x^2 - 25 = 0 \quad$ or $\; x^2 - 144 = 0$

$\qquad x^2 = 25 \qquad\qquad x^2 = 144$

$\qquad x = \pm 5 \qquad\qquad\; x = \pm 12$

Since x must be positive, $x = 5$ or $x = 12$.

If $x = 5$, then $y = \dfrac{60}{5} = 12$. If $x = 12$, then

$y = \dfrac{60}{12} = 5$. The dimensions of the bed of the

truck are 5 feet by 12 feet

Chapter 14 Cumulative Review Test

1. $4x - 2(3x - 7) = 2x - 5$

$$4x - 6x + 14 = 2x - 5$$
$$-2x + 14 = 2x - 5$$
$$-4x = -19$$
$$x = \frac{-19}{-4} = \frac{19}{4}$$

2. $2(x - 5) + 2x = 4x - 7$

$$2x - 10 + 2x = 4x - 7$$
$$4x - 10 = 4x - 7$$
$$-10 = -7$$

This is a contradiction, so the solution set is $\varnothing$.

3. $y = -2x + 2$

The slope is -2 and the y-intercept is $(0, 2)$.

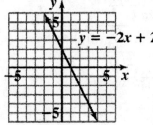

4. $\left(9x^2 y^5\right)\left(-3xy^4\right) = (9)(-3)x^{2+1}y^{5+4} = -27x^3 y^9$

5. $x^4 - x^2 - 42 \qquad$ let $u = x^2$

$$= u^2 - u - 42$$
$$= (u + 6)(u - 7)$$
$$= \left(x^2 + 6\right)\left(x^2 - 7\right)$$

6. Let x be the base of the sign. Then the height can be expressed as $x - 6$.

Area $= \dfrac{1}{2}$ (base $\times$ height)

$$56 = \frac{1}{2}(x - 6)(x)$$
$$112 = x^2 - 6x$$
$$0 = x^2 - 6x - 112$$
$$0 = (x - 14)(x + 8)$$
$$x - 14 = 0 \quad \text{or} \quad x + 8 = 0$$
$$x = 14 \qquad\qquad x = -8$$

Disregard the negative value. The base of the sign is 14 feet and the height is $14 - 6 = 8$ feet.

7. $\dfrac{3x^2 - x - 4}{4x^2 + 7x + 3} \cdot \dfrac{2x^2 - 5x - 12}{6x^2 + x - 12}$

$$= \frac{(3x - 4)(x + 1)}{(4x + 3)(x + 1)} \cdot \frac{(2x + 3)(x - 4)}{(3x - 4)(2x + 3)}$$

$$= \frac{(3x - 4)(x + 1)(2x + 3)(x - 4)}{(4x + 3)(x + 1)(3x - 4)(2x + 3)}$$

$$= \frac{x - 4}{4x + 3}$$

8. $\dfrac{x}{x + 3} - \dfrac{x + 5}{2x^2 - 2x - 24}$

$$= \frac{x}{x + 3} - \frac{x + 5}{2(x + 3)(x - 4)}$$

$$= \frac{2x(x - 4)}{2(x + 3)(x - 4)} - \frac{x + 5}{2(x + 3)(x - 4)}$$

$$= \frac{2x^2 - 8x - (x + 5)}{2(x + 3)(x - 4)}$$

$$= \frac{2x^2 - 9x - 5}{2(x + 3)(x - 4)}$$

9. $\dfrac{3}{x + 3} + \dfrac{5}{x + 4} = \dfrac{12x + 19}{x^2 + 7x + 12}$

$$\left(\frac{3}{x + 3} + \frac{5}{x + 4} = \frac{12x + 19}{(x + 3)(x + 4)}\right)(x + 3)(x + 4)$$

$$3(x + 4) + 5(x + 3) = 12x + 19$$
$$3x + 12 + 5x + 15 = 12x + 19$$
$$8x + 27 = 12x + 19$$
$$-4x = -8$$
$$x = 2$$

10. $f(x) = x^2 + 3x + 9$

$$f(10) = (10)^2 + 3(10) + 9$$
$$= 100 + 30 + 9$$
$$= 139$$

11. $\dfrac{1}{2}x - \dfrac{1}{3}y = 2 \overset{\times(-6)}{\Rightarrow} -3x + 2y = -12$

$\dfrac{1}{4}x + \dfrac{2}{3}y = 6 \overset{\times 12}{\Rightarrow} \underline{ 3x + 8y = 72}$

$ 10y = 60$

$ y = 6$

Substitute $y = 6$ into $-3x + 2y = -12$

$-3x + 2(6) = -12$

$-3x + 12 = -12$

$-3x = -24$

$x = 8$

The solution is $(8, 6)$.

12. $\qquad |3x + 1| > 4$

$3x + 1 > 4 \qquad 3x + 1 < -4$

$3x > 3 \qquad\quad 3x < -5$

$x > 1 \qquad\quad x < -\dfrac{5}{3}$

The solution set is $\left\{ x \;\middle|\; x < -\dfrac{5}{3} \ \text{ or } \ x > 1 \right\}$

13. $\left(\dfrac{18x^{1/2}y^3}{2x^{3/2}}\right)^{1/2} = \left(\dfrac{18}{2}x^{1/2-3/2}y^3\right)^{1/2}$

$= \left(9x^{-1}y^3\right)^{1/2}$

$= 9^{1/2}x^{-1/2}y^{3/2}$

$= \sqrt{9} \cdot x^{-1/2}y^{3/2}$

$= \dfrac{3y^{3/2}}{x^{1/2}}$

14. $\dfrac{6\sqrt{x}}{\sqrt{x}-y} = \dfrac{6\sqrt{x}}{\sqrt{x}-y} \cdot \dfrac{\sqrt{x}+y}{\sqrt{x}+y}$

$= \dfrac{6x + 6y\sqrt{x}}{x + y\sqrt{x} - y\sqrt{x} - y^2}$

$= \dfrac{6x + 6y\sqrt{x}}{x - y^2}$

15. $\qquad 3\sqrt[3]{2x+2} = \sqrt[3]{80x - 24}$

$\left(3\sqrt[3]{2x+2}\right)^3 = \left(\sqrt[3]{80x-24}\right)^3$

$27(2x+2) = (80x - 24)$

$54x + 54 = 80x - 24$

$54x = 80x - 78$

$-26x = -78$

$x = 3$

Check: $3\sqrt[3]{2(3)+2} \overset{?}{=} \sqrt[3]{80(3) - 24}$

$3\sqrt[3]{8} \overset{?}{=} \sqrt[3]{216}$

$3 \cdot 2 = 6 \ $ True

The solution is 3.

16. $3x^2 - 4x + 5 = 0$

$x = \dfrac{-b \pm \sqrt{b^2 - 4ac}}{2a}$

$= \dfrac{-(-4) \pm \sqrt{(-4)^2 - 4(3)(5)}}{2(3)}$

$= \dfrac{4 \pm \sqrt{-44}}{6}$

$= \dfrac{4 \pm 2i\sqrt{11}}{6}$

$= \dfrac{2 \pm i\sqrt{11}}{3}$

17. $\log(3x - 4) + \log(4) = \log(x + 6)$

$\log 4(3x - 4) = \log(x + 6)$

$\log(12x - 16) = \log(x + 6)$

$12x - 16 = x + 6$

$12x = x + 22$

$11x = 22$

$x = 2$

Check: $\log[3(2) - 4] + \log 4 \overset{?}{=} \log(2 + 6)$

$\log 2 + \log 4 \overset{?}{=} \log 8$

$\log(2 \cdot 4) = \log(8) \ $ True

The solution is 2.

18.
$$35 = 70e^{-0.3t}$$

$$\frac{1}{2} = e^{-0.3t}$$

$$\ln\frac{1}{2} = \ln e^{-0.3t}$$

$$\ln\frac{1}{2} = -0.3t$$

$$-\frac{1}{0.3}\ln\frac{1}{2} = t$$

$$t = -\frac{1}{0.3}\ln\frac{1}{2} \approx 2.31$$

19. $9x^2 + 4y^2 = 36$

$$\frac{9x^2}{36} + \frac{4y^2}{36} = 1$$

$$\frac{x^2}{4} + \frac{y^2}{9} = 1$$

Since $a^2 = 4$, $a = 2$.

Since $b^2 = 9$, $b = 3$.

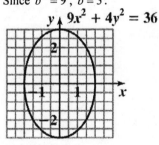

20. $\dfrac{y^2}{25} - \dfrac{x^2}{16} = 1$

Since $a^2 = 25$ and $b^2 = 16$, $a = 5$ and $b = 4$

The equations of the asymptotes are $y = \pm\dfrac{b}{a}x$,

or $y = \pm\dfrac{5}{4}x$. To graph the asymptotes, plot the

points (4, 5), (4, –5), (–4, 5), and (–4, –5).
The graph intersects the y-axis at (0, –5) and
(0, 5).

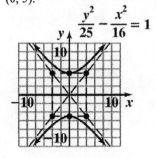

Chapter 15

Exercise Set 15.1

1. A sequence is a list of numbers arranged in a specific order.

3. A finite sequence is a function whose domain includes only the first n natural numbers.

5. In a decreasing sequence, the terms decrease.

7. A series is the sum of the terms of a sequence.

9. $\displaystyle\sum_{i=1}^{5}(i+4)$

The sum as i goes from 1 to 5 of $i+4$.

11. The sequence $a_n = 2n-1$ is an increasing sequence since the coefficient of n is positive.

13. Yes, $a_n = 1+(-2)^n$ is an alternating sequence since $(-2)^n$ alternates between positive and negative values as n alternates from odd to even.

15. $a_n = 6n$

$a_1 = 6(1) = 6$

$a_2 = 6(2) = 12$

$a_3 = 6(3) = 18$

$a_4 = 6(4) = 24$

$a_5 = 6(5) = 30$

The terms are 6, 12, 18, 24, 30.

17. $a_n = 4n-1$

$a_1 = 4(1)-1 = 3$

$a_2 = 4(2)-1 = 7$

$a_3 = 4(3)-1 = 11$

$a_4 = 4(4)-1 = 15$

$a_5 = 4(5)-1 = 19$

The terms are 3, 7, 11, 15, 19.

19. $a_n = \dfrac{7}{n}$

$a_1 = \dfrac{7}{1} = 7$

$a_2 = \dfrac{7}{2}$

$a_3 = \dfrac{7}{3}$

$a_4 = \dfrac{7}{4}$

$a_5 = \dfrac{7}{5}$

The terms are $7, \dfrac{7}{2}, \dfrac{7}{3}, \dfrac{7}{4},$ and $\dfrac{7}{5}$.

21. $a_n = \dfrac{n+2}{n+1}$

$a_1 = \dfrac{1+2}{1+1} = \dfrac{3}{2}$

$a_2 = \dfrac{2+2}{2+1} = \dfrac{4}{3}$

$a_3 = \dfrac{3+2}{3+1} = \dfrac{5}{4}$

$a_4 = \dfrac{4+2}{4+1} = \dfrac{6}{5}$

$a_5 = \dfrac{5+2}{5+1} = \dfrac{7}{6}$

The terms are $\dfrac{3}{2}, \dfrac{4}{3}, \dfrac{5}{4}, \dfrac{6}{5},$ and $\dfrac{7}{6}$.

23. $a_n = (-1)^n$

$a_1 = (-1)^1 = -1$

$a_2 = (-1)^2 = 1$

$a_3 = (-1)^3 = -1$

$a_4 = (-1)^4 = 1$

$a_5 = (-1)^5 = -1$

The terms are $-1, 1, -1, 1, -1$.

25. $a_n = (-2)^{n+1}$

$a_1 = (-2)^{1+1} = (-2)^2 = 4$

$a_2 = (-2)^{2+1} = (-2)^3 = -8$

$a_3 = (-2)^{3+1} = (-2)^4 = 16$

$a_4 = (-2)^{4+1} = (-2)^5 = -32$

$a_5 = (-2)^{5+1} = (-2)^6 = 64$

The terms are 4, -8, 16, -32, 64.

27. $a_n = 2n + 7$

$a_{12} = 2(12) + 7 = 24 + 7 = 31$

29. $a_n = \dfrac{n}{4} + 8$

$a_{16} = \dfrac{16}{4} + 8 = 4 + 8 = 12$

31. $a_n = (-1)^n$

$a_8 = (-1)^8 = 1$

33. $a_n = n(n+2)$

$a_9 = 9(9+2) = 9(11) = 99$

35. $a_n = \dfrac{n^2}{2n+7}$

$a_9 = \dfrac{9^2}{2(9)+7} = \dfrac{81}{18+7} = \dfrac{81}{25}$

37. $a_n = 3n - 1$

$a_1 = 3(1) - 1 = 3 - 1 = 2$

$a_2 = 3(2) - 1 = 6 - 1 = 5$

$a_3 = 3(3) - 1 = 9 - 1 = 8$

$s_1 = a_1 = 2$

$s_3 = a_1 + a_2 + a_3 = 2 + 5 + 8 = 15$

39. $a_n = 2^n + 1$

$a_1 = 2^1 + 1 = 2 + 1 = 3$

$a_2 = 2^2 + 1 = 4 + 1 = 5$

$a_3 = 2^3 + 1 = 8 + 1 = 9$

$s_1 = a_1 = 3$

$s_3 = a_1 + a_2 + a_3 = 3 + 5 + 9 = 17$

41. $a_n = \dfrac{n-1}{n+2}$

$a_1 = \dfrac{1-1}{1+2} = \dfrac{0}{3} = 0$

$a_2 = \dfrac{2-1}{2+2} = \dfrac{1}{4}$

$a_3 = \dfrac{3-1}{3+2} = \dfrac{2}{5}$

$s_1 = 0$

$s_3 = 0 + \dfrac{1}{4} + \dfrac{2}{5} = \dfrac{5}{20} + \dfrac{8}{20} = \dfrac{13}{20}$

43. $a_n = (-1)^n$

$a_1 = (-1)^1 = -1$

$a_2 = (-1)^2 = 1$

$a_3 = (-1)^3 = -1$

$s_1 = a_1 = -1$

$s_3 = a_1 + a_2 + a_3 = -1 + 1 + -1 = -1$

45. $a_n = \dfrac{n^2}{2}$

$a_1 = \dfrac{1^2}{2} = \dfrac{1}{2}$

$a_2 = \dfrac{2^2}{2} = \dfrac{4}{2} = 2$

$a_3 = \dfrac{3^2}{2} = \dfrac{9}{2}$

$s_1 = a_1 = \dfrac{1}{2}$

$s_3 = a_1 + a_2 + a_3 = \dfrac{1}{2} + \dfrac{4}{2} + \dfrac{9}{2} = \dfrac{14}{2} = 7$

47. Each term is twice the preceding term. The next three terms are 64, 128, 256.

49. Each term is two more than the preceding term. The next three terms are 17, 19, 21.

51. Each denominator is one more than the preceding one while each numerator is one. The next three terms are $\dfrac{1}{6}, \dfrac{1}{7}, \dfrac{1}{8}$.

53. Each term is -1 times the previous term. The next three terms are 1, -1, 1.

55. Each denominator is three times the previous one while each numerator is one. The next three terms are $\dfrac{1}{81}, \dfrac{1}{243}, \dfrac{1}{729}$.

57. Each term is $-\dfrac{1}{2}$ times the preceding term. The next three terms are $\dfrac{1}{16}, -\dfrac{1}{32}, \dfrac{1}{64}$.

59. Each term is 5 less than the preceding term. The next three terms are 17, 12, 7.

61. $\displaystyle\sum_{i=1}^{5}(3i-1) = \big[3(1)-1\big]+\big[3(2)-1\big]+\big[3(3)-1\big]+\big[3(4)-1\big]+\big[3(5)-1\big]$

$$= 2+5+8+11+14$$
$$= 40$$

63. $\displaystyle\sum_{i=1}^{6}\left(i^2+1\right) = \left(1^2+1\right)+\left(2^2+1\right)+\left(3^2+1\right)+\left(4^2+1\right)+\left(5^2+1\right)+\left(6^2+1\right)$

$$= (1+1)+(4+1)+(9+1)+(16+1)+(25+1)+(36+1)$$
$$= 2+5+10+17+26+37$$
$$= 97$$

65. $\displaystyle\sum_{i=1}^{4}\frac{i^2}{2} = \frac{1^2}{2}+\frac{2^2}{2}+\frac{3^2}{2}+\frac{4^2}{2} = \frac{1}{2}+\frac{4}{2}+\frac{9}{2}+\frac{16}{2} = \frac{1}{2}+2+\frac{9}{2}+8 = \frac{30}{2} = 15$

67. $\displaystyle\sum_{i=4}^{9}\frac{i^2+i}{i+1} = \frac{4^2+4}{4+1}+\frac{5^2+5}{5+1}+\frac{6^2+6}{6+1}+\frac{7^2+7}{7+1}+\frac{8^2+8}{8+1}+\frac{9^2+9}{9+1}$

$$= \frac{20}{5}+\frac{30}{6}+\frac{42}{7}+\frac{56}{8}+\frac{72}{9}+\frac{90}{10}$$
$$= 4+5+6+7+8+9$$
$$= 39$$

---

69. $a_n = n+8$

The fifth partial sum is $\displaystyle\sum_{i=1}^{5}(i+8)$.

71. $a_n = \dfrac{n^2}{4}$

The third partial sum is $\displaystyle\sum_{i=1}^{3}\frac{i^2}{4}$.

73. $\displaystyle\sum_{i=1}^{5}x_i = x_1+x_2+x_3+x_4+x_5$

$$= 2+3+5+(-1)+4$$
$$= 13$$

75. $\displaystyle\left(\sum_{i=1}^{5}x_i\right)^2 = \left(x_1+x_2+x_3+x_4+x_5\right)^2$

$$= \left(2+3+5+(-1)+4\right)^2$$
$$= 13^2$$
$$= 169$$

77. $\displaystyle\sum_{i=1}^{5}x_i^2 = x_1^2+x_2^2+x_3^2+x_4^2+x_5^2$

$$= 2^2+3^2+5^2+(-1)^2+4^2$$
$$= 55$$

79. $\bar{x} = \dfrac{15+20+25+30+35}{5} = \dfrac{125}{5} = 25$

81. $\bar{x} = \dfrac{72+83+4+60+18+20}{6} = \dfrac{257}{6} \approx 42.83$

83. **a.** Perimeter of rectangle: $p = 2l+2w$

$$p_1 = 2(1)+2(2\cdot1) = 2+4 = 6$$
$$p_2 = 2(2)+2(2\cdot2) = 4+8 = 12$$
$$p_3 = 2(3)+2(2\cdot3) = 6+12 = 18$$
$$p_4 = 2(4)+2(2\cdot4) = 8+16 = 24$$

b. $p_n = 2n+2(2n) = 2n+4n = 6n$

85 – 87. Answers will vary.

89. $\bar{x} = \dfrac{\sum x}{n}$

$$n\bar{x} = n\cdot\frac{\sum x}{n}$$
$$n\bar{x} = \sum x \text{ or } \sum x = n\bar{x}$$

91. Yes, $\sum_{i=1}^{n} 4x_i = 4\sum_{i=1}^{n} x_i$. Examples will vary.

93. a. $\sum x = x_1 + x_2 + x_3 = 3+5+2 = 10$

  b. $\sum y = y_1 + y_2 + y_3 = 4+1+6 = 11$

  c. $\sum x \cdot \sum y = 10 \cdot 11 = 110$

  d. $\sum xy = x_1 y_1 + x_2 y_2 + x_3 y_3$
  $= 3(4) + 5(1) + 2(6)$
  $= 12+5+12$
  $= 29$

  e. No, $\sum xy \neq \sum x \cdot \sum y$.

94. $2x^2 + 15 = 13x$
$2x^2 - 13x + 15 = 0$
$(2x-3)(x-5) = 0$
$2x-3 = 0 \quad \text{or} \quad x-5 = 0$
$x = \dfrac{3}{2} \qquad\qquad x = 5$

The solutions are $\dfrac{3}{2}$ and 5.

95. $6x^2 - 3x - 4 = 2$
$6x^2 - 3x - 6 = 0$
Use the discriminant to determine the number of real solutions:
$a = 6, b = -3, c = -6$.
$b^2 - 4ac = (-3)^2 - 4(6)(-6)$
$= 9+144$
$= 153$
Since $153 > 0$, the equation has two real solutions.

96. $\dfrac{x^2}{4} + \dfrac{y^2}{1} = 1$
$\dfrac{x^2}{2^2} + \dfrac{y^2}{1^2} = 1$
$a = 2$ and $b = 1$.

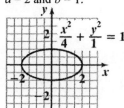

97. $x^2 + y^2 = 5$
$x = 2y$
Use the substitution method. Substitute $2y$ for x in the first equation.
$(2y)^2 + y^2 = 5$
$4y^2 + y^2 = 5$
$5y^2 = 5$
$y^2 = 1$
$y = \pm 1$
Now substitute 1 for y in the second equation and solve for x. Do this again for -1 .
$y = 1 \Rightarrow x = 2(1) = 2$
$y = -1 \Rightarrow x = 2(-1) = -2$
The solutions are $(1, 2)$ and $(-1, -2)$.

Note: Had we substituted $y = \pm 1$ into the first equation, we would have also obtained the solutions $(1, -2)$ and $(-1, 2)$. However, these are extraneous since they do not check when substituted into the second equation.

Exercise Set 15.2

1. In an arithmetic sequence, each term differs by a constant amount.

3. It is called the common difference.

5. The common difference, d, must be a positive number.

7. Yes. For example, $-1, -2, -3, -4, \ldots$ is an arithmetic sequence with $a_1 = -1$ and $d = -1$.

9. Yes. For example, $2, 4, 6, 8, \ldots$ is an arithmetic sequence with $a_1 = 2$ and $d = 2$.

11. $a_1 = 4$
$a_2 = 4 + (2-1)(3) = 4+3 = 7$
$a_3 = 4 + (3-1)(3) = 4+2(3) = 4+6 = 10$
$a_4 = 4 + (4-1)(3) = 4+3(3) = 4+9 = 13$
$a_5 = 4 + (5-1)(3) = 4+4(3) = 4+12 = 16$
The terms are $4, 7, 10, 13, 16$. The general term is $a_n = 4 + (n-1)3$ or $a_n = 3n+1$.

13. $a_1 = 7$

$a_2 = 7 + (2-1)(-2) = 7 - 2 = 5$

$a_3 = 7 + (3-1)(-2) = 7 + 2(-2) = 7 - 4 = 3$

$a_4 = 7 + (4-1)(-2) = 7 + 3(-2) = 7 - 6 = 1$

$a_5 = 7 + (5-1)(-2) = 7 + 4(-2) = 7 - 8 = -1$

The terms are 7, 5, 3, 1, –1. The general term is
$a_n = 7 + (n-1)(-2)$ or $a_n = -2n + 9$.

15. $a_1 = \dfrac{1}{2}$

$a_2 = \dfrac{1}{2} + (2-1)\left(\dfrac{3}{2}\right) = \dfrac{1}{2} + \dfrac{3}{2} = \dfrac{4}{2} = 2$

$a_3 = \dfrac{1}{2} + (3-1)\left(\dfrac{3}{2}\right) = \dfrac{1}{2} + 2\left(\dfrac{3}{2}\right) = \dfrac{1}{2} + \dfrac{6}{2} = \dfrac{7}{2}$

$a_4 = \dfrac{1}{2} + (4-1)\left(\dfrac{3}{2}\right) = \dfrac{1}{2} + 3\left(\dfrac{3}{2}\right) = \dfrac{1}{2} + \dfrac{9}{2} = \dfrac{10}{2} = 5$

$a_5 = \dfrac{1}{2} + (5-1)\left(\dfrac{3}{2}\right) = \dfrac{1}{2} + 4\left(\dfrac{3}{2}\right) = \dfrac{1}{2} + \dfrac{12}{2} = \dfrac{13}{2}$

The terms are $\dfrac{1}{2}$, 2, $\dfrac{7}{2}$, 5, $\dfrac{13}{2}$. The general term is

$a_n = \dfrac{1}{2} + (n-1)\dfrac{3}{2}$ or $a_n = \dfrac{3}{2}n - 1$.

17. $a_1 = 100$

$a_2 = 100 + (2-1)(-5)$

$\quad = 100 + (-5) = 100 - 5 = 95$

$a_3 = 100 + (3-1)(-5)$

$\quad = 100 + 2(-5) = 100 - 10 = 90$

$a_4 = 100 + (4-1)(-5)$

$\quad = 100 + 3(-5) = 100 - 15 = 85$

$a_5 = 100 + (5-1)(-5)$

$\quad = 100 + 4(-5) = 100 - 20 = 80$

The terms are 100, 95, 90, 85, 80. The general
term is $a_n = 100 + (n-1)(-5)$ or $a_n = -5n + 105$.

19. $a_n = a_1 + (n-1)d$

$a_4 = 5 + (4-1)3 = 5 + 3 \cdot 3 = 5 + 9 = 14$

21. $a_n = a_1 + (n-1)d$

$a_{10} = -9 + (10-1)(4) = -9 + 9(4) = -9 + 36 = 27$

23. $a_n = a_1 + (n-1)d$

$a_{13} = -8 + (13-1)\left(\dfrac{5}{3}\right)$

$\quad = -8 + 12\left(\dfrac{5}{3}\right) = -8 + 20 = 12$

25. $a_n = a_1 + (n-1)d$

$27 = 11 + (9-1)d$

$27 = 11 + 8d$

$16 = 8d$

$2 = d$

27. $a_n = a_1 + (n-1)d$

$28 = 4 + (n-1)(3)$

$28 = 4 + 3n - 3$

$28 = 1 + 3n$

$27 = 3n$

$9 = n$

29. $a_n = a_1 + (n-1)d$

$42 = 82 + (n-1)(-8)$

$42 = 82 - 8n + 8$

$42 = 90 - 8n$

$-48 = -8n$

$6 = n$

31. $s_{10} = \dfrac{10(a_1 + a_{10})}{2} = \dfrac{10(1+19)}{2} = 5(20) = 100$

$a_{10} = a_1 + (10-1)d$

$a_{10} = a_1 + 9d$

$19 = 1 + 9d$

$18 = 9d$

$2 = d$

33. $s_8 = \dfrac{8(a_1 + a_8)}{2} = \dfrac{8\left(\frac{3}{5} + 2\right)}{2} = 4\left(\dfrac{3}{5} + 2\right)$

$\quad = 4\left(\dfrac{3}{5} + \dfrac{10}{5}\right) = 4\left(\dfrac{13}{5}\right) = \dfrac{52}{5}$

$a_8 = a_1 + (8-1)d$

$a_8 = a_1 + 7d$

$2 = \dfrac{3}{5} + 7d$

$\dfrac{7}{5} = 7d$

$d = \dfrac{1}{7} \cdot \dfrac{7}{5} = \dfrac{1}{5}$

35. $s_6 = \dfrac{6(a_1 + a_6)}{2} = \dfrac{6(-5 + 13.5)}{2} = \dfrac{6(8.5)}{2} = 25.5$

$a_6 = a_1 + (6-1)d$

$a_6 = a_1 + 5d$

$13.5 = -5 + 5d$

$18.5 = 5d$

$3.7 = d$

37. $s_{11} = \dfrac{11(a_1 + a_{11})}{2} = \dfrac{11(7 + 67)}{2} = \dfrac{11(74)}{2} = 407$

$a_{11} = a_1 + (11-1)d$

$a_{11} = a_1 + 10d$

$67 = 7 + 10d$

$60 = 10d$

$6 = d$

39. $a_1 = 4$

$a_2 = 4 + (2-1)(3) = 4 + 3 = 7$

$a_3 = 4 + (3-1)(3) = 4 + 2(3) = 4 + 6 = 10$

$a_4 = 4 + (4-1)(3) = 4 + 3(3) = 4 + 9 = 13$

The terms are 4, 7, 10, 13.

$a_{10} = 4 + (10-1)(3) = 4 + 9(3) = 4 + 27 = 31$

$s_{10} = \dfrac{10(4 + 31)}{2} = \dfrac{10(35)}{2} = 175$

41. $a_1 = -6$

$a_2 = -6 + (2-1)(2) = -6 + 1(2) = -6 + 2 = -4$

$a_3 = -6 + (3-1)(2) = -6 + 2(2) = -6 + 4 = -2$

$a_4 = -6 + (4-1)(2) = -6 + 3(2) = -6 + 6 = 0$

The terms are $-6, -4, -2, 0$.

$a_{10} = -6 + (10-1)(2) = -6 + 9(2) = -6 + 18 = 12$

$s_{10} = \dfrac{10(-6 + 12)}{2} = \dfrac{10(6)}{2} = \dfrac{60}{2} = 30$

43. $a_1 = -8$

$a_2 = -8 + (2-1)(-5) = -8 - 5 = -13$

$a_3 = -8 + (3-1)(-5) = -8 + 2(-5) = -8 - 10 = -18$

$a_4 = -8 + (4-1)(-5) = -8 + 3(-5) = -8 - 15 = -23$

The terms are $-8, -13, -18, -23$.

$a_{10} = -8 + (10-1)(-5)$

$\quad = -8 + 9(-5) = -8 - 45 = -53$

$s_{10} = \dfrac{10\left[-8 + (-53)\right]}{2} = \dfrac{10(-61)}{2} = \dfrac{-610}{2} = -305$

45. $a_1 = \dfrac{7}{2}$

$a_2 = \dfrac{7}{2} + (2-1)\left(\dfrac{5}{2}\right) = \dfrac{7}{2} + 1\left(\dfrac{5}{2}\right) = \dfrac{7}{2} + \dfrac{5}{2} = \dfrac{12}{2} = 6$

$a_3 = \dfrac{7}{2} + (3-1)\left(\dfrac{5}{2}\right) = \dfrac{7}{2} + 2\left(\dfrac{5}{2}\right) = \dfrac{7}{2} + \dfrac{10}{2} = \dfrac{17}{2}$

$a_4 = \dfrac{7}{2} + (4-1)\left(\dfrac{5}{2}\right) = \dfrac{7}{2} + 3\left(\dfrac{5}{2}\right) = \dfrac{7}{2} + \dfrac{15}{2} = \dfrac{22}{2} = 11$

The terms are $\dfrac{7}{2}, 6, \dfrac{17}{2}, 11$.

$a_{10} = \dfrac{7}{2} + (10-1)\left(\dfrac{5}{2}\right)$

$\quad = \dfrac{7}{2} + 9\left(\dfrac{5}{2}\right) = \dfrac{7}{2} + \dfrac{45}{2} = \dfrac{52}{2} = 26$

$s_{10} = \dfrac{10(3.5 + 26)}{2} = \dfrac{10(29.5)}{2} = 147.5$

47. $a_1 = 100$

$a_2 = 100 + (2-1)(-7)$

$\quad = 100 + 1(-7) = 100 - 7 = 93$

$a_3 = 100 + (3-1)(-7)$

$\quad = 100 + 2(-7) = 100 - 14 = 86$

$a_4 = 100 + (4-1)(-7)$

$\quad = 100 + 3(-7) = 100 - 21 = 79$

The terms are 100, 93, 86, 79.

$a_{10} = 100 + (10-1)(-7)$

$\quad = 100 + 9(-7) = 100 - 63 = 37$

$s_{10} = \dfrac{10(100 + 37)}{2} = \dfrac{10(137)}{2} = 685$

49. $d = 4 - 1 = 3$

$a_n = a_1 + (n-1)d$

$43 = 1 + (n-1)(3)$

$43 = 1 + 3n - 3$

$43 = -2 + 3n$

$45 = 3n$

$15 = n$

$s_{15} = \dfrac{15(a_1 + a_{15})}{2} = \dfrac{15(1 + 43)}{2} = \dfrac{15(44)}{2} = 330$

51. $d = -5 - (-9) = -5 + 9 = 4$

$a_n = a_1 + (n-1)d$

$31 = -9 + (n-1)(4)$

$31 = -9 + 4n - 4$

$31 = -13 + 4n$

$44 = 4n$

$11 = n$

$s_{10} = \dfrac{11(a_1 + a_{10})}{2} = \dfrac{11(-9+31)}{2} = \dfrac{11(22)}{2} = 121$

53. $d = \dfrac{2}{2} - \dfrac{1}{2} = \dfrac{1}{2}$

$a_n = \dfrac{1}{2} + (n-1)\left(\dfrac{1}{2}\right)$

$\dfrac{17}{2} = \dfrac{1}{2} + \dfrac{1}{2}n - \dfrac{1}{2}$

$\dfrac{17}{2} = \dfrac{1}{2}n$

$17 = n$

$s_{17} = \dfrac{17(a_1 + a_{17})}{2}$

$= \dfrac{17\left(\frac{1}{2} + \frac{17}{2}\right)}{2} = \dfrac{17\left(\frac{18}{2}\right)}{2} = \dfrac{17(9)}{2} = \dfrac{153}{2}$

55. $d = 10 - 7 = 3$

$a_n = a_1 + (n-1)d$

$91 = 7 + (n-1)(3)$

$91 = 7 + 3n - 3$

$91 = 4 + 3n$

$87 = 3n$

$29 = n$

$s_{29} = \dfrac{29(a_1 + a_{29})}{2} = \dfrac{29(7+91)}{2} = \dfrac{29(98)}{2} = 1421$

57. $s_n = \dfrac{n(a_1 + a_n)}{2}$

$s_{50} = \dfrac{50(1+50)}{2} = \dfrac{50(51)}{2} = 1275$

59. $s_n = \dfrac{n(a_1 + a_n)}{2}$

$s_{50} = \dfrac{50(1+99)}{2} = \dfrac{50(100)}{2} = 2500$

61. $s_n = \dfrac{n(a_1 + a_n)}{2}$

$s_{30} = \dfrac{30(3+90)}{2} = \dfrac{30(93)}{2} = 1395$

63. The smallest number greater than 7 that is divisible by 6 is 12. The largest number less than 1610 that is divisible by 6 is 1608. Now find n in the equation $a_n = a_1 + (n-1)d$.

$1608 = 12 + (n-1)6$

$1596 = 6(n-1)$

$266 = n - 1$

$267 = n$

There are 267 numbers between 7 and 1610 that are divisible by 6.

65. $a_1 = 20,\ d = 2,\ n = 12$

$a_n = a_1 + (n-1)d$

$a_{12} = 20 + (12-1)(2) = 20 + 11(2) = 20 + 22 = 42$

$s_n = \dfrac{n(a_1 + a_n)}{2}$

$s_{12} = \dfrac{12(20+42)}{2} = \dfrac{12(62)}{2} = \dfrac{744}{2} = 372$

There are 42 seats in the twelfth row and 372 seats in the first twelve rows.

67. $26 + 25 + 24 + \cdots + 1$ or $1 + 2 + 3 + \cdots + 26$

$s_n = \dfrac{n(a_1 + a_n)}{2}$

$s_{26} = \dfrac{26(1+26)}{2} = \dfrac{26(27)}{2} = \dfrac{702}{2} = 351$

There are 351 logs in the pile.

69. $a_1 = 1,\ d = 2,\ n = 14$

$a_n = a_1 + (n-1)d$

$a_{14} = 1 + (14-1)(2) = 1 + 13(2) = 1 + 26 = 27$

$s_n = \dfrac{n(a_1 + a_n)}{2}$

$s_{14} = \dfrac{14(1+27)}{2} = \dfrac{14(28)}{2} = \dfrac{392}{2} = 196$

There are 27 glasses in the 14th row and 196 glasses in all.

71. $1 + 2 + 3 + \cdots + 100$

$= (1+100) + (2+99) + \cdots + (50+51)$

$= 101 + 101 + \cdots + 101$

$= 50(101)$

$= 5050$

73. $s_n = \dfrac{n(a_1 + a_n)}{2}$

$= \dfrac{n[1+(2n-1)]}{2} = \dfrac{n(2n)}{2} = \dfrac{2n^2}{2} = n^2$

75. a. $a_1 = 22, d = -\dfrac{1}{2}, n = 7$

$a_n = a_1 + (n-1)d$

$a_7 = 22 + (7-1)\left(-\dfrac{1}{2}\right) = 22 - 3 = 19$

Her seventh swing is 19 feet.

b. $s_n = \dfrac{n(a_1 + a_n)}{2}$

$s_7 = \dfrac{7(22+19)}{2} = 143.5$

She travels 143.5 feet during the seven swings.

77. $d = -6 \text{ in.} = -\dfrac{1}{2} \text{ ft}, a_1 = 6$

$a_n = a_1 + (n-1)d$

$a_9 = 6 + (9-1)\left(-\dfrac{1}{2}\right) = 6 + 8\left(-\dfrac{1}{2}\right) = 6 - 4 = 2$

The ball bounces 2 feet on the ninth bounce.

79. a. Note that if March 17th is day 1, then March 22nd is day 6.

$a_1 = 105, d = 10, n = 6$

$a_n = a_1 + (n-1)d$

$a_6 = 105 + (6-1)(10)$

$\quad = 105 + 5(10) = 105 + 50 = 155$

He can prepare 155 packages for shipment on March 22nd.

b. $s_n = \dfrac{n(a_1 + a_n)}{2}$

$s_5 = \dfrac{5(105+155)}{2} = \dfrac{5(260)}{2} = \dfrac{1300}{2} = 650$

He can prepare 650 packages for shipment from March 17th through March 22nd.

81. $s_n = \dfrac{n(a_1 + a_n)}{2}$

$s_{31} = \dfrac{31(1+31)}{2} = \dfrac{31(32)}{2} = 496$

On day 31, Craig will have saved $496.

83. a. $a_{10} = 42,000 + (10-1)(400) = 45,600$

She will receive $45,600 in her tenth year of retirement.

b. $s_{10} = \dfrac{10(42,000 + 45,600)}{2}$

$\quad = \dfrac{10(87,600)}{2}$

$\quad = 438,000$

In her first 10 years, she will receive a total of $438,000.

85. $360 - 180 = 180$

$540 - 360 = 180$

$720 - 540 = 180$

The terms form an arithmetic sequence with $d = 180$ and $a_3 = 180$.

$a_n = a_1 + (n-1)d$

$180 = a_1 + (3-1)180$

$180 = a_1 + 360$

$-180 = a_1$

$a_n = a_1 + (n-1)d$

$a_n = -180 + (n-1)(180)$

$\quad = -180 + 180n - 180$

$\quad = 180n - 360$

$\quad = 180(n-2)$

93. $A = P + Prt$

$A - P = Prt$

$\dfrac{A-P}{Pt} = r \ \text{ or } \ r = \dfrac{A-P}{Pt}$

94. $12n^2 - 6n - 30n + 15$

$= 3\left(4n^2 - 2n - 10n + 5\right)$

$= 3\left[(4n^2 - 2n) - (10n - 5)\right]$

$= 3\left[2n(2n-1) - 5(2n-1)\right]$

$= 3\left[(2n-1)(2n-5)\right]$

$= 3(2n-1)(2n-5)$

95. $y = 2x + 1$

$3x - 2y = 1$

Substitute $2x + 1$ for y in the second equation.

$\quad 3x - 2y = 1$

$\quad 3x - 2(2x+1) = 1$

$\quad\quad 3x - 4x - 2 = 1$

$\quad\quad\quad -x - 2 = 1$

$\quad\quad\quad\quad -x = 3$

$\quad\quad\quad\quad\quad x = -3$

Substitute -3 for x in the first equation.

$y = 2(-3) + 1 = -6 + 1 = -5$

The solution is $(-3, -5)$.

96. $(x+4)^2 + y^2 = 25$

$(x+4)^2 + y^2 = 5^2$

The center is $(-4, 0)$ and the radius is 5.

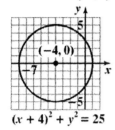

$(x+4)^2 + y^2 = 25$

Exercise Set 15.3

1. A geometric sequence is a sequence in which each term after the first is the same multiple of the preceding term.

3. To find the common ratio, take any term except the first and divide by the term that precedes it.

5. r^n approaches 0 as n gets larger and larger when $|r| < 1$.

7. Yes

9. Yes, s_∞ exists. $s_\infty = \dfrac{a_1}{1-r} = \dfrac{6}{1-\dfrac{1}{4}} = \dfrac{6}{\dfrac{3}{4}} = 6 \cdot \dfrac{4}{3} = 8$

This is true since $|r| < 1$.

11. $a_1 = 2$

$a_2 = 2(3)^{2-1} = 2(3) = 6$

$a_3 = 2(3)^{3-1} = 2(3)^2 = 2(9) = 18$

$a_4 = 2(3)^{4-1} = 2(3)^3 = 2(27) = 54$

$a_5 = 2(3)^{5-1} = 2(3)^4 = 2(81) = 162$

The terms are 2, 6, 18, 54, 162.

13. $a_1 = 6$

$a_2 = 6\left(-\dfrac{1}{2}\right)^{2-1} = 6\left(-\dfrac{1}{2}\right) = -3$

$a_3 = 6\left(-\dfrac{1}{2}\right)^{3-1} = 6\left(-\dfrac{1}{2}\right)^2 = 6\left(\dfrac{1}{4}\right) = \dfrac{3}{2}$

$a_4 = 6\left(-\dfrac{1}{2}\right)^{4-1} = 6\left(-\dfrac{1}{2}\right)^3 = 6\left(-\dfrac{1}{8}\right) = -\dfrac{3}{4}$

$a_5 = 6\left(-\dfrac{1}{2}\right)^{5-1} = 6\left(-\dfrac{1}{2}\right)^4 = 6\left(\dfrac{1}{16}\right) = \dfrac{3}{8}$

The terms are $6, -3, \dfrac{3}{2}, -\dfrac{3}{4}, \dfrac{3}{8}$.

15. $a_1 = 72$

$a_2 = 72\left(\dfrac{1}{3}\right)^{2-1} = 72\left(\dfrac{1}{3}\right) = 24$

$a_3 = 72\left(\dfrac{1}{3}\right)^{3-1} = 72\left(\dfrac{1}{3}\right)^2 = 72\left(\dfrac{1}{9}\right) = 8$

$a_4 = 72\left(\dfrac{1}{3}\right)^{4-1} = 72\left(\dfrac{1}{3}\right)^3 = 72\left(\dfrac{1}{27}\right) = \dfrac{8}{3}$

$a_5 = 72\left(\dfrac{1}{3}\right)^{5-1} = 72\left(\dfrac{1}{3}\right)^4 = 72\left(\dfrac{1}{81}\right) = \dfrac{8}{9}$

The terms are $72, 24, 8, \dfrac{8}{3}, \dfrac{8}{9}$.

17. $a_1 = 90$

$a_2 = 90\left(-\dfrac{1}{3}\right)^{2-1} = 90\left(-\dfrac{1}{3}\right) = -30$

$a_3 = 90\left(-\dfrac{1}{3}\right)^{3-1} = 90\left(-\dfrac{1}{3}\right)^2 = 90\left(\dfrac{1}{9}\right) = 10$

$a_4 = 90\left(-\dfrac{1}{3}\right)^{4-1} = 90\left(-\dfrac{1}{3}\right)^3 = 90\left(-\dfrac{1}{27}\right) = -\dfrac{10}{3}$

$a_5 = 90\left(-\dfrac{1}{3}\right)^{5-1} = 90\left(-\dfrac{1}{3}\right)^4 = 90\left(\dfrac{1}{81}\right) = \dfrac{10}{9}$

The terms are $90, -30, 10, -\dfrac{10}{3}, \dfrac{10}{9}$.

19. $a_1 = -1$

$a_2 = -1(3)^{2-1} = -1(3) = -3$

$a_3 = -1(3)^{3-1} = -1(3)^2 = -1(9) = -9$

$a_4 = -1(3)^{4-1} = -1(3)^3 = -1(27) = -27$

$a_5 = -1(3)^{5-1} = -1(3)^4 = -1(81) = -81$

The terms are $-1, -3, -9, -27, -81$.

21. $a_1 = 5$

$a_2 = 5(-2)^{2-1} = 5(-2)^1 = 5(-2) = -10$

$a_3 = 5(-2)^{3-1} = 5(-2)^2 = 5(4) = 20$

$a_4 = 5(-2)^{4-1} = 5(-2)^3 = 5(-8) = -40$

$a_5 = 5(-2)^{5-1} = 5(-2)^4 = 5(16) = 80$

The terms are $5, -10, 20, -40, 80$.

23. $a_1 = \frac{1}{3}$

$a_2 = \frac{1}{3}\left(\frac{1}{2}\right)^{2-1} = \frac{1}{3}\left(\frac{1}{2}\right) = \frac{1}{6}$

$a_3 = \frac{1}{3}\left(\frac{1}{2}\right)^{3-1} = \frac{1}{3}\left(\frac{1}{2}\right)^2 = \frac{1}{3}\left(\frac{1}{4}\right) = \frac{1}{12}$

$a_4 = \frac{1}{3}\left(\frac{1}{2}\right)^{4-1} = \frac{1}{3}\left(\frac{1}{2}\right)^3 = \frac{1}{3}\left(\frac{1}{8}\right) = \frac{1}{24}$

$a_5 = \frac{1}{3}\left(\frac{1}{2}\right)^{5-1} = \frac{1}{3}\left(\frac{1}{2}\right)^4 = \frac{1}{3}\left(\frac{1}{16}\right) = \frac{1}{48}$

The terms are $\frac{1}{3}, \frac{1}{6}, \frac{1}{12}, \frac{1}{24}, \frac{1}{48}$.

25. $a_1 = 3$

$a_2 = 3\left(\frac{3}{2}\right)^{2-1} = 3\left(\frac{3}{2}\right) = \frac{9}{2}$

$a_3 = 3\left(\frac{3}{2}\right)^{3-1} = 3\left(\frac{3}{2}\right)^2 = 3\left(\frac{9}{4}\right) = \frac{27}{4}$

$a_4 = 3\left(\frac{3}{2}\right)^{4-1} = 3\left(\frac{3}{2}\right)^3 = 3\left(\frac{27}{8}\right) = \frac{81}{8}$

$a_5 = 3\left(\frac{3}{2}\right)^{5-1} = 3\left(\frac{3}{2}\right)^4 = 3\left(\frac{81}{16}\right) = \frac{243}{16}$

The terms are $3, \frac{9}{2}, \frac{27}{4}, \frac{81}{8}, \frac{243}{16}$.

27. $a_6 = a_1 r^{6-1}$

$a_6 = 4(2)^{6-1} = 4(2)^5 = 4(32) = 128$

29. $a_9 = a_1 r^{9-1}$

$a_9 = -12\left(\frac{1}{2}\right)^{9-1} = -12\left(\frac{1}{2}\right)^8 = -12\left(\frac{1}{256}\right) = -\frac{3}{64}$

31. $a_{10} = a_1 r^{10-1}$

$a_{10} = \frac{1}{4}(2)^{10-1} = \frac{1}{4}(2)^9 = \frac{1}{4}(512) = 128$

33. $a_{12} = a_1 r^{12-1}$

$a_{12} = -3(-2)^{12-1} = -3(-2)^{11} = -3(-2048) = 6144$

35. $a_8 = a_1 r^{8-1}$

$a_8 = 2\left(\frac{1}{2}\right)^{8-1} = 2\left(\frac{1}{2}\right)^7 = 2\left(\frac{1}{128}\right) = \frac{1}{64}$

37. $a_7 = a_1 r^{7-1}$

$a_7 = 50\left(\frac{1}{3}\right)^{7-1} = 50\left(\frac{1}{3}\right)^6 = 50\left(\frac{1}{729}\right) = \frac{50}{729}$

39. $s_5 = \frac{a_1(1-r^5)}{1-r}$

$s_5 = \frac{5(1-2^5)}{1-2} = \frac{5(1-32)}{-1} = \frac{5(-31)}{-1} = \frac{-155}{-1} = 155$

41. $s_6 = \frac{a_1(1-r^6)}{1-r}$

$s_6 = \frac{2(1-5^6)}{1-5} = \frac{2(1-15,625)}{-4} = \frac{2(-15,624)}{-4} = 7812$

43. $s_7 = \frac{a_1(1-r^7)}{1-r}$

$s_7 = \frac{80(1-2^7)}{1-2}$

$= \frac{80(1-128)}{-1} = \frac{80(-127)}{-1} = \frac{-10,160}{-1} = 10,160$

45. $s_9 = \frac{a_1(1-r^9)}{1-r}$

$s_9 = \frac{-15\left[1-\left(-\frac{1}{2}\right)^9\right]}{1-\left(-\frac{1}{2}\right)}$

$= \frac{-15\left[1-\left(-\frac{1}{512}\right)\right]}{\frac{3}{2}}$

$= \frac{-15\left(1+\frac{1}{512}\right)}{\frac{3}{2}}$

$= \frac{-15\left(\frac{513}{512}\right)}{\frac{3}{2}}$

$= -15\left(\frac{513}{512}\right)\left(\frac{2}{3}\right)$

$= -\frac{2565}{256}$

47. $s_5 = \frac{a_1(1-r^5)}{1-r}$

$s_5 = \frac{-9\left[1-\left(\frac{2}{5}\right)^5\right]}{1-\frac{2}{5}} = \frac{-9\left(1-\frac{32}{3125}\right)}{\frac{3}{5}}$

$= \frac{-9\left(\frac{3093}{3125}\right)}{\frac{3}{5}} = -9\left(\frac{3093}{3125}\right)\left(\frac{5}{3}\right) = -\frac{9279}{625}$

49. $r = \frac{3}{2} \div 3 = \frac{3}{2} \cdot \frac{1}{3} = \frac{1}{2}$

$a_n = 3\left(\frac{1}{2}\right)^{n-1}$

51. $r = 18 \div 9 = 2$
$a_n = 9(2)^{n-1}$

53. $r = -6 \div 2 = -3$
$a_n = 2(-3)^{n-1}$

55. $r = \frac{1}{2} \div \frac{3}{4} = \frac{1}{2} \cdot \frac{4}{3} = \frac{2}{3}$
$a_n = \frac{3}{4}\left(\frac{2}{3}\right)^{n-1}$

57. $r = \frac{1}{2} \div 1 = \frac{1}{2}$
$s_\infty = \frac{1}{1-\frac{1}{2}} = \frac{1}{\frac{1}{2}} = 1\left(\frac{2}{1}\right) = 2$

59. $r = \frac{1}{5} \div 1 = \frac{1}{5}$
$s_\infty = \frac{1}{1-\frac{1}{5}} = \frac{1}{\frac{4}{5}} = 1\left(\frac{5}{4}\right) = \frac{5}{4}$

61. $r = 3 \div 6 = \frac{1}{2}$
$s_\infty = \frac{6}{1-\frac{1}{2}} = \frac{6}{\frac{1}{2}} = 6\left(\frac{2}{1}\right) = 12$

63. $r = 2 \div 5 = \frac{2}{5}$
$s_\infty = \frac{5}{1-\frac{2}{5}} = \frac{5}{\frac{3}{5}} = 5\left(\frac{5}{3}\right) = \frac{25}{3}$

65. $s_n = \frac{a_1\left(1-r^n\right)}{1-r}$
$93 = \frac{3\left(1-2^n\right)}{1-2}$
$93 = \frac{3\left(1-2^n\right)}{-1}$
$93 = -3\left(1-2^n\right)$
$-31 = 1-2^n$
$-32 = -2^n$
$32 = 2^n$
$n = 5$ since $2^5 = 32$

67. $s_n = \frac{a_1\left(1-r^n\right)}{1-r}$
$\frac{189}{32} = \frac{3\left[1-\left(\frac{1}{2}\right)^n\right]}{1-\frac{1}{2}}$
$\frac{189}{32} = \frac{3\left(1-\left(\frac{1}{2}\right)^n\right)}{\frac{1}{2}}$
$\frac{1}{2} \cdot \frac{1}{3} \cdot \frac{189}{32} = 1-\left(\frac{1}{2}\right)^n$
$\frac{63}{64} = 1-\left(\frac{1}{2}\right)^n$
$-\frac{1}{64} = -\left(\frac{1}{2}\right)^n$
$\frac{1}{64} = \left(\frac{1}{2}\right)^n$
$n = 6$ since $\left(\frac{1}{2}\right)^6 = \frac{1}{64}$

69. $r = 1 \div 2 = \frac{1}{2}$
$s_\infty = \frac{2}{1-\frac{1}{2}} = \frac{2}{\frac{1}{2}} = 2\left(\frac{2}{1}\right) = 4$

71. $r = \frac{16}{3} \div 8 = \frac{16}{3}\left(\frac{1}{8}\right) = \frac{2}{3}$
$s_\infty = \frac{8}{1-\frac{2}{3}} = \frac{8}{\frac{1}{3}} = 8\left(\frac{3}{1}\right) = 24$

73. $r = 20 \div -60 = \frac{20}{-60} = -\frac{1}{3}$
$s_\infty = \frac{-60}{1-\left(-\frac{1}{3}\right)} = \frac{-60}{1+\frac{1}{3}} = \frac{-60}{\frac{4}{3}} = -60\left(\frac{3}{4}\right) = -45$

75. $r = -\frac{12}{5} \div -12 = -\frac{12}{5}\left(-\frac{1}{12}\right) = \frac{1}{5}$
$s_\infty = \frac{-12}{1-\frac{1}{5}} = \frac{-12}{\frac{4}{5}} = -12\left(\frac{5}{4}\right) = -15$

77. $0.242424... = 0.24 + 0.0024 + 0.000024 + \cdots$
$\qquad\qquad = 0.24 + 0.24(0.01) + 0.24(0.01)^2 + \cdots$
$r = 0.01$ and $a_1 = 0.24$
$s_\infty = \frac{0.24}{1-0.01} = \frac{0.24}{0.99} = \frac{24}{99} = \frac{8}{33}$

79. $0.8888... = 0.8 + 0.08 + 0.008 + \cdots$

$\qquad = 0.8 + 0.8(0.1) + 0.8(0.1)^2 + \cdots$

$r = 0.1$ and $a_1 = 0.8$

$s_\infty = \dfrac{0.8}{1-0.1} = \dfrac{0.8}{0.9} = \dfrac{8}{9}$

81. $0.515151\cdots = 0.51 + 0.0051 + 0.000051 + \cdots$

$\qquad = 0.51 + 0.51(0.01) + 0.51(0.01)^2 + \cdots$

$r = 0.01$ and $a_1 = 0.51$

$s_\infty = \dfrac{0.51}{1-0.01} = \dfrac{0.51}{0.99} = \dfrac{51}{99} = \dfrac{17}{33}$

83. Consider a new series $b_1, b_2, b_3, \ldots$ where $b_1 = 15$ and $b_4 = 405$. Now $b_4 = b_1 r^{4-1}$ becomes

$405 = 15 r^3$

$\dfrac{405}{15} = r^3$

$27 = r^3$

$\sqrt[3]{27} = r$

$3 = r$

From the original series, $a_1 = \dfrac{a_2}{r} = \dfrac{15}{3} = 5$.

85. Consider a new series $b_1, b_2, b_3, \ldots$ where $b_1 = 28$ and $b_3 = 112$. Now $b_3 = b_1 r^{3-1}$ becomes

$112 = 28 r^2$

$\dfrac{112}{28} = r^2$

$4 = r^2$

so that $r = 2$ or $r = -2$. From the original series

$a_1 = \dfrac{a_3}{r^2} = \dfrac{28}{4} = 7$.

87. $a_1 = 1.40$, $n = 9$, $r = 1.03$

$a_n = a_1 r^{n-1}$

$a_9 = 1.4(1.03)^{9-1} = 1.4(1.03)^8 \approx 1.77$

In 8 years, a loaf of bread would cost $1.77.

89. $r = \dfrac{1}{2}$. Let a_n be the amount left after the nth

day. After 1 day there are $600\left(\dfrac{1}{2}\right) = 300$ grams

left, so $a_1 = 300$.

a. $37.5 = 300\left(\dfrac{1}{2}\right)^{n-1}$

$\dfrac{37.5}{300} = \left(\dfrac{1}{2}\right)^{n-1}$

$\dfrac{1}{8} = \left(\dfrac{1}{2}\right)^{n-1}$

$\left(\dfrac{1}{2}\right)^3 = \left(\dfrac{1}{2}\right)^{n-1}$

$n - 1 = 3$

$n = 4$

37.5 grams are left after 4 days.

b. $a_9 = 300\left(\dfrac{1}{2}\right)^{9-1} = 300\left(\dfrac{1}{256}\right) \approx 1.172$

After 9 days, about 1.172 grams of the substance remain.

91. After ten years will be at the beginning of the eleventh year. If the population increases by 1.1% each year, then the population at the beginning of a year will be 1.011 times the population at the beginning of the previous year.

a. Use $a_1 = 296.5$, $r = 1.011$, and $n = 11$.

$a_{11} = a_1 r^{11-1} = 296.5(1.011)^{10} \approx 330.78$

After ten years (the beginning of the eleventh year), the population is about 330.78 million people.

b. $a_n = 2(296.5) = 593$

Now, use $a_n = a_1 r^{n-1}$

$593 = 296.5(1.011)^{n-1}$

$\dfrac{593}{296.5} = (1.011)^{n-1}$

$2 = (1.011)^{n-1}$

Now, use logarithms.

$\log 2 = \log(1.011)^{n-1}$

$\log 2 = (n-1)\log 1.011$

$\dfrac{\log 2}{\log 1.011} = n - 1$

$63.4 \approx n - 1$

$64.4 \approx n$

The population will be double at the beginning of the 64.4 year, which is the end of the 63.4 year.

93. a. After 1 meter there is $\frac{1}{2}$ of the original light remaining, so $a_1 = \frac{1}{2}$.

$$a_1 = \frac{1}{2}$$

$$a_2 = \frac{1}{2}\left(\frac{1}{2}\right)^{2-1} = \left(\frac{1}{2}\right)\left(\frac{1}{2}\right) = \frac{1}{4}$$

$$a_3 = \frac{1}{2}\left(\frac{1}{2}\right)^{3-1} = \left(\frac{1}{2}\right)\left(\frac{1}{4}\right) = \frac{1}{8}$$

$$a_4 = \frac{1}{2}\left(\frac{1}{2}\right)^{4-1} = \left(\frac{1}{2}\right)\left(\frac{1}{8}\right) = \frac{1}{16}$$

$$a_5 = \frac{1}{2}\left(\frac{1}{2}\right)^{5-1} = \left(\frac{1}{2}\right)\left(\frac{1}{16}\right) = \frac{1}{32}$$

b. $a_n = \frac{1}{2}\left(\frac{1}{2}\right)^{n-1} = \left(\frac{1}{2}\right)^n$

c. $a_7 = \left(\frac{1}{2}\right)^7 = \frac{1}{128} \approx 0.0078$ or 0.78%

95. After 8 years will be the beginning of the ninth year. In any given year (except the first) there will be 106% of the previous amount in the account. Use $a_1 = 10{,}000$, $r = 1.06$, and $n = 9$

$$a_9 = 10{,}000\left(1.06\right)^{9-1}$$
$$\approx 10{,}000\left(1.5938481\right)$$
$$\approx 15{,}938.48$$

At the end of 8 years, there is $15,938.48 in the account.

97. a. $a_1 = 0.6(220) = 132$, $r = 0.6$

$$a_n = a_1 r^{n-1}$$
$$a_4 = 132(0.6)^{4-1} = 132(0.6)^3 = 28.512$$

The height of the fourth bounce is 28.512 feet.

b. $s_\infty = \frac{a_1}{1-r} = \frac{220}{1-0.6} = \frac{220}{0.4} = 550$

She travels a total of 550 feet in the downward direction.

99. a. $a_1 = 30(0.7) = 21$, $r = 0.7$

$$a_n = a_1 r^{n-1}$$
$$a_3 = 21(0.7)^{3-1} = 21(0.7)^2 = 10.29$$

The ball will bounce 10.29 inches on the third bounce.

b. $s_\infty = \frac{a_1}{1-r} = \frac{30}{1-0.7} = \frac{30}{0.3} = 100$

The ball travels a total of 100 inches in the downward direction.

101. Blue: $a_1 = 1$, $r = 2$

Red: $a_1 = 1$, $r = 3$

$$a_6 = a_1 r^{6-1} = a_1 r^5$$

Blue: $a_6 = 1(2)^5 = 32$

Red: $a_6 = 1(3)^5 = 243$

$243 - 32 = 211$

There are 211 more chips in the sixth stack of red chips.

103. Let a_n = value left after the nth year. After the first year there is $15000\left(\frac{4}{5}\right) = 12000$ of the value left, so $a_1 = 12000$, $r = \frac{4}{5}$.

$$a_2 = 12000\left(\frac{4}{5}\right)^{2-1} = 12000\left(\frac{4}{5}\right) = 9600$$

$$a_3 = 12000\left(\frac{4}{5}\right)^{3-1} = 12000\left(\frac{16}{25}\right) = 7680$$

$$a_4 = 12000\left(\frac{4}{5}\right)^{4-1} = 12000\left(\frac{64}{125}\right) = 6144$$

a. $12,000, $9,600, $7,680, $6,144

b. $a_n = 12000\left(\frac{4}{5}\right)^{n-1}$

c. $a_5 = 12000\left(\frac{4}{5}\right)^{5-1} = 12000\left(\frac{256}{625}\right) = 4915.20$

After 5 years, the value of the car is $4,915.20.

105. Each time the ball bounces it goes up and then comes down the same distance. Therefore, the total vertical distance will be twice the height it rises after each bounce plus the initial 10 feet. The heights after each bounce form an infinite geometric sequence with $r = 0.9$ and $a_1 = 9$.

$$s_\infty = \frac{9}{1-0.9} = \frac{9}{0.1} = 90$$

Total distance: 6

$$2\left(s_\infty\right) + 10 = 2\left(90\right) + 10 = 190$$

The total vertical distance is 190 feet.

107. a. y_2 goes up more steeply.

b.

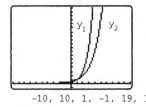

$$-10,\ 10,\ 1,\ -1,\ 19,\ 1$$

From the graph y_2 goes up more steeply, which agrees with our answer to part (a).

109. This is a geometric sequence with $r = \dfrac{2}{1} = 2$.

Also, $a_n = 1,048,576 = 2^{20}$. Using $a_n = a_1 r^{n-1}$ gives

$$2^{20} = 1(2)^{n-1}$$
$$2^{20} = 2^{n-1}$$
$$20 = n - 1$$
$$21 = n$$

Thus, there are 21 terms in the sequence.

$$s_{21} = \frac{a_1\left(1 - r^{21}\right)}{1 - r} = \frac{1\left(1 - 2^{21}\right)}{1 - 2}$$

$$= \frac{1 - 2,097,152}{-1} = \frac{-2,097,151}{-1} = 2,097,151$$

110. $(4x^2 - 3x + 6)(2x - 3)$
$$= 8x^3 - 12x^2 - 6x^2 + 9x + 12x - 18$$
$$= 8x^3 - 18x^2 + 21x - 18$$

111.
$$
\begin{array}{r}
8x - 15 \\
2x+5 \overline{\smash{\big)}\ 16x^2 + 10x - 18} \\
\underline{16x^2 + 40x} \\
-30x - 18 \\
\underline{-30x - 75} \\
57
\end{array}
$$

$$(16x^2 + 10x - 18) \div (2x + 5) = 8x - 15 + \frac{57}{2x + 5}$$

112. Let $x =$ the time it takes Mr. Donovan to load the truck by himself. Then $2x =$ the time it take Mrs. Donovan to load the truck by herself.

	Rate of Work	Time Worked	Part of Task
Mr. Donovan	$\dfrac{1}{x}$	8	$\dfrac{8}{x}$
Mrs. Donovan	$\dfrac{1}{2x}$	8	$\dfrac{8}{2x}$

$$\frac{8}{x} + \frac{8}{2x} = 1$$
$$2x\left(\frac{8}{x} + \frac{8}{2x}\right) = 2x(1)$$
$$16 + 8 = 2x$$
$$24 = 2x$$
$$12 = x$$

Mr. Donovan will need 12 hours to load the truck by himself.

113. $\left(\dfrac{9}{100}\right)^{-1/2} = \left(\dfrac{100}{9}\right)^{1/2} = \sqrt{\dfrac{100}{9}} = \dfrac{\sqrt{100}}{\sqrt{9}} = \dfrac{10}{3}$

114. $\sqrt[3]{9x^2 y}\left(\sqrt[3]{3x^4 y^6} - \sqrt[3]{8xy^4}\right)$
$$= \sqrt[3]{9x^2 y \left(3x^4 y^6\right)} - \sqrt[3]{9x^2 y \left(8xy^4\right)}$$
$$= \sqrt[3]{27x^6 y^7} - \sqrt[3]{72x^3 y^5}$$
$$= \sqrt[3]{27x^6 y^6 y} - \sqrt[3]{8 \cdot 9x^3 y^3 y^2}$$
$$= 3x^2 y^2 \sqrt[3]{y} - 2xy\sqrt[3]{9y^2}$$

115. $x\sqrt{y} - 2\sqrt{x^2 y} + \sqrt{4x^2 y}$
$$= x\sqrt{y} - 2x\sqrt{y} + 2x\sqrt{y}$$
$$= x\sqrt{y}$$

116. $\sqrt{a^2 + 9a + 3} = -a$
$$\left(\sqrt{a^2 + 9a + 3}\right)^2 = (-a)^2$$
$$a^2 + 9a + 3 = a^2$$
$$9a = -3$$
$$a = \frac{-3}{9} = -\frac{1}{3}$$

Check: $\sqrt{\left(-\dfrac{1}{3}\right)^2 + 9\left(-\dfrac{1}{3}\right) + 3} \overset{?}{=} -\left(-\dfrac{1}{3}\right)$

$$\sqrt{\frac{1}{9} - 3 + 3} \overset{?}{=} \frac{1}{3}$$
$$\sqrt{\frac{1}{9}} \overset{?}{=} \frac{1}{3}$$
$$\frac{1}{3} = \frac{1}{3} \quad \text{True}$$

The solution is $-\dfrac{1}{3}$.

Mid-Chapter Test: 15.1-15.3

1. $a_n = -3n + 5$

   $a_1 = -3(1) + 5 = -3 + 5 = 2$

   $a_2 = -3(2) + 5 = -6 + 5 = -1$

   $a_3 = -3(3) + 5 = -9 + 5 = -4$

   $a_4 = -3(4) + 5 = -12 + 5 = -7$

   $a_5 = -3(5) + 5 = -15 + 5 = -10$

   The terms are $2, -1, -4, -7, -10$.

2. $a_n = n(n + 6)$

   $a_7 = 7(7 + 6) = 7(13) = 91$

3. $a_n = 2^n - 1$

   $a_1 = 2^1 - 1 = 2 - 1 = 1$

   $a_2 = 2^2 - 1 = 4 - 1 = 3$

   $a_3 = 2^3 - 1 = 8 - 1 = 7$

   $s_1 = a_1 = 1$

   $s_3 = a_1 + a_2 + a_3 = 1 + 3 + 7 = 11$

4. Each term is 4 less than the previous term. The next three terms are $-15, -19, -23$.

5. $\displaystyle\sum_{i=1}^{5}(4i - 3)$

   $= [4(1) - 3] + [4(2) - 3] + [4(3) - 3]$

   $\quad + [4(4) - 3] + [4(5) - 3]$

   $= [4 - 3] + [8 - 3] + [12 - 3] + [16 - 3] + [20 - 3]$

   $= 1 + 5 + 9 + 13 + 17$

   $= 45$

6. $a_n = \dfrac{1}{3}n + 7$

   The fifth partial sum is $\displaystyle\sum_{i=1}^{5}\left(\frac{1}{3}i + 7\right)$.

7. $a_1 = -6$

   $a_2 = -6 + (2 - 1)(5) = -6 + 1(5) = -6 + 5 = -1$

   $a_3 = -6 + (3 - 1)(5) = -6 + 2(5) = -6 + 10 = 4$

   $a_4 = -6 + (4 - 1)(5) = -6 + 3(5) = -6 + 15 = 9$

   $a_5 = -6 + (5 - 1)(5) = -6 + 4(5) = -6 + 20 = 14$

   The terms are $-6, -1, 4, 9, 14$. The general term is $a_n = -6 + (n - 1)5$ or $a_n = 5n - 11$.

8. $a_n = a_1 + (n - 1)d$

   $-\dfrac{1}{2} = \dfrac{11}{2} + (7 - 1)d$

   $-\dfrac{1}{2} = \dfrac{11}{2} + 6d$

   $-\dfrac{12}{2} = 6d$

   $-6 = 6d$

   $-1 = d$

9. $a_n = a_1 + (n - 1)d$

   $-3 = 22 + (n - 1)(-5)$

   $-3 = 22 - 5n + 5$

   $-3 = 27 - 5n$

   $-30 = -5n$

   $6 = n$

10. $a_n = a_1 + (n - 1)d$

    $7 = -8 + (6 - 1)d$

    $7 = -8 + 5d$

    $15 = 5d$

    $3 = d$

    $s_6 = \dfrac{6(a_1 + a_6)}{2} = \dfrac{6(-8 + 7)}{2} = \dfrac{6(-1)}{2} = \dfrac{-6}{2} = -3$

11. $a_n = a_1 + (n - 1)d$

    $a_{10} = \dfrac{5}{2} + (10 - 1)\left(\dfrac{1}{2}\right) = \dfrac{5}{2} + 9\left(\dfrac{1}{2}\right) = \dfrac{5}{2} + \dfrac{9}{2} = \dfrac{14}{2} = 7$

    $s_{10} = \dfrac{10(a_1 + a_{10})}{2}$

    $= \dfrac{10\left(\frac{5}{2} + 7\right)}{2} = \dfrac{25 + 70}{2} = \dfrac{95}{2} = 47\dfrac{1}{2}$

12. $d = 0 - (-7) = 0 + 7 = 7$

    $a_n = a_1 + (n - 1)d$

    $63 = -7 + (n - 1)(7)$

    $63 = -7 + 7n - 7$

    $63 = 7n - 14$

    $77 = 7n$

    $11 = n$

    There are 11 terms in the sequence.

13. $16 + 15 + 14 + \cdots + 1$ or $1 + 2 + 3 + \cdots + 16$

    $s_n = \dfrac{n(a_1 + a_n)}{2}$

    $s_{16} = \dfrac{16(1 + 16)}{2} = \dfrac{16(17)}{2} = \dfrac{272}{2} = 136$

    There are 136 logs in the pile.

14. $a_1 = 80$

$$a_2 = 80\left(-\frac{1}{2}\right)^{2-1} = 80\left(-\frac{1}{2}\right) = -40$$

$$a_3 = 80\left(-\frac{1}{2}\right)^{3-1} = 80\left(-\frac{1}{2}\right)^2 = 80\left(\frac{1}{4}\right) = 20$$

$$a_4 = 80\left(-\frac{1}{2}\right)^{4-1} = 80\left(-\frac{1}{2}\right)^3 = 80\left(-\frac{1}{8}\right) = -10$$

$$a_5 = 80\left(-\frac{1}{2}\right)^{5-1} = 80\left(-\frac{1}{2}\right)^4 = 80\left(\frac{1}{16}\right) = 5$$

The terms are 80, –40, 20, –10, 5.

15. $a_7 = a_1 r^{7-1}$

$$a_7 = 81\left(\frac{1}{3}\right)^{7-1} = 81\left(\frac{1}{3}\right)^6 = 81\left(\frac{1}{729}\right) = \frac{1}{9}$$

16. $s_6 = \dfrac{a_1(1-r^6)}{1-r}$

$$s_6 = \frac{5(1-2^6)}{1-2} = \frac{5(1-64)}{-1} = \frac{5(-63)}{-1} = 315$$

17. $r = -\dfrac{16}{3} \div 8 = -\dfrac{16}{3}\cdot\dfrac{1}{8} = -\dfrac{2}{3}$

18. $r = 4 \div 12 = \dfrac{4}{12} = \dfrac{1}{3}$

$$s_\infty = \frac{a_1}{1-r} = \frac{12}{1-\frac{1}{3}} = \frac{12}{\frac{2}{3}} = 12\left(\frac{3}{2}\right) = 18$$

19. $0.878787\ldots = 0.87 + 0.0087 + 0.000087 + \cdots$

$$= 0.87 + 0.87(0.01) + 0.87(0.01)^2 + \cdots$$

$r = 0.01$ and $a_1 = 0.87$

$$s_\infty = \frac{0.87}{1-0.01} = \frac{0.87}{0.99} = \frac{87}{99} = \frac{29}{33}$$

20. a. A sequence is a list of numbers arranged in a specific order.

b. An arithmetic sequence is a sequence where each term differs by a constant amount.

c. A geometric sequence is a sequence in which each term after the first is the same multiple of the preceding term.

d. A series is the sum of the terms of a sequence.

Exercise Set 15.4

1. Answers will vary. One possibility follows. The first and last numbers in each row are 1 and the inner numbers are obtained by adding the two numbers in the row above (to the right and left).

```
            1
          1   1
        1   2   1
      1   3   3   1
    1   4   6   4   1
```

3. $1! = 1$

5. No. Factorials are only defined for nonnegative integers.

7. The expansion of $(a+b)^{13}$ has 14 terms, one more than the power to which the binomial is raised.

9. $\dbinom{5}{2} = \dfrac{5!}{2!\,(5-2)!} = \dfrac{5!}{2!\,3!} = \dfrac{5\cdot4\cdot3\cdot2\cdot1}{(2\cdot1)(3\cdot2\cdot1)} = \dfrac{20}{2} = 10$

11. $\dbinom{5}{5} = \dfrac{5!}{5!(5-5)!} = \dfrac{5!}{5!\cdot0!} = \dfrac{1}{0!} = \dfrac{1}{1} = 1$

13. $\dbinom{7}{0} = \dfrac{7!}{0!(7-0)!} = \dfrac{7!}{0!\cdot7!} = \dfrac{7!}{0!\cdot7!} = \dfrac{1}{0!} = \dfrac{1}{1} = 1$

15. $\dbinom{8}{4} = \dfrac{8!}{4!\,(8-4)!} = \dfrac{8!}{4!\,4!}$

$$= \frac{8\cdot7\cdot6\cdot5\cdot4\cdot3\cdot2\cdot1}{(4\cdot3\cdot2\cdot1)(4\cdot3\cdot2\cdot1)} = \frac{1680}{24} = 70$$

17. $\dbinom{8}{2} = \dfrac{8!}{2!\,(8-2)!} = \dfrac{8!}{2!\,6!}$

$$= \frac{8\cdot7\cdot6\cdot5\cdot4\cdot3\cdot2\cdot1}{(2\cdot1)\cdot(6\cdot5\cdot4\cdot3\cdot2\cdot1)} = \frac{56}{2} = 28$$

19. $(x+4)^3$

$$= \binom{3}{0}x^3 4^0 + \binom{3}{1}x^2 4^1 + \binom{3}{2}x^1 4^2 + \binom{3}{3}x^0 4^3$$

$$= 1x^3(1) + 3x^2(4) + 3x(16) + 1(1)64$$

$$= x^3 + 12x^2 + 48x + 64$$

21. $(2x-3)^3 = \binom{3}{0}(2x)^3(-3)^0 + \binom{3}{1}(2x)^2(-3)^1 + \binom{3}{2}(2x)^1(-3)^2 + \binom{3}{3}(2x)^0(-3)^3$

$= 1(8x^3)(1) + 3(4x^2)(-3) + 3(2x)(9) + 1(1)(-27)$

$= 8x^3 - 36x^2 + 54x - 27$

23. $(a-b)^4 = \binom{4}{0}a^4(-b)^0 + \binom{4}{1}a^3(-b)^1 + \binom{4}{2}a^2(-b)^2 + \binom{4}{3}a^1(-b)^3 + \binom{4}{4}a^0(-b)^4$

$= 1a^4(1) + 4a^3(-b) + 6a^2b^2 + 4a(-b^3) + 1(1)b^4$

$= a^4 - 4a^3b + 6a^2b^2 - 4ab^3 + b^4$

25. $(3a-b)^5 = \binom{5}{0}(3a)^5(-b)^0 + \binom{5}{1}(3a)^4(-b)^1 + \binom{5}{2}(3a)^3(-b)^2 + \binom{5}{3}(3a)^2(-b)^3 + \binom{5}{4}(3a)^1(-b)^4 + \binom{5}{5}(3a)^0(-b)^5$

$= 1(243a^5)(1) + 5(81a^4)(-b) + 10(27a^3)b^2 + 10(9a^2)(-b^3) + 5(3a)b^4 + 1(1)(-b^5)$

$= 243a^5 - 405a^4b + 270a^3b^2 - 90a^2b^3 + 15ab^4 - b^5$

27. $\left(2x+\dfrac{1}{2}\right)^4 = \binom{4}{0}(2x)^4\left(\dfrac{1}{2}\right)^0 + \binom{4}{1}(2x)^3\left(\dfrac{1}{2}\right)^1 + \binom{4}{2}(2x)^2\left(\dfrac{1}{2}\right)^2 + \binom{4}{3}(2x)^1\left(\dfrac{1}{2}\right)^3 + \binom{4}{4}(2x)^0\left(\dfrac{1}{2}\right)^4$

$= 1(16x^4)(1) + 4(8x^3)\left(\dfrac{1}{2}\right) + 6(4x^2)\left(\dfrac{1}{4}\right) + 4(2x)\left(\dfrac{1}{8}\right) + 1(1)\left(\dfrac{1}{16}\right)$

$= 16x^4 + 16x^3 + 6x^2 + x + \dfrac{1}{16}$

29. $\left(\dfrac{x}{2}-3\right)^4 = \binom{4}{0}\left(\dfrac{x}{2}\right)^4(-3)^0 + \binom{4}{1}\left(\dfrac{x}{2}\right)^3(-3)^1 + \binom{4}{2}\left(\dfrac{x}{2}\right)^2(-3)^2 + \binom{4}{3}\left(\dfrac{x}{2}\right)^1(-3)^3 + \binom{4}{4}\left(\dfrac{x}{2}\right)^0(-3)^4$

$= 1\left(\dfrac{x^4}{16}\right)(1) + 4\left(\dfrac{x^3}{8}\right)(-3) + 6\left(\dfrac{x^2}{4}\right)(9) + 4\left(\dfrac{x}{2}\right)(-27) + 1(1)(81)$

$= \dfrac{x^4}{16} - \dfrac{3x^3}{2} + \dfrac{27x^2}{2} - 54x + 81$

31. $(x+10)^{10} = \binom{10}{0}x^{10}(10)^0 + \binom{10}{1}x^9(10)^1 + \binom{10}{2}x^8(10)^2 + \binom{10}{3}x^7(10)^3 + \cdots$

$= 1x^{10}(1) + \dfrac{10}{1}x^9(10) + \dfrac{10\cdot9}{2\cdot1}x^8(100) + \dfrac{10\cdot9\cdot8}{3\cdot2\cdot1}x^7(1000) + \cdots$

$= x^{10} + 100x^9 + 4,500x^8 + 120,000x^7 + \cdots$

33. $(3x-y)^7 = \binom{7}{0}(3x)^7(-y)^0 + \binom{7}{1}(3x)^6(-y)^1 + \binom{7}{2}(3x)^5(-y)^2 + \binom{7}{3}(3x)^4(-y)^3 + \cdots$

$= 1(2187x^7)(1) + \dfrac{7}{1}(729x^6)(-y) + \dfrac{7\cdot6}{2\cdot1}(243x^5)(y^2) + \dfrac{7\cdot6\cdot5}{3\cdot2\cdot1}(81x^4)(-y^3) + \cdots$

$= 2187x^7 + 7(729x^6)(-y) + 21(243x^5)(y^2) + 35(81x^4)(-y^3) + \cdots$

$= 2187x^7 - 5103x^6y + 5103x^5y^2 - 2835x^4y^3 + \cdots$

35. $(x^2 - 3y)^8 = \binom{8}{0}(x^2)^8(-3y)^0 + \binom{8}{1}(x^2)^7(-3y)^1 + \binom{8}{2}(x^2)^6(-3y)^2 + \binom{8}{3}(x^2)^5(-3y)^3 + \cdots$

$= 1(x^{16})(1) + \frac{8}{1}(x^{14})(-3y) + \frac{8 \cdot 7}{2 \cdot 1}(x^{12})(9y^2) + \frac{8 \cdot 7 \cdot 6}{3 \cdot 2 \cdot 1}(x^{10})(-27y^3) + \cdots$

$= x^{16} + 8(x^{14})(-3y) + 28(x^{12})(9y^2) + 56(x^{10})(-27y^3) + \cdots$

$= x^{16} - 24x^{14}y + 252x^{12}y^2 - 1512x^{10}y^3 + \cdots$

---

37. Yes, $n! = n \cdot (n-1)!$
$4! = 4 \cdot 3 \cdot 2 \cdot 1$
$\quad = 4 \cdot (3 \cdot 2 \cdot 1)$
$\quad = 4 \cdot (3)!$
$\quad = 4 \cdot (4-1)!$

39. Yes, $(n-3)! = (n-3)(n-4)(n-5)!$ for $n \geq 5$.
Let $n = 7$:
$(7-3)! = (7-3)(7-4)(7-5)!$ or
$4! = 4 \cdot 3 \cdot 2! = 4 \cdot 3 \cdot 2 \cdot 1 = 4!$

41. $\binom{n}{m} = 1$ when either $n = m$ or $m = 0$.

43. $(x+3)^8$

First term is $\binom{8}{0}(x)^8(3)^0 = 1(x^8)(1) = x^8$.

Second term is $\binom{8}{1}(x)^7(3)^1 = 8(x^7)(3) = 24x^7$.

Next to last term is
$\binom{8}{7}(x)^1(3)^7 = 8(x)(2187) = 17,496x$.

Last term is $\binom{8}{8}(x)^0(3)^8 = 1(1)(6561) = 6561$.

45. $(a+b)^n = \sum_{i=0}^{n} \binom{n}{i} a^{n-i} b^i$

47. Let $x = 0$.
$2x + y = 10$
$2(0) + y = 6$
$\quad\quad y = 6$
The y-intercept is $(0, 10)$.

48. $x(x-11) = -18$
$x^2 - 11x + 18 = 0$
$(x-9)(x-2) = 0$
$x - 9 = 0 \quad$ or $\quad x - 2 = 0$
$\quad x = 9 \quad\quad\quad\quad x = 2$

49. $\frac{1}{5}x + \frac{1}{2}y = 4 \overset{\times(-10)}{\Rightarrow} -2x - 5y = -40 \quad (1)$

$\frac{2}{3}x - y = \frac{8}{3} \overset{\times 3}{\Rightarrow} \underline{\quad 2x - 3y = \quad 8 \quad (2)}$

$\quad\quad\quad\quad\quad\quad\quad\quad -8y = -32$
$\quad\quad\quad\quad\quad\quad\quad\quad\quad\quad y = 4$

Substitute $y = 4$ into equation (2):
$2x - 3(4) = 8$
$\quad\quad 2x = 20$
$\quad\quad\quad x = 10$
The solution is $(10, 4)$.

50. $\sqrt{20xy^4}\sqrt{6x^5y^7} = \sqrt{120x^6y^{11}}$

$= \sqrt{4x^6y^{10} \cdot 30y}$

$= 2x^3y^5\sqrt{30y}$

51. $f(x) = 3x + 8$
$y = 3x + 8$
$x = 3y + 8$
$3y = x - 8$
$y = \frac{x-8}{3}$

$f^{-1}(x) = \frac{x-8}{3}$

Chapter 15 Review Exercises

1. $a_n = n + 5$

$a_1 = 1 + 5 = 6$

$a_2 = 2 + 5 = 7$

$a_3 = 3 + 5 = 8$

$a_4 = 4 + 5 = 9$

$a_5 = 5 + 5 = 10$

The terms are 6, 7, 8, 9, 10.

2. $a_n = n^2 + n - 3$

$a_1 = 1^2 + 1 - 3 = -1$

$a_2 = 2^2 + 2 - 3 = 3$

$a_3 = 3^2 + 3 - 3 = 9$

$a_4 = 4^2 + 4 - 3 = 17$

$a_5 = 5^2 + 5 - 3 = 27$

The terms are -1, 3, 9, 17, 27 .

3. $a_n = \dfrac{6}{n}$

$a_1 = \dfrac{6}{1} = 6$

$a_2 = \dfrac{6}{2} = 3$

$a_3 = \dfrac{6}{3} = 2$

$a_4 = \dfrac{6}{4} = \dfrac{3}{2}$

$a_5 = \dfrac{6}{5}$

The terms are 6, 3, 2, $\dfrac{3}{2}$, $\dfrac{6}{5}$.

4. $a_n = \dfrac{n^2}{n + 4}$

$a_1 = \dfrac{1^2}{1 + 4} = \dfrac{1}{5}$

$a_2 = \dfrac{2^2}{2 + 4} = \dfrac{4}{6} = \dfrac{2}{3}$

$a_3 = \dfrac{3^2}{3 + 4} = \dfrac{9}{7}$

$a_4 = \dfrac{4^2}{4 + 4} = \dfrac{16}{8} = 2$

$a_5 = \dfrac{5^2}{5 + 4} = \dfrac{25}{9}$

The terms are $\dfrac{1}{5}, \dfrac{2}{3}, \dfrac{9}{7}, 2, \dfrac{25}{9}$.

5. $a_n = 3n - 10$

$a_7 = 3(7) - 10 = 21 - 10 = 11$

6. $a_n = (-1)^n + 5$

$a_7 = (-1)^7 + 5 = -1 + 5 = 4$

7. $a_n = \dfrac{n + 17}{n^2}$

$a_9 = \dfrac{9 + 17}{9^2} = \dfrac{26}{81}$

8. $a_n = (n)(n - 3)$

$a_{11} = (11)(11 - 3) = (11)(8) = 88$

9. $a_n = 2n + 5$

$a_1 = 2(1) + 5 = 2 + 5 = 7$

$a_2 = 2(2) + 5 = 4 + 5 = 9$

$a_3 = 2(3) + 5 = 6 + 5 = 11$

$s_1 = a_1 = 7$

$s_3 = a_1 + a_2 + a_3 = 7 + 9 + 11 = 27$

10. $a_n = n^2 + 8$

$a_1 = (1)^2 + 8 = 1 + 8 = 9$

$a_2 = (2)^2 + 8 = 4 + 8 = 12$

$a_3 = (3)^2 + 8 = 9 + 8 = 17$

$s_1 = a_1 = 9$

$s_3 = a_1 + a_2 + a_3 = 9 + 12 + 17 = 38$

11. $a_n = \dfrac{n + 3}{n + 2}$

$a_1 = \dfrac{1 + 3}{1 + 2} = \dfrac{4}{3}$

$a_2 = \dfrac{2 + 3}{2 + 2} = \dfrac{5}{4}$

$a_3 = \dfrac{3 + 3}{3 + 2} = \dfrac{6}{5}$

$s_1 = a_1 = \dfrac{4}{3}$

$s_3 = a_1 + a_2 + a_3$

$\quad = \dfrac{4}{3} + \dfrac{5}{4} + \dfrac{6}{5}$

$\quad = \dfrac{80}{60} + \dfrac{75}{60} + \dfrac{72}{60}$

$\quad = \dfrac{227}{60}$

12. $a_n = (-1)^n(n+8)$

$a_1 = (-1)^1(1+8) = (-1)(9) = -9$

$a_2 = (-1)^2(2+8) = 1(10) = 10$

$a_3 = (-1)^3(3+8) = -1(11) = -11$

$s_1 = a_1 = -9$

$s_3 = a_1 + a_2 + a_3 = -9 + 10 - 11 = -10$

13. This is a geometric sequence with $r = 4 \div 2 = 2$ and $a_1 = 2$.

$a_5 = 2(2)^{5-1} = 2 \cdot 2^4 = 32$

$a_6 = 2(2)^{6-1} = 2 \cdot 2^5 = 64$

$a_7 = 2(2)^{7-1} = 2 \cdot 2^6 = 128$

The terms are 32, 64, 128.

$a_n = 2(2)^{n-1} = 2^1 2^{n-1} = 2^n$

14. This is a geometric sequence with

$r = 9 \div (-27) = \dfrac{9}{-27} = -\dfrac{1}{3}$ and $a_1 = -27$.

$a_5 = -27\left(-\dfrac{1}{3}\right)^4 = -27\left(\dfrac{1}{81}\right) = -\dfrac{1}{3}$

$a_6 = -27\left(-\dfrac{1}{3}\right)^5 = -27\left(-\dfrac{1}{243}\right) = \dfrac{1}{9}$

$a_5 = -27\left(-\dfrac{1}{3}\right)^6 = -27\left(\dfrac{1}{729}\right) = -\dfrac{1}{27}$

The terms are $-\dfrac{1}{3}, \dfrac{1}{9}, -\dfrac{1}{27}$.

$a_n = -27\left(-\dfrac{1}{3}\right)^{n-1}$ or $a_n = (-1)^n\left(3^{4-n}\right)$

15. This is a geometric sequence with

$r = \dfrac{2}{7} \div \dfrac{1}{7} = \dfrac{2}{7} \cdot \dfrac{7}{1} = 2$ and $a_1 = \dfrac{1}{7}$.

$a_5 = \dfrac{1}{7}(2)^{5-1} = \dfrac{2^4}{7} = \dfrac{16}{7}$

$a_6 = \dfrac{1}{7}(2)^{6-1} = \dfrac{2^5}{7} = \dfrac{32}{7}$

$a_7 = \dfrac{1}{7}(2)^{7-1} = \dfrac{2^6}{7} = \dfrac{64}{7}$

The terms are $\dfrac{16}{7}, \dfrac{32}{7}, \dfrac{64}{7}$.

$a_n = \dfrac{1}{7}(2)^{n-1} = \dfrac{2^{n-1}}{7}$

16. This is an arithmetic sequence with $d = 9 - 13 = -4$ and $a_1 = 13$.

$a_5 = 13 + (5-1)(-4) = 13 - 16 = -3$

$a_6 = 13 + (6-1)(-4) = 13 - 20 = -7$

$a_7 = 13 + (7-1)(-4) = 13 - 24 = -11$

The terms are $-3, -7, -11$.

$a_n = a_1 + (n-1)d$

$\quad = 13 + (n-1)(-4)$

$\quad = 13 - 4n + 4$

$\quad = 17 - 4n$

17. $\displaystyle\sum_{i=1}^{3}(i^2 + 9) = (1^2 + 9) + (2^2 + 9) + (3^2 + 9)$

$\qquad = (1+9) + (4+9) + (9+9)$

$\qquad = 10 + 13 + 18$

$\qquad = 41$

18. $\displaystyle\sum_{i=1}^{4} i(i+5)$

$= 1(1+5) + 2(2+5) + 3(3+5) + 4(4+5)$

$= 1(6) + 2(7) + 3(8) + 4(9)$

$= 6 + 14 + 24 + 36$

$= 80$

19. $\displaystyle\sum_{i=1}^{5}\dfrac{i^2}{6} = \dfrac{1^2}{6} + \dfrac{2^2}{6} + \dfrac{3^2}{6} + \dfrac{4^2}{6} + \dfrac{5^2}{6}$

$\qquad = \dfrac{1}{6} + \dfrac{4}{6} + \dfrac{9}{6} + \dfrac{16}{6} + \dfrac{25}{6}$

$\qquad = \dfrac{55}{6}$

20. $\displaystyle\sum_{i=1}^{4}\dfrac{i}{i+1} = \dfrac{1}{1+1} + \dfrac{2}{2+1} + \dfrac{3}{3+1} + \dfrac{4}{4+1}$

$\qquad = \dfrac{1}{2} + \dfrac{2}{3} + \dfrac{3}{4} + \dfrac{4}{5}$

$\qquad = \dfrac{163}{60}$

21. $\displaystyle\sum_{i=1}^{4} x_i = x_1 + x_2 + x_3 + x_4 = 3 + 9 + 7 + 10 = 29$

22. $\displaystyle\sum_{i=1}^{4}(x_i)^2 = x_1^2 + x_2^2 + x_3^2 + x_4^2$

$\qquad = 3^2 + 9^2 + 7^2 + 10^2$

$\qquad = 9 + 81 + 49 + 100$

$\qquad = 239$

23. $\displaystyle\sum_{i=2}^{3}(x_i^2+1)=(x_2^2+1)+(x_3^2+1)$

$$=(9^2+1)+(7^2+1)$$
$$=(81+1)+(49+1)$$
$$=82+50$$
$$=132$$

24. $\displaystyle\left(\sum_{i=1}^{4}x_i\right)^2=(x_1+x_2+x_3+x_4)^2$

$$=(3+9+7+10)^2$$
$$=(29)^2$$
$$=841$$

25. a. perimeter of rectangle: $p=2l+2w$

$$p_1=2(1)+2(1+3)=2+8=10$$
$$p_2=2(2)+2(2+3)=4+10=14$$
$$p_3=2(3)+2(3+3)=6+12=18$$
$$p_4=2(4)+2(4+3)=8+14=22$$

  b. $p_n=2n+2(n+3)=2n+2n+6=4n+6$

26. a. area of rectangle: $a=l\cdot w$

$$a_1=1(1+3)=4$$
$$a_2=2(2+3)=10$$
$$a_3=3(3+3)=18$$
$$a_4=4(4+3)=28$$

  b. $a_n=n(n+3)=n^2+3n$

27. $a_1=5$

$$a_2=5+(2-1)(3)=5+1(3)=5+3=8$$
$$a_3=5+(3-1)(3)=5+2(3)=5+6=11$$
$$a_4=5+(4-1)(3)=5+3(3)=5+9=14$$
$$a_5=5+(5-1)(3)=5+4(3)=5+12=17$$

The terms are 5, 7, 9, 11, 13.

28. $a_1=5$

$$a_2=5+(2-1)\left(-\frac{1}{3}\right)=5+1\left(-\frac{1}{3}\right)=\frac{15}{3}-\frac{1}{3}=\frac{14}{3}$$
$$a_3=5+(3-1)\left(-\frac{1}{3}\right)=5+2\left(-\frac{1}{3}\right)=\frac{15}{3}-\frac{2}{3}=\frac{13}{3}$$
$$a_4=5+(4-1)\left(-\frac{1}{3}\right)=5+3\left(-\frac{1}{3}\right)=5-1=4$$
$$a_5=5+(5-1)\left(-\frac{1}{3}\right)=5+4\left(-\frac{1}{3}\right)=\frac{15}{3}-\frac{4}{3}=\frac{11}{3}$$

The terms are 5, $\dfrac{14}{3}$, $\dfrac{13}{3}$, 4, $\dfrac{11}{3}$.

29. $a_1=\dfrac{1}{2}$

$$a_2=\frac{1}{2}+(2-1)(-2)=\frac{1}{2}-2=-\frac{3}{2}$$
$$a_3=\frac{1}{2}+(3-1)(-2)=\frac{1}{2}-4=-\frac{7}{2}$$
$$a_4=\frac{1}{2}+(4-1)(-2)=\frac{1}{2}-6=-\frac{11}{2}$$
$$a_5=\frac{1}{2}+(5-1)(-2)=\frac{1}{2}-8=-\frac{15}{2}$$

The terms are $\dfrac{1}{2},-\dfrac{3}{2},-\dfrac{7}{2},-\dfrac{11}{2},-\dfrac{15}{2}$.

30. $a_1=-100$

$$a_2=-100+(2-1)\left(\frac{1}{5}\right)=-100+\frac{1}{5}=-\frac{499}{5}$$
$$a_3=-100+(3-1)\left(\frac{1}{5}\right)=-100+\frac{2}{5}=-\frac{498}{5}$$
$$a_4=-100+(4-1)\left(\frac{1}{5}\right)=-100+\frac{3}{5}=-\frac{497}{5}$$
$$a_5=-100+(5-1)\left(\frac{1}{5}\right)=-100+\frac{4}{5}=-\frac{496}{5}$$

The terms are $-100,-\dfrac{499}{5},-\dfrac{498}{5},-\dfrac{497}{5},-\dfrac{496}{5}$.

31. $a_9=a_1+(9-1)d$
$$a_9=6+(9-1)(3)=6+8(3)=6+24=30$$

32. $a_7=a_1+(7-1)d$
$$-14=10+6d$$
$$-24=6d$$
$$-4=d$$

33. $a_{11}=a_1+(11-1)d$
$$2=-3+10d$$
$$5=10d$$
$$d=\frac{5}{10}=\frac{1}{2}$$

34. $a_n=a_1+(n-1)d$
$$-3=22+(n-1)(-5)$$
$$-3=22-5n+5$$
$$-3=27-5n$$
$$-30=-5n$$
$$6=n$$

35. $a_8 = a_1 + (8-1)d$

$21 = 7 + 7d$

$14 = 7d$

$2 = d$

$s_8 = \dfrac{8(a_1 + a_8)}{2} = \dfrac{8(7+21)}{2} = \dfrac{8(28)}{2} = \dfrac{224}{2} = 112$

36. $a_7 = a_1 + (7-1)d$

$-48 = -12 + 6d$

$-36 = 6d$

$-6 = d$

$s_7 = \dfrac{7(a_1 + a_7)}{2} = \dfrac{7(-12-48)}{2} = \dfrac{7(-60)}{2} = -210$

37. $a_6 = a_1 + (6-1)d$

$\dfrac{13}{5} = \dfrac{3}{5} + 5d$

$\dfrac{13}{5} - \dfrac{3}{5} = 5d$

$\dfrac{10}{5} = 5d$

$2 = 5d$

$\dfrac{2}{5} = d$

$s_6 = \dfrac{6(a_1 + a_6)}{2}$

$= \dfrac{6\left(\frac{3}{5} + \frac{13}{5}\right)}{2} = \dfrac{6\left(\frac{16}{5}\right)}{2} = 6\left(\dfrac{16}{5}\right)\left(\dfrac{1}{2}\right) = \dfrac{48}{5}$

38. $a_9 = a_1 + (9-1)d$

$-6 = -\dfrac{10}{3} + 8d$

$-6 + \dfrac{10}{3} = 8d$

$-\dfrac{8}{3} = 8d$

$d = \dfrac{1}{8}\left(-\dfrac{8}{3}\right) = -\dfrac{1}{3}$

$s_9 = \dfrac{9(a_1 + a_n)}{2}$

$= \dfrac{9\left(-\frac{10}{3} - 6\right)}{2} = \dfrac{9\left(-\frac{28}{3}\right)}{2} = 9\left(-\dfrac{28}{3}\right)\left(\dfrac{1}{2}\right) = -42$

39. $a_1 = -7$

$a_2 = -7 + (2-1)(4) = -7 + 1(4) = -7 + 4 = -3$

$a_3 = -7 + (3-1)(4) = -7 + 2(4) = -7 + 8 = 1$

$a_4 = -7 + (4-1)(4) = -7 + 3(4) = -7 + 12 = 5$

The terms are $-7, -3, 1, 5$.

$a_{10} = -7 + (10-1)(4) = -7 + 9(4) = -7 + 36 = 29$

$s_{10} = \dfrac{10(-7+29)}{2} = \dfrac{10(22)}{2} = \dfrac{220}{2} = 110$

40. $a_1 = 4$

$a_2 = 4 + (2-1)(-3) = 4 + 1(-3) = 4 - 3 = 1$

$a_3 = 4 + (3-1)(-3) = 4 + 2(-3) = 4 - 6 = -2$

$a_4 = 4 + (4-1)(-3) = 4 + 3(-3) = 4 - 9 = -5$

The terms are $4, 1, -2, -5$.

$a_{10} = 4 + (10-1)(-3) = 4 + 9(-3) = 4 - 27 = -23$

$s_{10} = \dfrac{10[4+(-23)]}{2} = \dfrac{10(-19)}{2} = \dfrac{-190}{2} = -95$

41. $a_1 = \dfrac{5}{6}$

$a_2 = \dfrac{5}{6} + (2-1)\left(\dfrac{2}{3}\right) = \dfrac{5}{6} + \dfrac{2}{3} = \dfrac{9}{6} = \dfrac{3}{2}$

$a_3 = \dfrac{5}{6} + (3-1)\left(\dfrac{2}{3}\right) = \dfrac{5}{6} + 2\left(\dfrac{2}{3}\right) = \dfrac{5}{6} + \dfrac{4}{3} = \dfrac{13}{6}$

$a_4 = \dfrac{5}{6} + (4-1)\left(\dfrac{2}{3}\right) = \dfrac{5}{6} + 3\left(\dfrac{2}{3}\right) = \dfrac{5}{6} + \dfrac{6}{3} = \dfrac{17}{6}$

The terms are $\dfrac{5}{6}, \dfrac{3}{2}, \dfrac{13}{6}, \dfrac{17}{6}$.

$a_{10} = \dfrac{5}{6} + (10-1)\left(\dfrac{2}{3}\right) = \dfrac{5}{6} + 9\left(\dfrac{2}{3}\right) = \dfrac{5}{6} + 6 = \dfrac{41}{6}$

$s_{10} = \dfrac{10\left(\frac{5}{6} + \frac{41}{6}\right)}{2} = \dfrac{10\left(\frac{46}{6}\right)}{2} = 5\left(\dfrac{46}{6}\right) = 5\left(\dfrac{23}{3}\right) = \dfrac{115}{3}$

42. $a_1 = -60$

$a_2 = -60 + (2-1)(5) = -60 + 1(5) = -60 + 5 = -55$

$a_3 = -60 + (3-1)(5) = -60 + 2(5) = -60 + 10 = -50$

$a_4 = -60 + (4-1)(5) = -60 + 3(5) = -60 + 15 = -45$

The terms are $-60, -55, -50, -45$.

$a_{10} = -60 + (10-1)(5) = -60 + 45 = -15$

$s_{10} = \dfrac{10(-60-15)}{2} = 5(-75) = -375$

43. $d = 9 - 4 = 5$

$a_n = a_1 + (n-1)d$

$64 = 4 + (n-1)5$

$64 = 4 + 5n - 5$

$64 = 5n - 1$

$65 = 5n$

$13 = n$

$s_{13} = \dfrac{13(a_1 + a_{13})}{2} = \dfrac{13(4 + 64)}{2} = \dfrac{13(68)}{2} = 442$

44. $d = -4 - (-7) = -4 + 7 = 3$

$a_n = a_1 + (n-1)d$

$11 = -7 + (n-1)3$

$11 = -7 + 3n - 3$

$11 = 3n - 10$

$21 = 3n$

$7 = n$

$s_7 = \dfrac{7(a_1 + a_7)}{2} = \dfrac{7(-7 + 11)}{2} = \dfrac{7(4)}{2} = \dfrac{28}{2} = 14$

45. $d = \dfrac{9}{10} - \dfrac{6}{10} = \dfrac{3}{10}$

$a_n = a_1 + (n-1)d$

$\dfrac{36}{10} = \dfrac{6}{10} + (n-1)\dfrac{3}{10}$

$\dfrac{36}{10} = \dfrac{6}{10} + \dfrac{3}{10}n - \dfrac{3}{10}$

$\dfrac{36}{10} = \dfrac{3}{10} + \dfrac{3}{10}n$

$\dfrac{33}{10} = \dfrac{3}{10}n$

$n = \dfrac{10}{3}\left(\dfrac{33}{10}\right) = 11$

$s_{11} = \dfrac{11(a_1 + a_{11})}{2}$

$= \dfrac{11\left(\frac{6}{10} + \frac{36}{10}\right)}{2} = \dfrac{11\left(\frac{42}{10}\right)}{2} = 11\left(\dfrac{42}{10}\right)\left(\dfrac{1}{2}\right) = \dfrac{231}{10}$

46. $d = -3 - (-9) = -3 + 9 = 6$

$a_n = a_1 + (n-1)d$

$45 = -9 + (n-1)6$

$45 = -9 + 6n - 6$

$45 = -15 + 6n$

$60 = 6n$

$10 = n$

$s_{10} = \dfrac{10(a_1 + a_{10})}{2} = \dfrac{10(-9 + 45)}{2} = \dfrac{10(36)}{2} = 180$

47. $a_1 = 6$

$a_2 = 6(2)^{2-1} = 6(2) = 12$

$a_3 = 6(2)^{3-1} = 6(2)^2 = 6(4) = 24$

$a_4 = 6(2)^{4-1} = 6(2)^3 = 6(8) = 48$

$a_5 = 6(2)^{5-1} = 6(2)^4 = 6(16) = 96$

The terms are 6, 12, 24, 48, 96.

48. $a_1 = -12$

$a_2 = -12\left(\dfrac{1}{2}\right)^{2-1} = -12\left(\dfrac{1}{2}\right) = -6$

$a_3 = -12\left(\dfrac{1}{2}\right)^{3-1} = -12\left(\dfrac{1}{4}\right) = -3$

$a_4 = -12\left(\dfrac{1}{2}\right)^{4-1} = -12\left(\dfrac{1}{8}\right) = -\dfrac{3}{2}$

$a_5 = -12\left(\dfrac{1}{2}\right)^{5-1} = -12\left(\dfrac{1}{16}\right) = -\dfrac{3}{4}$

The terms are $-12, -6, -3, -\dfrac{3}{2}, -\dfrac{3}{4}$.

49. $a_1 = 20$

$a_2 = 20\left(-\dfrac{2}{3}\right)^{2-1} = 20\left(-\dfrac{2}{3}\right) = -\dfrac{40}{3}$

$a_3 = 20\left(-\dfrac{2}{3}\right)^{3-1} = 20\left(-\dfrac{2}{3}\right)^2 = 20\left(\dfrac{4}{9}\right) = \dfrac{80}{9}$

$a_4 = 20\left(-\dfrac{2}{3}\right)^{4-1} = 20\left(-\dfrac{2}{3}\right)^3 = 20\left(-\dfrac{8}{27}\right) = -\dfrac{160}{27}$

$a_5 = 20\left(-\dfrac{2}{3}\right)^{5-1} = 20\left(-\dfrac{2}{3}\right)^4 = 20\left(\dfrac{16}{81}\right) = \dfrac{320}{81}$

The terms are $20, -\dfrac{40}{3}, \dfrac{80}{9}, -\dfrac{160}{27}, \dfrac{320}{81}$.

50. $a_1 = -20$

$a_2 = -20\left(\dfrac{1}{5}\right)^{2-1} = -20\left(\dfrac{1}{5}\right) = -4$

$a_3 = -20\left(\dfrac{1}{5}\right)^{3-1} = -20\left(\dfrac{1}{25}\right) = -\dfrac{4}{5}$

$a_4 = -20\left(\dfrac{1}{5}\right)^{4-1} = -20\left(\dfrac{1}{125}\right) = -\dfrac{4}{25}$

$a_5 = -20\left(\dfrac{1}{5}\right)^{5-1} = -20\left(\dfrac{1}{625}\right) = -\dfrac{4}{125}$

The terms are $-20, -4, -\dfrac{4}{5}, -\dfrac{4}{25}, -\dfrac{4}{125}$.

51. $a_5 = a_1 \cdot r^{5-1} = 6\left(\frac{1}{3}\right)^{5-1} = 6\left(\frac{1}{3}\right)^4 = 6\left(\frac{1}{81}\right) = \frac{2}{27}$

52. $a_6 = a_1 \cdot r^{6-1} = 15(2)^{6-1} = 15(2)^5 = 15(32) = 480$

53. $a_4 = -8(-3)^{4-1} = -8(-3)^3 = -8(-27) = 216$

54. $a_5 = \frac{1}{12}\left(\frac{2}{3}\right)^{5-1} = \frac{1}{12}\left(\frac{2}{3}\right)^4 = \frac{1}{12}\left(\frac{16}{81}\right) = \frac{4}{243}$

55. $s_6 = \frac{6(1-2^6)}{1-2} = \frac{6(1-64)}{-1} = \frac{6(-63)}{-1} = 378$

56. $s_5 = \dfrac{-84\left[1-\left(-\frac{1}{4}\right)^5\right]}{1-\left(-\frac{1}{4}\right)}$

$= \dfrac{-84\left[1-\left(-\frac{1}{1024}\right)\right]}{1+\frac{1}{4}}$

$= \dfrac{-84\left(1+\frac{1}{1024}\right)}{\frac{5}{4}}$

$= -84 \cdot \left(\frac{1025}{1024}\right) \cdot \frac{4}{5}$

$= -\dfrac{4305}{64}$

57. $s_4 = \dfrac{9\left[1-\left(\frac{3}{2}\right)^4\right]}{1-\frac{3}{2}} = \dfrac{9\left(1-\frac{81}{16}\right)}{-\frac{1}{2}} = 9\left(-\frac{65}{16}\right)\left(-\frac{2}{1}\right) = \frac{585}{8}$

58. $s_7 = \dfrac{8\left[1-\left(\frac{1}{2}\right)^7\right]}{1-\frac{1}{2}} = \dfrac{8\left(1-\frac{1}{128}\right)}{\frac{1}{2}} = 8\left(\frac{127}{128}\right)\left(\frac{2}{1}\right) = \frac{127}{8}$

59. $r = 12 \div 6 = 2$
$a_n = 6(2)^{n-1}$

60. $r = -20 \div (-4) = 5$
$a_n = -4(5)^{n-1}$

61. $r = \dfrac{10}{3} \div 10 = \dfrac{10}{3} \cdot \dfrac{1}{10} = \dfrac{1}{3}$
$a_n = 10\left(\dfrac{1}{3}\right)^{n-1}$

62. $r = \dfrac{18}{15} \div \dfrac{9}{5} = \dfrac{18}{15} \cdot \dfrac{5}{9} = \dfrac{2}{3}$
$a_n = \dfrac{9}{5}\left(\dfrac{2}{3}\right)^{n-1}$

63. $r = \dfrac{5}{2} \div 5 = \dfrac{5}{2} \cdot \dfrac{1}{5} = \dfrac{1}{2}$

$s_\infty = \dfrac{5}{1-\frac{1}{2}} = \dfrac{5}{\frac{1}{2}} = 5\left(\dfrac{2}{1}\right) = 10$

64. $r = 1 \div \dfrac{5}{2} = 1\left(\dfrac{2}{5}\right) = \dfrac{2}{5}$

$s_\infty = \dfrac{\frac{5}{2}}{1-\frac{2}{5}} = \dfrac{\frac{5}{2}}{\frac{3}{5}} = \dfrac{5}{2}\left(\dfrac{5}{3}\right) = \dfrac{25}{6}$

65. $r = \dfrac{8}{3} \div (-8) = \dfrac{8}{3}\left(-\dfrac{1}{8}\right) = -\dfrac{1}{3}$

$s_\infty = \dfrac{-8}{1-\left(-\frac{1}{3}\right)} = \dfrac{-8}{\frac{4}{3}} = -8\left(\dfrac{3}{4}\right) = -6$

66. $r = -4 \div -6 = \dfrac{-4}{-6} = \dfrac{2}{3}$

$s_\infty = \dfrac{-6}{1-\frac{2}{3}} = \dfrac{-6}{\frac{1}{3}} = -6\left(\dfrac{3}{1}\right) = -18$

67. $r = 8 \div 16 = \dfrac{8}{16} = \dfrac{1}{2}$

$s_\infty = \dfrac{16}{1-\frac{1}{2}} = \dfrac{16}{\frac{1}{2}} = 16\left(\dfrac{2}{1}\right) = 32$

68. $r = \dfrac{9}{3} \div 9 = \dfrac{9}{3} \cdot \dfrac{1}{9} = \dfrac{1}{3}$

$s_\infty = \dfrac{9}{1-\frac{1}{3}} = \dfrac{9}{\frac{2}{3}} = 9\left(\dfrac{3}{2}\right) = \dfrac{27}{2}$

69. $r = -1 \div 5 = -\dfrac{1}{5}$

$s_\infty = \dfrac{5}{1-\left(-\frac{1}{5}\right)} = \dfrac{5}{\frac{6}{5}} = 5\left(\dfrac{5}{6}\right) = \dfrac{25}{6}$

70. $r = -\dfrac{8}{3} \div -4 = -\dfrac{8}{3}\left(-\dfrac{1}{4}\right) = \dfrac{2}{3}$

$s_\infty = \dfrac{-4}{1-\frac{2}{3}} = \dfrac{-4}{\frac{1}{3}} = -4\left(\dfrac{3}{1}\right) = -12$

71. $0.363636\ldots = 0.36 + 0.0036 + 0.000036 + \ldots$
$= 0.36 + 0.36(0.01) + 0.36(0.01)^2 + \cdots$
$a_1 = 0.36$ and $r = 0.01$

$s_\infty = \dfrac{0.36}{1-0.01} = \dfrac{0.36}{0.99} = \dfrac{36}{99} = \dfrac{4}{11}$

72. $0.621621... = 0.621 + 0.000621 + 0.000000621 + \cdots = 0.621 + 0.621(0.001) + 0.621(0.001)^2 + \cdots$

$a_1 = 0.621$ and $r = 0.001$; $\quad s_\infty = \dfrac{0.621}{1 - 0.001} = \dfrac{0.621}{0.999} = \dfrac{621}{999} = \dfrac{23}{37}$

73. $(3x + y)^4 = \dbinom{4}{0}(3x)^4(y)^0 + \dbinom{4}{1}(3x)^3(y)^1 + \dbinom{4}{2}(3x)^2(y)^2 + \dbinom{4}{3}(3x)^1(y)^3 + \dbinom{4}{4}(3x)^0(y)^4$

$\quad = 1(81x^4)(1) + 4(27x^3)(y) + 6(9x^2)(y^2) + 4(3x)(y^3) + 1(1)(y^4)$

$\quad = 81x^4 + 108x^3y + 54x^2y^2 + 12xy^3 + y^4$

74. $(2x - 3y^2)^3 = \dbinom{3}{0}(2x)^3(-3y^2)^0 + \dbinom{3}{1}(2x)^2(-3y^2)^1 + \dbinom{3}{2}(2x)^1(-3y^2)^2 + \dbinom{3}{3}(2x)^0(-3y^2)^3$

$\quad = 1(8x^3)(1) + 3(4x^2)(-3y^2) + 3(2x)(9y^4) + 1(1)(-27y^6)$

$\quad = 8x^3 - 36x^2y^2 + 54xy^4 - 27y^6$

75. $(x - 2y)^9 = \dbinom{9}{0}(x)^9(-2y)^0 + \dbinom{9}{1}(x)^8(-2y)^1 + \dbinom{9}{2}(x)^7(-2y)^2 + \dbinom{9}{3}(x)^6(-2y)^3 + \cdots$

$\quad = 1(x^9)(1) + 9(x^8)(-2y) + 36(x^7)(4y^2) + 84(x^6)(-8y^3) + \cdots$

$\quad = x^9 - 18x^8y + 144x^7y^2 - 672x^6y^3 + \cdots$

76. $(2a^2 + 3b)^8 = \dbinom{8}{0}(2a^2)^8(3b)^0 + \dbinom{8}{1}(2a^2)^7(3b)^1 + \dbinom{8}{2}(2a^2)^6(3b)^2 + \dbinom{8}{3}(2a^2)^5(3b)^3 + \cdots$

$\quad = 1(256a^{16})(1) + 8(128a^{14})(3b) + 28(64a^{12})(9b^2) + 56(32a^{10})(27b^3) + \cdots$

$\quad = 256a^{16} + 3072a^{14}b + 16,128a^{12}b^2 + 48,384a^{10}b^3 + \cdots$

---

77. This is an arithmetic series with $d = 1$, $a_1 = 101$, and $a_n = 200$.

$a_n = a_1 + (n - 1)d$

$200 = 101 + (n - 1)(1)$

$200 = 101 + n - 1$

$200 = n + 100$

$100 = n$

The sum is

$s_{100} = \dfrac{n(a_1 + a_{100})}{2}$

$\quad = \dfrac{100(101 + 200)}{2} = \dfrac{100(301)}{2} = 15,050$

78. $21 + 20 + 19 + \cdots + 1$ or $1 + 2 + 3 + \cdots + 21$

$s_n = \dfrac{n(a_1 + a_n)}{2}$

$s_{21} = \dfrac{21(1 + 21)}{2} = \dfrac{21(22)}{2} = \dfrac{462}{2} = 231$

There are 231 barrels in the stack.

79. This is an arithmetic sequence with $d = 1000$

a. $a_1 = 36,000$

$a_2 = 36,000 + (2 - 1)(1000)$

$\quad = 36,000 + 1000$

$\quad = 37,000$

$a_3 = 36,000 + (3 - 1)(1000)$

$\quad = 36,000 + 2000$

$\quad = 38,000$

$a_4 = 36,000 + (4 - 1)(1000)$

$\quad = 36,000 + 3000$

$\quad = 39,000$

Ahmed's salaries for the first four years are $36,000, $37,000, $38,000, and $39,000.

b. $a_n = 36,000 + (n - 1)(1000)$

$\quad = 36,000 + 1000n - 1000$

$\quad = 35,000 + 1000n$

c. $a_6 = 35,000 + 1000(6)$

$= 35,000 + 6,000$

$= 41,000$

In the 6th year, his salary would be \$41,000.

d. $a_{11} = 36,000 + (11-1)1000 = 46,000$

$$s_n = \frac{n(a_1 + a_n)}{2}$$

$$s_{11} = \frac{11(36,000 + 46,000)}{2}$$

$$= \frac{11(82,000)}{2}$$

$$= 451,000$$

Ahmed will make a total of \$451,000 in his first 11 years.

80. This is a geometric series with $r = 2$, $a_1 = 100$, and $n = 11$. (Note: There are 11 terms here since 200 represents the first doubling.)

$a_{11} = a_1 \cdot r^{11-1}$

$= 100(2)^{11-1} = 100(2)^{10} = 100(1024) = 102,400$

You would have \$102,400.

81. $a_1 = 1600$, $r = 1.04$, $a_n = a_1 r^{n-1}$

a. July is the 7th month, so $n = 7$.

$a_7 = 1600(1.04)^{7-1} = 1600(1.04)^6 \approx 2024.51$

Gertrude's salary will be \$2,024.51 in July.

b. December is the 12th month, so $n = 12$.

$a_{12} = 1600(1.04)^{12-1} = 1600(1.04)^{11} \approx 2463.13$

Gertrude's salary will be \$2,463.13 in December.

c. $s_n = \dfrac{a_1(1 - r^n)}{1 - r}$, $n = 12$

$$s_{12} = \frac{1600\left[1 - (1.04)^{12}\right]}{1 - 1.04} \approx 24,041.29$$

Gertrude will make \$24,041.29 in 2006.

82. Each year, the cost of the object will be 1.08 times greater than the previous year. After 12 years will be the 13th year. Therefore,

$a_{13} = 200(1.08)^{13-1} \approx 200(2.51817) \approx 503.63$

The item would cost \$503.63.

83. This is an infinite geometric series with $r = 0.92$ and $a_1 = 12$.

$$s_\infty = \frac{12}{1 - 0.92} = \frac{12}{0.08} = 150$$

The pendulum travels a total distance of 150 feet.

Chapter 15 Practice Test

1. A series is the sum of the terms of a sequence

2. a. An arithmetic sequence is one whose terms differ by a constant amount.

b. A geometric sequence is one whose terms differ by a common multiple.

3. $a_n = \dfrac{n-2}{3n}$

$a_1 = \dfrac{1-2}{3(1)} = \dfrac{-1}{3} = -\dfrac{1}{3}$

$a_2 = \dfrac{2-2}{3(2)} = \dfrac{0}{6} = 0$

$a_3 = \dfrac{3-2}{3(3)} = \dfrac{1}{9}$

$a_4 = \dfrac{4-2}{3(4)} = \dfrac{2}{12} = \dfrac{1}{6}$

$a_5 = \dfrac{5-2}{3(5)} = \dfrac{3}{15} = \dfrac{1}{5}$

The terms are $-\dfrac{1}{3}$, 0, $\dfrac{1}{9}$, $\dfrac{1}{6}$, $\dfrac{1}{5}$.

4. $a_n = \dfrac{2n+1}{n^2}$

$a_1 = \dfrac{2(1)+1}{1^2} = \dfrac{2+1}{1} = 3$

$a_2 = \dfrac{2(2)+1}{2^2} = \dfrac{4+1}{4} = \dfrac{5}{4}$

$a_3 = \dfrac{2(3)+1}{3^2} = \dfrac{6+1}{9} = \dfrac{7}{9}$

$s_1 = a_1 = 3$

$s_3 = a_1 + a_2 + a_3 = 3 + \dfrac{5}{4} + \dfrac{7}{9} = \dfrac{181}{36}$

5. $\displaystyle\sum_{i=1}^{5}(2i^2 + 3)$

$= [2(1)^2 + 3] + [2(2)^2 + 3] + [2(3)^2 + 3]$

$\qquad + [2(4)^2 + 3] + [2(5)^2 + 3]$

$= (2+3) + (8+3) + (18+3) + (32+3) + (50+3)$

$= 5 + 11 + 21 + 35 + 53$

$= 125$

6. $\displaystyle\sum_{i=1}^{4}(x_i)^2 = x_1^2 + x_2^2 + x_3^2 + x_4^2$

$\qquad = 4^2 + 2^2 + 8^2 + 10^2$

$\qquad = 16 + 4 + 64 + 100$

$\qquad = 184$

7. $d = \dfrac{2}{3} - \dfrac{1}{3} = \dfrac{1}{3}$

$\quad a_n = a_1 + (n-1)d$

$\qquad = \dfrac{1}{3} + (n-1)\left(\dfrac{1}{3}\right) = \dfrac{1}{3} + \dfrac{1}{3}n - \dfrac{1}{3} = \dfrac{1}{3}n$

8. $r = 10 \div 5 = \dfrac{10}{5} = 2$

$\quad a_n = a_1 r^{n-1} = 5(2)^{n-1}$

9. $a_1 = 15$

$\quad a_2 = 15 + (2-1)(-6) = 15 + 1(-6) = 15 - 6 = 9$

$\quad a_3 = 15 + (3-1)(-6) = 15 + 2(-6) = 15 - 12 = 3$

$\quad a_4 = 15 + (4-1)(-6) = 15 + 3(-6) = 15 - 18 = -3$

The terms are $15, 9, 3, -3$.

10. $a_1 = \dfrac{5}{12}$

$\quad a_2 = a_1 r^1 = \dfrac{5}{12}\left(\dfrac{2}{3}\right)^1 = \dfrac{5}{12}\left(\dfrac{2}{3}\right) = \dfrac{5}{18}$

$\quad a_3 = a_1 r^2 = \dfrac{5}{12}\left(\dfrac{2}{3}\right)^2 = \dfrac{5}{12}\left(\dfrac{4}{9}\right) = \dfrac{5}{27}$

$\quad a_4 = a_1 r^3 = \dfrac{5}{12}\left(\dfrac{2}{3}\right)^3 = \dfrac{5}{12}\left(\dfrac{8}{27}\right) = \dfrac{10}{81}$

The terms are $\dfrac{5}{12}, \dfrac{5}{18}, \dfrac{5}{27}, \dfrac{10}{81}$.

11. $a_{11} = a_1 + 10d = 40 + (10)(-8) = 40 - 80 = -40$

12. $s_8 = \dfrac{8(a_1 + a_8)}{2} = \dfrac{8[7 + (-12)]}{2} = \dfrac{8(-5)}{2} = -20$

13. $d = -16 - (-4) = -16 + 4 = -12$

$\quad a_n = a_1 + (n-1)d$

$\quad -136 = -4 + (n-1)(-12)$

$\quad -136 = -4 - 12n + 12$

$\quad -136 = 8 - 12n$

$\quad -144 = -12n$

$\qquad 12 = n$

14. $a_6 = a_1 r^5 = 8\left(\dfrac{2}{3}\right)^5 = 8\left(\dfrac{32}{243}\right) = \dfrac{256}{243}$

15. $s_7 = \dfrac{a_1\left(1 - r^7\right)}{1 - r}$

$\qquad = \dfrac{\frac{3}{5}\left[1 - (-5)^7\right]}{1 - (-5)}$

$\qquad = \dfrac{\frac{3}{5}[1 - (-78{,}125)]}{1 - (-5)}$

$\qquad = \dfrac{\frac{3}{5}(1 + 78{,}125)}{1 + 5}$

$\qquad = \dfrac{\frac{3}{5}(78{,}126)}{6}$

$\qquad = \dfrac{3(78{,}126)}{5 \cdot 6}$

$\qquad = \dfrac{78{,}126}{5 \cdot 2}$

$\qquad = \dfrac{39{,}063}{5}$

16. $r = 5 \div 15 = \dfrac{5}{15} = \dfrac{1}{3}$

$\quad a_n = a_1 r^{n-1} = 15\left(\dfrac{1}{3}\right)^{n-1}$

17. $r = \dfrac{8}{3} \div 4 = \dfrac{8}{3} \cdot \dfrac{1}{4} = \dfrac{2}{3}$

$\quad s_\infty = \dfrac{a_1}{1 - r} = \dfrac{4}{1 - \frac{2}{3}} = \dfrac{4}{\frac{1}{3}} = 4 \cdot \dfrac{3}{1} = 12$

18. $0.3939\cdots = 0.39 + 0.0039 + 0.000039 + \cdots$

$\qquad = 0.39 + 0.39(0.01) + 0.39(0.01)^2 + \cdots$

$\quad r = 0.01$ and $a_1 = 0.39$

$\quad s_\infty = \dfrac{0.39}{1 - 0.01} = \dfrac{0.39}{0.99} = \dfrac{39}{99} = \dfrac{13}{33}$

19. $\dbinom{8}{3} = \dfrac{8!}{3!(8-3)!}$

$\qquad = \dfrac{8!}{3!5!}$

$\qquad = \dfrac{8 \cdot 7 \cdot 6 \cdot \cancel{5 \cdot 4 \cdot 3 \cdot 2 \cdot 1}}{(3 \cdot 2 \cdot 1)(\cancel{5 \cdot 4 \cdot 3 \cdot 2 \cdot 1})}$

$\qquad = \dfrac{336}{6}$

$\qquad = 56$

20. $(x+2y)^4 = \binom{4}{0}(x)^4(2y)^0 + \binom{4}{1}(x)^3(2y)^1 + \binom{4}{2}(x)^2(2y)^2 + \binom{4}{3}(x)^1(2y)^3 + \binom{4}{4}(x)^0(2y)^4$

$\qquad = 1(x^4)(1) + 4(x^3)(2y) + 6(x^2)(4y^2) + 4(x)(8y^3) + 1(1)(16y^4)$

$\qquad = x^4 + 8x^3y + 24x^2y^2 + 32xy^3 + 16y^4$

---

21. $\bar{x} = \dfrac{\sum x}{n} = \dfrac{76+93+83+87+71}{5} = \dfrac{410}{5} = 82$

22. $13+12+11+\cdots+1$ or $1+2+3+\cdots+13$

$s_n = \dfrac{n(a_1+a_n)}{2}$

$s_{13} = \dfrac{13(1+13)}{2} = \dfrac{13(14)}{2} = \dfrac{182}{2} = 91$

There are 91 logs in the pile.

23. $a_1 = 1000$, $n = 20$

$a_n = a_1 + (n-1)d$

$a_{20} = 1000 + (20-1)(1000) = 20,000$

$s_{20} = \dfrac{20(1000+20,000)}{2} = \dfrac{20(21,000)}{2} = 210,000$

After 20 years, she will have saved $210,000.

24. $a_1 = 700$, $r = 1.04$, $a_n = a_1 r^{n-1}$

$a_6 = 700(1.04)^{6-1} = 700(1.04)^5 \approx 851.66$

She will be making about $851.66 in the sixth week.

25. $r = 3$, $a_1 = 500(3) = 1500$

$a_6 = 1500(3)^{6-1} = 1500(3)^5 = 364,500$

At the end of the sixth hour, there will be 364,500 bacteria in the culture.

Chapter 15 Cumulative Review Test

1. $A = \dfrac{1}{2}bh$, for b

$2A = bh$

$\dfrac{2A}{h} = \dfrac{bh}{h}$

$\dfrac{2A}{h} = b$

2. $m = \dfrac{y_2-y_1}{x_2-x_1} = \dfrac{9-(-2)}{1-4} = \dfrac{11}{-3} = -\dfrac{11}{3}$

Use the point-slope equation with $m = -\dfrac{11}{3}$ and $(4,-2)$.

$y - y_1 = m(x-x_1)$

$y - (-2) = -\dfrac{11}{3}(x-4)$

$y + 2 = -\dfrac{11}{3}x + \dfrac{44}{3}$

$y = -\dfrac{11}{3}x + \dfrac{44}{3} - 2$

$y = -\dfrac{11}{3}x + \dfrac{38}{3}$

3.

$\qquad\qquad 5x^3 + 4x^2 - 6x + 2$

$\qquad\qquad\qquad\qquad\qquad x + 5$

$\overline{\qquad 5x^4 + 4x^3 -\ 6x^2 +\ 2x\qquad}$

$\qquad\qquad 25x^3 + 20x^2 - 30x + 10$

$\overline{\qquad 5x^4 + 29x^3 + 14x^2 - 28x + 10}$

4. $x^3 + 2x - 5x^2 - 10 = (x^3 + 2x) - (6x^2 + 12)$

$\qquad\qquad\qquad\qquad\quad = x(x^2+2) - 6(x^2+2)$

$\qquad\qquad\qquad\qquad\quad = (x^2+2)(x-6)$

5. $(a+b)^2 + 8(a+b) + 16$ $\qquad$ [Let $u = a+b$.]

$= u^2 + 8u + 16$

$= (u+4)(u+4)$

$= (u+4)^2$ $\qquad$ [Back substitute $a+b$ for u.]

$= (a+b+4)^2$

6. $5 - \dfrac{x-1}{x^2+3x-10} = 5 - \dfrac{x-1}{(x+5)(x-2)}$

$\qquad = \dfrac{5(x+5)(x-2)}{(x+5)(x-2)} - \dfrac{x-1}{(x+5)(x-2)}$

$\qquad = \dfrac{5x^2+15x-50}{(x+5)(x-2)} - \dfrac{x-1}{(x+5)(x-2)}$

$\qquad = \dfrac{5x^2+15x-50-(x-1)}{(x+5)(x-2)}$

$\qquad = \dfrac{5x^2+15x-50-x+1}{(x+5)(x-2)}$

$\qquad = \dfrac{5x^2+14x-49}{(x+5)(x-2)}$

7. $y = kz^2$

$\quad 80 = k(20)^2$

$\quad 80 = 400k$

$\quad k = \dfrac{80}{400} = 0.2$

$\quad y = 0.2z^2$

Let $z = 50$: $\ y = 0.2(50)^2 = 0.2(2500) = 500$

8. $x + y + z = 1 \quad (1)$

$\quad 2x+2y+2z = 2 \quad (2)$

$\quad 3x+3y+3z = 3 \quad (3)$

Notice that if you multiply equation (1) by 2, the result is exactly the same as equation (2). This implies that this is a dependent system and therefore has infinitely many solutions.

9. $f(x) = 3\sqrt[3]{x-2}, \ g(x) = \sqrt[3]{7x-14}$

$\qquad f(x) = g(x)$

$\qquad 2\sqrt[3]{x-3} = \sqrt[3]{5x-15}$

$\qquad \left(2\sqrt[3]{x-3}\right)^3 = \left(\sqrt[3]{5x-15}\right)^3$

$\qquad 8(x-3) = 5x-15$

$\qquad 8x-24 = 5x-15$

$\qquad 3x = 9$

$\qquad x = 3$

10. $\sqrt{6x-5} - \sqrt{2x+6} - 1 = 0$

$\qquad \sqrt{6x-5} = 1 + \sqrt{2x+6}$

$\qquad \left(\sqrt{6x-5}\right)^2 = \left(1+\sqrt{2x+6}\right)^2$

$\qquad 6x-5 = 1 + 2\sqrt{2x+6} + 2x+6$

$\qquad 4x-12 = 2\sqrt{2x+6}$

$\qquad 2x-6 = \sqrt{2x+6}$

$\qquad (2x-6)^2 = \left(\sqrt{2x+6}\right)^2$

$\qquad 4x^2-24x+36 = 2x+6$

$\qquad 4x^2-26x+30 = 0$

$\qquad 2\left(2x^2-13x+15\right) = 0$

$\qquad 2(2x-3)(x-5) = 0$

$\qquad 2x-3 = 0 \quad \text{or} \quad x-5 = 0$

$\qquad x = \dfrac{3}{2} \qquad\qquad x = 5$

Upon checking, $x = \dfrac{3}{2}$ is an extraneous solution.

The solution is $x = 5$.

11. $x^2+2x+15 = 0$

$\qquad x^2+2x \qquad = -15$

$\qquad x^2+2x+1 = -15+1$

$\qquad (x+1)^2 = -14$

$\qquad x+1 = \pm\sqrt{-14}$

$\qquad x+1 = \pm i\sqrt{14}$

$\qquad x = -1 \pm i\sqrt{14}$

12. $x^2 - \dfrac{x}{5} - \dfrac{1}{3} = 0$

$\quad 15\left(x^2 - \dfrac{x}{5} - \dfrac{1}{3}\right) = 15(0)$

$\qquad 15x^2 - 3x - 5 = 0$

$\qquad x = \dfrac{-b \pm \sqrt{b^2-4ac}}{2a}$

$\qquad = \dfrac{-(-3) \pm \sqrt{(-3)^2 - 4(15)(-5)}}{2(15)}$

$\qquad = \dfrac{3 \pm \sqrt{9+300}}{30}$

$\qquad = \dfrac{3 \pm \sqrt{309}}{30}$

13. Let x be the number. Then
$$2x^2 - 9x = 5$$
$$2x^2 - 9x - 5 = 0$$
$$(2x+1)(x-5) = 0$$
$$2x+1 = 0 \quad \text{or} \quad x-5 = 0$$
$$x = -\frac{1}{2} \qquad x = 5$$

Since the number must be positive, disregard $x = -\frac{1}{2}$. The number is 5.

14. $y = 2^x - 1$

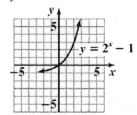

15. $\log_a \frac{1}{64} = 6$
$$a^6 = \frac{1}{64}$$
$$a^6 = \left(\frac{1}{2}\right)^6$$
$$a = \frac{1}{2}$$

16. $y = x^2 - 4x$
$$y = \left(x^2 - 4x + 4\right) - 4$$
$$y = (x-2)^2 - 4$$
The vertex is $(2, -4)$.

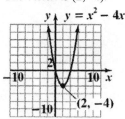

17. $(h, k) = (-6, 2), \quad r = 7$
$$(x-h)^2 + (y-k)^2 = r^2$$
$$\left(x-(-6)\right)^2 + \left(y-2\right)^2 = 7^2$$
$$(x+6)^2 + (y-2)^2 = 49$$

18. $(x+3)^2 + (y+1)^2 = 16$
$$(x+3)^2 + (y+1)^2 = 4^2$$
center: $(-3, -1)$ \quad radius: 4

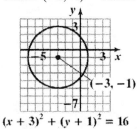

$(x + 3)^2 + (y + 1)^2 = 16$

19. $9x^2 + 16y^2 = 144$
$$\frac{9x^2}{144} + \frac{16y^2}{144} = 1$$
$$\frac{x^2}{16} + \frac{y^2}{9} = 1$$
$$\frac{x^2}{4^2} + \frac{y^2}{3^2} = 1 \implies a = 4 \text{ and } b = 3$$

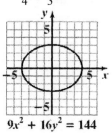

$9x^2 + 16y^2 = 144$

20. $r = 4 \div 6 = \frac{4}{6} = \frac{2}{3}$
$$s_\infty = \frac{a_1}{1-r} = \frac{6}{1-\frac{2}{3}} = \frac{6}{\frac{1}{3}} = 6 \cdot \frac{3}{1} = 18$$